ANNUAL REVIEW OF ASTRONOMY AND ASTROPHYSICS

ANNUAL REVIEW OF ASTRONOMY AND ASTROPHYSICS

GEOFFREY R. BURBIDGE, *Editor*
University of California, San Diego

DAVID LAYZER, *Associate Editor*
Harvard College Observatory

JOHN G. PHILLIPS, *Associate Editor*
University of California, Berkeley

VOLUME 13

1975

ANNUAL REVIEWS INC. 4139 EL CAMINO WAY PALO ALTO, CALIFORNIA 94306

ANNUAL REVIEWS INC.
Palo Alto, California, USA

International Standard Book Number: 0–8243–0913–8
Library of Congress Catalog Number: 63–8846

REPRINTS

The conspicuous number aligned in the margin with the title of each article in this volume is a key for use in ordering reprints. Available reprints are priced at the uniform rate of $1 each postpaid. Effective January 1, 1975, the minimum acceptable reprint order is 10 reprints and/or $10.00, prepaid. A quantity discount is available.

FILMSET BY TYPESETTING SERVICES LTD, GLASGOW, SCOTLAND
PRINTED AND BOUND IN THE UNITED STATES OF AMERICA

PREFACE

This volume of the Annual Review of Astronomy and Astrophysics was planned at a meeting of the Editorial Committee which was held on May 19, 1973, in Palo Alto, California.

It is a pleasure once again to thank the Associate Editors for carrying out the scientific editorial work very efficiently. I also wish to thank the Assistant Editor Rosalie West for the effective way in which she carried out her general editorial duties and for the considerable help she has afforded me in our dealings with the authors. She resigned in August 1974 and has been replaced by Marjorie Cutler who is doing an equally effective job. I would also like to acknowledge the help given to me by my secretary in La Jolla, Dawn Pedersen.

THE EDITOR

SOME RELATED ARTICLES APPEARING IN OTHER *ANNUAL REVIEWS*

ARTICLES OF DIRECT INTEREST TO OUR READERS

From the *Annual Review of Earth and Planetary Sciences,* Volume 3 (1975):

Walker, Robert M. Interaction of Energetic Nuclear Particles in Space with the Lunar Surface

From the *Annual Review of Nuclear Science,* Volume 25 (1975):

Baym, Gordon Neutron Stars
Wetherill, George W. Cosmochronology of the Solar System

ARTICLES OF PERIPHERAL INTEREST TO OUR READERS

From the *Annual Review of Earth and Planetary Sciences,* Volume 3 (1975):

Bullard, Edward The Emergence of Plate Tectonics: A Personal View

From the *Annual Review of Physical Chemistry,* Volume 26 (1975):

Johnston, Harold S. Pollution of the Stratosphere

CONTENTS

Computer Simulations of Stellar Systems, *Sverre J. Aarseth and Myron Lecar* 1

Radio Surveys and Source Counts, *D. L. Jauncey* 23

Neutrino Processes in Stellar Interiors, *Zalman Barkat* 45

Thermonuclear Reaction Rates, II, *William A. Fowler, Georgeanne R. Caughlan, and Barbara A. Zimmerman* 69

Chemical Composition of Extragalactic Gaseous Nebulae, *Manuel Peimbert* 113

Ultraviolet Studies of the Interstellar Gas, *Lyman Spitzer, Jr. and Edward B. Jenkins* 133

On-Line Computers for Telescope Control and Data Handling, *Lloyd B. Robinson* 165

Young Stellar Objects and Dark Interstellar Clouds, *S. E. Strom, K. M. Strom, and G. L. Grasdalen* 187

Stellar Populations in Galaxies, *Sidney van den Bergh* 217

High-Velocity Neutral Hydrogen, *Gerrit L. Verschuur* 257

Unseen Astrometric Companions of Stars, *Peter van de Kamp* 295

Equation of State at Ultrahigh Densities, II, *V. Canuto* 335

Astrophysical Processes near Black Holes, *Douglas M. Eardley and William H. Press* 381

Instrumental Technique in X-ray Astronomy, *Laurence E. Peterson* 423

On the Pulsar Emission Mechanisms, *V. L. Ginzburg and V. V. Zheleznyakov* 511

INDEXES

Author Index 537

Subject Index 546

Cumulative Index of Contributing Authors, Volumes 9–13 554

Cumulative Index of Chapter Titles, Volumes 9–13 555

ANNUAL REVIEWS INC. is a nonprofit corporation established to promote the advancement of the sciences. Beginning in 1932 with the *Annual Review of Biochemistry,* the Company has pursued as its principal function the publication of high quality, reasonably priced Annual Review volumes. The volumes are organized by Editors and Editorial Committees who invite qualified authors to contribute critical articles reviewing significant developments within each major discipline.

Annual Reviews Inc. is administered by a Board of Directors whose members serve without compensation.

Annual Reviews are published in the following sciences: Anthropology, Astronomy and Astrophysics, Biochemistry, Biophysics and Bioengineering, Earth and Planetary Sciences, Ecology and Systematics, Entomology, Fluid Mechanics, Genetics, Materials Science, Medicine, Microbiology, Nuclear Science, Pharmacology, Physical Chemistry, Physiology, Phytopathology, Plant Physiology, Psychology, and Sociology. The *Annual Review of Energy* will begin publication in 1976. In addition, two special volumes have been published by Annual Reviews Inc.: *History of Entomology* (1973) and *The Excitement and Fascination of Science* (1965).

COMPUTER SIMULATIONS OF STELLAR SYSTEMS

Sverre J. Aarseth
Institute of Astronomy, Cambridge, England CB3 0HA

Myron Lecar
Center for Astrophysics, Cambridge, Massachusetts 02138

INTRODUCTION

The subject of experimental stellar dynamics dates back only to 1960 when von Hoerner published results of the first *N*-body calculation. This new technique can be used to obtain self-consistent solutions for a variety of dynamical problems and to test the validity of theoretical considerations. Only a few such studies were available at the time of the previous review by Michie (1964) in this series, whereas current calculations embrace a wide range of self-gravitating systems.

In the present review of numerical simulations we concentrate mainly on *collisional* systems, that is, those whose age is greater than the time scale for significant orbital changes by encounters. Apart from the classical *N*-body problem itself, which is fundamental to the subject, the main topics are galactic and globular clusters as well as clusters of galaxies and interacting galaxies. We also include violent relaxation, which is collisionless, because it usually occurs during the early stages in collisional systems. At the other end of the spectrum, three-body calculations are only included if they relate to the general cluster problem or questions of numerical accuracy. Both *N*-body and galactic cluster calculations have been reviewed recently (Aarseth 1973a, Wielen 1974); however, these discussions primarily refer to the authors' own work. Some readers of this article will probably seek guidance on numerical techniques and we therefore begin by giving a brief summary of direct integration methods; the Monte Carlo and Boltzmann moment methods are discussed in the globular cluster section.

Although the standard *N*-body approach is entirely self-consistent, the time-consuming force calculation currently limits its practical usefulness to about 500 particles. Numerical difficulties due to close two-body encounters and persistent binaries may be avoided by making use of regularization methods that remove the dominant force term from the equations of motion. A comprehensive discussion of this powerful technique is given in the proceedings of the IAU Colloquium No. 10

(Lecar 1972); for a simplified version see Heggie (1973). The colloquium volume also contains detailed descriptions of the most popular integration methods for a variety of problems. In the recent approach by Ahmad & Cohen (1973a), direct N-body calculations can be speeded up significantly by evaluating the force field due to distant and neighboring particles on different time scales. This simplification makes it practical to study several thousand particles while retaining the essential nature of the problem. As regards the integration procedure itself, fixed-order polynomial methods with individual time steps (e.g. Wielen 1967, Aarseth 1971b) are still preferable for most purposes, although a recent alternative preserves the simplicity of using equal steps for all particles at the expense of variable individual orders (Janin 1974). If desired, critical triple encounters can now be treated by an efficient regularization method (Aarseth & Zare 1974), whereas close encounters of arbitrary multiplicity can be studied by introducing additional equations for each interaction (Heggie 1974a).

THE *N*-BODY PROBLEM

Validity of the Results

Numerical integrations of the gravitational N-body problem can in principle provide unique solutions for a given set of input data. Numerical errors or slightly perturbed initial conditions, however, produce entirely different individual particle orbits after a while. The growth rate of this divergence was first obtained by Miller (1964), who showed that the e-folding time is actually shorter than the crossing time.[1] Although this result has cast a shadow over N-body calculations ever since, it has not deterred subsequent investigations from being carried out in an optimistic spirit. As emphasized by Lecar (1968), the dispersion obtained by different investigators studying the same collapsing 25-body system is considerable. A similar disagreement has been noted for equilibrium-type initial conditions and different computers (Hayli 1970a). Again, the detailed solutions of two almost identical 32-body systems continued to diverge from each other when the first integrals of the motion were controlled by a stabilization technique (Miller 1971a,b). This instability is somewhat reduced when the Newtonian attraction is modified by a cutoff (Standish 1968a), indicating that it is driven by close 2-body encounters and hence should become less severe as N increases.

The numerical experiments show that gravitational interactions are strongly coupled in the sense that successive encounters lead to nonlinear error propagation. It is nevertheless possible to perform extremely accurate integrations of small systems by means of regularization methods. One such example is the general three-body problem where essentially complete solutions of considerable complexity have been obtained to a significant number of decimal places (Szebehely & Peters 1967). The question of the reproducibility of individual orbits was studied by Allen &

[1] The standard crossing time is defined as $t_{cr} = 2R/V = GM^{5/2}\ (-2E)^{-3/2}$, where R, V, M, and E are the virial theorem scale factor, rms velocity, total mass, and binding energy, respectively.

Poveda (1971) for systems with six particles initially at rest. Strong interactions often led to the ejection of one or two very energetic particles that were reproduced for a wide range of accuracies. Time reversal tests were most satisfactory for the high-velocity particles, thereby confirming the general validity of the escape phenomenon. Macroscopic quantities in such systems are not biased by truncation errors when the total energy is preserved to better than about 1%; however, large round-off errors do bias the results (Smith 1974). Hence in view of the intrinsic dispersion of results, high-accuracy integrations have no practical advantage when considering the system as a whole. It is also reassuring that the distributions of lifetimes, escape energies, and other quantities in triple systems are statistically similar for different accuracies (Valtonen 1974).

The results of single large calculations appear to be more reliable than indicated by earlier comparisons. Recent N-body calculations of three equal-mass Plummer models with $N = 250$ show similar time dependence for the radial structure (Aarseth, Hénon & Wielen 1974). These results are in general agreement with the Monte Carlo methods, confirming the correctness of the classical theory based on two-body encounters. In the following discussion we therefore put the intrinsic difficulties of the N-body problem on one side and examine the relationship between theory and experiment.

Relaxation Time

The basic theory of stellar dynamics covers many aspects that may be tested numerically. In the first place the nature of the particle motions are changed by encounters or departures from overall equilibrium. The relaxation time,[2] which is based on two-body encounters, assumes that consecutive changes in the orbital parameters are both small and independent, enabling the total contribution to t_{rh} to be obtained by a sum of squares. Because of difficulties in defining an unambiguous numerical procedure for the energy relaxation time, it is more convenient to compare the equivalent deflection time based on changes in the angular momentum. A stationary system of field particles may be used for this purpose; hence the N-dependence can be explored over a wide range. However, this procedure neglects "polarization" effects and such results may therefore not have complete dynamical validity. In the experiments of Standish & Aksnes (1969) about 100 orbits were integrated for selected values of $N \leqq 2500$, giving good agreement with the two-body relaxation formula. Similar calculations with pair correlations show that even well-separated pairs behave as single particles (Butterworth & Miller 1971). Finally, the predicted velocity dependence in small systems has been verified (Lecar & Cruz-Gonzalez 1971); these experiments also confirm the validity of obtaining the total orbital deflection by considering individual contributions as being statistically independent.

Numerical estimates of the energy relaxation time are less straightforward because the classical theory of Chandrasekhar (1942) assumes homogeneous systems. In general the binding energy of a particle is subject to both random and systematic

[2] The mean relaxation time is $t_{rh} = 0.033\, N t_{cr}/\ln(0.4\, N)$ (Spitzer & Hart 1971a).

variations, hence relaxation time estimates based on discrete changes depend somewhat on the adopted interval. Nevertheless, such determinations show approximate agreement with theory (Aarseth 1966, Wielen 1967). Some of these results are based on unequal masses that introduce additional complications because the standard prediction assumes equipartition of kinetic energy, whereas the numerical models exhibit mass segregation. Consequently, the light particles have longer relaxation times because their orbits are predominantly in low-density regions. An alternative procedure of relaxation time evaluation makes use of the analytical expression for the time derivative of individual binding energies (Wielen 1967). This expression also provides information about the relative importance of distant and close encounters. Analysis of an equal-mass system with $N = 250$ shows that in most cases the few nearest neighbors provide the dominant contribution (S. J. Aarseth 1970, unpublished).

The relaxation time concept may be approached more directly by considering the probability distribution of the random force field that modifies the particle motions. The classical stochastic description of Chandrasekhar & von Neumann (1942) has now been verified by numerical experiments with 1000 particles (Ahmad & Cohen 1973b). Further experiments show excellent agreement with theory for the probability distribution of the force derivative and dynamical friction, whereas the steeper time dependence of the force autocorrelation function may be due to the finite model (Ahmad & Cohen 1972, 1974a,b). According to a subsequent analytical calculation for large times and finite systems, the autocorrelation function decreases as $1/t^5$ in comparison with the classical $1/t$ dependence (Cohen 1974).

Equipartition and Mass Segregation

The modern theory of collisional relaxation shows directly that the two dominant terms are due to a fluctuating force field and a polarization drag (Gilbert 1968), confirming the fundamental analysis by Chandrasekhar (1942). These processes are competing because the former gives rise to a statistical acceleration independent of the mass, whereas the latter induces a retardation proportional to the mass. Consequently, the concept of dynamical friction implies mass segregation as well as a slower rate of populating the tail of the velocity distribution.

The application of the equipartition principle (cf Chandrasekhar 1942) has had unfortunate consequences for cluster dynamics. This assumption may readily be tested; computer studies have demonstrated that it does not apply in its simplest form (Aarseth 1966, Wielen 1967, van Albada 1968a). Different masses tend to occupy different volumes, however, and some degree of equipartition is consistent with a smaller rms velocity for all light particles if velocity comparisons are made locally. At the same time, equipartition at the center cannot be achieved beyond a mass ratio of about 4, corresponding to the velocity of escape (Aarseth 1973b).

Two-body encounters also lead naturally to mass segregation. This process is illustrated in a series of experiments with $N = 5$ and unequal masses (Baranov 1970). Moreover, the velocity distribution is isotropic in the central region and the velocity ellipsoid becomes elongated in the radial direction at greater distances, in qualitative agreement with results for larger systems (Aarseth 1968). In this connection we note that only the theoretical cluster model of Michie (1963a) emphasizes the velocity anisotropy. The largest N-body systems also exhibit significant mass segregation after a few crossing times, irrespective of whether an initial collapse takes place. Typically the average mass in spherical shells decreases by a factor of 3 from the innermost to the outermost 10% mass shell for $N = 500$ (Aarseth 1974a). A theoretical discussion by Spitzer (1969) considers the subsequent evolution of heavy particles that are assumed to be centrally concentrated. If their total mass exceeds a critical value, the corresponding mutual attraction requires such a large velocity dispersion that equipartition cannot occur. Instead, some heavy particles continue to lose kinetic energy and form a contracting core at the center, in qualitative agreement with the N-body results.

Binaries

The earliest calculations revealed the presence of temporary binaries (von Hoerner 1960, Aarseth 1963) but their full dynamical importance was not recognized for some time because of numerical difficulties. The continual loss of strongly bound particles from the core inevitably leads to the appearance of a close binary at the end of the collapse phase.[3] Typical time scales for $N = 250$ range from $t/t_{cr} \approx 40$ for equal masses to $t/t_{cr} \approx 5$ for a smooth mass spectrum (Aarseth 1971a). Both capture and exchange of companions play an important role in the subsequent evolution, which proceeds in the direction of increased mass (where appropriate) and energy; when the masses are unequal, the latter rapidly exceeds 50% of the total cluster energy (i.e. $\varepsilon > 0.5\ N$) even for $N = 500$. Most of the energy absorbed by the central binary goes into cluster expansion, which slows down the evolution rate. The shrinking cross section makes further interactions even less likely but high-velocity escapers are produced when they do occur (Aarseth 1973b). As first noted for $N \leqq 24$ (van Albada 1968a), stable hierarchical triples develop when the core becomes sufficiently depleted (Aarseth 1974a). The outer component absorbs energy from the field particles until its pericenter distance shrinks below the stability limit (cf Harrington 1972), whereupon the energy is transferred to the inner binary. Strong recoil effects may lead to escape in equal-mass systems, whereas the increased kinetic energy of a perturbed binary is converted into random motions by two-body encounters that expand the inner core.

Considering the dynamical importance of binaries in the computer models, it is surprising that this phenomenon was overlooked in earlier theoretical discussions. In the classical theory of binary formation by triple encounters, the third particle carries away the energy required to bind the two other components. The average hyperbolic excess energy increases with N, and binary formation at fixed ε therefore becomes less likely in large systems. Thus for a homogeneous system the number of *close* binaries formed per relaxation time is $\propto 1/N$ (Spitzer & Hart 1971a). Numerical three-body experiments show that the formation rate of *wide* binaries, due to

[3] In analogy with the close encounter definition $r \leqq 2G\bar{m}/\bar{v}^2$, we define the scaled binding energy of a close binary with mass components m_k, m_l and semimajor axis a by the condition $\varepsilon \equiv Gm_k m_l/\bar{m}\bar{v}^2 a \geqq 1$.

encounters with impact parameters below the mean particle separation, is approximately constant *per crossing time* for globular cluster conditions, in excellent agreement with new theoretical predictions (S. J. Aarseth and D. C. Heggie, in preparation). However, most of the wide binaries are not sufficiently energetic to survive beyond one crossing time.

The recent theoretical analysis by Heggie (in preparation) has clarified many important aspects of the binary problem. Detailed considerations of the formation rate indicate that the first stable binary ($\varepsilon \geqq 1.3$) in a uniform sphere with equal masses should occur on a time scale $t_b \approx 0.04\, N \log_{10} (0.4\ N)\ t_{rh}$ (Heggie 1974c). The actual formation rate is closely linked to the cluster structure, however, and it is now concluded that such binaries would form during the core collapse. In addition to mass segregation effects, the cross section for binary formation, "exchange," and "resonance" (i.e. disruption of bound triples) accounts for the preferential combination of heavy particles. Alternatively, the rapid evolution toward the most dynamically favorable configuration can be understood in terms of phase space arguments (Miller 1974).

The theory of binary evolution treats close and distant encounters as well as exchange events and distinguishes between "soft" and "hard" binaries defined by $\varepsilon < 1$ and $\varepsilon > 1$ (Heggie 1974b). Soft binaries reach an equilibrium distribution rather quickly but their lifetimes are short, whereas hard binaries survive and tend to become more energetic by "superelastic" encounters (cf Thüring 1968). This "watershed" effect has been confirmed by three-body calculations where the velocity of the third particle was selected from an appropriate Maxwellian (J. G. Hills, in preparation). The numerical evidence supports the prediction that the average rate of energy gain for hard binaries is approximately independent of ε, and the notion that single events provide the dominant contribution is also well established (Aarseth 1971a). To summarize, on the basis of numerical results and analytical considerations we anticipate that close binaries will become important in large systems as the core collapse phase is approached. The contracting core itself acts as an energy sink, hence only the last stages of the large N calculations may need modification.

We finally consider the possibility of initial binaries that cannot be ruled out from observations. Inspired by the fragmentation theory, Aarseth & Hills (1972) investigated a hierarchical configuration with 120 unequal-mass particles initially at rest inside eight subclusters. The subclusters lost their identity during the collapse but 5 surviving binaries became very energetic and prevented the core from shrinking. An equal-mass system with 10 initially close binaries has been studied recently (Aarseth 1974b). The early phase is characterized by mass segregation, indicating that a modest initial binary population may eventually dominate the central region if their average mass exceeds that of the single particles. One of these binaries is destroyed by collision but most of the others become more energetic, increasing the total energy stored in binaries, and the collapse phase is again halted by superelastic relaxation. Theoretical considerations suggest that internal energy changes of binaries near the most effective energy $\varepsilon \approx 10$ have a similar influence to the single particle relaxation if their number densities are comparable (Heggie 1974c). Collisions between binaries and elastic encounters with single particles have not been treated yet; this complex problem poses an exciting challenge for future work.

Close Encounters and Escape

Stellar systems lose their discreteness with increasing particle number. Close encounters may therefore be neglected at some stage, justifying a simpler treatment based on weak two-body encounters. According to the classical theory of escape, close and distant encounters contribute in the proportion $1/4.9 \ln (0.4\,N)$ (Spitzer & Thuan 1972). This ratio is comfortably small for globular clusters but it is still < 0.05 for $N = 250$ where close encounters dominate the escape rate even for equal masses (Aarseth 1974a, Wielen 1974). The presence of a general mass spectrum enhances the relative importance of close encounters because different particles contribute according to m^2 at least (Hénon 1969). Moreover, the high-density core can for some purposes be considered a nearly isolated subsystem. If associated with the encounter region, this implies that theoretical estimates of the escape energy should be based on a smaller effective particle number. Notwithstanding the possibility of stellar collisions and mass loss effects under realistic conditions, it appears likely that close encounters also play a significant role in the long-term evolution of relatively large systems.

Numerical integrations permit the escape phenomenon to be studied in considerable detail. Five main processes have been classified by van Albada (1968a) on the basis of calculations with $N = 10$ and $N = 24$: 1. two-body, triple, and multiple encounters; 2. encounters between binaries and field particles; 3. ejection from a bound triple system; 4. exchange of one binary component; and 5. small energy gains by halo orbits during central passages. Furthermore, initial mass motion effects (i.e. "violent relaxation") may also lead to a significant proportion of escapers (Standish 1968b). Analysis of the escaping particles in an unequal-mass system with $N = 500$ confirms all these processes (Aarseth 1974a); in addition, a few weakly bound halo orbits become unbound as the total mass decreases. The precise frequency distribution is not known but discrete events again account for the majority of the escapers.

Most predictions of the escape rate assume weak two-body encounters and are therefore not appropriate to the systems discussed here (cf Wielen 1974). An alternative description based on treating escape as a discontinuous random process has been adopted by Agekyan (1959) and Hénon (1960), who derived absolute escape rates due to single encounters. Subsequently the latter treatment was extended to include interactions between unequal masses in a formally consistent way (Hénon 1969). Even this simplified theory does not consider velocity anisotropy, mass segregation, and multiple encounters; furthermore, it has been derived only for the Plummer model. For comparison purposes we consider the relative escape rate per crossing time for a group of N_{i} particles, $L_{\mathrm{i}} = \Delta N_{\mathrm{i}} t_{\mathrm{cr}} / N_{\mathrm{i}}\, t$. Application to equal-mass Plummer models (which become more centrally concentrated later) with $N = 250$ give $L_1 = 5 \times 10^{-4}$ (Wielen 1974), 1.1×10^{-3}, and 9×10^{-4} (Aarseth 1974a), whereas $L_1 = 7 \times 10^{-4}$ for a similar model that developed from initial collapse conditions (Aarseth 1973a). These escape rates are significantly greater than the prediction by Hénon, that is, $L_1 = 1 \times 10^{-4}$. If anything, the discrepancy is greater than for systems with $N \leqq 25$ (von Hoerner 1963, Aleshin & Kaplan 1970, Hayli 1970b). The binding energies prior to escape frequently exceed the mean kinetic energy of field particles, whereas in the Monte Carlo methods escape can take place only for weakly bound orbits. This feature is essentially consistent with theory, however, because the predicted rms energy change for a central passage decreases as $1/(N)^{1/2}$ (see Spitzer 1974).

The introduction of a mass spectrum significantly increases the total escape rate (Wielen 1968). This feature is also reflected in the theoretical prediction which is now deficient by less than a factor of 2 for Plummer models in the range $N = 50$–500 (Wielen 1974). The corresponding relative escape rate is approximately independent of mass for all but the heaviest particles, which escape less frequently. The same conclusion applies to other small and large systems (van Albada 1968a, Aarseth 1974a), even though these results are influenced by close binaries that tend to discriminate against preferential mass effects. It is characteristic of such systems that an early active phase of escaper production is followed by a gradual slowing down (Aarseth 1973a); hence the agreement to within 30% for $N = 500$ is somewhat fortuitous. In contrast to the numerical results, the predicted escape probability increases slowly with decreasing mass. Given that the light particles spend most of their time outside the main escape region, however, it seems likely that the predicted mass dependence would be reduced if segregation were included.

The theory of Hénon also provides an expression for the mean energy of escaping particles. This prediction compares well with the experimental values for both equal and unequal masses when a few large contributions due to binary interactions are excluded (Aarseth 1974a). We may therefore conclude that, apart from critical encounters involving binaries, the discrete escape mechanism is essentially correct for small systems. Considering that the predicted escape rate depends on the cluster structure as well as the velocity distribution, the present underestimate for equal masses should not be interpreted as a decisive disagreement with the experiments.

In view of the detailed information provided by the N-body experiments, it is desirable to have a more comprehensive theory of stellar encounters. One promising approach is that of Kaliberda & Petrovskaya (1970, 1971), who have used the Kolmogorov-Feller equations to describe individual velocity variations as a purely discontinuous random process, taking into account the preferential dilution of multiple encounters proposed by Agekyan (1961). Although this treatment gives a more realistic description of the interactions, the solutions for the cluster structure are not yet self-consistent and numerical comparisons have therefore been omitted. It is hoped that further developments along these lines will establish the region of validity for methods based on the Fokker-Planck equation.

GALACTIC CLUSTERS

The galactic cluster problem is sufficiently well defined to form a natural extension of N-body simulations. Additional complications for a realistic approach are mainly due to the galactic tidal field, passing clouds, and mass loss from evolving stars. These effects are dissipative and, except for the clouds, may be included at little extra expense, thereby speeding up the calculations. Some aspects may also be studied by simple models. Allen & Poveda (1974) investigated the dynamical stability of trapezium systems by isolated N-body calculations, using newly determined initial conditions that do not support the expansion hypothesis. Although the integrations covered about 30 crossing times in the 10^6 yr available to the OB stars, two thirds of these unequal-mass systems survived as wide trapezia. Interpreting the Orion Trapezium as the most massive members of a cluster, Rybakov et al (1971) found that the addition of an observationally based restoring force can confine these stars within the cluster during 10^6 yr at least. Note that only the first model predicts an early formation of massive binaries. The limitations of mass point simulations of young compact groups were discussed by van Albada (1968b).

As emphasized in earlier theoretical discussions, the tidal field limits the spatial extent of dynamically old star clusters (von Hoerner 1957, King 1962). Appropriate force gradients for circular cluster orbits near the plane of symmetry can readily be included in the equations of motion (Aarseth 1967, Hayli 1967). This lowers the escape barrier and increases the disruption rate (Wielen 1968, Hayli 1970b). Because the vertical gradient acts as a restoring force, the probability of escape now depends on the direction of ejection as well as the energy. In order for escape to occur near the critical energy defined by the total mass and tidal radius, the motion must be within a small angle of the galactic plane; such stars leave the cluster via the equilibrium points (Hayli 1970b). On the other hand, high-energy stars with a significant vertical velocity remain trapped inside the partially closed zero velocity surface until they are deflected through the escape "window" (Aarseth 1973a). The tidal field also produces a significant flattening (Aarseth 1967, Hayli 1968); note, however, that the effect of rapid rotation is qualitatively similar (Aarseth 1969).

There are strong observational and theoretical reasons for including mass loss effects. In view of the uncertain status of this subject, only instantaneous supernova events have been considered (Wielen 1968, Aarseth 1973a). This simple model provides an interesting competition because massive binaries may develop at the center before the mass loss takes place. Both the total mass and cluster radius must now be specified in physical units and it becomes meaningful to discuss actual lifetimes. Whereas there is still some uncertainty about the precise definition of escape, alternative model calculations with 250 stars and the Salpeter mass function give typical half-lives of 1×10^8 yr (Wielen 1971a) or 2×10^8 yr (Aarseth 1973a), irrespective of whether mass loss is included. Although about half the observed galactic clusters disintegrate within 2×10^8 yr (Wielen 1971b), the present simulations cannot explain the existence of the oldest 10%. Many clusters have considerably more than 250 stars even now, however, and a corresponding scaling of the results shows that this would prolong the lifetime (Wielen 1971a). Furthermore, the parameter dependence is such that below a median radius of about 2 pc the reduced tidal effect is more than compensated for by the internal relaxation. There are some indications that old clusters also have favorable scale factors (van den Bergh & Sher 1960).

The cluster simulations are now approaching a suitable state for observational confrontation. It is very encouraging that mass segregation has been established observationally. Some examples are Pleiades (Limber 1962, Kholopov 1971), Hyades (Oort, Pels & Pels-Kluyver 1974), NGC 188, and M67 (van den Bergh & Sher 1960). There is some evidence for velocity anisotropy in the Pleiades where the velocity dispersion is approximately independent of the mass (Jones 1970a,b), and these features are consistent with the numerical results. Star counts by van den Bergh & Sher (1960) reveal a significant lack of faint members with respect to the standard luminosity function, in qualitative agreement with an extensive analysis of the nearby Hyades cluster (van Altena 1966). This deficiency may be caused by the preferential escape of low-mass stars predicted by dynamical theories (Spitzer & Härm 1958, Michie 1963b, Hénon 1969, Kaliberda & Petrovskaya 1971, King 1971), although Michie (1963c) had reservations about this. In contrast, the computer models show a constant escape rate for all but the heaviest stars (Aarseth & Woolf 1972). This problem was studied theoretically by Prata (1971a,b), who used a sequence of King's (1966) equilibrium models and included tidal effects, mass loss, and passing clouds. Allowing for preferential escape, the deficiency of low-mass stars in M67 could be explained only if the theoretical escape rates were increased by a factor of 5.

For technical reasons, the effect of HI clouds has not yet been included in the numerical cluster simulations and this could influence the results. A start on the solution to this problem was made by Bouvier & Janin (1970b), who studied the rate of energy increase in 25-body systems interacting with representative clouds. The corresponding disruption time is about four times shorter than predicted by Spitzer (1958). The theoretical time scale was later revised downwards by a factor of 4/9 (Bouvier 1971), whereas criticism by Hénon (1972) led to a reduction of the computed cloud disruption time from 4.5×10^8 to 3×10^8 yr (Bouvier 1972). Scaling of the results to 25 stars (cf Wielen 1971a) indicates that the tidal disruption time would be considerably shorter, in qualitative agreement with analytical considerations (Bouvier 1971). However, this conclusion is tied to the adopted cloud parameters, which are uncertain. The analogous problem of field binaries perturbed by passing stars was studied by Cruz-Gonzalez & Poveda (1971). Their dissolution times for "soft" binaries (cf Heggie 1974b) are in substantial agreement with the theory of Oort (1950) when Hénon's criticism is taken into account.

In conclusion, future simulations of galactic clusters, which should be extended to richer systems, require better descriptions of mass loss phenomena in evolving stars and improved cloud parameters. The neglect of a significant gas component can be justified theoretically (J. G. Hills 1973, personal communication). Likewise, the results are probably insensitive to the assumptions of simultaneous star formation and no initial mass segregation. Comparisons with the observations show substantial agreement on lifetimes. Computer models with 250 stars do not support the theoretical prediction of preferential escape, however. The most likely explanation for the relative deficiency of faint cluster stars is that either their initial distributions depart significantly from the Salpeter function, or tidal shocks from passing clouds (alternatively, galactic density waves) are mainly responsible for removing the halo stars.

GLOBULAR CLUSTERS

A very readable introduction to the observational and theoretical description of globular clusters (with pictures of clusters and projected density profiles) is given by King (1971). Illingworth (1974), with the aid of high-dispersion Coudé spectra, obtained velocity dispersions for a number of clusters that range from 10 to 20 km/sec (rms). The clusters had masses in the range 10^5–10^6 $M_\odot$, projected radii (containing one half the mass) of 2–3 pc, and M/L of 1.0–1.5. The qualitative evolution of globular clusters was predicted by Hénon (1961, 1965), who suggested that the central density would increase without limit, although the mass of the core would shrink to zero. Thus, in time, the core would have all the binding energy, whereas the halo would have all the mass. Further theoretical verification of the collapsing core came from Lynden-Bell & Wood (1968), who showed that a self-gravitating isothermal sphere (surrounded by a reflecting wall) is unstable to collapse if the density contrast exceeds 709. Because of the "negative specific heat" of self-gravitating systems, as the core collapses, its kinetic energy increases and therefore it has more heat (i.e. kinetic energy) to give away. The temperature of the core increases without limit, and the core runs away in a "gravothermal catastrophe."

Larson (1970a,b) developed a method of treating the Fokker-Planck equation by considering only the first four moments of the velocity distribution. His truncated moment equations provide a "fluid-dynamical" modeling of the Fokker-Planck equation. He immediately verified the gravothermal catastrophe. The central density became infinite in a finite time, t_0, which was about 16 t_{rh} [using Spitzer & Hart's (1971a) definition of relaxation time, t_{rh}, for the inner half mass]. Letting $\tau = t_0 - t$, the central density and velocity evolved as $\rho_c \propto \tau^{-1.34}$, $v_c^2 \propto \tau^{-0.22}$. The isothermal length scale, κ, scales as $\kappa^2 \propto v_c^2/\rho_c$, giving $\kappa^2 \propto \tau^{1.12}$. We may take for the core mass, $M_c \propto \rho_c \kappa^3 \propto \tau^{0.34}$. As τ goes to zero, M_c also goes to zero as predicted by Hénon. This dependence on τ is typical of several models treated by Larson, but t_0 depends on the initial conditions. Talpaert & Lefèvre (1972) extended the fluid-dynamical method to higher orders and replaced the Maxwellian assumption by a novel closure. Unfortunately, these equations are more prone to numerical instabilities.

Hénon (1967a) showed that the Fokker-Planck equation could be modeled by a Monte Carlo technique. For a homogeneous system the Monte Carlo method gave excellent agreement with exact solutions of the Fokker-Planck equation, and the velocity distribution evolved to a Maxwellian. This technique was later extended to globular cluster models (Hénon 1971a,b). As predicted, the core mass decreased to zero while the core absorbed nearly all the binding energy. In the core, the velocity distribution was isotropic. Progressing outward, the velocities decreased and developed a radial bias, giving $v_t^2 \propto r^{-2}$ (constant angular momentum), $v_r^2 \propto r^{-0.6}$. For systems with a mass distribution the evolution was qualitatively similar, but faster by a factor of about 4. As in the N-body calculations, the heavier masses segregated toward the center.

Meanwhile, Spitzer and his collaborators (Spitzer & Hart 1971a,b, Spitzer & Shapiro 1972, Spitzer & Thuan 1972) developed a somewhat different Monte Carlo technique. They found a density distribution corresponding to an isothermal sphere in the core, and a power law $r^{-3.5}$ in the halo. The velocity distribution in the core was given, to a good approximation, by a lowered Maxwellian, $f(E) \propto \exp(-\beta E) - 1$. The principal difference between the two methods is that Spitzer et al integrate the individual orbits of the test stars, whereas Hénon does not. There are also differences in the way the diffusion coefficients are calculated. Thus in Spitzer's formulation, finite velocity increments can result in the escape of halo stars, whereas in Hénon's treatment the stars diffuse toward zero energy but do not escape. Excellent descriptions and comparisons of the two methods are given by Hénon (1973) and Spitzer (1974). At first, the two procedures gave evolution rates that differed by a factor of 8. Comparison of the results led Hénon to introduce a variable time step that treated the core evolution more accurately, and Spitzer modified his diffusion coefficients. Their results are now in excellent agreement, which makes us confident that both methods are accurately modeling the Fokker-Planck equation. It is also encouraging that there is substantial agreement with the fluid-dynamical method on core collapse times.

The salient question is how well the Fokker-Planck equation models globular clusters. The main assumptions of the Fokker-Planck equation are that the one-particle distribution function suffices to describe the system (i.e. correlations, including energetic binaries, are neglected) and that two-body encounters dominate the relaxation. Some questions were answered by a comparison between N-body integrations of 100 and 250 bodies by Aarseth and Wielen, and Monte Carlo simulations by Hénon and Spitzer (Aarseth, Hénon & Wielen 1974). The main results of this comparison are presented in Figure 1, which shows the evolution of the radii containing 10, 50, and 90% of the mass. The agreement is almost better than could be expected. In particular, the time for infinite central density lies between 15 and 20 t_{rh} for all methods. Hénon (1974) also reexamined the relaxation time and included nondominant terms of relative size $1/\ln N$. The coefficient γ in $t_{rh} \propto N/\ln(\gamma N)$ is now reduced, giving $\gamma = 0.15$ for a system with equal masses and $\gamma = 0.06$ for a modest mass spectrum. With this correction to the time scale, the agreement between the Monte Carlo and N-body results is further improved. However, the agreement between the N-body experiments and the Monte Carlo calculations is not complete in the collapsing core. A close binary dominates the core during the final stages of the N-body calculations and, in addition to providing an energy sink, it has important dynamical effects. In particular, the binary prevents the complete collapse of the core by clearing out a space around itself, sometimes ejecting particles of very high energy. Furthermore, the importance of close encounters should decrease as $\ln(\gamma N)$, but as $\ln(10^5)/\ln(500)$ is only a factor of 2, that in itself would not account for the predominance of close-encounter escapes in the N-body experiments, as pointed out by Wielen (1974). On the other hand, Spitzer (1974) emphasized that even the classical theory predicts a significant escape energy when extrapolated to small N. Heggie (1974c) showed that the existence of initial binaries could influence the time scale of evolution. In particular, binaries observed in the field have about the most effective degree of "hardness" to be dynamically important in cluster evolution.

An additional dynamical effect is the tidal shock a cluster experiences if it passes rapidly through the galactic disk. Ostriker, Spitzer & Chevalier (1972) and Spitzer & Chevalier (1973) have considered the heating effect of "compressive shocks." The evolution of the core is affected only in extreme cases, but the shocks are efficient in enhancing the evaporation of the halo stars. As yet, the cluster simulations do not include the tidal field itself in the equations of motion; this effect is currently approximated by a radial cutoff. Recent calculations for a mass spectrum

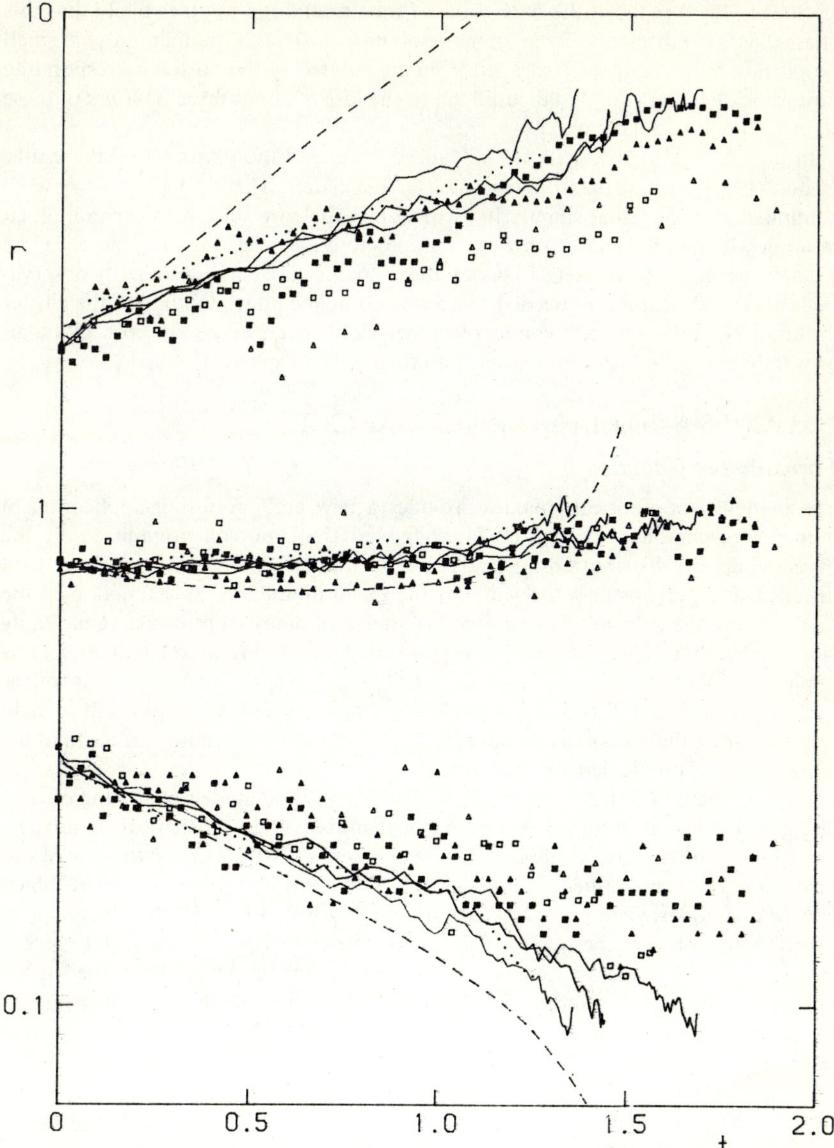

Figure 1 Radii containing 10, 50, and 90% of the total mass, plotted vs time ($t_{rh} = 0.09$) for a cluster with stars of equal masses. The data correspond to different methods: open triangles and squares (*N*-body, Wielen); filled triangles and squares (*N*-body, Aarseth); full lines (Monte Carlo, Hénon); dotted lines (Monte Carlo, Spitzer); dashed lines (fluid-dynamical, Larson). Note that Larson's model represents only 100 stars, whereas $N = 250$ in three of the *N*-body models.

(cf Spitzer 1974) reduced the core collapse time to about 3 t_{rh}; it is likely that this phase has already occurred in many globular clusters. Although only a small proportion of the light stars escaped from an isolated system in the corresponding time, shock heating and a tidal cutoff led to the loss of about three quarters of these stars (i.e. preferential escape).

In an attempt to reconcile the advanced core evolution with N-body results, Hénon (1974) introduced an energy sink at the center. The models could now be continued past the usual singularity, whereupon the core started to expand, albeit more slowly than the halo. Contrary to the N-body systems, however, there was no secondary collapse (cf Aarseth 1974a). It therefore appears that high-quality observations of central regions are needed to achieve complete understanding of the cluster evolution. Finally, one still wonders whether globular clusters were born globular, or whether initially they were flatter than their present spherical shape.

GALAXIES AND CLUSTERS OF GALAXIES

Formation of Galaxies

The modeling of elliptical galaxies provides a new and promising application of numerical techniques. Larson (1974) suggested that one can attempt to fit the observed density distributions, the variation of ellipticity and metal abundance with distance, and perhaps also the value of the metal abundance as a function of the total mass of the galaxies. The surface brightness of many ellipticals is remarkably close to Hubble's law, that is, $I/I_0 = (1+r/a)^{-2}$. Assuming a constant M/L, this implies the same law for the surface density, and a space density falling off as $1/r^3$ for $r \gg a$. Free fall collapse models (e.g. Larson 1969) typically result in halo space densities that fall off as the inverse 3.5–4.0 power of the radius rather than the more shallow Hubble density gradient.

Larson (1969, 1974) has computed a variety of two-fluid models consisting of stars and gas. The gas is assumed to be turbulent initially, and the critical parameters determining the density profile are the rate of gaseous dissipation and the rate of star formation. For some values of these parameters Larson was able to match Hubble's law. Many models yield a heavy element abundance that falls off by a factor of 10 over the inner 1 kpc; the halo distribution is more sensitive to model parameters. For a suitable choice of parameters he obtained an almost perfect fit to the surface photometry of NGC 3379 over a region of the galaxy in which the space density falls to 10^{-5} of its central value.

An orthogonal approach was adopted by Gott (1973, 1974), who assumed that elliptical galaxies can be modeled by a purely stellar system. There is no dissipation in his model and the relaxation is collisionless (i.e. due to variations in the mean field). For isolated galaxies he also obtains steep density gradients in the halo, where $\rho \propto r^{-3.5}$. He argues that the halo is largely populated by stars that initially partake in the cosmological expansion but are nevertheless weakly bound to the galaxy. As they fall in, they form a shallow halo with $\rho \propto r^{-2.8}$. Subsequently Gott simulated the scenario envisaged by Peebles (1971) where elliptical galaxies are initially nonrotating and acquire angular momentum by tidal torques from neighboring galaxies. Again including infall, he obtained an excellent fit to the elliptical galaxy NGC 4697, which obeys Hubble's law.

Thus at present we have the model by Larson of an initially *rotating gaseous* cloud, and the model by Gott of an initially *nonrotating stellar* cloud, both predicting density distributions that fit the observed surface photometry of elliptical galaxies. On the basis of stability arguments (including N-body calculations) and observed rotation curves, Ostriker & Peebles (1973) have suggested that galaxies may have even more extended halos. They propose an isothermal sphere that has $\rho \propto r^{-2}$ for large r. As yet, no numerical simulations have produced such a shallow density profile. It will be interesting to see whether a combination of Gott's cosmological infall and Larson's turbulent gas can yield an isothermal sphere.

Clusters of Galaxies

Among the earliest numerical simulations of self-gravitating systems were Aarseth's (1963, 1966, 1969) studies of the dynamical evolution of clusters of galaxies. A softened potential varying as $(r^2+\varepsilon^2)^{-1/2}$ was adopted to account for the appreciable size of galaxies relative to their mean separation. Clusters of 25, 50, and 100 galaxies were integrated for up to 20 crossing times. Even with the cutoff, this implied a considerable number of relaxation times and the influence of two-body encounters (mass segregation and binary formation) was therefore exaggerated. In the last study of the series a rotating cluster was modeled. The rotation induced flattening that persisted throughout the integration.

By now a number of rich clusters have been studied observationally, and the data seem sufficiently reliable to justify more intensive simulations. The density and velocity profiles of the Coma cluster have been determined by Rood et al (1972) out to 6 Mpc from the center (assuming $H = 50$ km/sec Mpc^{-1}). Bahcall (1973) finds that the projected density in the innermost 2 Mpc fits the surface density of an isothermal sphere with a core radius of 0.21–0.25 Mpc. According to Oemler (1974), the falloff in density is somewhat steeper, but with a pronounced minimum between 1 and 2 Mpc. Typical values for the virial mass inside 4 Mpc are $3–5 \times 10^{15}\ M_{\odot}$. The rms velocity is estimated at 1460 ± 30 km/sec and M/L falls in the probable range of 150–200. As an M/L value of 30 is typically assigned to an elliptical galaxy, this implies that about 80–85% of the mass is "hidden." There is evidence for mass segregation in "spiral-poor" clusters but not in "spiral-rich" clusters (Oemler 1974), indicating that collisional relaxation has been marginally important. This is approximately consistent with theory, because the rich clusters have probably undergone less than one relaxation time. Note, however, that the dynamical interpretation may be different for the "expanding-collapsing" cluster theory of Gunn & Gott (1972) (see Oemler 1974).

Peebles (1970) modeled the Coma cluster by integrating 300 equal-mass bodies over four crossing times, or about two mean relaxation times. Initially the galaxies were at rest or partaking in the cosmological expansion. The qualitative results for six models with different initial conditions (including one model with a highly asymmetrical density distribution) were similar and fitted the observational data reasonably well. In all cases, the final core resembled an isothermal sphere. The space density in the halo fell off with the inverse third to fourth power of the radius, perhaps somewhat more steeply than is implied by the observed projected density.

An ingenious scheme for building clusters of galaxies from already formed galaxies has been simulated by Press & Schechter (1974). Their model contained 1000 equal-mass objects uniformly distributed in an expanding universe. Gravitational clustering took place, and when a region reached 10 times the average density, they combined the objects in that region into a single object. After a few cycles, a mass spectrum developed that thereafter preserved its functional form. Below a cutoff mass (which increased with time), the probability of a given mass was proportional to the inverse 1.3–1.5 power of that mass, whereas the spectrum decreased exponentially above the cutoff mass. Depending on the fluctuations in the initial uniform density, the cutoff mass increased with the 1.5–2.0 power of the radius of the universe. Their mass spectrum fits reasonably well to the observed luminosity function of the Coma cluster.

To aid the interpretation of observations of groups and small clusters of galaxies, Aarseth & Saslaw (1972) evolved ensembles of small systems ($N \leqq 32$). They estimated the errors introduced in the mass determination using the virial theorem, due to projection effects, instantaneous departures from dynamical equilibrium, and most important, from selecting an incomplete sample of the most luminous galaxies for redshift determination. These errors caused the simulated observer to underestimate the mass by as much as a factor of 2. They also modeled initially unbound clusters and bound clusters that gradually became unbound by mass loss due to gravitational radiation. In both cases there is an evolution toward a flat density distribution, in contrast to the core-halo structure that develops in bound systems. Dearborn (1973), using 100-body systems, also simulated mass loss by gravitational radiation. He concluded that the flattened density profiles (after significant mass loss had taken place) were consistent with the observed data on the Virgo cluster.

It is somewhat surprising that no simulations have explicitly modeled the hidden mass. Lecar (1974) has suggested that the visible mass in luminous galaxies may be more centrally concentrated than the hidden mass, which is assumed to be intergalactic. If the galaxies are formed as isothermal spheres with low luminosity in extended halos (as suggested by Ostriker & Peebles 1973), reasonable estimates for the initial mass and radius are $5 \times 10^{12}\ M_\odot$ and 0.5 Mpc. If Coma initially contained 800 galaxies with the present density distribution, they would just be touching at about 2 Mpc from the center, and would be overlapping farther in. This suggests that most of the galaxies would have their halos torn off by tidal interactions with neighbors. Furthermore, the remaining cores would spiral into the center because of dynamical friction with the intergalactic matter. Assuming galaxy cores of $10^{12}\ M_\odot$, and taking a rms velocity of 1500 km/sec, the galaxies interior to 1 Mpc would halve their distance from the center of the cluster in 10^{10} yr.

Colliding Galaxies

Investigations of colliding galaxies date back to the work of Spitzer & Baade (1951). More recently, Alladin (1965) and his collaborators (Sastry & Alladin 1970, Alladin, Sastry & Ballabh 1974) conducted extensive calculations of collisions between spherical galaxies based on the impulsive approximation. Although the probability of binary formation by tidal capture is found to be small in rich clusters, the translational kinetic energy of galaxies may be significantly decreased by inelastic encounters, thereby reducing the efficiency of each two-body encounter. Gallagher & Ostriker (1972) estimated the mass loss resulting from hyperbolic collisions. A numerical simulation of colliding galaxies has recently been performed by Lauberts (1974), whose model comprised two nine-body systems (one heavy central body representing the core and eight lighter bodies modeling the halo). Mildly interpenetrating collisions at 1500 km/sec relative velocity result in mass loss of the order of 1% per collision for Hubble-type galaxies ($\rho \propto r^{-3}$). However, deeply penetrating collisions between isothermal galaxies ($\rho \propto r^{-2}$) would be expected to yield substantially greater mass loss. In a study of collisions between galaxies with extended halos, Biermann (1974) found that for encounter distances of one-fourth the galactic radius, as much as one-fourth of the mass was dispersed.

Interacting Galaxies

Many pairs of galaxies seem to be actively intertwined because either a "bridge" of luminous material connects them, or one or both galaxies have long spiral "tails." It has been argued that tidal interactions could not be responsible for such fine structures because the tails are sometimes very long and filamentary. Simulations of tidally interacting galaxies have now clarified the situation, and it appears that gravitational interactions can explain both the bridges and tails.

The model, based on a modified "restricted three-body problem," was introduced by Pfleiderer & Siedentopf (1961; also Pfleiderer 1963). The target galaxy is represented by a mass point surrounded by a disk of massless test particles, whereas the flyby is represented by a single mass point. It is hoped that the lack of self-consistency is justifiable. Pfleiderer and Siedentopf were primarily interested in inducing spiral arms that formed but wound up quickly. They observed tails, but were not motivated to pursue them further. A more extensive study along the same lines was recently conducted by Eneev, Kozlov & Sunyaev (1973). Contopoulos & Bozis (1964) investigated grazing and penetrating encounters at strongly hyperbolic velocities. They found that the flyby tore off a significant amount of mass (1 and 8%, respectively), but the escaping stars did not form a bridge. Yabushita (1971) showed that in order for the passing galaxy to capture stars, a close direct approach of nearly parabolic velocity was required. Wright (1972) established that long-lived (2×10^9 yr) tails could be produced by a flyby of approximately the same mass as the target galaxy, in a grazing, parabolic approach. In a billion years, the flyby galaxy can travel more than 100 kpc, and an observed tail can therefore be the remnant of an earlier close encounter. He also noticed that when the flyby has an inclination to the plane of the target galaxy, the tail develops in the plane of the latter. The target galaxy suffered significant mass loss for flyby velocities up to three times the parabolic value, and for closest approaches up to two galactic radii.

In a significant and comprehensive study, Toomre & Toomre (1972) pinned down the rules for forming bridges and tails. In both cases, a near parabolic (hyperbolic with eccentricity less than 2), near grazing (perigalacticon less than two galactic radii), direct approach is required. Tails result from a flyby galaxy comparable in mass to the target galaxy, whereas bridges are caused by the passage of a small companion. They speculated, convincingly, that such approaches were just what one would expect from binary galaxies with eccentric orbits.

COLLISIONLESS RELAXATION

Galaxies look relaxed in spite of the fact that their relaxation time by two-body encounters is much longer than the age of the universe. On the other hand, the N-body simulations show that the oscillations of an unstable cluster relax the system to a quasisteady state after an interval of only a few crossing times. Hénon (1964, 1966) simulated the collisionless relaxation of systems with spherical symmetry and showed that an encounterless relaxation took place in a few crossing times. He called this "dynamical mixing"; a combination of phase mixing and energy changes induced by the varying mean field. This process by itself tended to form a core-halo structure, with a fairly flat density profile in the core, and a halo where the density fell off as $r^{-3.5}$ to $r^{-4.0}$. He also found that the velocity dispersion in the halo had a radial bias. Subsequently Lecar (1966a,b) and Hohl & Feix (1967; also Hohl & Campbell 1968) simulated dynamical mixing in systems with plane symmetry (modeling the z-motion in the Galaxy).

A coherent theoretical explanation for these phenomena was given by Lynden-Bell (1967), who formulated a statistical mechanics of "violent relaxation." He assumed that interactions with the rapidly changing mean field would tend to populate all accessible volumes in phase space, and thus one is justified in looking for the most probable distribution function, subject to the usual constraints of the conserved quantities (mass, energy, and momentum). A novel feature of collisionless relaxation is that because the mass giving rise to the fluctuations in the mean field is much larger than any stellar mass, the final distribution is independent of the stellar mass, that is, there is no mass segregation. Furthermore, in collisionless relaxation, the microscopic phase space density (f) is invariant. This gives rise to a "degeneracy," and the most probable distribution function resembles Fermi-Dirac statistics. This is seen most easily by considering a system where initially the phase space density is either a constant (f_m) or zero. As time goes on, a macroscopic volume element will be interspersed with blobs with f_m and empty spaces. The average space density in this region ($\bar{f}$) will usually be less than f_m but in no case greater than f_m. If a given mass M initially occupies a volume $f_m r_0 v_0$ (in one dimension), at a later time M will occupy a volume $\bar{f}rv$. The condition $\bar{f} \leqq f_m$ implies that $rv \geqq r_0 v_0$. As $r_0 v_0$ is (an arbitrary) constant, we can label it "h." The analogy is clear; note, however, that "h" is not a constant of nature but depends on f_m (initial conditions) and the arbitrary choice of M.

The concept of violent relaxation prompted a series of numerical simulations. In spherical systems, the Lynden-Bell distribution leads to infinite mass (precisely as does the isothermal distribution that it reaches asymptotically), hence exact agreement with the theory is out of the question. However, in the interior regions (small radius, large binding energy), Hénon (1967b, 1968) and Bouvier & Janin (1970a) found good agreement. In the case of a one-dimensional system, the theory can be evaluated precisely (Lecar & Cohen 1971) and leads to a finite mass system, which in principle could be determined numerically. In general, however, the results were similar to those obtained for spherical systems. Cuperman, Goldstein & Lecar (1969), Goldstein, Cuperman & Lecar (1969), and Lecar & Cohen (1967) found that the core agreed with the prediction whereas the halo did not. Experiments by Cohen & Lecar (1967) verified that the final distribution function was independent of mass in a two-mass system. Cuperman, Harten & Lecar (1971a,b) showed that, as predicted by Lynden-Bell's theory, the final velocity dispersions were approximately inversely proportional to the initial phase space density, that is, an element with initially high phase space density would tend to a low velocity dispersion. Violent relaxation effects were also studied in N-body systems (Standish 1968b, Aarseth 1974a) and collisionless rotating systems (Gott 1973). The criterion for violent relaxation proposed by Saslaw (1970) remains untested, however.

Characteristically, in all the numerical investigations a halo tended to get ejected immediately, which thereafter only interacted weakly with the core. For this reason the discrepancies with the theoretical predictions were always more pronounced in the halo. That the theory predicted a Maxwellian tail (which leads to infinite mass in two or three dimensions) was thought to be one of its prime defects. However, if Ostriker & Peebles (1973) are correct in suggesting that real galaxies do have isothermal halos (in the end, severely reduced by neighboring galaxies), perhaps the Lynden-Bell mechanism is true to nature after all.

Literature Cited

Aarseth, S. J. 1963. *MNRAS* 126:223
Aarseth, S. J. 1966. *MNRAS* 132:35
Aarseth, S. J. 1967. *Bull. Astron.* 2:47
Aarseth, S. J. 1968. *Bull. Astron.* 3:105
Aarseth, S. J. 1969. *MNRAS* 144:537
Aarseth, S. J. 1971a. *Ap. Space Sci.* 13:324
Aarseth, S. J. 1971b. *Ap. Space Sci.* 14:118
Aarseth, S. J. 1973a. *Vistas in Astronomy,* ed. A. Beer, 15:13. Oxford: Pergamon
Aarseth, S. J. 1973b. *Recent Advances in Dynamical Astronomy,* ed. B. D. Tapley, V. Szebehely, 197. Dordrecht: Reidel
Aarseth, S. J. 1974a. *Astron. Ap.* 35:237
Aarseth, S. J. 1974b. *IAU Symp. No. 69, Besançon, France*
Aarseth, S. J., Hénon, M., Wielen, R. 1974. *Astron. Ap.* 37:183
Aarseth, S. J., Hills, J. G. 1972. *Astron. Ap.* 21:255
Aarseth, S. J., Saslaw, W. C. 1972. *Ap. J.* 172:17
Aarseth, S. J., Woolf, N. J. 1972. *Ap. Lett.* 12:159
Aarseth, S. J., Zare, K. 1974. *Celestial Mech.* 10:185
Agekyan, T. A. 1959. *Astron. Zh.* 36:41; *Sov. Astron.—AJ* 3:46
Agekyan, T. A. 1961. *Astron. Zh.* 38:1055; *Sov. Astron.—AJ* 5:809
Ahmad, A., Cohen, L. 1972. *Phys. Lett. A* 42:243
Ahmad, A., Cohen, L. 1973a. *J. Comput. Phys.* 12:389
Ahmad, A., Cohen, L. 1973b. *Ap. J.* 179:885
Ahmad, A., Cohen, L. 1974a. *Ap. J.* 188:469
Ahmad, A., Cohen, L. 1974b. *Lecture Notes in Mathematics,* ed. D. G. Bettis, 362:337. Berlin: Springer
Aleshin, V. I., Kaplan, S. A. 1970. *Astron. Zh.* 47:10; *Sov. Astron.—AJ* 14:6
Alladin, S. M. 1965. *Ap. J.* 141:768
Alladin, S. M., Sastry, K. S., Ballabh, G. M. 1974. *Galaxies and Relativistic Astrophysics,* ed. B. Barbanis, J. D. Hadjidemetriou, 129. Berlin: Springer
Allen, C., Poveda, A. 1971. *Ap. Space Sci.* 13:350
Allen, C., Poveda, A. 1974. *Proc. IAU Symp. No. 62,* 239
Bahcall, N. A. 1973. *Ap. J.* 183:783
Baranov, A. S. 1970. *Astrofizika* 6:261
Biermann, P. 1974. Habilitationsschrift. Univ. Göttingen, W. Germany
Bouvier, P. 1971. *Astron. Ap.* 14:341
Bouvier, P. 1972. *Astron. Ap.* 21:441
Bouvier, P., Janin, G. 1970a. *Astron. Ap.* 5:127
Bouvier, P., Janin, G. 1970b. *Astron. Ap.* 9:

461
Butterworth, E. M., Miller, R. H. 1971. *Ap. J.* 170:275
Chandrasekhar, S. 1942. *Principles of Stellar Dynamics.* Chicago, Ill.: Univ. Chicago Press
Chandrasekhar, S., von Neumann, J. 1942. *Ap. J.* 95:489
Cohen, L. 1974. *IAU Symp. No. 69, Besançon, France*
Cohen, L., Lecar, M. 1967. *NASA SP-153,* 309
Contopoulos, G., Bozis, G. 1964. *Ap. J.* 139:1239
Cruz-Gonzalez, C., Poveda, A. 1971. *Ap. Space Sci.* 13:335
Cuperman, S., Goldstein, S., Lecar, M. 1969. *MNRAS* 146:161
Cuperman, S., Harten, A., Lecar, M. 1971a. *Ap. Space Sci.* 13:411
Cuperman, S., Harten, A., Lecar, M. 1971b. *Ap. Space Sci.* 13:425
Dearborn, D. S. 1973. *Ap. J.* 179:45
Eneev, T. M., Kozlov, N. N., Sunyaev, R. A. 1973. *Astron. Ap.* 22:41
Gallagher, J. S., Ostriker, J. P. 1972. *Astron. J.* 77:288
Gilbert, I. H. 1968. *Ap. J.* 152:1043
Goldstein, S., Cuperman, S., Lecar, M. 1969. *MNRAS* 143:209
Gott, J. R. 1973. *Ap. J.* 186:481
Gott, J. R. 1974. *IAU Symp. No. 69, Besançon, France*
Gunn, J. E., Gott, J. R. 1972. *Ap. J.* 176:1
Harrington, R. S. 1972. *Celestial Mech.* 6:322
Hayli, A. 1967. *Bull. Astron.* 2:67
Hayli, A. 1968. *Bull. Astron.* 3:189
Hayli, A. 1970a. *Astron. Ap.* 7:17
Hayli, A. 1970b. *Astron. Ap.* 7:249
Heggie, D. C. 1973. See Aarseth 1973b, 34
Heggie, D. C. 1974a. *Celestial Mech.* 10:217
Heggie, D. C. 1974b. *Proc. IAU Symp. No. 62,* 225
Heggie, D. C. 1974c. *IAU Symp. No. 69, Besançon, France*
Hénon, M. 1960. *Ann. Ap.* 23:668
Hénon, M. 1961. *Ann. Ap.* 24:369
Hénon, M. 1964. *Ann. Ap.* 27:83
Hénon, M. 1965. *Ann. Ap.* 28:62
Hénon, M. 1966. *Colloq. Int. Ap., 14th, Liège,* 243
Hénon, M. 1967a. *Bull. Astron.* 2:91
Hénon, M. 1967b. *NASA SP-153,* 349
Hénon, M. 1968. *Bull. Astron.* 3:241
Hénon, M. 1969. *Astron. Ap.* 2:151
Hénon, M. 1971a. *Ap. Space Sci.* 13:284
Hénon, M. 1971b. *Ap. Space Sci.* 14:151
Hénon, M. 1972. *Astron. Ap.* 19:488
Hénon, M. 1973. *Dynamical Structure and Evolution of Stellar Systems,* ed. L. Martinet, L. Mayor, 183. Geneva: Observatory
Hénon, M. 1974. *IAU Symp. No. 69, Besançon, France*
Hohl, F., Campbell, J. W. 1968. *Astron. J.* 73:611
Hohl, F., Feix, M. R. 1967. *Ap. J.* 147:1164
Illingworth, G. 1974. *IAU Symp. No. 69, Besançon, France*
Janin, G. 1974. See Ahmad & Cohen 1974b, 291
Jones, B. F. 1970a. *Ap. J. Lett.* 159:L129
Jones, B. F. 1970b. *Astron. J.* 75:563
Kaliberda, V. S., Petrovskaya, I. V. 1970. *Astrofizika* 6:135
Kaliberda, V. S., Petrovskaya, I. V. 1971. *Astrofizika* 7:663
Kholopov, P. N. 1971. *Astron. Zh.* 48:529; *Sov. Astron.*—AJ 15:415
King, I. R. 1962. *Astron. J.* 67:471
King, I. R. 1966. *Astron. J.* 71:64
King, I. R. 1971. *Sky Telesc.* 41:139
Larson, R. B. 1969. *MNRAS* 145:405
Larson, R. B. 1970a. *MNRAS* 147:323
Larson, R. B. 1970b. *MNRAS* 150:93
Larson, R. B. 1974. *MNRAS* 166:585
Lauberts, A. 1974. *Astron. Ap.* 33:231
Lecar, M. 1966a. *The Theory of Orbits in the Solar System and in Stellar Systems,* ed. G. Contopoulos, 46. New York: Academic
Lecar, M. 1966b. *Colloq. Int. Ap., 14th, Liège,* 227
Lecar, M. 1968. *Bull. Astron.* 3:91
Lecar, M. 1972. *Gravitational N-Body Problem. IAU Colloq. No. 10.* Dordrecht: Reidel
Lecar, M. 1974. *IAU Symp. No. 69, Besançon, France*
Lecar, M., Cohen, L. 1967. *NASA SP-153,* 299
Lecar, M., Cohen, L. 1971. *Ap. Space Sci.* 13:397
Lecar, M., Cruz-Gonzalez, C. 1971. *Ap. Space Sci.* 13:360
Limber, D. N. 1962. *Ap. J.* 135:41
Lynden-Bell, D. 1967. *MNRAS* 136:101
Lynden-Bell, D., Wood, R. 1968. *MNRAS* 138:495
Michie, R. W. 1963a. *MNRAS* 125:127
Michie, R. W. 1963b. *MNRAS* 126:331
Michie, R. W. 1963c. *MNRAS* 126:499
Michie, R. W. 1964. *Ann. Rev. Astron. Ap.* 2:49
Miller, R. H. 1964. *Ap. J.* 140:250
Miller, R. H. 1971a. *J. Comp. Phys.* 8:449
Miller, R. H. 1971b. *J. Comp. Phys.* 8:464
Miller, R. H. 1974. *IAU Symp. No. 69, Besançon, France*
Oemler, A. 1974. PhD thesis. Calif. Inst. Tech., Pasadena
Oort, J. H. 1950. *Bull. Astron. Inst. Neth.* 11:91
Oort, J. H., Pels, G., Pels-Kluyver, H. A. 1974. *IAU Symp. No. 69, Besançon, France*

Ostriker, J. P., Peebles, P. J. E. 1973. *Ap. J.* 186:467
Ostriker, J. P., Spitzer, L., Chevalier, R. A. 1972. *Ap. J. Lett.* 176:L51
Peebles, P. J. E. 1970. *Astron. J.* 75:13
Peebles, P. J. E. 1971. *Astron. Ap.* 11:377
Pfleiderer, J. 1963. *Z. Ap.* 58:12
Pfleiderer, J., Siedentopf, H. 1961. *Z. Ap.* 51:201
Prata, S. W. 1971a. *Astron. J.* 76:1017
Prata, S. W. 1971b. *Astron. J.* 76:1029
Press, W. H., Schechter, P. 1974. *Ap. J.* 187:425
Rood, H. J., Page, T. L., Kintner, E. C., King, I. R. 1972. *Ap. J.* 175:627
Rybakov, A. I., Kalinina, E. P., Kholopov, P. N., Duboshin, G. N. 1971. *Astron. Zh.* 48:518; *Sov. Astron.—AJ* 15:406
Saslaw, W. C. 1970. *MNRAS* 150:299
Sastry, K. S., Alladin, S. M. 1970. *Ap. Space Sci.* 7:261
Smith, H. 1974. See Ahmad & Cohen 1974b, 360
Spitzer, L. 1958. *Ap. J.* 127:17
Spitzer, L. 1969. *Ap. J. Lett.* 158:L139
Spitzer, L. 1974. *IAU Symp. No. 69, Besançon, France*
Spitzer, L., Baade, W. 1951. *Ap. J.* 113:413
Spitzer, L., Chevalier, R. A. 1973. *Ap. J.* 183:565
Spitzer, L., Härm, R. 1958. *Ap. J.* 127:544
Spitzer, L., Hart, M. H. 1971a. *Ap. J.* 164:399
Spitzer, L., Hart, M. H. 1971b. *Ap. J.* 166:483
Spitzer, L., Shapiro, S. L. 1972. *Ap. J.* 173:529
Spitzer, L., Thuan, T. X. 1972. *Ap. J.* 175:31
Standish, E. M. 1968a. PhD thesis. Yale Univ., New Haven, Conn.
Standish, E. M. 1968b. *Bull. Astron.* 3:135
Standish, E. M., Aksnes, K. 1969. *Ap. J.* 158:519
Szebehely, V., Peters, C. F. 1967. *Astron. J.* 72:876
Talpaert, Y., Lefèvre, F. 1972. *Bull. Cl. Sci. Acad. R. Belg.* 58:759
Thüring, B. 1968. *Bull. Astron.* 3:177
Toomre, A., Toomre, J. 1972. *Ap. J.* 178:623
Valtonen, M. J. 1974. *Proc. IAU Symp. No. 62,* 211
van Albada, T. S. 1968a. *Bull. Astron. Inst. Neth.* 19:479
van Albada, T. S. 1968b. *Bull. Astron. Inst. Neth.* 20:57
van Altena, M. F. 1966. *Astron. J.* 71:482
van den Bergh, S., Sher, D. 1960. *Publ. David Dunlap Observ.* 2:203
von Hoerner, S. 1957. *Ap. J.* 125:451
von Hoerner, S. 1960. *Z. Ap.* 50:184
von Hoerner, S. 1963. *Z. Ap.* 57:47
Wielen, R. 1967. *Veroeff. Astron. Rechen Inst. Heidelberg, No. 19*
Wielen, R. 1968. *Bull. Astron.* 3:127
Wielen, R. 1971a. *Ap. Space Sci.* 13:300
Wielen, R. 1971b. *Astron. Ap.* 13:309
Wielen, R. 1974. *Proc. Eur. Astron. Meet., 1st,* ed. L. N. Mavrides, 2:326. Berlin: Springer
Wright, A. E. 1972. *MNRAS* 157:309
Yabushita, S. 1971. *MNRAS* 153:97

RADIO SURVEYS AND SOURCE COUNTS

D. L. Jauncey[1]
Division of Radiophysics, CSIRO, Sydney, Australia

INTRODUCTION

Radio source counts have been used extensively as a possible tool for probing the early history of the universe. Despite nearly two decades of improving observational results, however, there still appears to be a wide divergence of opinion on their interpretation and even on their value for such cosmological investigations. An accurate knowledge of the observed number-flux density distribution is certainly necessary in such an investigation, but of equal importance is an understanding of the significance of these results. It may not be possible to prove any one model correct, but if radio source counts are to contribute at all to cosmology it is certainly necessary to be in a position to disprove some models.

The last few years have seen a marked upsurge in the observational survey material that is becoming available for source counts. The Westerbork synthesis radio telescope is mapping more of the sky to millijansky [1 jansky (Jy) = 1 flux unit = 10^{-26} W m^{-2} Hz^{-1}] flux densities at frequencies of 610, 1415, and 5000 MHz (e.g. van der Laan & Le Poole 1974). One quarter of the sky accessible to the Arecibo reflector has been mapped at 611 MHz (Durdin et al 1973); at 408 MHz the Molonglo (Davies et al 1973) and Bologna (Colla et al 1970, 1972) surveys continue toward complete sky coverage. The Parkes 2700 MHz survey (Wall et al 1971) now covers much of the southern sky to flux densities between 0.1 and 0.35 Jy, while Jodrell Bank is extending the coverage at this frequency of the northern sky. At 5000 MHz the NRAO survey (Kellermann et al 1968) continues inexorably toward a complete coverage of the northern sky.

These observations are continuing; in some cases only preliminary results have been published. The present review aims at presenting a detailed examination of the published radio survey material, with the idea of constructing reliable counts from these surveys.

Observations over a wide range of flux densities are available at 178, 408, 1400,

[1] Part of the work for this review was carried out while the author was with the National Astronomy and Ionosphere Center at Cornell University, Ithaca, New York 14850. The NAIC is operated by Cornell University under contract to the National Science Foundation.

2700, and 5000 MHz, which are the finding frequencies considered here. The southern 86 MHz surveys (Mills & Slee 1957, Mills et al 1958, 1960) provided the first reliable counts for the stronger sources; however, they are not included here because no further low flux density surveys have been made to complement the original strong source counts.

All of the source count data are presented here in *differential* form rather than by the traditional integral or cumulative counts. Difficulties with the use of integral counts are not restricted to the underestimation of the uncertainties in the fitted statistics (Jauncey 1967, Crawford et al 1970); they also can lead to interpretative problems. Where changes in slope occur, a simple interpretation of the observed cumulative counts can be quite misleading. The problem is illustrated schematically: Figure 1*a* shows a power-law differential intensity distribution with a sudden change in slope at intensity S_0; Figure 1*b* is the corresponding integral count constructed from this distribution. It is clear that the sudden change at S_0 is lost in the integral counts and instead leaves a region of indeterminate slope from S_0 to some lower intensity where the integral count has flattened out. Consequently, slope changes, whatever their origin, are not readily determined, or often even recognizable, from integral counts.

There can be interpretative problems other than those associated with fitting cosmological models to observed counts. Near the lower intensity limit of a radio survey two strongly competing effects can introduce significant distortions of the true intensity distribution. One is the flux density overestimation problem, the so-called population law bias discussed by many authors. It occurs because there are many more weak sources to be scattered above a given intensity level than there are

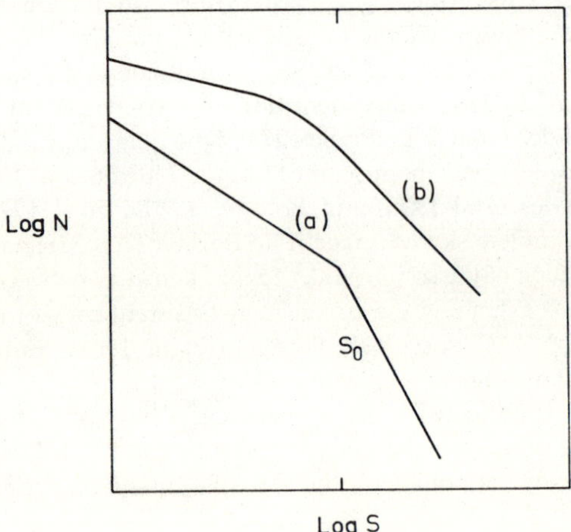

Figure 1 Differential and integral distribution comparison. *a*, schematic representation of a two-power-law *differential* distribution with a sharp change at S_0; *b*, the *integral* distribution derived from *a*: the sharp transition in the differential distribution is smoothed out and the slope of the integral distribution is undefined below S_0.

strong sources scattered below this level. This problem was first noted for radio sources by Bolton (1956) and examined later in more detail (Mills & Slee 1957, Bennett 1962a, Jauncey 1968). Murdoch et al (1973) have shown that significant corrections to the observed counts are necessary at an intensity of about five standard errors and that the corrections increase rapidly below this. The earliest radio surveys (e.g. Shakeshaft et al 1955) were of little value because of this effect. Their counts were extended almost to the rms background confusion and hence the reliability of most of the weakest sources was seriously impaired (Mills & Slee 1957).

The other problem is incompleteness in a survey as the relative flux density errors increase. Finding sources in any given survey is a complex problem, often critically dependent upon the details of the flux density error distribution, the actual source-finding techniques, and the method by which the survey is carried out. Consequently, while the usual criteria for setting the lower intensity limit of a survey (such as five times the standard error, or 20 beam areas per source) allow a reasonable estimate of the population law bias, they do not necessarily provide any completeness estimate.

In practice, this has proven a serious problem that could have easily been recognized had the raw *differential* numbers been plotted (see below). The wide divergence in the absolute luminosity of radio sources will ensure that any cosmological changes present in the source counts will be spread over a wide intensity range. By comparison, sharp changes in the counts over a small intensity range, particularly at the lower intensity levels of a survey, are more likely to indicate instrumental limitations in the survey.

As a corollary, the setting of the lower intensity limit above which reliable source counts can be constructed assumes considerable importance, particularly as the instrumental limitations of the surveying technique are approached. A full error analysis of both the individual source intensity measurements and the source-finding process is required. Where possible a detailed comparison with independent intensity measurements and survey measurements is highly desirable. Although a plot of the differential numbers may easily reveal incompleteness in a survey, it is not sufficient to determine the lower limit from such a plot alone. It is difficult to be truly objective in setting the lower limit when incompleteness is encountered, unless the limit is set well in excess of the region where the turndown in numbers is visible. The lower survey limit needs to be set *independently,* and the effects on the source counts of varying this limit need to be investigated.

COUNTS AT 178 MHz

The 3C revised survey (Bennett 1962a), the 4C survey (Pilkington & Scott 1965, Gower et al 1967), and the North Polar survey (Ryle & Neville 1962) form the basis for the counts at this frequency. Gower (1966) has presented an analysis of the *integral* counts from these surveys. By its sheer size the 4C survey is a good starting point from which to provide accurate source densities for comparison with the weaker and stronger, though smaller, North Polar and 3CR counts. The size of the 4C survey, however, is partially offset by the relatively small value of $S_0/\sigma = 5$, the ratio of the lower flux density limit of the survey to the rms error. It is now more widely appreciated (e.g. Murdoch et al 1973) that such low values of S_0/σ imply that large and often uncertain corrections are necessary in estimating the true $n(S)$ curve, particularly for confusion-limited surveys. The simple calculation below shows that the corrections used by Gower (1966) in his analysis of the 4C counts are insufficient to account for the full effects of confusion on the observed source densities.

The published error distribution for 2 Jy Monte Carlo sources (Pilkington & Scott 1965) shows a mean flux density overestimate of 0.23 Jy. The real lower flux density limit of the survey is therefore 1.77 Jy, rather than 2.0 Jy, so that this mean flux density overestimate results in a source density overestimate of an amount equal to $(2.0/1.77)^{\alpha}$, where α is the appropriate integral slope. For the fitted slope of 1.66 and for an incompleteness of 7% (Gower 1966) the resulting increase in source density due to the mean flux density overestimate is 14%. To this must be added the population law effects of the error distribution. By comparison, Gower's total correction, including the effects of the error distribution, is only 13%.

To determine the magnitude of the effects due to the flux density errors, the excess of weak sources inadvertently cataloged in the 4C survey was recalculated following Murdoch et al (1973). The published error distribution and, for convenience, an integral slope of 1.5 were used. In performing this calculation no sources were counted twice, that is, no confusion error was allowed that was greater than the source intensity. This provides a lower limit to the true effect, since this simple picture of confusion breaks down when three or more sources can contribute to the intensity of an observed "source."

This yielded a total source density overestimate of 19%. Increasing the slope to 1.66 [Gower's (1966) value], and hence increasing the contribution from the weaker sources, gave a value of 23%. Thus Gower's corrected source density above 2 Jy may be 5–9% too high, even without a correction for multiple blends. Such an error in total source density has most effect on the source density in the lowest flux density intervals. Decreasing the total density above 2 Jy by 5% decreases the source density between 2.0 and 3.0 Jy by 9%. Decreasing the total density by 9% produces a 16% decrease in the 2.0 to 3.0 Jy interval. A decrease of this order in the lowest flux density range of the 4C catalog will reduce substantially the slope in the 2–10 Jy range from the 1.66 found by Gower to about 1.55 or less. It is also important to realize, however, that what is significant is the uncertainty in the corrections, rather than just the corrected numbers of sources.

The reason for the discrepancy between Gower's (1966) calculation and that given above is not difficult to find: in the Monte Carlo calculations Gower (see his Figure 1) extends the simulated sources to only 1.25 Jy, whereas the error distribution shows that it should be extended to at least 1.0 Jy, and if multiple blending of sources exists, to even lower values (Murdoch et al 1973).

Gower's (1966) type of Monte Carlo technique is capable of yielding the appropriate corrections, provided that sources are injected to sufficiently low flux density levels. Indeed, repeating the above calculations with the lower limit increased to 5.0 Jy and using the appropriate published error distribution (Pilkington & Scott 1965) yields good agreement with Gower's quoted 10% correction.

The most practical solution at this stage is not to attempt corrections at 2 Jy but to set a higher value to the survey lower flux density limit, say 3 Jy. Then the source density above 3 Jy can be compared with the expected numbers of stronger and weaker sources. The very limited dynamic range of this restricted sample, 3–10 Jy, together with the fact that further corrections for the angular sizes of the stronger 4C sources are necessary (Longair 1973), precludes any reliable estimate of the slope being obtained from this limited sample.

A differential n/n_0 diagram has been constructed for the 178 MHz source counts and is given in Figure 2. Data from the 3CR, the restricted 4C, and North Polar surveys have been used. All n_0 were calculated with respect to a 5/2 (differential) slope. Gower's (1966) corrected source density of 302 sources sr^{-1} above 2.95 Jy (quoted by Refsdal 1969) was adopted for the normalization at this frequency. The plotted errors represent only the $(n)^{1/2}$ values and *do not* include any effects due to uncertainties in the corrections. The corrected 3CR source densities (Gower, quoted by Refsdal 1969) were used above 10 Jy, while the corrected 4C densities, excluding the 2–3 Jy interval, were also used. The dotted lines in this interval correspond to the two different corrections derived above and are included to indicate possible source densities in this range.

For the weaker sources, the Ryle & Neville (1962) North Polar survey was used, with an 8% increase in flux densities to align the flux density scales. In constructing Figure 2 only those sources from the restricted region declination $\geqq 87.5°$ were counted to avoid any possible problems arising from an incorrect radial correction of the type found for the 5C surveys (Condon & Jauncey 1973). This does not seriously affect the significance of the North Polar survey counts, because the central region contains most of the sources.

It should be noted that Refsdal's (1969) quoted figures for the North Polar survey are incorrect. The total number of sources in the central region above the lower

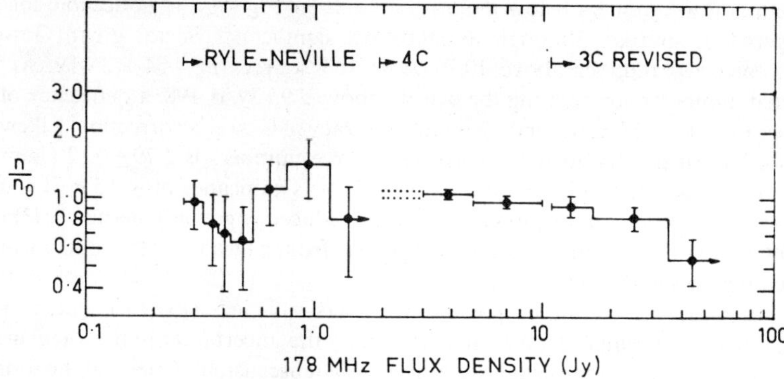

Figure 2 The normalized differential source counts at 178 MHz from the 3CR survey, the 4C survey, and the North Polar survey. The normalization is to a 5/2 differential slope and a total source density of 302 sources sr^{-1} with $S_{178} \geqq 2.95$ Jy. The dotted lines between 1.95 and 2.95 Jy indicate the approximate corrected source density calculated in the text.

flux density limit is 60 in an area of 0.006 sr. Consequently Refsdal's figures for the fitted slopes from this survey are also incorrect.

Plotting the differential North Polar counts reveals no obvious incompleteness, although the quoted rms background fluctuation level, including the ring sidelobes, is 0.06 Jy, compared with the 0.27 Jy lower limit. The effects of the flux density errors on the counts is difficult to assess because of this sidelobe contribution. Consequently no correction has been applied to the North Polar survey counts and the uncorrected source densities are plotted in Figure 2 where the error bars represent only the $(n)^{1/2}$ uncertainties.

On the basis of the integral counts at 178 MHz both Gower (1966) and Longair (1966) have claimed a dramatic change in slope over the range of flux densities covered by the North Polar survey. Examination of the differential counts in Figure 2 shows no evidence supporting this change; in fact, as shown below, there is no significant departure from a 5/2 slope (n/n_0 = constant) over the range 0.27 to about 10 Jy.

Any change in the form of the distribution at low flux densities can be tested by determining the slope from the North Polar survey source counts and also by extrapolation from the 4C survey to predict the total number of sources expected above 0.27 Jy in the North Polar survey. The maximum likelihood (integral) slope (Crawford et al 1970) for the uncorrected counts is 1.4 ± 0.2. The expected number in the central 19.6 square degrees for a 1.5 (integral) slope is 65 sources compared with an observed 60 sources. Thus, the North Polar survey counts reveal no reliable evidence for any turndown at low flux densities at 178 MHz.

Although the Ryle & Neville (1962) survey shows no evidence for any change, there is real evidence for a significant falloff in slope at low 178 MHz flux densities. This comes from the background sky deflection $P(d)$ analysis carried out by Hewish (1961) on the 4C survey records, and from considerations of the integrated background sky emission (Ryle & Clarke 1961).

The falloff in numbers below the $n/n_0 = 1$ line in Figure 2 is noticeable for the stronger 3CR sources, although its statistical significance is not great. Gower's (1966) corrected number above 10.70 Jy is 163 sources in 4.54 sr, whereas the expected number, extrapolating the density above 2.95 Jy, is 198, a deficiency of 35 sources, or only $\lesssim 2.5\,\sigma$, depending on the accuracy of Gower's correction. Likewise, the fitted integral slope to the uncorrected source numbers is 1.79 ± 0.12 (Jauncey 1967), once again differing from the simple 1.5 at a significance of $\sim 2.5\,\sigma$. It is only when an appropriate luminosity function is introduced (e.g. von Hoerner 1973) that the difference in both slope and source numbers from a steady state model assumes a much higher significance level.

At 178 MHz the 4C survey potentially covers sufficient sky area, but over its useful flux density range, 3–10 Jy, it is limited by the uncertainty in the angular size corrections caused by the partial resolution in right ascension of many of the sources (Longair 1973). The North Polar survey neither contains enough sources nor extends to low enough flux densities to establish whether any change from a 5/2 slope is present in the 0.27–2 Jy region. Between 1 and 5 Jy the peak that was formerly considered to be present in the 178 MHz source counts essentially disappears in Figure 2.

COUNTS AT 408 MHz

At 408 MHz the available surveys extend across four decades in flux density. An analysis of the integral counts of the data to 1968 was given by Pooley & Ryle (1968), but no detailed analysis of these data in differential form has been presented. Since 1968 more extensive surveys from Molonglo in the south (Davies et al 1973) and Bologna in the north (Colla et al 1970, 1972) have become available. Additionally, Robertson (1973) has examined the best available measurements to produce an all-sky catalog of the strong, $S_{408} \geqq 10.0$ Jy, sources to replace the earlier unpublished compilation used by Pooley & Ryle (1968). As important as the production of these more recent catalogs is the better appreciation of the shortcomings often present in the earlier radio surveys (Murdoch et al 1973).

In constructing the source counts at 408 MHz, Pooley & Ryle (1968) drew upon the work of Williams & Stewart (1967), Long et al (1966), and Williams et al (1968) for sources with flux densities between 4.0 and 10 Jy. Above 10 Jy the Parkes catalog (Shimmins et al 1966, Bolton et al 1964) was used south to declination −30°, while interpolation of the spectra from the 3CR source measurements made by Kellermann et al (1969) was used in the north.

The Parkes 408 MHz flux densities are present in these compilations both directly, $S_{408} \geqq 10$ Jy, and indirectly (Williams et al 1968). Recent measurements of

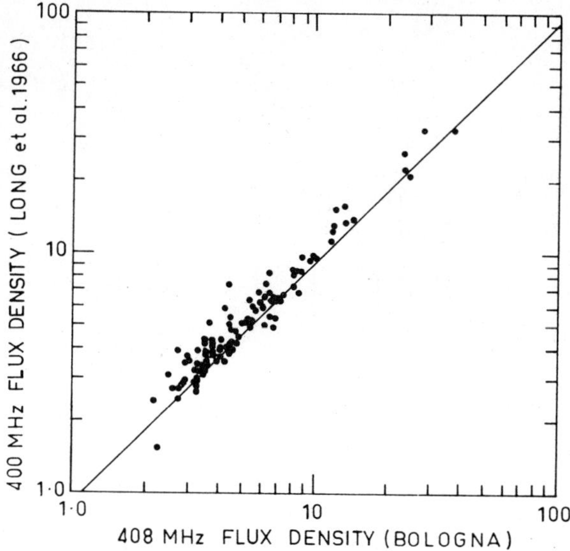

Figure 3 Comparison of the interpolated flux densities at 400 MHz (Long et al 1966) with measurements made with the Bologna Cross at 408 MHz (Colla et al 1970, 1972). The line at 45° shows the expected relationship, given the different flux density scales and the slight frequency difference. The large number of sources above this line indicates a significant overestimation on the part of the 400 MHz interpolated flux densities.

many of these sources with smaller antenna beams have revealed a systematic additive error in the original 408 MHz flux densities (Murdoch & Large 1968, Niell & Jauncey 1971). This bias is about 0.5 Jy and is less than the quoted confusion error, but it still has a significant effect on the source counts. The bias is still present, although diminished (Niell & Jauncey 1971) in the interpolated flux densities from Williams et al (1968), because the Parkes 408 MHz flux densities were used in this spectral interpolation. In compiling the spectral data for the 40–44° region of the 4C catalog, Williams & Stewart (1967) drew upon the VRO 610 MHz flux densities (MacLeod et al 1965). These too have been shown to suffer from a similar overestimation bias (Murdoch 1969), which must be reflected in the derived 400 MHz flux densities.

Pooley & Ryle (1968) compiled additional sources in the 6–10 Jy range based on the interpolated flux densities of Long et al (1966) from their investigation of the spectra of sources in the 20–40° region of the 4C catalog (Pilkington & Scott 1965). A comparison of the Bologna (Colla et al 1970, 1972) 408 MHz measurements with the interpolated S_{400} values of Long et al (1966) for the sources common to both lists is shown in Figure 3. Because of the slight frequency difference and the difference in flux density scales, the 45° line in Figure 3*a* does not pass through the origin. It is evident that there are many more sources above the line than below it, indicating that the interpolated flux densities of Long et al (1966) are systematically

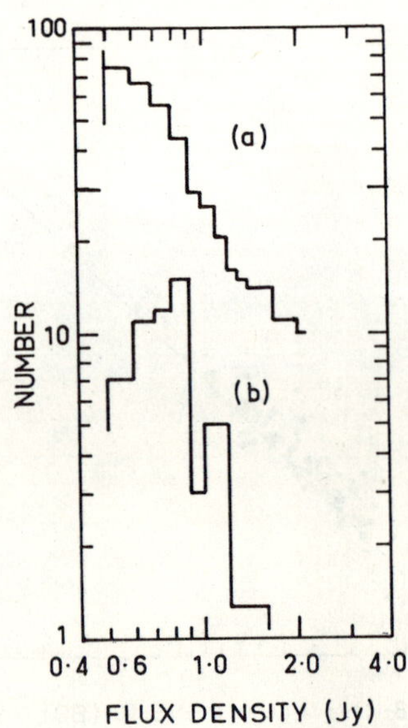

Figure 4 Comparison of the differential and the integral "staircase" source counts from the Cambridge fan-beam survey (Bailey & Pooley 1968) at 408 MHz. *a*, the cumulative numbers above each flux density level; *b*, the differential counts from the same survey, where the incompleteness below about 0.9 Jy becomes apparent.

high. In the interval 6.0 Jy $\leq S_{400} <$ 10 Jy the overestimates outnumber the underestimates by about three to one. Hence the source density in this flux density range based on the interpolated flux densities appears to be high by about 20%.

The presence of such systematic flux density errors in the compilations used by Pooley & Ryle (1968) renders them unsuitable for deriving reliable source counts in the 4–10 Jy flux density range. Unfortunately, no accurate measurements near 408 MHz of a large enough complete sample in this flux density range have since been published. This remains one of the main obstacles to determining accurate source counts at 408 MHz.

Between 0.5 and 4 Jy the sources from the fan-beam survey of Bailey & Pooley (1968) make up the sample used in compiling the earlier source counts. The direct translation from integral to differential counts reveals what is almost certainly a serious incompleteness in the central region of this survey where the catalog extends to the lowest flux densities. The differential numbers from the central $\pm 40'$ region plotted in Figure 4*b* show a dramatic falloff below $\sim$0.9 Jy: the expected number between 0.5 and 0.6 Jy is at least three times that in the interval 0.8–0.9 Jy, whereas the observed number is less than half. The "staircase" integral counts in Figure 4*a* hide this effect and illustrate very clearly the limitations of this form of analysis.

The reason behind this incompleteness is unknown, but it can possibly be traced directly to the effects of confusion on the fan-beam flux densities. At Arecibo, direct measurements of the confusion error with a comparable beam size (Condon 1974) indicate an expected rms error of about 0.13 Jy. This value is confirmed by a comparison with the Bologna (Colla et al 1970) flux densities and those from the later Cambridge fan-beam survey (Willson 1972). Because 0.5 Jy is less than four times the rms error, it is expected that confusion effects will be significant.

At flux densities below 0.5 Jy there has been essentially complete reliance on the Cambridge 5C aperture synthesis surveys (Pooley & Kenderdine 1968, Griffiths 1975), since these extend to flux density levels well below the level obtained with other instruments at this frequency. It was something of a surprise when an investigation of the spectra of the sources in the 5C survey regions uncovered a serious, radially dependent systematic bias in the aperture synthesis flux densities (Condon & Jauncey 1973). The bias is largest at the edges of the synthesis maps, where it is at least 30%. That it was detected at all is attributed directly to the accuracy and reliability of Maslowski's (1972) 1400 MHz survey measurements at nearly 3.5 times the frequency and with 45 times the beam area of the original 408 MHz surveys.

The nature of this bias is such that sources in the outer areas of these surveys (which are the stronger sources) have their flux densities overestimated. Until a complete reanalysis of the 5C surveys is undertaken, the flux densities and hence the source counts must be viewed with some caution.

Figure 5 shows the 408 MHz n/n_0 counts constructed from Robertson's (1973) all-sky survey, the Molonglo (Davies et al 1973) and Bologna (Colla et al 1972) surveys, and the revised 5C3 and 5C4 source numbers given by Griffiths (1975). The 5C counts have been included with the reservations given above. All have been normalized to the Molonglo flux density scale (Wyllie 1969), relying on the Arecibo measurements at 430 MHz (Niell & Jauncey 1971) and at 318 MHz (Condon et al

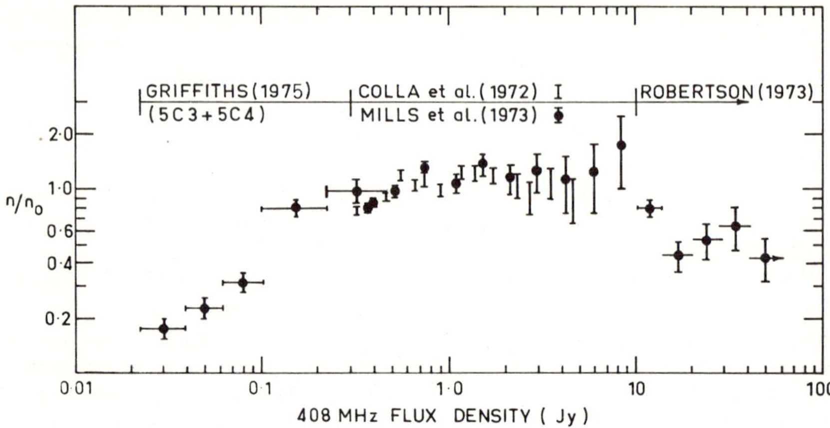

Figure 5 The normalized differential source counts at 408 MHz based on the 5C3 and 5C4 surveys, the Molonglo and Bologna surveys, and Robertson's (1973) all-sky catalog. The normalization is to a 5/2 differential slope and a total source density of 4725 sources sr^{-1} with $S_{408} \geqq 0.31$ Jy.

1971, Condon & Jauncey 1974a) to align the northern and southern hemisphere scales. Within the measurement error the Molonglo and Bologna scales are equal. To convert the 5C flux densities to the Molonglo scale they were increased by 12%; no correction was applied to the Bologna flux densities. Furthermore, Arecibo measurements of both strong and weak sources from Bologna (Condon & Jauncey 1974a) and Molonglo (Hunstead & Jauncey 1970) have established the linearity of both sets of measurements as well as the reliability of the error estimates.

The Molonglo and Bologna surveys are being carried out with similar instruments and reach comparable flux densities. Both are noise limited rather than confusion limited and hence are less likely to suffer from many of the difficulties encountered in the earlier surveys. In discussing the Molonglo counts, Mills et al (1973) have applied corrections to the observed numbers to account for the effects of noise and calibration errors, as well as the incompleteness caused by their criterion of selecting only those sources present in two observations. Above 0.3 Jy they find that these corrections are well defined, and so the counts here are restricted to sources above this lower limit. The Bologna sources have similarly been restricted to greater than 0.3 Jy.

In analyzing their counts, both groups present a careful error analysis, Mills et al (1973) in particular. Mills et al use integrated flux densities, eliminating the need for any corrections because of partial resolution. Furthermore, they present a detailed analysis of the effects of the error distributions for their particular source selection procedures. The Bologna counts (Colla et al 1972) have also been corrected for the population law bias, which is small at the present 0.3 Jy lower limit. They provide no correction for partial resolution in right ascension in the Bologna beam, which should have a small effect at these flux densities. Examination of the fraction of sources extended in right ascension in the Molonglo catalog (Davies et al 1973) shows that there should be an effect of about 1.5%, as does a similar estimate based on the observed fraction of extended sources in the 5C2 survey (Pooley & Kenderdine 1968). To this must be added a correction for the small differences in lower flux density limit before the Molonglo and Bologna source densities can be compared. Assuming a 3/2 integral slope for the Bologna counts to correct them from the Bologna 0.305 Jy limit to the 0.31 Jy Molonglo lower limit, the corrected Bologna source density becomes 4665 ± 240 sr^{-1}, which is in excellent agreement with 4725 ± 170 sr^{-1} for Molonglo.

Uncertainties in the flux density scale and resolution correction do become a limitation for anisotropy searches made by comparing source densities in different surveys and even in different parts of the same survey. The source density uncertainties have reached 2% while the usual ~2% relative scale error translates to a 3% error in source density. If the full power of these surveys is to be ultimately available for this type of investigation, then particular care must be paid to the scale uniformity within each survey as well as to the relative scales between the surveys.

Like the Pooley & Ryle (1968) compilation of the strongest sources, Robertson's (1973) 408 MHz flux densities in the north are based on interpolations from the Kellermann et al (1969) flux density measurements of the 178 MHz 3CR survey (Bennett 1962a,b). Robertson's catalog extends further south than Pooley and Ryle's. The sources were selected from the Parkes catalog (combined catalog, Ekers 1969) and most have been remeasured more accurately with the Molonglo fan beam. At the 10 Jy lower limit the fan-beam confusion and noise error is ± 0.3 Jy with an overall 5% calibration uncertainty, or a ratio of lower limit to rms uncertainty of 17. Consequently no significant corrections to the southern counts are necessary. In the north the reliability of the Kellermann et al (1969) flux densities is such that the quoted overall error of 8% in the interpolated flux densities is reached.

There are two reasons why some incompleteness may be expected in the north. Robertson (1973) uses a scale difference of 10% in relating the Wyllie (1969) to the Kellermann et al (1969) scales, whereas the measured difference is $12 \pm 2\%$. In the north this results in an incompleteness of $3 \pm 3\%$, or $1 \pm 1\%$ over the whole sky. To this must be added the $0.5 \pm 0.5\%$ arising from translation of the 178 MHz finding survey to a 408 MHz listing by Robertson.

Because they reach such low flux densities, the 5C3 and 5C4 counts (Griffiths 1975) have been included in Figure 5, despite the reservations expressed earlier about them. It is not clear whether Griffiths has applied any improved radial flux density corrections to the original survey flux densities; therefore, there may be systematic errors in the counts (Condon & Jauncey 1973). This effect is least for the central region of the surveys, and the weakest sources are thus less likely to be affected than the stronger sources nearer the edge of the maps.

The 5C2 sources are not included in Figure 5, although they agree closely below 0.1 Jy with 5C3 and 5C4. The 5C1 and 5C2 regions were preselected to contain no intense radio sources (Kenderdine et al 1966). The effects of this selection can be determined from Maslowski's (1971) published maps, which overlap both the 5C1 and 5C2 regions. The 1400 MHz survey (Maslowski 1972) was divided into 20 min intervals of right ascension starting at 0720 and the number of sources with $S_{1400} \geqq 0.30$ Jy were counted in each interval. The mean number of sources per region (excluding the 5C2 overlap regions) is 9.2, with a standard deviation of 3.6. The observed number in the 5C2 region was 3, with only 1 region out of the 25 having fewer sources above 0.30 Jy. From this it is clear that the 5C2 region cannot be regarded as *necessarily* a representative area of weak radio sources.

Machalski et al (1974) also draw attention to the apparent difference in population law between the four subregions of Maslowski's (1972) survey. It is most noticeable that their region II overlaps both the 5C1 and 5C2 survey areas and it is just this region that has the greatest deficit of sources with $0.34 \leqq S_{1400} < 1.43$ Jy. Repeating the contingency table test on Machalski's data, but excluding region II, results in a chi squared of 18.5 for 10 degrees of freedom. The difference between the regions I, III, and IV of Machalski et al (1974) is significant only at the 0.05 level. Thus, it appears that much of the evidence regarding anisotropy in the distribution of radio sources at source densities of about 10^3 sr^{-1} may be traced directly to the preselection of the 5C1 and 5C2 regions. This is not to say that anisotropy is not present, but rather that evidence based on the 5C1 and 5C2 regions must be used cautiously.

It is clear from the 408 MHz n/n_0 diagram in Figure 5 that there is close agreement between the Molonglo and Bologna counts; the change in slope near 0.9 Jy noted by Mills et al (1973) is clearly present in both surveys. Figure 5 was normalized to the Molonglo density of 4725 sources sr^{-1} with $S_{408} \geqq 0.31$ Jy. No single power law can be fitted to the observations. Following Mills et al (1973) three regions can be recognized: the high flux density region above ~ 10 Jy; an intermediate region from about 0.5 Jy to 10 Jy where the slope is indistinguishable from 5/2; and a region below 0.5 Jy where there is a pronounced flattening of the slope below 5/2. Because of the paucity of data in the regions between about 3 and 10 Jy and 0.05 and 0.3 Jy, the n/n_0 diagram is lacking in detail. It is in just these regions, however, that the slope changes appear and where accurate data are most desirable.

It is noticeable that both the 178 and 408 MHz counts exhibit slope changes that appear *between* different surveys rather than in the middle of any one survey. There is a need for a large area survey above 3 Jy, and for continued coverage in the 0.05 to 0.3 Jy range at 408 MHz. This last flux density region may be accessible to Molonglo because of the small beam size, provided that the instrumental noise level can be reduced.

The details of the slope changes are of considerable cosmological importance. The change around 10 Jy appears to occur over a very limited flux density range, whereas the large dispersion in the source luminosity distribution would tend to spread any cosmological change over a much wider range (Mills et al 1973). The situation around 0.1 Jy, the top point in the 5C surveys, is more pronounced because the change there appears to be more significant than that for the stronger sources. This highlights the need for more observations to cover these flux density ranges.

SURVEYS AND COUNTS AT 1400 MHz

At 1400 MHz many of the surveys have been made with the NRAO 300-ft (91-m) telescope where the low confusion level allows accurate and reliable flux density measurements. The data quality has been further improved through an appreciation of the instrumental limitations and a greater reliance on independent flux density measurement comparisons between observers. Although most of the measurements were made with the NRAO 300-ft telescope, the earlier low-resolution surveys—the DA survey of Galt & Kennedy (1968), and the OSU surveys of Scheer & Kraus (1967), Dixon & Kraus (1968), Fitch et al (1969), and Ehman et al (1970)—were made with smaller instruments. Both the DA and OSU surveys are seriously affected by systematic flux density errors and low measurement accuracy, so much so that their value for source counts is seriously impaired. Bridle et al (1972a), in compiling their complete sample above 2 Jy at 1400 MHz, remeasured many of the cataloged DA sources, as well as many of their marginally detected but uncataloged sources. The resultant Bridle et al/DA flux density comparison very clearly shows the extent of the DA flux density overestimation bias as well as the large uncertainties present in their measurements. This is a situation very similar to that encountered with the Parkes 408 MHz flux density measurements.

With the OSU survey, Bridle et al (1972a) confirmed the large errors in the stronger source measurements, $S_{1400} \gtrsim 1$ Jy. In addition, the weaker sources were found to be systematically underestimated in flux density (Jauncey & Niell 1971). This underestimation was found to be about 0.08 Jy and is about the same as the OSU confusion error. The combination of this bias and the large flux density errors makes the production of reliable source counts from this survey very difficult, although the survey has been important in uncovering many unusual radio sources.

Complete pencil-beam surveys with the NRAO 300-ft telescope have been carried out by Maslowski (1971, 1972) and M. M. Davis (unpublished data); a strong source catalog has been compiled by Bridle et al (1972a). The measured intensity-independent error due to confusion and noise is 0.035 Jy, with a proportional error of $\pm 2\%$ (Bridle et al 1972a). At much lower flux densities the Westerbork synthesis radio telescope provides the first 1415 MHz limited surveys to about 0.007 Jy (Katgert et al 1973). Consequently the available dynamic range both in terms of flux densities and source densities at 1400 MHz rival those at 408 MHz.

Bridle et al (1972a) have produced a complete catalog of sources with integrated flux densities stronger than 2.0 Jy in the area of sky $-5° < \delta < 70°$, $|b| > 5°$. By supplementing their 300-ft measurements with the 1410 MHz data from 3.77 sr of the Parkes survey south of $-5°$ declination (Gardner et al 1969) they have completed essentially full sky coverage above 6.0 Jy (Bridle et al 1972b). This source listing produces essentially correction-free source counts above 2.0 Jy with a claimed completeness of $98 \pm 2\%$.

In a separate investigation Maslowski (1971, 1972) used the NRAO 300-ft telescope to survey a 6° strip of sky centered on declination 48°45′ and covering the right ascension range 0717–1623. Integrated flux densities were measured and all sources stronger than 0.09 Jy were cataloged (Maslowski 1972), although the source counts (Maslowski 1973) extend only to 0.16 Jy, or five times the standard error. Detailed Monte Carlo calculations have been made with the survey data to determine accurately the errors in the flux density measurements and their effects on the source counts (Maslowski 1972, 1973). Even at the lowest flux density level the corrections are small, and an examination of the uncorrected source counts reveals no obvious incompleteness at 0.16 Jy.

While the Bridle et al (1972a) measurements were being submitted for publication, the source counts from their strong source catalog, together with the M. M. Davis (unpublished) survey above 0.5 Jy, were presented separately (Bridle et al 1972b). In comparing his source counts with the published Davis counts Maslowski (1973) found a significant discrepancy between the two surveys for source densities in the range 0.5–2 Jy. He concluded that this difference implied significant anisotropy on an angular scale of a few tenths of a steradian, contrary to the earlier analysis at 178 MHz of the 4C survey (Holden 1966). This apparent discrepancy has since been resolved. Fomalont et al (1974) remeasured sources from the Davis, Maslowski, and Westerbork surveys with the 300-ft telescope, using the same reduction procedures as Bridle et al (1972a). They confirmed the accuracy and linearity of the Maslowski measurements but found a significant underestimation bias in the Davis flux densities that was apparently due to his source-fitting techniques.

The first Westerbork synthesis survey was undertaken as an attempt to detect radio sources around optically selected quasistellar objects (Katgert et al 1973). Unfortunately, half of the fields surveyed did not have complete coverage of the u/v plane, because the observations were made before full operation of the WSRT was achieved. As a result, any effects of flux density errors and incompleteness are not uniform over all of the fields surveyed.

The source counts at 1400 MHz from the 2.0 Jy catalog (Bridle et al 1972a) to which three sources had been added, the corrected Davis survey (Fomalont et al 1974), Maslowski's (1973) survey, and the first Westerbork survey have been collected by

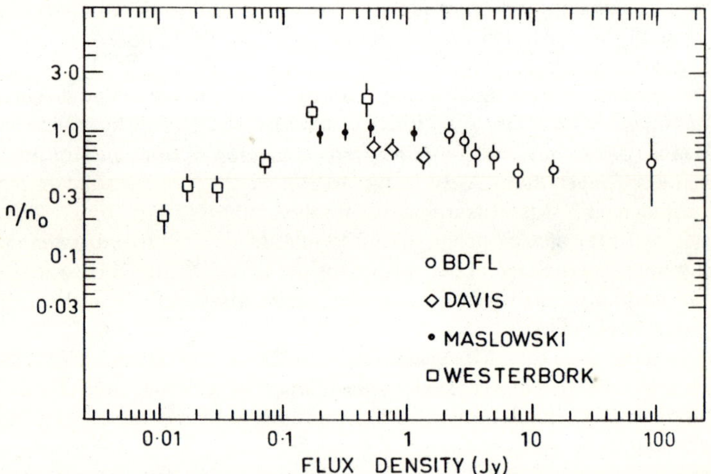

Figure 6 The normalized differential source counts at 1400 MHz, taken from Fomalont et al (1974), normalized to a 5/2 slope and a total source density of 52 sources sr^{-1} with $S_{1400} \geqq 1$ Jy.

Fomalont et al (1974) and are given in Figure 6. All of the NRAO 300-ft surveys have been normalized to the Kellermann et al (1969) scale, while the assumed flux densities for the Westerbork calibrators indicate that they are also within 2% of the same scale. As an additional check Fomalont et al (1974) measured a number of the most intense sources in the Westerbork survey. Within the measurement uncertainties they confirmed the equalities of the Westerbork and Kellermann et al (1969) scales and they found no evidence for any systematic trend in the Westerbork flux densities. This verifies the accuracy of the correction adopted by Katgert et al (1973) to correct observations of sources well beyond the half-power response of the Westerbork beam, although the scatter is still large for the strongest sources. In preparing the Westerbork counts Fomalont et al have renormalized the counts in the highest flux density range, $S_{1400} \geqq 0.465$ Jy, because Katgert et al chose a posteriori the flux density of the strongest source as their upper limit.

The correction to the Davis flux densities and the renormalization of the Westerbork counts above 0.465 Jy now remove much of the significance from the reported clustering (Katgert et al 1973) at source densities of around 500 sr^{-1} (Fomalont et al 1974). The Westerbork source density above 0.1 Jy is still higher than Maslowski's (1973) source density, although not significantly so in view of the small numbers involved in the synthesis survey.

The 1400 and 408 MHz curves of Figures 5 and 6 show good agreement. The 408 MHz falloff for the weak sources appears at about 0.5 Jy, or at an integral source density of $\sim 2.5 \times 10^3$ sources sr^{-1}. Although it is less well defined, the 1400 MHz falloff occurs below about the 0.16 Jy lower limit of Maslowski's (1973) counts, at about 3×10^3 sources sr^{-1}. At 1400 MHz there are as yet insufficient sources to bridge accurately the gap between the 2 Jy catalog and the 1 Jy level of the surveys, a situation very similar to that at 408 MHz noted earlier. The agreement for the strongest sources is expected, because both the 408 and 1400 MHz surveys contain many sources in common.

What is perhaps most noticeable in Figures 2, 5, and 6 is that the changes in slope occur between the surveys, rather than within any one survey, with the exception of the 408 MHz change near 0.5 Jy. The strong, medium, and weak surveys at 408 and 1400 MHz each reach comparable source densities. The changes in the intensity distribution occur by coincidence at just these source densities. Thus, the source counts at 1400 MHz provide little more quantitative information of the detailed shape of the $n(S)$ curve than is available at 408 MHz. This highlights once again the increasing need for more information over the relevant flux density intervals at both 408 and 1400 MHz.

Increasing the area to be surveyed with the 10′ arc resolution of the NRAO 300-ft telescope will lead to a solution regarding the slope change around 1–2 Jy but cannot lead directly to an accurate determination of the change near 0.1 Jy. Here this resolution leads to an rms confusion error of about 0.03 Jy, so that *reliable* source counting must stop at essentially the lower limit reached by Maslowski (1973). The gap may be bridged, however, by an analysis of the background deflection distribution (Scheuer 1957), for which the analysis techniques for pencil-beam observations now exist (Condon 1974). Fully computer-accessible surveys like

Maslowski's lend themselves very well to such analyses, which may provide independent confirmation of the synthesis results.

Higher resolution observations are also needed to define the shape of the $n(S)$ curve around 3×10^3 sources sr^{-1}, where the dropoff in numbers starts for the weak sources. Building up sufficient statistics at this level is slow and time consuming with the synthesis telescopes because of the small area surveyed and the time taken to achieve close to a full u/v coverage. The only pencil-beam instrument capable of these observations at 1400 MHz is the upgraded Arecibo reflector. This will have a tenfold improvement in angular resolution over the NRAO 300-ft telescope, which should lead to a similar decrease in survey lower limit.

SURVEYS AT 2700 AND 5000 MHz

The available surveys at these frequencies are the product of two groups: the 2700 MHz surveys in the southern hemisphere made at Parkes (Wall et al 1971, Shimmins 1971, Shimmins & Bolton 1972, Bolton & Shimmins 1973), and the 5000 MHz surveys of the NRAO group (Kellermann et al 1968, Davis 1971, Pauliny-Toth et al 1972, Pauliny-Toth & Kellermann 1972). At these frequencies the angular resolution available with filled aperture antennas is such that, except for the very weakest sources, the effects of confusion are negligible. Confusion is a problem only for the survey of the 5C1 and 5C2 regions by Kellermann et al (1968) and Pauliny-Toth et al (1972), respectively.

Much of the interest in high frequency surveys in recent years was prompted by the findings of Shimmins et al (1968) from their preliminary 2700 MHz survey. They found that the steep slope for their strongest sources appeared noticeably less pronounced at 2700 MHz than in the lower frequency surveys. Since then, the Parkes group has greatly extended its sky coverage at 2700 MHz, and at NRAO an appreciable fraction of the sky has already been surveyed at 5000 MHz.

While the confusion errors are usually less at the higher frequencies, the emergence of an increasing population of flat spectrum, optically thick, small diameter radio sources in the high frequency surveys (Kellermann et al 1968) introduces a complicating factor in using the source counts for cosmology. During the 10 years since the discovery of the first time variation in extragalactic radio sources (Dent 1965) an increasing proportion of these optically thick radio sources has been found to be variable (see e.g. Kellermann & Pauliny-Toth 1968). Even at 1400 MHz, intensity variations of up to 40% have been found, whereas at 5 GHz several sources have been found to vary by as much as a factor of 2. At 2.3 GHz, Nicholson (1973) found, from several years' observations of a complete sample at this frequency, that 31 out of 176 (17%) vary. He also found that at least 50% of the variables exhibit changes in flux density exceeding 30% in less than five years. At 8000 MHz the observed fraction of variables among the strongest sources has risen close to 60% (Brandie 1972).

This means that the flux densities and their distribution, and the sources present in a given survey, will change with time as a result of these intensity variations. This can be a serious complication, in view of the large flux density variations that have already been observed. When fitting to the observed number counts, the effects of the intrinsic intensity variations must be included. Although many individual sources have been studied in detail (e.g. Kellermann & Pauliny-Toth 1968), the necessary statistics for the complete surveys have not yet been assembled.

There is a further complication, namely, the change in the proportion of flat spectrum sources with flux density that has been found at 2700 MHz (Wall 1972, Condon & Jauncey 1974a) and at 5000 MHz (Condon & Jauncey 1974b). The strong correlation between flat spectrum and variability (Andrew et al 1972) implies that the fraction of variables also changes with flux density. Consequently, the effect of variability on the source counts needs to be determined as a function of both intensity and survey frequency. This variability problem is not entirely restricted to surveys above 1 GHz, since the possibility remains that the 408 MHz variability observed by Hunstead (1972) may be common at the lower frequencies.

A further aspect of the high frequency counts often not appreciated is that the luminosity distribution of the sources with "normal spectra" changes with finding frequency. The spectral index $\alpha = 0.75$ may be taken as an approximate dividing line for galaxies; those with a flatter spectral index tend to be intrinsically fainter than the steeper spectrum, strong radio galaxies (Kellermann et al 1969, Bridle et al 1972a). Because the median spectral index of the normal spectrum sources decreases with finding frequency, the high frequency surveys have proportionately more of the low luminosity galaxies and fewer of the high luminosity galaxies than the low frequency surveys (Condon & Jauncey 1974a). About 70% of the normal spectrum galaxies found in the NRAO 5 GHz survey are of low luminosity compared with only 30% of the 178 MHz galaxies (Condon & Jauncey 1974b). This implies that in going to high survey frequencies there is a change in the spatial distribution of the sources: there is an increase in the proportion of nearby galaxies accompanied by a decrease in the proportion of the more distant galaxies.

There is a similar selection among the quasars. The fraction of steep spectrum quasars of the type found mostly in the low frequency surveys decreases with increasing survey frequency. At the same time there is a rapid increase in the proportion of flat spectrum quasars.

These changes, with both finding frequency and flux density, must be incorporated in any physical description of the source counts from the high frequency surveys, as must the effects of variability.

The high angular resolution of the short wavelength surveys presents something of a problem in surveying the large areas of sky necessary to obtain significant source samples. A "fast scan" approach has usually been adopted (e.g. Kellermann et al 1968) where the two offset beams are scanned rapidly in a raster pattern across the sky. While confusion is not a problem, the survey completeness is determined by the receiver noise and stability, the tracking accuracy of the telescope, and the separation of the scans. In the first of the NRAO 5000 MHz surveys Kellermann et al claimed 100% completeness above 0.5 Jy. Their raw differential counts clearly show a rapid falloff below about 0.8 Jy, indicating incompleteness in the finding survey. This was confirmed in the later surveys (Pauliny-Toth et al 1972) and the declination separation between the right ascension scans was reduced to 3′.

The Parkes 2700 MHz surveys were made with declination scans at intervals of 6′ in right ascension (Wall et al 1971) so that no source should have been observed more than 0.75 of half the half-power beamwidth from the center of any scan. All of the sources above a predetermined intensity in the finding survey were reobserved and accurate position and intensity measurements made. The 100% completeness flux density level was chosen to be at least 1.5 times the finding survey cutoff. Subsequent comparison of the finding survey intensities with the more accurate measurements showed that few sources should have been missed with this technique (e.g. Shimmins 1971). In this way much of the southern sky has been surveyed to various flux density levels from 0.18 Jy (Shimmins & Bolton 1974) to 0.35 Jy (Wall et al 1971), and a complete survey to about 0.7 Jy of the whole sky south of declination 25° is in progress.

To extend the flux density range Wall et al (1971) conducted a special deep survey of the regions covered by six two-color plates made with the 48-inch Schmidt telescope of the Palomar Observatory. This survey covers 0.0753 sr and was designed to be complete to 0.10 Jy, although sources as weak as 0.06 Jy were cataloged. An examination of the differential counts from this survey reveals a rapid falloff in numbers below 0.10 Jy and suggests that incompleteness may still be a problem at flux densities slightly above the 0.10 Jy lower limit.

The increased area covered at 2700 MHz gives this survey much better statistics for total source densities above about 10^3 sources sr^{-1} than is available at lower frequencies. The n/n_0 diagram compiled by Wall & Cooke (1975) from the direct counts is shown in Figure 7. The dearth of sources above about 2 Jy is noticeable, but the transition from the intermediate flux density region, 0.1 to about 2 Jy at 2700 MHz, is smoother than at 408 MHz. In addition, the missing strong sources are fewer in number than at 408 MHz. This difference appears to be mostly due to

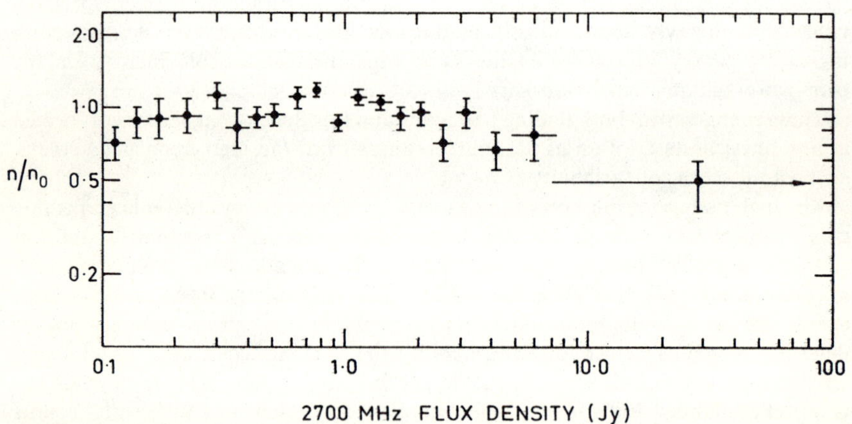

Figure 7 Normalized differential source counts at 2700 MHz collected by Wall & Cooke (1975) from the Parkes 2700 MHz survey.

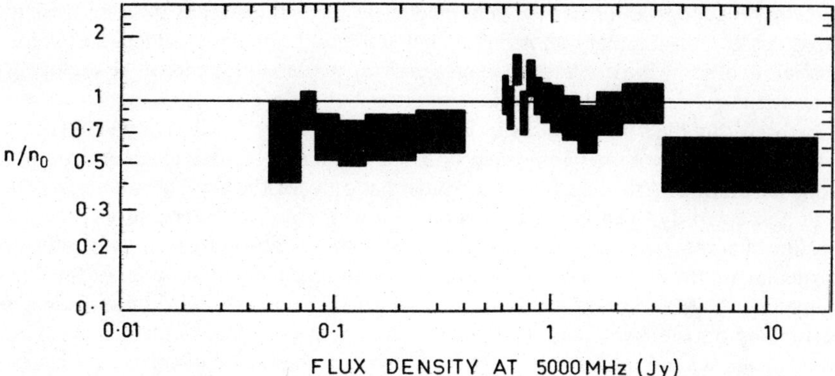

Figure 8 The normalized differential source counts at 5000 MHz taken from the results of Kellermann et al (1971).

the greater proportion of flat-spectrum sources at 2700 MHz and the much flatter $n(S)$ distribution for these sources as found in the Parkes $\pm 4°$ zone catalog (Condon & Jauncey 1974a). This difference between the high frequency and low frequency source counts shows clearly the importance of the dependence of radio luminosity function on finding frequency. Any satisfactory analysis of the source counts must take this into account.

The 0.10 Jy lower limit of the Parkes "selected areas" survey is set basically by the background sky confusion and the direct counting of sources is essentially impossible below this level. In the absence of any synthesis surveys at this frequency, Wall & Cooke (1975) have made measurements of the background deflection statistics and derived limits to the source counts using an extension of Scheuer's (1957) techniques. Although these limits are broad, nonetheless they show that the weak source falloff occurs at 2700 MHz and that the slope flattens below 5/2 in the 0.002–0.02 Jy range. With the sensitivity achievable with present-day receivers, this type of confusion analysis offers a potentially powerful alternative way of extending the source counts to millijansky levels as a valuable check on the synthesis surveys.

At 5.0 GHz the source counts extend to 0.05 Jy (Davis 1971) at a limiting source density little different from that reached in the 2.7 GHz deep survey (Wall et al 1971). Kellermann et al (1971) have assembled the source counts from the NRAO 5.0 GHz surveys and present a differential n/n_0 plot relative to a 5/2 slope; their results are reproduced in Figure 8. At the lower intensity limit of the Da vis high sensitivity deep survey, the signal-to-noise ratio drops to about 5 to 1. The counts from this survey have been corrected for the population law bias on the basis of Monte Carlo calculations with the original survey records. The sources were selected by visual inspection so that some incompleteness may be present near the lower limit of the survey. The "strong" and "intermediate" surveys extend to 0.6 and 0.25 Jy, respectively (Pauliny-Toth et al 1972), although the lower limit of the earlier 5000 MHz survey (Kellermann et al 1968) was increased to 0.8 Jy. Plotting the differential numbers from these catalogs confirms their incompleteness below the quoted lower limits. The rapid decrease in numbers immediately below these lower limits suggests that the possibility of some incompleteness even above the quoted limits cannot be excluded, although the effect should be small.

In discussing the source counts from the 5.0 GHz survey, Kellermann et al (1971) attack the problem of the significance, both statistical and cosmological, of the steep slope for the strong sources. They conclude that only for the very strongest sources with $S_{5.0} > 3.4$ Jy does the n/n_0 diagram show a significant departure from the Euclidean slope. Even then, the amount is only marginally statistically significant, particularly if the effects of source variability are to be included. Because of this low statistical significance, Kellermann (1972) rightly questions the wisdom of basing a cosmology on the 50 strongest sources. Kellermann et al (1971) further question whether the steep slope for the strongest sources can have *cosmological* significance because of the abrupt change in the n/n_0 diagram between the fast and the medium surveys. These same arguments are directly applicable to the low frequency counts, as has been seen earlier, although the number of strong sources missing over the whole sky is more significant at 400 than at 5000 MHz (Mills et al 1973).

CONCLUSION

After two decades of improving observations, many of the details of the log n-log S distribution of radio sources still need clarification. Three general regions have been recognized: a high flux density region above about 15 sources sr^{-1}; an intermediate region from about 1500 to 15 sr^{-1}, where the slope is indistinguishable from 5/2; and a low flux density region, at source densities above about 2000 sr^{-1}, where there is a pronounced flattening of the slope below 5/2. The details of the transitions between these regions are substantially missing because of the lack of adequate statistics. By coincidence, the survey flux density lower limits have usually been chosen in such a way that these changes occur between surveys rather than within any one survey. The instruments exist to bridge these gaps so that answers are possible.

More observations over these critical flux density ranges must be made, particularly at 408 and 1400 MHz, where the effects of intrinsic source variability are less than at the higher frequencies. The presence of unrecognized systematic errors in many of the earliest surveys emphasizes the need for both checks on the flux density measurements within a catalog and an independent evaluation of the completeness level of the finding surveys.

Before the high frequency 2700 and 5000 MHz source counts can be fully utilized for cosmological investigations, more needs to be known about the effects of source variability at all flux density levels. Additionally, high resolution observations are needed to extend the existing pencil-beam observations to lower flux density levels.

The value of radio source counts for cosmology is greatly enhanced as the details of the individual survey sources are incorporated. The optical identifications are most important in this regard, as is the observed flux density dependence of surface brightness and spectral index. Indeed, such information appears necessary if the full potential of radio source counts is to be realized for cosmology.

ACKNOWLEDGMENTS

I would like to thank A. H. Bridle, J. J. Condon, C. Hazard, K. I. Kellermann, and H. S. Murdoch for their many illuminating and valuable discussions over the years.

Literature Cited

Andrew, B. H., Medd, W. J., Harvey, G. A., Lock, J. L. 1972. *Nature* 236:447
Bailey, J. A., Pooley, G. G. 1968. *Mon. Notic. Roy. Astron. Soc.* 138:57
Bennett, A. S. 1962a. *Mon. Notic. Roy. Astron. Soc.* 125:75
Bennett, A. S. 1962b. *Mem. Roy. Astron. Soc.* 68:163
Bolton, J. G. 1956. *Observatory* 76:62
Bolton, J. G., Gardner, F. F., Mackey, M. B. 1964. *Aust. J. Phys.* 17:340
Bolton, J. G., Shimmins, A. J. 1973. *Aust. J. Phys. Ap. Suppl.* No. 30. 54 pp.
Brandie, G. W. 1972. *Astron. J.* 77:197
Bridle, A. H., Davis, M. M., Fomalont, E. B., Lequeux, J. 1972a. *Astron. J.* 77:405
Bridle, A. H., Davis, M. M., Fomalont, E. B., Lequeux, J. 1972b. *Nature Phys. Sci.* 235:123
Colla, G. et al 1970. *Astron. Ap. Suppl.* 1:281
Colla, G. et al 1972. *Astron. Ap. Suppl.* 7:1
Condon, J. J. 1974. *Ap. J.* 188:279
Condon, J. J., Jauncey, D. L. 1973. *Ap. J. (Lett.)* 184:L33
Condon, J. J., Jauncey, D. L. 1974a. *Astron. J.* 79:437
Condon, J. J., Jauncey, D. L. 1974b. *Astron. J.* 79:1220
Condon, J. J., Niell, A. E., Jauncey, D. L. 1971. *Bull. Am. Astron. Soc.* 3:25
Crawford, D. F., Jauncey, D. L., Murdoch, H. S. 1970. *Ap. J.* 162:405
Davies, I. M., Little, A. G., Mills, B. Y. 1973. *Aust. J. Phys. Ap. Suppl.* No. 28. 59 pp.
Davis, M. M. 1971. *Astron. J.* 76:980
Dent, W. A. 1965. *Science* 148:1458
Dixon, R. S., Kraus, J. D. 1968. *Astron. J.* 73:381
Durdin, J. M. et al 1973. *Nat. Astron. Ionos. Center Intern. Rep. 29*
Ehman, J. R., Dixon, R. S., Kraus, J. D. 1970. *Astron. J.* 75:351
Ekers, J. A., ed. 1969. *Aust. J. Phys. Ap. Suppl.* No. 7. 75 pp.
Fitch, L. T., Dixon, R. S., Kraus, J. D. 1969. *Astron. J.* 74:612
Fomalont, E. B., Bridle, A. H., Davis, M. M. 1974. *Astron. Ap.* 36:273
Galt, J. A., Kennedy, J. E. B. 1968. *Astron. J.* 73:135
Gardner, F. F., Morris, D., Whiteoak, J. B. 1969. *Aust. J. Phys.* 22:79
Gower, J. F. R. 1966. *Mon. Notic. Roy. Astron. Soc.* 133:151
Gower, J. F. R., Scott, P. F., Wills, D. 1967. *Mem. Roy. Astron. Soc.* 71:49
Griffiths, D. 1975. *Mon. Notic. Roy. Astron. Soc.* In press
Hewish, A. 1961. *Mon. Notic. Roy. Astron. Soc.* 123:167
Hoerner, S. von. 1973. *Ap. J.* 186:741
Holden, D. J. 1966. *Mon. Notic. Roy. Astron. Soc.* 133:225
Hunstead, R. W. 1972. *Ap. Lett.* 12:193
Hunstead, R. W., Jauncey, D. L. 1970. *Mon. Notic. Roy. Astron. Soc.* 149:91
Jauncey, D. L. 1967. *Nature* 216:877
Jauncey, D. L. 1968. *Ap. J.* 152:647
Jauncey, D. L., Niell, A. E. 1971. *Nature Phys. Sci.* 229:223
Katgert, P., Katgert-Merkelijn, J. K., Le Poole, R. S., Laan, H. van der. 1973. *Astron. Ap.* 23:171
Kellermann, K. I. 1972. *Astron. J.* 77:531
Kellermann, K. I., Davis, M. M., Pauliny-Toth, I. I. K. 1971. *Ap. J. (Lett.)* 170:L1
Kellermann, K. I., Pauliny-Toth, I. I. K. 1968. *Ann. Rev. Astron. Ap.* 6:417
Kellermann, K. I., Pauliny-Toth, I. I. K., Davis, M. M. 1968. *Ap. Lett.* 2:105
Kellermann, K. I., Pauliny-Toth, I. I. K., Williams, P. J. S. 1969. *Ap. J.* 157:1
Kenderdine, S., Ryle, M., Pooley, G. G. 1966. *Mon. Notic. Roy. Astron. Soc.* 134:189
Laan, H. van der, Le Poole, R. S. 1974. In *Proc. ESO/SRC/CERN Conf. Res. Programmes New Large Telesc.*, ed. A. Reiz, 259
Long, R. J., Smith, M. A., Stewart, P., Williams, P. J. S. 1966. *Mon. Notic. Roy. Astron. Soc.* 134:371
Longair, M. S. 1966. *Mon. Notic. Roy. Astron. Soc.* 133:421
Longair, M. S. 1973. Presented at IAU Symp. No. 63, Krakow
Machalski, J., Zieba, S., Maslowski, J. 1974. *Astron. Ap.* 33:357
MacLeod, J. M., Swenson, G. W. Jr., Yang, K. S., Dickel, J. R. 1965. *Astron. J.* 70:756
Maslowski, J. 1971. *Astron. Ap.* 14:215
Maslowski, J. 1972. *Acta Astron.* 22:227
Maslowski, J. 1973. *Astron. Ap.* 26:343
Mills, B. Y., Davies, I. M., Robertson, J. G.

1973. *Aust. J. Phys.* 26:417
Mills, B. Y., Slee, O. B. 1957. *Aust. J. Phys.* 10:162
Mills, B. Y., Slee, O. B., Hill, E. R. 1958. *Aust. J. Phys.* 11:360
Mills, B. Y., Slee, O. B., Hill, E. R. 1960. *Aust. J. Phys.* 13:676
Murdoch, H. S. 1969. *Proc. Astron. Soc. Aust.* 1:233
Murdoch, H. S., Crawford, D. F., Jauncey, D. L. 1973. *Ap. J.* 183:1
Murdoch, H. S., Large, M. I. 1968. *Mon. Notic. Roy. Astron. Soc.* 141:337
Nicholson, G. 1973. *Nature Phys. Sci.* 241:90
Niell, A. E., Jauncey, D. L. 1971. *Bull. Am. Astron. Soc.* 3:25
Pauliny-Toth, I. I. K., Kellermann, K. I. 1972. *Astron. J.* 77:797
Pauliny-Toth, I. I. K., Kellermann, K. I., Davis, M. M., Fomalont, E. B., Shaffer, D. S. 1972. *Astron. J.* 77:265
Pilkington, J. D. H., Scott, P. F. 1965. *Mem. Roy. Astron. Soc.* 69:183
Pooley, G. G., Kenderdine, S. 1968. *Mon. Notic. Roy. Astron. Soc.* 139:529
Pooley, G. G., Ryle, M. 1968. *Mon. Notic. Roy. Astron. Soc.* 139:515
Refsdal, S. 1969. *Ap. J.* 155:373
Robertson, J. G. 1973. *Aust. J. Phys.* 26:403
Ryle, M., Clarke, R. W. 1961. *Mon. Notic. Roy. Astron. Soc.* 122:349
Ryle, M., Neville, A. C. 1962. *Mon. Notic. Roy. Astron. Soc.* 125:39
Scheer, D. J., Kraus, J. D. 1967. *Astron. J.* 72:536
Scheuer, P. A. G. 1957. *Proc. Cambridge Phil. Soc.* 53:764
Shakeshaft, J. R., Ryle, M., Baldwin, J. E., Elsmore, B., Thompson, J. H. 1955. *Mem. Roy. Astron. Soc.* 67:106
Shimmins, A. J. 1971. *Aust. J. Phys. Ap. Suppl.* No. 21. 34 pp.
Shimmins, A. J., Bolton, J. G. 1972. *Aust. J. Phys. Ap. Suppl.* No. 26. 25 pp.
Shimmins, A. J., Bolton, J. G. 1974. *Aust. J. Phys. Ap. Suppl.* No. 32. 55 pp.
Shimmins, A. J., Bolton, J. G., Wall, J. V. 1968. *Nature* 217:818
Shimmins, A. J., Day, G. A., Ekers, R. D., Cole, D. J. 1966. *Aust. J. Phys.* 19:837
Wall, J. V. 1972. *Aust. J. Phys. Ap. Suppl.* No. 24:49
Wall, J. V., Cooke, D. J. 1975. *Mon. Notic. Roy. Astron. Soc.* In press
Wall, J. V., Shimmins, A. J., Merkelijn, J. K. 1971. *Aust. J. Phys. Ap. Suppl.* No. 19. 68 pp.
Williams, P. J. S., Collins, R. A., Caswell, J. L., Holden, D. J. 1968. *Mon. Notic. Roy. Astron. Soc.* 139:289
Williams, P. J. S., Stewart, P. 1967. *Mon. Notic. Roy. Astron. Soc.* 135:319
Willson, M. A. G. 1972. *Mon. Notic. Roy. Astron. Soc.* 156:7
Wyllie, D. V. 1969. *Mon. Notic. Roy. Astron. Soc.* 142:229

NEUTRINO PROCESSES IN STELLAR INTERIORS

Zalman Barkat
The Racach Institute of Physics, The Hebrew University of Jerusalem, Jerusalem, Israel

INTRODUCTION

The main driving force of stellar evolution is the continuous loss of energy into surrounding space. Photons are the carriers of the escaping energy during most of a star's lifetime. Struggling out of the depth of stars, photons suffer innumerable and varied collisions, where scattering, absorption, and reemission occur over and over again. A typical time scale for the leakage of electromagnetic radiation is of the order of 10^7 years. The astrophysicist who tries to describe the flow of this radiation must deal with complicated problems such as the calculation of opacities, etc. Since the work of Gamow & Schönberg (1941), it has been recognized that neutrinos play an important role in stellar evolution. But with the emergence of the theory of Universal Fermi interaction (Feynman & Gell-Mann 1958) physicists realized that there comes a time in at least some stars' lifetimes when the job of chief energy transporters is taken from the photons by neutrinos. Here, as well as elsewhere in this article, we use the term *neutrinos* to stand for both electron neutrinos and antineutrinos. Muon neutrinos (and antimuon neutrinos) are explicitly mentioned when needed. These elusive particles, whose interaction cross section with matter is of the order $\sim 10^{-44}\ x^2\ \text{cm}^2$ (where x is the neutrino momentum in MeV/c), have a mean free path, which in all but extreme conditions (see below) is much larger than stellar dimensions. Unlike photons, neutrinos thus once created behave essentially as local and instantaneous energy sinks. The astrophysicist, whose problem has been thus simplified, is now faced with another one, however, namely, the proper evaluation of the rate of production of neutrinos as a function of local conditions (temperature, density, composition). Much theoretical work concerning the description and quantitative formulation of possible physical processes where neutrinos are produced, scattered, or absorbed has been done through the years. Impressive experimental efforts to get precise information about and from these processes have been carried out and are still going on. Simultaneously, there have been numerous calculations of stellar evolution where the best up-to-date information has been incorporated to evaluate how evolution is affected by neutrinos. There have been attempts (see below) to present results of calculations of stellar evolution as astrophysical evidence for or against certain basic physical theories. Fowler & Hoyle (1964; see also Fowler 1967) have argued that "the terrestrial iron group isotopic abundance ratios strongly indicate the operation in massive stars of an energy loss mechanism having a loss rate of the same order of magnitude as that calculated for $e^{+}+e^{-} \rightleftarrows \nu+\bar{\nu}$ on the basis of the Universal Fermi interaction strength." A wider range of arguments has been presented by Stothers (1970b) who, by analyzing and comparing statistical astronomical data of various kinds with theoretical calculations sets lower and upper limits for the coupling constant of the electron-neutrino weak interaction. We also mention the conclusive ruling out of the suggested direct photon-neutrino coupling (Raychaudhuri 1970) on the basis of astrophysical evidence (Stothers 1970a, 1971).

Naturally there is room for scepticism. Most physicists probably would hesitate to accept evidence that relies only on the interpretation of observations through theoretical calculations, especially because these are quite complicated and involve a multitude of parameters for which at least a modest degree of uncertainty exists. We must mention, however, that astrophysical experts have become convinced that some results are highly invariant under the permissible variations of parameters, as strongly exemplified in the case of the missing solar neutrinos. In any case, although astrophysics must use basic physical results rather than produce them, attempts to produce physical results should go on, provided proper caution is taken.

Whereas behaving as cooling agents is the most common role of neutrinos, there is another role they are believed capable of. During catastrophic collapse, density and temperature may become so high that matter is no longer transparent; at the same time there is a tremendous flow of neutrinos from inner regions. The interaction of this flux with overlying matter may cause mass ejection.

This review is a discussion of the role of neutrinos in different relevant epochs of stellar evolution. We do not consider the interesting subject of neutrinos and cosmology, or the problem of the solar neutrinos that has been reviewed recently by Bahcall & Sears (1972). In Section 1 we outline the basic theoretical and experimental work. Section 2 is devoted to the effects of neutrino losses on late stages of stellar evolution in general. The evolution of stars in a special but important mass range $4 \lesssim M/M_{\odot} \lesssim 10$ (where pulsars might be formed) is discussed in Section 3. Section 4 deals with the dynamical collapse of stellar cores during which the special aspect of neutrinos' activity, that is, their interaction with outer layers, appears. The cooling of neutron stars is reviewed in Section 5. Concluding remarks are given in Section 6.

Earlier reviews on the present subject were given by Chiu (1966) and Ruderman (1965). We also mention the fundamental paper by Fowler & Hoyle (1964) and relevant chapters in the excellent book of Zel'dovich & Novikov (1971).

1 NEUTRINO PROCESSES

Neutrino processes can be divided into two categories: 1. neutrino (both electron and muon)-producing processes, and 2. processes where neutrinos interact with matter. Ordinary beta decays (including the Urca process) and interactions that rely on the direct electron-neutrino coupling predicted by the Universal Fermi interaction of Feynman & Gell-Mann (1958) or a similar model are discussed separately.

A *Neutrino-Producing Processes*

ORDINARY BETA PROCESSES Neutrinos (or antineutrinos) must participate in all processes where electrons (positrons) are either captured or emitted by nucleons or nuclei. The modern theory of beta decay that describes these processes has been applied and extended by Bahcall (1962, 1964) and by Peterson & Bahcall (1963) to cover conditions that are sometimes found in stellar interiors. In particular, allowance is made for the inhibition of decay due to the fact that final states might be occupied as a result of electronic degeneracy. Similarly one does expect that degeneracy can promote capture of electrons whose energy can become higher than the threshold for capture on a suitable nucleus. In high-temperature stellar interiors one can expect significant population of some low-lying excited states. Cameron (1959) has pointed out the importance of this situation ("photo-beta" processes) where a variety of intrinsically fast transitions can be introduced.

The basic equations relevant for these processes have been given in some of the works cited above. Useful analytical approximate expressions for the rates of the loss of energy due to the emitted neutrinos (supplemented by numerical tables) have been given by Hansen (1966, 1968), Beaudet et al (1972), and Mazurek (1973); for matter in nuclear statistical equilibrium by Tsuruta & Cameron (1965), Nadezhin & Chechetkin (1969), where references to other Russian works are given, and Barkat et al (1972). The processes treated are:

1. Interactions involving nuclei: Electron decay, positron decay, electron capture. Positron capture cannot occur in stellar interiors. It can occur only in neutron-rich matter (in the form of nuclei) at a high temperature and low density, where pairs can become abundant. As yet no one has been able to conceive of a scenario where such conditions can be met.

   The energy carried away by neutrinos produced in beta processes that accompany nuclear burning (e.g. positron emission in hydrogen burning either through the proton-proton chain or the CNO bi-cycle) is usually subtracted from the nuclear energy liberated. This type of neutrino loss is thus considered an integral part of the nuclear burning and not an independent sink, which is the situation in other types of neutrino loss.
2. Interactions with nucleons: Electron capture on free protons, and positron capture on free neutrons.
3. Urca process: In the Urca process a nucleus or nucleon alternatively captures an electron and undergoes a beta decay, emitting a neutrino and an antineutrino, respectively. The suggestion that such a process can take place in stellar interiors was first put forward by Gamow & Schönberg (1941), who have also coined the name "Urca" in memory of losers coming in and out of the gambling house of the Casino de Urca in Rio de Janeiro. The stellar conditions that promote

the operation of the Urca process are characterized by electron degeneracy such that the Fermi energy (E_F) is in the vicinity of the electron capture threshold of a candidate nucleus or nucleon. Phase space can become available for the Urca process by any of three mechanisms: (*a*) thermal rounding of the Fermi surface; (*b*) stellar vibrations that can periodically shift the E_F above and back below threshold; or (*c*) convective transportation of candidate nuclei across the level in the star where the E_F matches their respective threshold energies ("Urca shell").

Originally it was thought that thermal Urca losses could provide a dominant energy sink in stellar cores. However, at the conditions expected ($\rho \lesssim 10^9$ g/cm^3; $T \lesssim 6 \times 10^9$ °K), thermal Urca can hardly become significant. A most recent aspect of these losses is their contribution to delaying carbon ignition in the cores of stars of intermediate mass ($4 \leqq M/M_\odot \leqq 10$) (see Section 3). The revival of thermal Urca losses is associated with the theory of neutron stars where temperature and especially density are much higher than in evolving stellar cores.

Thermal rounding of the Fermi surface has been discussed by Chiu & Salpeter (1964), Finzi (1965), Ellis (1965), Bahcall & Wolf (1965a–c), and McFee & Schlitt (1971). Finzi (1966) discusses stellar vibrations, and Hansen (1966, 1968), and Tsuruta & Cameron (1970) discuss both mechanisms. Chiu & Salpeter (1964) introduced the "modified Urca process." The conditions expected in the interiors of neutron stars imply nonrelativistic degeneracy of the protons and neutrons, relativistic degeneracy of the electrons, and thermodynamic equilibrium. Under these conditions it is easy to show that the decay of a single neutron into a proton, electron, and antineutrino violates momentum conservation. The modified Urca process considers the collision of two neutrons at the top of the neutron Fermi sea, where one of them can decay. In this case momenta can be arranged in such a way as to conserve momentum.

The convective Urca mechanism was first suggested by Paczynski (1972), and later discussed and applied by Couch & Arnett (1973), Bruenn (1973b), Mazurek (1973), Paczynski (1973a, b), and Ergma & Paczynski (1974). All these works are concerned with the ability of this process to control carbon burning in high density degenerate cores (see Section 3).

For a purely thermal Urca process ($kT \gg E_F$) one would need a temperature in excess of 10^{10} °K. In dynamical collapse and hot neutron stars, however, such a high temperature accompanies a high density so that nevertheless $kT \ll E_F$.

DIRECT ELECTRON-NEUTRINO PROCESSES (LEPTONIC PROCESSES) The Universal Fermi interaction theory predicts a direct electron-neutrino coupling. On the basis of this theory, any time an electron changes its momentum state it is possible for a neutrino-antineutrino pair to be emitted. Several candidate processes have been considered by various authors over the years [see Fowler & Hoyle (1964), Chiu (1966), and Beaudet et al (1967)]. Of all processes considered, only five are actually significant:

1. Pair annihilation neutrinos. This process involves the formation of a neutrino-

antineutrino pair (rather than the usual photon) when an electron-positron pair annihilates

2. Photoneutrinos. The outgoing photon in Compton scattering is replaced by a neutrino-antineutrino pair
3. Plasma neutrinos. A photon propagates inside an electron gas ("plasmon") and transforms spontaneously into a neutrino-antineutrino pair
4. Bremsstrahlung neutrinos. Regular Bremsstrahlung is modified so that the radiated photon is replaced by a neutrino-antineutrino pair
5. Neutrino synchrotron. Regular photon radiation is replaced by a neutrino-antineutrino pair

The first three processes have been exhaustively discussed by Beaudet et al (1967), where references to earlier work are given (see also Dicus 1972).

The Bremsstrahlung case was quantitatively studied by Gandelman & Pinaev (1960) in the nonrelativistic limit and for nondegenerate electrons. Festa & Ruderman (1969) have extended the calculations into the realm of highly degenerate electrons. They take into account both screening of electrons on nuclei and the expected regular lattice arrangement of the nuclei at low temperatures. These authors give convenient formulas for the rates of energy loss. Cazzola, De Zotti & Saggion (1971) have studied the whole range of densities and temperatures relevant to astrophysics. Flowers (1973) has considered lattice effects and one-phonon corrections; De Zotti (1972) treats ion correlations.

The synchrotron neutrinos can become important only in the presence of a very strong magnetic field. Indeed, for the kind of fields expected in pulsars (see e.g. Gunn & Ostriker 1969), this process can become dominant (Canuto et al 1974).

The effects of strong magnetic fields on the various processes discussed above have been studied in recent years by Canuto and his associates (1974), Chou (1971), McFee & Schlitt (1971), and McFee (1971). Wolf (1966) and Itoh & Tsuneto (1972) have considered the effects of superfluidity.

At temperatures exceeding a few 10^{11} °K ($kT \gtrsim 50$ MeV) or densities $\gtrsim 7 \times 10^{14}$ g/cm^3 one can expect the formation of muon pairs that can decay into muon neutrino-antineutrino pairs. Some formulas for the rates of energy loss by muon neutrinos are given by Hansen (1968).

Bahcall & Wolf (1965a–c) have pointed out the importance of pions (if and when they appear) as candidates for modified Urca processes. They estimate that cooling rates can become larger by a factor $\sim 10^8$. This is due to the fact that pions are bosons and thus do not suffer from the exclusion principle. The latter introduces for fermions the restriction that only a small fraction, $\sim kT/E_F$, of them, those which are on the tail of the Fermi-Dirac distribution, can take part in the transitions implied by the Urca process.

Figure 1 shows (on the ρ-T plane) the regions of domination of the different neutrino-producing mechanisms. The schematic (isentropic sequences) evolutionary tracks (central conditions) of representative stellar cores are shown. The density temperature range covered here is relevant only for stellar evolution. Interiors of neutron stars fall beyond this range.

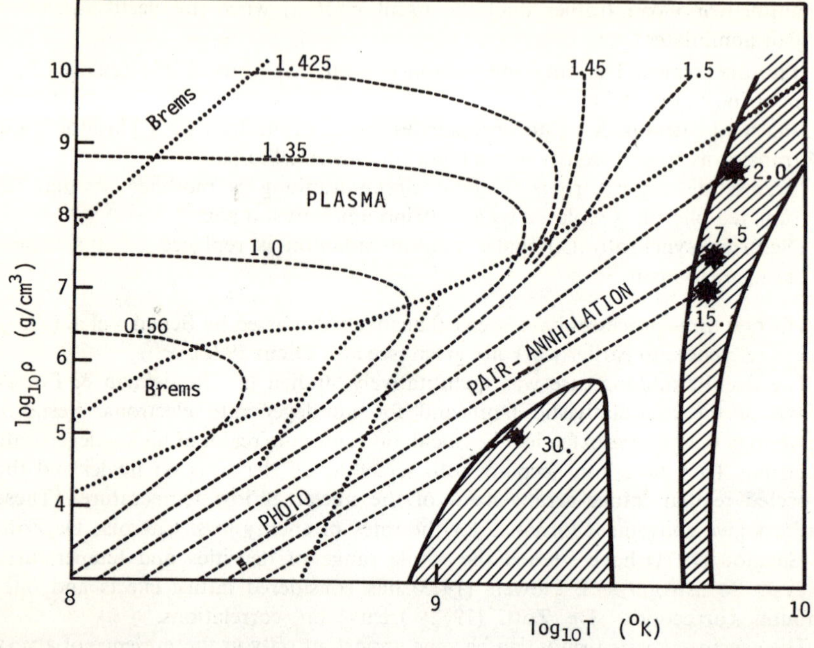

Figure 1 Regions of domination of four neutrino processes are displayed on the ρ-T plane. Also given are schematic (isentropic sequences) evolutionary tracks (representing central conditions) of stellar cores. These are labeled by their masses, which are assumed to be fixed. Regions where dynamical instabilities should be encountered are also shown. One is induced by pair formation ($e^+ + e^-$) and the other by iron-photodecomposition (Fe → He).

Whereas ordinary beta processes operate only under special conditions (in particular as far as composition is concerned), the processes discussed here are expected to be much more common because they depend almost entirely on temperature and density.

B *Interaction of Neutrinos with Matter*

The opacity of matter with respect to neutrinos has been investigated theoretically by Bahcall (1964), Bahcall & Frautschi (1964), where references to earlier work are given, and Bahcall & Wolf (1965a). These authors give estimates for the cross sections of the various possible interactions of neutrinos (both electronic and muonic) with matter: scattering off leptons and free nucleons, and absorption by bound nucleons. Hansen (1966, 1968) has given simple approximate formulas for these cross sections.

Theoretical work mainly in connection with gauge theories has been quite active recently. Weinberg's (1967, 1972) model has received most attention. The importance of the predictions of this model to astrophysical theories has been pointed out recently by Freedman (1973). It is obvious that his suggestions, as well as any other new theoretical developments, will receive attention. In fact Wilson (1974) has already looked into the possible significance of Freedman's findings in stellar dynamical collapse (see Section 4). Dicus (1972) has extended the work of Beaudet et al (1967) to include the effect of neutral currents that follow from Weinberg's theory.

A thorough review of the relevant experimental efforts cannot be given here. For further information and references the reader is referred to Bahcall & Sears (1972). Here we mention the continuous effort of Reines and co-workers, who were able to set quite stringent experimental limits on the ratio of the observed (σ exp) vs theoretical (σ_{VA}) cross sections: σ exp/$\sigma_{VA} = 1.1 \pm 0.9$. Further improvement on this is expected soon (F. Reines 1973, personal communication; see also Gurr et al 1972).

In the subsequent sections of this review we examine what the rates of energy loss and interaction cross sections imply for stellar structure and evolution. One of the most important problems in this regard is an evaluation of the sensitivity of results with respect to uncertainties in the data. The key problem here is to get a reasonably reliable estimate of the limits of uncertainty. A given margin for the coupling constant is equivalent to a scaling of rates. It is relatively easy to check what a scaling means in terms of stellar evolution, and we point this out when available. On the other hand, one must be prepared for more significant modifications than scaling as theory and experiment progress.

2 LATE STAGES OF EVOLUTION

In this section we discuss the role of neutrino losses in the late stages of stellar evolution, that is, the stages beyond which neutrino losses of direct coupling type (if they exist) become noticeable. This can happen only beyond helium core burning, that is, after a carbon-oxygen (C-O) core is formed.

A *General Properties*

Neutrinos produced in stellar interiors act as instantaneous and local energy sinks. To explain the nature of the influence of this sink on the late stages of stellar evolution, it is necessary to describe these stages. It is well known that stellar structure can be defined by the mass distribution of composition, $X_i(m)$, where X_i is the fractional density by mass of the ith nuclear component, together with the entropy density, $s(m)$, via the hydrostatic equilibrium condition (HEC)

$$\frac{\partial P}{\partial m} = -\frac{Gm}{4\pi r^4}, \qquad 2.1$$

where P is the pressure, r the radius within which the mass m is contained, G the gravitational constant, and t the time, together with the equation of mass conservation, expressed in the form:

$$\frac{\partial r}{\partial m} = \frac{1}{4\pi r^2 \rho}, \qquad 2.2$$

where ρ is the density. (All quantities are to be regarded as functions of m and t.) In addition one needs an equation of state $P = P(\rho, S;\ X_i)$ and a suitable boundary condition. Evolution is determined by the changes of these two mass distributions ("profiles"). For the rate of change of the entropy density profile one writes the energy equation [or, more properly, the heat conduction equation (HCE)]

$$T\frac{\partial s}{\partial t} = -\frac{\partial L}{\partial m} + \varepsilon + \sum_i \mu_i \frac{\partial X_i}{\partial t}, \qquad 2.3$$

where L is the energy flux. ε is the local net energy source (that is, sources minus sinks) and is written $\varepsilon = \varepsilon_{\text{nuc}} - \varepsilon_\nu$; ε_{nuc} is the heat source due to nuclear reactions and ε_ν is the energy sink due to neutrino losses. T is the temperature and μ_i is the chemical potential related to the ith nuclear component, defined thermodynamically as $\mu_i/T = \partial S/\partial X_i\ (\rho, u;\ X_i)$ where u is the energy density.

For the rate of change of the composition one writes:

$$\frac{\partial X_i}{\partial t} = R_i\,, \qquad 2.4$$

where R_i is the net rate at which the ith component is destroyed (or created) by nuclear reactions; R_i is calculated together with ε_{nuc}.

The term $\partial L/\partial m$, which appears in equation 2.3, is a troublemaker because of its nonlocal nature. In general L may be a combination of a radiative (and/or conductive) term (L_R) and a convective term (L_c). The first involves the opacity and/or conductivity and the temperature gradient, whereas special techniques are required for the evaluation of the latter. Note, however, that when convection is efficient one can actually avoid the calculation of L_c because such a convective zone can be looked upon as completely mixed (X_i independent of m) and isentropic (S independent of m); one needs to worry only about the integrated sources and sinks within it versus the radiative (and/or conductive) energy flux into and out of it (see Rakavy et al 1967). Late stages of evolution are characterized by the fact that in most cases $\partial L/\partial m$ can be neglected when compared to ε_ν. Entropy is then changed only by the nuclear (when and where nuclear reactions take place) and the neutrino terms, both of which are local. The correlation between different parts of the star is then provided only by the HEC (equation 2.1). A priori one would think that the difference between stellar evolution with and without neutrino losses would be very large. Indeed when one looks at the evolutionary tracks (it is convenient to describe evolution by drawing the path of the star's center on a ρ-T diagram) derived without regard to neutrino losses [see Figure 1 where we reproduce isentropic sequences of stellar-core models given by Rakavy (1967) and Rakavy & Shaviv (1968)], it is at once clear that at least the time scale of evolution, that is, the lifetime of a star at any given stage, can be cut down by a large factor because the conditions (ρ-T combinations) that occur call for very high rates of neutrino losses. The effect on the form of the track itself is not obvious, however. To induce a significant change in structure, neutrino losses would have to deform severely the otherwise typical entropy profiles. Indeed, the local nature of neutrino losses, together with their rather steep dependence on temperature and density, give rise to entropy profiles that are much steeper than otherwise. One might think that this should lead to significantly different evolutionary tracks; often this is not the case, however. In fact, the gross features of evolution can be derived even when sequences of isentropic models are constructed (see Rakavy & Shaviv 1968). We shall see later what important structural and evolutionary changes do occur in specific cases.

The criterion for stability against convection in the case of homogeneous composition (only!) can be expressed in the form.

$$\frac{\partial S}{\partial m} > 0, \tag{2.5}$$

whereas in a convective zone (in stellar interiors, where convection is always efficient) we have to high accuracy $\partial S/\partial m = 0$. It is thus clear that exoergic nuclear reactions always tend to induce convection outward whereas neutrino losses have a stabilizing effect outward and a destabilizing effect inward. Convection can be driven not only by nuclear reactions but also by neutrino losses.

The work of Schwarzschild & Harm (1965) has revealed that an evolutionary calculation starting from the main sequence gets into severe problems at the stage of double-shell (H, He) burning. In fact it is still not possible to proceed beyond this stage without employing special circumventing techniques. This situation led researchers to study late stages via a shortcut. One knows that stars develop inner cores composed of helium in the first stage and a mixture of oxygen and carbon in the second stage. Even though these cores are probably surrounded by outer layers of unburned helium and/or hydrogen ("envelope"), separated from the core by burning shells, there is a priori reason to believe that at least in the case of neutrino domination these elements have little if any influence on the core.

In fact the presence of outer layers is felt by the core at any given moment through the outer pressure, which appears in the HEC, and through the envelope's effect on the heat flux through the outer boundary, which appears in the HCE. The former is not expected to be significant at the relevant stages, because the envelope becomes very distended (giantlike structure). The modification of the heat flux set by the envelope cannot always be disregarded, however, especially if a burning shell exists there. Only in the case of neutrino domination, where heat fluxes are unimportant, can this modification be ignored. Cores might be affected by two other processes. First, it is possible that an outer convective zone will become deep enough so that it can invade outer layers of the core itself and mix them into the envelope, thus reducing the core's mass. Earlier investigation by Stothers & Chin (1969) indicated that extensive mixing of this type occurs for massive stars with $M \gtrsim 15\ M_\odot$. Subsequently it was shown by Sugimoto (1970a) that in the case of neutrino domination core-envelope mixing is rare. Paczynski (1970) finds core mass reduction only for $M \sim 7\ M_\odot$ (M_c reduced from 1.45 $M_\odot$ to 1.01 $M_\odot$). More work on this problem may be forthcoming.

Second, the mass of the core may be increased by the outward advance of a burning shell ("growing core"). The rate of growth must be compared to the rate of evolution of the core to see whether it can become significant. In Section 3 we present an example where the growth of the core is very important. Here we mention only that physically one must sometimes expect the existence of burning shells, which are associated with rather extensive convective zones, ahead of the thermonuclear active region. Such shells do not advance (masswise) in a continuous way. Instead they burn until extinction due to fuel consumption; meanwhile their convective zone may shrink. A new burning shell may form subsequently, but even if this happens we are faced with a jumpy behavior that must be distinguished from the continuous advance of thin burning shells.

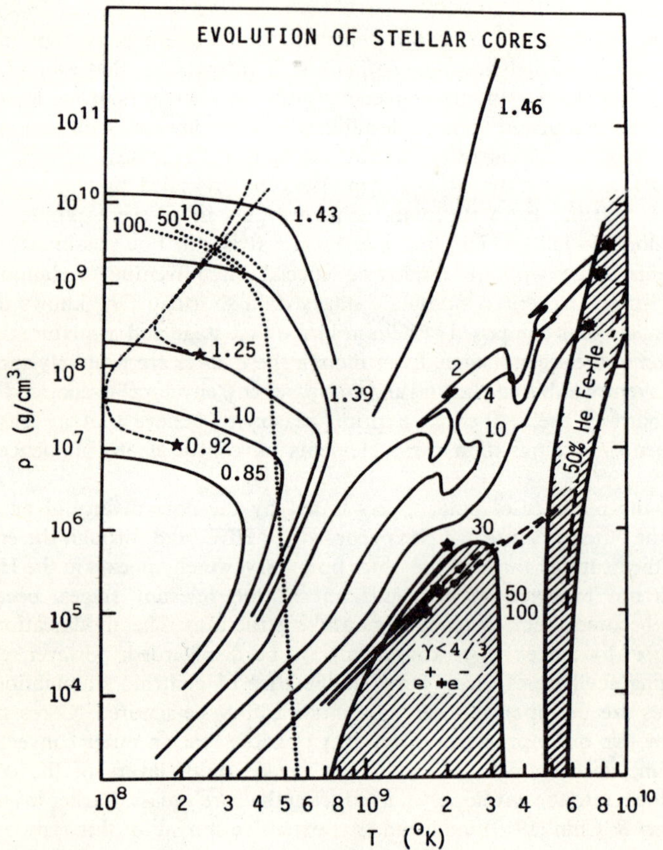

Figure 2 Evolutionary tracks of ^{16}O cores (solid lines). Masses label tracks and are assumed to be fixed. Neutrino losses are included. Stars at end of tracks represent dynamical instability. Dotted line is carbon ignition line which is given for three values of carbon mass fraction: 10, 50, and 100%. Dashed lines (*upper left*) represent evolution of growing carbon cores. Dashed lines (*lower right*) describe dynamical implosion-explosion.

Aside from the exception discussed in Section 3, evidence indicates that the description of inner cores as independent of their surroundings is justified (see Paczynski 1970, Arnett 1972). This does not mean, however, that the structure of the envelope does not depend on the core's behavior. On the contrary, the work of Stothers et al (see Stothers & Chin 1969 and references there) has shown that the ratio of expected observed red-to-blue supergiants may crucially depend on whether neutrino losses do occur in the cores of these stars (believed to be at stages of advanced core burning of C, O, Ne), and on the rate of loss. We discuss this problem in more detail in section 2F. In the rest of this section we discuss the results of the investigation of fixed mass cores. In the literature we find numerous calculations, all of which start from a certain He or C-O core. Qualitatively, the story of evolution can be classified according to the core's mass M_c. (In the following we use M_c to stand for the mass of a C-O core unless otherwise specified.) The results of the works described below must not be taken as referring to stellar masses on the main sequence. In fact only recently (see Arnett 1972a) has one been able to get a reasonable idea of the correlation between cores and their parent stars.

Figure 2 (taken mainly from Rakavy et al 1967, with some additions by the author) shows the evolutionary tracks of stellar cores of various masses. One immediately sees that a natural subdivision into mass regions is called for. We discuss separately the following cases: $M_c/M_\odot \lesssim 1.44$; $1.44 \lesssim M_c/M_\odot \lesssim 2$; $2 < M_c/M_\odot < 30$; and $30 < M_c/M_\odot$. The most important difference among these regions is the different instabilities that cores finally encounter. The boundaries of these regions are rather approximate and different authors do get somewhat different values.

B *Evolution in the Case* $M_c/M_\odot < 1.44$

A star whose mass satisfies the inequality $M < 5.75\ (Z/A)^2$ can end its life as a stable, cold, and degenerate configuration known as a "white dwarf." The right side is the famous "Chandrasekhar limit." The factor Z/A represents the composition, which in the simplest case is assumed to be homogeneous, and gives the ratio of nuclear charge Z to nuclear mass A. It is also true that actual stars do respect the Chandrasekhar limit, nearly as it stands, even when their composition is not necessarily homogeneous and they do not satisfy other simplifying assumptions. The factor Z/A is usually quite close to $\frac{1}{2}$ due to the fact that intermediate evolution leads relevant stars into regions where hydrogen must burn into helium.[1] Except for a minor neutronization (due to beta decays) at later stages, one does not expect the ratio Z/A to change. The approximate limiting mass thus becomes $M_{ch} \sim 1.44\, M_\odot$.

Any core whose mass is below M_{ch} (regardless of exact composition) evolves along a track that typically consists of three main regions: region I, where both temperature and density increase; region II, or the "Knee," where the temperature reaches a maximum and begins to drop while the density still increases; and region III, where the temperature eventually decreases at a constant density (see

[1] Very light stars can avoid hydrogen burning; such light stars are not of interest here, however, because neutrino losses never become significant in them.

Figure 1). In the literature one can find several calculations of the evolution of cores in this mass range where the role of neutrino losses is investigated. Probably the most recent work is that of Boozer et al (1973). Some earlier works are Vila (1966, 1967), Rakavy et al (1967), Ikeuchi et al (1972), where earlier work of C. Hayashi's group is cited, Beaudet & Salpeter (1969), Savedoff et al (1969), and Paczynski (1971). The characteristics of the effect of neutrino losses are: (*a*) neutrino losses become effective when $T \gtrsim 10^8$ °K ($\rho \sim 10^5$ g/cm^3); (*b*) the plasma neutrino rate dominates the photoneutrino rate for $\rho \gtrsim 10^6$ g/cm^3; and (*c*) neutrino losses become ineffective again along the decreasing temperature part of the evolutionary track due to their ρ-T dependence. Therefore, only for cores more massive than $\sim 0.3\, M_\odot$ do neutrino losses have any effect on evolution. Beyond this mass, however, they quickly become dominant.

Vila (1967) found that the shape of the evolutionary track does not change very much. It is flattened at the high temperature part so that T_{max} is reduced. This fact is important for considerations of the minimum mass for which carbon is still ignited in the course of evolution. This question was particularly investigated over the years by Salpeter, whose latest estimate is $M_{min} \sim 1.03\, M_\odot$ (Boozer et al 1973). Because the ignition line of carbon, defined as the location of points on the ρ-T plane where nuclear reactions supply energy at a rate that equals the rate of neutrino losses, happens to be almost parallel to the evolutionary tracks near T_{max}, one finds that small changes in T_{max} can lead to relatively large changes in M_{min}.

Much more pronounced are the following two effects of neutrino losses: As a core evolves toward the Knee the electrons become more and more degenerate and the thermal component of the pressure less and less important. At the same time plasma neutrino losses take over. That these losses may become higher toward the center (higher density) even if the temperature decreases, sets the stage for the unusual behavior where the temperature in the inner region not only can decrease faster than in outer layers but also can actually become lower, so that a positive temperature gradient is developed. This circumstance, which can be induced only by neutrino losses, leads to the possibility of nuclear fuel (e.g. C, O) ignition off center in a shell before it ignites in the center. This unique situation has recently been proposed by Iben (1972) as a possible resolution of the problem of pulsar formation (see Section 2), and may have other as yet unexplored consequences.

The rate of evolution along the Knee is accelerated significantly, especially for the more massive cores. For $M_c \sim 1\, M_\odot$ a typical time scale is shortened from 10^6 to 10^4 years. The same happens to the evolutionary track on the Hertzsprung-Russell (HR) diagram, which is traversed on a similarly short time scale. Another noticeable feature due to neutrino losses is an increase of the radiative luminosity (a factor of 10 for $1\, M_\odot$).

Stothers (1970b) has been able to use cooling times of bright white dwarfs to set an upper limit on the weak interaction's coupling constant by showing that a higher value would result in a deficiency of such objects, which is inconsistent with observation. These results offer an explanation for the observed behavior of the nuclei (central stars) of planetary nebulae (see Salpeter 1971 and other references therein). In particular the rapid evolution inferred observationally on the HR diagram is naturally understood.

In Section 4 we discuss some important effects of neutrinos where cores are growing (in mass) on account of an advancing burning shell.

C *Evolution in the Case* $1.44 \lesssim M_c/M_\odot \lesssim 2$

From Figures 1 and 2 one can clearly see that there is a region of stellar cores that is characterized by the achievement of high densities ($\rho \gtrsim 10^9$ g/cm^3) at relatively low temperature ($T \lesssim 4 \times 10^9$ °K). The significance of these conditions is due to the fact that they promote efficient electron capture (on the core's nuclei, i.e. ^{28}Si, ^{56}Fe) before the stage of iron decomposition can be realized (the latter is characteristic for the next mass range and is discussed below). Both processes induce a dynamical instability and it is not yet clear how different the two dynamical developments are (see Section 4). At this stage it is certainly possible that a significant qualitative difference does exist.

The upper boundary of the present region is somewhat arbitrary due to the existence of a transition zone where both electron capture (which appears first) and iron decomposition occur. It is clear that the effect of neutrino losses on this boundary is to increase its value. According to Ikeuchi et al (1972) the present region actually disappears if neutrino losses are omitted (upper boundary $\sim 1.4\,M_\odot$) and extends up to $\sim 2.6\,M_\odot$ when neutrino losses are included. A closer look at the reported results, however, does show that even without neutrino losses the width of the region is finite though indeed narrow (the upper boundary at $\sim 1.5\,M_\odot$).

This region has not been investigated as thoroughly as others for the following reasons: 1. On account of the cooling effect of neutrino losses and the resultant temperature inversion, fuel ignition (^{16}O) occurs in a shell off center. It is not easy to follow numerically the behavior of this shell as it converges toward the center. 2. The details of the electron capture process depend on the exact composition (e.g. small amounts of ^{32}S may be very important), which in turn is not easily and meaningfully determined.

In spite of these difficulties there have been a few calculations, not all of which have proceeded as far as the instability. We mention the works of Rakavy et al (1967), Kutter & Savedoff (1969), Ikeuchi et al (1972), and Paczynski (1971). One important result of these investigations, which is also valid for the following mass regions but not for the former, is that once neutrino losses dominate they never let down. In fact evolution is essentially, although not monotonically (burnings do cause temporary pauses), ever accelerated. The potential importance of this region for supernova and/or gravitational collapse theories will certainly be recognized in the near future.

D *Evolution in the Case* $2 \lesssim M_c/M_\odot \lesssim 30$

The work of Fowler & Hoyle (1964) emphasized the importance of massive cores ($M_c \sim 10\,M_\odot$) as candidates for type II supernovae via the famous mechanism of iron photodecomposition. Their work was followed by several evolutionary calculations by various authors [Rakavy et al (1967), Sugimoto (1970b), Ikeuchi et al

(1971), and Arnett (1972), where a comparison between these works appears]. Generally there is good agreement concerning the major features of evolution. In particular, evolution does lead cores in the present region along a track where temperature and density both increase (except for the appearance of loops related to fuel burnings) until the boundary of a region of dynamical instability (Fowler & Hoyle 1964) is crossed and soon afterward collapse ensues. This fact does not depend on the existence of neutrino losses, which do, however, change significantly a number of details:

1. The central density and the whole structure of the core at onset of instability are quite different (see Ikeuchi et al 1971, where the relevant density may differ by as much as two orders of magnitude).

2. The rate of evolution is highly accelerated. Typical time scales at the last stages are cut down from 10^3 yr to 10^4 sec. In fact Fowler & Hoyle (1964) have suggested that the time scale of evolution at the stage of ^{56}Ni beta decay into ^{56}Fe, which is terminated by the dynamical instability, and which thus determines the extent of this decay and thereby indirectly the ratios of isotopic iron group abundances, be used to indicate that neutrino losses do exist in the commonly predicted rate.

3. The size of the final (iron group) core, as well as the intermediate cores (^{16}O, ^{28}Si), is affected rather significantly. Rakavy et al (1967) find a narrow range of iron cores ($1.4 \lesssim M_{Fe}/M_{\odot} \lesssim 2.6$); Ikeuchi et al (1971) give similar results and emphasize the drastically different behavior of the relative mass fraction (M_c/M) of inner subsequent cores (i.e. C, O, Si, Fe), with respect to the total core mass: without neutrinos these ratios increase monotonically whereas with neutrinos the ratios decrease. Arnett (1972, 1973) gets generally similar results but he finds that with neutrinos the final M_c is nearly the same in all cases and rather close to $1.4\,M_{\odot}$. The fact that the imploding cores are very limited in size by neutrino losses is quite significant for the subsequent dynamical stage (see Section 4). The different chemical composition of the outer layers can be very important from the point of view of nucleosynthesis theory, that is, if it can be shown that these outer layers are ejected during the dynamical process (supernova).

E *Evolution in the Case* $30 \leqq M_c/M_{\odot}$

The evolutionary tracks of stellar cores whose mass is greater than $30\,M_{\odot}$ are known to carry the cores into a stage of dynamical instability due to pair formation (Ledoux 1965, Fowler & Hoyle 1964, Rakavy & Shaviv 1967).

Rakavy & Shaviv (1967) have shown that the effect of neutrino losses in this case as in other cases is mainly to accelerate evolution. Less pronounced are the structural effect and the resultant change in the value of the lower boundary of the present region ($30\,M_{\odot}$). More recent work by Barkat et al (1967), Fraley (1968), Ikeuchi et al (1971), and Arnett (1972) has produced essentially similar results. The investigation of the dynamics of these cores (Barkat et al 1967, Fraley 1968) has shown that neutrino losses do help to delay the bounce of the core during the implosion, which is rather mild. This delay is important inasmuch as it gives rise to a larger overshoot of the nuclear reactions (^{12}C and ^{16}O burning) and thereby a higher total energy release. The net result is that in all but the lighter cores the star is completely disrupted with no remnant left. These same studies have shown that there is an intrinsic negative feedback built into the dynamics, so that scaling up and down of neutrino losses (among other things) does not lead to a correspondingly large effect on the end result. In part, this negative feedback stems from the fact that a higher rate of energy release automatically leads to a shortening of the available time (earlier bounce) and thus the total energy integral does not vary too much.

Zel'dovich & Novikov (1971) have shown that neutrino radiation never becomes significant in supermassive stars ($M > 10^4 M_\odot$, if these exist), even when they collapse.

F *Envelope Structure*

One cannot overemphasize the importance of theoretical predictions concerning the envelope's structure that can be translated into observable quantities (i.e. luminosity L, effective temperature T_e). It is thus important to know the effect of neutrinos on the envelope's structure.

The temperature-density range for which neutrino losses are significant is such that stellar envelopes (He, H layers) can never be affected directly by them.

The envelope may be affected indirectly by neutrinos, however, through the neutrino's ability to change the time scale of evolution and the structure of the core. Much work on this problem has been carried out over the years by Stothers (1972) and others. The major issue involved is the establishment of a meaningful observational ratio of blue-to-red supergiants, and the interpretation of this ratio in terms of theoretical lifetimes of relevant stars at these stages. Stothers's (1972) latest important suggestion is that the observed paucity of red supergiants at very high mass confirms the conclusion that the carbon burning and later phases of evolution are very rapid, and consistent with the operation of neutrino losses.

3 THE EVOLUTION OF STARS IN THE RANGE $4 \lesssim M/M_\odot \lesssim 10$

Research has been directed in recent years toward finding an evolutionary link to the pulsars, where these are believed to be neutron star remnants of supernovae events. Circumstantial observational evidence suggests that the progenitors of such events are stars having $M \gtrsim 4M_\odot$ (see Shklovskii 1968, Sofia 1972, and Gunn & Ostriker 1970).

Evolutionary calculations starting from the main sequence encounter insurmountable technical difficulties at the stage of hydrogen-helium double shell burning (Weigert 1966, Schwarzschild & Harm 1967, Hoshi 1968). The difficulty is associated with a thermal instability that leads to a series of thermal relaxation oscillations.

Although it is impossible to ascertain whether the oscillations eventually cause mass loss and if so what amount is lost, one may reason as follows: If mass loss occurs and the envelope's mass is finally reduced to the extent that the total star's mass becomes less than M_{ch}, then one is left with a C-O white dwarf and there is not much of a story left. If, however, there remains enough of an envelope so that the total mass exceeds M_{ch}, then it becomes interesting to try and see what the fate of the star may be, using some special method to avoid the need of a detailed tracking of the oscillations. This has been done by Rose (1969), Arnett (1969), and Paczynski (1970).

It can be shown that essentially the core does not feel the presence of the envelope at all except through the burned out material it accretes at the wake of the advancing burning shell. It is reasonable to assume that the detailed and complicated burning can be translated and replaced by an effective, average rate of growth of the C-O core. Initially, that is, at helium exhaustion, which is approximately where neutrino losses take over, such a core is composed of a C-O mixture (where the exact proportion is still quite uncertain) and its mass turns out to be $\sim 0.5 M_{\odot}$ for a $3 M_{\odot}$ star, becomes $1 M_{\odot}$ for a $7 M_{\odot}$ star, and jumps to a value over $2 M_{\odot}$ for a star $\sim 10 M_{\odot}$ (Paczynski 1970). We feel that there is a real need for a more careful and systematic survey of the mass range $4 \lesssim M/M_{\odot} \lesssim 15$ where such effects as reduction of the core's mass due to convective (outer) mixing, and central (^{12}C) and off-central (^{16}O) ignition of fuel are expected to occur. Barkat (1971) has pointed out that the story of evolution in the present region can be told and analyzed in terms of the evolution of a growing core. Neutrino-induced entropy losses try to drive the core along its "natural" track, the track appropriate for its instantaneous mass (as discussed in Section 2A above), while the accretion of mass is shifting the track along an isentrope toward a higher mass natural track. As described above, cores satisfying $M_c \gtrsim M_{ch}$ evolve at an ever accelerated pace. It turns out that in this case the core does not have time to grow before it ends its evolution, that is, encounters an instability. On the other hand, if the core's mass soon after neutrino domination is still less than M_{ch}, we know that evolution must be slowed down beyond the Knee and in fact quite drastically so. Along the Knee itself evolution is so fast that no appreciable growth occurs. The nature of the transition zone of decelerating evolution and in particular its inclination on the ρ-T plane with respect to local isentropes favor a convergence of the evolutionary tracks of all growing cores. This analysis (Barkat 1971), which was offered as an explanation for the numerical results of Paczynski (1970) where such convergence had indeed been found, did open a way to parameterize evolution in the relevant stage. It thus became easy to get some quantitative estimates of variations of the unique (converged) track due to existing uncertainties (Barkat et al 1972a). Such estimates have become interesting for reasons that become clear later. Remember that the critical mass for ^{12}C ignition is $\sim 1 M_{\odot}$. It is clear that only if the core's mass at neutrino domination is smaller than $1 M_{\odot}$ (which, as mentioned above, corresponds to $7 M_{\odot}$) can the core preserve its original composition. The intermediate region $7 \lesssim M/M_{\odot} \lesssim 10$ has unfortunately not been explored yet; we do expect to find important results when it is. The work of Rose, Arnett, and Paczynski have shown that the unique track eventually leads to central carbon ignition at an approximate density of 2×10^9 g/cm^3. Here again *ignition* means equality of energy production rate (nuclear) and loss rate (neutrino). Arnett (1969) finds that ignition is followed by a nuclear runaway because there is no mechanism that can regulate burning in this case. He argues that a detonation (shock-induced burning that in turn reinforces the shock) is likely to be formed and sweep the whole core's mass, releasing the whole available thermonuclear energy on a dynamical time scale. The end result is a complete disruption of the whole star (core and envelope). If this is the case then obviously this is not where pulsars come from. Much of the work in recent years (for an overview see Buchler et al 1975) has been centered on the question of whether one could find a mechanism that will prevent disruption. Incidentally, it is not enough to show that a remnant is left; the remnant must be proven to evolve one way or another into a neutron star rather than a black hole. This means that collapse of the core (or part of it) must be accompanied by expulsion of outer layers. As will be shown in Section 4 these are not simple problems. Neutrino losses played an important role in the theoretical efforts to prevent disruption. For instance, electron capture on the nuclear products of burning behind the detonation front, accompanied by energy loss due to the emitted neutrinos, was suggested as a mechanism for removal of pressure that might lead to reimplosion (Colgate 1971, Barkat et al 1970). Quantitative examinations (Barkat et al 1970, 1971; Bruenn 1971, 1972; Buchler et al 1974a; and Mazurek et al 1974) have shown, however, that this can happen only when the central density at detonation is higher than a critical density, which is at least 10^{10} g/cm^3, compared to the value of 2×10^9 g/cm^3 that evolution suggests. This is where the question of whether it is possible to delay ignition beyond the critical density becomes interesting. The work of Barkat et al (1972) and of Bruenn (1972, 1973a, b) have shown that there is no way to delay ignition beyond $\rho \sim 6 \times 10^9$ g/cm^3 within resonable physical uncertainties. Among other factors examined were the inclusion of the hitherto neglected neutrino Bremsstrahlung losses (Festa & Ruderman 1969), simple scaling of total neutrino losses by factors of up to 100, and conceivable thermal neutrino Urca losses.

Still another suggestion was put forward by Paczynski (1972), who pointed out an important role that inevitable convective currents induced by the burning can play, well before runaway occurs: while nuclei are transported freely within a convective core, they see (as a function of time) different densities and thus different electronic Fermi energies (E_F). It becomes possible for a nucleus to be transported across a point (Urca shell) where E_F is equal to its own intrinsic threshold (E_{th}) for electron capture (emission). If this happens, convection sets up a new and interesting version of an Urca-shell cooling mechanism (Tsuruta & Cameron 1970). Paczynski could point to more than enough candidate nuclei that have their thresholds in the relevant density range and may be expected to be present even without relying on burning products of ^{12}C itself. Also, preliminary estimates have shown that the rate of neutrino losses could indeed become high enough to control burning (Paczynski 1972, Couch & Arnett 1973). However, Bruenn (1973b) and Mazurek (1973) have severely diminished the hope for the ability of convective Urca to control carbon burning: when electrons are captured from below E_F and if $kT \ll E_F$ (which is generally the case), electronic rearrangement will convert part of the degeneracy energy into thermal energy (γ rays absorbed locally). Because the convective current does transport nuclei into regions where E_F is considerably larger than E_{th}, it is expected that the net result will be heating rather than cooling. Although there has not yet been published any data that can definitely settle this problem, we feel that this conclusion will eventually be shown to be valid.

The crucial point in the evolution of stars in the range under discussion is that their cores manage to avoid carbon ignition at a relatively low density (where no explosion is expected if ignition does occur), but are then forced to ignite at a much higher density where explosion is possible. Neutrinos enable the cores to "turn the corner" while they are small enough and escape ignition. If neutrino losses had not occurred, cores would always grow big enough to ignite carbon non-explosively and then evolve peacefully as growing massive cores. No convergence would be expected.

4 DYNAMICAL COLLAPSE OF STELLAR CORES

Only stars whose mass is or can become smaller than M_{ch} can die peacefully as cold white dwarfs. It is possible and in some cases even suggested that heavy ($M > M_{ch}$) stars may lose mass at some stages of their evolution. However, even if and when this occurs, only outer layers, possibly a substantial part of the envelope, will be lost. The remaining stellar core may still be heavier than M_{ch}. Stellar evolution predicts that such stars must encounter a dynamical instability of one type or another. In the literature we find dynamical treatments of four different cases: 1. collapse of iron cores (belonging to massive, $M > 10 M_{\odot}$ stars) due to a dynamical instability induced by photodecomposition (Colgate & White 1966, Schwarz 1967, Arnett 1966, 1967, 1968, Ivanova et al 1969, Wilson 1971, Barkat et al 1974, and Wilson 1974); 2. collapse of a white dwarf due to (slow) electron capture (Hansen & Wheeler 1969, following a suggestion by Finzi & Wolf 1967, and Wheeler & Hansen 1971); 3. explosion of a stellar core due to a thermal (nuclear) runaway (Arnett 1969, Bruenn 1971, 1972, Buchler et al 1974a); and 4. collapse due to a dynamical instability set by pair formation of very massive cores $M_{core} \gtrsim 30 M_{\odot}$, which is turned into an explosion (Barkat et al 1967, Fraley 1968).

Of these four cases we have already partially discussed cases 3 and 4. In case 4 collapse is mild (accelerations are small in comparison with free fall) and a bounce definitely occurs. Case 3 does not involve collapse at all unless reimplosion occurs. Several suggestions for a mechanism to prevent disruption in this case, where neutrino losses are involved, have been discussed earlier (Section 3). If reimplosion does follow, it is expected that the physics will be very similar to that involved in cases 1 and 2. In fact, Colgate & White (1966) have shown that once collapse proceeds beyond a certain density (5×10^{10} g/cm^3), cooling due to the emission of neutrinos which accompany neutronization of matter (electron capture by protons) becomes efficient enough to ensure free fall. Once this stage is attained, why and how the collapse began is no longer important.

Therefore we concentrate on the collapse of iron cores (case 1 above), bearing in mind that other cases may behave similarly as far as the basic problems are concerned.

We have already mentioned the emission of neutrinos when protons capture electrons whose E_F becomes high enough to promote significant capture rate. At first only free protons, which have been knocked out of nuclei by hard photons, can capture electrons, but as the density increases, the E_F eventually becomes high enough so that protons in nuclei (mostly He) can capture electrons too. The energy carried away by the neutrinos helps to cool matter, but cooling becomes efficient only after its rate is higher than the rate of adiabatic compression (where $T \sim \rho^{1/3}$). According to Colgate & White's (1966) estimate, this happens at $\rho \sim 5 \times 10^{10}$ g/cm^3. In addition to cooling due to these (electron capture) neutrinos one also expects cooling due to the emission of neutrinos by direct processes. Colgate & White estimate, however, that the energy loss rate of pair-annihilation neutrinos, which is the second most important process here, will become higher than adiabatic compression only around $\rho \sim 2 \times 10^{11}$ g/cm^3. Sooner or later, depending on the exact form of the equation of state at about nuclear density, collapse of the inner regions must be halted and a core begins to form. The accumulation of falling matter on the inner core forms a shock ("core's shock"). It is expected that a large fraction of the shock heat created by the conversion of kinetic energy will be radiated away as a neutrino flux. It has been shown by Colgate & White that if this flux can leave the star unimpeded, collapse of the whole star (core and envelope) cannot be avoided. At this stage, however, the star is no longer transparent with respect to neutrinos. On this basis Colgate & White originally suggested that deposition of energy in outer layers due to interaction of flowing neutrinos with matter will create a powerful shock that may be effective enough to blow off large amounts of matter. This suggestion ["neutrino transport mechanism" (NTM)] seemed acceptable at first because of Colgate & White's illustrative calculations. As the developments of the theoretical calculations cited above have shown, one must be very careful in working with this problem, particularly in regard to the description of the neutrino flux. Both the source and the interaction of the flux with matter are determined by the ρ-T-composition profile (together with the relevant cross sections), but at the same time have a strong feedback effect on this profile. Among other questions concerning the source is whether muonic neutrinos are formed. If so, they must be carefully treated too. Wilson (1971) has emphasized the paramount importance of the correct treatment of the coupling of the neutrino flux (carrying $\sim 10^{53}$ ergs) with matter at the region where they are about to decouple (this region is gravitationally bound by only $\sim 10^{51}$ ergs). In fact, contrary to some earlier results, Wilson (1974) finds that the role of neutrinos in ejecting matter is actually negative, that is, they act as shock dampers by providing more cooling than heating. This result depends on the cross sections used, however, and, as recently indicated (Wilson 1974), if one is using opacities based on the Weinberg theory (see Freedman 1973) results can be quite different. Another important point was raised by Colgate & Chen (1972), who draw attention to the need for a study of the detailed physical shock structure, which is not reproduced by the common numerical codes (which use the pseudo-viscosity technique) and which may have a significant effect on the results.

Much of the published work has been concerned with heavy ($M > 3\,M_{\odot}$) collapsing iron cores. As evolutionary calculations have shown (see Section 2D) such cores are never formed.

Current estimates, based on the best but still quite uncertain physics available, seem to rule out the ability of NTM to eject matter off collapsing cores. However, the proponents of NTM have probably not had their last word (see e.g. Mazurek 1974, and Schramm & Arnett 1974).

5 COOLING OF NEUTRON STARS

There is as yet no satisfactory model for the formation of neutron stars. We do have a fair idea of the possible range of static neutron star models (see e.g. Cameron 1970) and know that, provided the mass does not exceed a certain critical mass, a neutron star can cool down to essentially zero temperature. Even in the absence of a formation model, and more properly because of this absence, it is interesting to see what can be said about the time scale of cooling of neutron stars. Such information may help to predict whether and how one can observe neutron stars. In fact, when the first few galactic X-ray sources were discovered, it was suggested that they might be neutron stars. It was concluded (Tsuruta & Cameron 1966) that on the basis of cooling alone one could not rule out this possibility. The discovery of pulsars, believed to be magnetic rotating neutron stars (Gold 1968), and more recently the indication that inner regions may be dominated by such phenomena as superfluidity and/or crystalline structure (see e.g. Pines 1972) have prompted a reexamination of the problem. The work on this subject has been closely related to the developments in both the theory of neutron star structure and the theory of neutrino processes. We do not wish to comment here on the former; we give a brief account of the latter.

The major share of cooling of the interiors of neutron stars (down to $T \sim 10^{9}\,°K$) belongs to neutrinos. At extremely high temperatures ($T > 10^{10}\,°K$), pair annihilation is the dominant loss mechanism and ensures cooling (down to $T < 10^{10}\,°K$) on a very short time scale (10^{-3} yr; Tsuruta 1972). As the temperature goes down, one finds (see Tsuruta et al 1972) that in the absence of a magnetic field and superfluidity, the Urca process is dominant. The Plasmon process competes with it at the higher temperature range and Bremsstrahlung at the lower range. It is shown (Canuto & Chou 1971) that a magnetic field (typical for pulsars) reduces the Urca neutrino luminosity as does superfluidity (Wolf 1966, Itoh & Tsuneto 1972). In this case neutrino Bremsstrahlung becomes the dominant cooling mechanism (see e.g. Flowers 1973). The cooling of magnetic superfluid neutron stars has recently been investigated by Tsuruta et al (1972) and Tsuruta (1972), as an extension of earlier work by Tsuruta & Cameron (1966) [see also Hansen & Tsuruta (1967), where vibrating neutron stars are investigated, and the review by Cameron (1970), where references to earlier work are given]. Typical cooling curves are given in Figure 3, which is taken from Tsuruta (1972). One observes significant earlier temperature drops due to both magnetic fields and superfluidity (a neutron star 10^6 yr old may be as cool as 10^3 °K as compared to 10^6 °K in the absence of these). This happens in spite of the reduction of Urca losses and is due to the effects of the magnetic field on the photon opacity (drastic reduction) on the one hand and the reduction of the heat capacity due to superfluidity on the other. Note also that differences appear only beyond 100 years. The current research activity in such fields as the equation of state at very high density (nuclear and subnuclear as well as ultradense densities) and neutron star structure will certainly be followed closely by an updating of our present knowledge and understanding of the subject.

6 CONCLUDING REMARKS

In this review we present an overall picture of the role of neutrinos in the evolution of stars. Our policy has been to include primarily the large amount of published

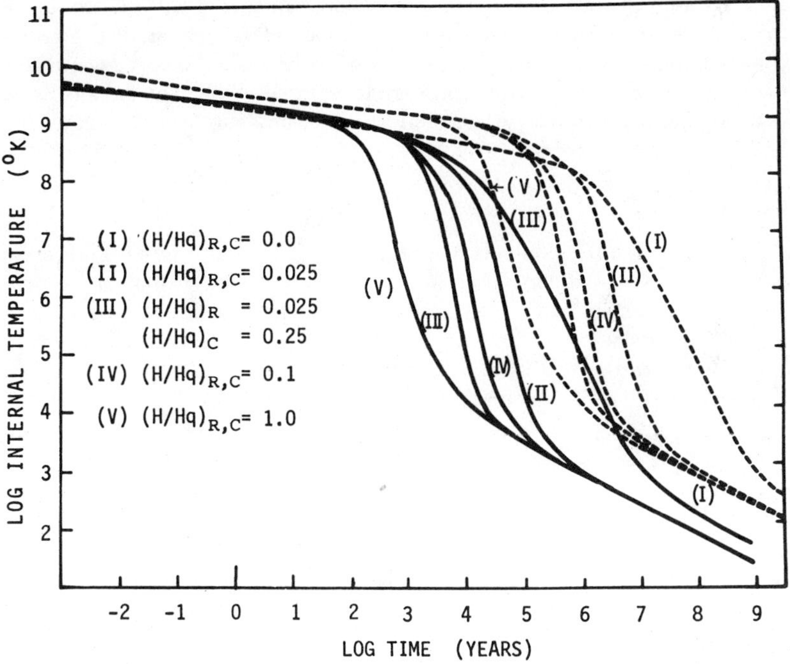

Figure 3 Cooling of neutron stars (internal temperature as function of time) as given by Tsuruta (1972). The dashed curves ignore superfluidity; the solid curves assume maximum superfluidity. Each family displays the effect of a magnetic field (assumed to be constant except in case III where the central field is larger than the surface field by a factor of 10). The magnetic field is measured in units of $H_q = m_e^2c^3/\hbar e = 4.41 \times 10^{13}$ gauss.

work based on the conventional charged current theory (besides ordinary beta decay). Recent attempts to check the implications of a neutral current theory have been mentioned only briefly. At the moment it does not seem that the latter are very significant, but perhaps one should not overstate this point at this time. It may not be out of place to express here our feelings, which we believe we share with a large number of workers in theoretical astrophysics, concerning the use of neutrino losses (of the direct coupling type). The main problem is that although neutrino losses are still not experimentally well established, researchers have become less and less hesitant to use these losses in their work. The most disturbing thing is that one does not have a satisfactory way to estimate probable errors or even extreme ones. Some authors still compare results "with" and "without" neutrino losses but this cannot be regarded as enough. The fact that the effects of conventional neutrino losses on stellar evolution are indeed found to be very important has led researchers to use astrophysical results as arguments in favor of their existence. It does remain true that a feeling of uneasiness is bound to remain with us until a more direct proof emerges. Therefore, when and if neutrino telescopes, which can efficiently look into the heart of stellar cores, can be built, one must expect a tremendous surge of interest in the present subject due to the possibility of having a more meaningful confrontation with observation. That Lande et al (1974) claim to have observed neutrino bursts may be the first sign of such a development (S. A. Bludman, personal communication). No one knows whether the current theory reviewed here will come out triumphant from such a confrontation or whether drastically new ideas will be needed. In any case, astrophysicists are no doubt waiting anxiously for such a development to occur.

ACKNOWLEDGMENTS

We would like to express our deep appreciation to the large number of authors who have sent us reprints and preprints of their work. Special thanks are due to Professors S. Treiman and F. Reines for very helpful comments. A long and interesting discussion with Professor S. Bludman is gratefully acknowledged. Finally it is a pleasure to thank Professor G. Rakavy and Dr. I. Reiss for their critical reading of the manuscript and many useful suggestions.

Literature Cited

Arnett, W. D. 1966. *Can. J. Phys.* 44:2553
Arnett, W. D. 1967. *Can. J. Phys.* 45:1621
Arnett, W. D. 1968. *Ap. J.* 153:335
Arnett, W. D. 1969. *Ap. Space Sci.* 5:180
Arnett, W. D. 1972. *Ap. J.* 176:681
Arnett, W. D. 1972a. *Ap. J.* 176:699
Arnett, W. D. 1973. *IAU Symp. No. 66, Warsaw, Poland*
Bahcall, J. N. 1962. *Phys. Rev.* 126:1143
Bahcall, J. N. 1964. *Ap. J.* 139:318
Bahcall, J. N., Frautschi, S. C. 1964. *Phys. Rev. B* 136:1547
Bahcall, J. N., Sears, R. L. 1972. *Ann. Rev. Astron. Ap.* 10:25
Bahcall, J. N., Wolf, R. A. 1965a. *Phys. Rev. B* 140:1452
Bahcall, J. N., Wolf, R. A. 1965b. *Ap. J.* 142:1254
Bahcall, J. N., Wolf, R. A. 1965c. *Phys. Rev. B* 140:1452
Barkat, Z. 1971. *Ap. J.* 163:433
Barkat, Z., Buchler, J. R., Wheeler, J. C. 1970. *Ap. Lett.* 6:117
Barkat, Z., Buchler, J. R., Wheeler, J. C. 1971. *Ap. Lett.* 8:21
Barkat, Z., Buchler, J. R., Wheeler, J. C.

1972. *Ap. J.* 173:183
Barkat, Z., Sack, N., Rakavy, G. 1967. *Phys. Rev. Lett.* 18:379
Barkat, Z., Wheeler, J. C., Buchler, J. R. 1972a. *Ap. J.* 171:651
Barkat, Z., Wheeler, J. C., Buchler, J. R., Rakavy, G. 1974. *Ap. Space Sci.* 29:267
Beaudet, G., Petrosian, V., Salpeter, E. E. 1967. *Ap. J.* 150:979
Beaudet, G., Salpeter, E. E. 1969. *Ap. J.* 155:203
Beaudet, G., Salpeter, E. E., Silvestro, M. L. 1972. *Ap. J.* 174:79
Boozer, A. H., Joss, P. C., Salpeter, E. E. 1973. *Ap. J.* 181:393
Bruenn, S. W. 1971. *Ap. J.* 168:203
Bruenn, S. W. 1972. *Ap. J.* 177:459
Bruenn, S. W. 1973a. *Ap. J.* 186:1157
Bruenn, S. W. 1973b. *Ap. J. Lett.* 183: L125
Buchler, J. R., Mazurek, T. J., Truran, J. W. 1975. *Comments Ap. Space Sci.* In press
Buchler, J. R., Wheeler, J. C., Barkat, Z. 1974a. *Ap. Space Sci.* 26:391
Cameron, A. G. W. 1959. *Ap. J.* 130:452
Cameron, A. G. W. 1970. *Ann. Rev. Astron. Ap.* 8:179
Canuto, V., Chou, C. K. 1971. *Ap. Space Sci.* 10:246
Canuto, V., Chiuderi, C., Chou, C. K., Fassio-Canuto, L. 1974. *Ap. Space Sci.* 28:145
Cazzola, P., De Zotti, G., Saggion, A. 1971. *Phys. Rev. D* 3:1722
Chiu, H.-Y. 1966. *Ann. Rev. Nucl. Sci.* 16:591
Chiu, H.-Y., Salpeter, E. E. 1964. *Phys. Rev. Lett.* 12:413
Chou, C. K. 1971. *Ap. Space Sci.* 10:291
Colgate, S. A. 1971. *Ap. J.* 163:221
Colgate, S. A., Chen, Y. 1972. *Ap. Space Sci.* 17:325
Colgate, S. A., White, R. H. 1966. *Ap. J.* 143:626
Couch, R. G., Arnett, W. D. 1973. *Ap. J. Lett.* 180:L101
De Zotti, G. 1972. *Mem. Soc. Astron. Ital.* 43:89
Dicus, D. A. 1972. *Phys. Rev. D* 6:941
Ellis, D. G. 1965. *Phys. Rev. B* 139:754
Ergma, E., Paczynski, B. 1974. *Acta Astron.* 24:1
Festa, G. C., Ruderman, M. A. 1969. *Phys. Rev.* 180:1227
Feynman, R. P., Gell-Mann, M. 1958. *Phys. Rev.* 109:193
Finzi, A. 1965. *Phys. Rev. B* 137:472
Finzi, A. 1966. *Phys. Rev. Lett.* 15:599
Finzi, A., Wolf, R. A. 1967. *Ap. J.* 150: 115
Flowers, E. 1973. *Ap. J.* 180:911
Fowler, W. A. 1967. *Proc. Int. Sch. Phys. Enrico Fermi* 35:367
Fowler, W. A., Hoyle, F. 1964. *Ap. J. Suppl.* 9:201
Fraley, G. 1968. *Ap. Space Sci.* 2:96
Freedman, D. Z. 1973. *NAL-Publ.-73/76-THY*
Gamow, G., Schönberg, M. 1941. *Phys. Rev.* 59:539
Gandelman, G. M., Pinaev, V. S. 1960. *Soviet Phys. JETP* 10:764
Gold, T. 1968. *Nature* 218:731
Gunn, J. E., Ostriker, J. P. 1969. *Nature* 221:454
Gunn, J. E., Ostriker, J. P. 1970. *Ap. J.* 169:972
Gurr, H. S., Reines, F., Sobel, H. W. 1972. *Phys. Rev. Lett.* 28:1406
Hansen, C. J. 1966. PhD thesis. Yale University, New Haven, Conn.
Hansen, C. J. 1968. *Ap. Space Sci.* 1:499
Hansen, C. J., Tsuruta, S. 1967. *Can. J. Phys.* 45:2823
Hansen, C. J., Wheeler, J. C. 1969. *Ap. Space Sci.* 3:464
Hoshi, R. 1968. *Progr. Theor. Phys.* 39: 957
Iben, I. Jr. 1972. *Ap. J.* 178:433
Ikeuchi, S., Nakazawa, K., Murai, T., Hoshi, R., Hayashi, C. 1971. *Progr. Theor. Phys.* 46:1713
Ikeuchi, S., Nakazawa, K., Murai, T., Hoshi, R., Hayashi, C. 1972. *Progr. Theor. Phys.* 48:1870
Itoh, N., Tsuneto, T. 1972. *Progr. Theor. Phys.* 48:1849
Ivanova, L. N., Imshennik, V. S., Nadezhin, D. K. 1969. *Sci. Inform. Astron. Counc. USSR Acad. Sci.* 13
Kutter, G. S., Savedoff, M. P. 1969. *Ap. J.* 156:1021
Lande, K. et al 1974. *Nature* 251:485
Ledoux, P. 1965. *Stars and Stellar Systems,* 3:527. Chicago: Univ. Chicago Press
Mazurek, T. J. 1973. PhD thesis. Yeshiva University, NY
Mazurek, T. J. 1974. "Chemical Potential Effects on Neutrino Diffusion in Supernovae." Preprint
Mazurek, T. J., Truran, J. W., Cameron, A. G. W. 1974. *Ap. Space Sci.* 27:261
McFee, R. W. 1971. *Ap. Space Sci.* 14:446
McFee, R. W., Schlitt, D. 1971. *Ap. Space Sci.* 12:4
Nadezhin, D. K., Chechetkin, V. M. 1969. *Soviet Astronomy AJ* 13:213
Paczynski, B. E. 1970. *Acta Astron.* 20:47
Paczynski, B. E. 1971. *Acta Astron.* 21:1
Paczynski, B. E. 1972. *Ap. Lett.* 11:53
Paczynski, B. E. 1973a. *Acta Astron.* 23:1
Paczynski, B. E. 1973b. *Ap. Lett.* 15:147
Peterson, V., Bahcall, J. N. 1963. *Ap. J.* 138:437

Pines, D. 1972. *Proc. IAU Symp. No. 53, Boulder, Colorado*
Rakavy, G. 1967. *High Energy Physics and Nuclear Structure,* ed. G. Alexander, 226. Amsterdam: North-Holland
Rakavy, G., Shaviv, G. 1967. *Ap. J.* 148: 803
Rakavy, G., Shaviv, G. 1968. *Ap. Space Sci.* 1:429
Rakavy, G., Shaviv, G., Zinamon, Z. 1967. *Ap. J.* 150:131
Raychaudhuri, P. 1970. *Can. J. Phys.* 48: 395
Rose, W. K. 1969. *Ap. J.* 155:491
Ruderman, M. A. 1965. *Rep. Progr. Phys.* 28:411
Salpeter, E. E. 1971. *Ann. Rev. Astron. Ap.* 9:127
Savedoff, M. P., Van Horn, H., Vila, S. C. 1969. *Ap. J.* 155:221
Schramm, D. N., Arnett, W. D. 1974. "The Weak Interaction and Gravitational Collapse." Preprint
Schwarz, R. A. 1967. *Ann. Phys.* 43:42
Schwarzschild, M., Harm, R. 1965. *Ap. J.* 142:855
Schwarzschild, M., Harm, R. 1967. *Ap. J.* 150:961
Shklovskii, I. S. 1968. *Supernovae.* New York: Wiley
Sofia, S. 1972. *Ap. J.* 172:53
Stothers, R. B. 1970a. *Phys. Rev. D* 2: 1417
Stothers, R. B. 1970b. *Phys. Rev. Lett.* 24: 10
Stothers, R. B. 1971. *Phys. Lett. A* 36:5
Stothers, R. B. 1972. *Ap. J.* 175:717
Stothers, R. B., Chin, C. W. 1969. *Ap. J.* 158:1039
Sugimoto, D. 1970a. *Progr. Theor. Phys.* 44: 375
Sugimoto, D. 1970b. *Progr. Theor. Phys.* 44:599
Tsuruta, S. 1972. *Proc. IAU Symp. No. 53, Boulder, Colorado*
Tsuruta, S., Cameron, A. G. W. 1965. *Can. J. Phys.* 43:2056
Tsuruta, S., Cameron, A. G. W. 1966. *Can. J. Phys.* 44:1863
Tsuruta, S., Cameron, A. G. W. 1970. *Ap. Space Sci.* 7:374
Tsuruta, S., Canuto, V., Lodenquai, J., Ruderman, M. 1972. *Ap. J.* 176:739
Vila, S. C. 1966. *Ap. J.* 146:437
Vila, S. C. 1967. *Ap. J.* 149:613
Weigert, A. 1966. *Z. Ap.* 64:395
Weinberg, S. 1967. *Phys. Rev. Lett.* 19:1264
Weinberg, S. 1972. *Phys. Rev. D* 5:1412
Wheeler, J. C., Hansen, C. J. 1971. *Ap. Space Sci.* 11:373
Wilson, J. R. 1971. *Ap. J.* 163:209
Wilson, J. R. 1974. *Phys. Rev. Lett.* 32:849
Wolf, R. A. 1966. *Ap. J.* 145:834
Zel'dovich, Ya. B., Novikov, I. D. 1971. *Relativistic Astrophysics.* Chicago: Univ. Chicago Press

THERMONUCLEAR REACTION RATES, II[1]

William A. Fowler,[2] *Georgeanne R. Caughlan,*[2,3] *and Barbara A. Zimmerman*[2]
California Institute of Technology, Pasadena, California 91125, and
Montana State University, Bozeman, Montana 59715

There is something fascinating about science. One gets such wholesale returns of conjecture out of such a trifling investment of fact.

Mark Twain, *Life on the Mississippi*

INTRODUCTION

It has been eight years since the publication of *Thermonuclear Reaction Rates* by Fowler, Caughlan & Zimmerman (1967) (hereafter referred to as FCZ I) in volume 5 of this series. In the interim an enormous amount of experimental data has been obtained in nuclear laboratories throughout the world and new theoretical and computer-programmed techniques for analyzing these data and applying them to astrophysical circumstances have been developed. It has been 101 years since Mark Twain spoke so forthrightly, so it is indeed time for another "investment of fact." In this article we revise and update the charged particle reactions involving nuclei with $A \lesssim 30$ in FCZ I, except for the $H^2(d,\gamma)He^4$ reaction whose reaction rate is very uncertain. Reactions involving neutrons are treated as neutron-producing reactions but factors for calculating the reverse reaction rate are provided as for all reactions. We have not revised the rates of the (n,γ) reactions given in FCZ I. To avoid duplication the reader is referred to FCZ I for the basic ideas underlying the determination of nuclear reaction rates under astrophysical conditions.[4]

[1] The survey of literature for this review was concluded in November 1974.

[2] Work supported in part by the National Science Foundation (GP-28027 and GP-36687X) at the California Institute of Technology.

[3] Work supported in part by the National Science Foundation (GP-9673) at Montana State University.

[4] We repeat here the units we have adopted: 10^9 °K as the unit of temperature, $T_9 = T/(10^9\,°K)$ and 1 MeV $= 1.6022 \times 10^{-6}$ erg $= 1.6022 \times 10^{-13}$ J as the unit of energy, $E_6 = E/(10^6$ eV). In these units Boltzmann's constant is given by $k = 8.6171 \times 10^{-2}$ MeV/10^9 °K = 1 MeV/(11.6049×10^9 °K). The atomic mass unit is given by $M_U = N_A^{-1} = 1.66053 \times 10^{-24}$ g = 931.4812 MeV/c^2, and Avogadro's number by $N_A = 6.0222 \times 10^{23}$ mole^{-1}. We did not think it necessary to adopt the small changes ($< 3 \times 10^{-5}$) in the fundamental constants recommended by Cohen & Taylor (1973) and Cohen (1974).

Reference to equations in that article will be preceded by I and definitions of many quantities used in equations in this article will be found only in FCZ I. The results of our analysis are given in Table 1. As in FCZ I we do not include factors for screening nor pycnonuclear effects (DeWitt, Graboske & Cooper 1973; Graboske et al 1973). Furthermore, we assume that the Maxwell-Boltzmann distribution holds for the relative velocities of the interacting particles, but see Clayton (1974).

We have in preparation a supplement to this article that will include a detailed description of the sources and analysis of the data used for each of the reactions given in Table 1. References to original material are not consistently included in the present article. A complete set of references for each reaction can be found for $A = 5$–10 in Ajzenberg-Selove & Lauritsen (1974); for $A = 11$–12 in Ajzenberg-Selove & Lauritsen (1968); for $A = 13$–15 in Ajzenberg-Selove (1970); for $A = 16$–17 in Ajzenberg-Selove (1971); for $A = 18$–20 in Ajzenberg-Selove (1972); and for $A = 21$–44 in Endt & Van der Leun (1973). The citations here refer in general to experimental, theoretical, or analytical techniques and not to data sources. Preprints of the supplement will be available in the near future and can be obtained by writing to us. In addition, we will accept inquiries for analysis of important and significant reactions ($A \leqq 30$) that have not been included. Comparison with experimental data for reactions involving nuclei with $A > 30$ will be included in a series of theoretical papers using Hauser-Feshbach calculations now in preparation by various permutations and combinations of Woosley et al (1975).

We frequently receive inquiries from nuclear physicists asking where they should direct their efforts to make significant new measurements or to improve old ones. The many uncertain entries in Table 1 preceded by the factor (0 to 1), which is defined below, should serve to answer these questions in a succinct manner. The promised supplement will give details.

We will be most grateful if those who find errors in Table 1 will let us know.

INSTRUCTIONS FOR THE USE OF TABLE 1

The first column of Table 1 lists nuclear reactions in standard notation insofar as possible within the limitations of computer printout. The chemical symbol for the "target" nucleus followed by the integral mass number is placed before the first parenthesis. The incident "particles," one or two in number, are placed inside the parentheses *before* the comma. The emitted "particles," one or two in number, are placed inside the parentheses *after* the comma. The symbols used inside the parentheses are G for gamma ray, E− for electron, E+ for positron, NU for neutrino, N for neutron, P for proton, D for deuteron, T for triton, HE3 for the helium-3 nucleus, and A for the alpha particle. The "residual" nuclei, one or two in number, are represented by their chemical symbols and integral mass numbers placed after the parentheses.[5] The particles and nuclei to the left of the comma

[5] In the notation of FCZ I, the reactions tabulated are of the type $0+1 \rightarrow 2+3$, $0+1 \rightarrow 2+\gamma$, $0+1 \rightarrow 2+3+4$, $0+1 \rightarrow 2+3+4+5$, $0+1+2 \rightarrow 3+\gamma$.

are the primary interacting bodies and those to the right are the product bodies. These products become the primary bodies in the reverse reaction. The three reactions between pairs of C^{12} and O^{16} nuclei are followed by the compound nucleus in parentheses; the branching ratios for breakup into the n, p, and α-channels are given at the end of Table 1.

The second column of Table 1 lists the Q-value of the reaction in MeV. In exoergic ($Q > 0$) reactions, this is the energy released to the thermal bath of particles and radiation per reaction; in endoergic ($Q < 0$) reactions, this is the energy drawn from the thermal bath per reaction. Implicit in these statements is the assumption that the reaction products are thermalized before reacting again. For a discussion of the problem see Gryzinski (1958, 1959).

In exoergic reactions that produce neutrinos the energy released *exclusive of neutrino energy* is also listed. Except in the "big bang" and the formation of neutron stars, the *difference* is carried away by the neutrino and is lost. In addition the neutrino carries away about half of the kinetic energy (CM) of the primary interacting particles when it is emitted with a positron and almost all of this energy when it is emitted without a positron. In endoergic reactions that produce neutrinos the absolute value of the negative Q listed is lost from the thermal bath and, in addition, the neutrino carries away the kinetic energy it receives. The energetics of neutrino processes are discussed in detail by Chiu (1968). In radioactive weak interactions such as beta decay ($\beta^{\pm}$) and electron capture (EC) the neutrino energy losses are

$$\langle E_{\nu}\rangle_{\beta} \approx \frac{W_0}{2}\left(1-\frac{1}{w_0^2}\right)\left(1-\frac{1}{4w_0}-\frac{1}{9w_0^2}\right) \qquad 1.$$

and

$$\langle E_{\nu}\rangle_{EC} \approx Q_{EC}, \qquad 2.$$

where $W_0 = w_0 m_e c^2 = Q_{\beta^-} + m_e c^2$ for electron decay and $W_0 = Q_{\beta^+} - m_e c^2$ for positron decay; the Q's are the *atomic* mass difference between the parent and the daughter in the radioactive process. The approximation given in equation 1 fits the values given in Table 6.5 of Chiu (1968) to better than $\pm 1.5\%$.

Total Q-values, exclusive of neutrino losses, for the proton-proton (pp) chain and the CN cycle are listed under H1(P, E+NU)H2 and N14(P, G)O15, respectively. The value for H1(P, E+NU)H2 holds only for the Sun. A general expression of good accuracy is $Q = 13.116[1+1.412\times10^{8}(1/X-1)\exp(-4.998/T_9^{1/3})]$ MeV, where X is the mass fraction of hydrogen. The energy release for the CNO tri-cycle is not much different from that for the CN cycle. These energy values can be used when the other relevant reactions are in equilibrium with these reactions. Under these conditions the N^{14} abundance should be taken approximately equal to the original C+N abundance if the temperature and density are such that O^{16} does not interact in the time scale involved. If the O^{16} does interact and the NO reactions come to equilibrium, then the N^{14} abundance should be taken approximately equal to the original C+N+O abundance.

Table 1 Reaction rates per second per (mole/cm**3)**($N-1$)

```
 13 DEC 74

H1(E-,NU)N1      Q=   -.782  1.55E-10*T932*EXP(-9.080/T9)*(1. + 0.084*T9)

H1(P,E+NU)H2     Q=   1.442  4.21E-15/T923*EXP(-3.380/T913)*(1.+0.123*T913+1.09*T923+0.938*T9)
                      1.192  EXCLUSIVE OF NU-ENERGY
                 Q=  13.811  PER PP-REACTION FOR ENTIRE PP-CHAIN IN SUN

H1(PE-,NU)H2     Q=   1.442  1.44E-20/T976*EXP(-3.380/T913)*(1.-0.729*T913+9.82*T923-2.58*T9)
                       .001  EXCLUSIVE OF NU-ENERGY

H2(P,G)HE3       Q=   5.494  2.65E+03/T923*EXP(-3.720/T913)
                              * (1.+0.112*T913+1.99*T923+1.56*T9+0.162*T943+0.324*T953)
                 REV RATIO   1.63E+10*T932*EXP(  -63.755/T9)

H2(P,N)2H1       Q=  -2.225  3.35E+07*EXP(-3.720/T913-25.817/T9)*(1.+0.784*T913+0.346*T923+0.690*T9)
                 REV RATIO   4.24E-10/T932*EXP(   25.817/T9)

H2(D,N)HE3       Q=   3.269  3.97E+08/T923*EXP(-4.258/T913)
                              * (1.+0.098*T913+0.876*T923+0.600*T9-0.041*T943-0.071*T953)
                 REV RATIO   1.73E+00*EXP(  -37.938/T9)

H2(D,P)H3        Q=   4.033  4.17E+08/T923*EXP(-4.258/T913)
                              * (1.+0.098*T913+0.518*T923+0.355*T9-0.010*T943-0.018*T953)
                 REV RATIO   1.73E+00*EXP(  -46.802/T9)

H3(P,G)HE4       Q=  19.815  2.20E+04/T923*EXP(-3.869/T913)
                              * (1.+0.108*T913+1.68*T923+1.26*T9+0.551*T943+1.06*T953)
                 REV RATIO   2.61E+10*T932*EXP( -229.947/T9)

 19 NOV 74

H3(P,N)HE3       Q=   -.764  7.07E+08*(1.-0.150*T912+0.098*T9)*EXP(-8.864/T9)
                 REV RATIO   9.98E-01*EXP(    8.864/T9)

H3(D,N)HE4       Q=  17.590  8.09E+10/T923*EXP(-4.524/T913-(T9/0.120)**2)
                              * (1.+0.092*T913+1.80*T923+1.16*T9+10.52*T943+17.24*T953)
                              + 8.73E+08/T923*EXP(-0.523/T9)
                 REV RATIO   5.54E+00*EXP( -204.130/T9)
```

```
H3(T,2N)HE4        Q= 11.332   1.67E+09/T923*EXP(-4.872/T913)
                                * (1.+0.086*T913-0.455*T923-0.272*T9+0.148*T943+0.225*T953)
                   REV RATIO   3.38E-10/T932*EXP( -131.512/T9)

HE3(E-,NU)H3       Q=  -.019   7.71E-12*T932*(1. + 6.48*T9 + 7.48*T9**2 + 2.91*T9**3)*EXP(-0.216/T9)

HE3(P,E+NU)HE4     Q= 19.796   8.78E-13/T923*EXP(-6.141/T913)
                      10.156   EXCLUSIVE OF NU-ENERGY

HE3(D,P)HE4        Q= 18.354   6.67E+10/T923*EXP(-7.181/T913-(T9/0.315)**2)
                                * (1.+0.058*T913-1.14*T923-0.464*T9+3.08*T943+3.18*T953)
                                + 4.36E+08/T912*EXP(-1.720/T9)
                   REV RATIO   5.55E+00*EXP( -212.994/T9)

HE3(T,D)HE4        Q= 14.321   5.46E+09*T9A56/T932*EXP(-7.733/T9A13)
                               T9A = T9/(1.+0.128*T9)
                   REV RATIO   1.60E+00*EXP( -166.193/T9)

HE3(T,NP)HE4       Q= 12.096   7.71E+09*T9A56/T932*EXP(-7.733/T9A13)
                               T9A = T9/(1.+0.115*T9)
                   REV RATIO   3.39E-10/T932*EXP( -140.376/T9)

HE3(HE3,2P)HE4     Q= 12.860   5.96E+10/T923*EXP(-12.276/T913)
                                * (1.+ 0.034*T913-0.199*T923-0.047*T9+0.032*T943+0.019*T953)
                   REV RATIO   3.39E-10/T932*EXP( -149.240/T9)

HE4(NN,G)HE6       Q=   .971   (0 TO 1)*4.04E-11/T9**2 * (1.+0.138*T9)*EXP(-9.585/T9)
                   REV RATIO   1.08E+20*T9**3*EXP( -11.269/T9)

HE4(NP,G)LI6       Q=  3.698   4.62E-06/T9**2 * (1.+0.075*T9)*EXP(-19.353/T9)
                   REV RATIO   7.22E+19*T9**3*EXP( -42.919/T9)

HE4(T,G)LI7        Q=  2.467   5.27E+05/T923*EXP(-8.080/T913)
                   REV RATIO   1.11E+10*T932*EXP(  -28.626/T9)

HE4(T,N)LI6        Q= -4.784   1.80E+08*EXP(-55.516/T9)*(1.-0.261*T9A32/T932)
                                + 2.72E+09/T932*EXP(-58.050/T9)
                               T9A = T9/(1.+49.18*T9)
                   REV RATIO   9.35E-01*EXP(   55.516/T9)
```

Table 1 *Continued*

```
HE4(HE3,G)BE7    Q=  1.586   6.33E+06/T923*EXP(-12.826/T913)
                              * (1.+0.033*T913-0.350*T923-0.080*T9+0.056* 943+0.033*T953)
                 REV RATIO   1.11E+10*T932*EXP(  -18.410/T9)

HE4(AN,G)BE9     Q=  1.573   2.59E-06/(T9**2*(1 + 0.344*T9))*EXP(-1.062/T9)
                 REV RATIO   5.84E+19*T9**3*EXP( -18.256/T9)

HE4(2A,G)C12     Q=  7.275   2.49E-08/T9**3*EXP(-4.4109/T9) + (0 TO 1)*1.35E-07/T932*EXP(-24.811/T9)
                 Q= 14.436   IF C12(A,G)O16 ALWAYS FOLLOWS
                 REV RATIO   2.00E+20*T9**3*EXP( -84.424/T9)

LI6(P,G)BE7      Q=  5.606   5.87E+05/T923*EXP(-8.413/T913)
                 REV RATIO   1.19E+10*T932*EXP(  -65.062/T9)

LI6(P,HE3)HE4    Q=  4.020   3.80E+10*T9A56/T932*EXP(-8.413/T9A13) + 5.97E+09/T932*EXP(-18.301/T9)
                             T9A = T9/(1.+0.095*T9)
                 REV RATIO   1.07E+00*EXP(  -46.652/T9)

LI6(A,G)B10      Q=  4.460   1.27E+07/T923*EXP(-18.790/T913-(T9/1.327)**2)
                              * (1.+0.022*T913+1.00*T923+0.156*T9+1.39*T943+0.549*T953)
                              + 3.18E+03/T932*EXP(-3.484/T9) + 1.54E+04/T9*EXP(-7.212/T9)
                 REV RATIO   1.58E+10*T932*EXP(  -51.759/T9)

LI7(P,N)BE7      Q= -1.644   6.77E+09*(1.-0.903*T912 + 0.218*T9)*EXP(-19.080/T9)
                 REV RATIO   9.98E-01*EXP(   19.080/T9)

LI7(P,A)HE4      Q= 17.348   8.04E+08/T923*EXP(-8.471/T913-(T9/30.068)**2)
                              * (1.+0.049*T913+0.230*T923+0.079*T9-0.027*T943-0.023*T953)
                              + 1.54E+06/T932*EXP(-4.479/T9) + 1.07E+10/T932*EXP(-30.443/T9)
                 REV RATIO   4.69E+00*EXP( -201.321/T9)
                             REV RATIO APPLIES TO BOTH LI7(P,A)HE4 AND LI7(P,G)2HE4

LI7(D,N)2HE4     Q= 15.123   2.92E+11/T923*EXP(-10.259/T913)
                 REV RATIO   9.95E-10/T932*EXP( -175.504/T9)

LI7(T,2N)2HE4    Q=  8.866   8.81E+11/T923*EXP(-11.333/T913)
                 REV RATIO   1.22E-19/T9**3*EXP(-102.886/T9)
```

```
LI7(HE3,NP)2HE4 Q=  9.630  1.11E+13/T923*EXP(-17.989/T913)
                REV RATIO  6.09E-20/T9**3*EXP(-111.750/T9)

LI7(A,G)B11     Q=  8.666  2.26E+08/T923*EXP(-19.161/T913-(T9/0.991)**2)
                            * (1.+0.022*T913+0.382*T923+0.058*T9+0.571*T943+0.221*T953)
                            + 1.51E+03/T932*EXP(-3.030/T9) + 3.45E+04/T912*EXP(-5.345/T9)
                REV RATIO  4.02E+10*T932*EXP( -100.563/T9)

LI7(A,N)B10     Q= -2.790  3.84E+08*EXP(-32.383/T9)
                REV RATIO  1.32E+00*EXP(   32.383/T9)

BE7(E-,NU+G)LI7 Q=   .862  1.34E-10/T912*(1.-0.537*T913+3.86*T923+1.20*T9+0.0027/T9
                            * EXP(2.515E-03/T9))
                     .049  EXCLUSIVE OF NU ENERGY
                           RATE MUST NOT EXCEED 1.51E-07/(RHO*(1.+X)/2.)
                           FOR T9 LESS THAN 0.001.

BE7(P,G)B 8     Q=   .137  4.35E+05/T923*EXP(-10.262/T913) + 3.30E+03/T932*EXP(-7.306/T9)
                REV RATIO  1.30E+10*T932*EXP(    -1.592/T9)

BE7(D,P)2HE4    Q= 16.768  1.07E+12/T923*EXP(-12.428/T913)
                REV RATIO  9.97E-10/T932*EXP( -194.585/T9)

BE7(T,NP)2HE4   Q= 10.510  2.91E+12/T923*EXP(-13.729/T913)
                REV RATIO  6.09E-20/T9**3*EXP(-121.966/T9)

BE7(HE3,2P)2HE4 Q= 11.274  6.11E+13/T923*EXP(-21.793/T913)
                REV RATIO  1.22E-19/T9**3*EXP(-130.830/T9)

BE7(A,G)C11     Q=  7.545  2.12E+08/T923*EXP(-23.212/T913-(T9/0.658)**2)
                            * (1.+0.018*T913+0.228*T923+0.029*T9+0.548*T943+0.175*T953)
                            + 7.27E+04/T932*EXP(-6.525/T9) + 1.23E+05/T9*EXP(-9.742/T9)
                REV RATIO  4.02E+10*T932*EXP(   -87.559/T9)

BE9(P,G)B10     Q=  6.585  1.33E+07/T923*EXP(-10.359/T913-(T9/0.846)**2)
                            * (1.+0.040*T913+1.52*T923+0.428*T9+2.15*T943+1.54*T953)
                            + 9.64E+04/T932*EXP(-3.445/T9) + 2.72E+06/T932*EXP(-10.620/T9)
                REV RATIO  9.73E+09*T932*EXP(   -76.422/T9)
```

Table 1 *Continued*

```
BE9(P,N)B 9      Q= -1.850   5.58E+07*(1.+0.042*T912+0.985*T9)*EXP(-21.466/T9)
                               + 1.02E+09/T932*EXP(-26.753/T9)
                 REV RATIO   9.98E-01*EXP(    21.466/T9)

BE9(P,D)2HE4     Q=   .651   2.11E+11/T923*EXP(-10.359/T913-(T9/0.520)**2)
                               * (1.+0.040*T913+1.09*T923+0.307*T9+3.21*T943+2.30*T953)
                               + 5.79E+08/T9*EXP(-3.046/T9) + 8.50E+08/T934*EXP(-5.800/T9)
                 REV RATIO   8.07E-11/T932*EXP(    -7.560/T9)

BE9(P,A)LI6      Q=  2.125   2.11E+11/T923*EXP(-10.359/T913-(T9/0.520)**2)
                               * (1.+0.040*T913+1.09*T923+0.307*T9+3.21*T943+2.30*T953)
                               + 4.51E+08/T9*EXP(-3.046/T9) + 6.70E+08/T934*EXP(-5.160/T9)
                 REV RATIO   6.18E-01*EXP(   -24.663/T9)

BE9(A,N)C12      Q=  5.702   4.62E+13/T923*EXP(-23.870/T913-(T9/0.049)**2)
                               * (1.+0.017*T913+8.57*T923+1.05*T9+74.51*T943+23.15*T953)
                               + 7.34E-05/T932*EXP(-1.184/T9) + 2.27E-01/T932*EXP(-1.834/T9)
                               + 1.26E+05/T932*EXP(-4.179/T9) + 2.40E+08*EXP(-12.732/T9)
                 REV RATIO   1.03E+01*EXP(   -66.167/T9)

B10(P,G)C11      Q=  8.691   4.61E+05/T923*EXP(-12.062/T913-(T9/4.402)**2)
                               * (1.+0.035*T913+0.426*T923+0.103*T9+0.281*T943+0.173*T953)
                               + 1.93E+05/T932*EXP(-12.041/T9) + 1.14E+04/T932*EXP(-16.164/T9)
                 REV RATIO   3.03E+10*T932*EXP( -100.862/T9)

B10(P,A)BE7      Q=  1.146   1.26E+11/T923*EXP(-12.062/T913-(T9/4.402)**2)
                               * (1.+0.035*T913-0.498*T923-0.121*T9+0.300*T943+0.184*T953)
                               + 2.59E+09/T9*EXP(-12.260/T9)
                 REV RATIO   7.54E-01*EXP(   -13.303/T9)

B11(P,G)C12      Q= 15.957   4.62E+07/T923*EXP(-12.095/T913-(T9/0.239)**2)
                               * (1.+0.035*T913+3.00*T923+0.723*T9+9.91*T943+6.07*T953)
                               + 7.89E+03/T932*EXP(-1.733/T9) + 9.68E+04/T915*EXP(-5.617/T9)
                 REV RATIO   7.01E+10*T932*EXP( -185.181/T9)

B11(P,N)C11      Q= -2.765   1.69E+08*(1.-0.048*T912 + 0.010*T9)*EXP(-32.084/T9)
                 REV RATIO   9.98E-01*EXP(    32.084/T9)
```

```
B11(P,A)2HE4     Q=  8.682  2.59E+11/T923*EXP(-12.095/T913-(T9/2.02)**2)
                             * (1.+0.035*T913+1.22*T923+0.295*T9+2.15*T943+1.32*T953)
                             + 7.40E+06/T932*EXP(-1.733/T9) + 8.14E+09/T932*EXP(-7.177/T9)
                             + 1.71E+09/T923*EXP(-12.696/T9)
                 REV RATIO  3.50E-10/T932*EXP( -100.758/T9)

C11(P,G)N12      Q=   .595  4.24E+04/T923*EXP(-13.658/T913-(T9/1.627)**2)
                             * (1.+0.031*T913+3.11*T923+0.665*T9+4.61*T943+2.50*T953)
                             + 8.84E+03/T932*EXP(-7.021/T9)
                 REV RATIO  2.33E+10*T932*EXP(   -6.910/T9)

C12(P,G)N13      Q=  1.944  2.04E+07/T923*EXP(-13.690/T913-(T9/1.500)**2)
                             * (1.+0.030*T913+1.19*T923+0.254*T9+2.06*T943+1.12*T953)
                             + 1.08E+05/T932*EXP(-4.925/T9) + 2.15E+05/T932*EXP(-18.179/T9)
                 REV RATIO  8.84E+09*T932*EXP(  -22.554/T9)

C12(A,G)O16      Q=  7.162  9.03E+07/T9**2*(1.+0.621*T923)**2/(1.+0.047/T923)**2
                            *EXP(-32.120/T913-(T9/5.863)**2) + 2.74E+07/T923*EXP(-32.120/T913)
                             + 1.25E+03/T932*EXP(-27.499/T9) + 1.43E-02*T9**5 * EXP(-15.541/T9)
                 REV RATIO  5.13E+10*T932*EXP(  -83.110/T9)

C12(A,N)O15      Q= -8.502  2.48E+07*(1.+0.188*T912 + 0.015*T9)*EXP(-98.663/T9)
                 REV RATIO  1.41E+00*EXP(   98.663/T9)

C13(P,G)N14      Q=  7.551  8.01E+07/T923*EXP(-13.717/T913-(T9/2.000)**2)
                             * (1.+0.030*T913+0.958*T923+0.204*T9+1.39*T943+0.753*T953)
                             + 1.35E+06/T932*EXP(-5.978/T9) + 2.66E+05/T932*EXP(-11.987/T9)
                             + 2.26E+06/T932*EXP(-13.463/T9)
                 REV RATIO  1.19E+10*T932*EXP(  -87.625/T9)

C13(P,N)N13      Q= -3.003  1.88E+08*(1.-0.167*T912 + 0.037*T9)*EXP(-34.848/T9)
                 REV RATIO  9.98E-01*EXP(   34.848/T9)

C13(A,N)O16      Q=  2.215  6.77E+15/T923*EXP(-32.329/T913-(T9/1.284)**2)
                             * (1.+0.013*T913+2.04*T923+0.184*T9)
                             + 3.82E+05/T932*EXP(-9.373/T9) + 1.41E+06/T932*EXP(-11.873/T9)
                             + 2.00E+09/T932*EXP(-20.409/T9) + 2.92E+09/T932*EXP(-29.283/T9)
                 REV RATIO  5.79E+00*EXP(  -25.707/T9)
```

Table 1 *Continued*

```
C14(P,G)N15     Q= 10.207  6.80E+06/T923*EXP(-13.741/T913-(T9/5.721)**2)
                            * (1.+0.030*T913+0.503*T923+0.107*T9+0.213*T943+0.115*T953)
                            + 5.36E+03/T932*EXP(-3.811/T9) + 9.82E+04/T913*EXP(-4.739/T9)
                REV RATIO  9.00E+09*T932*EXP( -118.456/T9)

C14(P,N)N14     Q=  -.626  7.21E+05*(1.+0.361*T912+0.502*T9)*EXP(-7.269/T9)
                            + 1.18E+08*EXP(-11.348/T9)
                REV RATIO  3.33E-01*EXP(     7.269/T9)

N13(P,G)O14     Q=  4.626  5.99E+07/T923*EXP(-15.202/T913-(T9/1.181)**2)
                            * (1.+0.027*T913+1.71*T923+0.329*T9+2.10*T943+1.03*T953)
                            + 8.36E+05/T932*EXP(-6.313/T9)
                REV RATIO  3.57E+10*T932*EXP(   -53.688/T9)

N14(P,G)O15     Q=  7.298  5.08E+07/T923*EXP(-15.228/T913-(T9/ 3.090)**2)
                            * (1.+0.027*T913-0.778*T923-0.149*T9+0.261*T943+0.127*T953)
                            + 2.28E+03/T932*EXP(-3.011/T9) + 1.65E+04*T913*EXP(-12.007/T9)
                Q= 24.970 PER N14(P,G)O15 REACTION FOR FULL CN-CYCLE.
                      USE N14 = ORIGINAL C+N+(0 TO 1)*O ABUNDANCE (SEE TEXT)
                REV RATIO  2.70E+10*T932*EXP(   -84.692/T9)

N14(P,N)O14     Q= -5.927  6.74E+07*(1.+0.658*T912 + 0.379*T9)*EXP(-68.785/T9)
                REV RATIO  2.99E+00*EXP(    68.785/T9)

N14(P,A)C11     Q= -2.922  2.63E+16*T9A56/T932*EXP(-31.883/T9A13 - 33.911/T9)
                           T9A = T9/(1.+4.78E-02*T9+7.56E-03*T953/(1.+4.78E-02*T9)**(2./3.))
                REV RATIO  2.72E-01*EXP(    33.911/T9)

N14(A,G)F18     Q=  4.416  7.78E+09/T923*EXP(-36.031/T913-(T9/0.881)**2)
                            * (1.+0.012*T913+1.45*T923+0.117*T9+1.97*T943+0.406*T953)
                            + 2.36E-10/T932*EXP(-2.798/T9) + 2.03E+00/T932*EXP(-5.054/T9)
                            + 1.15E+04/T923*EXP(-12.310/T9)
                REV RATIO  5.42E+10*T932*EXP(   -51.247/T9)
```

```
N15(P,G)O16     Q= 12.128  9.78E+08/T923*EXP(-15.251/T913-(T9/0.450)**2)
                            * (1.+0.027*T913+0.219*T923+0.042*T9+6.83*T943+3.32*T953)
                            + 1.11E+04/T932*EXP(-3.328/T9) + 1.49E+04/T932*EXP(-4.665/T9)
                            + 3.80E+06/T932*EXP(-11.048/T9)
                REV RATIO  3.62E+10*T932*EXP( -140.741/T9)

N15(P,N)O15     Q= -3.536  3.51E+08*(1.+0.452*T912-0.191*T9)*EXP(-41.032/T9)
                REV RATIO  9.98E-01*EXP(   41.032/T9)

N15(P,A)C12     Q=  4.966  8.16E+11/T923*EXP(-15.251/T913-(T9/0.522)**2)
                            * (1.+0.027*T913+6.74*T923+1.29*T9)
                            + 1.29E+08/T932*EXP(-3.676/T9) + 3.14E+08*EXP(-7.974/T9)
                REV RATIO  7.06E-01*EXP(  -57.631/T9)

N15(A,G)F19     Q=  4.013  2.54E+10/T923*EXP(-36.211/T913-(T9/0.616)**2)
                            * (1.+0.012*T913+1.69*T923+0.136*T9+1.91*T943+0.391*T953)
                            + 9.83E-03/T932*EXP(-4.232/T9) + 7.59E+01/T932*EXP(-6.319/T9)
                            + 1.52E+03*T9*EXP(-9.747/T9)
                REV RATIO  5.54E+10*T932*EXP(  -46.569/T9)

O16(P,G)F17     Q=   .601  1.50E+08/(T923*(1.+2.13*(1.-EXP(-0.728*T923)))) * EXP(-16.692/T913)
                REV RATIO  3.03E+09*T932*EXP(   -6.970/T9)

O16(P,A)N13     Q= -5.218  1.70E+19*T9A56/T932*EXP(-35.829/T9A13-60.556/T9)/(1.+5.*EXP(-2.396*T923))
                           T9A = T9/(1.+0.224*T9)
                REV RATIO  1.72E-01*EXP(   60.556/T9)

O16(A,G)NE20    Q=  4.730  5.49E+09/T923*EXP(-39.756/T913) + 4.09E+01/T932*EXP(-10.359/T9)
                            + 3.92E+02/T932*EXP(-12.243/T9) + 8.05*T9**2*EXP(-20.093/T9)
                REV RATIO  5.65E+10*T932*EXP(  -54.891/T9)

O17(P,G)F18     Q=  5.609  7.97E+07*T9A56/T932*EXP(-16.712/T9A13) + 1.51E+08/T923*EXP(-16.712/T913)
                            * (1.+0.025*T913-0.051*T923-8.82E-03*T9) + 1.56E+05/T9*EXP(-6.272/T9)
                            + (0 TO 1, SAME AS O17(P,A)N14 ) * 1.31E+01/T932*EXP(-1.961/T9)
                           T9A = T9/(1.+2.69*T9)
                REV RATIO  3.66E+10*T932*EXP(  -65.092/T9)
```

Table 1 *Continued*

```
O17(P,A)N14     Q=  1.193   1.53E+07/T923*EXP(-16.712/T913-(T9/0.565)**2)
                             * (1.+0.025*T913+5.39*T923+0.940*T9+13.5*T943+5.98*T953)
                             + 2.92E+06*T9*EXP(-4.247/T9)
                             +                        (0 TO 1) *
                            (4.81E+10*T9*EXP(-16.712/T913-(T9/0.040)**2) + 5.05E-05/T932*EXP(-0.723/T9))
                             + (0 TO 1, SAME AS O17(P,G)F18 ) * 1.31E+01/T932*EXP(-1.961/T9)
                REV RATIO   6.76E-01*EXP(  -13.845/T9)

O17(A,G)NE21    Q=  7.349   1.73E+17*FPT9A/GT9*T9A56/T932*EXP(-39.914/T9A13)
                             + 3.50E+15*FT9A/GT9*T9A56/T932*EXP(-39.914/T9A13)
                            T9A = T9/(1.+0.1646*T9)              GT9 = 1. + EXP(-10.106/T9)/3.
                            FT9A = EXP(-(0.786/T9A)**3.51)       FPT9A = EXP(-(T9A/1.084)**1.69)
                REV RATIO   8.63E+10*T932*EXP(  -85.281/T9)

O17(A,N)NE20    Q=   .588   1.03E+18/GT9*T9A56/T932*EXP(-39.914/T9A13)
                            T9A = T9/(1.+.0268*T9+.0232*T953/(1.+.0268*T9)**(2./3.))
                            GT9 = 1. + EXP(-10.106/T9)/3.
                REV RATIO   1.86E+01*EXP(   -6.819/T9)

O18(P,G)F19     Q=  7.993   2.39E+08/T923*EXP(-16.729/T913) + 1.80E+05/T923*EXP(-6.342/T9)
                             +                        (0 TO 1) *
                            3.96E+09/T923*EXP(-16.729/T913-(T9/0.138)**2)
                             * (1.+0.025*T913+5.88*T923+1.03*T9+29.80*T943+13.21*T953)
                             + 3.09E+02/T932*EXP(-1.662/T9)
                            DIVIDE ALL TERMS BY GT9 = 1. + 5.*EXP(-23.002/T9)
                REV RATIO   9.20E+09*T932*EXP(  -92.756/T9)

O18(P,A)N15     Q=  3.980   2.13E+11/T923*EXP(-16.729/T913-(T9/1.318)**2)
                             * (1. +0.025*T913+1.68*T923+0.292*T9+1.86*T943+0.824*T953)
                             + 9.64E+08/T9*EXP(-7.361/T9)
                             +                        (0 TO 1) *
                            7.88E+14/(T923*(0.439*(1.+5.18*T923)**2 + 0.561))
                             * EXP(-16.729/T913-0.534*T923)
                             + 1.99E+13/T923*EXP(-16.729/T913-(T9/0.138)**2)
                             * (1. + 0.025*T913+5.98*T923+1.04*T9+30.70*T943+13.61*T953)
                             + 1.40E+06/T932*EXP(-1.662/T9) + 2.34E+06/T932*EXP(-2.915/T9)
                            DIVIDE ALL TERMS BY GT9 = 1.+5.*EXP(-23.002/T9)
                REV RATIO   1.66E-01*EXP(  -46.187/T9)
```

```
O18(A,G)NE22     Q=   9.668  7.22E+17*FPT9A/GT9*T9A56/T932*EXP(-40.056/T9A13)
                              + 3.61E+14*FT9A/GT9*T9A56/T932*EXP(-40.056/T9A13)
                             T9A = T9/(1.+0.0483*T9+0.00569*T953 /(1.+0.0483*T9)**(2./3.))
                             GT9=1.+5.*EXP(-23.002/T9)            FT9A = EXP(-( 0.431/T9A)**3.89)
                             FPT9A = EXP(-(T9A/0.576)**1.87)
                 REV RATIO   5.85E+10*T932*EXP( -112.191/T9)

O18(A,N)NE21     Q=   -.698  7.22E+17*FT9A/GT9*T9A56/T932*EXP(-40.056/T9A13)
                              + 150.31/GT9*EXP(-8.101/T9)
                             T9A = T9/(1.+0.0483*T9+0.00569*T953 /(1.+0.0483*T9)**(2./3.))
                             GT9 = 1. + 5.*EXP(-23.002/T9)        FT9A = EXP(-(0.431/T9A)**3.89)
                 REV RATIO   7.84E-01*EXP(     8.101/T9)

F19(P,G)NE20     Q= 12.845   6.04E+07/T923*EXP(-18.113/T913-(T9/0.416)**2)
                              * (1. + 0.023*T913 + 2.06*T923 + 0.332*T9 + 3.16*T943 + 1.30*T953)
                              + 6.32E+02/T932*EXP(-3.752/T9) + 7.56E+04/T927*EXP(-5.722/T9)
                             DIVIDE ALL TERMS BY GT9=1.+4.*EXP(-2.090/T9)+7.*EXP(-16.440/T9)
                 REV RATIO   3.70E+10*T932*EXP( -149.063/T9)

F19(P,N)NE19     Q= -4.021   1.27E+08*(1.-0.147*T912 + 0.069*T9)*EXP(-46.659/T9)
                 REV RATIO   9.98E-01*EXP(    46.659/T9)

F19(P,A)O16      Q=   8.115  3.20E+11/T923*EXP(-18.113/T913-(T9/0.416)**2)
                              * (1.+0.023*T913 + 2.16*T923 + 0.348*T9 + 3.47*T943 + 1.42*T953)
                              + 5.23E+06/T932*EXP(-3.752/T9) + 3.88E+08/T918*EXP(-6.232/T9)
                             DIVIDE ALL TERMS BY GT9=1.+4.*EXP(-2.090/T9)+7.*EXP(-16.440/T9)
                 REV RATIO   6.54E-01*EXP(  -94.172/T9)

F19(A,P)NE22     Q=   1.675  4.50E+18/T923*EXP(-43.467/T913-(T9/0.637)**2)+7.98E+04*T932*EXP(-12.760/T9)
                 REV RATIO   6.36E+00*EXP(  -19.435/T9)

NE20(P,G)NA21    Q=   2.432  9.55E+06*EXP(-19.447/T913)/(T9**2*(1.+0.0127/T923)**2)
                              +2.05E+08/T923*EXP(-19.447/T913)*(1.+2.67*EXP(-SQRT(T9/0.210)))
                              + 1.80E+01/T932*EXP(-4.242/T9) + 1.02E+01/T932*EXP(-4.607/T9)
                              + 3.60E+04/T914*EXP(-11.249/T9)
                 REV RATIO   4.63E+09*T932*EXP(  -28.227/T9)
```

Table 1 *Continued*

```
NE20(A,G)MG24    Q=  9.315   4.11E+11/T923*EXP(-46.766/T913-(T9/2.219)**2)
                              * (1. + 0.009*T913 + 0.882*T923 + 0.055*T9 + 0.749*T943 + 0.119*T953)
                              + 5.27E+03/T932*EXP(-15.869/T9) + 6.51E+03*T912*EXP(-16.223/T9)
                              +                 (0 TO 1) *
                             4.21E+01/T932*EXP(-9.115/T9) + 3.20E+01/T923*EXP(-9.383/T9)
                             DIVIDE ALL TERMS BY GT9 = 1. + 5.*EXP(-18.960/T9)
                 REV RATIO   6.01E+10*T932*EXP( -108.094/T9)

NE21(P,G)NA22    Q=  6.740   3.42E+08/T923*EXP(-19.462/T913) + 1.56E+05/T932*EXP(-5.584/T9)
                              + 8.92E+05/T932*EXP(-5.810/T9) + 5.74E+06/T943*EXP(-8.131/T9)
                 REV RATIO   1.06E+10*T932*EXP(  -78.220/T9)

NE21(A,G)MG25    Q=  9.885   4.94E+15/GT9*T9A56/T932*EXP(-46.890/T9A13)       T9A = T9/(1.+0.0537*T9)
                             GT9 = 1. + 1.5*EXP(-4.068/T9) + 2.0*EXP(-20.258/T9)
                 REV RATIO   4.06E+10*T932*EXP( -114.717/T9)

NE21(A,N)MG24    Q=  2.553   4.94E+19*T9A56/T932*EXP(-46.890/T9A13) + 2.66E+07/T932*EXP(-22.049/T9)
                             T9A = T9/(1.+0.0537*T9)
                             DIVIDE ALL TERMS BY GT9 = 1.+1.5*EXP(-4.068/T9) + 2.0*EXP(-20.258/T9)
                 REV RATIO   1.29E+01*EXP(  -29.631/T9)

NE22(P,G)NA23    Q=  8.793   1.14E+09/T923*EXP(-19.475/T913) + 8.96E+03/T932*EXP(-4.840/T9)
                              + 6.52E+04/T932*EXP(-5.319/T9) + 8.33E+05/T912*EXP(-7.462/T9)
                 REV RATIO   4.67E+09*T932*EXP( -102.043/T9)

NE22(A,G)MG26    Q= 10.613   4.16E+19*FPT9A/GT9*T9A56/T932*EXP(-47.004/T9A13)
                              + 2.08E+16*FT9A/GT9*T9A56/T932*EXP(-47.004/T9A13)
                             T9A = T9/(1.+0.0548*T9)              GT9 = 1.+5.0*EXP(-14.791/T9)
                             FT9A = EXP(-(0.197/T9A)**4.82)       FPT9A = EXP(-(T9A/0.249)**2.31)
                 REV RATIO   6.15E+10*T932*EXP( -123.166/T9)

NE22(A,N)MG25    Q=  -.480   4.16E+19*FT9A/GT9*T9A56/T932*EXP(-47.004/T9A13)
                              + 1.44E-04/GT9*EXP(-5.574/T9)
                             T9A = T9/(1.+0.0548*T9)              GT9 = 1.+5.0*EXP(-14.791/T9)
                             FT9A = EXP(-(0.197/T9A)**4.82)
                 REV RATIO   5.44E-01*EXP(      5.574/T9)
```

```
NA23(P,G)MG24    Q= 11.692  2.93E+08/T923*EXP(-20.766/T913-(T9/0.297)**2)
                             * (1. + .020*T913 + 1.61*T923 + 0.226*T9 + 4.94*T943 + 1.76*T953)
                             + 9.34E+01/T932*EXP(-2.789/T9) + 1.89E+04/T932*EXP(-3.434/T9)
                             + 5.10E+04*T915*EXP(-5.510/T9)
                            DIVIDE ALL TERMS BY GT9 = 1. + 1.5*EXP(-5.105/T9)
                 REV RATIO  7.49E+10*T932*EXP( -135.679/T9)

NA23(P,N)MG23    Q= -4.839  9.29E+08*(1. - 0.881*T9A32/T932)*EXP(-56.157/T9)
                            T9A = T9/(1.+0.141*T9)
                 REV RATIO  9.98E-01*EXP(   56.157/T9)

NA23(P,A)NE20    Q=  2.377  6.79E+09/T923*EXP(-20.766/T913-(T9/0.131)**2)
                             * (1. + 0.020*T913 + 8.21*T923 + 1.15*T9 + 44.36*T943 + 15.84*T953)
                             + 2.75E+00/T932*EXP(-1.990/T9) + 1.10E+04/T943*EXP(-3.207/T9)
                             + 1.84E+06*T913*EXP(-5.170/T9)
                             + (0 TO 1)*3.06E-12/T932*EXP(-0.447/T9)
                            DIVIDE ALL TERMS BY GT9 = 1.+1.5*EXP(-5.105/T9)
                 REV RATIO  1.25E+00*EXP(  -27.585/T9)

MG24(P,G)AL25    Q=  2.270  9.79E+10/T923*EXP(-22.019/T913-(T9/0.168)**2 )
                             * (1. + 0.019*T913+6.50*T923+0.861*T9+29.67*T943+9.99*T953)
                             + 8.26E+02/T932*EXP(-2.492/T9) + 1.89E+03*T914*EXP(-4.261/T9)
                            DIVIDE ALL TERMS BY GT9 = 1. + 5.*EXP(-15.882/T9)
                 REV RATIO  3.13E+09*T932*EXP(  -26.346/T9)

MG24(A,G)SI28    Q=  9.985  4.78E+01/T932*EXP(-13.506/T9) + 1.97E+03/T932*EXP(-15.246/T9)
                             + 9.32E+02*T9*EXP(-17.425/T9)
                             +                (0 TO 1) *
                            1.72E-09/T932*EXP(-5.028/T9) + 1.25E-03/T932*EXP(-7.929/T9)
                             + 2.43E+01/T9*EXP(-11.523/T9)
                            DIVIDE ALL TERMS BY GT9 = 1. + 5.*EXP(-15.882/T9)
                 REV RATIO  6.27E+10*T932*EXP( -115.872/T9)
```

Table 1 *Continued*

```
MG25(P,G)AL26    Q=  6.307   2.70E+08/T923*EXP(-22.031/T913-(T9/0.678)**2)
                              * (1.+ 0.019*T913+3.05*T923+0.404*T9+7.12*T943+2.40*T953)
                              + 4.66E+03/T932*EXP(-3.529/T9) + 9.07E+04/T9*EXP(-5.006/T9)
                              +                 (0 TO 1) *
                             (7.40E+09/T923*EXP(-22.031/T913-(T9/0.129)**2)
                              * (1. + 0.019*T913+7.08*T923+0.937*T9+37.34*T943+12.57*T953)
                              + 3.54E+03/T932*EXP(-2.095/T9))
                             DIVIDE ALL TERMS BY GT9 = 1. + 10.*EXP(-13.180/T9)/3.
                 REV RATIO   1.03E+10*T932*EXP(   -73.187/T9)

MG25(A,G)SI29    Q= 11.127   8.97E+17/GT9*T9A56/T932*EXP(-53.410/T9A13)
                             T9A = T9/(1.+0.0963*T9)            GT9 = 1.+10.*EXP(-13.180/T9)/3.
                 REV RATIO   1.90E+11*T932*EXP( -129.125/T9)

MG25(A,N)SI28    Q=  2.653   3.59E+20/GT9*T9A56/T932*EXP(-53.410/T9A13)
                             T9A = T9/(1.+0.063*T9)             GT9 = 1.+10.*EXP(-13.180/T9)/3.
                 REV RATIO   2.00E+01*EXP(   -30.786/T9)

MG26(P,G)AL27    Q=  8.271   3.23E+08/T923*EXP(-22.042/T913-(T9/0.313)**2)
                              * (1. + 0.019*T913+3.55*T923+0.469*T9+9.45*T943+3.18*T953)
                              + 7.76E+02/T932*EXP(-3.265/T9) + 3.88E+04/T932*EXP(-3.781/T9)
                              + 1.80E+04*T934*EXP(-3.505/T9)
                              +                 (0 TO 1) *
                             (4.33E+10/T923*EXP(-22.042/T913-(T9/0.122)**2)
                              * (1. + 0.019*T913+7.31*T923+0.967*T9+40.04*T943+13.47*T953)
                              + 3.57E+03/T932*EXP(-2.010/T9))
                             DIVIDE ALL TERMS BY GT9 = 1. + 5.*EXP(-20.990/T9)
                 REV RATIO   3.14E+09*T932*EXP(   -95.982/T9)

MG26(A,G)SI30    Q= 10.643   4.18E+19*FPT9A/GT9*T9A56/T932*EXP(-53.505/T9A13)
                              + 4.71E+20*FT9A/GT9*T9A56/T932*EXP(-53.505/T9A13)
                             T9A = T9/(1.+0.248*T9)             GT9 = 1.+5.*EXP(-20.990/T9)
                             FT9A = EXP(-(0.664/T9A)**4.19)     FPT9A = EXP(-(T9A/0.870)**2.01)
                 REV RATIO   6.38E+10*T932*EXP( -123.509/T9)
```

```
MG26(A,N)SI29    Q=    .033  2.93E+20/GT9*T9A56/T932*EXP(-53.505/T9A13)
                             T9A = T9/(1.+0.0628*T9)             GT9 = 1.+5.*EXP(-20.990/T9)
                 REV RATIO   1.68E+00*EXP(     -.385/T9)

AL27(P,G)SI28    Q= 11.585   1.57E+08/T923*EXP(-23.261/T913-(T9/0.157)**2)
                              * (1. + 0.018*T913+5.68*T923+0.712*T9+26.45*T943+8.43*T953)
                              + 2.20E+00/T932*EXP(-2.269/T9) + 1.22E+01/T932*EXP(-2.517/T9)
                              + 4.14E+03*T985*EXP(-3.270/T9)
                              +                   (0 TO 1) *
                             3.34E-07/T932*EXP(-0.858/T9) + 5.60E-07/T932*EXP(-0.968/T9)
                             DIVIDE ALL TERMS BY GT9=1.+EXP(-9.792/T9)/3.+2.*EXP(-11.773/T9)/3.
                 REV RATIO   1.13E+11*T932*EXP( -134.446/T9)

AL27(P,A)MG24    Q=  1.601   1.10E+08/T923*EXP(-23.261/T913-(T9/0.157)**2)
                              * (1. + 0.018*T913 + 12.85*T923 + 1.61*T9 + 89.87*T943 + 28.66*T953)
                              + 1.29E+02/T932*EXP(-2.517/T9) + 5.66E+03*T972*EXP(-3.421/T9)
                              +                   (0 TO 1) *
                             2.00E-05/T932*EXP(-0.858/T9) + 2.81E-05/T932*EXP(-0.968/T9)
                             DIVIDE ALL TERMS BY GT9 = 1. + EXP(-9.792/T9)/3. + 2.*EXP(-11.773/T9)/3.
                 REV RATIO   1.81E+00*EXP(  -18.575/T9)

 20 NOV 74

C12+C12 (MG24)   Q= 13.931   1.26E+27*T9A56/T932*EXP(-84.165/T9A13)/(EXP(-0.010*T9A**4)
                              + 5.56E-03*EXP(1.685*T9A23))           T9A = T9/(1.+0.067*T9)

C12+O16 (SI28)   Q= 16.754   1.72E+31*T9A56/T932*EXP(-106.594/T9A13)/(EXP(-0.180*T9A**2)
                              + 1.06E-03*EXP(2.562*T9A23))           T9A = T9/(1.+0.055*T9)

O16+O16 (S32)    Q= 16.541   3.61E+37*T9A56/T932*EXP(-135.930/T9A13)/(EXP(-0.032*T9A**4)
                              + 3.89E-04*EXP(2.659*T9A23))            T9A = T9/(1.+0.067*T9)


                                                 BRANCHING RATIO (Q-VALUE)
                                          N-CHANNELS      P-CHANNELS     A-CHANNELS
                             C12+C12 T9 BELOW 3  0.00(-2.599)  0.50(2.240)  0.50(4.617)
                                     T9 ABOVE 3  0.05          0.30         0.65

                             C12+O16             0.10(-0.423)  0.50(5.169)  0.40(6.770)

                             O16+O16             0.14( 1.453)  0.80(7.677)  0.29(9.593)
```

Two-body Interactions

In the case of two-body primary interactions the third column of Table 1 gives $N_A\langle 01\rangle$ in reactions per sec per (mole/cm^3), or more simply in cm^3 sec^{-1} mole^{-1}.[6] As in FCZ I, $\langle 01\rangle$ is identical to $\langle\sigma v\rangle_{01}$, the Maxwellian averaged product of cross section times velocity for the two interacting nuclei, 0 and 1. This average is proportional to the Laplace transform of σE with $s = 1/kT$ and is given by

$$\langle\sigma v\rangle = \frac{(8/\pi)^{1/2}}{M^{1/2}(kT)^{3/2}}\int \sigma E \exp(-E/kT)\, dE, \qquad 3.$$

where M is the reduced mass of the interacting particles. In reference to the notation of FCZ I we also have $N_A\langle 01\rangle = [01]/\rho$. We tabulate $N_A\langle 01\rangle$ because it is only a function of temperature[7] and not of density and is most convenient for use in reaction chain differential equations that use X, the mass fraction, or $Y = X/A$, the number of moles g^{-1}. The number of moles cm^{-3} is given by ρY. Thus from equations I-3, I-4, and I-5

$$\frac{dY_0/dt}{\rho Y_0 Y_1} = \frac{dY_1/dt}{\rho Y_0 Y_1} = -N_A\langle 01\rangle \text{ cm}^3 \text{ sec}^{-1} \text{ mole}^{-1}. \qquad 4.$$

Thus, multiply $N_A\langle 01\rangle$ by $\rho Y_1 = \rho X_1/A_1$ in mole cm^{-3} to obtain the interaction rate per sec, or reciprocal of the mean lifetime for 0 interacting with 1, and permute for the reciprocal of the mean lifetime for 1 interacting with 0. Multiply $N_A\langle 01\rangle$ by $\rho N_A Y_0 Y_1/(1+\delta_{01}) = \rho N_A Y_0 Y_1/N_0!\,N_1!$ to obtain reactions g^{-1} sec^{-1}. N_I is the number of particles of type $I = 0, 1$ involved. Multiply this result by $1.6022\times 10^{-6}\ Q_6$ to obtain energy generation in erg g^{-1} sec^{-1}. To obtain reactions cm^{-3} sec^{-1} or erg cm^{-3} sec^{-1} multiply by an additional ρ.

For the rate of reverse reactions between *ground states* of the reacting nuclei multiply $N_A\langle 01\rangle$ by REV RATIO to obtain $N_A\langle 23\rangle$ or $N_A^2\langle 234\rangle$. For photodisintegrations induced by gamma radiation multiply $N_A\langle 01\rangle$ by REV RATIO to obtain $\lambda_\gamma(2) = 1/\tau_\gamma(2)$, the interaction rate per sec or the reciprocal of the mean lifetime to photodisintegration, for 2 from $\gamma+2\rightarrow 0+1$. The REV RATIOS listed are calculated using statistical weight factors for ground states only and are defined by equations I-10 and I-16 as given and equation I-20 modified to include a factor N_A on each side of the equation. When calculating reverse reaction rates at high temperature, $T_9 > 1$, see the section on reverse reaction rates and partition functions below. REV RATIOS are not listed for reactions that produce neutrinos.

[6] In general, as indicated at the top of Table 1, the reaction rates, direct or reverse, are given per second per (mole/cm^3)$^{N-1}$, where N is the number of interacting particles, including the target, *exclusive of photons* (gamma radiation). Recall that in the reverse reaction the residual nucleus of the direct reaction becomes the target.

[7] Fractional powers of temperature such as $T_9^{M/N}$, where M and N are integers, are given in Table 1 as $T9MN$, for example, $T923 \equiv T_9^{2/3}$. Integral powers of temperature are given in standard computer form, for example, $T9{*}{*}3 \equiv T_9^3$.

Three-Body Interactions

In the case of three-body primary interactions the third column of Table 1 gives $N_A^2\langle 012\rangle$ in reactions per sec per (mole/cm^3)^2 or, more simply, in cm^6 sec^{-1} mole^{-2}. Multiply $N_A^2\langle 012\rangle$ by $\rho^2 Y_1 Y_2(1+\delta_{01}+\delta_{20})/(1+\delta_{01}+\delta_{12}+\delta_{20}+2\delta_{012}) = \rho^2 Y_1 Y_2 N_0/N_0!\,N_1!\,N_2!$ to obtain the interaction rate per sec for 0 and permute for 1 and 2. N_I equals the number of particles of type $I = 0, 1, 2$ involved. For example, N_α equals 3 in $He^4(2\alpha,\gamma)C^{12}$ and $N_0!\,N_1!N_2!$ equals 6. Recall that O! equals 1. Multiply $N_A^2\langle 012\rangle$ by $\rho^2 N_A Y_0 Y_1 Y_2/N_0!\,N_1!\,N_2!$ for reactions gm^{-1} sec^{-1}. Multiply this result by $1.6022\times 10^{-6}\ Q_6$ to obtain the energy generation in erg gm^{-1} sec^{-1}. To obtain reactions cm^{-3} sec^{-1} or erg cm^{-3} sec^{-1} multiply by an additional ρ.

All of the reactions involving three interacting particles (012) included in Table 1 result in the production of a residual nucleus (3) plus a gamma ray. Thus multiply $N_A^2\langle 012\rangle$ by REV RATIO to obtain $\lambda_\gamma(3) = 1/\tau_\gamma(3)$, the interaction rate per sec or the reciprocal of the mean lifetime to photodisintegration, for 3 from $\gamma+3 \rightarrow 0+1+2$.

Reactions Producing Four Particles

Several two-body reactions in Table 1 result in the production of four particles. In these cases multiplication of the direct reaction rate by REV RATIO results in $N_A^3\langle 2345\rangle$ in reactions per sec per (mole/cm^3)^3 or, more simply, cm^9 sec^{-1} mole^{-3}. Multiply $N_A^3\langle 2345\rangle$ by $\rho^3 Y_3 Y_4 Y_5 N_2/N_2!\,N_3!\,N_4!\,N_5!$ to obtain the reaction rate per sec for 2 and permute for 3, 4, and 5. N_I is the number of particles of type $I = 2, 3, 4, 5$ involved. Multiply $N_A^3\langle 2345\rangle$ by $\rho^3 N_A Y_2 Y_3 Y_4 Y_5/N_2!\,N_3!\,N_4!\,N_5!$ to obtain reactions gm^{-1} sec^{-1} and so on as given previously.

Atomic Masses

The following atomic masses should be used in $Y = X/A$: $A_n = 1.008665$; $A_H = 1.007825$; $A_D = 2.014103$; $A_T = 3.016050$; $A_{He^3} = 3.016030$; and $A_{He^4} = 4.002603$ rounded off to the user's taste. All others are equal to the appropriate integral mass number within an error of no more than 3 parts in 1000. In our Q-value calculations we have used mass excesses given by Ajzenberg-Selove & Busch (1971) except for these revisions: Be^8, 4.94177 MeV; O^{15}, 2.8551 MeV; and Na^{21}, -2.18478 MeV.

Electron and Positron Capture

The expressions given in Table 1 for electron capture by H^1, He^3, and Be^7 are to be multiplied by n_{e^-}/N_A to obtain the reaction rate per sec. The electron density is designated by n_{e^-} and *at low temperature*

$$n_{e^-}/N_A = \rho Y_{e^-} = \rho/\mu_{e^-} = \rho[0.99935+0.98512X_H - \textstyle\sum X_{A,Z}\langle 0.99935-2Z/A\rangle]/2 \approx \rho(1+X_H)/2, \qquad 5.$$

where X_H is the mass fraction of hydrogen, $\sum X_{A,Z}$ is that of all atoms (A, Z) with

$A > 4$, and ρ is total density in g cm^{-3}. The last approximation must obviously not be used in neutron stars. The entry for the reaction $H^1(pe^-, \nu)H^2$ is to be multiplied by $\rho^2 N_A Y_e$- $Y_H^2/2 \approx \rho^2 N_A(1+X_H)X_H^2/4$ to obtain the number of reactions gm^{-1} sec^{-1}.

At higher temperatures ($T_9 > 1$) electron-positron pair formation sets in, the electron density is increased, and positron capture by nuclei becomes possible. Non-degenerate expressions for the electron and positron density are given in equations 10, 11, 12, and 17 of Fowler & Hoyle (1964). In their notation $n_0 \approx N_A \rho(1+X_H)/2$ and $n_\pm = n_{e\pm}$. Higher order terms in the temperature must also be included in addition to those given in Table 1 for electron capture reactions and these can be found in equations A20 and A20′ of Fowler & Hoyle (1964). Degenerate situations are discussed by Bahcall (1964) and Chiu (1968) and semidegenerate situations by Wagoner (1969).

Beta Decay

The mean decay rates for a number of electron and positron emitters are given in Table 15B of Wagoner (1969). Rates for other cases can be determined from half-lives given in *Nuclear "Wallet Cards"* (Ajzenberg-Selove & Busch 1971) by using $\lambda_\beta = \ln 2/t_{1/2}$. The modification of these rates under degenerate conditions is discussed by Chiu (1968) on p. 303 et seq, and under semidegenerate conditions by Wagoner (1969) in Table 14B.

Reverse Reaction Rates and Partition Functions

In Table 1 the listed REV RATIO gives the ratio of the rate of the reverse reaction to that quoted for the direct rate assuming that all the interacting nuclei are in their ground states. The appropriate equations for obtaining these ratios are equations I-10, I-16, and I-20. The generalization for the reactions in Table 1 in which four (or more) particles are produced is quite straightforward if it is recalled that a factor $N_I!$ must be introduced in the numerator of REV RATIO for each type of particle I, where N_I is the number of particles of type I produced in the direct reaction and a factor $N_K!$ must be introduced in the denominator for each type of particle K, where N_K is the number of particles of type K consumed in the direct reaction.

Under stellar circumstances the *bound* excited states of nuclei will be populated in general according to the appropriate statistical factor; the population in the ith state is given by

$$P^i = \frac{g^i}{G} \exp(-E^i/kT), \qquad 6.$$

where the partition function is given by

$$G = \sum_i g^i \exp(-E^i/kT) \qquad 7.$$

and $g^i = 2J^i + 1$ is the statistical weight of the state with spin J^i (in units $\hbar$) and E^i is the excitation energy of the state above the ground state. A nucleus is designated by the subscript I; its excited states by the superscript i. A quantity of considerable interest is the normalized partition function

$$\mathcal{G} = G/g^0 = 1/P^0, \tag{8.}$$

where g^0 is the statistical weight of the ground state because the reciprocal of $\mathcal{G}$ is just the fraction, P^0, of the nuclei populating the ground state ($E^0 = 0$) at equilibrium.

The population can be taken to be that for statistical equilibrium at temperature T because the *initial* rate for populating an excited state from the ground state by photoexcitation alone[8] can be shown to be

$$\lambda_i = \frac{1}{\tau^i} \equiv \frac{1}{n^i_{eq}} \frac{dn^i}{dt} = \frac{G}{g^0} \frac{\lambda^i_{\text{spon}}}{1 - \exp(-E^i/kT)} \geqq \lambda^i_{\text{spon}}, \tag{9.}$$

which is always larger than the spontaneous decay rate, λ^i_{spon}, of the excited state and at high temperature ($kT > E^i$) is larger by the factor GkT/g^0E^i. The exponential term in the denominator incorporates the effect of stimulated emission and assumes a black-body or Planck radiation spectrum. There are situations such as those associated with shock wave phenomena (Hoyle & Fowler 1973, Colgate 1974) where the time to equilibrate nuclei, ions, atoms, electrons, and radiation may be long and of considerable importance. Except for long-lived isomeric states of nuclei, $\tau^i \leqq 1/\lambda^i_{\text{spon}}$ is $< 10^{-9}$ sec for the excited states of the nuclei discussed in this paper. This is to be compared with the hydrodynamic time scale of explosive nucleosynthesis, $\tau_{\text{hyd}} \sim 10^3 \rho^{-1/2}$ sec, which is 10^{-3} sec even for $\rho = 10^{12}$ g cm^{-3}. Many nuclear processes can be slow compared to hydrodynamic time scales but photoexcitation is *not*, in the sense that the equilibrium population to be reached is low when the rate for photoexcitation is at a minimum ($E^i \gg kT$). Long-lived isomeric states, just like radioactive nuclei, must be treated as individual nuclear species in the handling of reaction chains under stellar circumstances.

Excited states of a nucleus that are not bound and do not decay primarily to the ground state can be taken as in equilibrium at a given time with those of their particle decay products, which are in greatest abundance at that time. The time to reach equilibrium after a sharp change in ρ and/or T, say, is of the order of $\hbar/\Gamma_c$, where Γ_c is the partial width for decay into the components of the appropriate channel (c). These times are nuclear times ($\sim 10^{-21}$ sec) divided by appropriate penetration factors, which can indeed be small, so that the equilibration time can be $\gg 10^{-21}$ sec; nonetheless, these times are usually very short compared to the "burning" times for transformation from the initial components to the final products. Toward the end of burning these excited states come into equilibrium with the final product or products. The case of one final product occurs in radiative capture and refers to the ground state or a *bound* excited state of the compound (final) nucleus in which case the unbound excited state does eventually reach equilibrium with the ground state and, in fact, with all the bound excited states. We emphasize that, during the initial stages of a radiative capture reaction, excited states that are not bound can be populated in much greater numbers than calculated from equilibrium

[8] Excitation by inelastic particle scattering (n, n'; p, p'; α, α'; etc) increases the rate.

with the ground state and the bound excited states. The opposite is of course true in the initial stages of photodisintegration, the reverse of radiative capture.

In calculating the abundance of a nucleus produced in nuclear burning one is interested in all those excited states of a nucleus that will decay to the ground state after the final cooling of any stellar event. Thus the appropriate upper limit on the summation over states for G or $\mathcal{G}$ is the excitation just above the point where the states become unbound and decay primarily by particle emission. For example, the nucleus N^{13} has only one bound state, its ground state with $J = 1/2$, and thus $G = g^0 = 2\mathcal{G} = 2$.

The above argument shows that the REV RATIOS needed in stellar reaction calculations should be based on the partition functions of the nuclei involved and not just on the statistical weights of the ground states. Thus a problem arises. In general, laboratory measurements yield σ and thus $N_A\langle\sigma v\rangle$ for the reaction from the *ground* state of the *target* nucleus (0) in the incident channel to *all bound* states of the *residual* nucleus (3) in the outgoing channel. Particles (1) and (2) are usually n, p, or α, which have excited states of such high excitation that we can neglect them for $T_9 \leqq 10$. Thus the REV RATIO in Table 1 should be corrected by dividing by $\mathcal{G}_3 = G_3/g_3^0$ for the residual nucleus. The reverse reaction-rate ratio as given in the table has the factor g_3^0 in the denominator, not the correct astrophysical factor G_3.

Certain of the heavier nuclei treated in this review such as F^{19}, Ne^{21}, Na^{23}, Mg^{25}, and Al^{27} have low-lying states between 0.1 and 1 MeV and these states will be significantly populated in the temperature range $1 \leqq T_9 \leqq 10$. Thus the target nuclei as well as the residual nuclei populate excited states under high temperature circumstances. In principle the interactions of these states can be determined by studying transitions to them in the reverse reaction. This often cannot be done, however, because the residual nucleus in the direct reaction (target nucleus in the reverse reaction) is radioactive with a short lifetime. We have attempted, where possible, to include the effects of these excited target states by using either inelastic scattering data as discussed by Bahcall & Fowler (1969a) or theoretical Hauser-Feshbach calculations following Michaud & Fowler (1970). For example, when the inelastic scattering-strength function, $S_{\text{inel}} = (2J_r+1)(\Gamma_1\Gamma_1'/\Gamma)_r$, has been measured at a resonance, r, as well as the reaction-strength function, $S_{\text{reac}} = (2J_r+1)(\Gamma_1\Gamma_2/\Gamma)_r$, then we increase the latter quantity by the factor

$$f = 1 + \left(\frac{\Gamma_1'}{\Gamma_1}\right)_r = 1 + \frac{S_{\text{inel}}}{(2J_r+1)\Gamma_r}\left(\frac{\Gamma}{\Gamma_1}\right)_r^2. \qquad 10.$$

Γ_r is the full width of the resonance, Γ_1 is the partial width for the incoming channel, Γ_1' is the partial width for the inelastic scattering channel, and Γ_2 is the partial width for the outgoing reaction channel. Γ_r is usually known but frequently $(\Gamma_1/\Gamma)_r$ is not. Fortunately in many of these unknown cases it is strongly indicated on theoretical grounds that $(\Gamma_1/\Gamma)_r \approx 1$ and f can be evaluated approximately.

In cases such as $Na^{23}(p, \alpha)Ne^{20}$, where we have been able to correct for the effects of excited states in the target nucleus, we express the results for $N_A\langle\sigma v\rangle$ in such a form that all terms are to be divided by $\mathcal{G}_0 = G_0/g_0^0$, the normalized partition function for the target nucleus. When this factor is properly included in the numerator of the reverse reaction-rate ratio and $\mathcal{G}_3 = G_3/g_3^0$ is properly included in the denominator, then, in the reverse reaction rate $\mathscr{G}_0$ cancels out and all terms are divided by $\mathscr{G}_3$ as expected from the reciprocity principle. To be explicit and completely general, in these cases multiply the REV RATIO in Table 1 by $\mathscr{G}_0\mathscr{G}_1/\mathscr{G}_2\mathscr{G}_3$. The reaction rates then involve the number densities of the interacting particles summed over all of their bound states. After "cooling" this will be the number density of the ground state.

When information concerning the interactions of the excited states of target nuclei is unavailable, one can in principle follow only the reactions of the ground state of such nuclei; the reaction rates involving their bound excited states, which will eventually decay to the ground state, are unknown. Under these circumstances we recommend the equal strength approximation of Bahcall & Fowler (1969a), which assumes that all states have the same $N_A\langle\sigma v\rangle$ as the ground state. Thus the rate for the correct statistical population is just that for the ground state. Once again, then, the REV RATIO in Table 1 should be corrected by $\mathscr{G}_0 = G_0/g_0^0$ and, in complete generality, by $\mathscr{G}_0\mathscr{G}_1/\mathscr{G}_2\mathscr{G}_3$. This means that in expressions like $n_0 n_1\langle\sigma v\rangle_{01}$ or $n_2 n_3\langle\sigma v\rangle_{23}$ the number densities, n_I, are the sums over all bound states. The extension to three or four interacting particles or three or four product particles should be obvious.

The partition functions involving the nuclei treated in this paper are defined and tabulated in Bahcall & Fowler (1970) (see their Table 2 and especially their note added in proof). The G listed in their Table 2 should be divided by the g^0 for the ground state listed in their Table 1 to obtain $\mathscr{G}$, the normalized partition function. Inspection of the table will reveal the temperatures below which the $\mathscr{G} \approx 1$ and our values for REV RATIO apply without substantial correction. We have used new measurements to yield the corrected values $\mathscr{G}(F^{19}) = 1+4\exp(-2.090/T_9)+7\exp(-16.440/T_9)$ and $\mathscr{G}(Ne^{21}) = 1+1.5\exp(-4.068/T_9)+2.0\exp(-20.258/T_9)$. Note that $\mathscr{G}(T_9)$ is designated by $GT9$ in Table 1.

Stimulated and Induced Reaction Rates

We have implicitly included stimulated gamma-ray emission in equilibrium and reverse reaction-rate considerations in equation 9. However, we have not included in Table 1 the increase in reaction rates, direct and reverse, due to stimulated gamma-ray emission and a number of induced processes. Here we can only catalog in a schematic way these various enhancement mechanisms that are significant at high temperature.

Given a black-body radiation environment, every radiation process is enhanced by $[1-\exp(-\hbar\omega_p/kT)]^{-1}$ where $\hbar\omega_p$ is the energy of the primary gamma ray emitted. Thus one needs to know in detail the various primary gamma-ray transitions that can result from the decay of resonant states in the compound nucleus. In nuclear processes it must be recalled that the excited states resulting from the primary gamma-ray emission may be unbound to particle emission and thus stimulated processes also occur here. For example, an initial two-particle channel may transform to a final two-particle channel via the primary emission of a gamma ray. Such a transformation will be enhanced under stellar circumstances by the factor given above.

At high temperature various "induced" processes occur. The compound nucleus produced as an intermediate state in a reaction can interact via inelastic and superelastic scattering with $\gamma, n, p, \alpha \ldots$ to produce the ground state of the compound nucleus, bound excited states that decay to the ground state, or excited states that break up into particle channels. All such processes can be treated as three reacting bodies in the incident channel and by considering the reactions in Table 1 as the first stage in these processes. For discussion of specific examples the reader is referred to Shaw & Clayton (1967) and to Truran & Kozlovsky (1969).

Remarks to Users

It should be appreciated at this point by the reader and user that our primary purpose is *not* to provide reaction rate expressions that can be employed most economically and efficiently in computer programs. Rather it is our primary purpose to provide expressions that accurately represent the available experimental data supplemented where necessary by accepted theoretical considerations, particularly in regard to extrapolation to low energies and thus low temperatures. In general we present expressions that represent the available data to $\lesssim 25\%$ accuracy over the temperature range 10^6 °K $\leqq T \leqq 10^{10}$ °K or $10^{-3} \leqq T_9 \leqq 10$. Space does not permit us to enter into the complex problem of the evaluation of the accuracy of primary data in this article.[9] In general the uncertainties range from $\sim 10\%$ for the pp reaction up to as much as a factor of 3 in some cases. The typical uncertainties are of the order of 50%. In cases where the available data only set upper limits on cross sections or resonance strengths, we have introduced the factor (0 to 1) in the expressions presented. Once the factor (0 to 1) appears it is to be included in *all* terms that follow it. Terms for which the factor must be the same are grouped together in brackets. We have also introduced this factor where resonance strengths have been evaluated by using dimensionless reduced widths $\theta_p^2 \leqq 1$ for protons and $\theta_\alpha^2 \leqq 0.1$ for $T = 0$ resonances, $\theta_\alpha^2 \leqq 10^{-4}$ for $T = 1$ resonances for alpha particles. T is the isotopic spin as defined by Chiu (1968), p. 318. When expressions that include the factor (0 to 1) are critical in determining ultimate results the user must refer to the original references cited and make his own judgment. Better still, make the calculations twice, once without the term in question and once with its full value. If a compromise is desired set (0 to 1) = 0.1. We will be happy to respond to inquiries in this regard. We will also be happy to provide numerical evaluations of all our expressions *but only* for our standard T_9 grid from 0.001(0.001) 0.016(0.002) 0.020(0.005) 0.030(0.01) 0.16(0.02) 0.20(0.05) 0.50(0.10) 1.00(0.25) 2.00(0.50) 4.00(1.00) to 10.00. A final word of advice: the user of Table 1 must be diligent in reading the parentheses correctly.

ANALYSIS OF THE EXPERIMENTAL DATA

Continuum Reaction Rates

The study of explosive nucleosynthesis on a short, *hydrodynamic* time scale (Fowler & Hoyle 1964) has created considerable interest in high-temperature reaction rates

[9] In any case we wish to remain on good terms with our hard-working colleagues in nuclear laboratories.

above $T_9 \sim 1$, which are in general related to reaction cross sections at energies above $E_6 \sim 1$. In this energy range nuclear reactions proceed through many resonances that are separated by intervals not much greater than their widths or that actually overlap to form a continuum. Frequently it is found that sharp resonances are superimposed on a continuum background. In any case the contribution to the reaction rate of the background and all but one or two of the low-lying resonances, which must be treated individually, can be summed and fitted to an expression of the form

$$\begin{aligned} N_A\langle\sigma v\rangle_{\text{cont}} &= \text{const}\, T^m \exp(-C/kT) \\ &= BT_9^m \exp(-11.605C_6/T_9)\ \text{cm}^3\ \text{sec}^{-1}\ \text{mole}^{-1}, \end{aligned} \qquad 11.$$

where B and C are free least square parameters; we have arbitrarily limited $|m|$ to integers, half integers, or rational fractions involving integers <10. In FCZ I we restricted ourselves to $m = 0$ but we have found many reactions require $m \neq 0$ for fits to $\lesssim 25\%$. The effective continuum threshold energy, C, can represent a true threshold in an endoergic reaction with $Q < 0$ in which case $C = |Q|$. In an exoergic reaction the cross section usually increases rapidly with energy as the penetration factor in the incident channel increases and flattens out either when the incident-channel partial width becomes equal to the sum of other partial widths or when the Coulomb barrier is reached in the incident channel. C is then roughly the energy at which this occurs.

Equation 11 is the rate that corresponds to a cross-section dependence on energy given by

$$\sigma(E) = 2\sigma(2C)\left(\frac{C}{E}\right)\left(\frac{E}{C} - 1\right)^{m+1/2} \qquad (E \geqq C), \qquad 12.$$

with $\sigma(2C)$ equal to the cross section at $E = 2C$ and $\sigma = 0$ for $E \leqq C$. From equation 3 one finds

$$\begin{aligned} N_A\langle\sigma v\rangle_{\text{cont}} &= N_A\Gamma(m+\tfrac{3}{2})\sigma(2C)\left(\frac{32C}{\pi M}\right)^{1/2}\left(\frac{kT}{C}\right)^m \exp(-C/kT) \\ &= 1.888\times 10^9\ \Gamma(m+\tfrac{3}{2})\sigma_b(2C_6)\left(\frac{C_6}{A}\right)^{1/2}\left(\frac{T_9}{11.605C_6}\right)^m \\ &\quad \times \exp(-11.605C_6/T_9)\ \text{cm}^3\ \text{sec}^{-1}\ \text{mole}^{-1}, \end{aligned} \qquad 13.$$

where Γ is the gamma function, $\sigma_b(2C_6)$ is the cross section at $2C_6$ in barns, and $A = MN_A$ is the reduced mass in atomic mass units. Note that for the limiting case of a single sharp resonance, $m = -\frac{3}{2}$, σ diverges at $E = C$ in such a way that σ must be treated as a δ-function in the customary manner (see equations I-62 and I-66).

Equation 12 for the cross section exhibits a maximum, characteristic of many reactions, for $-\frac{1}{2} \leqq m \leqq \frac{1}{2}$, at

$$E_{\max} = \frac{2C}{1-2m}. \qquad 14.$$

For $m < -\frac{1}{2}$, the cross section exhibits a singularity at $E = C$. For $m > \frac{1}{2}$, the cross section rises monotonically from the threshold at $E = C$. A case of special interest is that for $m = 0$, which applies to an endoergic reaction ($Q < 0$, $C = |Q|$) for s-wave neutrons (orbital angular momentum, $l = 0$). In this case the cross section rises as $(E-C)^{1/2}$ above threshold and drops off as $E^{-1/2}$ at high energy ($\sigma v \rightarrow$ const). For $m = -\frac{1}{2}$, σE exhibits a step function at $E = C$. The computer programmer will need definitions of T_9^m where $|m| = \frac{1}{2}, \frac{3}{2}, \frac{7}{2}, \frac{1}{3}, \frac{2}{3}, \frac{4}{3}, \frac{5}{3}, \frac{1}{4}, \frac{3}{4}, \frac{1}{5}, \frac{8}{5}, \frac{7}{6}, \frac{2}{7}$, and $\frac{1}{8}$ for complete coverage of the reactions listed in Table 1. In addition, powers of $T9A$ (see below) that should be defined are $\frac{3}{2}, \frac{1}{3}, \frac{2}{3}$, and $\frac{5}{6}$.

Sharp Resonance Reaction Rates

In the case of a sharp resonance at energy, E_r, for which the full width at half maximum, Γ_r, is much less than the resonance energy, that is, $\Gamma_r \ll E_r$, the reaction rate is given by equation 11 with $m = -\frac{3}{2}$ and $C = E_r$. Equation I-62 shows that the relevant experimental parameter is $(\omega\gamma)_r$, which is determined from a thin target experiment by equation I-66. In many experimental determinations thick targets are used, in which case the resonance results in a step in the excitation curve. The increase in yield per incident particle, Y_r, in crossing the resonance step determines $(\omega\gamma)_r$ through

$$Y_r = \int n\sigma\, dx = \int_r (\sigma/\varepsilon)\, dE \approx (1/\varepsilon_r) \int_r \sigma\, dE, \qquad 15.$$

so that

$$(\omega\gamma)_r = (ME_r/\pi^2\hbar^2)\varepsilon_r Y_r, \qquad 16.$$

where n is the number of *reacting* target nuclei per cm^3, dx is an element of target thickness, and $\varepsilon = (1/n)\, dE/dx$ is the stopping cross section in the target per reacting target nucleus. We assume that $\varepsilon \approx \varepsilon_r \equiv \varepsilon(E_r)$ over the interval of several Γ_r near E_r.

From equation I-62 one obtains

$$N_A\langle\sigma v\rangle = N_A \left(\frac{8}{\pi MkT}\right)^{1/2} \varepsilon_r Y_r (E_r/kT) \exp(-E_r/kT). \qquad 17.$$

This leads to an important consideration concerning the use of thick targets in determining upper limits on reaction rates due to low energy "near-threshold" resonances. The experimentalist uses a thick target and gradually decreases the voltage on his accelerator and thereby the energy of his incident beam of particles. If he detects a yield above background, he may be able to lower his beam energy until the step at E_r is reached and all is fine and dandy—equation 17 can be used to yield $N_A\langle\sigma v\rangle$ even though the standard deviation of the measurement may be substantial. In many cases, however, a limit is reached below which the accelerator cannot deliver a stable, well-focused beam of incident particles of sufficient intensity, and background effects make it possible to set only an upper limit on Y_r. The resulting upper limit on $N_A\langle\sigma v\rangle$ then depends critically, as equation 17 shows, on the still unknown E_r relative to kT where T is the temperature of interest. For example, there will be a factor of $\sim 10^3$ involved in the upper limit on $N_A\langle\sigma v\rangle$ for a given one on Y_r depending on whether $E_r = kT$ or 10 kT. (We neglect the slow variation of ε with E or ε_r with E_r at low energy.) Dwarakanath (1974) has discussed this matter extensively in connection with the bearing of his low-energy measurements on $He^3(He^3, 2p)He^4$ on the solar neutrino problem. The ultimate solution to the problem involves the determination of the resonance energy, if there is a resonance, by measuring the excitation of the resonant state in the compound nucleus in some other reaction or in showing that no resonance exists between the threshold and the lowest energy used in the direct measurements (Parker et al 1973). Indirect measurements of the properties of any state that does occur in this range may yield a value rather than an upper limit for $N_A\langle\sigma v\rangle$ (Rolfs et al 1975). The high-energy tails of subthreshold states can also be treated in this manner (Dyer & Barnes 1974).

Wide Resonance Reaction Rates

Wide resonances, $\Gamma_r > 0.1\,E_r$, cannot be treated accurately in the manner used for sharp resonances. Examples occur in the reactions $H^3(d, n)He^4$ where the standard tabulated values in the CM system are $\Gamma_r = 0.081$ MeV at $E_r = 0.064$ MeV and $Be^9(p, d)2He^4$ or $Be^9(p, \alpha)Li^6$, where $\Gamma_r = 0.120$ MeV at $E_r = 0.296$ MeV. For resonance energies below the Coulomb and/or centrifugal barrier, complications arise from the fact that the product, σE, is distorted from the symmetrical bell shape of the pure Lorentz factor $(\Gamma_r/2)^2/[(E-E_r)^2+(\Gamma_r/2)^2]$ by the rapid variation of the incoming and outgoing channel transition probabilities over the width of the resonance. The resonance shape is sharpened below the resonance energy and broadened above, and indeed the position of the maximum in σE may be shifted above E_r.

After a number of attempts we have found that the following procedure yields the most accurate results when compared with numerical integration over the resonance using equation 3. The low-energy tail of the resonance is treated as discussed in the next section on nonresonant reaction rates. For the resonant contribution we find that equation 11 is quite satisfactory if m is chosen somewhat greater than $-\frac{3}{2}$ and C is chosen as the centroid of the resonance area rather than the energy at the maximum in the product σE. Actually we determine B, m, and C as free parameters in a least squares fit of equation 11 to the cross section obtained by numerical integration using equation 3. In some cases such as $C^{12}(p, \gamma)N^{13}$ and $C^{13}(p, \gamma)N^{14}$ it is accurate enough to retain $m = -\frac{3}{2}$ and treat only B and C as free parameters.

Nonresonant Reaction Rates

As in FCZ I we have usually expressed the nonresonant *cross-section factor*, $S(E)$, as the first three terms of a truncated Maclaurin series in the center-of-momentum energy E according to the equation

$$\begin{aligned} S(E) &\equiv \sigma E \exp 2\pi\eta = \sigma E \exp\left(E_G^{1/2}/E^{1/2}\right) \\ &= S(0)+S'(0)E+\tfrac{1}{2}S''(0)E^2, \end{aligned} \qquad 18.$$

where $\eta = Z_0 Z_1 e^2/\hbar v$ and $E_G = (2\pi\alpha Z_0 Z_1)^2(Mc^2/2)$ is the "Gamow energy." Then $N_A\langle\sigma v\rangle$ is determined using equations I-51 (without the factor ρ)–I-59. A thorough discussion of the validity of this treatment of low-energy, charged-particle reactions has been given by Critchfield (1972). When low-energy experimental data are available the free parameters in equation 18, namely, $S(0)$, $S'(0)$, and $S''(0)$, can be determined by a least squares analysis.[10] In many cases treated in this paper such data are not available and it is necessary to determine the contribution of the tails of measured resonances to the low-energy resonant or, strictly speaking, off-resonance cross-section factor. In doing this we have found that it is necessary to include not only the resonance behavior, as was done in equations I-76 and I-77, but also the energy dependence of the width for the incoming and outgoing channels, namely, Γ_1 and Γ_2 of equation I-61. When the outgoing channel involves two particles the results are

$$S(0) = \frac{\pi\sigma_r E_r}{4K_{2l+1}^2(x)P_l(E_r)} \frac{P_L(Q)}{P_L(Q+E_r)} \frac{\Gamma_r^2/4}{E_r^2+\Gamma_0^2/4}, \qquad 19.$$

where P_l is the penetration factor as defined by Vogt (1968, equation 61) in the incoming channel for the appropriate orbital angular momentum, $l\hbar$, in that channel, and P_L is the penetration factor in the outgoing channel for the appropriate orbital angular momentum, $L\hbar$, in that channel. When several l's (or L's) were possible under the conservation of angular momentum and parity, we employed the minimum possible value for l (or L) except where available experimental evidence indicated that higher values make the major contribution to Γ_1 (or Γ_2). In the above expression for $S(0)$ we have used the asymptotic expression for P_l near zero energy given by

$$P_l(E) = \frac{\pi \exp\left[-(E_G/E)^{1/2} - \alpha_l E\right]}{4K_{2l+1}^2(x)}, \qquad E \to 0, \qquad 20.$$

where $K_{2l+1}(x)$ is the modified Bessel function of order $2l+1$ and $x = (8MR\ Z_0 Z_1 e^2/\hbar^2)^{1/2} = 0.525\ (AZ_0 Z_1 R_f)^{1/2}$ in the incoming channel. All other symbols in equations 19 and 20 are defined in FCZ I; α_l is defined in Burbidge et al (1957), p. 560. The radii, R_f in fermis, are taken from Michaud & Fowler (1970).

In equation 19 we have correctly used $\Gamma_0^2/4$ in the denominator rather than $\Gamma_r^2/4$ as in equation I-76. $\Gamma_0 \equiv \Gamma(E=0)$ is the sum of all partial widths at zero energy in the incoming channel. If this sum at resonance, $\Gamma_r = \Gamma(E_r)$, is mainly due to contributions from outgoing channels which do not vary rapidly with energy, then $\Gamma_0 \approx \Gamma_r$. However, if Γ_r is mainly due to the width of the incident channel that varies rapidly with energy in general and goes to zero at zero energy, then $\Gamma_0 \approx 0$ and the last part of the right-hand side of equation 19 becomes $\Gamma_r^2/4E_r^2$.

[10] We have not included the term $5/12\tau \propto T^{1/3}$ from equation I-52 in the entries in Table 1 unless the terms in $S'(0)$ and/or $S''(0)$ are known. The factor $1+5/12\tau$ can be calculated easily for any reaction using equations I-54 and I-55. Such factors are typically of the order $1+0.05T_9^{1/3}$.

For the derivatives of the cross-section factor at zero energy the results are

$$S'(0) = S(0)\left[\frac{2}{E_r} - \alpha_l - \alpha_L + \frac{1}{2}\left(\frac{E'_G}{Q^3}\right)^{1/2}\right], \tag{21}$$

where E'_G is the Gamow energy as defined in equation I-46 but for the outgoing channel and

$$S''(0) = S(0)\left[\frac{2}{E_r^2} + \left(\frac{S'(0)}{S(0)}\right)^2 - \frac{3}{4}\left(\frac{E'_G}{Q^5}\right)^{1/2}\right]. \tag{22}$$

The use of α_L in equation 21 is a poor approximation for Q equal to a substantial fraction of the Coulomb barrier in the outgoing channel, but its use is sufficiently accurate for our purposes.

When the outgoing channel involves gamma radiation the following replacements should be made: in equation 19 replace the ratio of the P_L's by $[Q/(Q+E_r)]^{2L+1}$, where L is now the multipole order of the radiation. Because electric or magnetic dipole radiation usually dominates, $L = 1$ and $2L+1 = 3$; in equation 21 set $\alpha_L = 0$ and replace the last term in the square brackets with $(2L+1)/Q$; in equation 22 replace the last term in the square brackets by $-(2L+1)/Q^2$.

In the customary random-phase approximations the contributions of each resonance to $S(0)$, $S'(0)$, and $S''(0)$ are added to obtain overall values. However, states of the same spin and parity can add coherently in either a constructive or destructive fashion at low energy. Such cases must be treated individually using the proper sum rules as given, for example, by Rolfs & Rodney (1974a).

In cases such as $Mg^{24}(\alpha, \gamma)$ no nonresonant term is included. This is because the nonresonant term made a substantial contribution to $N_A\langle\sigma v\rangle$ only at such low temperatures that the lifetime of Mg^{24}, even for $\rho X_{He} = 10^5$, was greater than the lifetime of the "universe." We dropped all terms that made a substantial contribution only at temperatures such that $N_A\langle\sigma v\rangle < 10^{-21}$ cm^3 sec^{-1} mole^{-1} for proton interactions and $< 10^{-23}$ for alpha-particle reactions.

Cutoff for Nonresonant Reaction Rates Used with Resonant or Continuum Rates

The truncated Maclaurin series we have used for $S(E)$ diverges positively or negatively (infrequent) at high energy and so $S_{eff}(T)$ defined by equation I-52 diverges at high temperature. For example, the expansion for $S(E)$ due to a resonance using the parameters given in equations 19, 21, and 22 falls below the true $S(E)$ at E somewhat less than E_r and becomes equal to $\sim 3\Gamma_r^2/2E_r^2$ times the true value at E_r. It then becomes equal to the true value at $\sim 4E_r/3$ and continues to rise while the true $S(E)$ decreases. In FCZ I we stipulated an upper limit on T for the use of the nonresonant rate. In the interim we have found that this limitation was not always applied by users, and therefore in this article we include cutoff factors in nonresonant reaction rates so that they can be used in combination with resonant or continuum rates at all temperatures. The resonant or continuum rates are automatically cut off at low temperatures by the factor $\exp(-E_r/kT)$ or $\exp(-C/kT)$.

It will be clear that the cutoff temperature for the nonresonant rate must be somehow related to the resonance energy, E_r, or the continuum threshold, C, because just below and above these energies the resonance or continuum terms adequately represent the cross section. This can be done quite straightforwardly through equation I-56 or, more explicitly, through equation I-60, which relates the temperature to the effective interaction energy below the Coulomb barrier for charged particles. Because the next term in the expansion for $S(E)$ would be the term in $E^3 \approx E_0^3 \propto T^2$, we have chosen the cutoff factor as

$$f_{\text{cutoff}} = \exp -(T_9/T_{9co})^2. \tag{23}$$

In certain cases the cutoff temperature, T_{9co}, can be determined by a detailed comparison of our analytical expressions with numerically integrated values for $N_A\langle\sigma v\rangle$. As a rough rule we have found that the following modification of equation I-60 is quite satisfactory:

$$\begin{aligned} T_{9co} &= \frac{23.46}{(W \ln 4)^{1/2}} E_{6r}^{3/2} \\ &= 19.92 W^{-1/2} E_{6r}^{3/2}, \end{aligned} \tag{24}$$

with $W = Z_0^2 Z_1^2 A$. In the continuum case replace E_{6r} by C_6. When the nonresonant cross section is due to the tails of many resonances, including those lumped in the continuum, the cutoff energy is taken as the energy of the resonance that makes the largest contribution at low energy, or the continuum threshold energy, if the contribution of the continuum resonances is the largest. Usually the first strong and not-too-narrow resonance determines T_{9co}.

In Figure 1, the deviation of the analytic expressions given in Table 1 from the numerically integrated values are shown in four cases where such detailed comparison is possible. In these cases T_{9co} has been chosen by trial and error to give a "balanced" plus-and-minus deviation near the temperature where the nonresonant term is cut off and the other terms take over. It is interesting that the maximum positive error occurs at T_{9co}. In the cases shown it has been possible to adjust the nonresonant term to be "exact" at low temperature and to adjust the resonant and/or continuum terms to be "exact" at high temperature. Figure 1 also shows for the sum of the reactions $Be^9(p, d)2He^4 + Be^9(p, \alpha)Li^6$ the deviation when we use our standard cutoff prescription given in equation 24. In this case the resonance is a wide one and our standard prescription does not yield a balanced deviation. However, the error is still only $\sim 40\%$ at a temperature rather high for the occurrence of Be^9-burning. Figure 1 includes the deviation for $C^{12}(p, \gamma)N^{13}$ when the cutoff factor is omitted, indicating the divergence at high temperature. The reader must recall that Figure 1 shows the deviation of our analytical expressions from the available cross-section data smoothed by integration into $N_A\langle\sigma v\rangle$ and does not illustrate the possible experimental error in that data. We claim only to fit the available data to $\pm 25\%$ in most cases. This is about as well as can be done for rates that may vary by 20 or more orders of magnitude over the range of temperature under consideration.

Alternative Nonresonant Reaction Rates

Calculation of the rates of nuclear processes at elevated temperatures ($T_9 > 1$) requires knowledge of cross sections at interaction energies in the MeV range. In this range the cross-section factor varies considerably and a truncated expansion intended for use just above zero energy, as given in equation 18, is not at all satisfactory. Because the major variation is usually an exponential decrease with increasing energy we have found it convenient to adjust the experimental data for $\sigma(E)$ and thus for $S(E)$ to expressions of the form

$$S(E) = S(0)q(E)\exp(-\alpha E), \tag{25.}$$

where the function $q(E)$ is normalized to $q(E=0)=1$. Frequently there is not enough low-energy data to determine α. In such cases we use an average over α_l

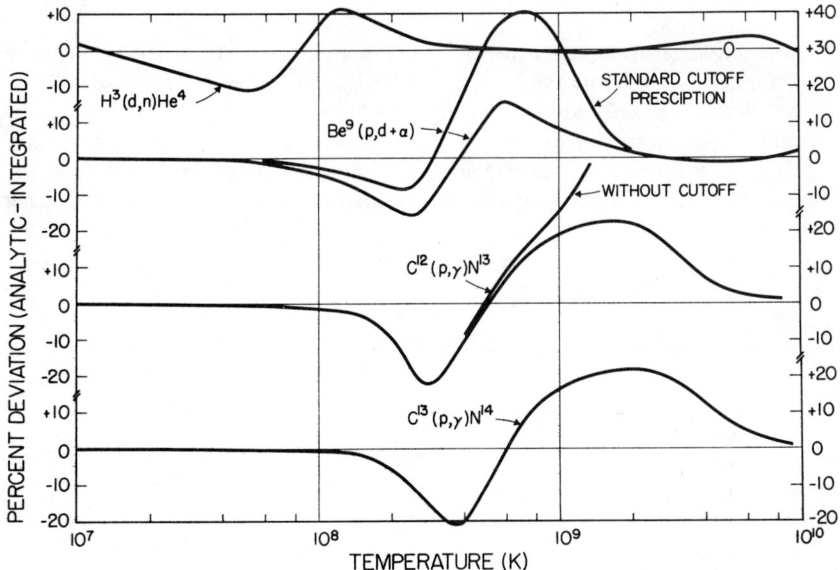

Figure 1 The percentage of deviation of the analytic expression for $N_A\langle\sigma v\rangle$ in Table 1 from the values given by numerical integration for the reactions indicated. In these cases it has been possible to choose the cutoff temperature for the nonresonant rate in such a way that a balanced plus-and-minus deviation is obtained. The standard cutoff prescription given by equation 24 does not always achieve this, as indicated for $Be(p,d)2He^4$ plus $Be^9(p,\alpha)Li^6$. If the cutoff is not applied, the result is disastrous, as indicated in the case of $C^{12}(p,\gamma)N^{13}$. The deviations shown are smaller than the uncertainty in the experimental data in most cases and must be considered in the light of the many orders of magnitude variation in the quantities being compared. For example, $N_A\langle\sigma v\rangle$ for $C^{12}(p,\gamma)N^{13}$ varies by 23 orders of magnitude over the range $10^7 \leqq T \leqq 10^{10}$ °K. This figure is to be compared with Figure 4 in FCZ I.

(see equation 20 et seq) for the first few partial waves believed to contribute to the reaction under consideration. Because $l = 0, 1, 2$, partial waves usually make the major contributions; because α_l decreases uniformly with increasing l we find that the use of $\alpha = \alpha_1$ is a good rule of thumb.

Consider the simplest case covered by equation 25, constant $q(E) = 1$. When $S(0)\exp(-E_G^{\frac{1}{2}}/E^{1/2} - \alpha E)$ replaces σE in the integrand in equation 3 it will be clear that an effective temperature

$$T_A = \frac{T}{1 + \alpha kT} \qquad 26.$$

replaces T in the temperature dependence of the integral of equation 3 but not of course in the factor $(kT)^{-3/2}$. The final result is

$$N_A\langle 01\rangle = 7.8327 \times 10^9\,(Z_0 Z_1/A)^{1/3} S(0)(T_{9A}^{5/6}/T_9^{3/2}) \times \exp(-4.2487\,Z_0^{2/3} Z_1^{2/3} A^{1/3}/T_{9A}^{1/3}), \qquad 27.$$

and the cutoff term, if any, involves T_{9A}. Equation 27 replaces $[01]/\rho$ given by equation I-51. An example of its use is found in the entry for $Mg^{26}(\alpha, n)Si^{29}$ in Table 1 where $\alpha = 0.729$ MeV^{-1} and $\alpha kT = 0.0628 T_9$.

A fairly frequent case, $q(E) = \exp(-\beta E^2)$, has been discussed in detail by Fowler (1974). Here a very good approximation results when one of the E-factors in βE^2 is replaced by E_o, the effective interaction energy, given by equation I-56. Then equation 26 becomes

$$T_A = \frac{T}{1 + \alpha kT + \beta E_o kT/(1 + \alpha kT)^{2/3}}. \qquad 28.$$

In full numerical detail

$$T_{9A} = \frac{T_9}{1 + 0.0862\alpha T_9 + 0.0105(Z_0^2 Z_1^2 A)^{1/3} \beta T_9^{5/3}/(1 + 0.0862\alpha T_9)^{2/3}} \qquad 29.$$

for α in MeV^{-1} and β in MeV^{-2}. An example of the use of equation 27 with T_{9A} given by equation 29 is found in the entry for $O^{17}(\alpha, n)Ne^{20}$ in Table 1 where $\alpha = 0.311$ MeV^{-1} and $\beta = 0.235$ MeV^{-2}.

In the general case a fair approximation is to set $q(E) = q(E_{oA})$ where E_{oA} is given by equation I-56 with T_9 replaced by T_{9A}. This occurs, for example, in $O^{16}(p, \alpha)N^{13}$. In cases such as $O^{16}(p, \gamma)F^{17}$, E is replaced by E_o since $\alpha = 0$.

In rare cases the nonresonant cross-section factor is best represented by a power-law expression, $S(E) \propto E^m$, as in $O^{17}(p, \alpha)N^{14}$ where $m = 5/2$. Again it is recommended that E be replaced by $E_o \propto T^{2/3}$ so that the final temperature dependence for the reaction rate is given by $N_A\langle 01\rangle \propto T^{2(m-1)/3} \times \exp(-\text{const}/T^{1/3}) \Rightarrow T\exp(-\text{const}/T^{1/3})$ for $m = 5/2$.

Direct Radiative Capture

In many of the reactions producing gamma rays we have used theoretical estimates for direct capture made for us by T. A. Tombrello or C. Rolfs. Fortunately, in many (p, γ) and (α, γ) reactions the direct capture is described by an approximately constant cross-section factor, although this is not always true. The estimates are made using the theory described in Christy & Duck (1961), Tombrello (1965), and Rolfs (1973). In general there is no cutoff for nonresonant direct-capture rates.

Reactions Involving C^{12} and O^{16} Pairs

Experimental measurements in the past few years have yielded a plethora of total cross-section data on reactions induced by $C^{12}+C^{12}$, $C^{12}+O^{16}$, and $O^{16}+O^{16}$. Information on the n, p, and α yield in these reactions has also become available.

The experimental data yield the cross sections that are directly relevant to the rates of these processes under explosive time scales at high temperature, $T_9 > 2$ for $C^{12}+C^{12}$, $T_9 > 4$ for $C^{12}+O^{16}$, and $T_9 > 6$ for $O^{16}+O^{16}$. However, to ascertain the rates under conditions of quasistatic equilibrium at lower temperatures it is necessary to parameterize the available data in terms of a reasonable theory of the reaction mechanisms. The problems raised are discussed in detail by Fowler (1974).

We have chosen a theoretical model that has the merit of great simplicity—a generalized "black-body" model (Michaud & Fowler 1970) incorporating a complex "square-well" potential with a repulsive real component. The standard black-body model assumes that the nuclear interaction vanishes beyond the nuclear radius, R_1, taken as an empirical parameter. R_1 is the radius of the square-well potential that gives results equivalent to those for a more "realistic" Woods-Saxon potential. Within this radius the wave function is taken as a traveling, ingoing wave described by $\phi \propto \exp(-iKr)$, $r \leqq R_1$, where K is a wave number appropriate to the nuclear interior, that is, $K^2 = 2M(V+E)/\hbar^2$ with M the reduced mass, E the interaction energy, and V the nuclear potential energy taken positive for an attractive potential and negative for a repulsive potential. We have taken $V = V_1 + iW_1$ and made a least squares analysis of the experimental data to determine R_1, V_1, and W_1 using transmission functions defined by Vogt (1968) in the Hauser-Feshbach theory described by Michaud & Fowler (1970). The "best fits" yield the following empirical parameters:

	R_1 (fermi)	V_1 (MeV)	W_1 (MeV)
$C^{12}+C^{12}$	7.50	−5.8	0.75
$C^{12}+O^{16}$	7.25	−5.0	1.0
$O^{16}+O^{16}$	7.96	−7.5	0.5

The smoothed-out, average cross-section factor for the sum of all reaction channels determined in this way is compared with the experimental data for $C^{12}+C^{12}$ in Figure 2. Similar results were obtained for $C^{12}+O^{16}$ and $O^{16}+O^{16}$. More detailed models of the reactions that assume $V_1 = V_1(r)$ usually incorporate a strong repulsive core at small r. Our constant V_1 is an average given by $V_1 \approx \int V_1(r) r dr / \int r dr$ and indeed our values for V_1 do represent the average value of the potentials used by Michaud (1973), for example.

The problem then reduces to fitting this average cross-section factor to a

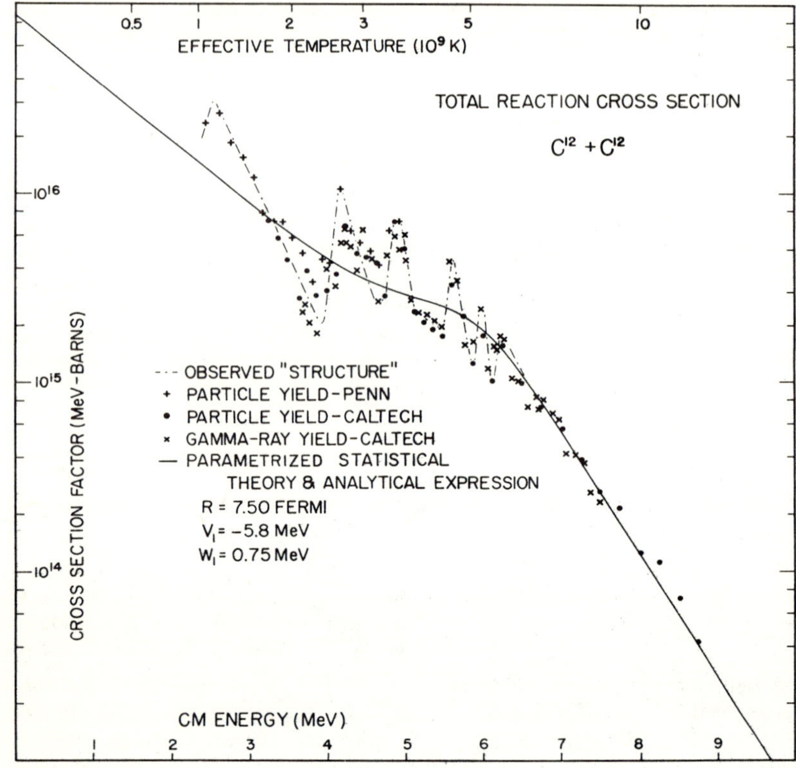

Figure 2 The cross-section factor for all reactions from $C^{12}+C^{12}$. The solid curve is a least squares adjustment to the available data of a parameterized statistical theory based on a generalized "black-body" model described in the text. This curve is also fitted to a few percent by the analytical expression in equation 30 with parameters given in the text. Even so, it goes without saying that the extrapolation to low energy given by a particular model is uncertain, probably by a factor of 3 either way.

reasonable analytical function of interaction energy. Almost perfect fits with deviations less than $\pm 2\%$ were obtained with $m > 0$, $\alpha > 0$, $\beta > 0$, $0 < b \ll 1$ in the expression

$$S(E) = \frac{S(0)\exp(-\alpha E)}{\exp(-\gamma E^m) + b\exp(+\beta E)}, \tag{30.}$$

which is asymptotic to $S(0)\exp(-\alpha E)$ at low energy and to $(1/b)S(0)\exp[-(\alpha+\beta)E]$ at high energy. The increase in the negative slope of $\ln S(E)$ vs E from α at low energy to $(\alpha+\beta)$ at high energy is characteristic of the observed cross-section factor as seen in Figure 2. The transition behavior around 5 MeV is determined by the empirical parameters γ and m in the term $\exp(-\gamma E^m)$. Our results are:

	$S(0)$ (MeV barns)	α (MeV^{-1})	b	β (MeV^{-1})	γ (MeV^{-m})	m
$C^{12}+C^{12}$	8.83×10^{16}	0.772	5.56×10^{-3}	0.697	5.01×10^{-5}	6
$C^{12}+O^{16}$	1.15×10^{21}	0.643	1.06×10^{-3}	0.837	6.26×10^{-3}	3
$O^{16}+O^{16}$	2.31×10^{27}	0.780	3.89×10^{-4}	0.681	9.05×10^{-6}	6

The total reaction rates given for these interactions in Table 1 are based on the substitution of the above parameters into equation 30 with E in the denominator terms replaced by E_{oA}. It must be emphasized that our extrapolation for $S(E)$ to low energy, $E < 2$ MeV for $C^{12}+C^{12}$ for example, is rather uncertain. This means that the quoted reaction rates are uncertain below the temperatures given in the second paragraph of this section. The problem has been discussed in some detail by Michaud (1973) and Fowler (1974).

The main products of the C^{12} and O^{16} pair reactions are n, p, and α, with the appropriate residual nuclei in both ground and excited states. After surveying the experimental literature we recommend the following fractional yields:

		n-channels	p-channels	α-channels
$C^{12}+C^{12}$	$T_9 < 3$	0.00(−2.599)	0.50(2.240)	0.50(4.617)
	$T_9 \geqq 3$	0.05	0.30	0.65
$C^{12}+O^{16}$		0.10(−0.423)	0.50(5.169)	0.40(6.770)
$O^{16}+O^{16}$		0.14(1.453)	0.80(7.677)	0.29(9.593)

The quantities in brackets are the ground state Q-values in MeV. The entries for $O^{16}+O^{16}$ add up to greater than unity because three-body channels such as $Si^{30}+2p$, $Al^{27}+p+\alpha$, and $Mg^{24}+2\alpha$ are open and deuteron production is included with the neutron production.

Neutron-producing Reactions

Table 1 includes a number of exoergic ($Q > 0$) neutron-producing reactions that have been analyzed in a straightforward manner using the techniques previously discussed: $H^2(d,n)He^3$; $H^3(d,n)He^4$; $H^3(t,2n)He^4$; $He^3(t,np)He^4$; $Li^7(d,n)2He^4$; $Li^7(t,2n)2He^4$; $Li^7(He^3,np)2He^4$; $Be^7(t,np)2He^4$; $Be^9(\alpha,n)C^{12}$; $C^{13}(\alpha,n)O^{16}$; $O^{17}(\alpha,n)Ne^{20}$; $Ne^{21}(\alpha,n)Mg^{24}$; $Mg^{25}(\alpha,n)Si^{28}$; and $Mg^{26}(\alpha,n)Si^{29}$.

A number of endoergic ($Q < 0$) neutron-producing reactions with thresholds at $E_{th} = |Q|$ have been analyzed by Bahcall & Fowler (1969b) in terms of a truncated Maclaurin series (three terms) in $E^{1/2}$ for the neutron cross-section factor $\mathscr{S}(E^{1/2})$. Their results are included in Table 1 as well as new results for $Li^7(\alpha,n)B^{10}$ for which $\mathscr{S}$ = const. Cases that cannot be treated in the manner of Bahcall & Fowler (1969b) and require more rapidly varying analytical expressions are discussed by Howard et al (1974) and Fowler (1974). In addition several reactions, $H^2(p,n)2H^1$, $He^4(t,n)Li^6$, and $Na^{23}(p,n)Mg^{23}$, have required treatment rather different from the other endoergic reactions but using analytical expressions that have already been described.

Two endoergic neutron-producing reactions, $O^{18}(\alpha,n)Ne^{21}-0.698$ MeV and

$Ne^{22}(\alpha, n)Mg^{25} - 0.480$ MeV, have required special treatment. In these reactions the energy dependence above the neutron-production threshold is determined by the barrier penetration factor in the alpha-particle channel. The neutron partial width exceeds the alpha-particle partial width within ~ 25 keV above threshold and so we can take the threshold as the energy above which the standard treatment for a charged particle reaction holds. The problem is discussed in Appendix C of Fowler & Hoyle (1964), who show that the usual reaction rate must be multiplied by the integral

$$F(T) = \frac{1}{\pi^{1/2}} \int_{x_t}^{\infty} \exp(-x^2)\, dx \Rightarrow 1/2 \quad \text{for} \quad x_t \Rightarrow 0. \qquad 31.$$

In equation 31 the threshold value for the variable in the integrand is

$$x_t = 2(E_t - E_o)/\Delta E_o \qquad 32.$$

where $E_t = |Q|$ is the threshold energy, $E_o \propto T^{2/3}$ is the effective interaction energy defined by equation I-56, and $\Delta E_o \propto T^{5/6}$ is the effective energy interval defined by equation I-58. These effective values are determined by the maximum and full width at $1/e$ times maximum in the product of the Gamow and Maxwell-Boltzmann exponentials in equation 3 with $S(E) \exp(-E_G^{1/2}/E^{1/2})$ from equation 18 replacing σE. In the standard treatment for exoergic charged-particle reactions, $E_t = 0$ and $2E_o > \Delta E_o$ over the temperature range of interest, so that the lower limit on the integral in equation 31 is large and negative and $F(T) \approx 1$.

It will be clear at low temperatures where $E_t \gg E_o$ and ΔE_o, that $F(T) \Rightarrow 0$ and at high temperatures where energies well above threshold are involved that $F(T) \Rightarrow 1$ even for endoergic reactions. These asymptotic values put constraints on the type of analytic function of T to be used to approximate $F(T)$. We are primarily interested in the rise near threshold from low temperature to high temperature as E_o increases through the value E_t and x_t decreases through zero. Evaluation of the integral about this point, where it is equal to 1/2, yields

$$F(T_9) = \exp - (\theta_9/T_9)^n, \qquad 33.$$

where

$$n = \frac{8}{3\pi^{1/2} \ln 2} \frac{\varepsilon_o}{\delta_o} \left(\frac{\varepsilon_o}{E_t}\right)^{1/4} = 2.171 \frac{\varepsilon_o}{\delta_o} \left(\frac{\varepsilon_o}{E_t}\right)^{1/4} \qquad 34.$$

and

$$\theta_9 = (\ln 2)^{1/n} \left(\frac{E_t}{\varepsilon_o}\right)^{3/2}, \qquad 35.$$

with

$$\varepsilon_o = E_o/T_9^{2/3} = 0.12204\,(Z_0^2 Z_1^2 A)^{1/3} \text{ MeV}/(10^9\,^\circ\text{K})^{2/3} \qquad 36.$$

and

$$\delta_o = \Delta E_o/T_9^{5/6} = 0.23682\,(Z_0^2 Z_1^2 A)^{1/6} \text{ MeV}/(10^9\,^\circ\text{K})^{5/6}. \qquad 37.$$

When necessary T_{9A} replaces T_9 in the above equations.

The threshold factor given by equation 33 holds only for T_9 near θ_9 and is much too small at low temperatures. If one thinks of the reactions in reverse, $Ne^{21}(n, \alpha)O^{18}$ and $Mg^{25}(n, \alpha)Ne^{22}$, which are exoergic, it will be clear, because there are no selection rules against s-wave neutrons, that $N_A\langle\sigma v\rangle_{\text{reverse}} = \text{const}$ and $N_A\langle\sigma v\rangle_{\text{direct}} = \text{const}\exp(-E_t/kT)$. Such a term has been evaluated theoretically and added to the rate of the reactions under discussion.

The onset of the neutron channel in the two interactions $O^{18}+\alpha$ and $Ne^{22}+\alpha$ has a profound effect on the two competing reactions $O^{18}(a,\gamma)Ne^{22}$ and $Ne^{22}(\alpha,\gamma)Mg^{26}$. Once the neutron channel is well above threshold the probability for neutron emission is large compared to gamma-ray emission or to alpha-particle reemission and the neutron channel cross section is approximately equal to the total reaction cross section for $E > E_t$. The gamma-ray cross section is a small fraction of this. However, below the threshold, $E < E_t$, the probability for neutron emission is zero, and because of the Coulomb barrier, the probability of alpha-particle reemission is small compared to gamma-ray emission. Thus the gamma-ray channel cross section is equal to the total reaction cross section, which can be obtained by extrapolating the observed neutron channel cross section to low energy, disregarding the threshold.

It will be clear that a cutoff factor corresponding to the threshold factor in equation 31 must be applied to the high-temperature reaction rate for the (α, γ) reactions when (α, n) becomes energetically possible. This cutoff factor is

$$F'(T) = \frac{1}{\pi^{1/2}}\int_{-\infty}^{x_t} \exp(-x^2)dx \Rightarrow 1/2 \quad \text{for} \quad x_t \Rightarrow 0. \qquad 38.$$

Now $F'(T) \Rightarrow 1$ at low temperature where $x_t \Rightarrow +\infty$ and $F'(T) \Rightarrow 0$ at high temperature where x_t is large and negative. Proceeding as in the case of $F(T)$ one finds

$$F'(T_9) = \exp-(T_9/\theta'_9)^{n'}, \qquad 39.$$

where

$$n' = \frac{8\ln 2}{3\pi^{1/2}}\frac{\varepsilon_o}{\delta_o}\left(\frac{\varepsilon_o}{E_t}\right)^{1/4} = 1.043\frac{\varepsilon_o}{\delta_o}\left(\frac{\varepsilon_o}{E_t}\right)^{1/4} = 0.4805n \qquad 40.$$

and

$$\theta'_9 = (\ln 2)^{-1/n'}\left(\frac{E_t}{\varepsilon_o}\right)^{3/2} = (\ln 2)^m\theta_9, \qquad 41.$$

with

$$m = -1/n - 1/n'. \qquad 42.$$

At high temperature the rate for gamma emission will be equal to a small fraction of the neutron emission rate including the $F(T)$ factor. In the case of the $O^{18}(\alpha,\gamma)Ne^{22}$ reaction the available experimental evidence indicates a rate $\sim 5\times10^{-4}$ that of the $O^{18}(\alpha,n)Ne^{21}$ reaction. We have used this same ratio in a number of cases, namely, $O^{17}+\alpha$, $Ne^{22}+\alpha$, $Mg^{25}+\alpha$, and $Mg^{26}+\alpha$. Because of the high Q of the $Ne^{21}(\alpha,n)Mg^{24}$ reaction we have used the ratio 10^{-4} from estimates made by S. E. Woosley (personal communication, 1974) for the ratio in this case.

In the cases of $O^{17}(\alpha,\gamma)Ne^{21}$, $Mg^{25}(\alpha,\gamma)Si^{29}$, and $Mg^{26}(\alpha,\gamma)Si^{30}$ we have noted in Hauser-Feshbach calculations that the cross sections are a substantial fraction of the corresponding (α,n) cross sections at low energy but drop to a fraction $\sim 5\times 10^{-4}$ at the onset of $O^{17}(\alpha,n_1)Ne^{20*}-1.046$ MeV, $Mg^{25}(\alpha,n_2)Si^{28**}-1.965$ MeV, and $Mg^{26}(\alpha,n_1)Si^{29*}-1.240$ MeV, respectively, where Ne^{20*} is the first excited state of Ne^{20} at 1.634 MeV, Si^{28**} is the second excited state of Si^{28} at 4.618 MeV, and Si^{29*} is the first excited state of Si^{29} at 1.273 MeV. The appropriate factors, $F'(T_9)$, have been applied in these cases.

Contribution of Bound States

In three reactions of considerable astrophysical importance, $C^{12}(\alpha,\gamma)O^{16}$, $O^{18}(p,\alpha)N^{15}$, and $Ne^{20}(p,\gamma)Na^{21}$, the tail of a bound state in the compound nucleus makes a major contribution to the low-energy, nonresonant cross section and thus to the cross-section factor. This contribution can be represented approximately as a function of energy by

$$S(E) = \text{const}\left|\frac{1}{E+E_b+i\Gamma/2}+\frac{1}{E_{\text{int}}}\right|^2 \Rightarrow \frac{\text{const}}{(E+E_b)^2+\Gamma^2/4} \quad \text{for} \quad E_{\text{int}}^{-1}=0, \tag{43}$$

where $E_b > 0$ is the binding energy (taken positive) of the state relative to the mass-energy equivalent of the components of the incident channel, Γ is the width of the state contributed by the outgoing and all competing channels, and E_{int}^{-1} is proportional to the amplitude, taken constant, of other states of the same spin and parity as the bound state in question. The contribution of these states can interfere coherently with that of the bound state. Again we replace E by E_o.

The simplest case is $Ne^{20}(p,\gamma)Na^{21}$, where the width of the bound state is $\Gamma=\Gamma_\gamma=0.31$ eV, whereas the binding energy is $E_b=7.1$ keV, and there seems to be no interference from other states (Rolfs et al 1975). Thus in this case $S\propto(0.5586T_9^{2/3}+0.0071)^{-2}$ because $E_o=0.5586T_9^{2/3}$ for $Ne^{20}+p$. Then, because $N_A\langle 01\rangle \propto S/T_9^{2/3}$, the factor that occurs in the denominator of the first entry for $Ne^{20}(p,\gamma)$ in Table 1 is $T_9^2(1+0.0127/T_9^{2/3})^2$. Because E_b is so small compared to the effective reaction energy for significant temperatures, we have $S(E)$ approximately proportional to E^{-2} for $E \gg E_b$ in this case. The question quite naturally arose concerning how accurate is the substitution of E_o for E, which was made to obtain the analytic formula we have employed. Accordingly we carried out a numerical integration to determine $N_A\langle 01\rangle$ accurately for $S(E)\propto(E+0.0071)^{-2}$ and found that our analytic expression was good to $\sim 1\%$ at low temperature ("noise" in the numerical integration) and was low by only 5% near $T_9=1$ and by 10% near $T_9=10$.

In the case of $O^{18}(p,\alpha)N^{15}$ the 7.90-MeV state in F^{19} is bound by 0.09282 MeV relative to $O^{18}+p$ for which $E_o=0.4805T_9^{2/3}$ and has a measured width given by $\Gamma=\Gamma_\alpha=0.210$ MeV. We neglect interference terms. The upshot is $S\propto[(0.4805T_9^{2/3}+0.09282)^2+0.0110]^{-1}$ and the factor that occurs in the denominator of the third term for $O^{18}(p, \alpha)N^{15}$ in Table 1 becomes $[0.439(1+5.18T_9^{2/3})^2+0.561]$ when normalized to unity for $T_9 = 0$. The reduced proton width for this state is not known so its contribution must perforce include the factor (0 to 1).

The importance of the bound state in O^{16} at 7.11867-MeV excitation ($J^\pi = 1^-$) just below the threshold for $C^{12}(\alpha, \gamma)O^{16}$ was discussed in detail in FCZ I. It is now known to be bound by 0.04295 MeV. In FCZ I it was necessary to estimate the nonresonant contribution by using a theoretical calculation for the reduced alpha-width of this state, $\theta_\alpha^2 = 0.085 \pm 0.040$. Measurement by Dyer & Barnes (1974) of the cross section of the $C^{12}(\alpha, \gamma)O^{16}$ reaction from just above the 3.160-MeV (LAB) resonance ($J^\pi = 1^-$) down to 1.88 MeV (LAB) has shown constructive interference of the two 1^- states and enabled them to extrapolate the cross section to low energy with some confidence using several theoretical methods that give substantially the same results (Koonin, Tombrello & Fox 1974; Humblet, Dyer & Zimmerman 1975). We have fitted their results below 1 MeV (LAB) to equation 43, neglecting $\Gamma \approx \Gamma_\gamma = 0.056$ eV. A good fit over this limited range of energy is obtained for $E_{\text{int}} = 1.4420$ MeV and the upshot, with $E_o = 0.9226T_9^{2/3}$ MeV, is the occurrence of the factors $(1+0.621T_9^{2/3})^2/[T_9^2(1+0.047/T_9^{2/3})^2]$, the first entry for $C^{12}(\alpha, \gamma)O^{16}$ in Table 1. The accuracy of this analytical approximation has been tested against numerical integration and found to deviate by no more than 1% over the range $0.01 \leqq T_9 \leqq 10$. Further discussion of the $C^{12}(\alpha, \gamma)O^{16}$ reaction rate will follow in due course. The Dyer & Barnes (1974) results yield $N_A\langle\sigma v\rangle$ equal to 0.62 that given in FCZ I for $E_{o6} \sim 0.3$ and $T_9 \sim 0.2$.

Three-Body Reactions

Table 1 includes four reactions induced by the interaction of three bodies, namely, $He^4(nn, \gamma)He^6$, $He^4(np, \gamma)Li^6$, $He^4(\alpha n, \gamma)Be^9$, and $He^4(2\alpha, \gamma)C^{12}$, all of which are of interest in connection with attempts (unsuccessful!) to bridge the mass gaps at $A = 5$ and 8 in "big bang" nucleosynthesis. The last reaction is of course important as the first stage in helium burning in red giant stars.

Direct or indirect experimental evidence is available on the reverse of all these reactions except $He^4(nn, \gamma)He^6$. When the photodisintegration cross section, σ_γ, is known as a function of gamma-ray energy, E_γ, the photodisintegration rate per nucleus per second is given by

$$\lambda_\gamma = \frac{c}{\pi^2(\hbar c)^3} \int_{E_t}^{\infty} \sigma_\gamma E_\gamma^2 \exp(-E_\gamma/kT)\, dE_\gamma \text{ sec}^{-1}, \qquad 44.$$

where E_t is the threshold energy. Both spontaneous and stimulated emission (in reverse) have been taken into account. This expression is to be compared with equation 3. Reciprocity through the use of the tabulated REV RATIOS makes it possible to determine the rate of the direct reactions involving three bodies that are under discussion. Equation 44 holds of course for photodisintegration into any number of bodies—2, 3, 4, ...

These reactions proceed through two stages, for example, $He^4 + He^4 \rightleftarrows Be^8$ and $Be^8(\alpha, \gamma)C^{12}$, and one or more excited states in the compound nucleus serve as resonances. For a single sharp resonance equation 44 yields

$$\lambda_\gamma \approx \frac{2J_r+1}{2J_0+1}\frac{\Gamma_{\text{rad}}(\Gamma-\Gamma_{\text{rad}})}{\hbar\Gamma}\exp(-E_r/kT)\ \text{sec}^{-1}, \qquad 45.$$

where E_r is the energy of the resonant state relative to the ground state in the compound nucleus, J_r is the spin of the resonant state, J_0 is the spin of the ground state, Γ_{rad} is the overall radiation width of the state, including gamma-ray transitions to *all* lower states and electron-positron pair emission, if it occurs, and Γ is the total width including radiation and particle emission. $\Gamma \gg \Gamma_{\text{rad}}$ is usually the case so $\lambda_\gamma \propto \Gamma_{\text{rad}}/\hbar$.

As discussed in FCZ I, equation 45 is used for the rate of $C^{12}(\gamma, 2\alpha)He^4$, which yields the rate of $He^4(2\alpha, \gamma)C^{12}$ and the latest discussion of this important case is given by Barnes & Nichols (1973). Experimental data are available on $\sigma_\gamma(E_\gamma)$ for $Li^6(\gamma, np)He^4$ and $Be^9(\gamma, n\alpha)He^4$. The excited states at 5.366 and 1.665 MeV, respectively, seem to establish the threshold energy in these reactions above which the logarithm of the cross section varies slowly and linearly with $(E_\gamma - E_t)$. The integral in equation 44 is thus easy to calculate. We have made a rough calculation for $He^6(\gamma, nn)He^4$ in analogy with $Li^6(\gamma, np)He^4$ and have been able to set a reasonable upper limit on the rate of this reaction and its reverse.

RINGING THE CHANGES SINCE FCZ I

We conclude with a few remarks concerning important processes of nucleosynthesis and energy generation, stressing the changes in reaction rates that have occurred since the publication of FCZ I.

"Big Bang" Nucleosynthesis

As noted previously, a number of three-body reactions involving neutrons, protons, and alpha particles have been analyzed because of their possible importance in bridging the mass gap at mass 5 and 8. It is known that Be^7 is produced in significant amounts during "big bang" nucleosynthesis and some C^{11} results from $Be^7(\alpha, \gamma)C^{11}$. Thus we have included a theoretical estimate for $C^{11}(p, \gamma)N^{12}(e^+\nu)C^{12}$ based on analogy with $C^{12}(p, \gamma)N^{13}$. In addition, $C^{11}(\alpha, p)N^{14}$ can be calculated from the rate for $N^{14}(p, \alpha)C^{11}$ in Table 1 which is based on experimental data. These reactions have been considered as more efficacious in reaching $A \geqq 12$ than $C^{11}(n, \gamma)C^{12}$ because the latter must compete with the much faster break-up reaction $C^{11}(n, \alpha)2He^4$.

It is well known that "big bang" nucleosynthesis results in the production of Li^7 in amounts comparable to the solar system abundance of this rare nucleus both directly as Li^7 and indirectly from the decay of Be^7. In studying this mechanism of synthesis it is necessary to include not only all reactions that produce Li^7 and Be^7 but also those that destroy them. Thus, in addition to H and He^4 reactions with these nuclei, Table 1 includes rates of their interactions with D, T, and He^3, which exist in the early stages of the "big bang" in significant amounts. For neutron interactions with these nuclei see FCZ I.

Many of our preliminary estimates for the reaction rates given in Table 1 were employed by Wagoner (1969) in his study of "big and little bangs." There is no reason to believe that our final results will change in any significant way his conclusion that no heavy elements beyond helium, with the exception perhaps of Li^7, are produced in the "big bang" in amounts of interest. "Little bang" synthesis does depend on the rates of these reactions but not in a critical manner.

Controlled Thermonuclear Fusion

Many of the reactions involving the use of D, T, He^3, and the lithium and boron isotopes in controlled thermonuclear fusion are also of interest under astrophysical circumstances. Thus we have taken some pains to fit the accurate experimental results for the D, T, and He^3 reactions with considerable care as illustrated for $H^3(d,n)He^4$ in Figure 1. Our study of the literature has revealed that much more precise measurements of reactions involving lithium and boron isotopes are warranted because these may well prove to be of interest in connection with laser-induced fusion.

The Proton-Proton Chain

The coefficient of the basic pp reaction, $H^1(p,e^+\nu)H^2$, given in Table 1 is 27% greater than that given in FCZ I. This results from use of the most recent value for the half lifetime of the neutron (12.5%), new radiative corrections (1.6%), and meson exchange effects with soft-core potentials (11.4%). This increase results in a significant decrease in the high-energy neutrino flux expected from the Sun but not enough to bring observation and theoretical expectations into agreement (Ulrich 1974). New measurements have been made on many of the pp chain reactions since the publication of FCZ I and these measurements yield the appropriate entries in Table 1. The result has been only small changes in the cross sections and reaction rates but considerable reduction in the experimental uncertainties. In a most important experiment Dwarakanath (1974) has shown that resonance (Fowler 1972) almost certainly does not occur in $He^3(He^3,2p)He^4$, so that the production of B^8 neutrinos via $He^4(He^3,\gamma)Be^7(p,\gamma)B^8(e^+\nu)2He^4$ should occur as calculated using the non-resonant rate for $He^3(He^3,2p)He^4$. The solar neutrino problem is still with us!

The CNO Tri-Cycle

Major changes have occurred in some of the reactions involving protons and the isotopes of carbon, nitrogen, and oxygen but fortunately not in the primary reactions of the CN cycle. However, Rolfs & Rodney (1974a) found that the cross-section factor for $N^{15}(p,\gamma)O^{16}$ is 2.5 times larger than that given in FCZ I. $N^{15}(p,\alpha)C^{12}$ has not been remeasured and thus remains unchanged. This means that leakage from the CN cycle to oxygen isotopes occurs in somewhat less than 1000 cycles under conditions in upper main-sequence stars.

In addition, Rolfs & Rodney (1974b) have found that the rate for $O^{17}(p,\alpha)N^{14}$ is probably not much faster than $O^{17}(p,\gamma)F^{18}(e^+\nu)O^{18}$, so that a *tri-cycle* occurs among the CNO isotopes involving the recycling reaction $O^{18}(p,\alpha)N^{15}$, as well as $O^{17}(p,\alpha)N^{14}$ and $N^{15}(p,\alpha)C^{12}$. Among other things this means that O^{17} survives hydrogen burning in greater abundance than heretofore thought and thus can serve as a source of neutrons in helium burning via $O^{17}(\alpha, n)Ne^{20}$.

Because of its importance in the fast CN cycle we have included an estimate for $N^{13}(p, \gamma)O^{14}$ based on analogy with $C^{12}(p, \gamma)N^{13}$—the extra proton in N^{13} is just a spectator in the reaction.

The NeNa and MgAl Cycles

Table 1 presents for the first time the rates of all the reactions involved in these cycles and the leakage from them. We assume in the MgAl cycle a time scale long compared to the mean lifetime of the ground state of Al^{26}, 1.04×10^5 yr. The most important new result is the experimental confirmation by Rolfs et al (1975) of a theoretical conjecture by Marion & Fowler (1957) that Ne^{21} is produced rapidly by $Ne^{20}(p, \gamma)Na^{21}(e^+ \nu)Ne^{21}$ through the off-resonant contribution of a bound state in Na^{21}. Ne^{21} may burn through $Ne^{21}(p, \gamma)Na^{22}$ more slowly than it is created so that it may survive hydrogen burning sufficiently to serve as a neutron source during helium burning via $Ne^{21}(\alpha, n)Mg^{24}$.

Helium Burning

Major experimental efforts have been made since the publication of FCZ I to elucidate the nature and rate of the basic helium-burning reactions, $3He^4 \rightarrow C^{12}$, $C^{12}(\alpha, \gamma)O^{16}$, $O^{16}(\alpha, \gamma)Ne^{20}$, ... The result of the work of many investigators [see references in Barnes & Nichols (1973)] has led to a change relative to FCZ I in the rate for $3He^4 \rightarrow C^{12}$ via the 7.655-MeV state in C^{12} by a factor $1.41 \exp(-0.117/T_9)$, which includes the change given in the notes added in proof. This is equal to 0.78 at the important red giant temperature, $T_9 = 0.2$—not a great change but the uncertainty in the new value is considerably less than before. The corresponding factor for the $C^{12}(\alpha, \gamma)O^{16}$ reaction rate, as noted previously, is 0.62. Significant changes have been made in $O^{16}(\alpha, \gamma)$, and new (α, γ) and (α, n) rates involving oxygen, neon, and magnesium isotopes have been included. A major decrease by several orders of magnitude is indicated for the important $N^{14}(\alpha, \gamma)F^{18}$ reaction so that it does not ignite before the onset of the $3\alpha \rightarrow C^{12}$ reaction in helium burning. The rate for $Ne^{20}(\alpha. \gamma)$ is still very uncertain at helium-burning temperatures.

Carbon, Oxygen, and Silicon Burning

Table 1 gives reliable rates for carbon burning, oxygen burning, and the first stages of silicon burning under explosive conditions. The importance of the photodisintegration rate of Mg^{24} is confirmed as determining the time scale for the freeing of neutrons, protons, and alpha particles in silicon burning. We emphasize once again that there is still considerable uncertainty (a factor of ~ 3 either way) in the reaction rates for quasistatic burning of carbon and oxygen and with this final hint to the experimental nuclear physicist we conclude this "Handbuch der Kernastrophysik."

NOTES ADDED IN PROOF The coefficient in the first term for $He^4(2A, G)C^{12}$ in Table 1 should be changed from $2.49E-08$ to $3.00E-08$. Questions have arisen concerning the rate of the $N^{15}(P, A)C^{12}$ reaction, and new measurements are under way in several laboratories.

ACKNOWLEDGMENTS

We acknowledge the advice, encouragement, and assistance of many people—Neta Bahcall, Charles Barnes, Bibiana Cujec, Keith Despain, Peggy Dyer, George Fox, George Fuller, Evaline Gibbs, Ralph Kavanagh, Frederick Mann, Frank Oppenheimer, Jan Rasmussen, William Rodney, Claus Rolfs, Fay Ajzenberg-Selove, Zygmunt Switkowski, Tom Tombrello, Cornelius van der Leun, Robert Wagoner, and Stanford Woosley. Many of those listed gave us a hard time, God bless them, but Frank Oppenheimer made it all worthwhile when he supplied the opening quotation by Mark Twain.

In addition we wish to express our sincere appreciation to the many experimentalists who generously supplied us with reaction data prior to publication. Without their interest and enthusiastic support our task would have been much harder.

Finally, we acknowledge the essential role played by the staff of Math and Computing, Lawrence Berkeley Laboratory, whose excellent facility made an otherwise difficult computing task simple and straightforward.

Literature Cited

Ajzenberg-Selove, F. 1970. *Nucl. Phys. A* 152:1

Ajzenberg-Selove, F. 1971. *Nucl. Phys. A* 166:1

Ajzenberg-Selove, F. 1972. *Nucl. Phys. A.* 190:1

Ajzenberg-Selove, F., Busch, C. L. 1971. *Nuclear "Wallet Cards."* Am. Phys. Soc.: Nucl. Phys. Div.

Ajzenberg-Selove, F., Lauritsen, T. 1968. *Nucl. Phys. A* 114:1

Ajzenberg-Selove, F., Lauritsen, T. 1974. *Nucl. Phys. A* 227:1

Bahcall, J. N. 1964. *Ap. J.* 139:318

Bahcall, N. A., Fowler, W. A. 1969a. *Ap. J.* 157:645

Bahcall, N. A., Fowler, W. A. 1969b. *Ap. J.* 157:659

Bahcall, N. A., Fowler, W. A. 1970. *Ap. J.* 161:119

Barnes, C. A., Nichols, D. B. 1973. *Nucl. Phys. A* 217:125

Burbidge, E. M., Burbidge, G. R., Fowler, W. A., Hoyle, F. 1957. *Rev. Mod. Phys.* 29:547

Chiu, H.-Y. 1968. *Stellar Physics,* Vol. 1. Waltham, Mass.: Blaisdell

Christy, R. F., Duck, I. 1961. *Nucl. Phys.* 24:89

Clayton, D. D. 1974. *Nature* 249:131

Cohen, E. R. 1974. *Phys. Today* 27(9):19, 80

Cohen, E. R., Taylor, B. N. 1973. *J. Phys. Chem. Ref. Data* 2:663

Colgate, S. A. 1974. *Ap. J.* 187:321

Critchfield, C. L. 1972. *Cosmology, Fusion and Other Matters,* ed. F. Reines, 186–91. Boulder, Colo.: Colorado Assoc. Univ. Press

DeWitt, H. E., Graboske, H. C., Cooper, M. S. 1973. *Ap. J.* 181:439

Dwarakanath, M. R. 1974. *Phys. Rev. C* 9:805

Dyer, P., Barnes, C. A. 1974. *Nucl. Phys. A* 233:495

Endt, P. M., van der Leun, C. 1973. *Nucl. Phys. A* 214:1

Fowler, W. A. 1972. *Nature* 238:24, 242:424

Fowler, W. A. 1974. *Quart. J. Roy. Astron. Soc.* 15:82

Fowler, W. A., Caughlan, G. R., Zimmerman, B. A. 1967. *Ann. Rev. Astron. Ap.* 5:525

Fowler, W. A., Hoyle, F. 1964. *Nucleosynthesis in Massive Stars and Supernovae.* Chicago: Univ. Chicago Press. This monograph was a reprint of *Ap. J.* (1960) 132:565 and *Ap. J. Suppl. No. 91* (1964) 9:201

Graboske, H. C., DeWitt, H. E., Grossman, A. S., Cooper, M. S. 1973. *Ap. J.* 181:457

Gryzinski, M. 1958. *Phys. Rev.* 111:900

Gryzinski, M. 1959. *Phys. Rev.* 115:1087

Howard, A. J., Jensen, H. P., Rios, M.,

Fowler, W. A., Zimmerman, B. A. 1974. *Ap. J.* 188:131
Hoyle, F., Fowler, W. A. 1973. *Nature* 241:384
Humblet, J., Dyer, P., Zimmerman, B. A. 1975. In preparation
Koonin, S. E., Tombrello, T. A., Fox, G. 1974. *Nucl. Phys. A* 220:221
Marion, J. B., Fowler, W. A. 1957. *Ap. J.* 125:221
Michaud, G. 1973. *Phys. Rev. C* 8:525
Michaud, G., Fowler, W. A. 1970. *Phys. Rev. C* 2:2041
Parker, P. D., Pisano, D. J., Cobern, M. E., Marks, G. H. 1973. *Nature Phys. Sci.* 241:106
Rolfs, C. 1973. *Nucl. Phys. A* 217:29
Rolfs, C., Rodney, W. S. 1974a. *Nucl. Phys. A* 235:450
Rolfs, C., Rodney, W. S. 1974b. *Ap. J. Lett.* 194:63
Rolfs, C., Rodney, W. S., Shapiro, M. H., Winkler, H. 1975. *Nucl. Phys. A*. In press
Shaw, P. B., Clayton, D. D. 1967. *Phys. Rev.* 160:1193
Tombrello, T. A. 1965. *Nucl. Phys.* 71:459
Truran, J. W., Kozlovsky, B.-Z. 1969. *Ap. J.* 158:1021
Ulrich, R. K. 1974. "Solar Neutrinos." In *Neutrinos—1974* (*Philadelphia*), ed. C. Baltay, 259–72. New York: Am. Inst. Phys.
Vogt, E. 1968. *Advan. Nucl. Phys.* 1:261
Wagoner, R. V. 1969. *Ap. J. Suppl. No. 162* 18:247. See also Wagoner, R. V., Fowler, W. A., Hoyle, F. 1967. *Ap. J.* 148:3
Woosley, S. E., Holmes, J. A., Fowler, W. A., Zimmerman, B. A. 1975. In preparation

CHEMICAL COMPOSITION OF EXTRAGALACTIC GASEOUS NEBULAE

Manuel Peimbert
Instituto de Astronomía, Universidad Nacional Autonoma de México,
Mexico 20, D.F., Mexico

1 INTRODUCTION

From the determination of chemical abundances in extragalactic gaseous nebulae it is possible to study the chemical evolution of galaxies and to obtain some clues about their formation. Moreover, by studying gaseous nebulae that have not been contaminated by stellar evolution, it is possible to determine pregalactic chemical abundances that might bear on the cosmological problem.

From optical observations of gaseous nebulae it is possible to derive the relative abundance of H, He, N, O, and Ne, which are among the six most abundant elements in the galactic neighborhood. Gaseous nebulae also show faint permitted lines of carbon in the optical region. These lines are difficult not only to detect but also to interpret, because they originate not only by recombination, but also by absorption of stellar radiation in spectrum lines (Seaton 1968); a detailed model of the nebula in question is necessary to derive the carbon abundance.

The study of chemical abundances from extragalactic gaseous nebulae is a very powerful tool compared with stellar observations, because, with the exception of the Magellanic Clouds, there have been no chemical abundance determinations of individual stars in other galaxies. On the other hand, with present equipment it is possible to derive helium-to-hydrogen (He/H) abundance ratios of reasonable accuracy from H II regions at many Mpc from us.

In this review we concentrate on normal H II regions, nuclear H II regions of normal galaxies, and planetary nebulae. We leave out quasars and Seyfert galaxy nuclei, because in these cases the results are strongly model dependent. Abundance determination methods have been reviewed in several places (Seaton 1960, Aller & Liller 1968, Osterbrock 1970, 1974); only a few particular problems associated with them are mentioned here.

Previous reviews on abundances in extragalactic gaseous nebulae have been presented by Osterbrock (1970), Spinrad & Peimbert (1975), Aller (1972), and Roberts (1972).

2 ABUNDANCE DETERMINATIONS

Chemical abundances are determined from emission-line intensity ratios. These ratios are affected by interstellar reddening and must be corrected; the usual procedure is to assume that the Balmer series is produced under case B and that the Whitford (1958) normal reddening law applies. Even if there are deviations from the normal reddening law, these will affect only the total absorption and not the relative line intensities in the 3500–7400 Å range, the range for which accurate line intensity determinations are available.

Three different types of line intensity ratios are used to derive relative ionic abundances: 1. permitted vs permitted, where the dependence on the electron temperature is very small; 2. forbidden vs forbidden, where in general the temperature dependence is not very strong; and 3. permitted vs forbidden, where the temperature dependence is strong and an accurate knowledge of the temperature is needed.

Unfortunately, for most atoms we do not observe all the stages of ionization and it is necessary to correct for the unobserved stages. This correction is usually estimated from the degree of ionization of other elements observed in more than one stage of ionization. Observations at different positions of a given nebula permit an empirical determination of the ionization correction factor, i_{cf}. It is also possible to estimate the i_{cf} from a theoretical model. A combination of both approaches should be used because each is subject to different errors.

a *Normal H II Regions*

By normal H II regions we mean those gaseous nebulae ionized mainly by main-sequence OB stars. A typical example of a normal H II region in the solar neighborhood is the Orion nebula. The H II regions observed in other galaxies are in general giant H II regions ionized by many more OB stars than the Orion nebula and of lower electron density.

HELIUM Previous reviews, devoted in particular to the He/H abundance ratio, have been presented by Danziger (1970), Seaton (1971), Searle & Sargent (1972b), and Churchwell et al (1974).

The total He/H abundance ratio is given by

$$y = y^0 + y^+ + y^{++}, \qquad 1.$$

where

$$y^i = \frac{\int N_e N(\mathrm{He}^i)\,dV}{\int N_e N(\mathrm{H}^+)\,dV}. \qquad 2.$$

y^+ and y^{++} can be determined directly from the following equations:

$$y^+ = 0.736 T_0^{0.11} \frac{\mathrm{I}(4472)}{\mathrm{I}(\mathrm{H}\beta)} = 0.522 T_0^{0.15} \frac{\mathrm{I}(5876)}{\mathrm{I}(\mathrm{H}\alpha)}, \qquad 3.$$

$$y^{++} = 2.08 \times 10^{-2} T_0^{0.15} \frac{I(4686)}{I(H\beta)}, \qquad 4.$$

where the effective recombination coefficients are derived from the computations by Brocklehurst (1971, 1972) interpolated in the 5000–10,000°K electron temperature range for $N_e = 10{,}000\ \text{cm}^{-3}$; T_0 is the average electron temperature and a mean square fluctuation of $t^2 = 0.055$ has been adopted. Equations 3 and 4 are almost independent of N_e and errors $\leqq 1000$°K in T_e produce errors smaller than 2% in the y^i determination. In equations 3 and 4 the effects of self-absorption and collisional excitation have been neglected. From observations of planetary nebulae (Peimbert & Torres-Peimbert 1971, Barker 1974) and from theoretical considerations (Brocklehurst 1972), it has been found that the contribution to the helium-line intensities λλ5876 and 4472 due to collisional excitation from the 2^3S level

Table 1 Helium-to-hydrogen abundance ratios by number

Object	He^+/H^+	i_{cf}(He)	He/H	Estimated Error	Observer
M82	0.065	2	0.130	0.03	Peimbert & Spinrad (1970b)
NGC 7679	0.078	1.16	0.090	0.02	Peimbert & Spinrad (1970a)
NGC 604	0.076	1.17	0.089	0.02	Mathis (1962)
NGC 604	0.132	1.15	0.152	0.02	Aller et al (1968)
NGC 604	0.094	1.19	0.112	0.02	Peimbert & Spinrad (1970a)
Orion nebula			0.101	0.005	Peimbert & Torres-Peimbert (1974)
η Car nebula			0.102	0.005	Peimbert & Torres-Peimbert (1974)
NGC 5461	0.082	1.10	0.090	0.015	Peimbert & Spinrad (1970a)
NGC 5455	0.066	1.10	0.073	0.02	Searle & Sargent (1972a)
NGC 4449	0.078	1.13	0.088	0.015	Peimbert & Spinrad (1970a)
NGC 6822	0.087	1.04	0.090	0.015	Peimbert & Spinrad (1970c)
⟨LMC⟩	0.082	1.08	0.089	0.010	Dufour (1974)
⟨LMC⟩	0.080	1.05	0.084	0.005	Peimbert & Torres-Peimbert (1974)
NGC 2070	0.081	1.04	0.084	0.02	Faulkner & Aller (1965)
NGC 2070	0.105	1.04	0.109	0.02	Mathis (1965)
NGC 2070	0.17		0.17	0.09	Mezger et al (1970)
NGC 5471	0.092	1.03	0.099[b]	0.015	Peimbert & Spinrad (1970a)
NGC 5471	0.082	1.03	0.088[b]	0.02	Searle & Sargent (1972a)
II Zw 40	0.056	1	0.056	0.02	Searle & Sargent (1972a)
⟨SMC⟩	0.077	1.05	0.081	0.015	Dufour (1974)
NGC 346	0.077	1.04	0.080	0.02	Aller & Faulkner (1962)
NGC 185-1[a]	0.14		0.14	0.04	Jenner et al (1973)
I Zw 18	0.081	1	0.085[b]	0.02	Searle & Sargent (1972a)

[a] Planetary nebula.

[b] A contribution of 0.004 due to He^{++} has been added.

has been overestimated by Cox & Daltabuit (1971) and that this effect for the temperatures encountered in normal H II regions can be neglected.

The He/H abundance ratios for well-observed H II regions are presented in Table 1. In most cases T_e was derived from the [O III] 4363/5007 line intensity ratio, whereas in a few cases it was assumed that $T_e = 8000$ or 10,000°K. The y^+ values for I Zw 18, II Zw 40, and the Magellanic Clouds were derived from the observations by Searle & Sargent (1972a) and Dufour (1974) by weighting the observed lines as described by Peimbert & Torres-Peimbert (1974).

Most bright H II regions are mainly ionized by early O-type main-sequence stars hot enough to ionize a substantial fraction of helium although not hot enough to doubly ionize it. Of the objects presented in Table 1, I Zw 18 and NGC 5471 have the highest y^{++} value, equal to 0.004; for the Orion nebula, $y^{++} < 4 \times 10^{-5}$ (Searle & Sargent 1972a, Peimbert & Goldsmith 1972). Therefore, for most bright H II regions $y = y^0 + y^+$, where $y^+ \gg y^0$. However, to derive very precise values of y it is necessary to estimate the amount of neutral helium inside H II regions.

To estimate the amount of neutral helium in the Orion nebula, Peimbert & Costero (1969) proposed that $i_{cf}(\mathrm{He}) = f(\mathrm{O}^+/\mathrm{O}^{++}, \mathrm{S}^+/\mathrm{S}^{++})$, where the dependence law was obtained from the condition that the y values were the same for all the observed points. Peimbert & Spinrad (1970a–c) and Dufour (1974) made use of the relations derived from the Orion nebula to estimate the y^0 values for other objects.

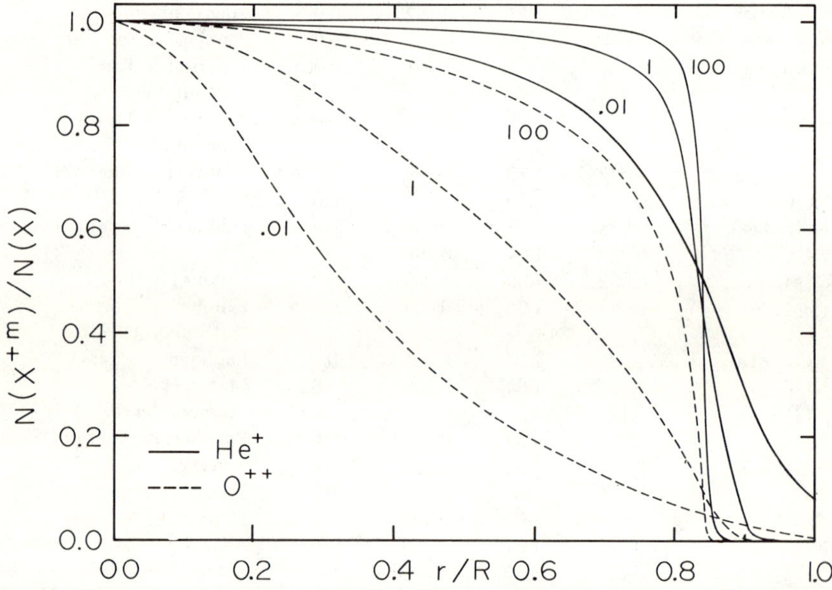

Figure 1 Ionization structure of oxygen and helium for a nebula excited by a main-sequence star of $T_* = 35{,}000$°K. Numbers on the graph correspond to hydrogen density, N(H) (Peimbert et al 1974).

From ionization structure models of H II regions (Peimbert et al 1974) it is found that y^0 depends not only on the ionization structure of the heavy elements but also on the electron density (see Figures 1 and 2). This dependence is in the sense that the higher the N_e, the larger the amount of neutral helium for a given $N(O^+)/N(O^{++})$. The root mean square electron density, N_e(rms), for the nucleus of M82 is similar to that of the Orion nebula; however, most giant H II regions have N_e(rms) values about two orders of magnitude smaller than those of the Orion nebula (Recillas-Cruz & Peimbert 1970, Peimbert & Spinrad 1970a–c). Consequently, the correction for the presence of neutral helium in giant H II regions based on the ionization structure of the Orion nebula overestimates y^0. Therefore we will adopt a helium ionization correction factor, for giant H II regions, given by $i_{cf}(He) = 1+\gamma/2$; where $1+\gamma$ is the $i_{cf}(He)$ derived from the dependence law used for the Orion nebula. From the previous discussion it follows that it is not possible to obtain a very precise helium abundance for those objects with a large correction factor unless observations at different points are made or a model including the spatial distribution of the electron density and the ionizing stars is computed.

The determinations of y^0 by Peimbert & Torres-Peimbert (1974) for Orion, η Car, and the Large Magellanic Cloud (LMC) nebulae are based on a combination of abundance determinations at different points and on theoretical models. In the cases of giant metal-poor H II regions such as I Zw 18 and II Zw 40, y^0 is negligible.

HEAVY ELEMENTS There are two types of abundance ratios: 1. a heavy element relative to another, and 2. a heavy element relative to hydrogen. In the latter case

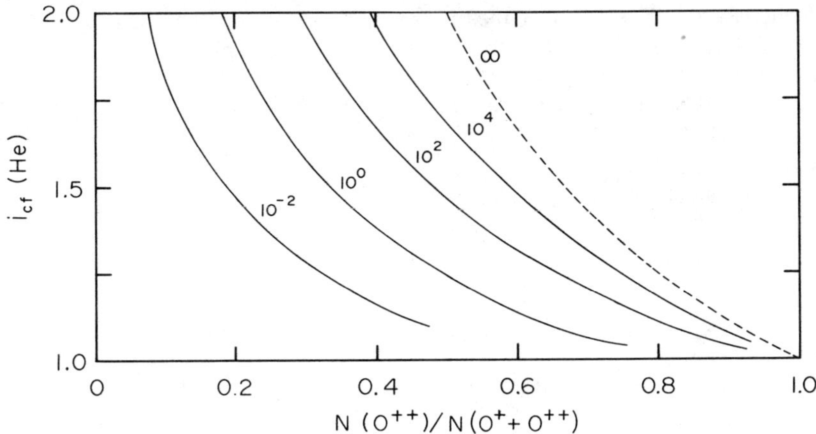

Figure 2 Helium ionization correction factor vs oxygen ionization degree. The upper part of the graph corresponds to models of nebulae ionized by single main-sequence stars of low T_*; the lower part is based on those ionized by stars of higher T_*. Hydrogen densities are indicated on the graph (Peimbert et al 1974).

a considerably better knowledge of temperature and ionization mechanisms is needed to determine the relative abundances.

Several methods for absolute abundance determinations have been attempted for the Orion nebula: 1. a constant temperature model where the temperature is in general provided by the [O III] line intensities; and 2. a simplified treatment of the variations in T, that considers the average temperature, T_0, and the mean square temperature fluctuation, t^2. The first model usually underestimates the heavy element abundances relative to that of hydrogen by factors of the order of 2, because in the presence of spatial temperature gradients or temperature fluctuations, $T_{[\mathrm{O\,III}]}$ is higher than the characteristic temperature associated with the formation of the forbidden and permitted lines (Peimbert & Costero 1969). The differences between the values derived from the two methods are smaller in the case of relative abundances than they are in the case of absolute abundances.

The key temperature for abundance determinations is obtained from the auroral-to-nebular line intensity ratios of O III. From the years 1966 to 1969 the collisional cross sections computed by Saraph et al (1966) were used to determine these electron temperatures. Eissner et al (1969) made approximate computations of the collisional cross sections including the presence of resonances. Their results, which yielded electron temperatures a few hundred degrees higher than those obtained previously, were used in most papers written during 1970–1974. More accurate calculations have now been made by Eissner & Seaton (1974) (see Table 2). Their results for $\Upsilon(2 \rightarrow 1)$ and $\Upsilon(3 \rightarrow 1)$ are in very close agreement with those of Saraph et al, although the most recent value for $\Upsilon(3 \rightarrow 2)$ is larger. These new results yield electron temperatures very similar to those derived from the computations by Saraph et al. Lower heavy element abundances correspond to higher adopted temperatures. Therefore the use of the most recent [O III] cross sections will increase the absolute abundance determinations based on the results by Eissner et al, by factors of about 0.1 in the log; the relative abundances are affected to a lesser degree.

The heavy element abundances in the gas provide a lower limit to the total Z value in H II regions because a fraction of the heavy elements is embedded in

Table 2 Collisional cross sections for O III[a]

T_e	$\Upsilon(^1D \rightarrow {}^3P)$[b]	$\Upsilon(^1S \rightarrow {}^3P)$	$\Upsilon(^1S \rightarrow {}^1D)$
5,000	2.17	0.276	0.807
7,500	2.26	0.305	0.859
10,000	2.36	0.325	0.856
12,500	2.44	0.339	0.834
15,000	2.49	0.348	0.807
17,500	2.53	0.353	0.779
20,000	2.55	0.356	0.752

[a] Eissner & Seaton (1974).

[b] $\Upsilon(i \rightarrow j) = \int_0^\infty \Omega(i,j) \exp(-\varepsilon_i/kT)\, d(\varepsilon_i/kT)$, $\varepsilon_i = \frac{1}{2} m v_i^2$.

Table 3 Heavy element abundances

Object	[O/H][a]	[N/O]	[Ne/O]	Observer
NGC 604	+0.11			Peimbert (1970)
Orion nebula	0.00	0.00	0.00	Peimbert & Costero (1969)
η Car nebula	−0.16	+0.10	−0.04	Peimbert & Torres-Peimbert (1974)
NGC 5455	−0.16			Searle & Sargent (1972a)
NGC 6822	−0.23	−0.55		Peimbert & Spinrad (1970c)
⟨LMC⟩	−0.28	−0.50	+0.04	Dufour (1974)
⟨LMC⟩	−0.31	−0.26	+0.11	Peimbert & Torres-Peimbert (1974)
⟨LMC⟩	−0.33	−0.36	+0.30	Aller et al (1974)
NGC 5471	−0.42		+0.17	Searle & Sargent (1972a)
II Zw 40	−0.70		+0.20	Searle & Sargent (1972a)
⟨SMC⟩	−0.66	−0.43	−0.04	Dufour (1974)
⟨SMC⟩	−0.82	−0.53	+0.36	Aller et al (1974)
NGC 185-1	−0.83			Jenner et al (1973)
I Zw 18	−1.52		+0.05	Searle & Sargent (1972a)

[a] $[A/B] \equiv \log(A/B) - \log(A/B)_{\text{Orion nebula}}$.

dust grains. In the case of the Orion nebula this fraction is not very large because the dust-to-gas ratio significantly varies from point to point (O'Dell & Hubbard 1965, Simpson 1973, Schiffer & Mathis 1974), whereas the most abundant heavy elements do not vary by values higher than the probable error of the abundance determination (Peimbert & Costero 1969, Simpson 1973). Also the dust-to-gas ratio in the Magellanic Clouds is smaller than in our galaxy (van den Bergh 1968). These arguments indicate that for H II regions similar to the Orion nebula and for metal-poor H II regions the fraction of the most abundant heavy elements embedded in dust grains is not considerable.

In Table 3 we list the heavy element abundances of extragalactic gaseous nebulae derived from photoelectric data and a direct determination of the electron temperature from the [O III] lines. The abundances for I Zw 18 and II Zw 40 have been recomputed and are slightly smaller than those derived by Searle & Sargent (1972a).

CHEMICAL ABUNDANCE GRADIENTS The following lines of evidence, independent of normal H II regions, indicate the presence of chemical abundance gradients with respect to the center of galaxies: 1. the distribution of cepheids according to their period (van den Bergh 1958); 2. the ratio of the most luminous blue stars to red giant stars (Walker 1964, van den Bergh 1968, Peimbert & Spinrad 1970c); 3. the strength of stellar absorption features close to and in the nuclei of galaxies (McClure 1969, Spinrad et al 1971, Spinrad et al 1972); 4. CN observations of K giant stars in the solar neighborhood (Janes & McClure 1972); and 5. the emission-line intensities in nuclear H II regions (Peimbert 1968). The stellar aspect has been reviewed by van den Bergh (1975) and the nuclear H II regions are considered in section 2b.

We concentrate now on the evidence based on normal H II regions. Aller (1942), from the study of the spectra of H II regions in M33, found a gradient in the [O III]/Hβ intensity ratio, this ratio being smaller for H II regions closer to the nucleus; changes in the [O II]/Hβ ratio are very small. Aller concluded from his observations that most of the oxygen is singly ionized and that the ionization degree of the outer H II regions is higher than those of the inner H II regions.

Searle (1971) made an extensive study of giant H II regions in several spiral galaxies and found that there is a gradient of three intensity ratios, [N II]/Hα, Hβ/[O III], and [N II]/[O II]; these ratios decrease from the center outward, forming a one-parameter family. The correlation is very good and holds from the innermost to the outermost H II regions. From simple theoretical models Searle established the presence of an N/O abundance gradient. Furthermore he suggested that the changes in the spectra of the H II regions could be caused by variations in one or more of the following parameters: 1. an increase in the O/H abundance ratio toward the nuclei of galaxies; 2. an increase in the H II region dust content; and 3. a decrease of the stellar effective temperature. Searle showed that even a very small change in the O/H ratio had a profound effect on the appearance of the spectra of H II regions and concluded that most of the change was probably caused by the presence of an O/H abundance gradient.

Representative values of the H, N, and O relative abundances across the disks of spiral galaxies are presented in Table 4 (Searle 1971). The few abundance determinations based on electron temperatures derived from [O III] line intensities (Table 3) support Searle's conclusions about the presence of O/H and N/O abundance gradients.

Shields (1974) computed more elaborate models than those used by Searle (1971) to interpret Searle's observations. In his models Shields considered variations in: (*a*) chemical composition; (*b*) dust-to-gas ratio; (*c*) clumpiness of the gas; and (*d*) effective temperature of the ionizing stars. From the analysis of his models, Shields supports Searle's conclusions that the variations of the spectra of H II regions are mainly caused by chemical abundance differences. According to Shields, however, the oxygen gradient varies from O/H $= 2 \times 10^{-4}$ for the outermost H II regions to O/H $= 30 \times 10^{-4}$ for the innermost regions; consequently the gradient is considerably steeper than that suggested by Searle.

Table 4 Representative abundances[a]

Region	[O/H]	[N/O]	[N/S]
Nucleus (0–20 pc)			1.23
Nucleus (0–50 pc)			0.73
Innermost H II region	0.14	0.71	
Inner H II region	0.07	0.23	0.23
Outer H II region	0.00	0.00	0.00
Outermost H II region	−0.09	−0.21	−0.60

[a] O/H and N/O from Searle (1971); N/S from Benvenuti et al (1973).

Rubin et al (1972), from photographic spectrophotometry of H II regions in M31, found line intensity gradients similar to those of Searle (1971). Moreover, based on the models by Searle, Rubin et al interpret the line-intensity variations between 5 and 15 kpc from the nucleus as caused by a decrease in the N/H and O/H abundance ratios by factors of 2 and 1.2, respectively.

Comte & Monnet (1974), from photometry with narrow-band interference filters combined with spectrophotometry of H II regions in M33, have also confirmed the excitation gradient found by Searle (1971).

Smith (1973; personal communication), using the image dissector scanner on the 120-inch telescope at Lick Observatory, is directing a very extensive program of spectrophotometric observations of extragalactic H II regions. He has been able to detect $\lambda 4363$ of O III in several objects, which permits a direct determination of the electron temperature and consequently of the chemical abundances. Smith finds an O/H gradient more pronounced than that derived by Searle, and a less steep N/H gradient.

Benvenuti et al (1973) found that in the inner H II regions of M33 the [N II] and [S II] lines originate in the same volume elements and consequently that

$$\frac{\mathrm{N(N)}}{\mathrm{N(S)}} \approx \frac{\mathrm{N(N^+)}}{\mathrm{N(S^+)}}, \qquad 5.$$

where

$$\frac{\mathrm{N(N^+)}}{\mathrm{N(S^+)}} = f(x, T_e)(1+0.14x)3.43 \exp{(5.0 \times 10^2/T_e)}\mathrm{I}(6584)/\mathrm{I}(6717+6731). \qquad 6.$$

In equation 6, $x = 10^{-2}N_e T_e^{-1/2}$ and f takes into account the collisional deexcitation of the D levels of the S^+ atom; for $N_e < 1000\ \mathrm{cm}^{-3}$, $f \sim 1$. Equation 6 is almost temperature and reddening independent; consequently at low densities the derived values of $\mathrm{N(N^+)/N(S^+)}$ ratios are very reliable. The results on the N/S abundance ratios by Benvenuti et al are presented in Table 4. From theoretical ionization structure models of H II regions Peimbert et al (1974) have found that in dilute radiation fields the N^+ and S^+ lines originate in the same volume elements. This requires low electron densities and/or large density fluctuations.

From observations of H II regions in M101 and M33, and from similar considerations to those of Benvenuti et al (1973), Comte (1975) found an N/S abundance gradient present in these galaxies. He did not, however, find any gradient over a considerable fraction of the disk of M51. His results are presented in Figure 3. Also presented in Figure 3 is the extremely low N/S ratio of NGC 5471 in M101, derived by Benvenuti et al from the observations by Searle & Sargent (1972a).

b *Nuclear H II Regions*

The nuclei of many spiral and elliptical galaxies exhibit emission-line intensities that are different from normal H II regions. Moreover, it is not clear what the relative importance of the different mechanisms and sources of ionization is: The ionization could be partly radiative and partly collisional; the ionizing photons might originate in OB stars or horizontal branch stars, or could be due to

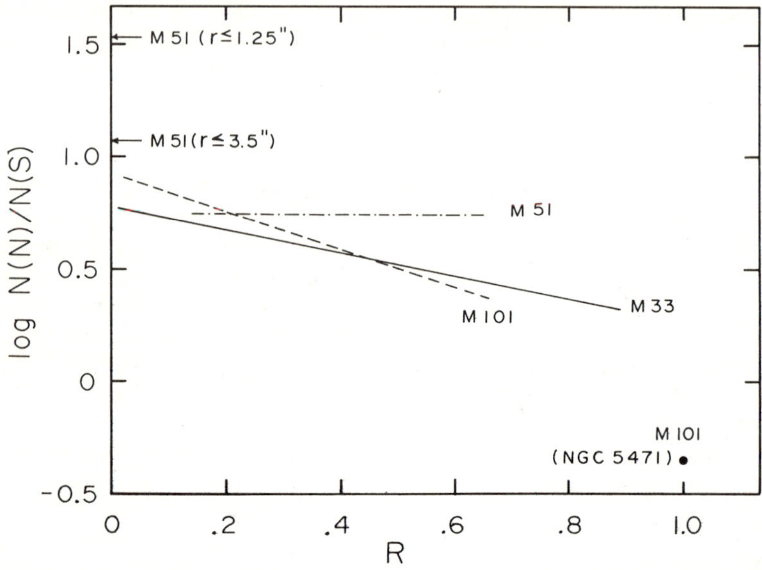

Figure 3 The observed N/S abundance gradient for different galaxies (Comte 1975, Benvenuti et al 1973). The abscissa is the distance normalized to the apparent optical radius.

recombination of the ionized gas, whereas the collisional ionization might arise from shock waves. Therefore to derive chemical abundances one should consider the different ionization mechanisms.

The first very important result in this field is that a considerable number of spiral galaxies exhibit a variation of the I[N II]/I(Hα) intensity ratio. The ratio varies from about $\frac{1}{3}$ in normal H II regions, to larger than 1 in the nuclei (Burbidge & Burbidge 1962, 1965; Burbidge et al 1963).

From photoelectric observations of Hα, Hβ, and the nebular lines of [O I], [O II], [O III], and [N II] Peimbert (1968) concluded that the chemical composition in the nuclei of M51 and M81 was not solar and that the change in the [N II]/Hα line intensity ratio was most probably caused by an overabundance of nitrogen in the nuclei of galaxies relative to normal H II regions.

Alloin (1973) has made an extensive study of spiral galaxies with a nuclear radio component. Out of the 89 objects of her survey she selected M81, NGC 2903, NGC 3351, and NGC 4569 to carry out a spectrophotometric study of the emission lines. Alloin considered five possible ionization mechanisms for the nuclear H II regions: 1. collisional ionization; 2. ionization by relativistic electrons; 3. ionization by energetic protons; 4. photoionization by hard uv radiation; and 5. photoionization by normal stars. She concluded that the last mechanism is the most likely for the objects studied. Under the assumption of radiative ionization by normal stars, Alloin finds that the four nuclei are overabundant in nitrogen by factors from 6 to 17. A nitrogen overabundance would also be obtained if the ionization were collisional.

An abnormally high N/O abundance ratio has also been obtained for the peculiar galaxy NGC 5253 (Welch 1970), which has been classified as an irregular. Welch finds an underlying distribution of stellar surface brightness, however, that resembles that of an elliptical galaxy. Bohuski et al (1972) found that the nucleus of the irregular galaxy NGC 5408 shows very similar line intensities to those of the nucleus of NGC 5253, with the exception of the nebular nitrogen lines, which are considerably weaker. Bohuski et al conclude that the nitrogen abundance is probably higher in NGC 5253 and suggest that the higher nitrogen abundance might indicate a later evolutionary stage for NGC 5253. This would agree with the distribution of stellar surface brightness found by Welch.

Benvenuti et al (1973), assuming equation 5 and basing their study on observations by Peimbert (1968, 1971a) and Warner (1973), concluded that the N(N)/N(S) abundance ratio increased considerably toward the center of M51 and M81 (see Table 4 and Figure 3). In the presence of a strong radiation field, that is, for a large electron density, equation 5 is no longer valid, because sulfur becomes ionized twice in regions where nitrogen is ionized singly. In the cases of M51 and M81 the electron densities are high and increase toward the nucleus (Warner 1973); the density fluctuations are very large (Peimbert 1968), however, and dilute the radiation field, which compensates in part for the high value of the density.

It is also possible that a fraction of the ionization in the nuclei of M51 and M81 is caused by shock waves. Evidence in favor of this mechanism is based on the large [O I] line intensities (Peimbert 1968, Alloin 1970) and on the presence of noncircular motions (Burbidge & Burbidge 1964, Goad 1974). In the case of collisonal ionization, equation 5 is a good approximation, regardless of the electron density (Peimbert et al 1974).

c *Planetary Nebulae*

Jenner et al (1973) have obtained the helium and oxygen abundances relative to hydrogen in NGC 185-1, a planetary nebula in NGC 185. The O/H ratio is lower than that of the Orion nebula by about an order of magnitude while the He/H ratio is similar to those of planetary nebulae in our galaxy (see Tables 1 and 3).

Sanduleak et al (1972) found that the [N II]/Hα ratios in the planetary nebulae, and probably also in the H II regions, of the SMC are very low; from these results they conclude that nitrogen must be deficient in the planetary nebulae of the SMC. This conclusion is in agreement with the H II region abundance determinations in the SMC by Dufour (1974) and Aller et al (1974).

3 DISCUSSION

In Table 1 we have presented the best He/H abundance ratios available in the literature, in order of decreasing metallicity (O/H abundance ratio). According to Peimbert & Torres-Peimbert (1974) the differences in the y values between the Orion nebula and the LMC are real and indicate a general trend for other H II regions, in the sense that the higher the metal abundance, the higher the helium abundance. From Figure 4 it can be seen that the extremely metal-poor regions are helium deficient relative to the Orion and η Car nebulae; this result is not due to the presence of neutral helium, because in general (see Figure 4 and Table 1) the ionization degree of the extremely metal-poor H II regions is higher and the amount of neutral helium is correspondingly smaller than that of the Orion nebula. By disregarding the three determinations with highest probable errors in Table 1, it is found that $\langle y \rangle = 0.092$. The 9 determinations based on H II regions with nearly normal abundances (including NGC 4449) yield $\langle y \rangle_{\text{normal}} = 0.1100$, whereas the 11 determinations based on the most metal-poor objects yield $\langle y \rangle_{\text{poor}} = 0.086$. It is clear that the differences are very small. We think that they are significant, however.

Oxygen, neon, and sulfur are correlated and their high abundance makes them excellent tracers of the z value. Nitrogen, on the other hand, is underabundant relative to oxygen in metal-poor H II regions and overabundant relative to oxygen

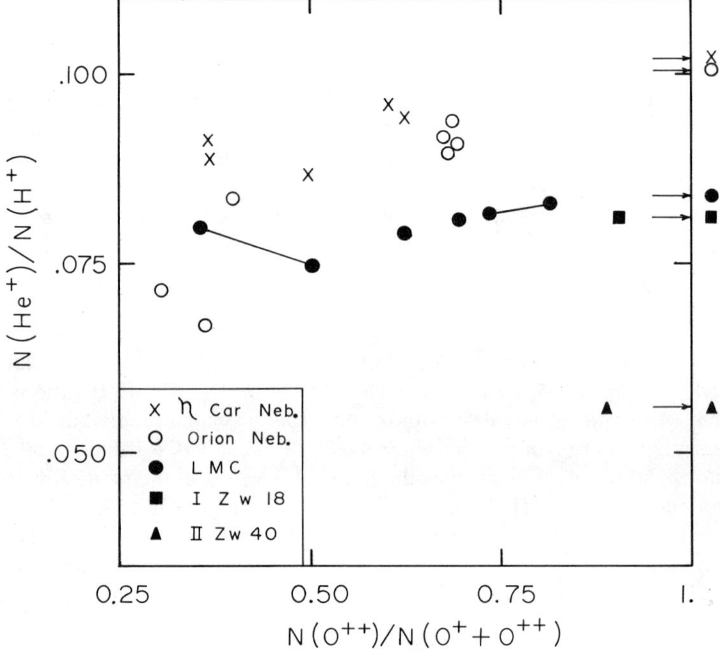

Figure 4 Comparison of the observed helium ionic abundance with the observed degree of ionization of oxygen (Peimbert & Torres-Peimbert 1974). From the observations by Searle & Sargent (1972a) we have added the points corresponding to I Zw 18 and II Zw 40. The estimate of y^0 is made from the extrapolated value at $N(O^{++})/N(O^{+}+O^{++}) = 1$.

in metal-rich H II regions. This observational result provides a tool to evaluate the relative efficiency for enrichment of the interstellar medium by explosive burning that produces oxygen, neon, and sulfur, and by hydrostatic burning, which is responsible for the production of nitrogen. The observed nitrogen abundance gradients extend along the spiral arm regions and become steeper toward the nuclei of galaxies (see Table 4).

Those H II regions located in systems with high ratios of interstellar-to-stellar matter are poorer in metal content than those H II regions located in systems of low gas-to-star mass ratios. This result is to be expected if the enrichment of the interstellar medium is caused by mass loss from objects such as planetary nebulae, red giants, and supernovae. The same explanation applies to the abundance gradients found across the disks of spiral galaxies, because the star density, as well as the ratio of the mass in stars to the mass in the form of gas, decreases from the center outward; thus the fraction of the enriched product material diminishes from the center outward.

There is observational evidence from our galaxy for enrichment of the interstellar medium by supernova remnants: y enrichment by the Crab nebula (Woltjer 1958; J. S. Miller, personal communication) and z enrichment by Cas A (Peimbert & van den Bergh 1971, Peimbert 1971b). Even if most planetary nebulae show y values in the 0.09–0.13 range with $\langle y \rangle = 0.11$ (Peimbert & Torres-Peimbert 1971, Barker 1974), there is a small fraction of them (5–10%), presumably the most massive, that shows values about twice as high. Examples of these are NGC 6302, NGC 6445, and PB6 (Danziger et al 1973, Aller et al 1973, Torres-Peimbert & Peimbert 1975). Consequently, planetary nebulae might be mainly responsible for the helium enrichment of the interstellar medium.

Because apparently metal-rich regions are also helium rich, the existence of a metal gradient across the disks of spiral galaxies implies the presence of a y gradient. The expected y gradient is small and difficult to detect because the degree of ionization of H II regions decreases and y^0 increases with increasing metal abundance. The [S II]/Hα line intensity ratios, which generally increase toward the nuclei of galaxies (Peimbert 1971a, Rubin & Ford 1972, Rubin et al 1972, Warner 1973), provide observational evidence in favor of a y^0 gradient (Rodríguez et al 1974).

The decrease of the ionization degree with the increase of the metal abundance in H II regions might be caused by dust particles absorbing more efficiently those photons able to ionize helium than those able to ionize hydrogen, or to a slightly varying stellar initial mass function (IMF) in the sense that the more metal rich the H II region, the lower the mass of the most massive object (Rodríguez et al 1974).

The very similar y values obtained from H II regions in different galaxies indicate that the pregalactic values of y were approximately the same, if not the same, and that the slightly higher values for the metal-rich H II regions could be the product of galactic chemical evolution.

From arguments presented in this section, in what follows we adopt the view that there is a pregalactic helium-to-hydrogen ratio, y_p, which corresponds to

$z = 0$, and that the chemical evolution of the interstellar matter in galaxies is characterized by the parameter $\alpha \equiv \Delta y/\Delta z$.

In Figure 5 we present the best abundance determinations from which to obtain values of α and y_p. Searle (1972) has given evidence in favor of a constant IMF in space and time; if this is the case, and if the y_p values are the same for all the objects, then all the H II region abundances in Figure 5 should lie on the same curve. The large observational errors do not permit us to test the previous statement, however. The straight line with $\alpha = 3.3$ was derived from the LMC and Orion nebula values (Peimbert & Torres-Peimbert 1974) and corresponds to $y_p = 0.069 \pm 0.008$ or $y_p = 0.216 \pm 0.02$. The extrapolation of α to $z = 0$ to derive y_p is valid if the stars that produce the helium and metal enrichment are the same, or if the lifetime of the stars that produce the z and of the stars that produce the y enrichment is short relative to the age of the galaxy. By considering the different uncertainties in the y and z determinations we conclude that for the range $z = 0$–$z = 0.02$, $1.5 < \langle\alpha\rangle < 4.5$.

To study the cosmological implications of the pregalactic helium abundance we will make use of the computations by Wagoner (1973) for a Friedmann universe with cosmological constant $\Lambda = 0$, and no neutrino degeneracy (standard big-bang model). In Figure 6 we present the average density of the universe, ρ_0, the primeval $x(D)_p$ and $x(^3\text{He})_p$ mass fractions for different y_p values from the computations by Wagoner; and in Table 5 the ρ_0, $x(D)_p$, and $x(^3\text{He})_p$ for the y_p value derived

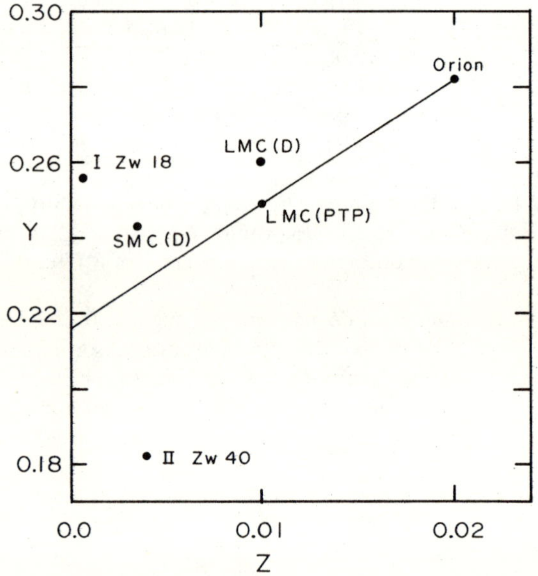

Figure 5 The observed chemical composition (by mass). The straight line corresponds to $\Delta y/\Delta z = 3.3$.

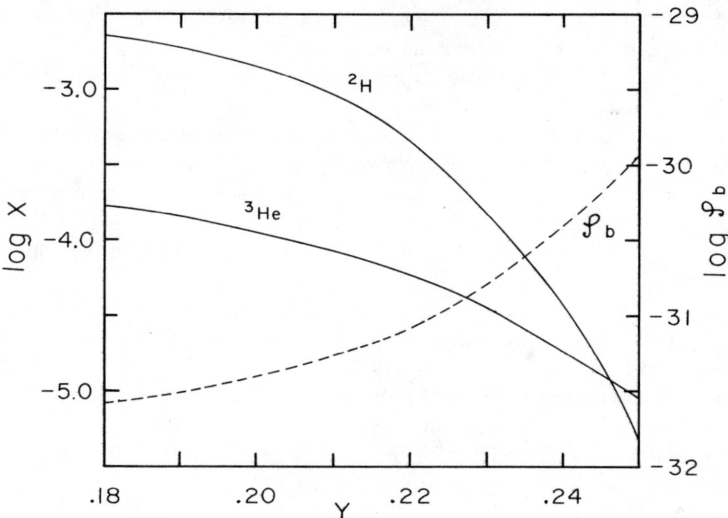

Figure 6 The dependence of the primeval mass fraction of D and ^{3}He as well as that of the present density of the universe on the primeval ^{4}He mass fraction for the standard big-bang model (Wagoner 1973).

above, as well as for the limiting primeval helium abundance corresponding to the estimated errors. From Figure 5 it can be seen that the results by Dufour (1974) indicate a slightly higher y_p value than that derived by Peimbert & Torres-Peimbert (1974). However, from Figure 5 and Table 1 it follows that y_p is smaller than the helium abundance of the LMC. Therefore, in Table 5 we have added this value as a conservative upper limit to the primeval helium abundance.

The values for the deceleration parameter, q_0, and the age of the universe given in Table 5 are obtained by adopting $H_0 = 55$ km sec^{-1} Mpc^{-1}, the Hubble constant derived by Sandage (1972) from Sc galaxies. It is clear from the results presented in

Table 5 Primeval abundances, age, and present density of the Universe[a]

y	log $x(D)$	log $x(^3\text{He})$	log ρ_0 (g cm^{-3})	q_0[b]	t_0[b] (10^9 yr)
0.196	−2.80	−3.92	−31.44	0.003	17.5
0.216	−3.21	−4.15	−31.17	0.006	17.4
0.235	−4.10	−4.58	−30.62	0.021	16.8
0.249[c]	−5.25	−5.02	−30.00	0.088	15.2

[a] Based on the computations by Wagoner (1973).
[b] For $H_0 = 55$ km sec^{-1} Mpc^{-1}.
[c] y value for the LMC from Peimbert & Torres-Peimbert (1974).

Table 5 that the limiting y_p value, together with the adopted H_0 and the standard big-bang hypothesis, imply an open universe with $q_0 < 0.09$. In Figure 7 we present the (H_0, q_0) diagram, where it can be seen that for $H_0 > 25$ km sec^{-1} Mpc^{-1} the y_p value derived from the LMC indicates an open universe.

Gott et al (1974) find a very narrow region of permitted values in the (H_0, q_0) diagram. This permitted region also corresponds to an open universe and implies $0.235 < y_p < 0.241$, in good agreement with the LMC upper limit on y_p, but slightly higher than the y_p derived by Peimbert & Torres-Peimbert (1974). Gott et al derived the permitted region from the following three restrictions: 1. that q_0 is larger than the best estimate of the smoothed-out density in galaxies, best q_{0*}; 2. that t_0 is less than 18 billion years; and 3. that the density of the universe is smaller than that derived from twice the value of the deuterium-to-hydrogen (D/H) ratio obtained by Rogerson & York (1973) (broken line in Figure 7). The lower limit on the D/H primeval ratio is obtained by assuming that the fraction of astrated material, that is, the fraction of interstellar matter in the solar neighborhood that has been inside stars, where presumably the primeval deuterium is destroyed, is at least 50%.

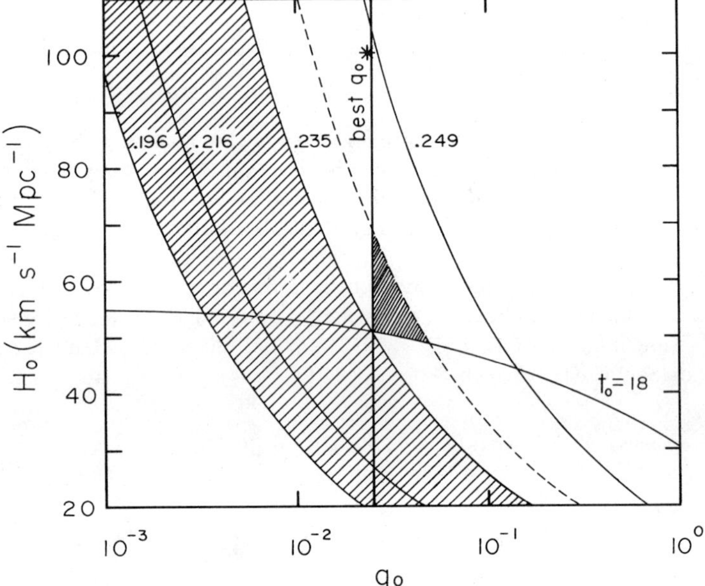

Figure 7 Cosmological restrictions. Ordinate is the Hubble constant and abscissa is the deceleration parameter. The large shaded area is the permitted region adopting the standard big-bang model (Wagoner 1973) and the primeval helium mass fraction (Peimbert & Torres-Peimbert 1974). The numbers on the graph correspond to primeval helium mass fractions. The small triangle corresponds to the possible values of H_0 and q_0 from the work by Gott et al (1974).

It is very important to study the assumptions used in the determination of ρ_0 from the D/H value. In addition to the assumption of the standard big-bang model it is assumed that 1. the fraction of deuterium tied-up in molecules is larger than that of hydrogen; 2. deuterium has been produced mainly by the big bang; and 3. the fraction of astrated material is known. Hoyle & Fowler (1973) and Colgate (1973, 1974) have suggested that deuterium can be produced not only by the big bang but also by spallation in envelopes of supermassive objects or supernovae. However, Epstein et al (1974), from a study of light elements produced in a hydrodynamic model of a type II supernova, argue that it is unlikely that the interstellar deuterium in the solar neighborhood was produced in this way.

4 SUMMARY: THE NEED FOR MORE OBSERVATIONS

One of the most interesting results of extragalactic abundance determinations is the correlation found between the degree of ionization and metallicity of giant H II regions. More observations should be attempted, including direct determinations of the electron temperature, to confirm and possibly extend the correlation to H II regions of higher metallicity. Infrared observations, particularly of metal-rich H II regions, will help us decide whether the difference in degree of ionization is caused by the presence of dust or a varying IMF.

To study the chemical evolution of galaxies it is important to estimate from observations the amount and chemical composition of the enriched material that is being added to the interstellar medium. From this estimate it will be possible to attempt an explanation of the different y and z values observed in H II regions.

It is important to determine chemical abundances by observations at different distances from the center of metal-rich H II regions, because they will provide an empirical estimate of the ionization correction factor of helium and consequently a more accurate y value. Moreover, the determination of abundances at different projected distances from the center of a given H II region will allow us to estimate the fraction of heavy elements embedded in dust grains.

Additional determinations of the pregalactic y value are needed to confirm the conclusion reached by Peimbert & Torres-Peimbert (1974) that the y_p coupled with the standard big-bang model computations by Wagoner (1973) implies an open universe. To reach this goal, high accuracy determinations of the y value in metal-poor H II regions should be attempted. These determinations may be achieved from observations of the integrated volume because their ionization correction factors are very close to 1. Furthermore the accuracy of the y determination in metal-poor H II regions can be higher than in metal-rich H II regions because the former are almost dust free, that is, the amount of heavy elements locked up in grains is negligible and the reddening correction is very small.

It will be possible to obtain the fraction of astrated material from accurate values of y_p and the deuterium mass fraction under the assumption of the standard big-bang model.

Independent determinations of q_0, H_0, or ρ_0 are needed to test the assumptions on which the standard big-bang model is based.

Literature Cited

Aller, L. H. 1942. *Ap. J.* 95:52
Aller, L. H. 1972. *Ann. NY Acad. Sci.* 194:45
Aller, L. H., Czyzak, S. J., Craine, E., Kaler, J. B. 1973. *Ap. J.* 182:509
Aller, L. H., Czyzak, S. J., Keyes, C. D., Boeshaar, G. 1974. *Proc. Nat. Acad. Sci. USA* 71:4496
Aller, L. H., Czyzak, S. J., Walker, M. F. 1968. *Ap. J.* 151:491
Aller, L. H., Faulkner, D. J. 1962. *Publ. Astron. Soc. Pac.* 74:219
Aller, L. H., Liller, W. 1968. In *Nebulae and Interstellar Matter*, ed. B. M. Middlehurst, L. H. Aller, 483. Chicago: Univ. Chicago Press
Alloin, D. 1970. *Astron. Ap.* 9:45
Alloin, D. 1973. *Astron. Ap.* 27:433
Barker, T. 1974. PhD thesis. Univ. California, Santa Cruz
Benvenuti, P., D'Odorico, S., Peimbert, M. 1973. *Astron. Ap.* 28:447
Bohuski, T. J., Burbidge, E. M., Burbidge, G. R., Smith, M. G. 1972. *Ap. J.* 175:329
Brocklehurst, M. 1971. *MNRAS* 153:471
Brocklehurst, M. 1972. *MNRAS* 157:211
Burbidge, E. M., Burbidge, G. R. 1962. *Ap. J.* 135:694
Burbidge, E. M., Burbidge, G. R. 1964. *Ap. J.* 140:1445
Burbidge, E. M., Burbidge, G. R. 1965. *Ap. J.* 142:634
Burbidge, G. R., Gould, R. J., Pottasch, S. R. 1963. *Ap. J.* 138:945
Churchwell, E. B., Mezger, P. G., Huchtmeier, W. 1974. *Astron. Ap.* 32:283
Colgate, S. A. 1973. *Ap. J. Lett.* 181:L53
Colgate, S. A. 1974. *Ap. J.* 187:321
Comte, G. 1975. *Astron. Ap.* In press
Comte, G., Monnet, G. 1974. *Astron. Ap.* 33:161
Cox, D. P., Daltabuit, E. 1971. *Ap. J.* 167:257
Danziger, I. J. 1970. *Ann. Rev. Astron. Ap.* 8:161
Danziger, I. J., Frogel, J. A., Persson, S. E. 1973. *Ap. J. Lett.* 184:L29
Dufour, R. J. 1974. PhD thesis. Univ. Wisconsin, Madison
Eissner, W., Martins, P. de A. P., Nussbaumer, H., Seaton, M. J. 1969. *MNRAS* 146:63
Eissner, W., Seaton, M. J. 1974. *J. Phys. B.* 7:2533
Epstein, R. I., Arnett, W. D., Schramm, D. N. 1974. *Ap. J. Lett.* 190:L13
Faulkner, D. J., Aller, L. H. 1965. *MNRAS* 130:393
Goad, J. W. 1974. *Ap. J.* 192:311
Gott, J. R. III, Gunn, J. E., Schramm, D. N., Tinsley, B. M. 1974. *Ap. J.* 194:543
Hoyle, F., Fowler, W. A. 1973. *Nature* 241:384
Janes, K. A., McClure, R. D. 1972. In *IAU Colloq. No. 17*, ed. G. Cayrel de Strobel, A. M. Delplace, Sect. 28. Paris: Meudon Observ.
Jenner, D. C., Ford, H. C., Epps, H. W. 1973. *Bull. Am. Astron. Soc.* 5:13
Mathis, J. S. 1962. *Ap. J.* 136:374
Mathis, J. S. 1965. *Publ. Astron. Soc. Pac.* 77:189
McClure, R. D. 1969. *Astron. J.* 74:50
Mezger, P. G., Wilson, T. L., Gardner, F. F., Milne, D. K. 1970. *Ap. Lett.* 5:117
O'Dell, C. R., Hubbard, W. B. 1965. *Ap. J.* 142:591
Osterbrock, D. E. 1970. *Quart. J. Roy. Astron. Soc.* 11:199
Osterbrock, D. E. 1974. *Astrophysics of Gaseous Nebulae.* San Francisco: Freeman
Peimbert, M. 1968. *Ap. J.* 154:33
Peimbert, M. 1970. *Publ. Astron. Soc. Pac.* 82:636
Peimbert, M. 1971a. *Bol. Observ. Tonantzintla Tacubaya* 6:97
Peimbert, M. 1971b. *Ap. J.* 170:261
Peimbert, M., Costero, R. 1969. *Bol. Observ. Tonantzintla Tacubaya* 5:3
Peimbert, M., Goldsmith, D. W. 1972. *Astron. Ap.* 19:398
Peimbert, M., Rodríguez, L. F., Torres-Peimbert. S. 1974. *Rev. Mex. Astron. Astrof.* 1:129
Peimbert, M., Spinrad, H. 1970a. *Ap. J.* 159:809
Peimbert, M., Spinrad, H. 1970b. *Ap. J.* 160:429
Peimbert, M., Spinrad, H. 1970c. *Astron Ap.* 7:311
Peimbert, M., Torres-Peimbert, S. 1971. *Ap. J.* 168:413
Peimbert, M., Torres-Peimbert, S. 1974. *Ap. J.* 193:327
Peimbert, M., van den Bergh, S. 1971. *Ap. J.* 167:223
Recillas-Cruz, E., Peimbert, M. 1970. *Bol. Observ. Tonantzintla Tacubaya* 5:247
Roberts, M. S. 1972. In *IAU Symp. No. 44*, ed. D. E. Evans, 12. Dordrecht: Reidel
Rodríguez, L. F., Torres-Peimbert, S., Peimbert, M. 1974. *Rev. Mex. Astron. Astrof.* 1:161
Rogerson, J. B. Jr., York, D. G. 1973. *Ap. J. Lett.* 186:L95
Rubin, V. C., Ford, W. K. Jr. 1972. In *IAU Symp. No. 44*, ed. D. E. Evans, 49. Dordrecht: Reidel
Rubin, V. C., Kumar, C. K., Ford, W. K. Jr. 1972. *Ap. J.* 177:31

Sandage, A. 1972. *Quart. J. Roy. Astron. Soc.* 13:282

Sanduleak, N., MacConnell, D. J., Hoover, P. S. 1972. *Nature* 237:28

Saraph, H. E., Seaton, M. J., Shemming, J. 1966. *Proc. Phys. Soc.* 89:27

Schiffer, F. H. III, Mathis, J. S. 1974. *Ap. J.* 194:597

Searle, L. 1971. *Ap. J.* 168:327

Searle, L. 1972. In *IAU Colloq. No. 17,* ed. G. Cayrel de Strobel, A. M. Delplace, Sect. 52. Paris: Meudon Observ.

Searle, L., Sargent, W. L. W. 1972a. *Ap. J.* 173:25

Searle, L., Sargent, W. L. W. 1972b. *Comments Ap. Space Phys.* 4:59

Seaton, M. J. 1960. *Rep. Progr. Phys.* 23: 313

Seaton, M. J. 1968. *MNRAS* 139:129

Seaton, M. J. 1971. In *Highlights of Astronomy,* ed. C. de Jager, 2:288. Dordrecht: Reidel

Shields, G. A. 1974. *Ap. J.* 193:335

Simpson, J. P. 1973. *Publ. Astron. Soc. Pac.* 85:479

Smith, H. E. 1973. *Bull. Am. Astron. Soc.* 5:448

Spinrad, H., Gunn, J. E., Taylor, B. J., McClure, R. D., Young, J. W. 1971. *Ap. J.* 164:11

Spinrad, H., Peimbert, M. 1975. In *Galaxies and the Universe,* ed. A. R. Sandage, J. Kristian. Chicago: Univ. Chicago Press. In press

Spinrad, H., Smith, H. E., Taylor, D. J. 1972. *Ap. J.* 175:649

Torres-Peimbert, S., Peimbert, M. 1975. *Rev. Mex. Astron. Astrof.* In press

van den Bergh, S. 1958. *Astron. J.* 63:492

van den Bergh, S. 1968. *The Galaxies of the Local Group. David Dunlap Observ. Commun. No. 195*

van den Bergh, S. 1975. *Ann. Rev. Astron. Ap.* 13:217

Wagoner, R. V. 1973. *Ap. J.* 179:343

Walker, M. F. 1964. *Astron. J.* 69:744

Warner, J. W. 1973. *Ap. J.* 186:21

Welch, G. A. 1970. *Ap. J.* 161:821

Whitford, A. E. 1958. *Astron. J.* 63:201

Woltjer, L. 1958. *BAN* 14:39

ULTRAVIOLET STUDIES OF THE INTERSTELLAR GAS

ⅹ2078

Lyman Spitzer, Jr. and Edward B. Jenkins
Princeton University Observatory, Princeton, New Jersey 08540

1 INTRODUCTION

Observations from space have had a singularly important effect on interstellar research. Interstellar atoms are normally in the ground state, and most cannot absorb photons longward of 3000 Å. Ground-based observations of interstellar absorption lines have been restricted to minor constituents in relatively scarce ionization stages. Such data have given fundamental information on the velocities of interstellar clouds (Adams 1949, Hobbs 1969, 1973, 1974; Marschall & Hobbs 1972), but have been less useful for studying the composition and physical state of the gas.

Small telescopes on sounding rockets first detected ultraviolet interstellar absorption lines from atoms (Morton & Spitzer 1966) and from H_2 molecules (Carruthers 1970b, Smith 1973). The first successful NASA Orbiting Astronomical Observatory (OAO-2), with a low-resolution (12 Å) spectrometer, provided important information on the Lα line of interstellar H (Savage & Jenkins 1972). Abundances of Mg, Mn, and Fe have been determined from ultraviolet measurements with the ESRO TD-1A satellite (de Boer et al 1972b). The later OAO-3 satellite, *Copernicus,* has provided extensive new data on many interstellar lines, observed with a grating spectrometer (Rogerson, Spitzer et al 1973) giving a spectral resolution of 0.05 Å from 950 to 1500 Å; other *Copernicus* data channels, giving high resolution at longer ultraviolet wavelengths and lower resolution throughout the ultraviolet, have also been used for some interstellar studies, but are not described here.

The *Copernicus* scanning spectrometer accumulates data slowly, moving one wavelength step of about 0.023 Å every 16 sec. Because observations are possible only about one third of the time, a week is required to scan from 1000 to 1300 Å in one star. Figure 1 shows a high resolution scan of ζ Ophiuchi in the vicinity of the H_2 (4, 0) Lyman band. In most stars data have been restricted to a number of wavelength bands about 0.8 Å wide, each centered on an interstellar line.

The photomultiplier tubes used provide high relative photometric accuracy, limited only by the photon statistics. In the region of peak efficiency, between 1025 and 1200 Å, about 1000 counts are accumulated during each stepping interval for an unreddened B1 star of magnitude 5.0, with a corresponding accuracy of 3%

in the relative intensities. With multiple scans, summed by computer on the ground, higher accuracies can be achieved even on fainter stars, although observing times become impractically long for stars with $E(B-V)$ appreciably greater than 0.40. Outside this wavelength region of peak response the sensitivity drops steadily, reaching about one thirtieth of its peak value at 950 and 1450 Å, where observations are feasible only on the brighter unreddened O or B stars.

Apparent line profiles must be corrected for the instrumental function of the spectrometer, that is, the recorded profile for a monochromatic emission line. From multiple scans of the weaker H_2 features in the spectrum of δ Per (Spitzer & Morton 1975) the instrumental profile at about 1060 Å can be well fitted by a Gaussian with a full width at half maximum equal to 0.051 Å. In addition there is scattered light from the grating that, from measures of the apparent central intensities of wide, saturated interstellar lines, amounts to about 6% of the adjacent continuum at 1060 Å. The apparent scattered light level varies less rapidly with wavelength than does the stellar spectrum, and at 950 and 1450 Å contributes about half the observed signal. The large additional stray light noted in the early *Copernicus* observations (Rogerson, Spitzer et al 1973) has generally been eliminated from later high-resolution observations by using a movable mirror (part of the intermediate resolution assembly) to occult the outgassing aperture in the far ultraviolet phototube.

The observed line scans can be used to determine line profiles or equivalent widths. The stronger H I and H_2 lines are more than 1 Å in width (see Figure 1) and fitting the observations with Voigt profiles is the best way to determine column densities [Spitzer, Drake et al 1973, Rogerson & York 1973, Drake (in preparation)]. Profiles of the weaker H_2 lines have been determined (Spitzer & Morton 1975) from tenfold scans to determine the column densities in the several clouds that

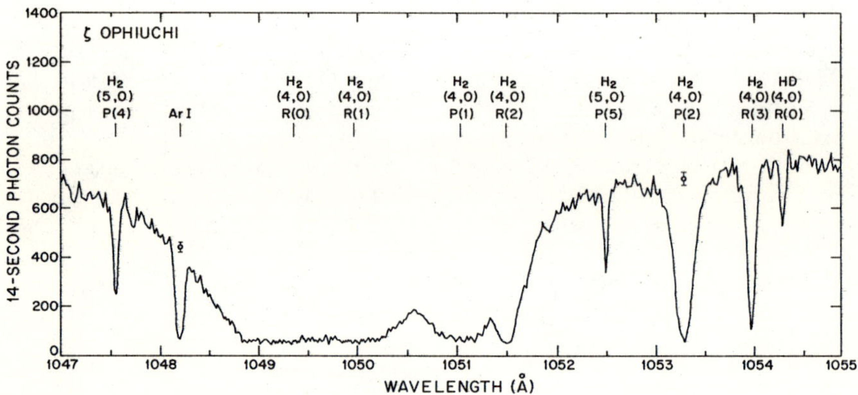

Figure 1 High-resolution scan of the O9.5 V star ζ Oph [$m_V = 2.56$, $E(B-V) = 0.32$] over an 8-Å interval. Error bars show the dispersion in photon counts expected from statistical fluctuations. The wavelengths of an Ar I line and of various rotational features in the (4, 0) and (5, 0) vibrational Lyman bands of H_2 and HD are shown by vertical lines.

produce components with different radial velocities. The absence of damping wings has also been used to determine upper limits on the column densities of O I, N I, N II, C II, and similar abundant species in the line of sight to ζ Oph (Morton 1975).

Most of the *Copernicus* observations have been used for determinations of the equivalent width, W_λ. In the best photometric cases, with high photon counts and a straight continuum, W_λ can be measured with a probable error of about 10^{-3} Å (Spitzer et al 1975). To determine from W_λ the corresponding column density N per cm^2 in the line of sight requires use of the curve of growth, the plot of log W_λ/λ against log $Nf\lambda$ (Spitzer 1968), where f is the oscillator strength [see the extensive compilation for interstellar lines by Morton & Smith (1973), also Morton (1975)]. If the distribution of radial velocities for the atoms in the line of sight is Maxwellian, with a dispersion $b/2^{1/2}$, one obtains the familiar theoretical curves that have been used for many years in the interpretation of the interstellar lines of Na I and Ca II, using the doublet ratio to determine the degree of saturation and the column densities N(Na I) and N(Ca II).

In fact the assumption of a Maxwellian velocity distribution for interstellar atoms is unrealistic, because Adams (1949) and others have shown that most interstellar absorption features are composed of several components, each attributable to a separate absorbing region or cloud. Theoretical calculations for two identical clouds, with nonoverlapping Maxwellian velocity distributions, show that the curve of growth is virtually indistinguishable from that for a single such cloud, except that the effective b value is increased (Strömgren 1948). For clouds with widely differing values of N and b, on the other hand, the curve of growth can have quite a different shape from that computed for a single cloud, and use of the usual doublet-ratio method can give large errors in N (Nachman & Hobbs 1973). While unsaturated lines remain unaffected by these problems, permitting reliable determinations of N, column densities for stronger lines must be found from empirical curves of growth.

In constructing such empirical curves, one would expect that atoms whose abundances vary in similar ways from cloud to cloud should produce similar curves and could be combined together. Curves obtained in this way for ζ Oph are shown in Figure 2, where the circles represent data for atoms that are in their dominant stage of ionization in an H I cloud (such as N I, Ar I, Mg II, and S II), whereas the triangles represent neutral atoms (such as C I, Na I, S I) of elements that are predominantly singly ionized in such clouds (Morton 1975); the crosses represent the H_2 lines from excited rotational levels (Spitzer et al 1975, Morton 1975). One would expect the circles all to lie on the same curve of growth, provided that the chemical composition of the gas is the same in the different clouds. For the triangles, the fraction of each atom present in neutral form should vary rapidly with the gas density, and if variations of the ionizing radiation between clouds are ignored, the relative contribution of different clouds should be the same for all the triangles, giving a common curve of growth for all. The crosses should define a single curve of growth if the rotational excitation of the H_2 molecules is assumed to be similar in all the contributing clouds. The small scatter shown in each of these curves suggests that these assumptions are nearly correct for ζ Oph, and that the derived column densities are reliable. The large differences between the three curves indicate that care must be taken in using generally a curve of growth obtained for some particular atom or molecule.

For other lines of sight different groupings of lines are appropriate. For example, in λ Scorpii, where absorption in both H I and H II regions apparently contributes importantly to C II and Si II lines (York 1975a), different curves of growth have been found for different atoms depending on the relative contribution of these two types of regions.

Radial velocity data are helpful in these analyses, because lines with significantly different radial velocities must originate in different regions. The *Copernicus* spectrometer does not yield precise absolute radial velocities, because no wavelength calibration source shortwards of 1600 Å could be provided, and the wavelength calibration of the instrument changes with time by up to ± 7 km sec^{-1}. Nevertheless differences of radial velocity between adjacent lines have been measured with a dispersion of about 1 km sec^{-1}. In ζ Oph the three groups of lines plotted in Figure 2 show systematic velocity differences between each other by some 5 km sec^{-1} (Morton 1975). The velocity data for λ Sco in Figure 3 (York 1975a) show clearly the difference in location between lines of different ionization potential, with

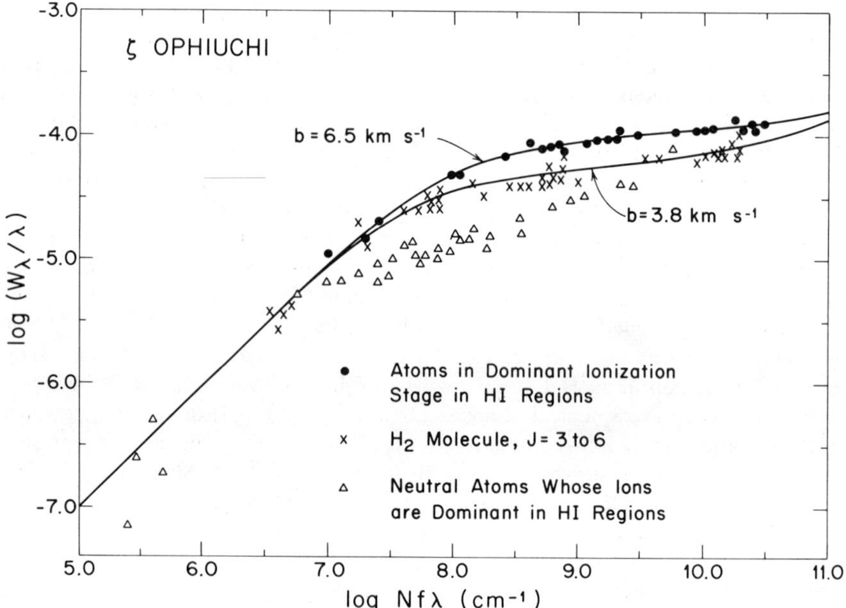

Figure 2 Curves of growth for different groups of interstellar lines in ζ Oph. The filled circles represent lines produced by N I, Ar I, Mg II, Si II, S II, and Fe II; the triangles show C I, Na I, Mg I, S I, K I, and Fe I. The crosses represent H_2 Lyman lines from the rotational levels $J = 3$–6.

velocity differences between the H I and H II regions and even between different parts of the H II zone. Even for lines all formed with the same curve of growth, relative velocity shifts can arise with changing line strength if the overall velocity distribution is not symmetric about its mean value.

Column densities based on detailed curves of growth have been obtained from values of W_λ for some 50 lines of H_2 observed in 28 stars (Spitzer et al 1975). A similar analysis for the atomic lines in ζ Oph has led to column densities for 16 elements in various stages of ionization, in addition to the 5 determined from ground observations (Morton 1975). Column densities have also been obtained from detailed analyses of λ Sco and α Vir (York 1975a, York & Kinahan 1975). Specific observational programs for the deuterium lines and for the doublets from highly ionized species (N V and O VI) are discussed in Sections 4.2 and 5.2, respectively. In addition, relative f values have been determined from observed equivalent widths of interstellar Fe II and Mn II (de Boer et al 1974) and C I (de Boer & Morton 1975).

2 ATOMIC HYDROGEN

The most abundant element, hydrogen, produces the strongest of all the interstellar absorption lines at the wavelength of Lα (1215.67 Å), a feature typically about 10 Å wide for stars only 200 pc away. Consequently, the measurement of hydrogen column densities has been attainable with only moderate wavelength resolutions. Also, this feature is so strong that for all but the closest stars radiation damping

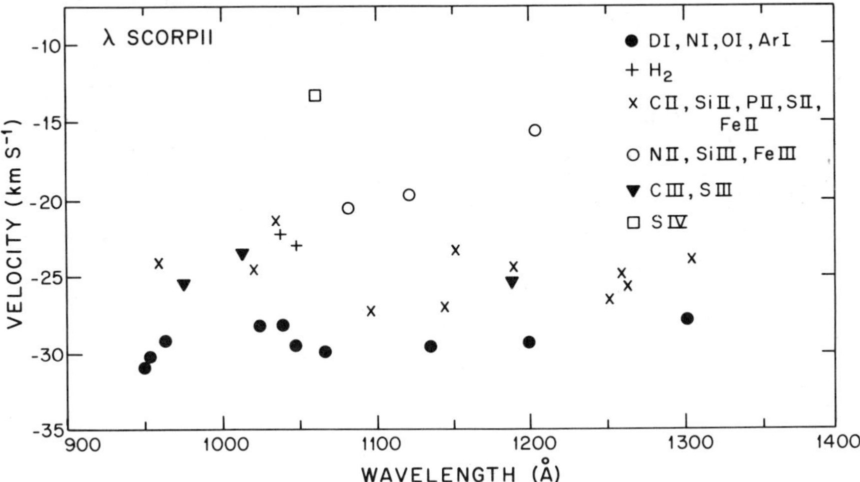

Figure 3 Heliocentric radial velocities of interstellar lines in the B1.5 IV star λ Sco [$m_V = 1.63$, $E(B-V) = 0.03$]. The x's represent singly ionized atoms for which the ionization potential, I, required for their production is less than 13.6 V. For the open circles, triangles, and squares, I is about 15, 24, and 35 V, respectively.

is the principal source of line broadening, which makes the derivations of column densities from equivalent widths or profile shapes rather straightforward, although information on the behavior of radial velocities is absent.

The earliest observations of Lα absorption were carried out with spectrographs on sounding rockets (Morton 1967, Carruthers 1970a, Jenkins 1970 and references cited). At most, however, only a few stars could be observed on each flight. Only after the survey by Savage & Jenkins (1972), based on observations by the spectrometer aboard OAO-2 for 69 stars of type B2 or hotter, was there a large enough data base for deriving generalized conclusions on the behavior of the gas. This study was later extended by Jenkins & Savage (1974) to include 26 new stars. At the wavelength resolution of 12 Å provided by the OAO-2 instrument, serious blending of nearby stellar features with the broad, shallow wings of the Lα damping profile made straight equivalent-width measurements untrustworthy, which resulted in a number of erroneously high column densities reported initially by Savage & Code (1970). Savage & Jenkins (1972) overcame this difficulty by using higher resolution rocket data, where the stellar features (except stellar Lα) were well distinguished, as a guide for unraveling the instrumentally blended features for stars of various spectral types. However, observations of 10 stars at 7 Å resolution by the spectrograph aboard the *Mariner IX* Mars orbiter suggested the OAO-2 column densities may in some cases be too large by a factor of 1.4 (Bohlin 1973).

For all stars more than 10° away from the galactic equator, Savage & Jenkins (1972) compared the Lα measures of hydrogen column density with those derived from observations of 21-cm emission in approximately the same directions. The correlation between the two was exceptionally poor (see their Figure 7*b*), with the Lα column densities usually lower than those from 21 cm, even after a simple estimate for the general background emission behind the stars had been subtracted. This poor correspondence is probably a consequence of early-type stars being within or in front of rich concentrations of interstellar material, although some exotic physical processes for anomalously enhancing the 21-cm radiation have also been proposed (Fischel & Stecher 1967, Kovach 1972). No improvement in the correlation seems to result if one substitutes more recent 21-cm data (Grayzeck & Kerr 1974, Heiles & Habing 1974) taken with smaller beamwidths centered more closely on the actual star positions.

The 95 stars observed by OAO-2 had distances ranging from 100 to 1000 pc, a scale considerably smaller than the space occupied by the large gas complexes identified in 21-cm emission surveys. C. Heiles and E. B. Jenkins (in preparation) make use of the Lα measurements over the well-defined paths to the stars to place constraints on the distances of principal features appearing in their photographic display of H I data at high galactic latitudes. For instance, conspicuous concentrations of material projected through Taurus and Orion must be more than 500 pc away, because the Orion stars show comparatively low column densities, whereas much of the gas in the Scorpius-Ophiuchus region must be in front of the stars at 200 pc. A striking feature of the H I distribution shown by Jenkins & Savage (1974) is a pronounced deficiency of gas over a sector of galactic longitude from 180 to 270°. Here the average volume density $\langle n(\text{H I})\rangle$ out to 1 kpc was found to be only 0.13 atoms cm^{-3}, compared with 0.72 atoms cm^{-3} for the other stars in the survey. Stars closer than 140 pc showed somewhat less hydrogen than average, but a reliable figure for the density of nearby gas could not be attained because relatively few stars were very close, and many of those were classified as type B2, indicating a strong likelihood that stellar Lα absorption could make a significant contribution.

For sampling the interstellar gas in the more immediate vicinity of the sun, Lα absorption appears in the *Copernicus* scans of chromospheric emissions from late-type stars. For instance, Moos et al (1974) found an average density within the range 0.02–0.1 atoms cm^{-3} over the 11 pc toward α Boötis. This range compares rather favorably with the estimates of 0.06 atoms cm^{-3} by Fahr (1970) and from 0.08 to 0.16 atoms cm^{-3} by Bertaux et al (1972), who interpreted the diffuse Lα glow in the sky (not the geocoronal component) as being produced by resonant backscatter of the solar Lα emission by interstellar H I impinging on our solar system [see a recent review by Fahr (1974)]. Similar measurements with helium emission lead to $0.009 < n(\text{He I}) < 0.024$ (Weller & Meier 1974); the helium abundance ratio in our galaxy suggests n(H I) should be roughly 10 times n(He I).

Early comparisons of 21-cm emission with optical obscuration (or reddening) demonstrated that interstellar dust was associated with H I [see references cited in Jenkins (1970)]. However, differences in sampling geometry of the radio beams and optical paths, coupled with possible self-absorption of the 21-cm radiation by cold gas, made the correlations somewhat indistinct. Occasionally, inverse correlations were discovered for more dense regions, an effect attributed to either low temperatures or the formation of molecular hydrogen.

The Lα measurements allow us to compare accumulations of dust, as revealed by $B-V$ color excesses, with temperature-independent measurements of H I in exactly matching volumes of space. By plotting N(H I) vs $E(B-V)$ for the stars in the OAO-2 survey Jenkins & Savage (1974) demonstrated that these two quantities were well correlated, except that many stars with somewhat more than the normal reddening per unit distance $E(B-V)/r$ exhibited, to varying degrees, lower than average $N(\text{H I})/E(B-V)$. The *Copernicus* results for a few of these stars (Spitzer, Drake et al 1973) showed that the conversion of H I to H_2 was an explanation consistent with the observed departures. After eliminating, on the basis of large $E(B-V)/r$, those stars for which $N(H_2)/N(\text{H I})$ was thought to be appreciable, Jenkins and Savage found for the remaining stars

$$\Sigma N(\text{H I})/\Sigma E(B-V) = 6.2 \times 10^{21} \text{ atoms cm}^{-2} \text{ mag}^{-1}. \qquad 1.$$

The observed rms scatter of individual N(H I) values from those computed with this ratio and the observed $E(B-V)$ was $\pm 2 \times 10^{20}$ cm^{-2}. In addition to this result, Jenkins & Savage demonstrated that there was a close correspondence between excesses and deficiencies of both H I and reddening per unit distance, showing that equation 1 was not just a consequence of viewing stars at different distances, but that gas and dust are likely to be physically associated. For hydrogen in all forms Jenkins and Savage found

$$\langle N_H/E(B-V)\rangle = 7.5 \times 10^{21} \text{ atoms cm}^{-2} \text{ mag}^{-1}, \qquad 2.$$

after applying a rough estimate of 20% for the upward correction to the H I column densities for gas within each observed star's Strömgren sphere.

Ryter et al (1975) and Gorenstein (1975) have compared reddening with the attenuation, primarily by interstellar C, N, and O (Brown & Gould 1970), of soft X rays from supernova remnants. Both investigations find the constant in equation 2 to be 6.8×10^{21} atoms cm^{-2} mag^{-1}, assuming that the ratios of heavy elements to hydrogen equal the cosmic values and that the shielding of X-ray absorption for material in dust grains (Fireman 1974) is unimportant. Knapp & Kerr (1974) arrived at 5.1×10^{21} atoms cm^{-2} mag^{-1} for this constant in their study of the 21-cm H I column densities and reddening toward globular clusters.

3 INTERSTELLAR MOLECULES

3.1 H_2 *Molecules*

Systematic measures of the rich spectrum of H_2 in the far ultraviolet, including Lyman vibrational bands from (0,0) up to (5,0) and above, give detailed values of $N(J)$, the column density of H_2 molecules in the rotational level J of the ground vibrational and electronic state. The sum of these gives $N(H_2)$, which may be compared with predictions of theories for H_2 formation and disruption. The rotational excitation evident in Figure 1 may be measured by the ratios among the various $N(J)$ along a particular line of sight. This excitation depends on physical conditions in the clouds and may be used to determine the gas temperature T, the volume density n(H I) of neutral H, and β_0, the probability per unit time that an H_2 molecule at the cloud surface absorbs a photon in any of the Lyman or Werner lines. Evidently β_0 is a measure of the ultraviolet radiation flux incident on a cloud.

Values of $N(H_2)$ obtained from observed column densities [Spitzer et al (1975), York (1975a,b), Drake (in preparation)] are shown in Figure 4, plotted against $E(B-V)$ for each star. For stars with $E(B-V) \leqq 0.03$, $N(1)$ generally much exceeds $N(0)$, and measures of $N(1)$ and $N(3)$ (York 1975a,b) have replaced earlier upper limits on $N(0)$ (Spitzer, Drake et al 1973). The dashed line in Figure 4 is obtained from equation 1 if all hydrogen within the cloud is assumed molecular. The plot shows that in some of the stars, including most of those with $E(B-V) \geqq 0.1$, from 5 to 50% of the hydrogen is in molecular form, whereas in the others the molecular fraction, f, equal to $2N(H_2)/N_H$, where N_H is the total column density of hydrogen nuclei in the cloud, is less than 2×10^{-4} and generally between 10^{-5} and 10^{-6}. No values of f have been observed between 2×10^{-4} and 5×10^{-2}, a gap that appears to be statistically significant.

The enormous variation of f shown in Figure 4 is generally consistent with H_2 dissociation by photons in the Lyman bands; a molecule cascading back from the excited electronic state may enter an unbound vibrational level ($v'' > 14$) of the ground electronic state, a process proposed by Solomon (see Field et al 1966). When clouds are optically thick in these lines, much of the hydrogen can accumulate as H_2, with f approaching unity, whereas in the more transparent clouds f can be extremely small.

To present the quantitative theory of H_2 equilibrium, including this effect, we consider first the rate of H_2 formation on the surface of grains, the most likely formation mechanism in the cool clouds. When H atoms strike a grain, a fraction γ will be assumed to come off in H_2 molecules; at the gas temperature of 80° typical of H I clouds, γ is estimated as about 0.3 (Hollenbach & Salpeter 1971). The rate of H_2 formation per cm^3 may be written as $Rn(\text{H I})n_{\text{H}}$, where $n(\text{H I})$ and n_{H} again represent the particle volume densities of neutral H atoms and of all H nuclei, respectively, and where

$$R = \frac{\gamma n_{gr} \sigma_{gr} \langle v_{\text{H}} \rangle}{2n_{\text{H}}}. \qquad 3.$$

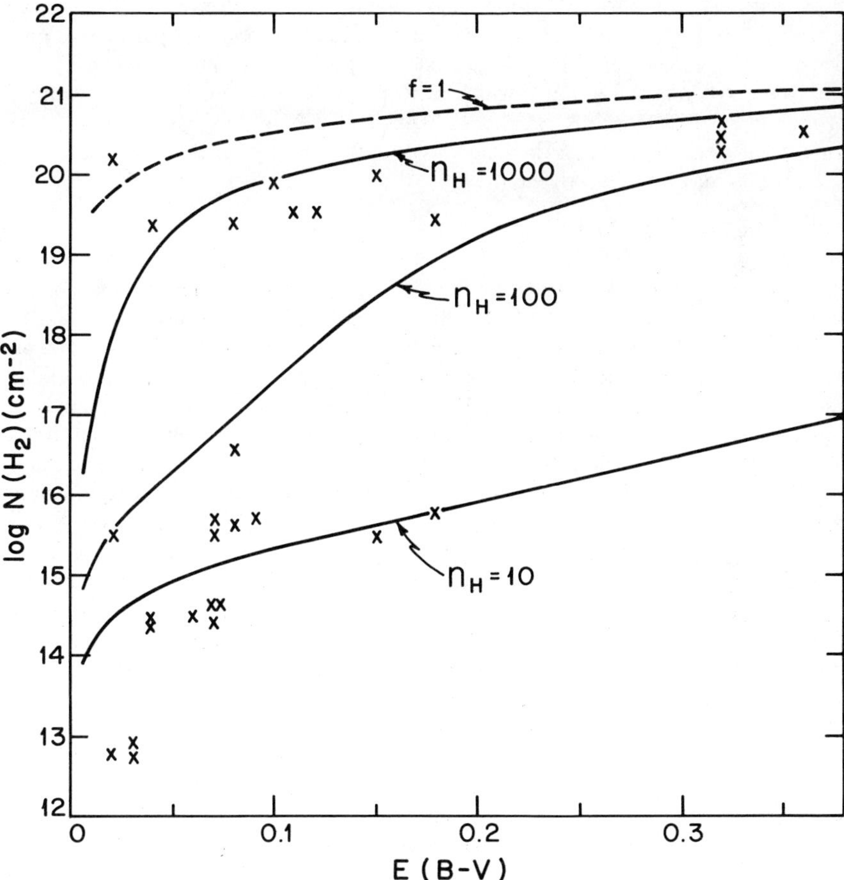

Figure 4 Observed H_2 column densities $N(H_2)$, plotted against color excesses, $E(B-V)$. If N_{H} is strictly proportional to $E(B-V)$ (see equation 1), the dashed curve results if $2N(H_2) = N_{\text{H}}$, whereas the three solid curves result from the theory of Hollenbach et al (1971) applied to clouds with three values of n_{H} (in units of cm^{-3}), the overall particle density of hydrogen.

If $n_{gr}\sigma_{gr}/n_{\rm H}$, the total cross-sectional area of the grains per H nucleus in the gas, is evaluated from measures of selective extinction in the visible (Spitzer 1968), and if $\langle v_{\rm H}\rangle$, the mean random H-atom velocity, is evaluated at 80°, we obtain

$$R = 2.5 \times 10^{-17}\gamma\,{\rm cm}^3\,{\rm sec}^{-1}. \qquad 4.$$

The increase of extinction recently observed in the ultraviolet (Bless & Savage 1972, York et al 1973) indicates that very small grains contribute a large additional cross-sectional area, perhaps tripling the numerical value in equation 4.

The photodissociation rate may be expressed in terms of $\beta(J)$, the probability of an upward electronic transition for a molecule in rotational state J of the ground vibrational and electronic states; $n(J)$ denotes the corresponding particle volume density of H_2 molecules present in this state. Of all the transitions in the Lyman and Werner bands a fraction k will lead to photodissociation. From the detailed probabilities of Dalgarno & Stephens (1970), Jura (1974b) has computed $k = 0.11$ for typical stellar spectra; k does not depend appreciably on J. The equation of equilibrium may now be written

$$Rn({\rm H\ I})n_{\rm H} = k\sum_J \beta(J)n(J). \qquad 5.$$

In the absence of self-shielding by the cloud, $\beta(J)$ equals β_0, nearly independent of J. Summing the known ultraviolet stellar fluxes (Jura 1974b) gives β_0 equal to 5×10^{-10} sec^{-1} near the Sun. Within a cloud, $\beta(J)$ is reduced by absorption in the H_2 lines, especially for low J, as well as by grain extinction. Detailed solutions of equation 5 have been obtained by Hollenbach et al (1971) by assuming constant $n_{\rm H}$ and T within a cloud, and considering radiative transfer in eight typical lines. More detailed computations by Glassgold & Langer (1974), considering all relevant H_2 transitions (60 Lyman lines) and determining $n_{\rm H}$ and T in thermal equilibrium (see Section 6.2) at constant pressure, give accordant results. Typical curves of $N(H_2)$ as a function of $N_{\rm H}$, as measured by $E(B-V)$, are plotted in Figure 4 for three values of $n_{\rm H}$, with R set equal to 10^{-17} cm^3 sec^{-1}, and the effects of shielding computed (J. H. Black and A. Dalgarno, personal communication) for a plane cloud. With increasing density the fraction f of hydrogen in molecular form increases markedly, because the formation rate varies as $n_{\rm H}^2$ (see equation 5). The large variation in f shown in Figure 4 for a fixed $E(B-V)$ may be attributed to variations of $n_{\rm H}$ between clouds; variations of β_0 and R may also contribute.

The three parameters $n_{\rm H}$, β_0, and R may be determined from the measured rotational excitations, together with the observed N(H I). The populations for each J involve collisional excitation and deexcitation, depending on $n_{\rm H}$ and the gas temperature T, as well as on radiative cascading following molecule formation or photon absorption (optical pumping). The analysis of rotational excitation becomes particularly simple for clouds with strong saturated lines from the $J = 0$ and 1 levels; the strong self-shielding drastically reduces $\beta(0)$ and $\beta(1)$, and collisions of molecules with protons (Dalgarno et al 1973) outweigh other processes in determining the relative population in these two levels. Under these conditions $N(0)/N(1)$ equals its Boltzmann value at the kinetic temperature of the gas. Averaging T determined in this way over 13 clouds with $N(H_2)$ exceeding 10^{19} cm^{-2} (Spitzer & Cochran 1973) yields

$$T = 81° \pm 13°K \text{ (av dev)}, \qquad 6.$$

in good agreement with the values obtained from a comparison of 21-cm absorption and emission data.

Consideration of the higher rotational levels as well as the equilibrium between formation and disruption has been carried through in detail by Jura (1975a,b), obtaining R, n_H, and β_0 separately for each cloud; the low gas temperature obtained in equation 6 was used throughout. Essentially β_0 is determined from the ratio of $N(J)$ in the highest rotational levels to the total $N(H_2)$, corrected as necessary for self-shielding; optical pumping excites these levels, which are then depopulated by downward quadrupole radiative transitions. The population of an intermediate level such as $J = 3$, for example, depends not only on $\beta(J)$ but also on n(H I), because collisional deexcitation can increase the downward transition probability significantly above the quadrupole radiative probability $A_{3,1}$, which decreases dramatically with decreasing J (Dalgarno & Wright 1972). Thus a measurement of $N(5)/N(3)$ can yield a determination of n(H I) if β_0 is known. The detailed analysis depends critically on the results for optical pumping obtained by Black & Dalgarno (1975), who computed the necessary probabilities in the downward cascading through various vibrational and rotational states. Because in equilibrium formation balances disruption (see equation 5), the rotational excitation of newly formed molecules is also of importance (see Jura 1975a). Jura's resultant value of β_0 was about equal to the expected value of 5×10^{-10} sec^{-1} in four out of eight clouds with unsaturated H_2 lines; in the remaining four, β_0 was about an order of magnitude greater. The values of n(H I) were between 10 and 30 cm^{-3} in the clouds with the expected β_0, and were generally about an order of magnitude higher in the others. The value of R obtained was about 3×10^{-17} cm^3 sec^{-1}, in general agreement with equation 4 and the subsequent discussion. Analysis of nine clouds showing strong H_2 lines gave about the same distribution of β_0, with n_H mostly between 20 and 200 cm^{-3}, about the same range obtained for the more transparent clouds. These clouds of high β_0 are presumably relatively close to the stars in whose spectra they are seen. Some evolutionary connection between the stars and such circumstellar clouds seems likely, as suggested by Steigman et al (1975).

It is probably significant that the clouds characterized by high β_0 are not only of higher n_H but must also be relatively thin and sheetlike, because their value of N(H I) is in many cases not very large. In the most extreme case, the high-velocity cloud in ζ Orionis with a velocity of -20 km sec^{-1} in the local standard of rest, shows $\beta_0 = 40 \times 10^{-10}$ sec^{-1}, and $n_H = 1000$ cm^{-3} (Jura 1975a). As pointed out by Drake & Pottasch (1975), a high density of either H atoms or electrons is indicated in this cloud by the excitation of the fine-structure levels in C I and Si II, but N(H I) is less than 5×10^{18} cm^{-2} according to the 21-cm emission data. These results suggest that this particular region is a sheet less than 0.002 pc thick!

These determinations of n_H and β_0 from optical pumping theory for a cold gas are plausible, but more cross checks are needed for confirmation. Further consideration of multiple clouds and of inhomogeneities within clouds could modify the results somewhat. Excitation by nonthermal electrons has been proposed by Jura (1975b) to explain some discrepancies between his models and the observations. Moreover, collisional excitation at temperatures of 1000°K or higher (Aannestad & Field 1973, Spitzer & Zweibel 1974) may be important in certain clouds with relatively low $N(H_2)$. It is even possible that in regions at these higher temperatures, where H atoms bounce off grains with very little formation of H_2 (Hollenbach & Salpeter 1970, 1971), the molecules may be produced instead by one of the reactions $H + H^+ \rightarrow H_2^+ + h\nu$ or $H + H^- \rightarrow H_2 + e$, whose rate coefficients have been evaluated by Bates (1951) and by Browne & Dalgarno (1969), respectively. These reactions evidently require rather special ionization conditions that could occur, for example, in the transition layers between H I and H II regions. The large difference in radial velocity shown between the H_2 lines and the lines of neutral atoms in Figure 3, and observed also in τ Sco (Spitzer et al 1975), suggests that the H_2 molecules present in these stars may be formed by one of the above processes rather than on grains, although a separate rather dense H I cloud provides an alternative interpretation.

An acceptable theory of H_2 formation should explain the conspicuous horizontal gap in $N(H_2)$ evident in Figure 4. Perhaps two distinct types of clouds must be assumed; alternatively, an instability of thermal or other origin may be responsible for this gap.

The measured widths of H_2 line profiles (Spitzer & Cochran 1973, Spitzer et al 1975) increase with increasing J in some cases; similar results are indicated by the curves of growth. More detailed profile measures (Spitzer & Morton 1975) have established that this effect is due in each case to a second cloud with a slightly different mean velocity, with low $N(0)$ and $N(1)$ but with strong rotational excitation (presumably caused by a high β_0); the composite profile at low J is little affected by this second component, which becomes conspicuous at high J, broadening the profile markedly.

3.2 *Other Molecules*

Equivalent widths of two HD lines have been measured in several stars (Spitzer et al 1975). The observed presence of these lines can be explained, consistently with the D/H ratio discussed in Section 4.2 below, if a small fraction of the H and D atoms in the cloud are ionized, leading to the exothermic exchange reaction $D^+ + H_2 \rightarrow HD + H^+$. The proton density required is about 10^{-2} cm^{-3} (Black & Dalgarno 1973, Jura 1974b; O'Donnell & Watson 1974).

Toward several reddened stars absorptions by CO have been registered (Jenkins et al 1973). Radio observations of the $J = 0$ to 1 transition of CO at 115 GHz apply chiefly to very dense clouds ($n_H > 10^4$), whereas the ultraviolet measurements sample regions of lower density, with a substantial increase in sensitivity for small column densities $N(CO)$. In spite of some instrumental blending of the band structure, the *Copernicus* scans allow direct measures of rotation temperature T_R, provided $N(J)$ is characterized by a single temperature for all J levels and there is no saturation in any of the line components. Once we obtain a better understanding of CO excitation cross sections and a solution of the equations of rotational equilibrium, we should be able to derive combinations of density and kinetic temperature for the regions containing CO. Jenkins et al (1973) found $T_R = 5°K$ for three stars that had $N(CO)/N_H \approx 3 \times 10^{-8}$, whereas several other stars showed substantially less CO. The largest abundance of CO yet observed in the ultraviolet has been toward ζ Oph (Morton 1975), with $N(CO)/N_H$ somewhere between 8×10^{-7} and 4×10^{-6}. Even this range is substantially lower than that found in dense molecular clouds observed by radio astronomers, where much of the available carbon is in the form of CO (Rank, Townes & Welch 1971).

An intensive search for absorptions by a Rydberg transition of OH near 1220 Å, identified by Douglas (1974), has given upper limits of about 5 mÅ for several stars with $E(B-V) > 0.1$ [Larach 1973, E. B. Jenkins (unpublished)]. The f value for this transition is unknown, although Douglas notes the lines are much stronger than those near 3080 Å, which can be covered by ground-based observations (Herbig 1968). If the f value for the Rydberg transition turns out to be as large as 0.1 (suggested by Jenkins et al 1973) the conclusion that $N(OH) < 2 \times 10^{12}$ cm^{-2} for these reddened stars may lead to very stringent limits for the proton densities inside clouds (Black & Dalgarno 1973).

Molecules other than H_2, HD, and CO have yet to be observed by *Copernicus*, although Jenkins et al (1973) searched for a variety of possible molecular features. Many spectra, however, show prominent lines that remain unidentified (see Morton 1975). Because finding lists for atomic lines may be reasonably complete, while there are none yet compiled for molecules in the far ultraviolet, it seems likely that these unidentified features are molecular in origin. Correlations of these features with the abundances of H_2 and CO might test this possibility.

4 ABUNDANCES OF ATOMS IN H I REGIONS

4.1 *General Abundances*

It is reasonable to assume that the composition of early-type stars reflects that of the present-day interstellar medium in all its forms—atoms, molecules and dust grains—because these stars have condensed recently out of this material. [For a suggestion to the contrary, however, see a proposal by Talbot & Arnett (1973) that stars form out of selectively metal-rich clouds of gas.] Thus a solar abundance scale[1] (Withbroe 1971), which is nearly indistinguishable from that of the O and B stars (Pagel 1974), may be adopted as a standard with which we may compare ultraviolet observations of various elements in atomic form.

It has long been suspected that, relative to hydrogen, the populations of many heavy atoms in space fall short of the respective cosmic (or solar) ratios. Although faced with moderately large uncertainties in ionization corrections and comparisons with hydrogen, investigators who interpreted visual absorption data concluded that calcium and titanium, and to a lesser extent sodium, exhibited a noticeable amount of depletion (Strömgren 1948, Howard et al 1963, Habing 1969, Hobbs 1971, Wallerstein & Goldsmith 1974). The fact that some heavier elements could be relatively scarce was an attractive possibility to theorists concerned with the heat

[1] Except for D and Li (and possibly Be and B), which are consumed by nuclear burning in the Sun.

balance of the interstellar gas, because it was difficult to propose ways of generating enough heat in H I regions to balance the loss by radiation from a cosmic abundance of carbon (Field et al 1969, Goldsmith et al 1969, Dalgarno & McCray 1972). It is clear that a nonnegligible fraction of the heavier elements must be in the form of dust grains: A rough estimate combining equation 2 with a model for grain extinction (Gilra 1971) suggests that around 0.6% by mass of all interstellar material (excluding He) exists as solid particles (Jenkins & Savage 1974). The same ratio for cosmic abundances is 2%. When we look ahead to the ultraviolet results (see Figure 5), it appears that the depletion of most of the more abundant elements is more severe than the factor $2/(2-0.6)$, a point emphasized recently by Greenberg (1974). Although this discrepancy is not yet conclusive, perhaps a significant amount of matter is in the form of larger particles or aggregates that do not cause appreciable extinction.

The first discussions of element abundances measured with *Copernicus* were presented by Morton et al (1973) and Rogerson, York et al (1973) for the material toward nine stars. For a perspective on the pattern of depletion for various elements, we consider here the more recent analysis by Morton (1974) of the interstellar lines, both visual and ultraviolet, appearing in the spectrum of ζ Oph. This star is situated behind a dense cloud producing a rich variety of atomic and molecular lines (Herbig 1968). The very fact that ζ Oph is one of the few bright stars in the sky to display such a profusion of various lines means, of course, that it represents an atypical sample of interstellar material. Nonetheless, ζ Oph is perhaps the most intensively studied star for absorption lines in the visible and was selected for a complete coverage of the available ultraviolet spectrum by the high-resolution detectors aboard *Copernicus*.

Morton (1974) compared column densities (or their upper limits) of 25 elements to the total hydrogen N_H in H I regions along the line of sight to ζ Oph. His compilation showed that the relative abundances of various elements range anywhere from 2×10^{-4} to nearly 1.0 times the respective cosmic ratios. The pattern of relative depletions from one element to another offers some clues on the nature of the condensation process. For instance, if the accretion on grains occurs at high temperatures ($\sim 10^3$ °K) and high pressures (~ 100 dyne cm^{-2}), such as in the outer layer of a star's atmosphere or in a surrounding nebula, chemical equilibrium is rapidly established between free atoms and condensable compounds. Depending upon the actual values for temperature and pressure, a selective depletion of some elements occurs when the more refractory compounds form, whereas other atoms remain in the gas phase because their compounds are too volatile. An alternate possibility is that the grains grow slowly in the denser interstellar clouds (van de Hulst 1949); this is essentially a nonequilibrium process where we expect no appreciable selectivity according to volatility in the material that condenses out. Instead, atoms should deplete at a rate proportional to their mean speed in the gas, whereas ions deplete at roughly 3.5 times this rate (Mészáros 1972) if the grains are negatively charged (Spitzer 1968), and not at all if the grains have a positive charge (Watson 1972).

Figure 5 displays the information we need to decide on the relative importance

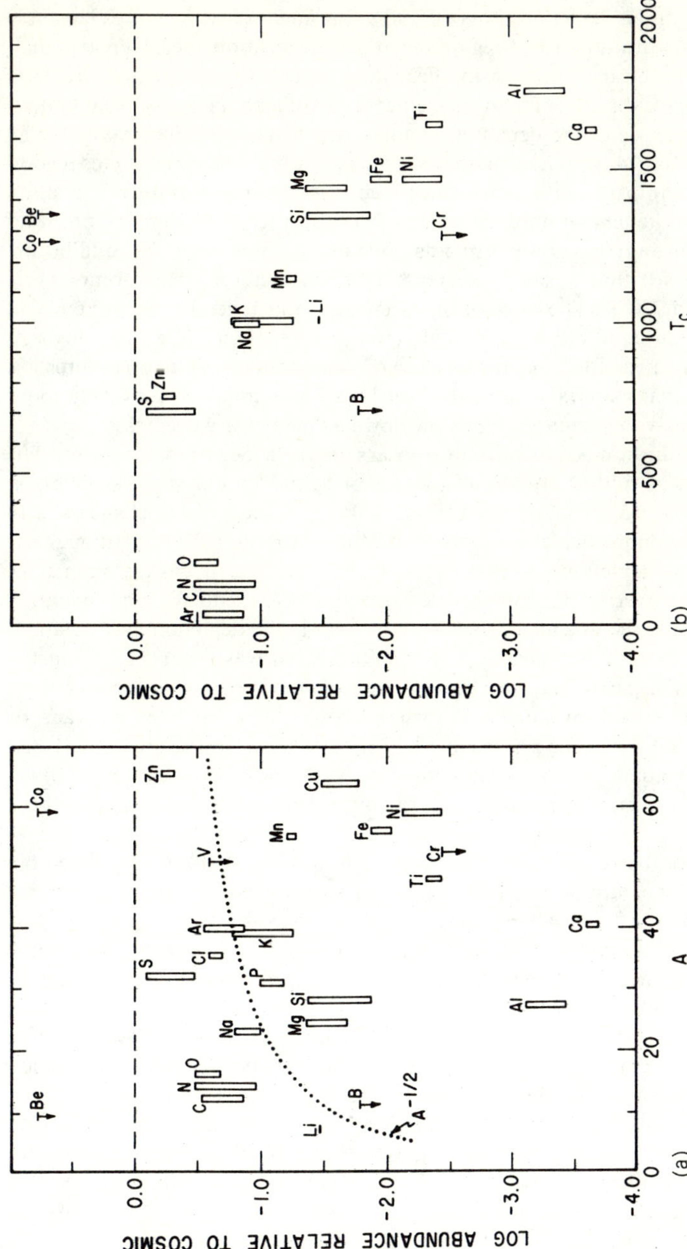

Figure 5 Depletion below solar abundances for elements in their atomic form in the H I gas toward ζ Oph, plotted against atomic mass A in (*a*) and condensation temperature T_c in (*b*). The vertical width of each bar represents the experimental error arising from uncertainties in the respective element's curve of growth. For grains and atoms of a given charge, the nonequilibrium accretion rate in cool interstellar clouds should be proportional to $A^{-1/2}$, illustrated by the dotted line in (*a*). All elements shown here except N, O, and Ar should be predominantly ionized in H I regions.

of the two methods of grain formation. If nonequilibrium processes dominate, we would expect to find a depletion proportional to (atomic mass)$^{-1/2}$ if the element is predominantly neutral in an H I region (i.e. has an ionization energy greater than 13.6 eV), whereas the modification in effective grain cross section by electrostatic attraction or repulsion holds for those elements existing as ions. No clear pattern emerges when one views the depletion factors on a scale of atomic mass. On the other hand, Field (1974a) pointed out that the behavior of depletions exhibited a good regularity when compared with each element's condensation temperature T_c, a quantity defined to be the temperature where significant transfer of an element into solid compounds occurs as a gas gradually cools while maintaining chemical equilibrium. In Figure 5 we see that a steady decrease in relative abundances occurs as T_c increases beyond 700°K. The slope of this decrease is much gentler than one would expect for a gas in equilibrium at a single temperature. Hence one must envision contributions from condensing regions over a distribution of temperatures or, alternatively, that as material expands outward from a region it progressively cools, and simultaneously the reaction rates slow down as the density decreases.

Some nonequilibrium condensation appears to have occurred, however. The depletions of C, N, and O are about a factor of 5, and they do not fit an extrapolation of the trend at higher temperatures. In fact, these three elements with $T_c < 200$°K have more depletion than S or Zn at around 700°K. Although Ar ($T_c = 30$°K) has a depletion similar to that of C, N, and O, its cosmic abundance is somewhat controversial (i.e. compare de Boer et al 1972a with Walker et al 1974).

One element that seems conspicuously out of place in the depletion vs T_c relationship is boron, whose abundance upper limit for ζ Oph was reported by Morton et al (1974). This disparity may be a consequence of an error in the adopted cosmic abundance of B, based on studies of carbonaceous chondrites; recent scans by Boesgaard et al (1974) of the photospheric 1362 Å absorption feature of B II in α Lyr give a B/H abundance roughly consistent with the interstellar upper limit. Alternatively, there may be some refractory compound involving B that has been overlooked.

Under his hypothesis of equilibrium condensation, Field (1974b) has pointed out that for ζ Oph the ratio of high T_c depletions to those at low T_c suggests that less than 2% of the material in our galaxy has escaped processing in the hot, dense regions. Furthermore, he argues that most of the processing must have occurred in dense nebulae rather than in stellar atmospheres. The reasoning here is based on the premise that, in the course of their evolution, stars return only about 10% of the matter initially captured, when one considers the average effect over all masses in a Salpeter mass function. To explain the almost complete processing of gas, much of the gas would have to recycle several times through stars, and the relative fraction of galactic mass locked up in stars would have to be significantly larger than what is now observed. Hence there is reason to favor the hypothesis that significant condensation occurs in nebular regions that can be easily dissipated and returned to the general medium.

A possible alternative to Field's theory is that the behavior of depletions with T_c is not brought about by grain formation but rather by destructive processes. For instance, mutual collisions between grains may heat them enough to cause some evaporation (Oort & van de Hulst 1946), or sputtering may deplete volatile substances more easily than it does nonvolatiles (Salpeter 1974).

When one searches the literature to study variabilities in relative depletions from one star to the next, special care must be taken to ensure that apparent differences are not the result of the following: 1. differing contributions from H II regions, an effect that can alter the n_H comparison standard without appreciably influencing the amount of such ions as C II, Si II, or Mg II (see Section 5.1); 2. inaccuracies in defining the curves of growth for strong lines, particularly applicable to C II and O I; and 3. changes ("improvements") in assumed transition f values. With these considerations in mind, however, some differences and similarities stand out. For instance, York's (1974a,b, 1975a) analyses of the unreddened stars α Virginis, β Centauri, and λ Sco show that the general pattern of abundances reproduces the one seen for ζ Oph (see Figure 5), but the average depletions are less. On the other hand, two stars within the Gum nebula that also have little reddening, γ^2 Velorum (Grewing et al 1973) and ζ Puppis (de Boer & Pottasch 1973), exhibit depletions of Mg and Fe which differ little from those found for ζ Oph. For a number of elements toward the star σ Sco, which has far-ultraviolet extinction properties markedly different from ζ Oph (Bless & Savage 1972), the depletions (E. B. Jenkins, unpublished) are nearly the same as those shown in Figure 5.

4.2 *Deuterium*

Measurements of the atomic deuterium abundances in the interstellar gas have a special significance. As stressed by Schramm & Wagoner (1974), deuterium is formed only in regions where low densities and high particle energies are present. Unlike all elements heavier than boron, deuterium is destroyed rather than manufactured during nuclear burning in stars. Although deuterium has the distinction of tracing a specialized history of astrophysical events, reliable measurements of its general abundance have been elusive: 1. it is not seen in the atmospheres of stars (which is not surprising); 2. terrestrial, meteoritic, and interstellar molecular abundances are anomalously high because of chemical fractionation effects; and 3. interstellar D atoms are only marginally detectable with present-day radio receivers (Pasachoff & Cesarsky 1974).

Rogerson & York (1973) found that in the nearby unreddened star β Cen the absorption by interstellar atomic hydrogen in the higher Lyman lines (Lβ–Lε) was weak enough that the neighboring lines from deuterium were not obliterated. Because the H_2 lines were weak and no HD could be seen, the relative atomic abundance could be considered a reliable indicator, and they derived a D/H number ratio of $1.4 \pm 0.2 \times 10^{-5}$.

This determination could have far-reaching cosmological significance. If one assumes all of the deuterium was produced in the early stages of formation of the universe and accepts the analysis by Wagoner (1973) of light-element production, an abundance measurement permits us to derive a value for the present mass density of the universe, ρ_0. Of all the light elements, deuterium appears to be the best discriminant for ρ_0, because the density ρ during an intermediate phase of the initial expansion governs how thoroughly the deuterium is removed by burning into helium. After correcting the β Cen result upward by a factor of 6 [an admittedly uncertain number from Truran & Cameron (1971)] to allow for the consumption of D by stars during the age of our galaxy, one finds $\rho_0 = 1.5 \times 10^{-31}$ g cm^{-3}, considerably below the closure density of 4.7×10^{-30} g cm^{-3} if $H_0 = 50$ km sec^{-1} Mpc^{-1}. Even without the upward correction, the β Cen abundance value implies $\rho_0 = 4.7 \times 10^{-31}$ g cm^{-3}, so that the universe will continue to expand. More recent measurements by D. G. York and J. B. Rogerson (unpublished) for other stars indicate that abundances are comparable to or slightly greater than the measurement toward β Cen, thus reinforcing the earlier conclusion.

One way to reconcile the large deuterium abundance with a closed universe, without challenging the basic assumptions in Wagoner's (1973) analysis, is to invoke the proposal by Hoyle & Fowler (1973) and Colgate (1973, 1974) that the observed deuterium may originate from spallation reactions within shocks produced by supernova explosions. This suggestion has been controversial: Weaver & Chapline (1974) have concluded that radiation diffusion is a significant dissipative mechanism in a supernova shock wave, with the effect that there is less high-energy material present for the deuterium production assumed by Colgate. In addition, Epstein et al (1974) have shown that with Colgate's shock model there would be a large overproduction of ^{7}Li, ^{9}Be, and ^{11}B, a result these same authors (in preparation) consistently obtain even with a wide variation of shock parameters. The overproduction of these light elements would be limited to only ^{7}Li if the explosion occurred in an object lacking the normal abundance of C, N, and O (i.e. extreme population II).

5 PROPERTIES OF IONIZED GAS

5.1 *Normal H II Regions*

Of all the abundant species that may be observed, the N II ion is perhaps the most straightforward tracer of H II regions. (In speaking of H II regions, we refer to gas with $10^3 \leqq T \leqq 10^4$°K, which is ionized by starlight, as opposed to collisionally ionized, low-density gas with $T \approx 10^6$ °K discussed in Section 5.2.) Because the ionization potential of neutral nitrogen (14.53 V) is only slightly higher than that of hydrogen (13.60 V), the response of N to ionizing radiation beyond the Lyman limit should be very similar to that of H, and the recombination coefficients are not markedly different. Virtually no N II should be found in H I regions, even if cosmic-ray or X-ray ionizing fluxes are large (which is unlikely; see Section 6.2), because the exothermic reaction $N^+ + H_2 \rightarrow NH^+ + H$ and perhaps also $N^+ + H \rightarrow N + H^+$ should accelerate the disappearance of N^+ ions (Steigman 1975c). Since N III and higher stages of ionization are usually far less abundant than N II, we may conclude that the integrated electron density from H II regions penetrated by the line of sight is directly related to N(N II)/ξ(N), where ξ(N) is the elemental abundance of nitrogen relative to hydrogen.

There are some drawbacks, however, in a simple application of N(N II) measurements to a study of H II regions. As a rule, errors in the column densities are large because of the availability of only one line, which is usually very strongly saturated. One must therefore apply a curve of growth derived indirectly from other species (recall from Figure 2 that we may need to select from several possibilities here). In addition, N II measures the average electron density $\langle n_e \rangle$ through the region, whereas we would often prefer to know $\langle n_e^2 \rangle$. For instance, the size of an H II region excited by a star of a given spectral type, as well as its emission measure, are both governed by $\langle n_e^2 \rangle$.

Fortunately, the ground electronic state of N II is split into three fine-structure levels. Absorptions from the two upper levels are not as saturated as that from the lower level. These levels are populated primarily by collisions with electrons (Bahcall & Wolf 1968), and the balancing of these collisions against radiative decay establishes relative populations that, in the limit of low densities, are proportional to $n_e T^{-0.5}$. Deviations from the simple proportionality occur when the occupation of the levels approaches their relative degeneracies, or when the gas is close enough to a star that optical pumping through ultraviolet transitions assists the electrons in populating the upper states. To express such deviations we define the parameter

$$z = \log(n_{\text{excited}}/n_{\text{total}}) - \log n_e T^{-0.5} \qquad (n_e \text{ in cm}^{-3},\ T \text{ in °K}), \qquad 7.$$

which is a sum of two terms: z_c from collisions alone, and a correction term z_p, which allows for the effect of pumping by a star's continuum radiation integrated along the line of sight through the Strömgren sphere (of uniform density). Actual computations of z_p for an O9.5 V and a B1 IV star (E. B. Jenkins, unpublished) show a good fit to the equation

$$z_p = (b - m \log n_e T^{-0.5})^2. \qquad 8.$$

Values of z_c, b, and m are given in Table 1 for the two levels of N II, as well as C II and Si II, which also have fine-structure levels [collision rates and transition f values are from Bahcall & Wolf (1968) and Morton & Smith (1973), respectively]. Except for Si II, with $\log n_e T^{-0.5} > -3.0$, z_p is of order 0.1 or less for an O9.5 V or hotter star, but increases substantially for the cooler star (B1 IV) because the Strömgren radius shrinks dramatically, with only a modest decrease in pumping photons. In reality the true pumping corrections can be significantly smaller if the photons are redistributed in frequency by absorption and reemission in the H II region, or if stellar absorption features from these same ions reduce the fluxes to well below the average continuum values.

Table 1 Values for z_c (as a function of $\log n_e T^{-0.5}$), m, and b, for excited levels found in H II regions[a]

Fine Structure Level	z_c for Indicated Log $n_e T^{-0.5}$					b for Spectral Type		m for Spectral Type	
	$\leqq -2.0$	-1.5	-1.0	-0.5	0.0	O9.5 V	B1 IV	O9.5 V	B1 IV
C II $J = 3/2$	0.41	0.37	0.27	0.07	-0.28	0.03	0.24	0.118	0.185
N II $J = 1$	0.39	0.35	0.27	0.06	-0.32	0.02	0.17	0.103	0.185
N II $J = 2$	-0.67	-0.65	-0.63	-0.62	-0.70	0.13	0.77	0.118	0.150
Si II $J = 3/2$	-0.83	-0.83	-0.83	-0.86	-0.91	0.34	0.42	0.155	0.172

[a] See equations 7 and 8 and accompanying text for explanation.

In most cases practically all of the absorption from the excited levels of N II should come only from the observed star's own H II region because foreground H II regions are likely to be much less dense. If the large-scale distribution of gas around the object is uniform (but with or without small-scale clumpiness), one can derive, from a measurement of the column density of excited material $N(\text{N II*})$, an expected radius r_S of the star's H II region from the expression

$$r_S = [(3/4\pi)N_L \alpha_0^{-1} T^{0.2} \xi(\text{N}) 10^z N(\text{N II*})^{-1}]^{1/2} \qquad 9.$$

where the constant $\alpha_0 = 1.64 \times 10^{-10}$ cm^3 sec^{-1} °K$^{0.7}$ equals the recombination coefficient to the $n = 2$ level of hydrogen, multiplied by $T^{0.7}$. N_L is the total flux of photons beyond the Lyman limit for the star. Except for a difference in the temperature dependence, one finds a similar relation with the emission measure E_m from the H II region, as deduced from the observed intensity of Hα, high-level recombination lines, or radio continuum fluxes. Assuming that one half of E_m arises from material behind the star, we expect

$$r_S = [(3/2\pi)N_L \alpha_0^{-1} T^{0.7} E_m^{-1}]^{1/2}, \qquad 10.$$

with E_m expressed in cm^{-5}.

If we consider $N(\text{N II*})$ toward λ Ori A (Morton et al 1973), ζ Oph (Morton 1975), and δ Sco (E. B. Jenkins, unpublished), and then evaluate r_S using equations 7–9 with $T = 10^4$ °K, log ξ(N) equal to the solar value of -3.94, and N_L values from Panagia (1973), we obtain a very good agreement (to within a factor of 1.5) with the apparent sizes of the H II regions obtained from the Hα photographs of Sivan (1974) and the distances of Lesh (1968). When emission measures are available, the computations of r_S using equation 10 also show a favorable correspondence, indicating ξ(N) is not far from the cosmic abundance ratio in these H II regions. From the data for ξ Per given by Gómez-González & Lequeux (1975), the values for r_S derived using equations 9 and 10 agree well, but the observed size is lower by a factor of 4; a possible explanation is that in some directions dust attenuates some of the Lyman limit flux from the star.

As long as changes in ξ(N) and z are small along the entire line of sight, one may obtain a measure of $\langle n_e^2 T^{-0.5}\rangle/\langle n_e\rangle$ through the use of equations 7 and 8 and a comparison of $N(\text{N II*})$ with the total column density $N(\text{N II})$ for all levels. Various stars show the logarithm of this quantity to vary between -3.0 and -1.4, corresponding to a uniform n_e between 0.1 and 4 cm^{-3}; however, we reemphasize that the amount of unexcited N II is often uncertain. (Comparing only one excited level with the other is usually not productive because one must work with the differences of the levels' z values, which remain fixed near 1.0 over a wide range of realistic densities.) If we believe that all of the N II is associated with the star's Strömgren sphere we can compare the expected column density

$$N(\text{N II}) = \xi(\text{N})[(3/4\pi)N_L \alpha_0^{-1} T^{0.7} r_S^{-1} \langle n_e\rangle^2/\langle n_e^2\rangle]^{1/2} \qquad 11.$$

with the observed value to obtain a crude measure of the clumpiness of the gas $\langle n_e\rangle^2/\langle n_e^2\rangle$ inside. The difference between the two actually gives only an upper limit for this quantity inside the Strömgren sphere, because some of the observed unexcited

N II could actually come from intervening H II regions of low density, which are unrelated to the star. Most stars give results consistent with little or no clumpiness ($\langle n_e \rangle^2 / \langle n_e^2 \rangle = 1$); only ξ Per (for which r_S or N_L is probably untrustworthy for the reasons stated earlier) and λ Sco (York 1975a) appear to give limits below 1.0. The observed N(N II) for λ Ori A, on the other hand, is more than 10 times the value found from equation 11, though the line wings attributed to damping could in fact be due to high-velocity gas.

In Table 1 the two other ions, C II and Si II, can be produced by photons with energies less than 13.6 eV and may reside in either H I or H II regions. We can argue, however, that absorption features produced by Si II* must come only from H II regions. If they came from H I regions, we would expect column densities of the two excited levels of O I to surpass those of Si II* by factors of 45 and 30, figures that are really lower limits because we assume an optimistically large value for n_e/n_H of 5×10^{-4} (Glassgold & Langer 1975) and no depletion of silicon below its cosmic ratio to oxygen. In reality, virtually no O I* has been observed, and hence the Si II* must come from H II regions where most of the oxygen is ionized. We therefore may derive

$$\log N(\text{Si II}) = \log N(\text{Si II*}) + \log N(\text{N II}) - \log N(\text{N II*}) - z_{\text{Si II}} + z_{\text{N II}} \qquad 12.$$

for the H II regions, and deduce abundances of Si relative to N after higher stages of ionization have been included. If we recall from the earlier discussion that the abundance of N in H II regions is approximately cosmic, we find that the Si/N results for the H II regions toward ζ Oph (Morton 1975), δ Sco (E. B. Jenkins, unpublished), and β Cen (D. G. York, unpublished) indicate Si is depleted relative to hydrogen by a factor comparable to the amount found for the H I gas. Recalling our reconstruction of grain formation in Section 4.1, we may suppose that the volatile nitrogen-bearing compounds deposited on the outside of the grains during non-equilibrium accretion are evaporated off in H II regions, leaving behind the more durable silicate cores that formed in the dense nebulae.

Although H II regions should also contain C II*, a roughly comparable contribution to the absorption should come from intervening H I clouds, which effectively prevents our applying the aforementioned analysis to carbon, unless there is a good velocity separation between the H I and H II regions.

For H II regions surrounding stars of different temperatures, Thuan (1975) and Steigman et al (1975) have computed expected column densities for various stages of ionization of several abundant elements. With a few exceptions, their results generally show satisfactory agreement with the observations, after one has allowed for the relative depletions noted earlier, and the contributions from H I regions.

5.2 *High-Temperature, Low-Density Phase of Gas*

In the expectation of finding highly ionized interstellar atoms produced by cosmic rays or X rays in the normal gas, absorptions from such ions as N V, O VI, Si IV, and S IV were scanned in the early study of interstellar lines by *Copernicus*. Except for some weak lines of S IV and Si IV in ξ Oph and γ Arae (Morton et al 1973), no features with the characteristic narrowness of the usual interstellar absorptions could be detected. On the other hand, broad, shallow features of O VI were seen in some stars, and a more extensive follow-up by Jenkins & Meloy (1974) revealed that a majority of the stars on the *Copernicus* survey program exhibited detectable O VI lines with a high-velocity dispersion ($b = 20$ to 50 km sec^{-1}).

The fact that these lines were so broad precluded the interpretation they were formed in the normal H I gas clouds or H II regions, and yet they were substantially narrower than the expected widths of photospheric features for stars with reasonably large projected rotational velocities. If one assumes the lines are broadened purely by thermal Doppler motions of the atoms, the applicable temperatures from 2×10^5 to 3×10^6 °K are consistent with the range where the ratio of O VI to all forms of oxygen $[n(\text{O VI})/n_{\text{O}}]_T$ reaches a maximum in collisional equilibrium. Although the origin of the O VI absorptions might be traced to circumstellar shells existing around hot stars, statistical evidence from radial velocities given by Jenkins & Meloy (1974) seems to favor the interpretation that this coronal gas is situated in the general interstellar space.

Reporting on a rather intensive search for highly ionized ions in the spectra of six stars, York (1974c) could find significant absorptions only for O VI. Because such ions as S IV and N V pass through their maximum population ratios at temperatures below that favored by O VI, York was able to use his upper limits for $N(\text{N V})$ and $N(\text{S IV})$ to derive a characteristic lower limit to the temperature, again assuming collisional equilibrium and cosmic abundances in the gas. This lower limit was on the order of 2×10^5 °K, which is in accord with the prediction by Cox & Smith (1974) that at normal interstellar pressures gas below this temperature cools very rapidly and disappears from the high-temperature region.

From their overall average density $\langle n(\text{O VI})\rangle = 1.7 \times 10^{-8} \text{ cm}^{-3}$ over the various lines of sight, Jenkins & Meloy (1974) calculated a representative pressure for the coronal gas, given by

$$p/k = \langle n(\text{O VI})\rangle [n(\text{O VI})/n_{\text{O}}]_T^{-1} (n_{\text{H}}/n_{\text{O}})_{\text{cosmic}}\, 2.3 f^{-1}\, T \text{ (cm}^{-3}\text{ °K)}, \qquad 13.$$

to range between 1.4×10^2 to $1.6 \times 10^4 \text{ cm}^{-3}$ °K for $2 \times 10^5 \leqq T \leqq 1 \times 10^6$ °K and a volume-filling factor of the hot gas $f = 1.0$. Thus the O VI regions could occupy a substantial fraction of space and be in pressure equilibrium with the normal interstellar gas at $10^3 \lesssim p/k \lesssim 10^4$ °K. The large uncertainty in p/k (and also n_{H}) within the O VI regions results from the rapid change in $[n(\text{O VI})/n_{\text{O}}]_T$ as T is allowed to vary.

Independent evidence for the existence of a hot, low-density phase of the interstellar gas at large distances from the galactic plane has been discussed by Spitzer (1956). The *Copernicus* results suggest this coronal material could be rather evenly distributed in our galaxy, a finding that may be supported by the possibility that the widespread diffuse X-ray background observed at energies around 0.2 keV (Williamson et al 1974) may come from this same component of gas. A plausible origin for the material is the interior volumes of supernova remnants, where gas is predicted to have $n_{\text{H}} \approx 10^{-2} \text{ cm}^{-3}$ and $T \gtrsim 2 \times 10^5$ °K (Chevalier 1974). A discussion of the global effect of contributions from supernovae in our galaxy has been presented by Cox & Smith (1974).

6 IONIZATION AND THERMAL EQUILIBRIUM

Observational information in the far ultraviolet has led to many theoretical analyses of the physical state of the gas, especially the kinetic temperature and ionization. Some of the results obtained are summarized here.

6.1 *Ionization by Energetic Radiation*

A central problem is the extent to which unobserved energetic radiation, including cosmic-ray particles and X-ray photons, produces ionization in the interstellar gas. Cosmic-ray particles with low MeV energies have been proposed (see the surveys by Spitzer 1968 and Dalgarno & McCray 1972) to account for the gas temperatures of roughly 100°K observed in 21-cm studies and for the mean electron density of about 0.03 cm^{-3} deduced from pulsar dispersion measures. Soft X rays with energies of about 100 eV have also been proposed (Silk & Werner 1969) as heating and ionizing agents. The hydrogen ionization rate, ζ_H, associated with these effects depends on the model assumed, but is generally about 10^{-15} sec^{-1}.

An attractive general picture of the interstellar gas, based on such processes, has been developed by Field and his colleagues (Field et al 1969, Field 1973). They point out that a medium heated by low-energy cosmic rays has two thermally stable phases, the conspicuous dense clouds at low temperature and a hot, low-density region ($T = 7500$°K, $n_{\rm H} = 0.24$ cm^{-3}) or intercloud medium (ICM). Ten percent of the hydrogen in this ICM is ionized, consistent with the pulsar data; the corresponding value of $\zeta_{\rm H}$ is 1.2×10^{-15} sec^{-1}. A temperature above 1000°K for the ICM is indicated by comparison of the absorption and emission measures at 21 cm. Other interstellar observations are also consistent with this sweeping picture, which with some modifications can be adapted to ionization and heating by soft X rays. The extensive discussions of such two-phase models of the interstellar gas are summarized by Dalgarno & McCray (1972). However, as shown below, the *Copernicus* observations cast considerable doubt on these theories.

In addition, the assumption that low-energy cosmic rays permeate the galactic plane seems in contradiction with present theories of cosmic-ray propagation (Wentzel 1974). Energetic particles of sufficient particle density are believed to excite Alfvén waves, which then scatter the particles and limit their streaming velocity to about the Alfvén velocity of some 30 km sec^{-1} in the interstellar gas. Because the lifetime of a 2 MeV proton, for example, is only about 5×10^4 years in H gas of density 0.5 cm^{-3}, the distance traveled from the source (presumably a supernova) is only about 2 pc. The particle density of 2 MeV protons required for the assumed $\zeta_{\rm H}$ is more than enough to excite the Alfvén waves (Kulsrud & Cesarsky 1971). It does not seem to have been realized that these theoretical results are inconsistent with uniform ionization and heating by low-energy cosmic rays.

6.2 *Cool H I Clouds*

The evidence from ultraviolet observations on ionization and heating mechanisms is presented first for cool H I clouds, defined as those in which the R(0) lines of

H_2 are saturated. The temperature of such clouds is known to be low, averaging some 80°K (see Section 3), and our knowledge of their state seems somewhat less incomplete than it is for other regions. Very soft X rays cannot be of much importance in clouds with appreciable column densities of H I, because the mean free path of a 100-eV photon corresponds to $N(\text{H I}) = 2.6 \times 10^{19}$ cm^{-2}. However, the range of a 2-MeV proton in hydrogen is $N_H = 1.8 \times 10^{21}$ cm^{-2}, equal to the column density of a cloud for which $E(B-V) = 0.24$ (see equation 2). Although in principle the magnetic lines of force within a cloud could be unconnected with the surrounding field, thus excluding cosmic-ray particles, such a configuration would create difficulties in explaining either the confinement of cosmic rays to the galaxy (Parker 1966) or the intensity of galactic γ rays (Bignami & Fichtel 1974), and seems somewhat unlikely. The propagation problems discussed in the preceding paragraph raise doubts as to whether low-energy cosmic rays can ever reach clouds from a possible source; if these particles somehow manage to permeate the intercloud medium they may penetrate the clouds. Hence the determination of ζ_H in a cool H I cloud gives some information on the likely level of low-energy cosmic rays in interstellar space.

A specific value for ζ_H in clouds containing H_2 can be obtained from the ratios $N(\text{HD})/N(H_2)$ and $N(\text{OH})/N(H_2)$. As shown in Section 3.2, measures of HD give a proton density of about 10^{-2} cm^{-3}, and the corresponding value for ζ_H is between 10^{-17} and 10^{-16} sec^{-1} (O'Donnell & Watson 1974, Jura 1974b), one to two orders of magnitude below the value normally assumed for the nonthermal two-phase models of the interstellar gas, but somewhat greater than the value 7×10^{-18} sec^{-1} produced by the measurable primary cosmic-ray flux (Spitzer 1968). The upper limit on $N(\text{OH})$ obtained by Herbig (1968) for ζ Oph has been used by Black & Dalgarno (1973) to give an upper limit on ζ_H that is again markedly less than 10^{-15} sec^{-1}. Whereas the molecular rate coefficients involved in these calculations of ζ_H are somewhat uncertain, the results strongly suggest that the ionization produced by cosmic rays is significantly less than had previously been believed.

An independent upper limit of 4×10^{-16} sec^{-1} for ζ_H is obtained (Glassgold & Langer 1974) from the large ratio of $N(H_2)/N_H$ observed in ζ Oph, because more intense ionization dissociates H_2 molecules and reduces their relative abundance; with reasonable changes in the uncertain parameters, however, this upper limit could presumably be increased above 10^{-15} sec^{-1}, and hence does not significantly strengthen the evidence against cosmic-ray heating. An upper limit of 3×10^{-16} sec^{-1} for ζ_H is also indicated (Brown 1973) by the absence of H recombination lines from conspicuous H I clouds.

Atoms whose ionization potential is less than that of H can be ionized by starlight in H I regions; the interstellar radiation field is observable at wavelengths longwards of 912 Å and is gradually becoming well known (Witt & Johnson 1973). The ionization equilibria for such atoms have been considered by many workers (see Morton 1975, and references cited). Estimates of the absorption coefficient for ionizing radiation and of the radiative recombination coefficient, as well as of the mean ultraviolet radiation field in interstellar space, make it possible to determine n_e from such observed ratios as $N(\text{C II})/N(\text{C I})$, $N(\text{S II})/N(\text{S I})$, $N(\text{Mg II})/N(\text{Mg I})$,

N(Ca II)/N(Ca I), and N(Fe II)/N(Fe I). The most complete analysis has been for ζ Oph (Morton 1975), where these ratios, except that for Fe, yield $n_e = 0.7$ cm^{-3} within a factor of 2. This analysis gives a lower limit for n_e in the region where the neutral atoms are concentrated, derived on the assumption that all the observed ions are concentrated in this region; the difference in curves of growth shown in Figure 2 indicates that in fact an appreciable fraction of the ions are in quite different regions from these neutrals, with the H_2 molecules in still different regions. If the electrons come from carbon only, assumed to be depleted by a factor of 5 (see Section 4.1), the minimum n_H in the region containing the neutral atoms equals about 10^4 cm^{-3}. The observed N_H limits the thickness of such a high-density region to at most 0.05 pc. The maximum value of ζ_H found above for cool H I clouds would not increase n_e appreciably at these high densities, which exceed by an order of magnitude the density found in the region containing the rotationally excited H_2 molecules (see Section 3) in the line of sight to ζ Oph.

The denser portions of H I regions should be responsible for absorptions from C I, for which populations of two excited fine-structure levels can be compared with that of the ground level. To explain the C I level populations toward ζ Oph, de Boer & Morton (1975) concluded that if $n_H = 10^4$, T must be as low as 20°K. Although this value is at variance with the 56° temperature found from the ratio of $J = 1$ to $J = 0$ populations of H_2, Figure 2 gives evidence that C I and H_2 have somewhat different spatial distributions, as we have already noted.

These conclusions based on ionization equilibrium are altered if the region containing the neutral atoms is assumed to be about 15 pc from ζ Oph, at the boundary of the observed H II region. The large stellar radiation flux then increases the computed n_e by a factor of 20 (Morton 1975) to a value 14 cm^{-3}. The corresponding overall hydrogen density is also increased, unless the H atoms are partially ionized by some process in this boundary layer between the H II zone and the surrounding H I region.

To explain the observed mean temperature of 70–80°K in cool H I clouds (see Section 3) requires a variety of rather extreme assumptions. If the relative abundance of carbon, the chief radiating agent, has its standard cosmic value of about 4×10^{-4} relative to hydrogen, by number, no mechanism seriously proposed will maintain the temperature at 70°K if n_H is as great as 10 cm^{-3}. If carbon is depleted by a factor of 5, temperatures of 60–70°K can be achieved by cosmic-ray heating if $\zeta_H = 1.2 \times 10^{-15}$ sec^{-1} (Field 1973). If this process is excluded by the analysis of the HD and OH measures, such temperatures can be produced (Glassgold & Langer 1974) if two other heating processes are assumed present: 1. photoelectric emission from grains with a quantum efficiency of 10% for ultraviolet light, a mechanism proposed by Watson (1972); and 2. a kinetic energy of about 2 eV imparted to each newly formed H_2 molecule when it leaves its parent grain. Each of these processes is quantitatively uncertain. If, as seems probable in dense clouds, ice mantles form around grains, the photoelectric yield would be expected to be much less than 10%. An appreciable kinetic energy of H_2 molecules at formation was suggested by Spitzer & Cochran (1973) in a preliminary interpretation of H_2 linewidths, a theoretical suggestion now superseded by more accurate line profiles (see Section 3). The other heating processes reviewed by Silk (1973) and the heating produced by the reaction chain involving HCl^+ and Cl (Jura 1974a) are less powerful than these two. It is entirely possible that the chief heating mechanism in H I clouds remains to be discovered.

6.3 *Intercloud Medium* (*ICM*)

Observations of unreddened stars with $E(B-V) \leq 0.03$ are of particular interest, because, in the absence of conspicuous obscuring clouds in the line of sight, absorption lines produced by the ICM might be detectable. As first pointed out by Silk (1970) and Silk & Brown (1971), a large flux of low-energy cosmic rays or soft X rays would be expected to produce appreciable numbers of highly ionized atoms. In particular, for a given ζ_H, X rays should be more effective than charged particles in producing highly charged heavy atoms. The Auger effect, in which a second electron is ejected whenever an electron is removed from the K shell, contributes importantly to the relative number of highly ionized atoms (Weisheit & Dalgarno 1972, Weisheit 1975).

The relative populations of different ionization stages for C, N, and O atoms in H I regions have been computed by Weisheit (1973). In his computations for the ICM, n_H, n_e, and T are set equal to 0.1 cm^{-3}, 0.02 cm^{-3}, and 10^4°K, respectively; the cosmic-ray heating is attributed to 2 MeV protons, whereas the X ray heating is attributed to 0.1 keV X rays, with the intensity of each adjusted to give $\zeta_H = 2 \times 10^{-15}$ sec^{-1}. His results for C II, C IV, N I, and N V are given in Table 2, which includes for comparison the corresponding relative ion abundances if only the observed X-ray flux at energies above 0.3 keV is present; this observed flux, which is also assumed present in the calculations for the other two heating sources shown in the table, fails by several orders of magnitude to produce the ionization of H assumed. Each value in the table gives the fraction of each element in the indicated ionization stage.

The early observations on the bright, well-observed object, λ Sco (Rogerson, York et al 1973), have been extensively compared with theoretical predictions similar to those in Table 1 (Weisheit & Tarter 1973, Mészáros 1973, 1974; Kafatos et al 1974, Hill & Silk 1975, Steigman 1975a). Later, more precise observations (York 1975a) have substantially modified the earlier values of the column densities. In

Table 2 Relative abundances in the intercloud medium with ionization by X rays and cosmic rays

	Fractional Abundance with Heating by:		
Ion	Observed X Rays	X Rays at 0.1 keV	Protons at 2 MeV
C II	1.00	0.31	0.96
C IV	6.8×10^{-4}	0.25	2.0×10^{-3}
N I	0.97	8.0×10^{-3}	0.60
N V	2.6×10^{-7}	0.21	1.4×10^{-5}

particular, better knowledge and control of scattered light in the vicinity of the strong C III line at 977 Å, and an analysis of relatively narrow stellar features, possibly due to an atmospheric shell or to a secondary component, have increased N(C III) in this star by a factor of at least 40. Also, the earlier identification of the C I line at 1657 Å has now been replaced by an upper limit on N(C I) based on lines at shorter wavelength. As a result, some of the detailed conclusions concerning electron density and ζ_H are no longer valid.

To test for the presence of ionization by energetic radiation we may use the upper limit of 6×10^{12} cm^{-2} for N(C IV) obtained from the spectrum of α Vir (York & Kinahan 1975) at a distance of 86 pc. The measured value of N(H I) is 1.0×10^{19} cm^{-2}, consistent with the assumed ICM extending over about 40 % of the line of sight. If carbon has one fifth its cosmic abundance relative to H,[2] then N_C, the total column density of C in all stages of ionization, equals 7×10^{14} cm^{-2}; the observed value of N(C II) is 1.0×10^{16} cm^{-2}, presumably mostly in the H II region. Thus the overall ratio of C IV to C II in the H I gas is at most equal to 9×10^{-3}; this value is consistent with ionization by the observed X rays or by 2-MeV protons, with recombination at the assumed n_e and T, but is inconsistent with heating by 0.1-keV X rays. Reasonable changes in the assumed density or temperature can probably not eliminate this discrepancy. Similar results may be obtained from N V, for which the column density in α Vir associated with the H I gas is less than 2.4×10^{11} cm^{-2}, as compared with 4.6×10^{14} cm^{-2} observed for N I [corresponding to a depletion factor of 2.5 relative to the measured N(H I)].

Charge-exchange processes, if sufficiently rapid, could appreciably reduce the relative population of highly charged ion states and possibly reconcile soft X rays with the observations. According to Steigman (1975b) the rate coefficient k for charge exchange between H and multiply charged ions may be reasonably high, in the range 10^{-13}–10^{-10} cm^{-3} sec^{-1}. The reaction $C^{3+} + H \rightarrow C^{2+} + H^+$ will reduce the relative population of C IV, dividing the ratio N(C IV)/N(C II) in Table 2 by roughly the factor $kn_H/\alpha n_e$. The observed upper limit on N(C IV)/N(C II) in α Vir is about 1% of the high value shown in Table 2 for heating by 0.1-keV X rays. To obtain so marked a reduction with n_e/n_H equal to 0.2 and α for C IV equal to 7.5×10^{-12} cm^3 sec^{-1} at 10^4 °K (Aldrovandi & Pequignot 1973) requires that $k = 1.5 \times 10^{-10}$ cm^3 sec^{-1}, at the upper limit of the possible range. Although these approximate results are not conclusive, especially since charge exchange of H with C III might indirectly reduce N(C IV), the *Copernicus* data seem somewhat unfavorable to the hypothesis that these very soft X rays can ionize the assumed ICM. Ionization by 2-MeV cosmic rays also appears unlikely in view of the interpretation of the HD and OH abundances in H I clouds, described above. Hence the ultraviolet data cast grave doubt on current theories for producing a warm intercloud gas, assumed to account for half the 21-cm emission and most of the electron density deduced from pulsar dispersion measures.

[2] It may be noted that York & Kinahan deduce a solar abundance ratio of C to H from their α Vir data. If this inference is accepted, the ratio N(C IV)/N(C II) is decreased by a factor 1/5, and becomes even more difficult to reconcile with ionization by soft X rays.

6.4 *Alternatives to a Warm, Partly Ionized ICM*

One alternative to the steady state ICM is the time-dependent model (Dalgarno & McCray 1972) in which X rays from frequent supernovae are assumed to ionize and heat the gas, with subsequent cooling followed by recombination. Monte Carlo computations by Gerola et al (1974) give mean values of T, T^{-1}, n_e, $n_e^2 T^{-3/2}$, etc in reasonable agreement with various observations. Computations for highly ionized atoms (Schwarz 1973, Kafatos et al 1974) show that the ionization changes steeply with time and the ultraviolet absorption measures for any single star can readily be fitted; the *Copernicus* data are not sufficiently extensive to permit the statistical comparisons needed to test this theory.

Another alternative is to assume that the free electrons and the warm H I gas, which were found together in the two-phase model of the ICM, usually occur in different regions, with most free electrons in H II zones. As noted earlier, the pulsar dispersion measures yield $\langle n_e \rangle \approx 0.03\ \text{cm}^{-3}$, whereas the diffuse galactic emission gives $\langle n_e^2 \rangle \approx 0.05\ \text{cm}^{-6}$ (Reynolds et al 1973), if $T = 6000°\text{K}$. Comparison between the observed spatial variation of this emission (Reynolds et al 1974) and the distribution of early-type stars shows (Elmergreen 1975) that the Hα observations can apparently be explained by emission from the usual Strömgren spheres around O and B stars with values of n_e ranging from 0.2 up to 2 cm^{-3}, and the pulsar measures can perhaps also be accounted for in this way. Extended H II regions around dying hot stars (Hills 1972), including the central stars of density-bounded planetary nebulae (Terzian 1974), must also be considered, although Mészáros (1973) has shown that with the effective temperature of 150,000°K proposed by Hills (1972) the stellar radiation produces about the same distribution of ions as do the 0.1-keV X rays.

Various possibilities may be listed for the nature of the gas some distance from the O and B stars. Much of the volume between the stars may be filled with the hot coronal gas discussed in Section 5.2 (Cox & Smith 1974). Indeed the simplest picture theoretically is to assume that supernova shock waves have engulfed all low-density regions (J. P. Ostriker, personal communication), filling most of the space with coronal gas, leaving only the dense H I or H II clouds with perhaps some additional H II regions produced in gas ejected from the stars themselves. However, it is also possible that an appreciable fraction of the galactic volume is occupied with low density H I or H II gas at temperatures of 1000–10,000°K.

An extended low-density H II gas could be maintained, perhaps, by ultraviolet photons from O and B stars outside dense H II regions. Torres-Peimbert et al (1974) have shown that the number of photons radiated from such objects, often runaway stars, are sufficient to maintain nearly full ionization in an extended uniform gas of low density. Radiation from the dying hot stars discussed above may also contribute. Intervening clouds of dust and neutral H will tend to absorb ultraviolet photons, however, and it is not yet clear how much of the volume of the galaxy can be kept ionized in this way.

A warm low-density H I medium might conceivably be kept hot by one of the various mechanisms listed by Silk (1973), though little study has been given to this problem. However, there is some question as to whether the observations do in fact require such a medium. The chief evidence for warm H I gas has been the weakness of the 21-cm absorption produced by some clouds that are seen in emission at 21 cm (Hughes et al 1971, Radhakrishnan et al 1972). Additional support for this view is provided by statistical studies of 21-cm emission (Radhakrishnan 1975, Baker & Burton 1975), which show that the observations are best fitted with a two-component model, containing a uniformly distributed medium characterized by a high-velocity dispersion, in addition to the usual cool clouds. The observed distribution of ultraviolet line emission from interstellar neutral H and He passing near the Sun also suggests $T \approx 5000°K$ (Fahr 1974, Weller & Meier 1974) as well as low densities [see Section 2 (but Wallis 1974 has criticized this conclusion)]. Countervailing evidence is provided by Greisen (1973), who finds very small-scale structure in 21-cm absorption, and concludes that the absorption-emission comparison is often misleading and that there is no firm evidence for warm H I gas. In addition, no H_2 molecules should form on grains at temperatures greater than 1000°K, and $N(H_2)$ should be below the observational limit for all warm H I clouds. In fact the R(1) lines of H_2 have been detected in the four unreddened stars within 100 pc for which the error in W_λ has been pushed significantly below 1 mÅ by tenfold scans (York 1975b). Whether or not much warm H I gas exists in the galaxy is unclear.

6.5 *Summary*

Ultraviolet line observations by *Copernicus* offer new insight into the composition and physical properties of both H I clouds and circumstellar H II regions. In particular, for those H I regions with strong H_2 lines the data indicate mean temperatures of about 80°K (confirming earlier 21-cm results), particle densities between 10 and 1000 cm^{-3}, and a depletion of heavy elements that becomes greatly enhanced with increasing condensation temperature. About half these clouds seem relatively close to early-type stars, and many of them seem very thin and presumably sheetlike. The more transparent H I clouds have been less well studied, but their densities and compositions appear comparable to those of the more opaque clouds. For almost all these H I clouds the H_2 content appears too great to be consistent with the temperature exceeding 1000°K indicated by 21-cm data and the studies of interplanetary H and He. Significant ionization of such clouds by radiation other than photons from hot stars seems in conflict with present tentative evidence. Hence pulsar dispersion measures and diffuse $H\alpha$ emission must both require a widespread distribution of H II regions; conventional Strömgren regions around most of the O and B stars may suffice. A high-temperature coronal gas, at a temperature of about a million degrees and a very low density, seems rather conclusively indicated by the observations of wide O VI lines; this gas could result from successive supernova explosions, and may occupy much of interstellar space. Clearly ultraviolet spectroscopy has provided a powerful new tool for studies of the interstellar gas.

ACKNOWLEDGMENTS

We are grateful for helpful suggestions from our Princeton colleagues, especially D. C. Morton and D. G. York, and from A. Dalgarno, G. B. Field, and M. Jura. This work has been supported in part by the National Aeronautics and Space Administration under contract NAS5-1810 with Princeton University.

Literature Cited

Aannestad, P. A., Field, G. B. 1973. *Ap. J. Lett.* 186:L29
Adams, W. S. 1949. *Ap. J.* 109:354
Aldrovandi, S. M. V., Pequignot, D. 1973. *Astron. Ap.* 25:137
Bahcall, J. N., Wolf, R. A. 1968. *Ap. J.* 152:701
Baker, P. L., Burton, W. B. 1975. *Ap. J.* 198:281
Bates, D. R. 1951. *MNRAS* 111:303
Bertaux, J. L., Ammar, A., Blamont, J. E. 1972. *Space Res.* 12:1559
Bignami, G. F., Fichtel, C. E. 1974. *Ap. J. Lett.* 189:L65
Black, J. H., Dalgarno, A. 1973. *Ap. J. Lett.* 184:L101
Black, J. H., Dalgarno, A. 1975. *Ap. J.* In press
Bless, R. C., Savage, B. D. 1972. *Ap. J.* 171:293
Boer, K. S. de, Olthof, H., Pottasch, S. R. 1972a. *Astron. Ap.* 16:417
Boer, K. S. de, Hoekstra, R., van der Hucht, K. A., Kamperman, T. M., Lamers, H. J., Pottasch, S. R. 1972b. *Astron. Ap.* 21:447
Boer, K. S. de, Morton, D. C. 1975. *Astron. Ap.* In press
Boer, K. S. de, Morton, D. C., Pottasch, S. R., York, D. G. 1974. *Astron. Ap.* 31:405
Boer, K. S. de, Pottasch, S. R. 1973. *Astron. Ap.* 28:155
Boesgaard, A. M., Praderie, F., Leckrone, D. S., Faraggiana, R., Hack, M. 1974. *Ap. J. Lett.* 194:L143
Bohlin, R. C. 1973. *Ap. J.* 182:139
Brown, R. H., Gould, R. J. 1970. *Phys. Rev. D* 1:2252
Brown, R. L. 1973. *Ap. J.* 184:693
Browne, J. C., Dalgarno, A. 1969. *J. Phys B* 2:885
Carruthers, G. R. 1970a. *Space Sci. Rev.* 10:459
Carruthers, G. R. 1970b. *Ap. J. Lett.* 161:L81
Chevalier, R. A. 1974. *Ap. J.* 188:501
Colgate, S. A. 1973. *Ap. J. Lett.* 181:L53
Colgate, S. A. 1974. *Ap. J.* 187:321
Cox, D. P., Smith, B. W. 1974. *Ap. J. Lett.* 189:L105
Dalgarno, A., Black, J. H., Weisheit, J. C. 1973. *Ap. Lett.* 14:77
Dalgarno, A., McCray, R. A. 1972. *Ann. Rev. Astron. Ap.* 10:375
Dalgarno, A., Stephens, T. L. 1970. *Ap. J. Lett.* 160:L107
Dalgarno, A., Wright, E. L. 1972. *Ap. J. Lett.* 174:L49
Douglas, A. E. 1974. *Can. J. Phys.* 52:318
Drake, J. F., Pottasch, S. R. 1975. In preparation
Elmergreen, B. S. 1975. *Ap. J. Lett.* 198:L31
Epstein, R. I., Arnett, W. D., Schramm, D. N. 1974. *Ap. J. Lett.* 190:L13
Fahr, H. J. 1970. *Nature* 226:435
Fahr, H. J. 1974. *Space Sci. Rev.* 15:483
Field, G. B. 1973. *Molecules in the Galactic Environment*, ed. M. A. Gordon, L. E. Snyder, 21. New York: Wiley
Field, G. B. 1974a. *Ap. J.* 187:453
Field, G. B. 1974b. *The Dusty Universe*, ed. A. G. W. Cameron, G. B. Field. Washington DC: Smithsonian Inst. Press
Field, G. B., Goldsmith, D. W., Habing, H. J. 1969. *Ap. J. Lett.* 155:L149
Field, G. B., Somerville, W. B., Dressler, K. 1966. *Ann. Rev. Astron. Ap.* 4:207
Fireman, E. L. 1974. *Ap. J.* 187:57
Fischel, D., Stecher, T. P. 1967. *Ap. J. Lett.* 150:L51
Gerola, H., Kafatos, M., McCray, R. 1974. *Ap. J.* 189:55
Gilra, D. P. 1971. *Nature* 229:237
Glassgold, A. E., Langer, W. D. 1974. *Ap. J.* 193:73
Glassgold, A. E., Langer, W. D. 1975. *Ap. J.* Submitted for publication
Goldsmith, D. W., Habing, H. J., Field, G. B. 1969. *Ap. J.* 158:173
Gómez-Gonzáles, J., Lequeux, J. 1975. *Astron. Ap.* 38:29
Gorenstein, P. 1975. *Ap. J.* 198:95
Grayzeck, E. J., Kerr, F. J. 1974. *Astron. J.* 79:368
Greenberg, J. M. 1974. *Ap. J. Lett.* 189:L81
Greisen, E. W. 1973. *Ap. J.* 184:379
Grewing, M., Lamers, H. J., Walmsley, C. M., Wulf-Mathies, C. 1973. *Astron. Ap.* 27:115
Habing, H. J. 1969. *Bull. Astron. Inst. Neth.* 20:177
Heiles, C., Habing, H. J. 1974. *Astron. Ap. Suppl.* 14:1
Herbig, G. H. 1968. *Z. Ap.* 68:243

Hill, J. K., Silk, J. 1975. *Ap. J.* 198:299
Hills, J. B. 1972. *Astron. Ap.* 17:155
Hobbs, L. M. 1969. *Ap. J.* 157:135
Hobbs, L. M. 1971. *Ap. J.* 166:333
Hobbs, L. M. 1973. *Ap. J.* 181:79
Hobbs, L. M. 1974. *Ap. J. Lett.* 188:L67
Hollenbach, D. J., Salpeter, E. E. 1970. *J. Chem. Phys.* 53:79
Hollenbach, D. J., Salpeter, E. E. 1971. *Ap. J.* 163:155
Hollenbach, D. J., Werner, M. W., Salpeter, E. E. 1971. *Ap. J.* 163:165
Howard, W. E. III, Wentzel, D. G., McGee, R. X. 1963. *Ap. J.* 138:988
Hoyle, F., Fowler, W. A. 1973. *Nature* 241: 384
Hughes, M. P., Thompson, A. R., Colvin, R. S. 1971. *Ap. J. Suppl.* 23:323
Hulst, H. C. van de. 1949. *Rech. Astron. Observ. Utrecht* 11:Pt. 2
Jenkins, E. B. 1970. *Ultraviolet Stellar Spectra and Related Ground-Based Observations,* ed. L. Houziaux, H. E. Butler, 281. Dordrecht: Reidel
Jenkins, E. B. et al 1973. *Ap. J. Lett.* 181:122
Jenkins, E. B., Meloy, D. A. 1974. *Ap. J. Lett.* 193:L121
Jenkins, E. B., Savage, B. D. 1974. *Ap. J.* 187:243
Jura, M. 1974a. *Ap. J. Lett.* 190:L33
Jura, M. 1974b. *Ap. J.* 191:375
Jura, M. 1975a. *Ap. J.* 197:575
Jura, M. 1975b. *Ap. J.* 197:581
Kafatos, M., Gerola, H., Hatchett, S., McCray, R. 1974. *Ap. J. Lett.* 187:L113
Knapp, G. R., Kerr, F. J. 1974. *Astron. Ap.* 35:361
Kovach, W. S. 1972. *Ap. J.* 173:287
Kulsrud, R. M., Cesarsky, C. J. 1971. *Ap. Lett.* 8:189
Larach, D. R. 1973. Junior paper. Princeton Univ., Princeton, NJ
Lesh, J. R. 1968. *Ap. J. Suppl.* 17:371
Marschall, L. A., Hobbs, L. M. 1972. *Ap. J.* 173:43
Mészáros, P. 1972. *Ap. J.* 177:79
Mészáros, P. 1973. *Ap. J.* 183:469
Mészáros, P. 1974. *Ap. J.* 191:79
Moos, H. W., Linsky, J. L., Henry R. C., McClintock, W. 1974. *Ap. J. Lett.* 188: L93
Morton, D. C. 1967. *Ap. J.* 147:1017
Morton, D. C. 1974. *Ap. J. Lett.* 193:L35
Morton, D. C. 1975. *Ap. J.* 197:85
Morton, D. C. et al 1973. *Ap. J. Lett.* 181:L103
Morton, D. C., Smith, W. H. 1973. *Ap. J. Suppl.* 26:333, No. 233
Morton, D. C., Smith, A. M., Stecher, T. P. 1974. *Ap. J. Lett.* 189:L109
Morton, D. C., Spitzer, L. 1966. *Ap. J.* 144:1
Nachman, P., Hobbs, L. M. 1973. *Ap. J.* 182:481
O'Donnell, E. J., Watson, W. D. 1974. *Ap. J.* 191:89
Oort, J. H., Hulst, H. C. van de. 1946. *Bull. Astron. Inst. Neth.* 10:187
Pagel, B. E. J. 1974. Presented at NATO Advan. Study Inst. Origin Abundance Chem. Elements, Cambridge, England
Panagia, N. 1973. *Astron. J.* 78:929
Parker, E. N. 1966. *Ap. J.* 145:811
Pasachoff, J. M., Cesarsky, D. A. 1974. *Ap. J.* 193:65
Radhakrishnan, V. 1975. *Galactic Radio Astronomy. IAU Symp. No. 60,* ed. F. J. Kerr. Dordrecht: Reidel
Radhakrishnan, V., Murray, J. D., Lockhart, P., Whittle, R. P. J. 1972. *Ap. J. Suppl.* 24:15, No. 203
Rank, D. M., Townes, C. H., Welch, W. J. 1971. *Science* 174:1083
Reynolds, R. J., Scherb, F., Roesler, F. L. 173. *Ap. J.* 185:869
Reynolds, R. J., Roesler, F. L., Scherb, F. 1974. *Ap. J. Lett.* 192:L53
Rogerson, J. B., Spitzer, L., Drake, J. F., Dressler, K., Jenkins, E. B., Morton, D. C., York, D. G. 1973. *Ap. J. Lett.* 181:L97
Rogerson, J. B., York, D. G. 1973. *Ap. J. Lett.* 186:L95
Rogerson, J. B., York, D. G. et al 1973. *Ap. J. Lett.* 181:L110
Ryter, C., Cesarsky, C. J., Andouze, J. 1975. *Ap. J.* 198:103
Salpeter, E. E. 1974. *H. N. Russell Lecture.* Presented at Meet. Am. Astron. Soc., 144th, Gainesville, Fla.
Savage, B. D., Code, A. D. 1970. *Ultraviolet Stellar Spectra and Related Ground-Based Observations. IAU Symp. No. 36,* ed. L. Houziaux, H. E. Butler, 302. Dordrecht: Reidel
Savage, B. D., Jenkins, E. B. 1972. *Ap. J.* 172:491
Schramm, D. N., Wagoner, R. V. 1974. *Phys. Today,* December 1974, p. 40
Schwarz, J. 1973. *Ap. J.* 182:449
Silk, J. 1970. *Ap. Lett.* 5:283
Silk, J. 1973. *Publ. Astron. Soc. Pac.* 85:704
Silk, J., Brown, R. L. 1971. *Ap. J.* 163: 495
Silk, J., Werner, M. 1969. *Ap. J.* 158:185
Sivan, J. P. 1974. *Astron. Ap. Suppl.* 16:163
Smith, A. M. 1973. *Ap. J. Lett.* 179:L11
Spitzer, L. 1956. *Ap. J.* 124:20
Spitzer, L. 1968. *Diffuse Matter in Space.* New York: Wiley Interscience
Spitzer, L., Cochran, W. D. 1973. *Ap. J. Lett.* 186:L23
Spitzer, L., Cochran, W. D., Hirshfeld, A.

1975. *Ap. J. Suppl.* 28:373, No. 266
Spitzer, L., Drake, J. F. et al 1973. *Ap. J. Lett.* 181:L116
Spitzer, L., Morton, W. A. 1975. In preparation
Spitzer, L., Zweibel, E. G. 1974. *Ap. J. Lett.* 191:L127
Steigman, G. 1975a. *Ap. J. Lett.* 195:L39
Steigman, G. 1975b. *Ap. J.* 199: In press
Steigman, G. 1975c. Preprint
Steigman, G., Strittmatter, P. A., Williams, R. E. 1975. *Ap. J.* 198: In press
Strömgren, B. 1948. *Ap. J.* 108:242
Talbot, R. J., Arnett, W. D. 1973. *Ap. J.* 186:69
Terzian, Y. 1974. *Ap. J.* 193:93
Thuan, T. X. 1975. *Ap. J.* 198:307
Torres-Peimbert, S., Lazcano-Araujo, A., Peimbert, M. 1974. *Ap. J.* 191:401
Truran, J. W., Cameron, A. G. W. 1971. *Ap. Space Sci.* 14:179
Wagoner, R. V. 1973. *Ap. J.* 179:343
Walker, A. B. C., Rugge, H. R, Weiss, K. 1974. *Ap. J.* 188:423
Wallerstein, G., Goldsmith, D. 1974. *Ap. J.* 187:237
Wallis, M. K. 1974. *MNRAS* 167:103
Watson, W. D. 1972. *Ap. J.* 176:103, 271
Weaver, T. A., Chapline, G. F. 1974. *Ap. J. Lett.* 192:L57
Weisheit, J. C. 1973. *Ap. J.* 185:877
Weisheit, J. C. 1975. *Ap. J.* 190:735
Weisheit, J. C., Dalgarno, A. 1972. *Ap. Lett.* 12:103
Weisheit, J. C., Tarter, C. B. 1973. *Ap. J. Lett.* 186:L33
Weller, C. S., Meier, R. R. 1974. *Ap. J.* 193:471
Wentzel, D. G. 1974. *Ann. Rev. Astron. Ap.* 12:71
Williamson, F. O., Sanders, W. T., Kraushaar, W. L., McCammon, D., Borken, R., Bunner, A. N. 1974. *Ap. J. Lett.* 193:L133
Withbroe, G. D. 1971. *The Menzel Symposium. NBS Spec. Publ. 353,* ed. K. B. Gebbie, 127. Washington DC: GPO
Witt, A. N., Johnson, M. W. 1973. *Ap. J.* 181:363
York, D. G. 1974a. *Bull. Am. Astron. Soc.* 6:225
York, D. G. 1974b. *Proc. Congr. I.A.F., 24th, Baku, USSR*
York, D. G. 1974c. *Ap. J. Lett.* 193:L127
York, D. G. 1975a. *Ap. J. Lett.* 196:L103
York, D. G. 1975b. *Ap. J.* Submitted for publication
York, D. G., Kinahan, B. F. 1975. In preparation
York, D. G. et al 1973. *Ap. J. Lett.* 182:L1

ON-LINE COMPUTERS FOR TELESCOPE CONTROL AND DATA HANDLING[1]

Lloyd B. Robinson
Board of Studies in Astronomy and Astrophysics, Lick Observatory,
University of California, Santa Cruz, California 95064

INTRODUCTION

The use of small computers in almost every field of science and technology has been expanding rapidly for more than a decade. As applications and demand have increased, the technology has improved and prices have fallen dramatically, leading to even faster growth and more new applications. The applications are so varied and new developments are so numerous that formal up-to-date documentation is rare; new techniques usually are disseminated at conferences, by word of mouth, and in trade journal advertisements. Formal papers in the scientific journals often mention the use of on-line computer equipment, but rarely describe it.

The phenomenal speed of development of computer hardware, software, and methodology contributes to the difficulty experienced by those involved in decision-making processes regarding data acquisition and control systems. The enormous range of possibilities offered by the programmability of the computer almost has turned the design of computer-aided control and data-acquisition systems into an art form, reflecting the personality and experience of the designer, as well as the limitations of economics and technology. The use of computers for on-line control and data acquisition adds complexity to the problem; computer functions must not only be carried out correctly but also be synchronized to operations outside the computer. Finally, the necessity that the finished system be unobtrusive, easy to operate, and even aesthetically pleasing adds to the intricacy of the design.

This article is written with the hope that it will be interesting and helpful to some of the astronomers who will specify, design, and use on-line digital computing equipment for ground-based data acquisition, data reduction, and telescope control. Because only on-line applications are considered, the characteristics of small "mini" computing systems are emphasized.

[1] *Contributions from the Lick Observatory,* No. 404.

HISTORICAL SUMMARY

The use in science of on-line digital computing techniques was pioneered in the 1950s by nuclear chemists using pulse-height analyzers. These early machines had from 100- to 1000-word ferrite magnetic core memories, and an arithmetic unit that could add unity to the content of any word. They provided a cathode-ray tube display, graph plotter, typewriter, punched paper tape for output, and a switch panel control.

The ferrite-core memory was the most expensive part of these machines, and it was argued that greater arithmetic capabilities should be built into the pulse-height analyzers to utilize the expensive memory. By 1962, however, the cost of a small digital stored-program computer was similar to that of some pulse-height analyzers. Computer manufacturers offered pulse-height analyzing systems, while analyzer manufacturers designed digital computers. The small computers could be programmed to control parts of the data-taking experiments, as well as to process the data (Broude 1964). These small machines were soon used in many fields of science and technology, and competed with slightly less expensive hard-wired special purpose controllers by virtue of superior flexibility and wider range of usefulness (Fulbright 1969).

A decade later, technical development and greatly reduced prices have combined to make the small computer ubiquitous in science and industry. For example, a single model of a small general purpose digital computer in the PDP 8 series is sold at the rate of over 2000 per month. A large number of different manufacturers are offering various types and sizes of such machines for sale. The machines are used not only in the laboratory, but also to control and analyze such diverse operations as medical procedures, industrial processes, typesetting, and city traffic. Central processors can be fabricated on a few semiconductor chips and can sell for only a few hundred dollars.

Observational astronomy has, of course, benefited from the explosive growth of this technology. Telescopes are now in operation that can be controlled almost totally by computer (Hollis 1974); some digital data-acquisition systems are fully automatic; and the on-line reduction of data during an observing run is commonplace. Fortunately, much of the experience gained in nuclear physics was available to those who wished to develop on-line computer-aided systems for astronomy (Strand 1971), and the introduction of computers into the observatory has been accomplished at relatively low cost in both time and money.

TERMINOLOGY

The computer vocabulary is replete with jargon and unfamiliar words. A short list and explanation of some commonly used terms follows.

Central Processor Unit (*CPU*): The arithmetic, logical, and sequencing circuitry; the "brain" of the computer

Minicomputer: A term used to describe small computer systems with typically 4K words of 16-bit internal memory, teletype, and possibly magnetic tape and disk

Microcomputer: All the circuits for a 4- or 8-bit processor can be built on a single semiconductor wafer. Several such "chips," combined with a small memory, often are used to form a small special purpose computer, at very low cost

Register: An array of flip-flop circuits capable of storing one computer word. Arithmetic is generally done on the contents of registers by specially wired links between the registers

Interrupt: An external electrical signal can cause the processor to interrupt a program and activate a special program located at some particular memory location. At the end of the special program, the interrupted program is resumed

Hardware arithmetic: Allows the computer to do multiplication and division in microseconds rather than milliseconds

Internal memory: Fast-access "main" memory. Ferrite magnetic cores allow any "word" to be accessed in a "memory cycle" of about one microsecond. Semiconductor memory devices now cost about the same as ferrite cores (the order of one cent per bit), and can provide access times of small fractions of a microsecond

Semiconductor memory: Fast-access memory using solid-state integrated-circuit technology. This technology appears to be replacing ferrite-core memories because of lower cost and higher speed. Unlike the ferrite core memory, most semiconductor memories lose all data when a power failure occurs

Bit: A single information element, of value zero or one

Byte: An 8-bit segment of information

Word: A number of bits, usually used for one instruction, read from memory in one memory cycle. Typical word sizes range from 12 to 32 bits

K: 1024 words ($2^{10} = 1024$), e.g. a 4K memory has 4096 words

Memory field or page: A memory segment small enough in size to be addressed by a single instruction. Used for computers whose word length is too small to address the whole memory with a single instruction, i.e. a 12-bit word can address a field of not more than 2^{12}, or 4096 words

Disk memory: Rotating magnetic platter about the size of a phonograph record, on which data are written and read by a single "moving-head" from one of perhaps 256 "tracks," or by a number of "fixed-heads." Access time to a selected item varies from 15 msec with fixed heads to a few tenths of a second for a moving head. A single disk may hold from a few thousand to a million computer data words. The cost per bit of disk memory is typically one tenth that of core memory

Floppy disk: A movable-head disk made of flexible material similar to magnetic tape. The disks are removable and cost only a few dollars each. Reliability appears to be very good. Access time is typically a few milliseconds. One disk will hold 10^5 words or so

IBM or *industry-compatible magnetic tape:* $\frac{1}{2}$-inch wide tape on which "blocks" or "records" of data are written, usually with a density of 800 bits per inch. Seven or 9 "tracks" are written in parallel, one track providing a "parity" check on each 6- or 8-bit "character" written. The IBM standard changed several years ago from 7 to 9 track, but 7-track tape is still used on many other computer systems

Parity bit: A bit added to a data word to provide a simple error check. Odd parity means that the number of bits set to "one" in a correct word or character will always be an odd number

DECtape: A miniature, highly reliable magnetic tape, manufactured by Digital Equipment Corporation. Timing and address marks allow absolute identification by its location on the tape of each data item on the tape. High reliability is achieved at the expense of lower data density. Access time is less than 30 sec. One tape holds over 10^5 data words

Cathode-ray tube display (CRT) or *visual display unit* (VDU): Graphic and/or alphanumeric information is often presented on a CRT or TV display. These displays must be continually refreshed, either directly from the computer, or from a dedicated core, disk, or semiconductor memory; or the CRT may be built with internal memory

Light pen: A small penlike device used to mark a location on a CRT display. Seemingly continuous CRT displays are really a series of bright flashes that occur as the electron beam is moved about the tube. The flashes can be detected by a photosensitive diode in the light pen, which can interrupt the computer just as a specific display element is displayed. If the computer has been generating the display, it can relate the time of interrupt to a specific element of information and take appropriate action

Joystick or *cursor:* A computer-generated pointer can be moved about on a display under program control by using sense switches or a joystick, which the computer interrogates. The computer can then relate the position of the pointer to a particular item of displayed data. Programming is easier than it is for a light pen and the results are very similar

Machine language: The numerical code used to program a computer. The code is specific to a particular model of computer

Assembler: A bookkeeping program that converts symbolic terms into numerical code to make up a machine-language computer program. Usually generates one instruction per line of code

Compiler: Generates machine-language code by interpreting high-level symbolic terms (much more powerful than an assembler). Programs written in symbolic languages such as FORTRAN may be compiled on almost any computer, provided a compiler and suitable operating system are available. Usually generates many instructions per line of code

Editor: An interactive computer program that aids a programmer in preparing error-free symbolic text for input to assemblers and compilers

Loader: Loads the machine-language numerical code into the required locations of core memory. The loader may link separate programs at "load-time"

Debug program: An aid to test new programs. It allows registers and memory locations to be interrogated at various points during the operation of a program

Operating system (OS): A program that facilitates interaction between the operator, the computer, and peripheral devices such as printers, tapes, and disks. It may control the loading and running of other programs. It may include assemblers, compilers, and special subroutines needed by languages such as FORTRAN

Real-time operating system (RTOS): A program that facilitates and controls access of one or more machines or users to the computer's resources, but with special provisions for fast response to external demands. The RTOS often will have to save one program on the disk, and load another one into core in response to a request from an external device. Such necessities can lead to unexpected difficulties in high-speed on-line operations. Real-time operating systems require considerable amounts of memory and efficient interrupt to handle hardware, plus complex hardware and software devices to protect parts of the program from being changed by user's programs

High-order language: A language whose symbols are compiled or interpreted by the computer, so that a single symbol can result in a sizeable sequence of machine-language code

COMPUTER HARDWARE

Although minicomputers are built of standardized main components, a bewildering number of configurations are often offered for a single type of small computer. A minimum-sized system will usually consist of a CPU, 4K of internal memory, and a teletype that includes 10 characters per second paper-tape equipment. The next step in expense may include a 60 characters per second paper-tape reader and punch, and a 8K memory.

The inconvenience, wasted time, and unreliability of paper-tape equipment is often avoided by use of cassette magnetic tape, DECtape, or industry-compatible ½-inch tape. The recently developed floppy disk is another good alternative to paper tape.

Multiplication and division customarily have been done in small machines by special subroutines that use only addition, subtraction, and shifting of data. The time required to divide two floating-point numbers by such subroutines may be many milliseconds, but can be reduced to a few microseconds by the use of special "fast-arithmetic" or "floating-point" hardware.

The need for fast response to external stimuli is met by the use of "interrupt" hardware that can switch the program to a specific memory location, in response to an external voltage level.

The desire to make the computer operation more automatic, and to allow the machine to protect the user against various mistakes or possible hardware faults has led to the development of complex "operating," "executive," or "monitor" systems. Internal memory of 16K or more is required to run most operating systems and still leave some space for useful programs; particularly if a high-order language such as FORTRAN is used.

Although the cost of high-speed core and semiconductor memory for computers has fallen by an order of magnitude during the past decade, small on-line computer power is still chiefly limited by memory capacity. Core memory is from 8K to 32K words in most on-line systems. Both magnetic disk memory, providing 10^5 to 10^6 words and access time of a few tens of milliseconds, and magnetic tape, providing 10^7 words with access times of up to a minute or so, are essential for all but the simplest on-line computer systems.

The addition of devices that display data and allow direct communication with the computer makes possible interactive data reduction or direct step-by-step human control of computer aided hardware. Such items as a digital or electrostatic plotter, cathode-ray tube display, joystick, light pen, keyboard, or other interactive devices have been found invaluable by many workers who wish to retain control of their experiment or interact with their data-reduction program at many points in the process.

There is little consensus on the relative value of many peripheral devices or the optimum size or cost of an on-line computer system. The only universal complaint is that the size of the available disk, tape, and internal memory is too small.

COMPUTER SOFTWARE

Computer programs consist of sequences of binary numbers stored in the memory. Each number, usually stored in one word of memory, represents a single elementary operation such as add, store, fetch, etc.

Program preparation is greatly aided by using the bookkeeping abilities of the computer itself to make new programs. The computer is used to convert words in symbolic code into one or more computer instructions. A language defines the way the computer converts symbolic command words, entered by a human operator, into numerical codes that can control the machine. Programming languages are divided roughly into assembly languages, interpretive languages, and compiler languages.

1. *Assembly languages* are unique to each particular computer, usually generating a single word of machine language code for each symbolic instruction. Assembler language uses the basic computer instructions and provides maximum operating efficiency, at the cost of having to program a large number of steps for each simple operation. Different assemblers may provide few or many diagnostic aids, Boolean operations, automatic program linkages, etc. Some assemblers include "macro" capability, allowing a standard instruction sequence to be generated from a single code-word. Programs for very small computers often can be assembled on a larger general purpose computer, but most minicomputers are provided with a stand-alone assembler; logistical problems can be avoided by assembling the program on the computer where it will be used. Assembly-language programming is tedious and expensive, but often necessary for control of special devices or for computers with small memories, especially if very efficient operation is needed. The rules and difficulties of machine-language programming vary greatly from one computer to the next. The quality and convenience of the assembler language and the skill of the programmer may be crucial to successful development of an on-line system, especially in a small computer.
2. *Compiler languages* are exemplified by FORTRAN, in which "high-level" commands are automatically converted to machine-language instructions. The compiler will generate many words of machine-language "object" code for a single FORTRAN instruction. Idiosyncrasies of the hardware and the intricacies of variable storage and multiple-precision or floating-point arithmetic are handled automatically by the compiler program. Special machine-language subroutines often are accessed by a FORTRAN instruction to operate special hardware

devices. Because FORTRAN language instructions are largely independent of the computer type, a program written for one computer system may run on another one with only minor modifications. Interactive and real-time versions of FORTRAN exist which make the language suitable for on-line use. Although versions of FORTRAN have been written that can run on a 4K computer with only a teletype, generally a minimum of 16K memory and a mass storage device are needed for a useful system. Currently, most manufacturers offer a version of FORTRAN IV.

3. *Interpretive languages* are represented by BASIC and FOCAL. Interactive versions of FORTRAN may use an interpreter rather than a compiler, and if so, will run more slowly than ordinary FORTRAN. Simple commands in these languages can initiate and control complex computer operations. Unlike a compiler, an interpreter does not convert the program into machine language before it is executed. Instead, the text is stored and interpreted (or converted to machine code), statement by statement, as the program is executed. Because the original program text is held in the computer, these languages are ideal for *interactive* programming; the text can be modified and used at once with no intermediate compilation or assembly. Because each instruction must be interpreted as it is executed, a program in FOCAL will run much more slowly than an equivalent program in FORTRAN. Some versions of BASIC do a certain amount of preprocessing for faster program execution. Whereas it can take weeks to master assembly language, interpretive languages usually can be mastered after a few hours of practice. Additional commands can be put in these languages by simply inserting additional machine-language subroutines. These languages are easier to use than FORTRAN, but much slower in execution for "number-crunching" types of operations.
4. *Special languages.* Computer systems have been developed using special commands designed especially for the particular job to be done. These systems may be written in assembler language and made to execute selected subroutines when a particular switch is set or when a key on the keyboard is depressed.

   A hierarchical language, FORTH (Moore & Rather 1973), has been used at several computer installations for astronomy, particularly at Kitt Peak and Steward Observatories (Hollis 1974). This language appears to be extremely flexible, combining many of the advantages of both assembly and compiler languages. It allows command words to be defined in terms of a string of other commands, down to the level of single machine-language instructions. This allows the definition of a special language, custom tailored to a specific application. The speed of operation is said to be close to that of an assembler language.

   As a rule, special purpose languages such as FORTH, not widely used in the computer community, create problems of documentation, debugging, and training of programmers and users. Astronomers who have used the FORTH language are enthusiastic about its possibilities, however, particularly the rapidity with which a new program sequence can be developed. The most serious objection is that it is somewhat less convenient than FORTRAN or BASIC for evaluation of mathematical expressions.

There are a number of other more or less standard computer languages not as well known for scientific application as BASIC or FORTRAN. Lasker (1972) reported on the development of a special control language, but suggested that the ALGOL language could have been used efficiently.

UTILITY PROGRAMS

The complexity and multiplicity of possible operations by even a very small computer has led to development of various programs that serve as aids to the user.

1. *Diagnostic programs* can prove that the computer and peripheral hardware are able to operate correctly. Often a hardware failure can be detected and pinpointed by a well-designed diagnostic program.
2. *Debug programs* allow a program to be tested in segments, or may provide the status of registers and variables at selected points in a failing program. Debug programs are especially valuable when developing and testing programs in machine or assembly language.
3. *Assemblers* and *editors* are essential for the bookkeeping needed for assembly-language programming. They also aid greatly in making the changes and corrections invariably required to get an assembler-language program working.
4. *Operating systems* can be simple or sophisticated but are as essential as the hardware for satisfactory operation of a computer. An operating system allows both hardware devices to be controlled by simple commands and interaction with external devices. It sequences operations and reduces the amount of special knowledge needed to operate the computer.

Unfortunately, the quality and usefulness of a particular software package is difficult to judge until it has been used for some time. Wide discrepancies commonly occur between expectation and reality in this area.

A common error made by first-time purchasers of a minicomputer is the assumption that all necessary software can be developed easily and quickly by the users. Software available for some small computers has cost the manufacturer many man-years of effort. Expenditures of this kind are not made needlessly by manufacturers in the competitive cost-conscious computer industry, and the necessity for this kind of effort by either the manufacturer or the user should not be ignored. While some extremely talented programmers are able to create satisfactory software for a specific application with only a few weeks of intensive effort, note should also be taken of small on-line systems that have required several years of programming and debugging to achieve satisfactory performance.

SOFTWARE MODULES

Small, plug-in electronic circuit modules with standardized rules for input and output have been used to great advantage by computer-hardware designers. Design and debugging are facilitated because any standard module can be modified or replaced without affecting other modules.

Ideally, one would like software to be constructed in the same way, so that any segment of a program could be replaced or modified at will. The software designer has a much richer set of possibilities to work with than does the hardware designer, however; the temptation to gain efficiency at the cost of nonstandard program modules or with special linkages between subroutines has often proved overwhelming. The result is sometimes a highly efficient program, but so "cast in concrete" that it cannot be modified or corrected except by the original programmer. The desire to bring software costs more into line with falling hardware costs has led to increased emphasis on standardization and modularization of software for small computers.

DATA ACQUISITION BY ON-LINE COMPUTERS IN ASTRONOMY

The use of a computer to handle data in electronic form at the telescope is fairly commonplace, but not yet universal. Most major telescopes are now equipped with one or more computers that can be used on line for data taking and controlling photoelectric or other electronic detectors. The following brief description of several such systems gives a sample of the current situation.

1. McDonald Observatory (Nather 1971, Wells 1969) has a total of seven NOVA computers, each equipped with a teletype and at least 4K of 16-bit memory. The computers used for on-line work have a CRT display and a data-acquisition interface built by observatory personnel. Two of these on-line minicomputers also are connected directly to an IBM 1800 process-control computer, equipped with three disk memories, a line printer, card-punch equipment, and an incremental plotter. The IBM computer is located in the 107-inch telescope building and runs in a time-shared mode, with a link to a teletype in each telescope dome. The data-acquisition interface of the NOVA computers includes a 50 MHz counter and a number of digital input and output connections, which are used to sense microswitches and drive stepping motors, relays, etc. A display interface can control a CRT monitor. The NOVA computer is controlled from the keyboard, and display parameters are set from the computer sense switches. Programs for the NOVA are written in assemby language, often by an astronomer. A full-time programmer handles the IBM 1800 control programs.

The IBM 1800 also precesses coordinates, and can provide access to star catalogs. It provides flexure and refraction corrections for pointing the 107-inch telescope and can be used to automatically offset the telescope.

2. The Hale Observatories (Dennison 1971a) have installed a "comprehensive" single computer system used for both control and data handling at the telescope, in contrast to the "distributed" multicomputer system described above. They use a Raytheon 703 minicomputer equipped with 16K of core memory, an ASR 33 teletype, a CRT display with a character generator, a 9-track IBM-compatible magnetic-tape transport, and a strip printer. The computer communicates with other equipment by means of a "universal input/output controller." The input/output controller can connect the computer to as many as 250 different devices, and issue up to 256

different commands to each device. Data are transmitted serially, eight bits at a time. Specially designed pushbutton boxes provide only the controls deemed necessary to allow the observer to control a data-collection program. This computer can read telescope, dome, and focus settings and can issue commands to control these telescope functions. The CRT display provides a video output, so that any information available to the computer can be displayed on a simple TV monitor at any remote location. A large number of machine-language subroutines are required to handle special input/output requirements, whereas computation of tracking rates, and other telescope control parameters are done in FORTRAN.

It should be emphasized that this case of a single computer handling both telescope control and data acquisition is made easier by the existence of sophisticated external hardware that can maintain telescope tracking or continue data acquisition for an interval of time while the computer handles another task.

3. An on-line data-taking system of intermediate complexity is illustrated by the Lick Observatory data-acquisition computer (Robinson 1971). Lick Observatory owns three PDP 8/I computers, each equipped with 8K of memory, 32K magnetic disk, dual DECtapes, a 9-level IBM tape transport, a memory CRT, a teletype, a labeled sense-switch panel, a joystick control, and an incremental plotter. One of these machines is installed in the 120-inch telescope building on Mount Hamilton where it is coupled alternately to one of several data-acquisition systems at the Cassegrain and Coudé foci. The other computers are located on the Santa Cruz campus where they control photographic data-reduction equipment (a microphotometer and a measuring engine), and allow on-campus data reduction and program development without traveling to the observing station (Klemola et al 1974). Serial digital multiplexers allow the computers to communicate with equipment via a pair of coaxial cables, at a distance of several hundred feet, with transmit and response times of a few microseconds per computer word. Software for the Lick systems is written in an expanded version of the Digital Equipment Corporation's interpretive FOCAL language. Fifty or so additional commands have been added to the language, so that single commands can be used to read or write magnetic tape, read data from an image-tube-scanner memory, etc. Long FOCAL language programs can be chained together from the magnetic tape, and machine-language subroutines that support the FOCAL commands are overlaid into core memory from the disk memory, so that very long and complex program sequences are used, even with the very small core memory. This system is quite flexible; it is programmed routinely by a number of astronomers and students for data taking and data reduction. The major application of the system has been to control and handle data for a multichannel image-tube scanner (Robinson & Wampler 1972). Programs are usually controlled by switches on the specially labeled sense-switch panel, so that little or no computer knowledge or learning of special keyboard response formats is required of observers who use the system.

The most serious limitation of this system is its small core and disk memory. This has caused programming restrictions and reduced speed of program execution. The initial lack of IBM-compatible tape led to some wasted time, because data-reduction jobs done on the small machine could have been done more economically at the campus computer center. Memory was much more expensive when the equipment was purchased, but at today's prices, a 16K core memory and 256K disk would easily make up their own cost by savings of many man-months of programming time. On the other hand, by having several inexpensive but identical machines, much of the initial development cost of the programming system has been shared among three computers. It has proven very convenient to be able to run identical data-reduction programs on campus or at the telescope, and to prepare and test data-acquisition programs away from the telescope. The ability to do interactive data reduction on these systems has been very valuable.

COMPUTER DATA ACQUISITION: ADVANTAGES

The introduction into the quiet austerity of a telescope dome of an expensive, complex, and noisy electronic computer, with its attendant train of engineers and programmers, has some unattractive aspects. What are the reasons that lead more and more astronomers to accept this potential disruption?

The primary reason is sheer necessity. The human mind cannot cope with the large amount of raw data produced by electronic detection systems. Immediate preliminary data reduction on the computer allows an observer to estimate quickly whether useful information is being obtained, instruments are behaving properly, and the object being observed has been correctly identified. Proper adjustment of complex instruments is greatly facilitated by use of the computer. The ability to change the observing agenda during the night, because of incoming data, has led to significant improvements in the efficiency with which an astronomer can use a large telescope. During extended runs, it is often possible to partially reduce data during the night or next day to allow choice of an optimum observing program.

An additional advantage of computer-aided data acquisition is the ease with which a computer-controlled data-taking system may be modified. An improved version of a working system can be tested by merely loading a new program, while the astronomer retains the option of returning to the old system. With a computer-aided system, individual astronomers who will be sharing the use of any major astronomical instrument can, by programming, tailor the system to individual needs without interfering with other applications.

COMPUTER DATA ACQUISITION: REQUIREMENTS

The hardware for a data-acquisition system can be as simple as a 4K 12-bit computer with teletype, scaler, and clock, or may include multiple disk and magnetic-tape transports, line printers, CRT and other visual display, plotters, keyboards, light pens, or joystick equipment that make it easy for the astronomer to interact with the computer.

As a general rule, programming costs during the useful life of a system will far exceed the cost of hardware for a data-acquisition computer, particularly if on-line data reduction is desired. Such hardware items as extra memory, magnetic disk and tape, line printers, etc, can actually save enough on programming costs to pay for themselves, besides adding to the usefulness of the computer. The ability to operate in a high-order language such as BASIC, FORTRAN, or FORTH can greatly simplify programming; such languages usually require extra core memory and a mass storage device, but reduce programming costs.

Data-acquisition systems may be designed for single purpose instruments, as in the McDonald Observatory system, or for general purpose, as in the Hale Observatory system. There is no clear verdict on which type is most satisfactory; the decision is based usually on the particular application, personal preferences, and on the money and personnel available. The use of a general purpose interface such as CAMAC (described below in the section on computer interfacing) and general purpose real-time operating system programs may make it easier to add new functions, but may increase the initial cost and complexity over that of a small single purpose system by a large factor.

Fulbright (1969) has discussed experiences and requirements for on-line computing systems in nuclear physics. Much of the material in that report has direct application to on-line systems in astronomy. A large number of computer systems used for nuclear physics work were discussed at the Skytop Conference (1969).

TELESCOPE CONTROL BY ON-LINE COMPUTER

The tasks that should be performed by a computer that controls a large telescope have been outlined by Wehner (1971). The telescope must be pointed, with corrections for telescope flexure and atmospheric refraction. The dome and windscreen positions must be synchronized with the telescope position. The telescope tracking speed has to be controlled at the correct sidereal rate, with allowance for guiding at various rates by the observer. Automatic offsetting or chopping between star and sky should be possible. The safety of the telescope must be assured, with due regard for limit conditions, oil-bearing pressure, high humidity, etc. It is vital also to provide not only good communication with the observer or night assistant, but also position readouts, warning of approaching limit conditions, and indications as to the general status of the system. Because large telescopes have been operated satisfactorily for many years with simple analog readouts and oscillator-controlled drive systems, the installation of a complex digital-computer control system should involve a significant improvement in reliability and efficiency of telescope use.

The 88-inch telescope on Mauna Kea in Hawaii provides an example of a major optical telescope operating under computer control. An IBM 1800 computer with 16K words of 16-bit core memory is fitted with a 5×10^5 word disk memory, card reader, typewriter, and keyboard. The computer can read telescope position with 1″ precision by means of Datex absolute encoders. Tracking rates can be continually updated by the computer, to precision of 0.001″/sec. Tracking is controlled by an up-down counter that compares the number of pulses from a digital tachometer connected to the telescope drive (20 pulses per seconds of arc) with the number of pulses demanded by the computer. To avoid errors, the tracking system requires the computer to interact at 10 msec intervals, although an alternative hardware system can be used to control the telescope without the computer. A modified system is being developed that will maintain a computer-initiated tracking rate indefinitely (Wolff 1974). The computer is interrupted 100 times per second by a high-priority clock to handle telescope control tasks. It updates track rates every second (10 times per second if guiding), checks the dome position every second, and corrects the dome position every 100 sec. Dome and windscreen positions are computed exactly from trigonometric equations describing the geometry of the telescope configuration (Harwood 1974).

Display of telescope coordinates is by digital display lamps monitored by a TV camera. Hour angle and declination are displayed directly from the encoders, but the computer must continually calculate and update the right ascension display.

The software for this system uses 10K words of core memory for a core-resident executive program consisting of IBM programs that handle interrupts, perform error analysis, and do other housekeeping functions. The executive also contains subroutines associated with telescope control functions that must be executed immediately; reading in from the disk would be too slow.

The remaining 6K words of memory are available for overlay programs from the disk. Assembling, compiling, and running data-reduction programs are done by reading punched cards that cause specific programs to be loaded from the disk. Keyboard requests from the operator also load special programs into core from the disk.

The computer can be used simultaneously for telescope control and data acquisition. For example, an optical multichannel analyzer is controlled by the computer while the computer is also controlling the telescope.

An interesting example of the problems inherent in such on-line time sharing occurred when it was discovered that data taking was interrupted each time the dome control operated. This was because of time lost while swapping the data-taking and dome-control programs on and off the disk. The problem was cured by making the dome-control program part of the core-resident executive.

CATEGORIES OF ON-LINE COMPUTER SYSTEMS

Computer systems for telescope control and data acquisition can be divided roughly into three classes:

1. Special purpose computer with hard-wired peripherals to carry out well-defined tasks on a permanent basis, for example, a nonshared telescope controller
2. Programmable computer system, with a few instruments connected, but for which a special program, usually in assembly language, is written for each new observational problem
3. Programmable system, with general purpose interface such as CAMAC, and with a real-time operating system or "monitor," with a command structure that allows a large number of observational and data-reduction tasks to be selected with little or no reprogramming

Class 3 can be further subdivided into dedicated and shared machines. In some cases a single computer is used simultaneously to control a telescope and acquire or reduce data. For example, the NRAO 36-foot radio telescope (Hollis 1974) allows data acquisition, telescope tracking, data analysis, and program monitoring with a single PDP 11/40 computer. More commonly, dual computers are used, one for controlling the telescope and one for data handling. The latter alternative provides greater protection for the telescope if users are to have unrestricted use of the computer for data acquisition.

It is also very easy for a time-shared computer to become overloaded, even with simple tasks. For example, at the automated MIT Wallace Observatory, it was found desirable to build a hard-wired dome controller to reduce the burden on the computer (Brookes 1974).

The costs of the three approaches increase from (1) through (3), and proponents for each present strong arguments covering cost, reliability, complexity, flexibility, and efficiency. All three types of system are used successfully; the author has not found that any working system has been replaced by one using a different approach.

COMPUTER INTERFACING: CAMAC

A large part of the cost for installation of an on-line computer lies in the need to connect special devices and instruments to the computer. Because each computer has its own interfacing rules, and each instrument has special requirements, each new system requires a large amount of costly one-of-a-kind electronic and software development.

An interfacing technique called CAMAC (Mack 1972, Van Breda 1972), which has become a standard in Europe and America for nuclear physics instrumentation, has been proposed as a standard for astronomy (Stephens & Van Breda 1972). It has been adopted for use by Kitt Peak National Observatory (Hoag 1972); at the Anglo-Australian 150-inch telescope (Bothwell 1971); at Cerro Tololo (Aikens & Lasker 1973); and at the Isaac Newton telescope (Stephens & Van Breda 1972).

The CAMAC system consists essentially of an instrumentation "crate" with power supply, into which up to 24 modular instruments can be plugged. Power supply and mechanical standards are such that modules built by a manufacturer are guaranteed to fit into crates built by any other manufacturer. A special control module is required to connect the crate to any particular computer, otherwise the system is the same for all computers. The crate supplies a "data highway," by which 24-bit numbers can be transferred between the computer and any module. Expansion to a large number of crates is possible, and a serially linked CAMAC system (where data are transmitted in serial form along a few wires, instead of in parallel form along many wires) also has been defined (CAMAC 1973).

CAMAC seems to be the only possible standard in an industry characterized by deliberate incompatibility between manufacturers. A large amount of effort by the nuclear physics community has gone into its development, and it is now supported on a number of different computers (Costrell 1974). Some computer manufacturers offer CAMAC controllers as an option. Work is underway to develop standard program rules and languages (Hooton & Hagan 1974) so that CAMAC instruments can be controlled by FORTRAN or BASIC programs that can be moved easily from one computer to another. Several instrument manufacturers now sell hardware (scalers, displays, tape controllers, analog-to-digital converters, and bread-board items) that conform to CAMAC standards.

The minimum CAMAC system costs several thousand dollars, and requires considerable special effort for initial program development. As an alternative, some observatories (Dennison 1971b, Nather 1971) have developed their own standard interfacing system. Others (Rodgers et al 1973) have determined that for a particular installation, most of the benefits of CAMAC can be obtained at lower cost using interfacing hardware provided by certain computers. The cost of CAMAC may be prohibitive for many special-purpose on-line computers; its adoption as a standard for astronomy is not yet unanimous.

GENERAL COMMENTS

The list in Table 1 of on-line computer installations at various observatories leaves little doubt that these machines are valuable in astronomy. There are still pitfalls for the unwary, however. Although most owners of on-line systems show considerable pride of ownership, and present cogent arguments for the approach taken, the admission is occasionally made that some difficulties have existed or still exist.

One expected source of trouble has been almost absent. Early installations often planned for a back-up computer so that observing could continue even if a computer failed. In fact, computers have proven to be among the most reliable parts of the telescope. Developments in digital electronics during the past decade have made the conventional picture of cranky unreliable electronic devices quite obsolete. The central processor of a small computer can be expected to run for several years with no maintenance other than routine cleaning of dust filters. Magnetic-tape and disk units may give equally good service, although mechanical printers, typewriters, and plotters will be less reliable.

The problem that is most serious, and rarely solved in a very satisfactory way, is the cost and difficulty of providing adequate, easily used software. The programmable computer *must* be programmed; all the much advertised flexibility of a computer-aided system is only achieved if the *program* structure is flexible. Improvements in programming costs and techniques have not kept pace with the advances in hardware, so that the cost of providing adequate software has in most cases exceeded the cost of the hardware.

The expense of software development for even a small computer system can be incredibly large. Also, the problems caused by peripheral devices that do not always work, and by software that has not been carefully designed and adequately debugged, can easily prevent a system from ever giving satisfactory service. Unfortunately, the world is not always fair to the small and weak, even in the computer world. Only those computers that for several years have sold in large quantities are likely to be well supported by reliable peripherals and good software.

One way to judge the probable support available for a particular species of

Table 1 A list of some of the small on-line computer systems in use at astronomical observatories

Observatory	Computer, Memory	Peripherals	Application	Reference
Algonquin Radio Observatory	SEL 840A 20K 24 bit	Three teletypes, paper tape, two 7-track mag tapes		Bradt et al 1972
Anglo-Australian 150-in telescope	Interdata 70 32K 16 bit	Dual disk, VDU, printer, paper tape	Telescope control	Wehner 1971
Anglo-Australian 150-in telescope	Interdata 70 24K 16 bit	Dual disk, VDU, mag tape, CAMAC, plotter, paper tape	Data acquisition	Bothwell 1971
Cerro Tololo	Nova 1220 64K 16 bit	9-track mag tape, CRT terminal, light pen, CAMAC	Vidicon control	Aikens & Lasker 1973
Cerro Tololo	Nova 16K 16 bit	Paper tape	Pulse counting	Lasker 1971 Lasker 1972
Cerro Tololo	Nova	Dual disk, 2 mag tapes, CAMAC	"Console" computer	Aikens & Lasker 1973
Copenhagen University	PDP 8/I 4K 12 bit	Teletype	Pulse counting	Nielsen 1972
ESO 3.6 m	HP 2100 32K 16 bit	2 disks, mag tape, CRT	Data acquisition	ESO Annual Report 1972
ESO 2.2 m	HP 2100 8K 16 bit	2 disks, mag tape, CRT	Telescope control	Bahner & Solf 1971
Effelsburg 100-m Radio Telescope	Ferrante-Argus 24K	Disk, 7-level mag tape, printer, Calcomp plotter, visual display	Telescope control, data handling	Wielebinski 1971

ESO 50-cm Telescope Copenhagen	Nova 4K 16 bit	Mag tape, punched cards	Telescope control	Nielsen 1971
Fleurs Radio Telescope	PDP 11 16 bit	30-mile land line to telescope	Data handling	Christiansen 1973
Hale Observatory	Raytheon 703 16K 16 bit	Mag tape, strip printer, "universal I/0 controller," CRT	Telescope control, data acquisition	Dennison 1971a,b
Hawaii University	PDP 11/45		Coronal spectrophotometer, SIT vidicon camera	Landman et al 1974
Hamburg 60-cm Refractor	PDP 8S	Teletype	Photoelectric scanning micrometer	Høg 1971
High Altitude Observatory, Boulder	PDP 8L 8K 12 bit	7-level mag tape	Solar eclipse studies	Lee et al 1970
Isaac Newton	Interdata 5			Ring et al 1972
Jodrell Bank	Argus 400 12K 24 bit	CRT, magnetic drum	Telescope control, data processing	Davies 1971
Kitt Peak	16 bit (10 systems)	Mag tape, dual disk CAMAC, I/0 connectors, teletype or CRT terminals	Telescope control, data acquisition	Hoag 1972 Crawford 1972
	Datacraft 6024/3	7-track mag tapes, disk, teletype, printer, wired link to Varian machines	On-line data reduction	Slaughter 1974

Table 1 (*continued*)

Observatory	Computer, Memory	Peripherals	Application	Reference
Lick	PDP 8/I 8K 12 bit (3 identical)	Disk, 9-track mag tape, DECtape, CRT, plotter, joystick, teletype	Image tube scanner, microphotometer, measuring engine	Robinson & Wampler 1972 Robinson 1971 Klemola et al 1974
Lindheimer	PDP 8/S 12 bit	Teletype	Scanning spectrophotometer	Bahng 1971
Lowell	PDP 11 16K PDP 11 12K PDP 11 4K 16 bit	CRT, teletype DECtape, paper tape	Coudé spectrum scanner, data reduction	Albrecht et al 1971 Boyce et al 1973
MacDonald	Nova 4K 16 bit (7 of) IBM 1800 16 bit		Pulse counting, telescope control	Nather 1971 Wells 1969
Mauna Kea (Hawaii)	IBM 1800 12K 16 bit	Disk, typewriter	Telescope control, data acquisition	Wolff 1974 Harwood 1974
Max-Planck Institute for Astronomy (2.2 m)	Honeywell H316 8K 16 bit	Card reader, tape reader, display unit	Telescope control	Bahner & Solf 1971
McMath Solar Telescope	XDS 910 16K 24 bit	Disk	Magnetograph control, photometry	Livingston et al 1971
MIT	Nova 16 bit Datacraft 24 bit		Telescope control, data handling	McCord et al 1972 Brookes 1974

Mt. Stromlo	PDP 11 HP 2100A	VDU, disk, mag tape	Data acquisition	Rodgers et al 1973
New Mexico Tech	IBM 360	Printer, microwave link display, etc	Remote control of supernova detector	Colgate & Moore 1971
NRAO-Tucson 36-ft Telescope	PDP 11/40	Disk, TV screen, card reader, CRT terminal, 9-track mag tape	Data acquisition, data reduction, telescope control	Hollis 1974
Onsala Observatory	Linc 8	CRT, teletype, line-tape	Data acquisition	Winnberg 1971
Parkes Radio Telescope	PDP 9 8K 18 bit	Dual DECtape	Telescope control, data acquisition	
Michigan State	Raytheon 706	Teletype, CRT	Automated photometric telescope	Hill & Linnell 1974
Sacramento Peak Vacuum Solar Telescope	XDS Sigma 2 24K 16 bit	Disk, card reader, plotter, CRT, TV display, data link	Altazimuth telescope control, time shared	Dunn 1971
St. Andrews	H316 8K 16 bit Varian 620F	Fast paper tape, mag tape, CAMAC	Data acquisition, microdensitometer	Stephens & Van Breda (P514) 1972
Stanford Linear Array	HP 2114B 8K 16 bit	CRT, cassette mag tape, phone line to computer center	Data acquisition	Bracewell et al 1973
Washburn	PDP 8 4K 12 bit	Cassette mag tape, teletype		McNall et al 1968, 1972

computer is to scan the advertisements of independent firms supplying add-on memory, magnetic disks and tapes, and other special devices. A computer for which several independent firms are offering peripherals is unlikely to end up as an unwanted orphan.

Although individual observatories have standardized on one or another model of computer, there have been few attempts to standardize on models between observatories and fewer attempts to standardize on software. This is partly on account of the rapid growth in the computer industry, which has tended to make each new machine obsolete almost before it is delivered. The well-known independence and individuality of scientists in general, and astronomers in particular, also may be a contributing factor.

The adoption of two or three computer models as standard, plus some interfacing standard such as CAMAC, and a standard language such as BASIC, FORTRAN, or FORTH would be of value to astronomy. Then new on-line computer systems may be able to ride on the shoulders of their predecessors and not have to be treated as completely new development projects.

Literature Cited

Aikens, R. S., Lasker, B. M. 1973. *Vancouver Symp. Astron. Observ. Telev. Type Sensors,* ed. J. W. Glaspey, G. A. H. Walker, 65
Albrecht, P., Boyce, P., Chastian, J. 1971. *Publ. Astron. Soc. Pac.* 83:683
Bahng, J. A. R. 1971. *Publ. Astron. Soc. Pac.* 83:327
Bahner, K., Solf, J. 1971. *IAU Colloq.* 11:33
Bothwell, G. W. 1971. *ESO/CERN Conf. Large Telesc. Des.,* 437
Boyce, P. B., White, N. M., Albrecht, P., Slettebak, A. 1973. *Publ. Astron. Soc. Pac.* 85:91
Bradt, D. J., Higgs, L. A., Jones, S. G., O'Neill, T. G., Wolfe, J. L. 1972. *Publ. Astron. Soc. Pac.* 84:194
Bracewell, R. N. et al 1973. *Proc. IEEE* 61:1249
Brookes, M. 1974. *Rev. Sci. Instrum.* 46:1372
Broude, C. 1964. *EANDC Conf. Automat. Acquis. Reduct. Nucl. Data, Karlsruhe,* 143
Christiansen, W. N. 1973. *Proc. IEEE* 61:1266
Colgate, S. A., Moore, E. P. 1971. *Bull. Am. Astron. Soc.* 3:450
Costrell, L. 1974. *IEEE Trans. Nucl. Sci.* 1:870
Crawford, D. L. 1972. *IAU Colloq.* 11:14
Davies, J. G. 1971. *IAU Colloq.* 11:42
Dennison, E. W. 1971a. *ESO/CERN Conf. Large Telesc. Des.,* 363
Dennison, E. W. 1971b. *Science* 174:240
Dunn, R. B. 1971. *IAU Colloq.* 11:37
European-American Nuclear Data Committee. 1969. *Proc. Skytop Conf. Comput. Syst. Exp. Nucl. Phys., 1969.* Conf. 6903-01, 121u. Washington DC: EANDC (U.S.)
European Southern Observatory. 1972. *Ann. Rep.*
Fulbright, H. W. 1969. *On-line Data Acquisition Systems in Nuclear Physics.* Washington DC: Nat. Acad. Sci.
Harwood, J. 1974. Personal communication
Hill, S. J., Linnell, A. P. 1974. *Bull. Am. Astron. Soc.* 6:218
Hoag, A. A. 1972. *Conf. Auxiliary Instrum. Large Telesc.,* 39
Høg, E. H. 1971. *Astrophys. Space Sci.* 11:22
Hollis, J. M. 1974. *NRAO Comput. Div. Intern. Rep. 18. 36 Foot Telescope Computer System Manual*
Hooton, I. N., Hagan, P. J. 1974. *IEEE Trans. Nucl. Sci.* 1:903
Klemola, A. R., Robinson, L. B., Vasilevskis, S. 1974. *Publ. Astron. Soc. Pac.* 86:820
Landman, D. A., Orrall, F. Q., Zane, R. 1974. *Bull. Am. Astron. Soc.* 6:290
Lasker, B. M. 1971. *IAU Colloq.* 11:12
Lasker, B. M. 1972. *Publ. Astron. Soc. Pac.* 84:207
Lee, R. H., McQueen, R. M., Mankin, W. G. 1970. *Appl. Opt.* 9:2653
Livingston, W., Harvey, J., Slaughter, C. 1971. *IAU Colloq.* 11:53
Mack, D. A. 1972. *Publ. Astron. Soc. Pac.* 84:167
McCord, T. B., Snellen, G., Paavola, S. 1972. *Publ. Astron. Soc. Pac.* 84:220
McNall, J. F., Michalski, D. E., Miedaner, T. L. 1972. *Publ. Astron. Soc. Pac.*

84:145
McNall, J. F., Miedaner, T. L., Code, A. D. 1968. *Astron. J.* 73:756
Moore, C. H., Rather, E. D. 1973. *Proc. IEEE* 61:1346
Nather, R. E. 1971. *IAU Colloq.* 11:205
Nielsen, R. F. 1971. *ESO/CERN Conf. Large Telesc. Des.*, 415
Nielsen, R. F. 1972. *IAU Colloq.* 11:60
Ring, J., Stephens, C. L., Wayte, R. C. 1972. *Conf. Auxiliary Instrum. Large Telesc.*, 357
Robinson, L. B. 1971. *IAU Colloq.* 11:198
Robinson, L. B., Wampler, E. J. 1972. *Publ. Astron. Soc. Pac.* 84:161
Rodgers, A. W., Roberts, R., Rudge, P. T., Stapinski, T. 1973. *Publ. Astron. Soc. Pac.* 85:268
Slaughter, C. D. 1974. Personal communication
Stephens, C. L., Van Breda, I. G. 1972. *Conf. Auxiliary Instrum. Large Telesc.*, 499
Strand, R. C. 1971. *Conf. Photo-Astron. Tech.*, ed. H. Eichhorn, 15. *NASA CR 1825*
U.S. Atomic Energy Commission. 1973. *CAMAC: Serial System Organ.*, TID 26488. Washington DC: U.S. At. Energy Comm.
Van Breda, I. G. 1972. *Publ. Astron. Soc. Pac.* 84:212
Wehner, H. 1971. *ESO/CERN Conf. Large Telesc., Des.* 9
Wells, D. C. 1969. *Bull. Am. Astron. Soc.* 1:210
Wielebinski, R. 1971. *Naturwissenshaften* 58:109
Winnberg, A. 1971. *Sky Telesc.* 42:274
Wolff, R. 1974. Personal communication

YOUNG STELLAR OBJECTS AND DARK INTERSTELLAR CLOUDS

×2080

S. E. Strom, K. M. Strom, and G. L. Grasdalen
Kitt Peak National Observatory,[1] Tucson, Arizona 85726

Ah, what a dusty answer gets the soul
When hot for certainties in this our life!

George Meredith, *Modern Love*

If, as they say, some dust thrown in my eyes
Will keep my talk from getting overwise,
I'm not the one for putting off the proof.
Let it be overwhelming.

Robert Frost, *Dust in the Eyes*

1 INTRODUCTION

During the last several years, there has been considerable renewed activity directed toward achieving an understanding of the star formation process and the early stages of stellar evolution. This activity has been stimulated primarily by technical advances that have permitted astronomers to probe the dark interstellar clouds, the current birthplaces of young stars in our own galaxy.

In this review we attempt a synthesis of our current observational knowledge concerning pre-main-sequence (PMS) evolution and the interaction of PMS stars with their dark cloud environment.

Our outlook is no doubt prejudiced by our greater familiarity with recent optical and infrared observations, both those circulated in the underground literature as well as our own. Moreover, we have chosen quite deliberately to use this forum to raise questions of current interest and to share with our colleagues thoughts regarding the direction of future fruitful research. To the extent that we have sacrificed the archival for the provocative, we offer our apologies.

2 THE EARLY PHASES OF STELLAR EVOLUTION

The last burst of activity in the study of young stellar objects, nearly two decades ago, was stimulated by the theoretical work of Henyey, LeLevier & Levée (1955)

[1] Operated by the Association of Universities for Research in Astronomy, Inc., under contract with the National Science Foundation.

and others. These authors provided the first estimates of the expected evolutionary tracks for the PMS stars. The tracks suggested that the lower mass stars in young clusters containing main-sequence (MS) O and early B stars would still be in the PMS phase if stars at all masses were formed coevally. Walker (1956, 1957, 1959, 1961, 1969) surveyed a number of young clusters and indeed confirmed the presence of stars located above and to the right of the MS, apparently in the domain filled by the Henyey tracks. Until recently, a considerable amount of work has centered on attempts to make detailed comparisons of observations with computed PMS tracks.

Various investigators have attempted to use the statistical distribution of PMS in the Hertzsprung-Russell (HR) diagram in order to effect such a comparison. Despite the initial encouragement provided by the pioneering work of Hayashi, Hōshi & Sugimoto (1962) and Hayashi (1970), however, such attempts have not been fruitful, primarily because the cluster members of differing mass are apparently formed over a period of time comparable to their PMS lifetimes.

2.1 *Current Theoretical Picture*

It is historically inaccurate to state that current interest in the early phases of stellar evolution originated with more sophisticated theoretical insight; rather, advances in radio and IR techniques have been the driving force.

Nevertheless, contemporaneously with these advances, Hayashi et al (1962), Hayashi (1970), and Larson (1969, 1972) have published several seminal theoretical papers on PMS evolution. These authors have attempted the first hydrodynamical computations following stars of different masses as they collapse from dark cloud material until they reach the MS. These calculations provide an excellent framework for structuring our discussion of the early phases of stellar evolution.

Of necessity, the theoretical computations are greatly simplified. For example, Larson assumes spherical collapse with (*a*) no rotation, and (*b*) no magnetic field; therefore, we at best expect from these computations only a qualitative guide

Table 1 Expected appearance of PMS stars

	$1M_\odot$ T(years)	Qualitative Appearance	$5M_\odot$ T(years)	Qualitative Appearance
1.	10^2	infalling envelope, $T_{env} \sim 60°K$	10^2	infalling envelope, $T_{env} \sim 30°K$
2.	10^3	infalling envelope, $T_{env} \sim 90°K$	10^3	infalling envelope, $T_{env} \sim 60°K$
3.	10^4	infalling envelope, $T_{env} \sim 125°K$	10^4	infalling envelope, $T_{env} \sim 300°K$
4.	10^5	infalling envelope, $T_{env} \sim 300°K$	10^5	infalling envelope, near-rapid luminosity increase, transition from hydrodynamic to equilibrium radiative evolution, $T_{env} \sim 1000°K$
5.	10^6	envelope optically thin	3×10^5	near MS, infalling envelope, $T_{env} \sim 3000°K$

to the major features of PMS evolution. The essential conclusions of Larson's computations that bear directly on observable tests of his picture are:

1. The hydrodynamic phases of evolution constitute a significant fraction of a star's PMS lifetime
2. The stellarlike core of a PMS star is surrounded by an infalling cloud of optically opaque circumstellar (CS) material. One expects to observe stars during these phases only at IR wavelengths; the dust in the CS envelope absorbs short wavelength photons from the stellar core and radiates this absorbed energy in the IR

Figure 1 illustrates the tracks computed by Larson for stars of mass $5M_{\odot}$ and $1M_{\odot}$. We have indicated by tick marks those points along the tracks that are of particular interest. Listed in Table 1 are the time scales appropriate to each track segment and the expected, qualitative appearance of the star. We wish to emphasize a characteristic of the $5M_{\odot}$ track, namely, the rapid rise in luminosity prior to the star's reaching the quasistatic equilibrium phase of evolution (equilibrium radiative or Henyey track in the case of the $5M_{\odot}$ stars, and the final phases of fully convective equilibrium, the Hayashi track, in the case of the $1M_{\odot}$ stars).

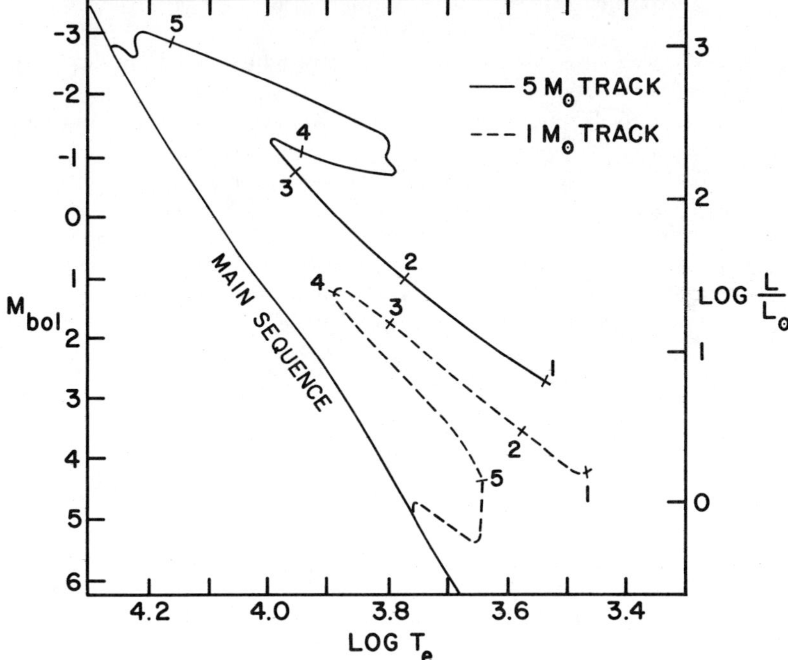

Figure 1 The evolutionary tracks computed by Larson for stars of $1M_{\odot}$ and $5M_{\odot}$. Table 1 lists the qualitative appearance of these PMS stars at each point on the tracks denoted by a tick mark.

3 THE RENEWED SEARCH FOR YOUNG STELLAR OBJECTS

To make comparisons with the major features of Larson's tracks, we must locate a sample of stars with ages considerably less than 10^6 years. Walker's work was directed at clusters with minimum ages of several million years. It is certainly possible that some members of these groups are younger; however, there is no obvious way to select such objects even if they exist. A more profitable approach is to locate "signposts" indicating the presence of very young stellar objects analogous to the O and B stars that guided Walker's surveys.

3.1 *The Herbig Ae and Be Stars*

Herbig (1960) was motivated by similar interests to select a group of stars that he hoped to identify with massive stars still approaching the MS. Because many of the PMS objects associated with the Walker clusters were late-type emission variables (of the T Tauri class), Herbig reasoned that young, massive objects could be selected from lists of early-type emission objects. To exclude evolved stars of the rapidly rotating class, he imposed the further criterion that such stars be associated with reflection nebulosity, an indication that these objects were still closely associated with their presumed birthplaces, the dark clouds. Most of the Ae and Be stars associated with nebulosity (henceforth called Herbig emission stars or HES) were also found to be irregular variables, adding further strength to his identification by analogy with the T Tauri stars. Unfortunately, Herbig's spectroscopic study did not provide convincing confirmation of the PMS character of these objects.

More recently, Strom et al (1972) carried out a spectrophotometric study of these objects and succeeded in locating the HES in the HR diagram. Their results provided quantitative confirmation of Herbig's intuition: the HES as a class appear to be approaching the MS along equilibrium radiative tracks; if so, they have ages of between 10^5 and 10^6 years. We note that these ages are based on equilibrium evolution time scales and may not necessarily provide an accurate estimate of the time since the "initial" collapse (see Appenzeller & Tscharnuter 1974).

The next logical step is to survey these regions to delineate the nature of the associated stellar population. The prejudice implicit in the approach is that this population will have an age similar to that of the HES. There is always the danger (as is the case with the Walker clusters) that star formation takes place continuously or in bursts over long periods of time within the same cloud complex (Strom et al 1975).

Herbig (1960) had already reported the presence of Hα-emission objects, some of which can be identified as T Tauri stars (see Herbig & Rao 1972). Furthermore, a survey of the Palomar Observatory Sky Survey (POSS) plates by Strom, Grasdalen & Strom (1974c) revealed the presence of several Herbig-Haro (H-H) objects (Herbig 1951, Haro 1952) in the vicinity of the HES; a more complete list can be found in Herbig (1974). However, it may be that a significant fraction of the stellar population is still embedded within the dark cloud material; the younger the stellar group is the more likely is this to occur. Searches for these embedded stars must be undertaken at IR wavelengths to minimize the effects of dark cloud and CS extinction.

Strom et al (1972) attempted such searches at a wavelength of 1 μ using an S-1 image intensifier. The advantage of searching at 1 μ is quite significant because $\tau(1\ \mu) \sim 0.5\tau(0.5\ \mu)$. They discovered a large number of very red stellar sources, most of which appear to be ordinary stars embedded in the dark cloud material. Unfortunately, only a limited number of regions have been explored in further detail and the nature of these sources has not been thoroughly examined. Additional searches, complemented by IR photometry and near-IR spectroscopy, would be of considerable interest.

That visible region impressions of the dark cloud populations are deceptive is even more dramatically illustrated by the 2-μ maps published by Grasdalen, Strom & Strom (1973), Strom et al (1975), and Vrba et al (1975). These authors report the presence of dense young clusters hidden within the dark cloud material.

3.2 *Young Stellar Objects Near HII Regions*

Other searches for young stellar objects have been carried out implicitly in the course of mapping HII regions in the IR (Wynn-Williams, Becklin & Neugebauer 1972, Frogel & Persson 1974). In retrospect, the presence of the HII region appears to have been an indirect signpost indicating an active region of star formation, that is, the HII region indicates that a massive molecular cloud has, in one region, given birth to massive stars. Star formation is no longer active in the ionized region, however, but rather in the associated molecular cloud, if at all. The sources discovered in the IR surveys of HII regions seem to be of three classes: 1. those associated with small, high-emission measure "compact" HII regions; 2. those embedded within massive molecular clouds associated with the HII region; and 3. those associated with OH and/or H_2O maser sources. The sources of class 1 appear to be massive stars that have just reached the MS with a likely characteristic age of several times 10^4 years; the compact ionized regions probably have just begun to expand. Sources of class 2 may be similar to those found in dark dust clouds such as Lynds 1630 (Strom et al 1975): representatives of the upper MS of clusters embedded within the clouds. Class 3 sources may represent (*a*) high-density concentrations destined to become massive stars or stellar clusters, or (*b*) concentrations of gas and dust surrounding already formed objects; these might best be termed dense placental clouds. The fact that H_2O and OH maser sources, compact HII regions, and IR sources in some cases are coincident render suggestion *b* almost certainly confirmed (Morris et al 1974a).

3.3 *T-associations*

T Tau stars associated with dark cloud complexes [Ambartsumian's (1955) T-associations] may provide another signpost of recent star-forming activity. It is not clear, however, whether such groups represent (*a*) the precursors of OB associations having representatives at all stellar masses, or (*b*) a heterogeneous set of clusters, some of which may be relatively old and destined never to contain massive stars. It may be possible (see Section 4.5) to define a statistical "age" for a T-association from the observed IR and visual properties of the stars. Nevertheless, we believe that current evidence best supports the view that T-associations have a wide range of ages and do not provide unique signposts for young stellar objects.

3.4 *Radio Molecular Searches*

Surveys of dark clouds in various radio molecular transitions offer great promise in isolating active star-forming regions. Because recently formed stars are likely to be surrounded by dense placental material, searches of dark clouds in lines enhanced in high-density regions provide an excellent guide for locating young stellar objects. Recent work by Morris et al (1974a) and Loren et al (1973) has begun to suggest the power of this procedure. The major emphasis to date has been placed on study of molecular clouds associated with HII regions, however, rather than on the dense cloud complexes associated with other signposts of recent star formation. Moreover, work in this area has been directed toward studies of overall molecular cloud properties rather than purposeful searches for placental clouds. This situation has been considerably improved during the last few years as combined optical, IR, and molecular studies have converged to provide a more coherent framework for this research.

In Table 2 we list the signposts characteristic of regions that differ in "age." By "age" we mean to convey not a single number appropriate to a given region, but rather the time elapsed since the *last* discernible star-forming event; the same region may contain representatives from previous star-forming events. Also listed in the table is the range of objects found in close proximity to the signpost.

In attempting a comparison with the current theoretical picture of PMS evolution, we are unfortunately forced at present to depend primarily on those objects for which optical spectroscopy provides some guidelines concerning their probable location in the HR diagram. Consequently, the discussion which follows centers on a detailed analysis of objects in categories II and III in Table 2: 1. the Ae and Be stars, 2. the H-H objects, 3. the T Tauri stars, and 4. the non-emission, nonvariable PMS stars. For each class of object, we have summarized available information concerning their evolutionary state and the characteristics of their CS envelopes. A synthesis of these summaries leads to the evolutionary "scenarios" presented in Section 7.

4 THE HERBIG Ae AND Be STARS ASSOCIATED WITH NEBULOSITY

A discussion of the optical and IR properties of these objects can be found in Strom et al (1972). As a class, they share the following characteristics:

1. Optical emission spectra ranging in complexity from those with only Hα through the rich emission spectrum of V380 Ori (Herbig 1960). In many

Table 2 A pocket guide to stellar ages

	Signpost	Age	Radio Appearance	Associated Objects
1.	Strong H_2O and OH masers, compact radio continuum source	10^4 to 10^5 years	High excitation molecules such as CS, HCN; $n \gg 10^4$ cm^{-3}	Infrared source associated with radio source; embedded clusters of point sources; often found near optical HII regions that contain OB stars and members of the Orion population. The latter probably represents another, older region of the massive molecular cloud
2.	Herbig Ae and Be stars	$\sim 10^5$ to 10^6 years	A mix of high and low excitation molecules; n between 10^2–10^5 cm^{-3}	Herbig-Haro objects (some possibly associated with H_2O masers); T Tauri stars, and in some cases objects already on the MS with likely ages greater than 10^6 years; embedded clusters of point IR sources
3.	Young clusters with massive stars already on MS	$>10^6$ years	Basically CO and a few high-density. "spots"	T Tauri stars; occasional HES; nonvariable, nonemission line PMS stars; IR clusters embedded in the associated dark cloud

cases the emission spectrum is remniscent of that found in T Tauri stars; this accounts for references to some of these objects in the older literature as "early-type" T Tauri stars

2. An underlying absorption spectrum ranging from type B1 to F8 (composite). The surface gravities derived from analysis of the wings of the hydrogen lines locate many of these objects above and to the right of the zero-age main sequence (ZAMS)
3. P Cygni profiles. With the exception of RR Tau, for which Herbig (1960) suggests the presence of an inverse P Cygni profile on occasion, the HES characteristically show P Cygni profiles in the Balmer series and in some cases in other lines as well
4. Irregular variability. Most stars of the class have shown episodes of apparently irregular variability with amplitudes of up to four magnitudes. Many of them have shown long periods of remarkable stability, however
5. Optical polarization. Breger (1974) has demonstrated that these objects have both polarizations discernibly different from those of nearby objects and rotation in position angle of the electric vector with wavelength; both characteristics provide certain identification of the observed polarization as intrinsic to the objects

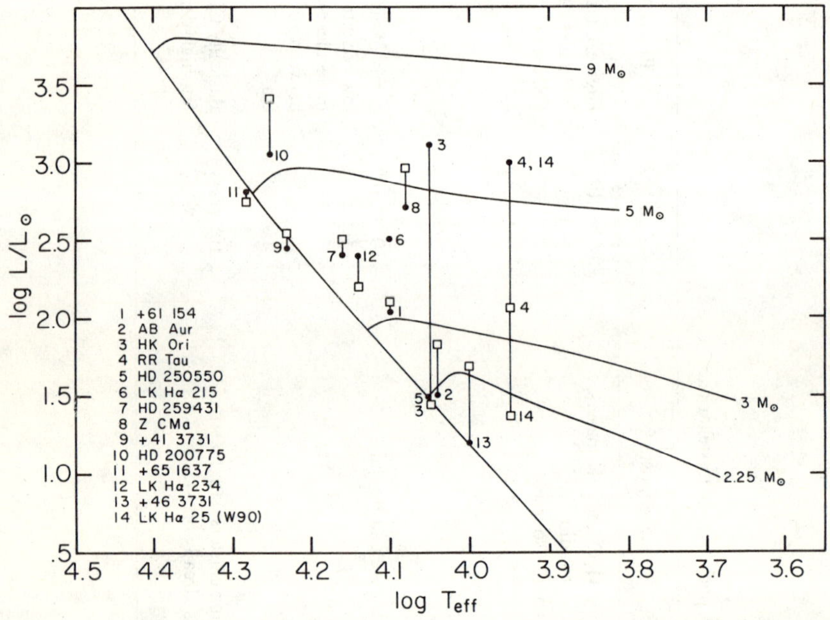

Figure 2 The location of the Ae and Be stars in the L-T_{eff} plane. The spectroscopic luminosity determinations are plotted as solid dots; bolometric luminosities are plotted as open squares.

6. Infrared excesses. Gillett & Stein (1971) and Geisel (1970) first reported strong IR excesses for these stars. Strom et al (1972) published a more complete IR survey of objects in this class; these results provide the basis for a more detailed analysis of HES envelope characteristics. Some stars in the group (AB Aur, LkHα 198) show evidence for the 10.2-μ "bump" often identified with emission from silicate material (Cohen 1973b; S. E. Strom, K. M. Strom, G. L. Grasdalen, unpublished observations)

Strom et al (1972) combined the optical and IR material to locate these objects in the log L-log T_{eff} plane (see Figure 2). In an attempt to understand the CS envelope properties of the HES, they compared the "spectroscopic" luminosity, L_{sp}, determined from the hydrogen line profiles and a "bolometric" luminosity, L_{bol}, derived from the observed total (optical plus IR) flux. In most cases, these quantities compare well (to within the errors); however, in a few cases (HK Ori, RR Tau, LkHα 25), L_{bol} is considerably smaller than L_{sp}. These authors were heavily influenced by Larson's theoretical picture and chose to interpret their data in terms of a dust envelope model. The IR excesses were attributed to re-emission by CS grains of absorbed visual radiation from the underlying stars; the grains were identified with Larson's remnant infalling envelopes. In the cases of stars with $L_{\mathrm{bol}} \ll L_{\mathrm{sp}}$, they argued that the dust is arranged in a disk, optically thick in the IR, but observed nearly edge on so that the amount of IR radiation emitted in the observer's direction is small.

Subsequent observations supplemented by more careful thought suggest several flaws with this interpretation:

1. In some cases, the ratio of envelope IR flux to photospheric flux exceeds unity; if the dust is arranged in a disk, it is difficult to imagine how such a large flux ratio could result given the necessarily small solid angle intercepted by the disk
2. It is difficult to understand how a *remnant* dust envelope could survive a stellar wind with outflow velocities 200–300 km/sec and mass-loss rates of 10^{-6} to $10^{-8} M_{\odot}$/yr (as suggested by the observed P-Cyg profile) unless the remnant material is stabilized in a disk or in optically thick "clumps"; however, for the above reason, such a disk would not explain the observed IR properties. Even at ages as small as 10^5 years, outflow of material rather than infall from remnant envelopes characterizes the HES
3. Observations of Hα–emission line strength have been made for most of these stars. In Figure 3, we plot the Hα flux against the measured flux at 10 μ. The strong correlation between these quantities is difficult to explain on the basis of a dust model because: (*a*) Increasing IR flux implies increasing absorption by the CS envelope of optical radiation. It is remotely possible that dust is created in a stellar wind and that Hα intensity provides a crude measure of the amount of gas (and dust) introduced by the wind into the CS environment. This seems a tortuous way to account for the observed correlation. (*b*) There is no evidence of increased visual reddening for these stars with the largest IR fluxes

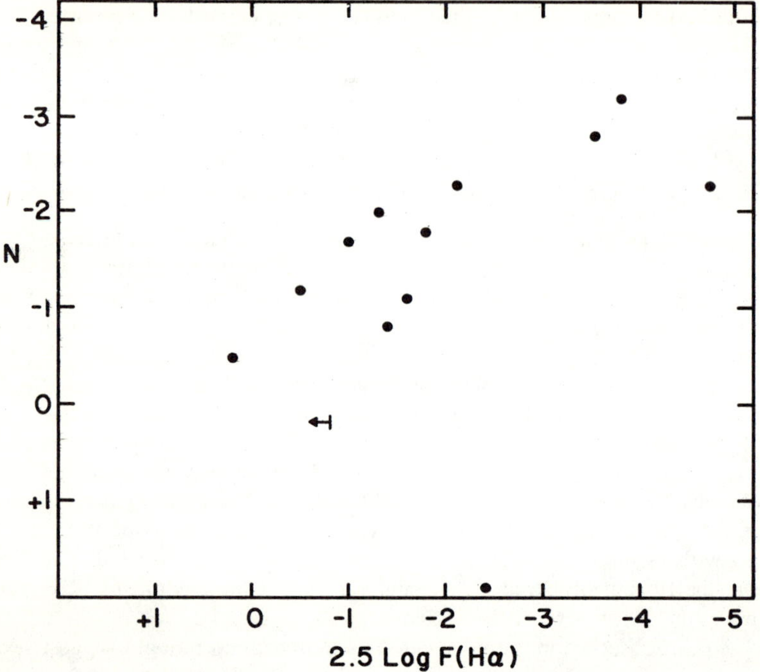

Figure 3 A plot of the observed Hα strengths for the HES against the observed broad-band 10-μ magnitude, *N*.

The observations in some cases of 10-μ bumps suggest the presence of some dust in the CS envelope. It seems more reasonable, however, that the IR excess shortward of 10 μ results from free-free emission in the same envelope region that produces the Hα flux (see Dyck & Milkey 1972, Milkey & Dyck 1973); ion-proton, free-free radiation seems most likely.

Even in the absence of a totally satisfying description of the envelope IR emission, it is possible to argue from their location in the HR diagram that the majority of these objects are approaching the MS along equilibrium radiative tracks.

5 THE T TAURI STARS

Objects of this class are characterized by:

1. Balmer line emission
2. Emission in the H and K lines of CaII as well as FeI, FeII, TiII, and [SII] in varying degrees
3. Underlying spectral types between late F and middle M; often the photospheric

lines are broad. The broadening of the lines has usually been interpreted as arising from rotation (Herbig 1957)
4. P Cygni profiles. With the exception of the class of "YY Ori" stars discussed by Walker (1972), the T Tauri stars exhibit P Cygni profiles in the Balmer line and in the H and K lines of CaII
5. An "ultraviolet excess" and in some cases a "blue continuum." The uv excess appears best interpreted as Balmer continuum radiation. However, at wavelengths shortward of ~4200 Å, many T Tauri stars are observed to exhibit a steep rise in their spectral energy distributions. This is often accompanied by the apparent weakening or disappearance of photospheric lines; the washed out spectral appearance is described as "veiling." In some cases, no photospheric absorption features are visible in the red, indicating that the veiling phenomenon extends at least to 6500 Å. No entirely satisfactory explanation of this effect has been proposed
6. Optical polarization. Hunger & Kron (1957) first reported optical polarization for T Tauri. Since then, little work had been done on this question until Serkowski (1969a,b) and Breger (1974) surveyed a few objects of this class. The polarization appears to be variable with time, and from some of our own unpublished data can achieve values as high as 12%. The observation of variable polarization indicates an intrinsic origin
7. IR excesses. Mendoza V. (1966, 1968) was the first to report the unusual IR characteristics of the T Tauri stars. Some, though not the majority, of these stars appear to show evidence of a 10-μ emission bump (K. M. Strom and S. E. Strom, unpublished observations) and a sharp rise at 20 μ (Cohen 1973a,b; K. M. Strom and S. E. Strom, unpublished observations)
8. Irregular variability at optical wavelengths. Some stars have amplitudes as large as five magnitudes; more typically, ranges of one to two magnitudes are observed. No evidence for periodicity has been found (Plagemann 1970). In one case (V1057 Cygni), a six-magnitude rise in a period of <100 days has been observed. This rapid rise in luminosity is similar to that observed for FU Orionis (Herbig 1966), a previously discovered star presumably analogous to V1057 Cyg

Because T Tauri stars are found in regions with signposts of categories II and III (see Table 2), it is reasonable to assume they range in age from $\sim 10^5$ to 10^7 years. The problem of locating them in the HR diagram and determining their evolutionary state is considerably more complex than it is for the HES. There are no accurate spectroscopic luminosity criteria for these late-type stars and, in many cases, we are not at all certain that the underlying photosphere is observed. Previous arguments regarding the evolutionary state of T Tauri stars have been sharply prejudiced by their presence in OB-associations.

Because O and B stars in these clusters already lie on the MS, it seems most natural to assume that the T Tau stars are objects of relatively low mass (0.2–$2M_\odot$) approaching the MS along equilibrium radiative tracks. This conclusion is based, however, on the belief that formation of stars at all masses is virtually simultaneous within a cluster. Nevertheless, the observations of star-forming complexes suggest that the formation times may extend over periods as long as 10^7 years (see Strom et al 1975). Therefore, we cannot logically exclude the possibility that the T Tauri phenomenon can extend over a wide range of stellar masses and ages. It appears necessary to scrutinize carefully the accepted assumptions concerning the evolutionary state of these stars.

5.1 *The Masses and Evolutionary State of T Tauri Stars*

At present, the only determination of the mass of a T Tauri star rests on Grasdalen's (1973) derived mass for V1057 Cygni. Prior to its FU Orionis-like outburst, this star was classified by Herbig (1958) as a normal member of the T Tauri class (possibly of underlying type K). Quite fortunately, its postoutburst spectrum was of early A-type so that a surface gravity estimate from the Balmer line profiles was possible. The mass derived from this estimate, the observed bolometric luminosity, and the derived surface temperature was $8M_\odot$.

This result is most surprising and suggests that the T Tauri phenomenon extends to considerably higher mass PMS stars. Further investigation of T Tauri masses depends on calibrating an accurate surface gravity index. These efforts could center on one or more of the following possibilities:

1. The size of the feature at 1.6 μ. For stars later than type M0, this bump results from a minimum in the H^-, H_2O, and CN opacity in the photosphere and is surface gravity dependent (Lee 1970). The difficulty in applying this procedure lies in the unknown contribution of overlying CS IR emission
2. The MgH band. Spinrad & Wood (1965) have demonstrated the sensitivity of this feature to luminosity; application to the T Tauri stars awaits spectrophotometric observations with the level of sensitivity achievable, for example, with an image-dissecting scanner (Robinson & Wampler 1972)
3. Observation of the CO first overtone band at 2.3 μ. The luminosity sensitivity of this feature has already been exploited by Baldwin et al (1973) in their study of stellar populations in galactic nuclei. The growing sensitivity of 2-μ detectors should render such a measurement feasible if CS IR emission does not overwhelm the underlying photosphere

The importance of observing coeval groups of T Tauri stars can be appreciated by referring to Figure 1. Suppose one observes a pair of stars of differing mass born at identical times. If one or both stars are in hydrodynamic evolutionary phases, then in general no correlation is expected between the spectral types and luminosities of the pair. If, however, both stars lie on equilibrium radiative tracks, the more massive star will be further along on its evolutionary path; hence, its luminosity will be greater and its spectral type earlier than that of its low-mass companion. An optical and IR survey of T Tauri close pairs selected from the list of Herbig (1962) was recently completed by A. E. Rydgren, S. E. Strom, and K. M. Strom (unpublished observations). Of six pairs with disparate spectral type, in all cases the earlier type companion was found to be more luminous; in the case of a pair of closely identical type, the luminosities of the two stars did not differ by more than 0.1 in the log.

This evidence leads to the conclusion that the majority of T Tauri stars lie on equilibrium radiative tracks approaching the MS. If so, then their apparent location on the log L-log $T_{\rm eff}$ plane can be used to deduce the mass range appropriate to a selected group. Formal mass determinations for 25 T Tauri stars located in the Taurus complex suggest $M \leqq 3.0M_{\odot}$.

We must emphasize, however, the contradictory evidence presented by V1057 Cygni for which a mass of $8M_{\odot}$ is derived. We believe that a conservative interpretation of available data suggests the following:

1. A majority of T Tauri stars observed in cloud complexes are probably low-mass (0.2 to $2M_{\odot}$) stars currently approaching the MS along equilibrium radiative tracks
2. The class also contains more massive stars that for a time share the spectral characteristics of T Tauri stars and then undergo a rapid increase in luminosity (the FU Ori phase)
3. The range of ages may lie between $\sim 10^5$ to 10^7 years

Equally important as the determination of gross evolutionary state is a description of the physical state of the T Tauri photosphere and CS envelope.

5.2 *Mass Loss or Inflow?*

Central to a comparison with Larson's tracks is the location of stars for which remnant, infalling envelopes can be certified from observation. As is the case for the HES, however, the overwhelming majority of carefully observed T Tauri stars appears to have P Cygni profiles indicative of mass outflow (Kuhi 1964, Herbig 1962). Walker (1972) has discussed a class of T Tauri stars that appear to possess inverse P Cygni profiles, the so-called YY Ori stars. The data appear at first glance convincing. We wish to raise a few points, however, that render this interpretation more doubtful. Walker has noted that inverse P Cygni profiles can disappear and in some cases become normal P Cygni profiles in a time scale of a few days. It seems inconceivable that a mass inflow can reverse itself in this short a time. To avoid this difficulty, Walker favors a model in which there is isotropic mass outflow, but irregular inflow from a disk or ring of material. While this model is somewhat ad hoc, it does offer a possible explanation of the data. A. E. Rydgren (unpublished observations) has discovered several stars, however, that exhibit inverse P Cygni profiles in the upper Balmer lines, but that at Hα show an absorption feature flanked both by a strong violet emission peak and weaker redward emission. The profile is strikingly similar to that of a Be star with variable violet-to-red (V/R) emission ratios. It is tempting to argue that the variation from normal to inverse P Cygni profiles in the YY Ori stars results from V/R variations rather than from changes in the character of the flow. Observations of the radial velocities of each of these features will be decisive in checking this hypothesis.

The evidence seems conservatively interpreted as supporting mass outflow for the majority of the T Tauri stars with the possibility that some may have infalling material.

5.3 *The Variability*

A central question in understanding the stability of these stars is, What fraction of the light variation results from photospheric changes as opposed to CS envelope changes? This question is motivated by the observation of both a strong, blue continuum and "veiling" in many T Tauri stars—features that would appear to be produced in an envelope region. Does the variability result primarily from appearance and disappearance of the superposed continuum? An almost total ignorance concerning the photospheres of these stars prevents us from answering this question. Until recently, it has not been possible technically to achieve sufficient spectral resolution to carry out a reliable differential abundance analysis of a T Tauri star and a star of more normal characteristics. Such an analysis would serve to delineate the range in optical depths over which the T Tauri photosphere can be considered normal; it also may provide a test of Herbig's stimulating suggestion (Herbig 1970a) that a chromospheric region extending well into the photospheric layers could explain a number of the observed characteristics of these stars.

Statistical work by Rydgren (unpublished observations) suggests that relatively small amplitude ($\leqq 1.0$ mag) variability is well correlated with the uv excess; the larger the excess the greater the range of variability. This fact, combined with the tendency of the emission line spectrum to be more prominent at light maximum (Herbig 1962), suggests that some of the observed variation indeed results from changes in the envelope emission characteristics. It is doubtful that all brightness changes are a result of envelope variations, however, particularly in view of the observed irregular variations of the HES, which show no veiling and weak, if any, uv excesses, as do a few classical T Tauri stars.

5.4 *The Infrared Excesses*

Until recently, these excesses were believed to arise from reemission by CS dust of absorbed stellar radiation. Support for this view comes from the observation of 10-μ emission bumps and the strong 20-μ rise, possibly associated with a secondary peak in silicate emission spectra (see Cohen 1973a for an exposition of this view). Strom (1972) reported a strong correlation, however, similar to that found for the HES, between Hα-emission strength and IR flux. The arguments offered in Section 4 for the HES stars should apply to the T Tauri stars as well; we therefore prefer to interpret the IR radiation observed shortward of 10 μ in terms of a free-free model. Additional support is supplied by a plot of F (3650 Å) against F (3.5 μ) (Strom 1972). Again, from the tightness of the correlation, it appears as if the uv and IR excesses are most simply interpreted as arising from gas emission processes in a CS envelope (see Smith 1972 and Breger 1972 for further supporting arguments).

The IR excess may also provide indirect evidence bearing on the evolutionary status of these stars. Knacke et al (1973) first argued that the IR excesses were greater in the R-CrA T-association than in the Ophiuchus complex. Because the Ophiuchus T Tauri stars are associated with an old (10^7 years) OB-association, it seemed plausible to suggest that the IR excess might decrease with the age of the stellar group.

An index of the IR excess is provided by the difference, ΔL, between the CS envelope and underlying stellar continuum at 3.5 μ. A. E. Rydgren, S. E. Strom, and K. M. Strom (unpublished data) have observed a significant sample of T Tauri stars in the Ophiuchus and Taurus dark cloud complex. For the Ophiuchus T Tauri stars, $\Delta L \lesssim 0.5$; for the Taurus T Tauri stars, $0.2 \lesssim \Delta L \lesssim 1.5$, with $\Delta L \sim 1.0$ most representative. Because the Taurus dark clouds contain no known OB star members, it is plausible to conjecture that the stellar population is so young that the massive stars have yet to reach their equilibrium radiative tracks; if so, then ΔL is a decreasing function of increasing age as suggested by Knacke et al (1973). Of course it can be argued that T-associations may never contain massive stars and that the age of a given group is indeterminate.

Additional arguments suggest, however, that the observed changes in ΔL may be related to evolutionary effects. For groups of large ΔL, both the Hα fluxes and the observed Hα velocity widths are substantially larger. A naive interpretation of these data suggests to us that stars with large Hα fluxes and widths have large mass-loss rates. We consequently interpret the changes in optical and IR emission properties as arising from differences in mass-loss rate and believe the data indicate a decreasing mass-loss rate with increasing age. Within each group, there is a large dispersion of optical and IR emission strength. This probably indicates a dispersion in age, mass, and in whatever initial conditions (e.g. angular momentum) control the rate of mass loss for stars of a given mass.

5.5 *The Nonemission, Nonvariable PMS Stars*

The presence of "normal" stars occupying the same region of the HR diagram as the T Tauri variables is often overlooked. Nearly 50% of the apparently late-type PMS stars in the Walker clusters, however, show no evidence of variability or emission lines. Some of these stars may be field interlopers. Others may be stars in which Hα was too weak to detect, was temporarily absent (Herbig 1962), or for which sufficiently accurate or long-term searches for variability were not available. The case for a continuum of hydrogen-line emission strengths is quite plausible in view of the work by Strom et al (1971). They found among the A and F PMS stars in NGC 2264 a number of stars that have weak Hβ and Balmer continuum radiation; this was unrecognized in previous surveys. If, moreover, variability and envelope emission strength are correlated, the weak Hα stars may also be variables of smaller amplitude.

The question of membership of these stars in the Walker clusters is a vexing one and statistical arguments based on field luminosity functions are not satisfying. Several stars lacking variability or Balmer-line emission were classified by Walker as giants or subgiants, some with broad lines. These are almost certainly PMS stars. An alternative approach is to take advantage of the increased strength of the LiI line in young stellar objects (Zappala 1972). A spectroscopic search by S. E. Strom and K. M. Strom (unpublished observations) provides weak evidence for Li-line enhancement in a few members of this group in Orion. The latter approach appears to provide a simple criterion for selecting young, late-type PMS stars and should be pursued further.

If we accept the reality of this group, we must face the evolutionary implications. In our view, they fall into the following categories:

1. Older PMS stars of relatively low mass that have already undergone a less dramatic episode similar to the FU Ori phenomenon. We know from observations of V1057 Cygni that the T Tauri emission spectrum characteristic of the preoutburst star has virtually disappeared. The loss of emission characteristics might indicate the end of the hydrodynamic phase and the beginning of quasistatic evolution toward the MS along equilibrium radiative tracks (Grasdalen 1973)
2. Stars similar in mass and evolutionary state to the T Tauri stars for which the mass-loss rate and consequent envelope emission is considerably lower. This hypothesis demands a range of mass-loss rate dependent on some unknown, probably initial condition

A more radical suggestion is to view the T Tauri stars as atypical, and the "normal" stars as representative of "ordinary" PMS evolution. Perhaps the T Tauri stars represent unrecognized PMS spectroscopic binaries in which the emission line features are produced in an envelope common to the two stars. What little reliable radial velocity work is available tends to contradict this unsettling hypothesis; the four well-observed T Tauri stars (T Tau, RY Tau, UX Tau, and SU Aur) seem to have mean radial velocities relatively close to the Taurus dark cloud velocities determined from molecular line studies (Herbig 1962; S. E. Strom and K. M. Strom unpublished observations). The radial velocity data are few, however. Moreover, their reliability is questionable because of the complexities of measurement introduced by the inherent line width and the overlying emission. More work at higher spectral resolution is clearly necessary. The binary origin hypothesis for T Tauri stars cannot be ignored, despite its unpleasant consequence, because we have not yet identified PMS binaries. Furthermore, we expect 30–50% of stars in the mass range 0.5–1.5$M_{\odot}$ to be binaries; this figure is hauntingly close to the observed fraction of T Tauri stars relative to the total number of PMS objects occupying the same region of the HR diagram.

The coexistence of T Tauri and "normal" stars in the same part of the HR diagram argues against identifying isolated PMS stars having weak or absent emission with later stages in the evolution of a T Tauri star. Statistically, this seems to be an acceptable way of ordering *groups* of T Tauri stars in an age sequence. To suggest on the basis of weak emission lines and/or weak IR excess that a particular star is "post-T Tauri," however, seems dangerous to us (see Herbig 1973 for another view).

5.6 *Line Widths and Polarization*

The broad lines observed for several T Tauri stars have been introduced as evidence to support the contention that these stars are rotating at velocities between 20 and 100 km/sec, considerably greater velocities than are typical for their spectral type. Herbig (1962) has argued that such rotational velocities are consistent with those expected when the stars in question reach the MS at late A- or early F-type. We note in passing that this argument rests heavily on the belief that T Tauri stars are approaching the MS along Henyey tracks.

The observation of significant visual polarization, however, raises some questions concerning the rotational broadening picture. The polarization most plausibly results from either electron scattering or scattering by dust particles in an inhomogeneous envelope.

If the envelope is expanding and $\tau_{scat} \sim 1$, it is possible that the photospheric line profile, observed both directly and scattered by the envelope material, may appear broadened. Herbig (1970b) suggested such a picture for VY CMa. In this picture, the line would appear asymmetrical, with the red wing broader than the wing on the blue side of line center. Moreover, on low resolution spectra, the asymmetry may introduce a bias so that the radial velocity of the absorption line spectra will be estimated somewhat positive relative to the true stellar velocity. However, if a dark cloud velocity is available from molecular line studies, it may be possible to test indirectly the importance of this effect by searching for a systematic, positive shift of the observed stellar velocity relative to the dark cloud value.

6 THE HERBIG-HARO OBJECTS

Until quite recently, the H-H objects were believed to represent a very early stage of stellar or possibly protostellar evolution. This view was based on (*a*) the morphological characteristics of the objects—they are defined as "semistellar" patches of nebulosity characterized by a low excitation spectrum dominated by [SII] and [OII] and the Balmer lines; (*b*) their location in dark cloud complexes bearing other representatives of a young stellar population; and (*c*) their spectroscopic similarity to CS envelopes surrounding a few T Tauri stars. Various proposals have been offered to account for the observed characteristics of these objects (Herbig 1969, Magnan & Schatzman 1965); all require the presence of a stellar or protostellar object obscured from view at optical wavelengths and located within the optical nebulosity. Strom et al (1974a,c) have suggested that the H-H objects are actually reflection nebulae, whose illuminating stars are extremely young objects located within the dense, dark cloud material. Variations on this theme have been offered recently by others (e.g. Schwartz 1974). Their arguments rest heavily on the following observations:

1. The discovery of single IR sources associated with multiple-component optical H-H objects; these sources are typically displaced relative to the optical objects
2. The measurement of optical polarization for the individual knots in a complex of H-H objects (Strom et al 1974b); the polarization is large and the **e**-vector orientation for the knots is consistent with illumination by a single source coincident in location with the IR object. The H-H objects are often located near the HES and are not found in older regions. A characteristic age

of several times 10^5 years or less is suggested by these observations. These authors argue that the illuminating H-H stars represent the immediate precursors of T Tauri stars. Consistent with this hypothesis, the IR spectral

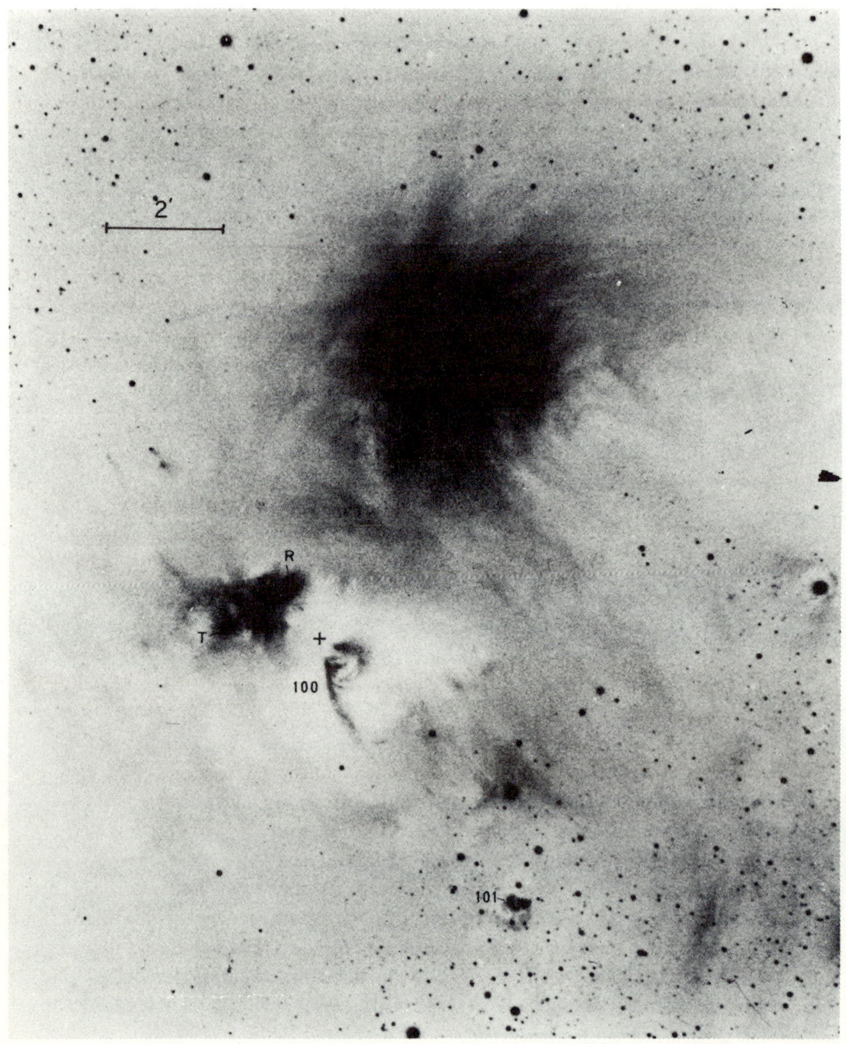

Figure 4 A photograph of the region near H-H 100 in the CrA dark cloud (*top,* north; *left,* east). A 098-02 plate plus RG 610 filter define the effective bandpass. Note the nearly circular "white patch" in which H-H 100 is located. We believe that this region represents the high-density, placental cloud from which the H-H star was formed. We thank Dr. T. Gull for allowing us to reproduce this photograph.

energy distribution of the H-H stars seems well represented by a typical T Tauri distribution reddened by 20–30 magnitudes of visual extinction

These reddening values, combined with the sizes of the associated dark clouds, permit the minimum dark cloud gas densities to be crudely estimated as 10^4–10^5 cm^{-3}. These estimates agree with the densities required to excite CS and HCN, both of which are typically observed in the vicinity of H-H objects (Morris et al 1974a, Lada et al 1974).

A rather dramatic visual demonstration of the apparent association of H-H objects with dense, dark cloud material is provided by the red photograph of H-H 100 in CrA, reproduced in Figure 4. The H-H star (cross) appears located near the center of a dark, circular region of dimension $\sim$0.1 pc located within the CrA cloud. It is tempting to identify this region with the remnant placental region out of which the H-H star was formed.

The reflection-nebula hypothesis renders the H-H phenomenon somewhat less exotic than heretofore believed. If accepted, however, it suggests an uncharacteristic kindness on nature's part, in that we are permitted, albeit by reflected light, to observe a very early phase of stellar evolution. Spectroscopic observations of the optical H-H knots reveal a further important characteristic: the measured radial velocities of the emission lines are negative (by up to 140 km/sec) relative to the dark cloud velocities. If the emission lines are formed in an expanding envelope, small in dimension with respect to the star, these observations can be interpreted in terms of a mass outflow. Strom et al (1974c) and Schwartz (1974) argue that the mass-loss rate for these objects may be a factor of 10 or more greater than the rates of $10^{-8} M_\odot$ $years^{-1}$ characteristic of the T Tauri stars (Kuhi 1964). Hence, we face the surprising result that mass outflow, rather than inflow, is observed even for stars as young as 10^5 years or less. This is at variance with the details of Larson's PMS tracks and should be recognized as an important, unsolved problem in understanding the early phases of stellar evolution. It appears to add support, however, to the explanation proposed in Section 5.4 for the apparent decrease with time of optical emission line strength and the IR excess for objects of the T Tauri class. The H-H stars, viewed as the precursors of T Tauri stars, would be expected to have higher mass-loss rates and consequently more prominent emission lines, blue continua, and IR excesses. The presence of strong stellar winds at ages of 10^5 years or less will no doubt affect any conclusions regarding the survival of proto–solar system nebulae. By this time, the material in such nebulae must be stabilized in a disk or in self-gravitating bodies.

7 A SCENARIO FOR PMS EVOLUTION

Despite the serious disagreement in detail between Larson's computations and available observations of PMS stars, it appears possible to use his evolutionary tracks as a qualitative guide to early stellar evolution.

From our previous discussion, it seems logical to propose the following scheme:

1. Stars first become visible in their hydrodynamic phase of evolution

2. For stars of moderate mass ($15 \gtrsim M/M_{\odot} \gtrsim 2$), the transition between hydrodynamic and quasistatic equilibrium evolution is indicated by the rapid rise in luminosity observed as the FU Ori phenomenon. This rapid rise is already implicit in Larson's tracks; moreover, the computed luminosity change is larger for the higher mass stars. Observationally, we suppose this evolutionary path to result in a transition from H-H stars to T Tauri stars to a HES. Once on its radiative equilibrium track, the star begins a relatively stable evolution toward the MS
3. For lower-mass stars, the transition phase between hydrodynamic and quasistatic evolution is less dramatic. Stars either join equilibrium radiative tracks or join the lower luminosity region of a Hayashi track. Observationally, these stars evolve from H-H stars to T Tauri stars with high amplitude variability, to more stable objects of somewhat higher luminosity than their immediate predecessors
4. Mass loss characterizes observable PMS stars of all masses. Despite a large "cosmic scatter" of mass-loss rates for stars of identical mass (possibly resulting from differences in initial angular momentum), the mass-loss rate decreases as the star approaches the MS. Infall of material is probably important only at ages considerably smaller than 10^5 years; moreover, the supporting evidence for an infall picture is to date not entirely convincing
5. Stars of all masses are increasingly more difficult to observe the younger they are. The optical and radio observations of the H-H stars (Strom et al 1974a–c, Lada et al 1974) provide compelling evidence that suggests that these extremely young objects are still associated with the dense, placental, dark material out of which they are formed. It would appear that this material as opposed to dense CS envelope material, is the factor that limits the "visibility" of young stellar objects for a majority of their PMS lifetime. The mechanism by which this material is dissipated is not yet well understood, although the stellar wind is a likely candidate (see Section 8.5).

The PMS evolution of stars with $M \gtrsim 15M_{\odot}$ is more difficult to delineate from available data. In part, this results from the short evolution time to the MS and the consequently limited time the object has to emerge from its placental cloud. As a result, the signpost for the appearance of a massive star on the MS, a radio compact HII region, an H_2O or OH maser often leads to the discovery of a luminous, heavily obscured, IR object. These objects are characterized by strong ice and silicate absorption features and an extremely steep spectral energy distribution. The interpretation of the observed spectral energy distribution is complicated by our partial ignorance of the range of extinction laws possible in dense dark clouds. We do know that the grains in such dense clouds are likely to be larger than is typical of the lower density interstellar medium, possibly as a result of the growth of ice mantles (Carrasco et al 1973, Grasdalen 1974, Vrba et al 1975). How large they grow and how the shape of the reddening law is influenced is currently under investigation. Further observations of the optical depths of the ice and silicate absorption features as a function of deduced particle size will be a critical first step in improving our understanding. With an inadequate appreciation of the possible reddening law variations, it is easy to interpret mistakenly the observed spectral energy distribution of a highly reddened luminous star as one arising from a low temperature "protostellar" object. More vexing still are the problems of the temperature structure of the dark cloud material surrounding the star and the likely importance of an intrinsic, stellar IR excess.

As examples of these difficulties of interpretation, we mention the various discussions of the Becklin-Neugebauer (BN) object (Gillett & Forrest 1973, Penston, Allen & Hyland 1971, Grasdalen 1974) and of Haro 13a (Allen et al 1975).

Equally vehement arguments have been offered in defense of a "protostellar" hypothesis and a heavily obscured PMS star hypothesis for the BN object. We feel that the most consistent interpretation of this object can be made by carefully scaling observed reddening laws from well-understood, less obscured objects located in similar dark cloud complexes and thereby deriving an extinction law appropriate to BN; the BN object therefore appears to be a massive star currently on its equilibrium radiative track approaching the MS, obscured by 50 magnitudes of visual extinction. It appears to have a small, intrinsic IR excess beyond 5 μ.

The case of Haro 13a is illustrative of the role played by a larger intrinsic IR excess. By following a similar procedure to that used in interpreting BN, it is possible to understand the observed spectral energy distribution from this object from 8000 Å to 3.5 μ if we assume that it is obscured by material giving rise to 20 magnitudes of visual extinction. Longward of 3.5 μ, the object appears to show both an IR excess typical of the HES described earlier and silicate emission features at 10 and 20 μ. If this interpretation is correct, it would appear to be an analog of the HES with $M_{\text{bol}} \sim -5$. If so, it will arrive on the MS as a late O or early B star. It is significant that this massive young (several 10^4 years) object shows indirect evidence (if the IR excess arises from free-free emission) for mass outflow, as is the case for the HES.

8 THE INTERACTION OF YOUNG STARS AND DARK CLOUDS

The role played by young stars in affecting the physical conditions within dark clouds has largely been ignored. In making this statement, we have not overlooked the extensive literature concerned with ionized hydrogen regions. Rather, we note that until recently the focus of attention has rarely been on their association with massive, dark cloud complexes.

Because of the paucity of existing literature, this section is of necessity somewhat brief and speculative. We view this area of research, however, as one that will grow considerably in importance over the next few years.

8.1 *Embedded Stellar Population*

It is somewhat surprising that dark cloud interiors were thought to be regions generally devoid of objects other than "protostars." Herbig (1962) had speculated that the visual impressions of dark cloud regions may provide a very deceptive view of the extent of star formation. In retrospect, it is almost a logical necessity that a cloud must pass through the following stages:

1. Production of stars in selected regions; the stars are hidden from view by placental dark cloud material
2. Emergence of a few T Tauri stars or an HII region
3. Emergence of visible clusters of stars

Speculation regarding the internal stellar content of dark clouds, however, received very little attention until the application of IR techniques. Maps of dark cloud regions by Grasdalen et al (1973), Wynn-Williams et al (1972), and Vrba et al (1975) have provided rather dramatic evidence that suggests that massive clusters are very likely embedded within these complexes.

In Figure 5, we reproduce the IR (2 μ) map of the Ophiuchus cloud complex made by Vrba et al (1975). Sixty-seven sources were detected in the course of their survey; the large majority appear to be heavily obscured ($A_V \sim$ 10–40 mag)

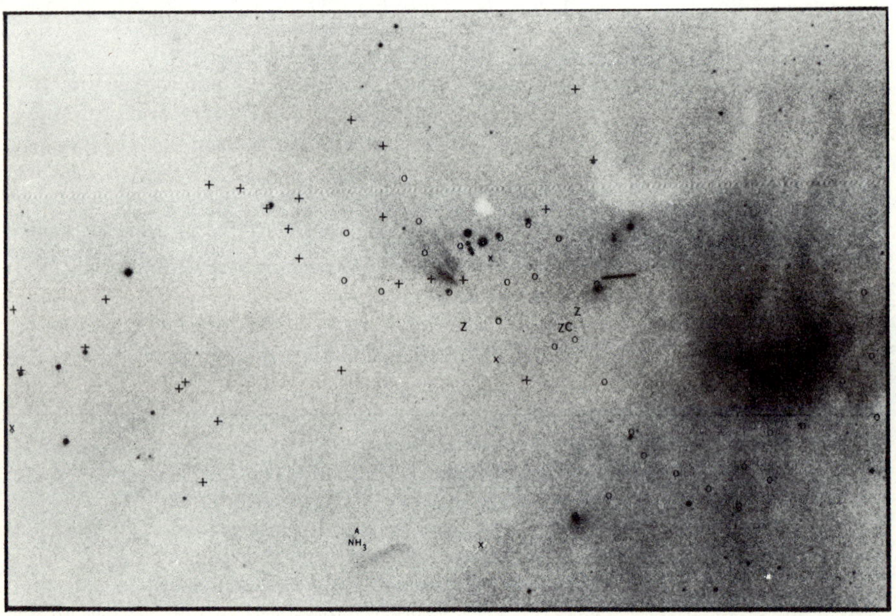

Figure 5 A plot of the sources observed in the course of the Grasdalen et al (1973) and Vrba et al (1975) 2-μ surveys of the Ophiuchus dark cloud (*top,* north; *left,* east). The underlying photograph is the POSS red plate. The scale can be estimated from the long dimension of the inscribed rectangle (36′). The bright star associated with nebulosity is HD 147889.

stars embedded in the dark cloud. These authors argue that the IR sources represent the upper MS of a young cluster. In support of their contention, they compared the luminosity function of this putative cluster with the one observed by Walker (1956) for NGC 2264; they are virtually identical. *This observation strongly suggests that dark clouds can contain embedded clusters even several million years after the cluster forming event(s).*

8.2 *Cloud Dissipation*

Because stars and clusters are observed outside the confines of dark clouds, an obvious question presents itself: how is the cloud material dissipated? *Locally,* HII regions can dissipate cloud material. However, HII regions typically comprise only a small fraction of the dark cloud mass [e.g. the ionized gas and stars in the Orion nebula complex contribute less than 10% of the mass of OMC1 and OMC2 (Gatley et al 1974, Morris et al 1974b, Werner et al 1974)]. Moreover, clouds associated with T-associations do not now and may not ever contain stars massive enough to form an HII region. A supernova explosion could provide the mechanism of cloud destruction. Another possibility is that a combination of cloud-cloud collisions and shearing through differential galactic rotation may eventually destroy the clouds; both these mechanisms, however, require time scales of $\sim 10^8$ years. The fact that dark clouds may hide developed clusters from view for times as long as 10^8 years may be of some significance in understanding the distribution of some spiral tracers. Whereas the current spiral arms are delineated by OB stars and HII regions, objects such as M supergiants and cepheids could be found well away from the currently observed spiral arms, if their precursors were embedded within dark cloud complexes for 10^7–10^8 years. How and when the complexes are dissipated are questions that clearly require more attention.

8.3 *The Radiation Field in Cloud Interiors*

The presence of embedded clusters may also affect significantly the atomic and molecular equilibrium within the clouds. The assumption that the uv radiation field is negligibly small within dark clouds can no longer be taken for granted. An embedded B star can (*a*) create a small ionized hydrogen region; (*b*) create a larger ionized carbon region; (*c*) disassociate CO and many more complex molecular species; and (*d*) provide the energy to release complex molecular species that may form on grain surfaces. Continuum emission (B. Zuckerman, personal communication) and carbon recombination lines (Brown et al 1974) have already been detected in the direction of Ophiuchus; their presence has been attributed to the ionizing radiation provided by early-type members of the embedded cluster discovered by Grasdalen et al (1973).

The importance of the uv radiation field on a global scale within the cloud will be determined by the uv albedo and the scattering phase function of the grains. In the case of high albedo and strong forward scattering, the uv radiation field within the cloud may be very significant.

8.4 *Local Heating of Dark Cloud Material*

The embedded stars may also provide a mechanism for locally heating the cloud material. Absorbed stellar radiation can heat the grains; collisions between the grains and H_2 molecules can then heat the gas. The correspondence of CO "hotspots" with partially embedded stars or stellar clusters (see Loren et al 1973 for examples) possibly arises from this effect. The heated grains may also be observed in the submillimeter range. The detection of 350-μ radiation from the Ophiuchus complex by Simon et al (1973) is probably yet another manifestation of the effect of the embedded stars on the cloud environment.

8.5 *Interaction of Stellar Winds and Dark Cloud Material*

The fact that many, if not all, young stars possess stellar winds may be very significant in understanding velocity fields observed in cloud complexes. We know from observations of HES and T Tauri stars that stellar winds characterized by mass-loss rates of 10^{-7}–$10^{-8} M_{\odot}$/yr, and velocities of a few hundred km/sec persist over time scales of 10^5–10^6 years; the winds may be even stronger during the H-H star phase. These winds are capable of clearing out a significant region within the cloud complex (see Strom et al 1975). A crude static analysis suggests that in a cloud of density 10^3 cm^{-3} regions of order 0.1 pc in size can be swept clean by stellar winds of this magnitude.

Hence, the wind from a single star may significantly affect the region around a PMS star. This may account for our ability to glimpse a number of partially obscured clusters in the vicinity of young, carly-to-middle B stars (Strom et al 1975). Furthermore, it may represent the mechanism by which a star dissipates its placental cloud. Some further evidence in favor of this view may be found in the optical appearance of H-H objects. The "holes" in the clouds through which light from the H-H stars reaches the scattering patches of bright nebulosity may result from the interaction of stellar winds with inhomogeneous, dark cloud material. The stellar winds are highly supersonic and should produce shocks in the dark cloud medium. Schwartz (1974) argues that the forbidden line radiation in the optical H-H objects may be produced in a shock region; the similarity of the observed spectra of H-H objects to that of supernova remnants lends some credence to this suggestion.

The observations of Loren et al (1974) of the dark cloud regions surrounding selected HES provide evidence that suggests that the stellar winds from these objects may affect the cloud velocity field. Enhanced CO line widths and the presence of multiple velocity components find attractive explanation in terms of a wind-cloud interaction.

If an extensive cluster of PMS objects is contained within a cloud complex, it may be difficult to disentangle (with the available resolution of mm-radio dishes) the velocity dispersion produced by wind-cloud interaction from (*a*) the intrinsic cloud velocity dispersion produced by internal motions of gas not associated with stars and (*b*) differential motion on a larger scale. This possibility cannot be ignored in an analysis of mm-line observations of dark clouds; careful and extensive mapping provide the only possible means to distinguish the various possibilities.

8.6 *HII Regions*

EVOLUTION Observable HII regions associated with recently formed stellar complexes range in character from low surface brightness regions such as that found near NGC 2264, to higher surface brightness regions such as the Orion nebula, to the extremely high emission measure, optically obscured "compact" HII regions. It is important to note that HII regions of all types in our own galaxy are almost invariably found in association with large molecular cloud complexes that are usually considerably more massive than the ionized and stellar material in the region. This association is not surprising because the age of these regions probably ranges from 10^4 to 10^6 years. This is just the range observed for other young stellar objects still confined near their place of origin in the dark cloud complexes. The compact and low surface brightness HII regions no doubt represent the extremes of an evolutionary sequence. A plausible scenario for their evolutionary history probably includes the following steps:

1. Arrival of an O star or cluster of early-type stars on the MS. The grains near the star are heated and evaporated and a small HII region forms near the star. The propagation of the ionization front is controlled by the absorption by the grains of Lyman continuum photons
2. Heating of the grains sufficiently expands the dark cloud material surrounding the star so that absorption of Lyman continuum radiation by the gas dominates over absorption by the dust grains
3. The ionization front propagates in the direction of low density regions. Higher density regions act as "walls." As the HII region tries to propagate in the direction of the walls, material flows from the walls in the direction of the lower density cloud material

Many features of this picture have been discussed at length by Zuckerman (1973), Grasdalen (1974), and Grasdalen & Carrasco (1975). Unfortunately, detailed computations are not available. Previous computations of HII region evolution (e.g. Mathews 1969) were made before molecular line observations forced on us the realization that these regions begin their lifetimes in inhomogeneous regions of extremely high density. Computations would be of considerable importance in determining the likely time scales, particularly for the early evolutionary stages. This problem may be of more than academic interest because it is possible that in regions of high density within a dark cloud the time scale for evolution of an HII region may be comparable to or greater than the MS lifetime of the ionizing star(s). A statistical study of HII regions and dark cloud velocities covered at radio frequencies suggests that HII regions are found principally near the periphery of dark clouds. It is conceivable that, whereas massive stars do form near the dark cloud centers, they fail to produce observable HII regions during their MS lifetime.

HII regions are among the strongest known sources at wavelengths between

10 and 100 μ. At these wavelengths, the observed radiation exceeds by a large factor the radiation expected from extrapolating the free-free emission measured at cm wavelengths. This excess was originally postulated to arise from heated dust contained within the HII region (Krishna-Swamy & O'Dell 1967). The heating is thought to arise from a combination of (*a*) nebular Lyman α photons, (*b*) nebular Lyman continuum photons, and (*c*) stellar photons longward of the Lyman limit. If heating arises from (*a*) and/or (*b*) then there should be a direct relation between the observed IR flux and the number of ionizing photons. A good correlation between the 10-μ and radio fluxes for selected HII regions has been found by several investigators (see Wynn-Williams & Becklin 1974 for details).

Strom et al (1974d) have compared the IR fluxes from the "Air Force Cambridge Research Laboratory Infrared Sky Survey" (Walker & Price 1974) with measured cm fluxes; the comparison with this more objective selection of HII regions confirms the previous result. Despite these observed correlations, however, the accepted picture of the origin of IR radiation in HII regions is faced with serious observational contradictions. From the theoretical work of Krishna-Swamy & O'Dell (1967), one expects a correlation between the grain temperatures and the mean intensity of the Lyman α and Lyman continuum radiation fields; the latter quantities in turn are related to the observed cm continuum surface brightness and ultimately to the electron density and size of the region. Estimates of grain temperature depend on the assumed emissivity of the dust at IR wavelengths; $T_{gr} \sim 150°K$ is probably representative. Near this value, small changes in T_{gr} should be reflected as large changes in the observed flux ratio between 10 and 20 μ; this flux ratio in turn should be related to the mean electron density in the HII region. However, the flux ratio not only fails to show the expected correlation with N_e, but also is remarkably constant over a wide range of HII region densities.

Another conundrum is provided by the attempt to observe extragalactic HII regions in Sc galaxies at 10 μ (Strom et al 1974d). The observed radio fluxes and the relation between 10 μ and cm flux for HII regions in our own galaxy suggested that the extragalactic regions could be detected; with one possible exception, none was. It is important to note that the extragalactic HII complexes observed are greater in size and lower in mean N_e than most galactic regions; nevertheless, 10-μ radiation from galactic HII regions of comparable surface brightness has been observed. One observable property of the extragalactic regions may hold the key to the conundrum: no molecular clouds have been observed associated with the HII regions despite a few attempts at detection. Moreover, the mass of ionized material ($10^6 M_\odot$) in these extragalactic HII complexes is comparable to the mass contained in a molecular cloud complex in our own galaxy; it is possible that these regions have formed a sufficient number of massive stars to ionize all the original cloud material. These observations, if confirmed by more diligent attempts at detection of molecular line radiation, suggest to us that the 10–20 μ IR radiation may arise not within ionized regions, but at the dark cloud-HII region boundary.

It has already been suggested (see Harper 1974 for a discussion) that the submillimeter IR radiation arises from a region more extended and somewhat displaced relative to the radio continuum boundaries of the HII regions. Harper interprets this observation as an indication that the long wavelength IR radiation is produced at the cloud-ionized region interface. Unfortunately, shorter wavelength (10–20 μ) maps of these regions have not been made at sufficiently low spatial resolution to determine both their global emission properties and their spatial relationship with the dark cloud material.

The strongest argument in support of an internal origin for the short wavelength IR radiation comes from the detailed correlation of IR and radio flux maxima. This correlation could be produced equally well, however, if both the IR and radio radiation were produced at the cloud-HII region interface. At this point, critical observations aimed at understanding the origin of HII region IR radiation should be directed at (*a*) a detailed attempt to correlate grain temperature estimates with surface brightness, and (*b*) mapping HII regions using large beam systems so that the emission properties in the low surface brightness regions can be studied.

We also note the importance of studying the detailed IR spectral energy distributions of the compact HII regions. The presence of a bright IR source within a dense, presumably placental, cloud offers a unique opportunity to probe the properties of interstellar material under high density conditions. Gillett & Capps (1975) have recently observed enormously strong ice and silicate absorption features in the compact source W33A. A comparison of these feature strengths with the deduced wavelength dependence of extinction may provide important clues in understanding the details of grain growth in dense regions.

8.7 *Infrared Sources Associated with H_2O and OH Masers*

Several bright IR sources have been found associated with H_2O and OH masers located within dark cloud complexes (Wynn-Williams & Becklin 1974, Wynn-Williams, Becklin & Neugebauer 1972). In some cases, these maser sources appear to be spatially coincident with radio continuum peaks (compact HII regions); in others, no continuum source is seen. In the former case, the maser is probably produced in the high density placental region surrounding a newly formed, early-type star. In the latter case, the maser may be indicating a high density, protostellar condensation, or the high density placental cloud surrounding a massive star that has not yet reached the MS or in which an HII region has not yet propagated. It is difficult to distinguish among these alternatives (see Section 7). Observations at sub-mm wavelengths are essential in providing an estimate of the temperature range and total luminosity of these sources. Furthermore, observations of cloud complexes in high excitation molecular lines should allow an a priori selection of regions likely to be conducive to masering.

The observations made by Dickinson et al (1974) and by Lo (1974) of H_2O masers near H-H stars suggest that the velocity of the maser may differ significantly from the dark cloud velocity; therefore searches need not be restricted to frequencies near the cloud velocity. The observed maser-line velocities may provide important indications of the presence of a stellar wind and its possible importance in affecting the dark cloud environment.

9 PROSPECTS FOR THE FUTURE

The preceding discussion illustrates the considerable progress that has been made in understanding the early phases of stellar evolution. Although significant gaps in observational knowledge and insight abound, the framework for future progress seems reasonably well established.

We expect that the focus of activity will shift increasingly toward the study of dark clouds rather than stellar evolution.

The problems of cloud collapse and fragmentation into stars are blessed with considerable theoretical speculation but little empirical insight. Our data and understanding are sufficient to rule out a situation where a massive molecular cloud undergoes spherical collapse and where a single burst of star formation takes place in the central dense core of the collapsing cloud. Optical and radio studies clearly suggest that star formation can take place in dark clouds over periods perhaps as long as several times 10^7 years. Dark cloud complexes seem to contain representatives of extremely young objects (10^4–10^5 years old) found near dense placental material and much older, well-distributed stars formed during previous episodes of star formation. A critical question that must be answered is, What mechanism prevents the global collapse of massive cloud complexes? It is becoming increasingly apparent that magnetic fields may hold the key to understanding this question (see Lada et al 1974). Unfortunately, attempts to estimate the internal magnetic fields of dark clouds through observation of Zeeman splitting in OH and SO have so far proved disappointing. Obtaining reliable results from these methods demands much higher spatial resolution mapping so that the Zeeman splitting can be separated from the effects of dark cloud velocity fields. Optical and IR polarization measurements of partially and heavily obscured stars within the clouds offer the hope of determining field geometry, but little hope of deducing field strengths.

Nevertheless, an analysis of the field geometry may provide the basis for understanding both the large-scale morphology of the clouds and, indirectly, the role played by the fields during the collapse phase.

An increasing amount of effort is presently aimed at determining the run of temperature, density, and velocity within cloud complexes. Maps of selected dark clouds in several molecular transitions that sample a variety of density ranges will be critical in understanding the fragmentation process. We cannot overemphasize the importance of combining these maps with IR surveys of the same regions. As discussed in Section 8.5, the presence of embedded sources may affect not only cloud temperatures, but also the velocity fields. We believe that considerable effort should be spent in attempting to distinguish differences between clouds associated with T-associations and those containing HII regions. If the T-associations are destined to produce only stars of relatively low mass, then the study of the associated cloud complex should provide considerable insight into the mechanism(s) responsible for production of the observed stellar mass spectrum. Further optical study of regions of recent star formation is also essential for determining the luminosity functions in these regions.

We suspect that the much-sought-after "protostellar" objects have already been observed; the H_2O and OH masers are probably the signposts for likely candidates. Confirmation of this speculation awaits the availability of higher spatial resolution studies at submillimeter (continuum) and mm (molecular-line) wavelengths. If mm-line interferometers and large mm-dishes at ultradry sites become available, we will almost certainly be able to witness and understand the process of stellar birth.

While our enthusiasm and optimism concerning the future application of IR and radio molecular techniques to studies of the star formation process is patently obvious, other approaches must not be ignored. In particular, we believe that efforts directed at studies of star formation in external galaxies may prove critical. These systems provide us with ranges of chemical composition, dust-to-gas ratio, and dynamical histories far greater than those available for study in our own galaxy. Perhaps an educated compilation of differences in star formation characteristics on a galactic scale will free us from some of the blind prejudices we no doubt possess.

Literature Cited

Allen, D. A., Strom, K. M., Grasdalen, G. L., Strom, S. E., Merrill, K. M. 1975. *MNRAS*. In press

Ambartsumian, V. A. 1955. *Mem. Soc. Roy. Sci. Liège, Ser. 4* 14:293

Appenzeller, I., Tscharnuter, W. 1974. *Astron. Ap.* 30:423

Baldwin, J. R., Danziger, I. J., Frogel, J. A., Persson, S. E. 1973. *Ap. J.* 184:427

Breger, M. 1972. *Ap. J.* 171:539

Breger, M. 1974. *Ap. J.* 188:53

Brown, R. L., Gammon, R. H., Knapp, G. R., Balick, B. 1974. *Ap. J.* 192:607

Carrasco, L., Strom, S. E., Strom, K. M. 1973. *Ap. J.* 182:95

Cohen, M. 1973a. *MNRAS* 161:97

Cohen, M. 1973b. *MNRAS* 161:105

Dickinson, D. F., Kojoian, G., Strom, S. E. 1974. *Ap. J. Lett.* 194:L93

Dyck, H. M., Milkey, R. W. 1972. *Publ. Astron. Soc. Pac.* 84:597

Frogel, J. A., Persson, S. E. 1974. *Ap. J.* 192:351

Gatley, I. et al. 1974. *Ap. J. Lett.* 191:L121

Geisel, S. L. 1970. *Ap. J. Lett.* 161:L105

Gillett, F. C., Capps, R. 1975. *Ap. J. Lett.* In press

Gillett, F. C., Forrest, W. J. 1973. *Ap. J.* 179:483

Gillett, F. C., Stein, W. A. 1971. *Ap. J.* 164:77

Grasdalen, G. L. 1973. *Ap. J.* 182:781

Grasdalen, G. L. 1974. *Ap. J.* 193:373

Grasdalen, G. L., Carrasco, L. 1975. *Astron. Ap.* In press

Grasdalen, G. L., Strom, K. M., Strom, S. E. 1973. *Ap. J. Lett.* 184:L53

Haro, G. 1952. *Ap. J.* 115:572

Harper, D. A. 1974. *Ap. J.* 192:557

Hayashi, C. 1970. *Mem. Soc. Roy. Sci. Liège, Ser. 5* 19:127

Hayashi, C., Hōshi, R., Sugimoto, D. 1962. *Progr. Theor. Phys. Suppl. 22*

Henyey, L. G., LeLevier, R., Levée, R. D. 1955. *Publ. Astron. Soc. Pac.* 67:154

Herbig, G. H. 1951. *Ap. J.* 113:697

Herbig, G. H. 1957. *Ap. J.* 125:612

Herbig, G. H. 1958. *Ap. J.* 128:259

Herbig, G. H. 1960. *Ap. J. Suppl.* 4:337

Herbig, G. H. 1962. *Advan. Astron. Ap.* 1:47

Herbig, G. H. 1966. *Vistas Astron.* 8:109

Herbig, G. H. 1969. *Non-periodic Phenomena in Variable Stars,* 75. Budapest: Academic

Herbig, G. H. 1970a. *Mem. Soc. Roy. Sci. Liège, Ser. 5* 19:13

Herbig, G. H. 1970b. *Ap. J.* 162:577

Herbig, G. H. 1973. *Ap. J.* 182:129

Herbig, G. H. 1974. *Lick Obs. Bull. 658*

Herbig, G. H., Rao, N. K. 1972. *Ap. J.* 174:401

Hunger, K., Kron, G. E. 1957. *Publ. Astron.*

Soc. Pac. 69:347
Knacke, R. F., Strom, S. E., Strom, K. M., Young, E. 1973. *Ap. J.* 179:847
Krishna-Swamy, K. S., O'Dell, C. R. 1967. *Ap. J.* 147:529
Kuhi, L. V. 1964. *Ap. J.* 140:1409
Lada, C. J., Gottlieb, C. A., Litvak, M. M., Lilley, A. E. 1974. *Ap. J.* 194:609
Lar on, R. 1969. *MNRAS* 145:271
Larson, R. 1972. *MNRAS* 157:121
Lee, T. A. 1970. *Ap. J.* 162:217
Lo, K. Y. 1974. PhD thesis. MIT, Cambridge, Mass.
Loren, R. B., Vanden Bout, P. A., Davis, J. H. 1973. *Ap. J. Lett.* 185:L67
Loren, R. B., Peters, W. L., Vanden Bout, P. A. 1974. *Ap. J. Lett.* 194:L103
Magnan, C., Schatzman, E. 1965. *Contrib. Roy. Acad. Sci. Paris* 260:6289
Mathews, W. G. 1969. *Ap. J.* 157:583
Mendoza, E. E. 1966. *Ap. J.* 143:1010
Mendoza, E. E. 1968. *Ap. J.* 151:977
Milkey, R. W., Dyck, H. M. 1973. *Ap. J.* 181:833
Morris, M., Palmer, P., Turner, B. E., Zuckerman, B. 1974a. *Ap. J.* 191:349
Morris, M., Zuckerman, B., Turner, B. E., Palmer, P. 1974b. *Ap. J. Lett.* 192:L27
Penston, M. V., Allen, D. A., Hyland, A. R. 1971. *Ap. J. Lett.* 170:L33
Plagemann, S. 1970. *Mem. Soc. Roy. Sci. Liège, Ser. 5* 19:331
Robinson, L. B., Wampler, E. J. 1972. *Publ. Astron. Soc. Pac.* 84:161
Schwartz, R. 1974. *Ap. J.* 191:419
Serkowski, K. 1969a. *Ap. J. Lett.* 156:L55
Serkowski, K. 1969b. *Ap. J. Lett.* 158:L107
Simon, M., Righini, G., Joyce, R. R., Gezari, D. Y. 1973. *Ap. J. Lett.* 186:L127
Smith, M. 1972. *Ap. J.* 176:617
Spinrad, H., Wood, D. B. 1965. *Ap. J.* 141:109
Strom, K. M., Strom, S. E., Carrasco, L., Vrba, F. J. 1975. *Ap. J.* 196:489
Strom, K. M., Strom, S. E., Grasdalen, G. L. 1974a. *Ap. J. Lett.* 187:L83
Strom, K. M., Strom, S. E., Kinman, T. D. 1974b. *Ap. J. Lett.* 191:L93
Strom, K. M., Strom, S. E., Yost, J. 1971. *Ap. J.* 165:479
Strom, S. E. 1972. *Publ. Astron. Soc. Pac.* 84:745
Strom, S. E., Grasdalen, G. L., Strom, K. M. 1974c, *Ap. J.* 191:111
Strom, S. E., Strom, K. M., Grasdalen, G. L., Capps, R. W. 1974d. *Ap. J. Lett.* 193:L7
Strom, S. E., Strom, K. M., Yost, J., Carrasco, L., Grasdalen, G. 1972. *Ap. J.* 173:353
Vrba, F. J., Strom, K. M., Strom, S. E., Grasdalen, G. L. 1975. *Ap. J.* 197:77
Walker, M. F. 1956. *Ap. J. Suppl.* 2:365
Walker, M. F. 1957. *Ap. J.* 125:636
Walker, M. F. 1959. *Ap. J.* 130:57
Walker, M. F. 1961. *Ap. J.* 133:438
Walker, M. F. 1969. *Ap. J.* 155:447
Walker, M. F. 1972. *Ap. J.* 175:89
Walker, R., Price, S. D. 1974. *AFCRL Infrared Sky Survey.* Preliminary version. Personal communication
Werner, M. W., Elias, J. H., Gezari, D. Y., Westbrook, W. E. 1974. *Ap. J. Lett.* 192:L31
Wynn-Williams, C. G., Becklin, E. E. 1974. *Publ. Astron. Soc. Pac.* 86:5
Wynn-Williams, C. G., Becklin, E. E., Neugebauer, G. 1972. *MNRAS* 160:1
Zuckerman, B. 1973. *Ap. J.* 183:863
Zappala, R. R. 1972. *Ap. J.* 172:57

STELLAR POPULATIONS IN GALAXIES

Sidney van den Bergh
David Dunlap Observatory, University of Toronto, Richmond Hill, Ontario, Canada

1 INTRODUCTION

The concept of stellar populations was introduced by Baade (1944), who noted that galaxies could be thought of as consisting of two distinct population components: population I and population II.[1] On Baade's picture the Hubble (1936) classification sequence E-Sa-Sb-Sc-Ir may be understood in terms of a variation in the relative importance of populations I and II. According to Baade elliptical galaxies consist of pure population II whereas irregulars constitute pure population I. Baade defined population II as consisting of stars that lie on color-magnitude diagrams similar to those of galactic globular clusters, and population I as comprising stars that are located on color-magnitude diagrams similar to those of young or old metal-rich open clusters. It should be emphasized that in Baade's original definition stellar populations were defined in terms of the characteristics of their color-magnitude diagrams.

According to Baade (1944) the stellar population in the central bulge of M31 resolves at the same magnitude level as the globular clusters associated with that galaxy. Baade therefore concluded that the nuclear bulge of M31 consists of globular clusterlike stars. He suggested that this idea could be tested by studying the stellar content of the nuclear bulge of our own Galaxy, which presumably contains stars similar to those that occur in the nuclear bulge of M31. In particular it should be possible to observe the RR Lyrae variables in the nuclear bulge of the Galaxy in low-absorption windows such as that centered on the globular cluster NGC 6522 at $l = 1°, b = -4°$. This expectation was apparently confirmed when Baade (1951) was able to show that the low-absorption window centered on NGC 6522 contains large numbers of RR Lyrae variables. A histogram of the frequency distribution of these variables shows a rapid rise to a maximum $B \approx 17.5$ and then a decline as the line of sight passes beyond the nuclear bulge into the sparsely populated halo region of the Galaxy.

The first doubts about this population picture, in which elliptical galaxies and the cores of spiral nebulae consist of a globular clusterlike population, arose in 1957.

[1] Previous reviews of this subject are by King (1971) and by Spinrad (1966).

In that year Morgan & Mayall (1957) were able to show that the integrated spectrum of the nuclear region of the Andromeda nebula is dominated by cyanogen giants. This conclusion was subsequently confirmed by photoelectric spectrum scans obtained by van den Bergh & Henry (1962). The fact that the cyanogen band $\lambda 4216$ is so strong in the nucleus of M31 shows that light from this region is dominated by strong-lined metal-rich stars.

At first sight the observation that the surface density of RR Lyrae stars in the direction of NGC 6522 was ~ 1000/square degree appeared to give overwhelming support to Baade's hypothesis that the galactic nucleus consisted of globular cluster-like population II stars. Nassau & Blanco (1958) were able to show, however, that the RR Lyrae stars in this field are outnumbered two to one by M-type giants. Because M giants do not occur in halo-type globular clusters it follows that the dominant stellar population in the nuclear bulge cannot be pure population II. This conclusion is confirmed by detailed color-magnitude diagrams of the nuclear bulge (van den Bergh 1971a, 1974a) and by studies of the integrated light of the Sagittarius star cloud by Morgan (1959), which show that the light from this region is dominated by strong-lined cyanogen giants.

Observations of the color-magnitude diagram of the nuclear bulge at $b = -4°$ by van den Bergh (1971a) show that the dominant stellar population at $Z = -0.6$ kpc resembles that in the old metal-rich cluster NGC 188. Metal-poor halo stars account for at most a few percent of all giant stars. At $Z = -1.2$ kpc the nuclear bulge is observed (van den Bergh 1974a) to contain a larger fraction of population II giants than it does at $Z = -0.6$ kpc.

UBV photometry of star-free sky patches in the nuclear bulge (van den Bergh 1971a) (see Table 1) also shows that most of the brightest main-sequence stars in the nuclear bulge must also be metal rich.

The model that emerges from these studies is that the dominant stellar population in the nuclear bulges of the Galaxy and M31 consist of old metal-rich stars (similar to those in the old open cluster NGC 188) and that these regions contain only a sprinkling of true globular cluster stars. A similar conclusion was first reached by Arp (1965).

Baade's 1944 population picture was revised and extended by Oort (1958), who suggested that the population of the Galaxy might be thought of as consisting of five population subtypes: 1. halo population II, 2. intermediate population II, 3. disc population, 4. intermediate population I, and 5. extreme population I.

Table 1 Colors of galactic nuclear bulge

	B-V[a]	*U-B*[a]
M3	0.54	0.0:
Nuclear bulge	0.77 ± 0.04	0.33 ± 0.06
NGC 188	0.80	0.37

[a] Integrated color indices (corrected for reddening) of all stars with $M_V > +2.5$.

2 STELLAR POPULATIONS IN THE GALAXY

Classification Type and Luminosity of the Galaxy

A number of lines of evidence, summarized in Table 2, suggest that the Galaxy has a Hubble-type intermediate between Sb and Sc. Two of the entries in this table require more detailed comment. It has been shown (van den Bergh 1971b) that there exists a surprisingly tight correlation between the maximum rotational velocity in spiral galaxies and their intrinsic B-V color. According to Plaut & Oort (1975), the circular velocity near the sun, which is very close to V(max), is $\gtrsim 200$ km sec^{-1}. Substitution of this value into the observed relation

$$V(\max) = -53 + 424(B\text{-}V)_0 \qquad 1.$$

yields $(B\text{-}V)_0 \gtrsim 0.60$. This value compares with $\langle B\text{-}V\rangle_0 = 0.56$ for galaxies of type Sc I-II and $\langle B\text{-}V\rangle = 0.64$ for galaxies of type Sb I-II. The value $(B\text{-}V)_0 \gtrsim 0.60$ is consistent with the work of Schmidt-Kaler & Schlosser (1973) who, on the basis of a study of the galactic nuclear bulge, conclude that the Galaxy has an intrinsic color $(B\text{-}V)_0 = 0.61 \pm 0.13$.

One of the most remarkable features of the Galaxy discovered from surveys at radio wavelengths is the concentration of H II regions situated within 1°.2 (200 pc) of the galactic nucleus. This region of violent star-forming activity is apparently embedded in a much larger zone that is almost completely devoid of star-forming activity. Inspection of photographs of external galaxies shows that this kind of distribution of star-forming activity is frequently (but not exclusively) observed in barred spirals.

In Table 3 (van den Bergh 1972b) the Galaxy is compared with M31 and M33. The data in this table show that the Galaxy is intermediate between M31 and M33 in mass and in size; it appears to be somewhat closer to M31 ($M_V = -21.1$) than it is to M33 ($M_V = -18.8$). This suspicion is strengthened by the observation that the number of galactic globular clusters (Hogg 1973) is more similar to that in M31 (Vetešnik 1965) than it is to that in M33 (Hiltner 1960). On the basis of these results it will be assumed that the Galaxy has an absolute visual magnitude $M_V = -20.5 \pm$

Table 2 The Hubble type of the galaxy

Type of information	Favors Sb	Favors Sc
Fraction of mass in H I	x	
Shape of 21-cm spiral arms	?	
Shape of optical spiral arms		?
Rate of star formation		x
Existence of giant H II regions		?
Presence of a central bulge	?	
H II regions near nucleus		x
$V_{rot}(\max) > 200$ km sec^{-1}	x	

Table 3 Comparison of the Galaxy, M31, and M33

	M31	Galaxy	M33
Absolute magnitude M_V	−21.1	?	−18.8
No. globular clusters	~200	~130	~6
Total mass ($\mathcal{M}_\odot$)	3.1×10^{11}	1.3×10^{11}	3.9×10^{10}
Distance to $\sigma = \sigma_\odot$ (kpc)[a]	12	9	4

[a] $\sigma_\odot$ is the surface brightness of the Galaxy near the Sun as seen from the direction of the galactic pole [$\sigma_\odot \approx 23.4$ mag (arc sec)$^{-2}$].

0.3 corresponding to $L = (1.4 \pm 0.4) \times 10^{10} L_\odot$. With this value the visual mass-to-light ratio of the Galaxy (in solar units) is ~9.

Stellar Populations and Composition Gradients in the Galaxy

AGE GRADIENTS The color-magnitude diagram of stars near the Sun shows that these objects constitute a mixed population that contains both young stars of population I, large numbers of old disk stars, and a sprinkling (Bond 1970, Sturch & Helfer 1972) of metal-poor stars. The radial variation in the relative importance of old and young stellar populations in the Galaxy is dramatically illustrated on spectra published by Morgan & Osterbrock (1969). Their results show that the spectrum of the integrated light of the nuclear bulge of the Galaxy, observed through a transparent window near the globular cluster NGC 6522 ($l - 1°$, $b = -4°$), is similar to that of the nuclear bulge of M31. The spectral type in the blue region is observed to be near K0 in both cases and the metallic line indicator near $\lambda 4390$ is strong, indicating that most of the light in the blue region of the spectrum in both galaxies is contributed by metal-rich stars. A spectrum obtained in a partially obscured region about 2° from NGC 6522 shows a strengthening of $H\gamma$ and $H\delta$ relative to the G-band yielding an overall integrated spectral type near F8 for the galactic disk. This observation shows that the integrated spectral type in the Galaxy exhibits a radial variation in the same sense as that which is observed in the Andromeda nebula.

COMPOSITION GRADIENTS In addition to this evidence for a radial variation in the relative importance of old and young stellar populations, there is evidence for a metallicity gradient within the galactic disk. The first evidence for such a gradient was obtained by van den Bergh (1958), who showed that Cepheids with periods in the range 2–4 days are predominantly located in the anticenter direction whereas Cepheids with periods of 7–9 days mainly occur in the direction toward the galactic center. Furthermore Hartwick (1970) finds that the ratio of the number of red to blue supergiants increases with increasing distance from the galactic center as it does in M33 (Walker 1964, Madore 1971). Finally Janes & McClure (1972) find that K giants with unusually strong cyanogen absorption predominantly enter the solar vicinity on orbits that connect with the inner part of the galactic disk. These results are entirely consistent with those of Peimbert (1968) and Searle (1971), who find strong evidence for a radial composition gradient in the interstellar gas in spiral nebulae. It would be particularly interesting to see whether such a composition gradient could be detected in the galactic disk by intercomparison of H II regions in the Sagittarius and Perseus spiral arms.

Evolution of Stellar Population Components in the Galaxy

INTRODUCTION It appears highly probable that the evolutionary history of galaxies is very complex and that both the rate of star formation and the mass spectrum by which stars are formed depends on physical conditions in the star-forming regions. Only our ignorance of the past and our desire for simplicity lead us to hope that the mass spectrum of star formation and the rate of star formation are independent of position and time.

A number of lines of observational evidence show that these hopes are not fulfilled. The integrated light of elliptical galaxies is dominated by old stars, whereas the light of irregular galaxies is dominated by young ones. This observation shows that the mean rate of star formation differs in different kinds of galaxies. (In the central bulges of spiral nebulae the rate of star formation has obviously declined much more rapidly than it has in the outer spiral arm regions.) Furthermore the mass-to-light ratio in giant elliptical galaxies (Dickens & Peach 1972) is obviously much greater ($\mathscr{M}/L \sim 60$) than it is in globular clusters (Illingworth 1973) for which $\mathscr{M}/L \sim 1.5$. Furthermore there are strong indications (Ostriker & Peebles 1973) that the mass-to-light ratio in the halos of galaxies may be much greater than it is in their cores and disks.

Models for the evolution of stellar populations in the Galaxy are subject to the following principal observational constraints:

1. They must be able to account for the observed frequency distribution of metal abundance that is observed in main-sequence stars near the Sun. In particular such models should be able to account for the fact that very metal-poor stars are exceedingly rare
2. Evolutionary models should be able to account for the radial abundance gradient that is observed within the disks of other galaxies and that is also inferred to be present in our own Milky Way system
3. Such models must be able to account for the fact that stars in the galactic halo are generally much more metal poor than are those in the galactic disk
4. Any theory of galactic evolution must account for the fact that the heavy element abundances in disk stars have not changed by more than a factor of 2 during the last $\sim 1 \times 10^{10}$ years (Hearnshaw 1972)
5. Models must be consistent with the stellar luminosity function that is observed in the solar neighborhood
6. Among external galaxies models for stellar evolution must be able to account for the fact that the average metallicity of stars in luminous galaxies is greater than it is in dwarfs
7. Such models must be consistent with the observation (McClure & van den Bergh 1968b) that elliptical galaxies of a given luminosity have the same metallicity regardless of whether or not they are located in clusters

MODELS WITH A TIME-DEPENDENT MASS SPECTRUM OF STAR FORMATION The first attempt to account for the stellar luminosity function near the Sun was made by Salpeter (1955), who assumed both the rate of star formation and the mass spectrum of star formation to be time independent. Subsequently van den Bergh (1957) and Schmidt (1959) elaborated on Salpeter's work by allowing a variable (decreasing) rate of star formation but retaining an invariant mass spectrum of star formation. It was first noted by van den Bergh (1962) and Schmidt (1963) that such simple models have great difficulty in accounting for the paucity of metal-poor stars in the vicinity of the Sun. A number of interesting suggestions have been made to overcome this difficulty. Schmidt (1963) proposed that the first generation of stars contained more massive objects than did subsequent generations. Because heavy elements are formed predominantly by massive stars, such a model would lead to both a very rapid, initial, heavy-element enrichment in the Galaxy and an embarrassing overproduction of heavy elements. This difficulty can be avoided (Truran & Cameron 1971) by assuming that high-mass stars were formed only during a brief initial phase corresponding to the formation of the galactic halo.

Variations in the relative numbers of massive stars will alter the relative production rates of different elements. According to Arnett & Schramm (1973), excellent agreement between observed and theoretical abundances is obtained from the presently observed mass spectrum of star formation. This suggests (Tinsley 1974) that only the low-mass end ($\mathscr{M} \lesssim \mathscr{M}_{\odot}$) of the mass spectrum of star formation varies as a function of time. On this hypothesis the initial spectrum of stellar masses contained very few stars with $\mathscr{M} < \mathscr{M}_{\odot}$. As Tinsley (1974) points out, this idea is not entirely ad hoc, because it is physically plausible that the lower rate of cooling in metal-poor gas would inhibit the formation of low-mass stars. Some tentative observational support for this speculation may be provided by the inference by Searle & Sargent (1972) that only stars with masses greater than $\sim 2\mathscr{M}_{\odot}$ are currently forming in two metal-poor dwarf irregular galaxies that they have studied. If this conclusion is correct, then the mass spectrum of star formation in the Small Magellanic Cloud (SMC) may also turn out to be deficient in faint stars. On this hypothesis it would, however, be difficult to account for the existence of vast numbers of low-mass stars in (presumably metal-poor) galactic halos (Ostriker & Peebles 1973).

GAS INFLOW MODELS Larson (1972, 1974a) has proposed models for the formation of galaxies in which metal-poor gas is continually streaming in to the star-forming regions of galaxies. In such models evolving stars can keep the metallicity of the interstellar gas quite high in the star-forming regions *if* the influx of metal-poor gas is sufficiently low. A difficulty with Larson's inflow model is that it can only rearrange the frequency distribution of stellar metallicity but it cannot change the overall mean metallicity of stars. This conclusion is, of course, dependent on the assumption that a galaxy can be treated as a closed system in which the effect of its internal dynamics is mainly to redistribute the heavy elements without significantly changing the amount produced.

To explain the observed dependence of metallicity on galaxy luminosity, one would have to invoke differences in the mass spectrum of star formation. The high metallicity of giant galaxies would, on this hypothesis, require a luminosity function that is deficient in faint stars. One might therefore expect giant ellipticals to have a lower mass-to-light ratio than dwarfs. This is the exact opposite of what is actually observed. Alternatively Larson (1974a, b) suggests that small galaxies may tend to lose their processed gas and that massive galaxies may tend to accrete such metal-enriched gas from their surroundings. Although this hypothesis may account for the fact that the more massive galaxies tend to have higher metal abundancies, it would appear to be inconsistent with the observation (McClure & van den Bergh 1968b) that ellipticals in clusters have a metallicity similar to field ellipticals of the same brightness. (This objection could of course be met by assuming that all field ellipticals are escapees from clusters.)

CHEMICALLY INHOMOGENEOUS MODELS Searle (1972) and Talbot & Arnett (1973) have pointed out that evolving galaxies are likely to be chemically inhomogeneous. If it is assumed that star formation takes place preferentially in those interstellar clouds that have the highest metallicity (Talbot 1974), then it may be possible to understand the paucity of metal-poor stars near the Sun. Against this picture it may be argued that the SMC, which is moderately metal poor, seems to have no particular trouble forming stars. This point is significant because stars with $Z \approx Z_{\odot}/4$ similar to those in the SMC are observed to be relatively rare near the Sun.

METAL-POOR DWARF STARS MOSTLY IN HALO It might be assumed that the deficiency of metal-poor stars in the Galaxy is a figment of the observations, which results from the fact that most low-mass stars are situated in the Ostriker-Peebles (1973) halo and are therefore not observed. This explanation appears implausible because the low frequency of metal-poor stars has been derived for objects with $\mathcal{M} \approx \mathcal{M}_{\odot}$. The very low frequency with which planetary nebulae occur in the galactic halo (van den Bergh 1973b) shows that stars of $\sim 1\ \mathcal{M}_{\odot}$ are in fact quite rare in the galactic halo.

Enrichment in Disk or Halo?

Galaxies consist of three distinct population components: 1. a central spheroid embedded in a spherical halo, 2. a disk component, and 3. a nucleus. The relative importance of these three components differs from galaxy to galaxy. Nuclei are observed only in galaxies that are more luminous than $M_V \approx -15$ (van den Bergh 1972c). Dwarf spheroidals contain only a spherical component whereas giant ellipticals contain both a spherical component and a nucleus. Spiral galaxies and galaxies of type S0 usually contain all three population components. The relative importance of the disk and spherical components does, however, vary greatly from object to object. For example, Freeman (1973) has shown that the spheroidal component is much more important relative to the disk in NGC 4459 than it is in NGC 598 (M33). In general, galaxies with well-developed spherical components such as the Galaxy and M31, are accompanied by large numbers of globular clusters, whereas galaxies similar to M33, in which the disk component is dominant, contain but few globulars.

In a model such as that advocated by Schmidt (1963) or in the prompt initial enrichment model (Truran & Cameron 1971) the first generation of heavy-element-producing massive stars were members of the halo population. On this picture one would therefore expect that galaxies such as M33, which did not experience a vigorous burst of star formation during the halo phase of their evolution, may be metal poor. In fact M33 probably has a relatively normal heavy-element abundance (Smith 1974) with O and Ne down by only a factor of ~ 2 in the outer H II regions. This suggests that heavy-element abundance enrichment most probably took place within the disk of M33 itself. If this is true in M33 then it may equally well be argued that heavy-element enrichment in the Galaxy was also a result of the formation of massive stars in the galactic disk. In fact it seems rather unreasonable to assume that the galactic halo simultaneously formed both very massive stars, which produced most of the galactic heavy elements, and a population of faint M dwarfs, which account for the missing halo mass that is required to stabilize the galactic disk.

3 STELLAR POPULATIONS IN THE MAGELLANIC CLOUDS

Introduction

Because of their proximity to the Galaxy the Magellanic Clouds are ideally suited for studies of their population content. From the wealth of observational data accumulated during recent years it appears that there are numerous important differences between the Galaxy and the Magellanic Clouds. Most of these differences can be understood in terms of differences in the evolutionary histories of the Galaxy and the Clouds:

1. In the Magellanic Clouds the present rate of star formation appears to be not very different from its average rate during the last $\sim 1 \times 10^{10}$ years, whereas in the Galaxy the present rate of star formation is much lower than it was in the past. As a result, the average age of galactic stars is much greater than the average age of stars in the Magellanic Clouds
2. Stars of any age in the Magellanic Clouds have a lower heavy-element abundance than do stars of similar age in the Galaxy. For young stars of population I Z (Galaxy) ≈ 0.03, Z [Large Magellanic Clouds (LMC)] ≈ 0.025, and Z(SMC) ≈ 0.01
3. The history of cluster formation in the Galaxy and in the Clouds is radically different

Star Clusters in the Magellanic Clouds

The formation of star clusters in the Galaxy may naturally be divided into two quite distinct eras. During the first of these, which coincided with the collapse of the protogalaxy, very massive globular clusters were formed in the galactic halo. During the second phase of galactic evolution low-mass star clusters were formed in the galactic disk. For reasons that are not understood the masses of open clusters formed in the galactic disk are $\sim 10^2$ times smaller than those of the globular clusters formed during the halo phase of galactic evolution. In the Magellanic Clouds (perhaps because they are irregular galaxies) no such dichotomy exists. Not only are there massive old globular clusters such as NGC 121 (SMC) and NGC 1466 (LMC), but there are also massive young clusters such as NGC 330 (SMC) and NGC 1866 (LMC). Furthermore the clusters NGC 419 and K3 are almost certainly massive, intermediate age clusters. These moderately metal-poor objects of intermediate age are of particular importance because they do not have a galactic counterpart. [A possible exception is the galactic open cluster NGC 2420 (McClure, Forrester & Gibson 1974, Keenan & Innanen 1974).]

To avoid confusion it appears desirable to avoid the term "globular cluster" for the massive young and intermediate age clusters in the Magellanic Clouds. Following Hodge we have adopted the nomenclature "populous clusters" for these objects.

All available data on the color-magnitude diagrams of populous old clusters in the Magellanic Clouds are assembled in Table 4. These data are discussed below.

POPULATION GRADIENT ALONG THE HORIZONTAL BRANCH All of the old LMC clusters studied have blue horizontal branches whereas those in the SMC have red horizontal branches.

THE REDDEST GIANT STARS The reddest giant stars in the SMC clusters are redder than the reddest stars in the LMC clusters. The reddest LMC giants are, however, still redder than the stars at the tips of the giant (asymptotic?) branches of galactic globular clusters.

MAGNITUDE LEVEL OF THE HORIZONTAL BRANCH For the five LMC clusters studied down to the horizontal branch (HB), $19.1 \leq V(\mathrm{HB}) \leq 19.6$. This dispersion could

Table 4 Observational data on old populous clusters in the Magellanic Clouds

Cluster	V(HB)	$V_{1.4}$[a]	Δ	$(B\text{-}V)$(max)[b]	HB	Reference
SMC:						
NGC 121[c]	19.5	16.85	2.65	1.6	Red	Tifft (1963)
NGC 339	—	16.05:	—	1.3	—	Gascoigne (1966)
NGC 361	—	16.50	—	2.2	—	Arp (1958)
NGC 419	19.3	16.80	2.5:	4.1	Red	Walker (1972b)
L1	—	16.95	—	2.2	Red?	Gascoigne (1966)
K3	19.1	16.65	2.45	2.3	Red	Walker (1970)
LMC:						
NGC 1466[c]	19.1	16.30	2.8	1.5	Blue	Gascoigne (1966)
NGC 1783	19.2	16.80	2.4	1.8	Blue	Gascoigne (1962)
NGC 1841	19.6	16.60	3.0	1.7	Blue	Gascoigne (1966)
NGC 1978[c]	—	16.45	—	2.0	—	Hodge (1960)
NGC 2209	19.6	17.40	2.2	1.9	Blue	Gascoigne (1966)
NGC 2231	—	16.5:	—	1.6:	—	Gascoigne (1966)
NGC 2257[c]	19.4	16.10	3.3	1.8	Blue	Gascoigne (1966)

[a] Magnitude of giant branch at $(B\text{-}V)_0 = 1.40$.
[b] In some clusters the reddest star may be a field object.
[c] Cluster contains RR Lyrae variables.

be accounted for *if* 1. all the LMC cluster horizontal branches have the same absolute magnitude and 2. the LMC cluster system has a depth of ~10 kpc along the line of sight. Alternatively the observed dispersion in V(HB) may be due to a real spread in M_V(HB) of the LMC clusters. For the Large Cloud, $\langle V(\mathrm{HB})\rangle = 19.4 \pm 0.1$. With $(m\text{-}M)_V = 18.6 \pm 0.1$ for the LMC, this yields $\langle M_V(\mathrm{HB})\rangle = +0.8$.

There is marginal evidence that the two LMC clusters that contain RR Lyrae variables have brighter horizontal branches than do those in which no RR Lyrae stars have been found. In NGC 1466, V(HB) = 19.1 (Gascoigne 1966); and in NGC 2257, V(HV) = 19.4 (Gascoigne 1966) or V(HB) = 19.2 (Walker 1972a).

For the SMC globular NGC 121, we find that V(HB) = 19.5 (Tifft 1963). With $(m\text{-}M)_V = 18.9$ for the Small Cloud, this yields M_V(HB) = +0.6. A significantly brighter value is found for the cluster K3, which is probably of intermediate age. For this cluster V(HB) = 19.1 (Walker 1970), and hence M_V(HB) = +0.2. These results suggest (but do not prove) that: 1. clusters containing RR Lyrae variables have brighter horizontal branches than do clusters that do not; and 2. intermediate age clusters have brighter horizontal branches than do true globulars.

MAGNITUDE OF BRIGHTEST CLUSTER GIANTS Harris (1975) has introduced a parameter, δV_5, which he defines as the magnitude difference between the fifth brightest cluster red giant star and the horizontal branch. He finds that δV_5 is the same in *all* galactic globular clusters after a small correction for cluster richness has been applied. In particular δV_5 is independent (or almost independent) of cluster metallicity. With M_V(HB) = +0.5, Harris's data yield $M^*_{V5} = -2.3$ for the fifth brightest red giant. (M^*_{V5} denotes the value of M_{V5} *corrected* to the value that would be observed if the cluster had 100 giants brighter than the horizontal branch.) The value $M^*_{V5} = -2.3$ is in reasonably good agreement (see Table 5) with the values

Table 5 Magnitude of fifth brightest cluster star

Cluster	V_5	n_g	V_5[a]	M^*_{V5}	Reference
SMC:					
NGC 121[b]	16.8	50	16.5	−2.4	Tifft (1963)
NGC 339	17.5	56	17.2	−1.7	Gascoigne (1966)
NGC 361	16.4	109	16.4	−2.5	Arp (1958)
NGC 419	16.2	70	16.0	−2.9	Walker (1972b)
L1	16.6	93:	16.6	−2.3	Gascoigne (1966)
K3	16.6	37	16.1	−2.8	Walker (1970)
LMC:					
NGC 1466[b]	16.6	39	16.1	−2.5	Gascoigne (1966)
NGC 1783	16.5	80:	16.4	−2.2	Gascoigne (1962)
NGC 1841	16.6	74	16.5	−2.1	Gascoigne (1966)
NGC 2209	18.0:	19	17.2:	−1.4:	Gascoigne (1966)
NGC 2231	17.4	60:	17.2	−1.4	Gascoigne (1966)
NGC 2257[b]	16.2	71	16.0	−2.6	Gascoigne (1966)

[a] Corrected to $n_g = 100$ using the precepts given by Harris (1975).
[b] Cluster contains RR Lyrae variables.

$M_{V5}^* = -2.4$ (NGC 121), $M_{V5}^* = -2.5$ (NGC 1466), and $M_{V5}^* = -2.6$ (NGC 2257) that are found for the clusters known to contain RR Lyrae variables. The most luminous stars in possible intermediate age clusters, such as NGC 419 ($M_{V5}^* = -2.9$) and K3 ($M_{V5}^* = -2.8$), are significantly *brighter* than those in galactic globular clusters.

Strong observational data by Walker (1970) yield δV_5 (observed) = 2.5 for the SMC cluster K3. This value is in excellent agreement with δV_5 (predicted) = 2.34 that is obtained from Harris's work. *This suggests that the magnitude of the brightest cluster stars and the magnitude level of the horizontal branch shift up or down together.*

On the basis of their ΔV values most of the *old* clusters in the Magellanic Clouds appear to be of intermediate metallicity. This conclusion is confirmed by the spectroscopic work of Andrews & Lloyd-Evans (1971). According to Ford (1971), the Fe I lines are stronger in NGC 419 (SMC) and in NGC 1783 (LMC) than they are in the other clusters.

The fact that massive clusters have been forming more or less continuously in the Small Magellanic Cloud suggests that there may not have been a great burst of cluster formation when the SMC collapsed (cf Freeman & Munsuk 1972). If this conclusion is correct, then with the small sample of massive clusters that have so far been studied, there is no a priori reason to believe that any of the populous clusters so far investigated in detail are in fact as old as the SMC itself. This hypothesis leads to a natural explanation of the observation that all SMC globulars, for which color-magnitude diagrams are available, have stubby red horizontal branches despite the fact that they appear to be relatively metal poor. According to Rood (1973) this is exactly what one would expect if these clusters are (1 or 2) $\times 10^9$ years younger than the globular clusters in the Galaxy. Recent work on mass loss by red giant stars seem to suggest (Fusi-Pecci & Renzini 1975) that the observed differences in globular cluster horizontal branch morphology might be accounted for by age differences only half as large as those found by Rood.

The evolution of the SMC appears to differ somewhat from that of the LMC. The oldest Large Cloud clusters studied have blue horizontal branches. This suggests that these clusters were formed during the collapse of the Large Cloud and hence might have ages that are almost as great as those of galactic globular clusters. The fact that a few such old clusters are observed in the Large Cloud suggests that the LMC (which is more massive than the SMC) started its evolution with a more violent burst of star (and cluster) formation than did the Small Cloud. In this connection it is interesting to note that M33 contains only half a dozen clusters, which, on the basis of their colors (Hiltner 1960), may be regarded as globular clusters. This observation suggests that the globular cluster forming era in M33 was even less active than it was in the Magellanic Clouds. In this respect the Clouds and M33 differ from the Galaxy and the Andromeda nebula, in which large numbers of halo clusters were formed.

Additional support for the notion that clusters such as NGC 419 constitute an intermediate age population is provided by the observation that field stars similar to those on the giant branch of NGC 121 occur throughout the halo of the SMC (Tifft 1963), whereas bright red giants like those in NGC 419 are observed only in the vicinity of the SMC Bar.

Both old and young star clusters in the Magellanic Clouds differ from their galactic counterparts in a number of interesting ways: 1. No galactic globular clusters are as highly flattened as NGC 121 in the Small Cloud and NGC 1978 in the Large Cloud. (Both of these clusters contain RR Lyrae stars and are therefore true globulars.) 2. Cloud globular clusters appear to have a lower mean surface brightness than do their galactic counterparts.

The color-magnitude diagrams of the globular clusters in the Magellanic Clouds and of old populous clusters such as K3 and NGC 419 differ from those of galactic globular clusters in that many of them contain red (asymptotic?) giant branch stars with B-$V > 2.0$ (van den Bergh 1968). Such very red stars are not known to occur in galactic globular clusters.

These observations suggest that physical conditions prevailing in the Magellanic Clouds during the era of globular cluster formation differed systematically from those prevailing during the early evolution of the Galaxy.

Young clusters in the Galaxy and in the Magellanic Clouds (Hagen & van den Bergh 1974) also exhibit a number of differences. Figure 1 shows a comparison between the color-magnitude diagrams of the young populous SMC cluster NGC 330 and stars in galactic clusters with ages of $\sim 15 \times 10^6$ years. The figure shows that the red giants in NGC 330 are brighter and bluer than are their galactic counterparts. A similar but slightly less pronounced difference between galactic clusters and the populous young LMC cluster NGC 1866 has been noted by Hagen (1974). Intercomparison of these results with theoretical evolutionary tracks strengthens and confirms the conclusion that the observed differences between

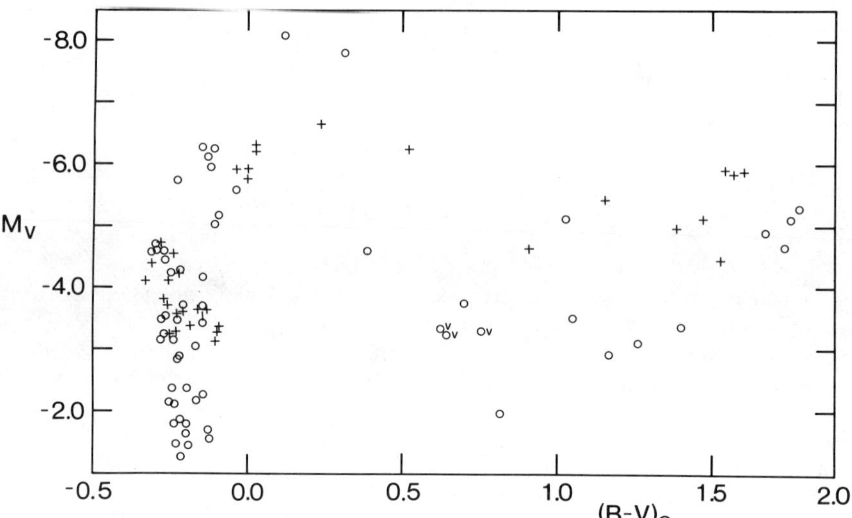

Figure 1 Comparison between the color-magnitude diagrams of the young populous SMC cluster NGC 330 and stars in galactic open clusters with ages $\sim 15 \times 10^6$ years (Hagen & van den Bergh 1974). ○ galaxy, + NGC 330.

clusters in the Galaxy and in the Clouds can be accounted for in terms of metal abundance differences between the Clouds and the Galaxy.

A color-magnitude diagram for the brightest stars in the Clouds (Walraven & Walraven 1971) shows a deficiency of supergiants with spectral types later than F5 in the SMC that is not present in the LMC. Probably this effect is due to the fact that $Z(\mathrm{SMC}) < Z(\mathrm{LMC})$.

Distribution of Stellar Populations in the Clouds

Observations by Hogg (1955) and by Elsässer (1958) show that the core of the SMC is much bluer than its outer regions. The work by Searle, Sargent & Bagnuolo (1973) shows that the integrated color $(B\text{-}V) \approx 0.2$ observed for the region within 0.5° (0.6 kpc) of the center of the Small Cloud can be explained only if the present rate of star formation in the core of the Small Cloud is greater than its average rate during the last $\sim 1 \times 10^{10}$ years.

The composite nature of the Small Cloud is also strikingly evident when photographs taken in the blue and near infrared (Johnson 1961, Walker, Blanco & Kunkel 1969a,b) are intercompared. Such a comparison shows that the young blue stellar population, which is concentrated near the center of the Small Cloud, is superimposed on a much more widely distributed old red population.

The conclusion that a burst of star formation has recently taken place in the SMC Bar is confirmed by observations of Cepheids (Payne-Gaposchkin & Gaposchkin 1966). The Cepheids of longest period in the Small Cloud are more strongly concentrated in the core of the SMC than are those of shorter period. In this respect the Small Cloud differs radically from the Large Cloud. In the LMC regions of active star formation, as outlined by associations, H II regions, long-period Cepheids, and Wolf-Rayet (WR) stars are currently distributed more or less uniformly over the face of the Large Cloud. Using the Cepheid age calibration of Tammann (1969) it is found that star formation, distributed more or less uniformly over the face of the LMC, has been taking place for $\sim 3 \times 10^7$ years. A totally different pattern emerges for the distribution of Cepheids with periods in the range $0.7 < \log P < 0.9$ which have ages in the range $(4\text{–}5) \times 10^7$ years. Cepheids in this period range are found to be very strongly concentrated near the center of the LMC Bar. The conclusion that the Bar was an active center of star formation until fairly recently is confirmed by Tifft & Snell (1971), who find that the color-magnitude diagram of the Bar contains a narrow vertical blue sequence of relatively young stars that extend up to $V = 14.5$ ($M_V = -4.2$). Associated with this population are a few bright red giant stars with magnitudes in the range $14 < V < 16$. The relative paucity of supergiants in the region of the Bar confirms the impression that the LMC Bar is not presently a site of very active star formation. The remarkable shift from a concentration of star formation in the Bar to star formation predominantly outside the Bar that took place only 4×10^7 years ago remains entirely unexplained.

The work of Tifft & Snell (1971) shows that the character of the stellar population in the LMC Bar changes quite dramatically at $V \approx 16$ where numerous red giants with colors up to $B\text{-}V \approx 1.8$ suddenly occur. A few very red stars with $B\text{-}V > 2.0$, which are almost certainly associated with a globular cluster or old populous cluster asymptotic branch population, are also seen near $V = 16$. It would be particularly important to see whether the Bar of the Large Cloud contains large numbers of normal population I red giants with $M_V \approx +0.5$ ($V = 19.2$). Due to severe crowding problems in the LMC Bar the answer to this question must await the completion of the present generation of 4-m telescopes in the southern hemisphere. Nevertheless, available population studies appear to favor the view that the LMC Bar consists predominantly of what, in our own Galaxy, would be referred to as old population I.

On the basis of this picture it is, however, rather puzzling that the 11 novae observed in the Large Cloud do not exhibit a very noticeable concentration toward the Bar. The fact that planetary nebulae (Westerlund & Smith 1964) do not exhibit a concentration to the Bar of the LMC may be accounted for by selection effects. In this connection it should be emphasized, however, that the planetaries in the Small Cloud are observed to be concentrated toward the SMC Bar. This puzzle may be resolved by future observations of the LMC Bar with one of the new southern 4-m telescopes. Such observations should reveal planetary nebulae with $V > 19$ as small faint disks with diameters $> 1''$.

Variable Stars

CEPHEIDS A number of significant differences are observed between the variable stars in the Galaxy and in the Magellanic Clouds. Perhaps the best documented of these (see Figure 2) is that the mean periods of the Cepheids in the Large and Small Clouds are significantly shorter than the mean period of Cepheid variables in the Galaxy. The median values of the periods of Cepheids in the Magellanic Clouds (Payne-Gaposchkin 1971) are listed in Table 6. The difference in period frequency distribution of Cepheids in the LMC and in the SMC has been known for a long time and is certainly real. Payne-Gaposchkin has emphasized the fact that the Harvard material, on which differences in period frequency between the two Clouds is found, is based on plates for the two Clouds that are strictly comparable because they were obtained with the same instruments. Furthermore this material was studied by the same investigators using very similar methods.

Within the Galaxy (van den Bergh 1958) the mean period of Cepheids is a function of distance from the galactic center. A number of investigations, such as those by McClure (1969), Peimbert (1968), and Searle (1971), suggest the existence of a metal abundance gradient within the disks of spiral galaxies. It therefore seems reasonable to assume that the mean periods of Cepheids are related to the metal abundance of the interstellar gas in the regions in which they are formed. The fact that both the mean period and the metal abundance increase toward the nuclei of spirals suggests that a long mean Cepheid period goes with a high metal abundance. If this conclusion is correct, then the Small Cloud has a lower metal abundance than does

Table 6 Period (in days) of Cepheid variables

	Galaxy	LMC	SMC
Median log P	0.785:	0.635	0.422

the Large Cloud. By the same token the metal abundance in the LMC should be intermediate between that in the Galaxy and the Small Cloud.

Gascoigne (1969) has found that the SMC Cepheids are 0.1 mag bluer in *B-V* than the Cepheids in the Galaxy. This conclusion is strengthened and confirmed by the recent *UBV* photometry of Cloud Cepheids by Madore (1974). According to Bell & Parsons (1972) the observed color difference between the Cepheids in the Galaxy and in the Small Cloud may be accounted for in terms of a difference in line blanketing resulting from a metal abundance four times lower in the SMC Cepheids than in their galactic counterparts.

NOVAE Observations of novae in the Galaxy and in M31 (van den Bergh 1975b) show a relatively tight correlation between the absolute magnitude of a nova at maximum light and its rate of decline. Application of this relation to the novae in the LMC yields an apparent distance modulus, $(m\text{-}M)_V = 19.3$. This value is significantly greater than the mean distance modulus, $(m\text{-}M)_V = 18.6$, obtained from all distance indicators. This observation suggests that the novae in the Magellanic Clouds may differ systematically in maximum luminosity from their galactic counterparts. The plausability of this hypothesis cannot be adequately assessed because the existence of

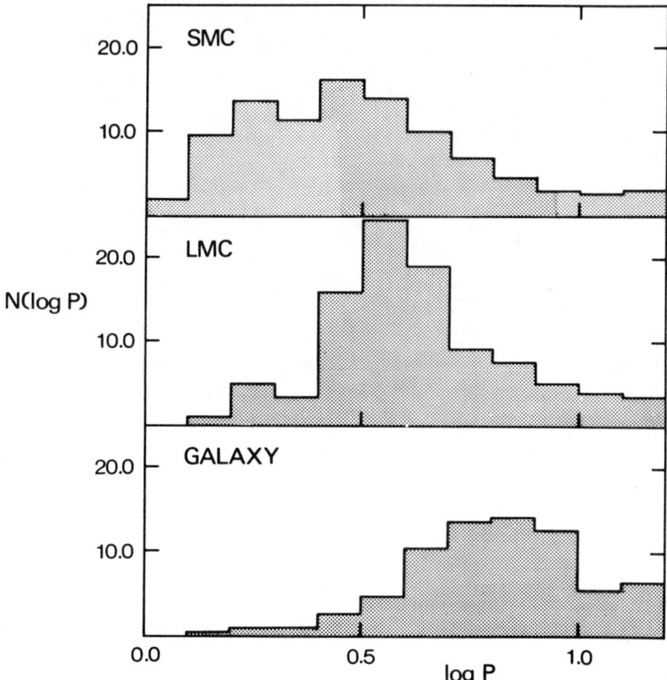

Figure 2 Comparison of the frequency distribution of Cepheid periods in the SMC, the LMC, and the Galaxy.

a relationship between the maximum brightness of novae and their rate of decline is not understood theoretically.

The discrepancy between the observed and the expected magnitudes of novae is particularly glaring in the case of Nova Doradus 1971a for which excellent photoelectric observations are available (Ardeberg & de Groot 1973). From the period vs rate of decline relationship derived from novae in the Galaxy and in M31 (van den Bergh 1975b) this nova (for which $t_2 = 16.8$ days) would be expected to have $M_B(\max) = -8.1$. With $B(\max) = 11.86$ this yields $(m\text{-}M)_B \approx 20.0$. The blue color of Nova Doradus 1971a near maximum light shows that this discrepancy cannot be accounted for by interstellar absorption.

As in the case of the LMC, the SMC novae also give a larger distance modulus than do the other distance indicators. For SMC novae $(m\text{-}M)_V = 19.4$ compared to $(m\text{-}M)_V = 18.9$ for the mean of all distance determinations to the Small Cloud. It is therefore concluded that the novae in the Magellanic Clouds are similar to each other but differ from those in the Galaxy and in M31.

VARIABLE M-TYPE SUPERGIANTS Lloyd-Evans (1971) finds that $\sim 15\%$ of the M-type supergiants in the SMC are large amplitude variables. Large amplitude M supergiant variables such as S Persei are quite rare in the Galaxy and in the LMC. The fact that an object of similar type [Sandage's (1971) variable no. 19] has been found in IC 1613, suggests that low metal abundance may favor the occurrence of large amplitude red supergiant variables.

Planetary Nebulae

Sanduleak et al (1972) have pointed out that λ6584 of [N II] is remarkably faint in the spectra of the planetary nebulae in the SMC. Sanduleak et al find that [N II] λ6584 is absent in 31 out of 32 planetary nebulae that they have observed in the Small Cloud. This contrasts with the situation in the Large Cloud where λ6584 is observed in 40% of the 110 known planetaries. A comparison of the intensity ratio 6584 : Hα is given in Table 7. The table shows that the 6584 : Hα ratios in the Galaxy and in the LMC are similar but differ radically from those obtained in the SMC. The data in this table strongly suggest that nitrogen is deficient in the Small Cloud planetary nebulae. The low nitrogen abundance in the SMC may be understood if it is assumed (Peimbert & Torres-Peimbert 1971) that the interstellar gas in the Small Cloud has

Table 7 Percentage distribution of I(6584)/I(Hα) in planetary nebulae (Sanduleak et al 1972)

System	I(6584/I(Hα) <0.1 (%)	0.1–0.5 (%)	>0.5 (%)	n
SMC	97	0	3	32
LMC	61	15	24	110
Galaxy	55	22	22	49

Table 8 Abundance determinations in the Magellanic Clouds

Star	Sp	[Fe/H][a]	Reference
LMC:			
HD 32034	B9Ie	-0.8	Przybylski (1971)
HD 33579	A3Ia-O	-0.2 ± 0.2	Przybylski (1968)
		~ 0.0	Wares et al (1968)
		-0.3[b]	Wolf (1972)
SMC:			
HD 5045	B3Ia	$\lesssim -0.6$	Osmer (1973)
HD 7099	B2.5I	$\lesssim -0.6$	Osmer (1973)
HD 7583	A0Ia-O	-1.0	Przybylski (1972)
		-0.3[b]	Wolf (1973)

[a] $[Fe/H] \equiv \log(Fe/H)_{star} - \log(Fe/H)_{Galaxy}$.
[b] From best determined elements (Mg, Si, Ca, Ti, Cr, and Fe) relative to α Cyg.

been less enriched in nitrogen by evolving stars than is the case for interstellar gas in the Large Cloud and the Galaxy.

Abundances in Stars

Data on the abundance determinations of stars in the Magellanic Clouds are summarized in Table 8. Taken at face value the data in this table suggest that the stars in the LMC are possibly slightly deficient in heavy elements whereas those in the SMC show a substantial heavy element deficiency. These conclusions are entirely consistent with the observations of Cepheids and star clusters, which also suggest that LMC objects are intermediate between those in the Galaxy and in the SMC.

Gas and Dust in the Magellanic Clouds

The dust clouds in the Magellanic Clouds have recently been studied by van den Bergh (1974b) and by Hodge (1972, 1974). These investigations show that both the LMC and the SMC contain dark nebulae. The dust clouds in the Large Cloud are much more prominent, however, than those in the Small Cloud. On the basis of these observations it is concluded that the interstellar gas in the LMC has been able to produce dust much more efficiently than has the gas in the Small Magellanic Cloud. Because interstellar dust grains are largely composed of heavy elements, it seems quite probable that the observed difference between the LMC and the SMC is due to a lower heavy element abundance in the Small Cloud than in the Large Cloud. Strong support for this speculation is provided by the observations discussed by Peimbert (1973), who finds the oxygen abundance in the SMC emission nebulae to be ~ 5 times lower than it is in the LMC emission nebulae.

Intercomparison of the 21-cm maps of the Large (McGee & Milton 1964) and Small Cloud (Hindman 1967) clearly shows that the gas in the LMC has a much clumpier distribution than the gas in the SMC. The Large Cloud contains 52 clearly defined gas concentrations, the SMC only 3. It seems highly probable that this difference is in some way related to the fact that the gas in the Small Cloud (which contains little dust and few heavy elements) cools less efficiently than the gas in the LMC.

4 GLOBULAR CLUSTERS

Globular clusters are of interest because they are the oldest entities that can be studied in the Universe. They therefore give us information on the earliest phase in the evolution of galaxies. As a group, globular clusters are metal poor and thus confirm the view that heavy elements were formed after star formation was initiated in collapsing protogalaxies. The metal abundance in individual globular clusters may be determined in three ways: 1. from studies of the integrated colors or integrated spectra of clusters, 2. from the morphology of cluster color-magnitude diagrams, or 3. from the colors or spectra of individual stars.

Studies employing the integrated cluster light are particularly powerful because globular clusters as a whole are $\sim 10^2$ times more luminous than the brightest individual cluster stars. The principal difficulty with investigations based on an interpretation of the integrated characteristics of clusters is that it is difficult to distinguish differences that result from differing metallicity of cluster stars from those that result from differing morphology of the horizontal branch. A blue integrated color may, for example, be due to either low metallicity of all of the cluster stars or to the presence of a strongly developed blue horizontal branch in the color-magnitude diagram.

Studies of individual galactic globular clusters show that they range in metallicity from $Z \approx Z_{\odot}/300$ to $Z \approx Z_{\odot}/3$. Certain characteristics such as ΔV [the magnitude difference between the giant branch at $(B\text{-}V)_0 = 1.4$ and the horizontal branch] and the UV excess of individual stars correlate quite strongly with metal abundance. Other features of globular cluster morphology, such as the population gradient along the horizontal branch, do *not* appear to correlate closely with metallicity. This observation shows (van den Bergh 1961, 1967; Sandage & Wildey 1967; Hartwick 1968) that globular clusters do not form a one-parameter family. In addition to metallicity a second parameter, which has not yet been identified with certainty, appears to determine the morphology of globular cluster color-magnitude diagrams. Van den Bergh (1961) originally proposed that this second parameter might be a variation in helium abundance. Some support for this view is provided by recent observations (Peimbert 1975, Peimbert & Torres-Peimbert 1975) that appear to show that the helium abundance in the LMC is slightly lower than it is in the Galaxy. A second possibility (Iben & Rood 1968, van den Bergh 1973a, 1975a) is that globular clusters may differ from each other in age. Theoretical work by Rood (1973) suggests that age differences of the order of $\sim 1 \times 10^9$ years may be sufficient to account for the differences in horizontal branch gradients that occur between galactic globular clusters of similar metallicity.

Perhaps the most striking difference between globular clusters associated with different galaxies is that the mean metallicity of globular cluster families correlates with the luminosity (or mass) of their parent galaxies. All of the globulars associated

Table 9 Relation between mass, line strength, and color of globular clusters (van den Bergh 1975a)

Object	$\mathcal{M}/\mathcal{M}_{\odot}$	$\langle L \rangle$	$\langle B\text{-}V \rangle$
Fornax[a]	1.3×10^7	0	0.67
NGC 205[b]	1.6×10^9	—	0.65
Galaxy	1.3×10^{11}	4	—
M31	3.1×10^{11}	8	0.73
M87	$\sim 1 \times 10^{13}$	—	0.8

[a] $\dfrac{\mathcal{M}/\mathcal{M}_{\odot}}{L_V/L_{\odot}} = 1$ assumed.

[b] $\dfrac{\mathcal{M}/\mathcal{M}_{\odot}}{L_V/L_{\odot}} = 5$ assumed.

with the Fornax dwarf spheroidal system are metal poor. The certain globular clusters (i.e. those that contain RR Lyrae stars) in the Magellanic Clouds such as NGC 121, 1466, 1978, and 2257 all appear to be moderately metal poor. The globular clusters associated with the Galaxy exhibit the entire range from extreme metal deficiency to metal abundances close to that of the Sun. The globular clusters near the Andromeda nebula exhibit the same range in line strengths as do those associated with the Galaxy. The mean metallicity of the M31 globulars is considerably greater, however, than that for the clusters associated with the Galaxy. (This effect may be even more striking if some of the metal-poor globulars near M31 were originally associated with but subsequently stripped from M32.) Recent observations by Ables, Newell & O'Neil (1974) appear to show that the globular clusters associated with the giant elliptical galaxy M87 are even redder (and hence presumably more metal rich) than those in M31. The data on the correlation between the mass of the parent galaxy and the mean *B-V* color of the globular clusters associated with them is summarized in Table 9.

Within the Galaxy, metal-poor globular clusters appear to have formed predominantly in the low-density halo regions, whereas metal-rich clusters are mainly found in the high-density inner disk and central bulge of the Galaxy. Taken at face value these observations suggest that the metal abundance of globular clusters was mainly determined by the density of the regions where they were formed. It is not yet clear (van den Bergh 1969) if the same relationship holds in M31. Discovery and observation of globular clusters in the outer halo of M31 is planned to look into this question in more detail.

5 STELLAR POPULATIONS IN M31 AND ITS COMPANIONS

The Brightest Stars

According to Baade (1963) the disk population of the Andromeda nebula and *all* of the M31 globular clusters are resolved at the same magnitude. This statement appears improbable for two reasons: 1. The absolute magnitudes of cluster red giants are expected to depend on metal abundance (Arp 1955, Sandage & Wallerstein 1960). 2. Different clusters are dimmed by differing amounts of interstellar absorption.

Through the kindness of Miss Henrietta Swope, I have been permitted to examine Baade's original plates in the files of the Hale Observatory. These plates show that the globular clusters associated with M31 remain unresolved on those plates which only barely resolve the stellar population of the Andromeda nebula. This observation does not imply that the red giants in the globular clusters are fainter than the red giants in M31 itself. It simply means that 1. it is more difficult to resolve stars in very crowded clusters than in the general field, and 2. a larger total population is sampled in the field than in any one globular cluster. Only on Baade's best plate (PH834B), which is centered between M31 and NGC 205, are all of the bright clusters located within the coma-free field resolved. Baade's two next best photovisual plates (PH525B and PH1173B) seem to show some clusters resolved and others unresolved. The relevant data are summarized in Table 10, which is taken from van den Bergh (1969). All of the clusters in this table are located outside the main body of M31 and are therefore presumably not greatly affected by reddening within the Andromeda nebula.

The data in the table may be interpreted in two ways: 1. The red giants in clusters 73 and 76 are brighter than those in the other clusters because they are quite metal poor (van den Bergh 1969). 2. The brightest stars in these two clusters are more luminous than those in the other clusters because the total stellar population in these clusters is greater than it is in the others (Harris 1975). The choice between alternatives 1 and 2 is important because it is related to the question of whether or not the absolute magnitude level of the RR Lyrae stars is itself dependent on metallicity.

If both the absolute luminosity of RR Lyrae stars *and* the magnitude difference between the horizontal branch and the tip of the giant branch are independent of metal abundance, then the absolute magnitude of the brightest stars of population II must itself be independent of metallicity. This conclusion appears to be in conflict with observations of the companions of the Andromeda nebula. Inspection of Baade's plates shows that NGC 205 is much more easily resolved than is M32. This difference in resolvability of the two closest companions to M31 is far greater than can be explained from the effects of crowding, which are more severe in M32 than they are in NGC 205. I would estimate that the brightest red giants in NGC 205 are ~ 1 mag brighter than those in M32. It would be important to strengthen this conclusion using a wide-field Ritchey-Chrétien telescope with which both M32 and

Table 10 Resolution of globular clusters on PH525B and PH1173B

No.	V	B-V	Q	L	Resolution
57	15.9	0.98	−0.20	—	Unresolved
69	17.1	—	—	—	Unresolved
73	15.4	0.95	−0.37	8	Incipient resolution
76	15.6	0.66	−0.43	1	Resolved
83	16.1	0.78	−0.21	—	Unresolved

NGC 205 could be photographed on the same plate. Inspection of Baade's plates suggests that the stars in the inner halo of M31 have luminosities that are intermediate between those of the red giants in NGC 205 and in M32. Because of the high background density of unresolved stars in the nuclear bulge of M31 it is not possible to make any statement regarding the luminosity of the brightest red giants in the central lens of the Andromeda nebula.

It may of course be argued that the resolved stars in NGC 205 are bright because they are *intermediate age* metal-poor objects similar to those in the SMC cluster NGC 419. The fact that some star formation is still taking place in NGC 205 at the present time (Baade 1951, Hodge 1973) lends some plausibility to this view.

According to van den Bergh & Racine (1967) $V = 21.8 \pm 0.3$ for the brightest stars in NGC 205. Because the brightest stars in M32 are approximately 1 mag fainter than those in NGC 205, it follows that $V \approx 22.8$ for the brightest stars in M32. According to Harris (1975) the brightest stars in very rich clusters are located 3.1 mag above the cluster horizontal branch. This suggests that the horizontal branch stars in M32 are located at $V \approx 25.9$. With $(m\text{-}M)_V = 24.6$ for M31 it follows that $M_V(\text{HB}) \simeq +1.3$ in M32. The uncertainty of this value is probably ~ 0.5 mag. It follows that the horizontal branch in M32 is significantly fainter than the value $M_V(\text{HB}) \approx +0.4$ that is obtained (Sandage 1969) for metal-poor globular clusters. This result is expected because: 1. Hartwick & Hesser (1974) find $M_V(\text{HB}) = +0.85$ or $+0.9$ by fitting the main sequence of the strong-lined globular cluster 47 Tucanae to the main sequence of moderately metal-deficient stars for which trigonometric parallaxes are available, and 2. McClure & van den Bergh (1968a) find that the integrated intermediate and wide-band colors of M32 indicate that this object contains a stellar population resembling that of a dwarf-enriched, metal-rich, galactic globular cluster.

On the basis of these results it is tentatively concluded that the brightness of the horizontal branch for population II stars depends on metallicity in such a way that metal-poor horizontal branches are more luminous than are metal-rich horizontal branches. As we have seen in Section 3 the observations of old populous clusters in the Magellanic Clouds give marginal support for this view.

The Integrated Light Studies of M31, M32, and NGC 205

THE CENTRAL BULGE OF M31 A long slit photograph of the central region of M31 taken by Babcock (Morgan & Osterbrock 1969) dramatically illustrates the variation in the stellar population along the major axis of M31. At 30′ (6 kpc) from the nucleus the strength of the Balmer lines indicates a strong contribution to the integrated light by F-type stars. As the nucleus is approached the early-type spectrum fades and the integrated light becomes dominated by giants in the spectral range G8–K3. The presence of TiO absorption in the red region of the spectrum indicates that M stars are important contributors to the light at longer wavelengths.

Inspection of the spectrograms published by Morgan & Mayall (1957) shows that the blue region of the spectrum in the central part of M31 is dominated by cyanogen giants. This conclusion is confirmed by the photoelectric spectrum scans of van den Bergh & Henry (1962). The great observed strength of cyanogen absorption in the central region of the Andromeda nebula may be accounted for by one of two hypotheses: 1. The light from the central region of M31 is dominated by strong-lined giants, or 2. If dwarf stars contribute to the integrated light, and hence dilute the strength of cyanogen, then the giants must have super strong cyanogen absorption. Work by Faber (1972) shows that intermediate and broad-band photometric measurements in themselves are insufficient to establish the fractional dwarf contribution to the integrated light. From 38 color data on the nuclear region of M31 Faber finds that an acceptable fit to the observations is provided by models with $15 < \mathcal{M}/L < 60$. This result is consistent with the value $\mathcal{M}/L \approx 13$ that Morton & Thuan (1973) obtained from application of the virial theorem to velocity dispersion observations in the nucleus of M31. A relatively low mass-to-light ratio (i.e. a giant-dominated model) for the central region of M31 is also favored by Whitford's (1972) observations of the strength of the Wing-Ford molecular feature at λ9910. This unidentified molecular band is a useful population discriminant because it is much stronger in dwarfs than it is in giants. Whitford's scanner observations of the nucleus of M31 at a resolution of 30 Å give no positive indication of a dwarf contribution to the total light. This conclusion is strengthened and confirmed by Baldwin et al (1973), who have measured the strength of a CO band at 2.3 μm in the central region of M31. Their observations yield a photometric CO index of 0.098 ± 0.013 for the nucleus of M31. This value may be compared with a calculated CO index of 0.079 for a giant-dominated model with $\mathcal{M}/L = 15$ and 0.028 for a dwarf-

Table 11 Population model for M31 with $\mathcal{M}/L_V = 15$ (Faber 1972, 1974)

Spectral type	No. stars[a]	Percentage of mass	Percentage of V light
G0V–G4V	1.22E+3	0.77	11.56
G5V–K0V	1.34E+3	0.76	5.10
K1V–K2V	8.95E+2	0.40	2.29
K3V–K4V	1.91E+3	0.78	3.07
K5V–K7V	2.98E+3	1.12	1.24
M0V–M2V	2.89E+3	0.73	0.27
M3V–M4V	5.46E+4	10.3	1.09
M5V–M6V	3.64E+4	4.6	0.15
M7V	7.34E+5	69.4	1.74
M8V	1.63E+5	10.2	0.03
G0IV–G4IV	4.65E+2	0.35	11.88
G5IV–G9IV	3.44E+2	0.26	8.79
SMR K0–K1IV	1.66E+2	0.13	6.74
SMR K2 III–IV	1.65E+2	0.12	26.57
SMR K3 III	4.61E+1	0.03	12.23
SMR K4–5 III	1.48E+1	0.01	5.98
M5III–M6III	3.02E 0	0.003	1.22
HB	9.59E−2	0.000	0.05

[a] For a total population of 1,000,000 stars.

dominated model with $\mathscr{M}/L = 44$. It is concluded that the observations of the Wing-Ford band and the 2.3 μm CO absorption feature confirm Morgan's conclusion that the integrated light in the blue green region of the spectrum of the nucleus of M31 is dominated by giant stars.

A population model for the nuclear region of M31 (Faber 1972) is given in Table 11. In this model, for which $\mathscr{M}/L_V = 15$, giants and subgiants contribute 33 and 41% of the visual light, respectively. It should be emphasized that a good fit to the observed photometric parameters for the nuclear region of M31 can be achieved only if the giants in M31 are very strong lined and hence presumably "super metal rich."

THE HALO OF M31 It has recently been shown by Ostriker & Peebles (1973) that the disks of spiral galaxies are unstable to barlike deformations unless the mass in the halo interior to some radius R is comparable to that situated in the disk out to distance R. Some observational support for the conclusion that very large amounts of mass are located well outside the optical images of galaxies is provided by the 21-cm observations of M31 by Roberts & Rots (1973). These observations, together with similar results for the supergiant Sc galaxy M83 by Rogstad et al (1974), suggest that the luminosity function in the outer regions of spiral galaxies is dominated by very faint M dwarfs. Such a population of faint late-type stars could produce an exceedingly high mass-to-light ratio in the outer regions of galaxies. If a similar population is present in the outer regions of giant elliptical galaxies it may account for the very high total masses of clusters of galaxies that appear to be indicated by the virial theorem (Rood 1974). If this interpretation is correct, the luminosity function of star formation in galaxies must depend quite critically on distance from the galactic nucleus. On the picture outlined above, the luminosity function of star formation would be dominated by massive stars near the centers of galaxies, whereas low-mass stars dominate the luminosity functions in the outer parts of galaxies. Such a situation may arise if the mass spectrum of star formation depended on density (van den Bergh 1973a, 1975a) in such a way that high-mass stars are mainly produced in regions of high density whereas low-mass stars are predominantly formed in those regions of a protogalaxy in which the gas density is low.

COMPOSITION GRADIENT IN THE DISK OF M31 Observations by McClure & van den Bergh (1968a), McClure (1969), Spinrad et al (1971), and Spinrad, Smith & Taylor (1972) have shown that the strength of cyanogen absorption increases sharply toward the nucleus of M31. McClure (1969) has observed a similar increase in the cyanogen strength over the inner 90″ (1.4 kpc) of the central disk of M81. Spectroscopic observations by Joly & Andrillat (1973) also show that the line intensities of NaI $\lambda\lambda$5890, 5896, Mg I $\lambda\lambda$5167, 5173, 5184, Fe I $\lambda\lambda$4383, 3746, 3734, 3647, and Ca II λ3933 strengthen as the distance from the center of M31 decreases. Similar radial variations in the strengths of the cyanogen bands $\lambda\lambda$3883, 4216, the G band, the Mg I + Mg H blend at λ5175, and the Na D blends in the giant E galaxy NGC 4472 and in the S0 galaxy NGC 3115 have been observed by Welch & Forrester (1972). The most straightforward interpretation of these results is that the mean heavy-element abundance in stars increases as the nuclei of these galaxies are approached. Strong support for this conclusion is provided by the observation that the heavy-element abundance in the interstellar gas (Peimbert 1968, Searle 1971, Shields 1974) also increases sharply toward the nuclei of spiral galaxies. This result can be understood in one of two ways: 1. The luminosity function of star formation near the nuclei of spirals may have been weighted in favor of high-mass stars that subsequently evolved into heavy-element-producing supernovae, or 2. Heavy elements produced by supernovae in the outer regions of these galaxies may have settled down toward the nucleus (Larson 1974b) so that the stars in the nuclear region were formed from more highly enriched material than those that formed farther out.

NGC 205 AND NGC 185 A detailed discussion of the stellar content of the dwarf elliptical galaxy NGC 205 is given by Hodge (1973). Inspection of yellow or red plates of this galaxy shows that the overwhelming majority of the stars in this object are evolved red giants. On blue and UV plates, a second and much less conspicuous population of blue stars is observed. These blue stars, which are mainly concentrated in the central part of NGC 205, have a clumpy distribution. This suggests that the blue stars have recently been formed from the interstellar gas in this galaxy. The existence of interstellar gas in this object is indicated by the presence of a number of conspicuous dust patches. The fact that the nuclear region of NGC 205 is much bluer (B-$V = 0.52$) than its outer region (B-$V = 0.72$) is due to the fact that the population I component is concentrated toward the center of this galaxy. Hodge (1973) estimates that the total mass of population I stars in NGC 205 lies in the range 2×10^5–$2 \times 10^6 \mathcal{M}_\odot$. The brightest blue star has an apparent magnitude $B = 19.4$ corresponding to $M_B = -5.3$. A few young blue stars similar to those in NGC 205 are also found in NGC 185, which appears to be generally similar to (but less luminous) than NGC 205. In NGC 185 (Hodge 1963) the nuclear region is also bluer (B-$V = 0.95$) than the outer portion (B-$V = 1.15$) of this galaxy. As in the case of NGC 205 the reason for this color gradient appears to be that the young blue stars are concentrated near the center of NGC 185. It is, however, rather puzzling that NGC 185 is 0.4 mag redder in B-V than is NGC 205. Inspection of direct plates of NGC 185 shows that this galaxy resolves at approximately the same magnitude level as does NGC 205. Because both of these objects are presumably at almost the same distance, the observed color difference cannot be due to reddening of NGC 185, which would have made the giants in that galaxy ~ 1 mag fainter than those in NGC 205. It follows that there must be some as yet not understood fundamental difference in the stellar content of these two galaxies.

According to Spinrad (1966) the integrated spectrum of NGC 205 resembles that of the moderately metal-poor globular cluster M5. A somewhat lower metal abundance seems to be indicated by image-tube spectra of the disk and nucleus of NGC 205 obtained by van den Bergh (1969). Some caution should be exercised, however, in interpreting this result. At a dispersion of 50 Å mm^{-1} the lines in NGC 205 have a much greater width than do those in globular clusters. This makes a direct intercomparison of the spectrum of NGC 205 with those of globular clusters difficult.

Christensen (1972) has observed the nuclear region of NGC 205 with a spectrum scanner at the Cassegrain focus of the 200-inch telescope. He finds that the dominant stellar populations in the nucleus consists of a mild population II similar to that in the M31 globular clusters H12 and H140. According to van den Bergh (1969) both of these clusters have line-strength indices $L = 8$, indicating intermediate metal abundance.

M32 = NGC 221 Stellar population models for M32 have been constructed from integrated light observations by McClure & van den Bergh (1968a), Spinrad & Taylor (1971), and Faber (1972). Observations by Penston (1973) at 2.2 μm show that M32 is too blue in the infrared to fit the population model of Spinrad and Taylor. McClure & van den Bergh (1968a) find that a dwarf-enriched, metal-rich, globular cluster model gives an acceptable fit to the integrated color of M32. There appears to be no evidence for blue stars similar to those in NGC 205 associated with M32. This result is uncertain, however, because part of M32 is projected on an outer spiral arm of M31. As a result it is not possible to make any strong statements on the complete absence of bright blue stars in M32.

STAR FORMATION IN DWARF ELLIPTICALS Apparently the rate at which star formation takes place in dwarf elliptical galaxies varies greatly from object to object. Possibly the star formation currently taking place in such dwarf ellipticals as NGC 185 and NGC 205 is due to the fact that gas ejected by evolving stars is not able to leave these systems and accumulates until its density becomes high enough for star formation to take place. The fact that the rate of star formation is so different from system to system may be accounted for if such star formation occurs in bursts. Searle & Sargent (1972) have previously suggested that star formation in dwarf *irregular* galaxies may also take place in bursts.

Examples of dwarf ellipticals in various states of star-forming activity are:

1. NGC 147. There is no evidence either for any star formation taking place in this galaxy or for the presence of dust patches in this object
2. NGC 185. A single prominent dust patch is present in this galaxy. Blue and UV photographs reveal the presence of a sprinkling of young blue stars
3. NGC 205. This galaxy contains a number of prominent dust patches and star formation in it is proceeding more vigorously than it is in NGC 185
4. NGC 5253. A huge burst of star formation is presently taking place in this peculiar galaxy. Inspection of the glass copies of the Palomar Sky Survey plates shows that the underlying luminosity distribution of NGC 253 resembles that of an E or S0 galaxy. Superimposed on this smooth distribution of stars is a conspicuous concentration of knots grouped around the center of this galaxy. About a dozen such knots are recognizable on a plate taken in excellent seeing with the University of Toronto 24-inch telescope on Las Campanas in Chile. A widened image-tube spectrum of the brightest of these knots has recently been published by van den Bergh (1972a). It shows an early-type absorption-line spectrum resembling that of the bright knots (super star clusters) near the nucleus of M82. This observation suggests that NGC 5253 has recently experienced a violent burst of star formation. The rich emission-line spectrum in this galaxy has been described by Welch (1970)

Another relatively nearby dwarf elliptical galaxy that contains prominent dust patches, strong emission lines, and knots of young stars is NGC 3077. Possibly NGC 5195 is also related to the class of objects that has been discussed above. According to Morgan & Osterbrock (1969) it seems probable that the spectrum of NGC 5195 is of the weak-lined variety. This conclusion has been disputed by Spinrad (1973), however. The dust observed in NGC 5195 may have been "borrowed" from NGC 5194 = M51. A similar explanation may of course be invoked to account for the dust patches in NGC 205, which is quite close to M31. Such an explanation could not, however, account for the dust observed in NGC 185, which is quite far removed from the Andromeda nebula.

LUMINOSITY DEPENDENCE OF COLORS AND SPECTRA OF E GALAXIES Evidence for the fact that luminous early-type galaxies have stronger lines than objects of lower luminosity was first given by Spinrad (1961) and Deutsch (1964). This observation was confirmed by McClure & van den Bergh (1968a), who were able to show that the strength of the cyanogen band $\lambda 4216$ increases with increasing galaxy luminosity. The most detailed study of the dependence of absorption-line strengths on luminosity among elliptical galaxies has been made by Faber (1973b). Her observations show that elliptical galaxies form a one-parameter family after their colors have been corrected for interstellar reddening. Faber also finds that the strengths of the absorption features due to CN and Mg"b" + MgH increase monotonically with

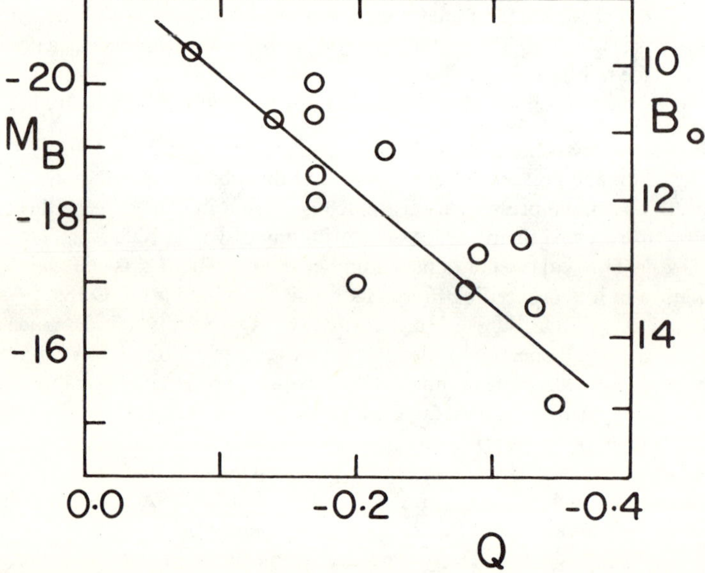

Figure 3 Plot of the reddening-free parameter $Q = (U\text{-}B) - 0.72\ (B\text{-}V)$ vs luminosity for E and S0 galaxies in the Virgo cluster (McClure & van den Bergh 1968b).

Table 12 Colors of cluster and noncluster E galaxies (McClure and van den Bergh 1968b)

Location	$\langle Q \rangle$[a]	n
In rich clusters	-0.16 ± 0.01	23
In poor clusters	-0.18 ± 0.01	10
Field objects	-0.18 ± 0.01	27

[a] reduced to $\langle M_B \rangle = -19.6$.

increasing luminosity. She finds that a reddening-free absorption-line index monitoring CN and Mg can be used to determine the absolute magnitude of an elliptical galaxy with an accuracy of ± 0.56 mag. The difference in color and line strength between M32 and the most luminous elliptical galaxies is consistent with an increase in [Fe/H] from the solar value in M32 to twice the solar value in luminous ellipticals.

Baum (1959), De Vaucouleurs (1961), Rood (1969), and Tifft (1969) have noted that intrinsically faint elliptical galaxies tend to be bluer than the most luminous ones. This effect is illustrated in Figure 3, which shows a plot (McClure & van den Bergh 1968b) of the reddening-free parameter $Q = (U\text{-}B) - 0.72\ (B\text{-}V)$ vs brightness for E and S0 galaxies in the Virgo cluster. The figure shows that the most luminous E galaxies are much redder than the fainter ellipticals. The data in Table 12 show that galaxies of the same absolute magnitude have similar Q values in clusters and in the general field. This result may be interpreted in one of two ways: 1. The evolutionary history of an elliptical galaxy is independent of the environment in which it is formed, that is, the galaxy is essentially insulated from its neighbors after star formation and heavy-element enrichment begins, or 2. Field E galaxies have escaped from clusters.

In a recent paper Sandage (1972a) has drawn attention to the fact that the dwarf elliptical galaxies in the Local Group appear to be redder than their counterparts in the Virgo and Coma clusters. Data on the $U\text{-}B$ colors of the Local Group dwarfs are summarized in Table 13. The intrinsic colors of the companions of the Andromeda nebula were calculated by assuming them to have the same reddening $E_{B\text{-}V} = 0.11 \pm 0.02$ (van den Bergh 1975b) and hence $E_{U\text{-}B} = 0.09$ as M31 itself. M32 was not plotted because both its colors (Faber 1973a) and the absence of globular clusters

Table 13 Integrated colors and magnitudes of local group dwarfs

System	M_V	$U\text{-}B$	$(U\text{-}B)_0$	Reference
Fornax	-13.0	0.08	0.08	De Vaucouleurs & Ables (1968)
NGC 147	-14.9	0.32	0.23	Sandage (1972a)
NGC 185	-15.2	0.36[a]	0.25[a]	Hodge (1963), Sandage (1972a)
NGC 205	-16.4	0.18	0.09	De Vaucouleurs (1961)

[a] Colors refer to nuclear region only.

associated with it indicate that this galaxy has been reduced in size and luminosity by tidal stripping during close encounters with M31.

Figure 4 shows a plot of V vs U-B for E and S0 galaxies in the Coma cluster, the Virgo cluster, and the Local Group.

The data in Figure 4 may be interpreted in one of three ways. Either 1. $H > 100$ km sec^{-1} Mpc^{-1}, or 2. Local Group dwarfs are, in the mean, redder than those in the Virgo and Coma cluster, or 3. the reddening values of NGC 147 and NGC 185 have been underestimated. If the red colors of NGC 147 and NGC 185 are to be accounted for in terms of reddening, then it is difficult to see how the brightest stars in these galaxies could have approximately the same V magnitude as the brightest stars in NGC 205 and the dwarf spheroidal companions to M31 (van den Bergh 1972d).

NUCLEI OF GALAXIES Faber (1973a) shows that M32 occupies a peculiar position in a plot of CN + Mg strength vs M_V. On the basis of its CN + Mg index, M32 would be expected to be 2 mag more luminous than it is actually observed to be. Faber suggests that this observation may be accounted for if M32 has lost its outer envelope as a result of tidal stripping during a close encounter with M31. Additional support for this conclusion is provided by inspection of photographs of M32 that show this object to be unusually compact. Furthermore, M32 does not have any globular

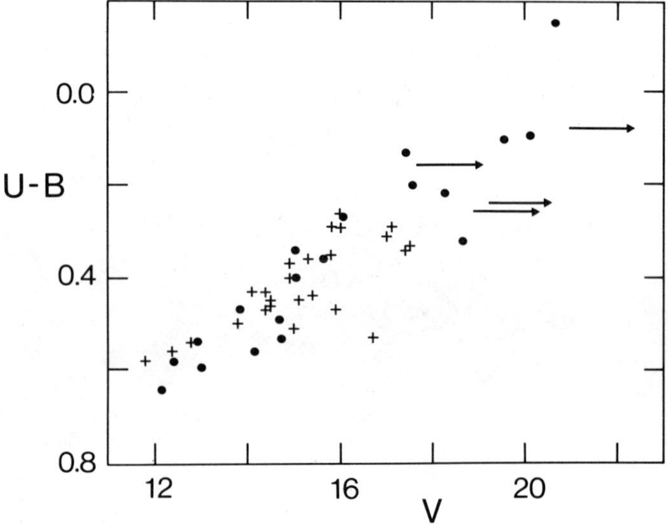

Figure 4 Plot of U-B (corrected to zero redshift) vs apparent magnitude V for galaxies in the Coma cluster (crosses). Also plotted in the figure are galaxies in the Virgo cluster (dots) with 3.66 added to their V values. Arrows show how changing the Hubble constant H from 100 to 50 km sec^{-1} Mpc^{-1} would shift the positions of the Fornax system, NGC 147, NGC 185, and NGC 205 in the diagram.

clusters associated with it whereas NGC 205, which has approximately the same absolute magnitude, has six. (On the basis of both its radial velocity and its strong-lined spectrum NGC 205: III is regarded as a physical member of the M31 cluster system.) Presumably the outermost globular clusters initially associated with M32 were also lost as a result of tidal stripping. From the general correlation between the mean metallic line strength of globular clusters and the luminosity of their parent galaxy (see Table 9) it is expected that the average metallicity of the globulars that once belonged to M32 should be quite low. Possibly a considerable fraction of the small number of metal-poor globulars that now accompany M31 might therefore originally have been associated with M32. Tidal stripping cannot, however, account for the lack of globular clusters in the inner region of M32. Possibly the absence of such clusters near the core of M32 could be accounted for (Ostriker 1974) by the assumption that such objects have spiraled into the nucleus of M32 due to the effects of dynamical friction (Chandrasekhar & von Neumann 1943). This speculation receives some support from the observation (Spinrad et al 1972) that the semistellar nucleus of M32 has a lower cyanogen strength than the core of this galaxy. A difficulty with this view, however, is that the luminosity of the nucleus of M32 is so high. According to Walker (1962) the region within 3".1 of the center of M32 has $V = 11.3$, which with $(m\text{-}M)_V = 24.6$, yields $M_V = -13.3$. Not only is this value twice as great as that of the nucleus of M31, but it is also almost 100 times greater than the luminosity of a typical globular cluster. It does not seem reasonable to assume that so large a number of globulars disintegrated by spiraling into the center of M32.

Additional evidence against the assumption that the semistellar nuclei of galaxies have been produced by the spiraling in of globular clusters due to dynamical friction is provided by observations of the semistellar nucleus of M33. This nucleus, which has a diameter of 1".5, has $M_V \approx -10$, $B\text{-}V \approx 0.65$, and $U\text{-}B \approx 0.00$. I have taken a widened spectrum of this nucleus through the "moonlight eliminator" at the Cassegrain focus of the 200-inch telescope. For these image-tube observations a $2'' \times 2''$ entrance slot was used. This spectrogram, which has a dispersion of 49 Å mm^{-1}, shows Hγ and Hδ, the H and K lines of Ca II, and λ4226 of Ca I. The G band, and $\lambda\lambda$4045, 4325 of Fe I are weakly present. According to Walborn (1972), comparison of this image-tube spectrum with main-sequence standards observed with the same instrument yields the following spectral types: Late A from K/H + Hε; F2–F4 from 4226/Hγ; F3–F4 from CH/Hγ. These observations suggest that the light from the semistellar nucleus of M33 is dominated by relatively young stars similar to those that contribute most of the light in the disk of the Triangulum nebula. If this conclusion is correct, then the nucleus of M33 must contain a substantial young population component indicating that it was the site of relatively recent star formation. This observation could be understood if the semistellar nucleus in M33 was formed from gas that accumulated at the center of this galaxy. If this conclusion is correct, then star formation during an early epoch in the history of M31 and M32 may also be invoked to account for the formation of the semistellar nuclei that are observed in those galaxies. The only alternative to this suggestion is that the relatively early spectral type of the nucleus of M33 is due to the fact that it is composed of very metal-poor stars similar to those occurring in globular clusters. More detailed spectroscopic observations would be required to rule out this possibility, although the fact that Fe I $\lambda\lambda$4045, 4325 is weakly present would appear to militate against this suggestion.

6 DWARF SPHEROIDAL GALAXIES

The first two dwarf spheroidals were discovered by Shapley (1938). Subsequent discoveries have raised the number of these low-luminosity systems known within the Local Group to 9. The total number of known members of the Local Group now stands at 20. Because additional nearby dwarf spheroidals probably remain to be discovered, it appears likely that these objects are intrinsically the most numerous class of galaxies in the Universe. The distribution of dwarf spheroidals in the Local Group and in the nearby M81 and South Pole clusters suggests that the majority of them are distant satellites of spiral galaxies.

Data on the color-magnitude diagrams of dwarf spheroidals are available for the Sculptor system (Hodge 1965), the Leo II system (Swope 1967), the Ursa Minor system (van Agt 1967), and the Draco system (Baade & Swope 1961). These color-magnitude diagrams show that the stellar population contained in dwarf spheroidal galaxies is very similar to that which occurs in globular clusters. This conclusion is strengthened by the observation that the three dwarf spheroidals located close to M31 resolve at approximately the same magnitude level as do the population II stars in the elliptical companions to the Andromeda nebula.

Available observations of dwarf spheroidal systems show that the old metal-poor population in these objects differs from that in galactic globular clusters in a number of important respects:

1. No W Virginis stars ($P > 10$ days) have been discovered in any dwarf spheroidals
2. The BL Herculis ($P < 10$ days) variables in dwarf spheroidals (van Agt 1967, 1973) (see Figure 5) are significantly brighter than the BL Her stars in galactic globular clusters
3. Hodge (1965) finds that the two reddest stars at the tip of the giant (asymptotic?) branch in the Sculptor system have $B\text{-}V > 2.0$. Such red stars are not known to occur in galactic globular clusters

The peculiarities noted above are in some respects reminiscent of those observed among the population II stars in the Magellanic Clouds: 1. W Vir stars ($P > 10$ days) are exceedingly rare in the Clouds, 2. The 1.43 day BL Her star that Tifft (1963) has observed near the globular cluster NGC 121 in the Small Cloud is brighter than stars of similar period in galactic globular clusters, and 3. Many of the old populous red clusters in the Magellanic Clouds (see Table 4) contain stars with $B\text{-}V > 2.0$.

For three dwarf spheroidal galaxies, information is available on the population gradient along the horizontal branch. The Draco system and the Leo II system have red horizontal branches whereas the Ursa Minor system has a blue horizontal branch. On the basis of Rood's (1973) computations these results would appear to suggest that

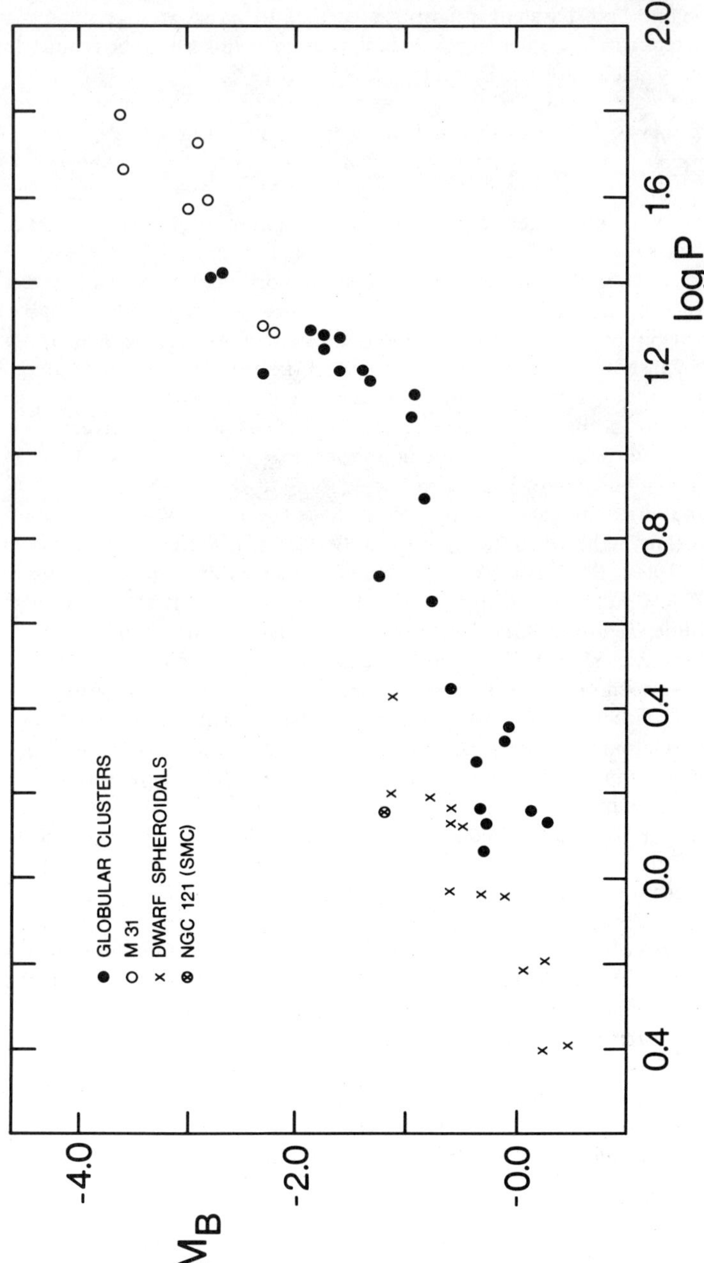

Figure 5 Period luminosity relation for W Virginis stars.

the Draco and Leo II systems might be (1 or 2) $\times 10^9$ years younger than the Ursa Minor system. (An age difference only half as large as this would be required by the evolutionary calculations of Fusi-Pecci & Renzini 1975.)

7 THE MISSING MASS IN GALAXY HALOS AND IN CLUSTERS OF GALAXIES

The best population models for galaxies suggest that most of their light is provided by giant stars whereas most of their mass is locked up in dwarfs. Because the mass spectrum of star formation may be dependent on position within a galaxy, it follows that the distribution of light in a galaxy does not necessarily provide a reliable guide to the distribution of mass. In fact, observations such as those of Roberts & Rots (1973), which show that the rotation curve of M31 hardly drops in the outer regions of that galaxy, suggest that a major fraction of the total mass of spirals is situated beyond the optical images of these objects. If this conclusion is correct, then some galaxies may be thought of as resembling icebergs in which most of the mass is situated in regions that escape optical detection.

Within individual galaxies there is clear-cut evidence for metal abundance gradients. The highest metallicity invariably occurs in the dense cores of galaxies (McClure 1969). Progressively lower heavy-element abundances are observed as regions of lower density are approached. The correlation between mean metallicity and absolute magnitude (and hence presumably mean density) that is observed in elliptical galaxies (Faber 1973b) also seems to hold for spirals and irregulars. It would be very important to gain some theoretical understanding of the reason for the correlation between the metallicity of a stellar system and its density. Possibly such a correlation could be understood if the luminosity function of star formation (Salpeter function) in dense regions favored the formation of high-mass stars that evolved to produce heavy elements. The assumption that low-density regions favor the formation of low-mass stars may also account for the large mass (Roberts & Rots 1973) that appears to occur in the outer region of M31. Similarly the high mass-to-light ratio of the halo regions of spiral galaxies, which is required to stabilize the disks of spirals (Ostriker & Peebles 1973), may be accounted for if the halos of these objects consist primarily of low-mass stars. Finally such low-mass stars may account for the well-known fact that the sum of the masses of (the inner regions) of all galaxies making up a cluster is insufficient to account for the total mass of such a cluster, which is derived by applying the virial theorem to the observed velocity dispersion of cluster galaxies.

It should be emphasized that a stellar luminosity function in which all, or almost all, stars have $\mathcal{M} \ll \mathcal{M}_{\odot}$ will not give rise to the formation of any heavy elements. The opposite conclusion would, of course, be arrived at if the missing halo mass is in the form of black holes formed from very massive stars that exploded during the halo phase of galactic evolution.

If the mass spectrum of star formation in the halo was very heavily weighted toward the formation of low-mass stars, the question arises why globular clusters, which were also formed in the galactic halo, have mass-to-light ratios (in solar units) in the range 1–2 (Illingworth 1973). Possibly this difference in $\mathcal{M}/L$ is due to the fact that globular clusters were formed from the densest condensations in the protogalactic halo. Some support for such a picture is possibly provided by the observation (van den Bergh & Sher 1960) that the luminosity functions of open clusters in the galactic disk contains far fewer faint low-mass stars than does the luminosity function of field stars in the galactic disk.

8 THE EVOLUTION OF ELLIPTICAL GALAXIES

The integrated light of giant elliptical galaxies is dominated by old strong-lined giants. As these giants die and become planetary nebulae or supernovae they eject matter back into interstellar space. In a giant elliptical the total amount of ejected matter is probably a few $\mathcal{M}_{\odot}$ per year. If this amount of gas were allowed to build up for $\sim 10^9$ years, the amount of gas in ellipticals would become comparable to that observed in spirals. With the exception of a few unusual objects, such as NGC 1275 and NGC 5128, giant elliptical galaxies give no evidence for the existence of *large* quantities of interstellar material. This apparent lack of gas and dust could be accounted for in a number of different ways: 1. Mathews & Baker (1971) have suggested that a high supernova rate may set up a "galactic wind" that could sweep gas out of a galaxy and into intergalactic space. 2. Gunn & Gott (1972) have shown that elliptical galaxies may be swept clean of gas as they move through intracluster gas at high velocity. Obviously this mechanism would not work for galaxies outside of clusters or for the massive central galaxies in clusters that are at rest relative to the center of mass of a cluster. 3. A significant fraction of the gas ejected by stars may eventually collapse into the nucleus of a giant elliptical, giving rise to a violent explosion. Such an explosive event may sweep galaxies clean of any remaining gas (van den Bergh 1963, 1972a).

A number of lines of evidence are now beginning to emerge that tend to favor the view that bursts of star formation may accompany violent explosive events in the nuclei of supergiant elliptical galaxies. New observations of the peculiar elliptical galaxy NGC 5128 (= Centauras A), recently obtained with the 1.5-m telescope of the Cerro Tololo Observatory, show that the integrated colors of the regions near the equatorial dark band in this galaxy are somewhat bluer than the areas near its poles. This observation suggests that the observed integrated colors of NGC 5128 are affected by a younger stellar population superimposed on the old metal-rich population that constitutes the dominant population component in giant elliptical galaxies (McClure & van den Bergh 1968a). Some support for this view is provided by the observation that a number of bright knots occur in the bluest parts of NGC 5128 that are adjacent to its prominent equatorial dust band. *UBV* observations of the brightest of these knots through a 10-sec diaphragm yield $V = 14.17$ ($M_V \approx -13$), $B\text{-}V = 0.41$, and $U\text{-}B = -0.58$. These colors suggest that the brightest knot in NGC 5128 is either a reddened OB cluster or a knot of emission nebulosity. The idea that some star formation is still taking place in or near the dark belt in NGC 5128 receives powerful support from a photograph of this object taken in blue light by Rickard (see Figure 6) and from recent y–m plates by the author which show huge

Figure 6 Blue exposure of NGC 5128 (= Centaurus A) showing evidence for recent star formation.

associations of OB stars within the dark band. Possibly the presence of large amounts of neutral hydrogen gas in NGC 5128 can be accounted for by the fact that this is one of those rare giant elliptical galaxies that is situated outside a rich cluster and hence cannot have its interstellar gas swept out in the manner suggested by Gunn and Gott.

Possibly NGC 4594 (The Sombrero) is similar to NGC 5128. One might imagine that the present chaotic regions of star formation associated with the dark band in NGC 5128 might after a few times 10^8 years be smeared out by differential rotation into a spirallike configuration resembling that which is presently observed in NGC 4594.

Additional evidence for the occurrence of bursts of star formation in giant elliptical galaxies is provided by observations of NGC 1275 (= Perseus A). The nucleus of this object emits a strong continuous spectrum that is presumably due to synchrotron radiation. Broad emission lines are superimposed on this continuum. An image-tube spectrum of a region located 5″ to the south of the nucleus of NGC 1275 has been published by van den Bergh (1972a). This spectrum shows a well-developed A-type absorption-line spectrum. This unexpected result confirms a previous observation by Minkowski (1968). The most straightforward interpretation of this observation is that the active nucleus of NGC 1275 is surrounded by a region in which a burst of star formation has recently taken place. The suspicion that some star formation may still be taking place in NGC 1275 is strengthened by the observation of spiral armlike strings of bright knots located in the outer parts of this galaxy.

It may of course be argued that NGC 1275 is not an elliptical galaxy at all. The two principal arguments against this point of view are: 1. The core of the Perseus cluster, of which NGC 1275 is a member, consists only of E and S0 galaxies. The brightest members of such early-type clusters are almost invariably found to be E or S0 galaxies. 2. On direct plates the overall luminosity distribution in NGC 1275 appears to resemble that of typical giant ellipticals.

Possibly the gas that was ejected from evolving stars in NGC 1275 was not swept out of this galaxy by intracluster gas because NGC 1275 has little or no net motion relative to the center of mass of the Perseus cluster. This view is consistent with the observation by Chincarini & Rood (1971) that NGC 1275 has a radial velocity of 5291 km sec^{-1}, which does not differ significantly from the mean cluster velocity of 5460 ± 200 km sec^{-1}.

Some support for the notion that violent explosive events may trigger bursts of star formation is provided by the observation that the nucleus of the posteruptive galaxy M82 (Kronberg et al 1972) is surrounded by a dozen young "super star clusters," each of which has $M_V \approx -15$.

According to Sandage (1972b) (see Figure 7) giant elliptical galaxies that are radio sources are often bluer than those that are radio quiet. This observation can be accounted for by recent bursts of star formation and/or by the presence of emission lines.

It may be assumed that the observation of a young stellar population component in certain giant elliptical galaxies would require an agonizing reappraisal of our ideas

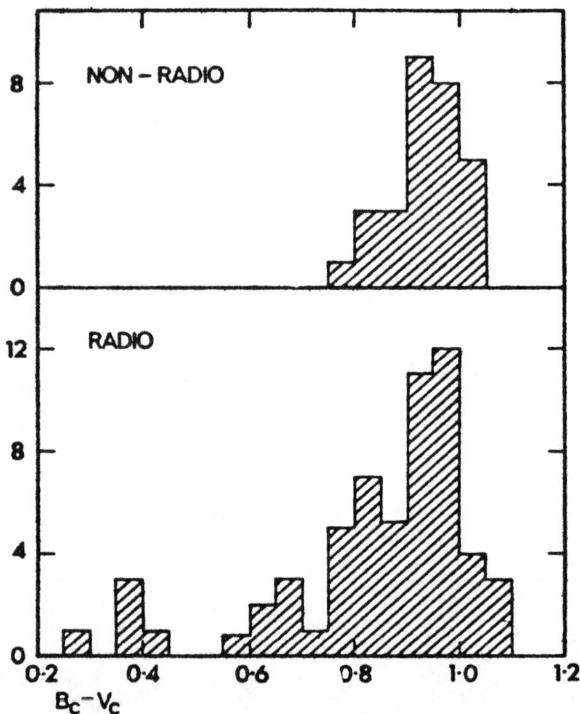

Figure 7 Colors of the brightest E galaxies in clusters. The figure shows that E galaxies that are radio sources frequently have colors bluer than radio-quiet ellipticals. (From Sandage 1972b.)

on the evolution of galaxies. Perhaps this view is unduly pessimistic because young stars (even though they may momentarily contribute significantly to the integrated light of an elliptical galaxy) probably account for only a negligible fraction of the total mass of such an object. Possibly elliptical galaxies with relatively strong hydrogen lines in their nuclei such as NGC 3379 (Spinrad 1972) are objects in which a burst of star formation took place within the last $\sim 1 \times 10^9$ years. If this view is correct, elliptical galaxies with bluer than average nuclei might be regarded as fossil radio sources.

Literature Cited

Ables, H. D., Newell, E. B., O'Neil, E. J. 1974. *Publ. Astron. Soc. Pac.* 86:311
Andrews, P. J., Lloyd-Evans, T. 1971. *The Magellanic Clouds,* ed. A. B. Muller, 88. Dordrecht: Reidel
Ardeberg, A., de Groot, M. 1973. *Astron. Ap.* 26:53
Arnett, W. D., Schramm, D. N. 1973. *Ap. J. Lett.* 184:L47
Arp, H. C. 1955. *Astron. J.* 60:317
Arp, H. C. 1958. *Astron. J.* 63:487
Arp, H. C. 1965. *Ap. J.* 141:43
Baade, W. 1944. *Ap. J.* 100:137
Baade, W. 1951. *Publ. Univ. Michigan Observ.* 10:7
Baade, W. 1963. *Evolution of Stars and*

Galaxies, 101. Cambridge, Mass.: Harvard Univ. Press
Baade, W., Swope, H. H. 1961. *Astron. J.* 66:300
Baade, W., Swope, H. H. 1963. *Astron. J.* 68:435
Baldwin, J. R. et al 1973. *Ap. Lett.* 14:1
Baum, W. A. 1959. *Publ. Astron. Soc. Pac.* 71:106
Bell, R. A., Parsons, S. B. 1972. *Ap. Lett.* 12:5
Bond, H. E. 1970. *Ap. J. Suppl.* 22:117
Chandrasekhar, S., von Neumann, J. 1943. *Ap. J.* 97:1
Chincarini, G., Rood, H. J. 1971. *Ap. J.* 168: 321
Christensen, C. G. 1972. PhD thesis. Calif. Inst. Technol., Pasadena, Calif.
Deutsch, A. J. 1964. *Ap. J.* 139:532
De Vaucouleurs, G. 1961. *Ap. J. Suppl.* 5:233
De Vaucouleurs, G., Ables, H. D. 1968. *Ap. J.* 151:105
Dickens, R. J., Peach, J. V. 1972. *External Galaxies and Quasi-Stellar Objects,* ed. D. S. Evans, 89. Dordrecht: Reidel
Elsässer, H. 1958. *Z. Ap.* 45:24
Faber, S. M. 1972. *Astron. Ap.* 20:361
Faber, S. M. 1973a. *Ap. J.* 179:423
Faber, S. M. 1973b. *Ap. J.* 179:731
Faber, S. M. 1974. Private communication
Ford, H. C. 1971. *Bull. Am. Astron. Soc.* 3:19
Freeman, K. C. 1973. Presented at IAU Symp. No. 58, Canberra, Australia
Freeman, K. C., Munsuk, C. 1972. *Proc. Astron. Soc. Aust.* 2:151
Fusi-Pecci, F., Renzini, A. 1975. In preparation
Gascoigne, S. C. B. 1962. *MNRAS* 124:201
Gascoigne, S. C. B. 1966. *MNRAS* 134:59
Gascoigne, S. C. B. 1969. *MNRAS* 146:1
Gunn, J. E., Gott, J. R. 1972. *Ap. J.* 176:1
Hagen, G. L. 1974. PhD thesis. Univ. Toronto, Canada
Hagen, G. L., van den Bergh, S. 1974. *Ap. J. Lett.* 189:L103
Harris, W. E. 1975. In preparation
Hartwick, F. D. A. 1968. *Ap. J.* 154:475
Hartwick, F. D. A. 1970. *Ap. Lett.* 7:151
Hartwick, F. D. A., Hesser, J. E. 1974. *Bull. Am. Astron. Soc.* 6:216
Hearnshaw, J. B. 1972. *MNRAS* 77:55
Hiltner, W. A. 1960. *Ap. J.* 131:163
Hindman, J. V. 1967. *Aust. J. Phys.* 20:147
Hodge, P. W. 1960. *Ap. J.* 132:346
Hodge, P. W. 1963. *Astron. J.* 68:691
Hodge, P. W. 1965. *Ap. J.* 142:1390
Hodge, P. W. 1972. *Publ. Astron. Soc. Pac.* 84:365
Hodge, P. W. 1973. *Ap. J.* 182:671
Hodge, P. W. 1974. *Publ. Astron. Soc. Pac.* 86:263
Hogg, A. R. 1955. *MNRAS* 115:473
Hogg, H. S. 1973. *Publ. David Dunlap Observ.* 3:6
Hubble, E. 1936. *The Realm of the Nebulae,* 45. New Haven: Yale Univ. Press
Iben, I., Rood, R. T. 1968. *Ap. J.* 154:215
Illingworth, G. 1973. PhD thesis. Aust. Nat. Univ. Mt. Stromlo Observ., Canberra, Australia
Janes, K. A., McClure, R. D. 1972. *Bull Am. Astron. Soc.* 4:241
Johnson, H. M. 1961. *Publ. Astron. Soc. Pac.* 73:20
Joly, M., Andrillat, Y. 1973. *Astron. Ap.* 26:95
Keenan, D. W., Innanen, K. A. 1974. *Ap. J.* 189:205
King, I. R. 1971. *Publ. Astron. Soc. Pac.* 83:377
Kronberg, P. P., Pritchet, C. J., van den Bergh, S. 1972. *Ap. J. Lett.* 173:L47
Larson, R. B. 1972. *Nature Phys. Sci.* 236:7
Larson, R. B. 1974a. *MNRAS* 166:585
Larson, R. B. 1974b. *MNRAS* 169:229
Lloyd-Evans, T. 1971. In *Supergiant Stars. Trieste Colloq. Ap., 3rd,* ed. M. Hack, 125. Trieste Observ., Italy
Madore, B. F. 1971. MSc thesis. Univ. Toronto, Canada
Madore, B. F. 1974. PhD thesis. Univ. Toronto, Canada
Mathews, W. G., Baker, J. C. 1971. *Ap. J.* 170:241
McClure, R. D. 1969. *Astron. J.* 74:50
McClure, R. D., Forrester, W. T., Gibson, J. 1974. *Ap. J.* 189:409
McClure, R. D., van den Bergh, S. 1968a. *Astron. J.* 73:313
McClure, R. D., van den Bergh, S. 1968b. *Astron. J.* 73:1008
McGee, R. X., Milton, J. A. 1964. In *The Galaxy and the Magellanic Clouds. IAU Symp. No. 20,* ed. F. J. Kerr, A. W. Rogers, 289. Canberra: Aust. Acad. Sci.
Minkowski, R. 1968. *Astron. J.* 73:842
Morgan, W. W. 1959. *Astron. J.* 64:432
Morgan, W. W., Mayall, N. U. 1957. *Publ. Astron. Soc. Pac.* 69:291
Morgan, W. W., Osterbrock, D. E. 1969. *Astron. J.* 74:515
Morton, D. C., Thuan, T. X. 1973. *Ap. J.* 180:705
Nassau, J. J., Blanco, V. M. 1958. *Ap. J.* 128: 46
Oort, J. H. 1958. In *Stellar Populations,* ed. D. J. K. O'Connell, 507. Amsterdam: North-Holland
Oort, J. H. 1974. Presented at Workshop Cent. Region Galaxy, Charlottesville, NC
Osmer, P. S. 1973. *Ap. J. Lett.* 184:L127
Ostriker, J. P. 1974. Presented at Workshop

Cent. Region Galaxy, Charlottesville, NC
Ostriker, J. P., Peebles, P. J. E. 1973. *Ap. J.* 186:467
Payne-Gaposchkin, C. 1971. *The Magellanic Clouds,* ed. A. B. Muller, 34. Dordrecht: Reidel
Payne-Gaposchkin, C., Gaposchkin, S. 1966. *Smithsonian Contrib. Ap.,* Vol. 9
Peimbert, M. 1968. *Ap. J.* 154:33
Peimbert, M. 1973. Presented at IAU Symp. No. 58, Canberra, Australia
Peimbert, M. 1975. *Ann. Rev. Astron. Ap.* 13:113–131
Peimbert, M., Torres-Peimbert, S. 1971. *Bol. Observ. Tonantzintla Tucabaya* 6:101
Peimbert, M., Torres-Peimbert, S. 1975. *Ap. J.* In preparation
Penston, M. V. 1973. *MNRAS* 162:359
Plaut, L., Oort, J. H. 1975. *Astron. Ap.* In press
Przybylski, A. 1968. *MNRAS* 139:313
Przybylski, A. 1971. *MNRAS* 152:197
Przybylski, A. 1972. *MNRAS* 159:155
Roberts, M. S., Rots, A. H. 1973. *Astron. Ap.* 26:483
Rogstad, D. H., Lockhart, I. A., Wright, M. C. H. 1974. *Ap. J.* 193:309
Rood, H. J. 1969. *Ap. J.* 158:657
Rood, H. J. 1974. *Ap. J.* 188:451
Rood, R. T. 1973. *Ap. J.* 184:815
Salpeter, E. E. 1955. *Ap. J.* 121:161
Sandage, A. R. 1969. *Ap. J.* 157:515
Sandage, A. R. 1971. *Ap. J.* 166:13
Sandage, A. R. 1972a. *Ap. J.* 176:21
Sandage, A. R. 1972b. *Ap. J.* 178:25
Sandage, A. R., Wallerstein, G. 1960. *Ap. J.* 131:598
Sandage, A. R., Wildey, R. 1967. *Ap. J.* 150:469
Sanduleak, N., MacConnell, D. J., Hoover, P. S. 1972. *Nature* 237:28
Schmidt, M. 1959. *Ap. J.* 129:243
Schmidt, M. 1963. *Ap. J.* 137:758
Schmidt-Kaler, T., Schlosser, W. 1973. *Astron. Ap.* 29:409
Searle, L. 1971. *Ap. J.* 168:327
Searle, L. 1972. In *IAU Colloq. No. 17,* ed. G. Cayrel de Strobel, A. M. Delplace, Sect. 52. Paris: Meudon Observ.
Searle, L., Sargent, W. L. W. 1972. *Ap. J.* 173:25
Searle, L., Sargent, W. L. W., Bagnuolo, W. G. 1973. *Ap. J.* 179:427
Shapley, H. 1938. *Nature* 142:715
Shields, G. A. 1974. *Ap. J.* 193:335
Smith, H. E. 1974. PhD thesis. Univ. Calif., Berkeley, Calif.
Spinrad, H. 1961. *Publ. Astron. Soc. Pac.* 73:336
Spinrad, H. 1966. *Publ. Astron. Soc. Pac.* 78:367
Spinrad, H. 1972. *Ap. J.* 177:285
Spinrad, H. 1973. *Ap. J.* 182:381
Spinrad, H., Gunn, J. E., Taylor, B. J., McClure, R. D., Young, J. W. 1971. *Ap. J.* 164:11
Spinrad, H., Smith, H. E., Taylor, D. J. 1972. *Ap. J.* 175:649
Spinrad, H., Taylor, B. J. 1971. *Ap. J. Suppl.* 22:445
Sturch, C. R., Helfer, H. L. 1972. *Astron. J.* 77:730
Swope, H. H. 1967. *Publ. Astron. Soc. Pac.* 79:439
Talbot, R. J. 1974. *Ap. J.* 189:209
Talbot, R. J., Arnett, W. D. 1973. *Ap. J.* 186:69
Tammann, G. A. 1969. *Astron. Ap.* 3:308
Tifft, W. G. 1963. *MNRAS* 125:199
Tifft, W. G. 1969. *Astron. J.* 74:354
Tifft, W. G., Snell, C. M. 1971. *MNRAS* 151:365
Tinsley, B. M. 1974. Preprint
Truran, J. W., Cameron, A. G. W. 1971. *Ap. Space Sci.* 14:179
van Agt, S. L. T. J. 1967. *Bull. Astron. Inst. Nether.* 19:275
van Agt, S. L. T. J. 1973. In *Variable Stars in Globular Clusters and in Related Systems,* ed. J. D. Fernie, 35. Dordrecht: Reidel
van den Bergh, S. 1957. *Z. Ap.* 43:236
van den Bergh, S. 1958. *Astron. J.* 63:492
van den Bergh, S. 1961. *Publ. Astron. Soc. Pac.* 73:135
van den Bergh, S. 1962. *Astron. J.* 67:486
van den Bergh, S. 1963. *Publ. Astron. Soc. Pac.* 75:498
Ibid. 1967. 79:460
van den Bergh, S. 1968. *The Galaxies of the Local Group. David Dunlap Observ. Commun. No. 195*
van den Bergh, S. 1969. *Ap. J. Suppl.* 19:145
van den Bergh, S. 1971a. *Astron. J.* 76:1082
van den Bergh, S. 1971b. *Publ. Astron. Soc. Pac.* 83:663
van den Bergh, S. 1972a. *RASC J.* 66:237
van den Bergh, S. 1972b. *Astron. Ap.* 20:469
van den Bergh, S. 1972c. In *External Galaxies and Quasi-Stellar Objects,* ed. D. S. Evans, 1. Dordrecht: Reidel
van den Bergh, S. 1972d. *Ap. J. Lett.* 178:L99
van den Bergh, S. 1973a. *Astron. Ap.* 28:469
van den Bergh, S. 1973b. In *Variable Stars in Globular Clusters and in Related Systems,* ed. J. D. Fernie, 26. Dordrecht: Reidel
van den Bergh, S. 1974a. *Astron. J.* 79:603
van den Bergh, S. 1974b. *Ap. J.* 193:63
van den Bergh, S. 1975a. *IAU Symp. No. 58,* ed. J. Shakeshaft. Dordrecht: Reidel
van den Bergh, S. 1975b. In *Stars and Stellar Systems,* ed. A. R. Sandage, M. Sandage, J. Kristian, 9:509. Chicago: Univ. Chicago Press

van den Bergh, S., Henry, R. C. 1962. *David Dunlap Observ. Publ.* 2:281
van den Bergh, S., Racine, R. 1967. *Astron. J.* 72:69
van den Bergh, S., Sher, D. 1960. *David Dunlap Observ. Publ.* 2:203
Vetešnik, M. 1965. *Publ. Univ. Brno No. 5*
Walborn, N. 1972. Personal communication
Walker, M. F. 1962. *Ap. J.* 136:695
Walker, M. F. 1964. *Astron. J.* 69:744
Walker, M. F. 1970. *Ap. J.* 161:835
Walker, M. F. 1972a. *MNRAS* 156:459
Walker, M. F. 1972b. *MNRAS* 159:379
Walker, M. F., Blanco, V. M., Kunkel, W. E. 1969a. *Astron. J.* 74:44
Ibid. 1969b. 74:964
Walraven, T., Walraven, J. H. 1971. In *The Magellanic Clouds,* ed. A. B. Muller, 117. Dordrecht: Reidel
Wares, G. W., Ross, J. E., Aller, L. H. 1968. *Ap. Space Sci.* 2:344
Welch, G. A. 1970. *Ap. J.* 161:821
Welch, G. A., Forrester, W. T. 1972. *Astron. J.* 77:333
Westerlund, B. E., Smith, L. F. 1964. *MNRAS* 127:449
Whitford, A. E. 1972. *Bull. Am. Astron. Soc.* 4:230
Wolf, B. 1972. *Astron. Ap.* 20:275
Wolf, B. 1973. *Astron. Ap.* 28:335

HIGH-VELOCITY NEUTRAL HYDROGEN

Gerrit L. Verschuur
Department of Astro-Geophysics, University of Colorado, Boulder, Colorado 80302

INTRODUCTION

General Comments

The existence of neutral hydrogen emission, often in the form of clouds or concentrations which exhibit a velocity not readily explainable by a simple model of galactic structure, has been an enigma for many years. The phenomena have usually been referred to as high-velocity clouds or intermediate-velocity clouds because most of the early observational work stressed the existence of structures at anomalous velocities. (We exclude the galactic center region from these discussions.) Looking back on the progress in surveying the neutral hydrogen emission away from the general galactic hydrogen, whether in position or velocity, we are struck by the incompleteness of the data available to the astronomer in any epoch. Historically, the early surveys in the Netherlands covered part of the northern sky over a limited velocity range. Those observations showed an illusory preponderance of negative-velocity matter away from the galactic plane. Later surveys of some of the southern sky, still made by northern hemisphere observers, revealed positive-velocity matter with large velocities also unexpected for their location in the sky.

Studies of the random motions of gas within particular small areas of the sky, combined with information about the motion of stars, gave some indication that the so-called random motions of matter in interstellar space were of the order of, at most, 10 km sec^{-1}. Available data suggested that the disk of the Milky Way itself was about 200 pc thick. It should be borne in mind that such data referred to the solar neighborhood and regions within the solar circle.

Large apparent radial velocities ($\sim$100–200 km sec^{-1}) for galactic neutral hydrogen were expected on the basis of models for the rotation of our Galaxy, in which we have matter moving in circular orbits about the nucleus. Differential rotation effects, therefore, produced observed radial velocities for hydrogen in the galactic plane of up to 200 km sec^{-1} in some directions in space, depending on the longitude and distance of the gas.

When the first surveys of gas away from the Milky Way band itself were made (with reasonably sensitive equipment), the observers were very surprised to find that emission was detected from gas which appeared to have a radial motion of many tens of km sec^{-1} and in some directions well away from the galactic plane clouds of gas were found to have velocities as high as -150 to -200 km sec^{-1} (Muller et al 1966). Such velocities were clearly much greater than the previously observed random motions. In fact, they simply could not be random motions in the local gas because they would then be supersonic.

Neutral hydrogen emission arising from gas in distant spiral arms was also not expected at latitudes much above 10°, because it was assumed that we lie in a flat galaxy. The emission was, therefore, anomalous on two counts: 1. A considerable amount of gas seemed to exist well away from the galactic plane in parts of the sky where one was expecting to observe only local gas, and 2. The velocities of this gas were very much higher than could be account for by random motions in a uniformly rotating galaxy.

In addition, most of the clouds originally found had negative velocities with respect to the local standard of rest. Thus the picture grew that the skies were filled with H I clouds coming toward us with velocities of 50–200 km sec^{-1}.

This concept, that the high-velocity clouds were actually coming toward us, stuck for many years, and models involving infall of matter from extragalactic space or the falling of matter initially ejected from some other region of the galaxy back down into the plane were widely touted. This was despite the cautions aired by some who pointed out that we were, after all, seeing only one component of the true space velocity of the "high-velocity" clouds. The true motions might be anything but toward us. Most of the clouds were seen in that range of galactic longitude (the second quadrant) in which distant gas, whether parts of distant spiral arms, intergalactic clouds, or clouds in orbit about our galaxy, would always show a negative velocity. Some of the clouds in the anticenter direction were, however, expected to show zero velocity for their longitude, no matter what their distance, provided they were in circular orbit about the galactic center. Those clouds were not adequately explained on any simple model.

To stress the point that negative velocities could be produced equally well in other ways than by assuming local infall of matter, two papers were written suggesting the possible extragalactic nature of the clouds. Kerr & Sullivan (1969) considered the clouds as objects in orbit about our galaxy, and found that the observed velocities of the clouds in the line of sight could be explained by this model. Verschuur (1969) suggested that the clouds were members of the Local Group of galaxies, based on their distribution in the sky and their velocities with respect to the galactic center which showed the same distribution as the Local Group galaxies. Also, the clouds would be virially stable at the distances of M31, that is, about 500 kpc (Burke 1967).

The basic problem in understanding the high-velocity clouds is the question of the distance. There is no information on this parameter for the highest-velocity clouds and there is some evidence that some of the so-called intermediate-velocity clouds are at distances of about 1 kpc, although some of these data are at present circumstantial, as discussed below.

To a certain extent the validity of at least one of the extragalactic models has been supported by the discovery of the Magellanic Stream, which places one of the originally discovered high-velocity clouds at the distance of the Magellanic Clouds.

A small workshop was held early in September 1971 at the National Radio Astronomy Observatory in Green Bank on the topic of high-velocity hydrogen. It is interesting that the models proposed by Verschuur (1971b, 1972, 1973a) and Davies (1972a,b), which are similar to one another in many respects, were not actually considered at that meeting. The present author did realize, on listening to the tape recordings of the meeting, that although many other possibilities (both wild and serious) were discussed, the concept that the clouds were no more or less than parts of distant spiral arms distorted with respect to our normal view of the Galaxy was not seriously discussed.

A Summary of the Present Models

The so-called high-velocity clouds in the Galaxy include two basic groups of clouds and cloud complexes sometimes referred to as high-velocity clouds and intermediate-velocity clouds. A review of the properties and distributions of these clouds shows that there are at least three clearly distinct phenomena presently being observed.

1. The intermediate-velocity clouds around $l = 210°$, $b = +70°$, whose velocities are around -40 to -50 km sec^{-1} with respect to the local standard of rest, are undoubtedly relatively close to the sun. Their distance is of the order of several hundred parsecs. They are associated with a distinct lack of zero-velocity matter and they appear to have been accelerated to their present negative velocities by some force. Two alternatives exist to explain this acceleration: either a supernova has swept away the local matter which is now seen at the velocities of -40 to -50 km sec^{-1}, or some large cloud has fallen in from above the galactic plane and swept up the matter which would otherwise have been at zero velocity. (Both models are discussed below.) The supernova model is favored over the infall model. In any event this localized infall model is not to be confused with the general infall model used to explain other aspects of the highest-velocity cloud phenomenon.
2. One class of the highest-velocity clouds is clearly extragalactic at distances of the order of those of the Magellanic Clouds and greater. The clouds that are included in this category are the highest velocity features near the south galactic pole.
3. The third category of high-velocity clouds is almost certainly associated with distant spiral structure in some way. One model attaches the highest-velocity clouds around $l = 120°$ to the outer arms of the galaxy whereas another model gives the highest-velocity clouds an independent existence beyond these most distant clearly defined arms. They are then independent spiral features. The difference between these two models is small and it seems difficult to decide firmly between them. They do have the common property that the clouds at the highest velocities are placed at great distances from the sun, at least 10 kpc, and the observed velocities are primarily the result of differential galactic rotation effects. It is also possible that the highest-velocity clouds are as much as 20 kpc away so that they form part of a

tail to our Galaxy much in the same way that the cloud near the south pole forms a tail on the far side of the Magellanic Clouds.

A subclass of clouds in this category are the intermediate positive- and negative-velocity clouds which are at latititudes below about 50°. These are parts of nearby spiral arms reaching to high-z distances.

The general infall hypothesis to account for any of the high-velocity clouds, whether with positive or negative velocities, is not necessary to explain these clouds since they basically have velocity and spatial distributions consistent with one of the above models. This hypothesis is critically reviewed below.

OBSERVATIONS OF HIGH-VELOCITY CLOUDS

The Discovery of Anomalous-Velocity Hydrogen away from the Milky Way

The earliest detailed papers on observations of the so-called high-velocity clouds, and here we lump the intermediate- and high-velocity clouds together, were published in 1966. Observations had been in progress since the early 1960s in the Netherlands to study the distribution of H I emission at intermediate and high galactic latitudes, and the observers at Gröningen and Leiden divided the velocity range to be studied. The former concentrated on emission between ± 70 km sec^{-1} and the latter studied the emission at velocities more negative than 70 km sec^{-1}. This distinction led to the original classification of "high-velocity clouds" as separate entities from "intermediate-velocity clouds," but it now appears that the distinction, although there are differences between the two types of clouds, was mainly geographical! If anything, a velocity demarcation should be made around 90 km sec^{-1} (Wesselius & Fejes 1973).

In the early surveys Hulsbosch & Raimond (1966) reported that a number of large clouds, later called "cloud complexes" by Hulsbosch (1968), were found in the northern skies around longitudes 100° and 160° whose velocities were of the order -150 km sec^{-1} and which were located up to latitude $+50°$. This was far too high a latitude for the material to be thought of as belonging to any normal population of matter in the Galaxy. At the same time, Blaauw & Tolbert (1966) reported that the intermediate-velocity survey had revealed a preponderance of negative-velocity matter at high positive latitudes. Associated with this excess of intermediate-velocity matter there appeared to be a relative absence of low- and zero-velocity material, almost as if the local matter had been disturbed in some way with the result of hydrogen gas being accelerated to some negative velocity of the order of 50 km sec^{-1}.

There also appeared a rough correlation between the areas of high-velocity matter and the distribution of intermediate-velocity clouds as well as the local absence of zero-velocity matter. This seemed to argue that all three phenomena were associated, and Oort (1966) argued just this point. He suggested that we were seeing an infall of matter into the galactic plane, matter that had either been thrown out of the plane and was now falling back in, or matter that was still falling into the galaxy as a whole—perhaps part of an intergalactic wind.

Before these extensive early surveys at intermediate and high latitudes were done, Dieter (1965) surveyed the hydrogen gas near the galactic poles and noted the presence of an excess of negative-velocity hydrogen toward both poles. At the south galactic pole matter with a velocity as high as -134 km sec^{-1} was reported by Hulsbosch & Raimond (1966).

The grouping of many formerly discovered clouds into large complexes was reported by Hulsbosch (1968), who found that the complexes appeared as coherent entities in this survey, with sizes of many tens of degrees long, by several degrees wide, in some cases. Within these complexes small concentrations could be seen, and recently Giovanelli et al (1973) have mapped a number of the complexes in detail (see below).

For several years it was believed that the high-velocity clouds had mainly negative velocities with only one or two exceptions. However, a series of high-sensitivity observations made by the Bell Telephone Laboratories group (Wannier & Wrixon 1972, Wannier et al 1972) showed that there are also extensive positive-velocity, high-velocity clouds at longitudes around 270°. They suggested that these clouds might well be at the distance of the Magellanic Clouds and if so, they would be at very large z distances because they also appeared to partake of galactic rotation.

A notation for referring to high-velocity clouds has evolved over the years. The abbreviation HVC followed by the longitude, latitude, and velocity of the cloud are used as follows: HVC 132+24−210 is the high-velocity cloud at $l = 132°$, $b = +24°$, whose velocity is -210 km sec^{-1} with respect to the local standard of rest.

The Distribution of High-Velocity Clouds

The history of the types of surveys that have been made show that the coverage of the sky is patchy and far from complete. The earliest surveys used very coarse grids at which emission profiles were obtained and even the best surveys, which used grid spacings of 2°, often missed some of the smaller high-velocity concentrations completely. Until 1974 there was virtually no data at all available for high-velocity hydrogen in the southern skies.

The various surveys that have been performed during the last 10 years are listed in Table 1. It can be seen that most of them had velocity cutoffs around -250 km sec^{-1} and only very coarse surveys have covered velocities as great as $+250$ km sec^{-1}. Various observers have searched at velocities beyond this range, but have not published any results.

In the lower part of Table 1 some information on the main mapping programs which studied individual cloud parameters are listed.

As an introduction to the subsequent discussions in this review we now present several maps of the distribution of the anomalous-velocity gas on the plane of the sky. In Figure 1 are indicated the so-called intermediate-velocity clouds as summarized by Verschuur (1973a). The cutoff in velocity was 100 km sec^{-1} for this map, although it is later suggested that a better cutoff might be around 90 km sec^{-1}. Different shadings have been used to indicate the locations of clouds taken from different surveys.

Table 1 Recent high-velocity cloud surveys

Observer	Region	Velocity range (km sec^{-1})	Beam-width	Grid	Band-width (kHz)
Hulsbosch & Raimond (1966)	$\|b\| > 20°$	-250 to -50	40′	10°	50
Hulsbosch (1968)	All sky	-250 to -50	40′	5°	50
		$+50$ to $+250$	40′	10°	50
Meng & Kraus (1970)	$\|b\| > 10°$	-250 to $+60$	$10' \times 40'$	Drift scans $\Delta\delta = 20'$	95
Tolbert (1971) (analyzed by Wesselius & Fejes 1973)	$\|b\| > 15°$	-250 to $+60$ (sometimes $+160$)	40′	5°	16
van Kuilenburg (1972)	$\|b\| > 15°$ 0° to 72° Dec.	-270 to -60	40′	Drift scans $\Delta\delta = 2.°5$	50
	0° to $-30°$ Dec.	$+60$ to $+270$	40′	$\Delta\delta = 2.°5$	50
Dieter (1972a)	$\|b\| < 15°$	negative; variable	35′	1° or 2°	10
		positive; variable	35′	2° or 10°	10
Dieter (1972b)	$\|b\| > 15°$	negative; variable	35′	2° × 5°	10
				2° × 30°	10
Heiles & Habing (1974)	$\|b\| > 10°$	-92 to $+75$	35′	$18' \times 36'$	10
Wannier et al (1972)	$b = 252°$ to 322°	$+18$ to $+334$	120′	2°	75
	$b = 10°$ to 30°	$+300$ to $+600$	120′	5°	75
Mathewson et al (1974)	$\delta < -375°$	-70 to $+300$	50′	Drift scans $\Delta\delta = 2.°5$	33
		-340 to $+110$	50′	$\Delta\delta = 5.°0$	33
Limited or Individual Cloud Mapping Programs					
Hulsbosch (1968)	Individual clouds		40′	0.°5 or 1°	50
Meng & Kraus (1970)	See above				
Wannier et al (1972)	See above				
Wesselius & Fejes (1973)	See Tolbert (1971), above				
Giovanelli et al (1973)	Individual clouds		10′	10′	3.5

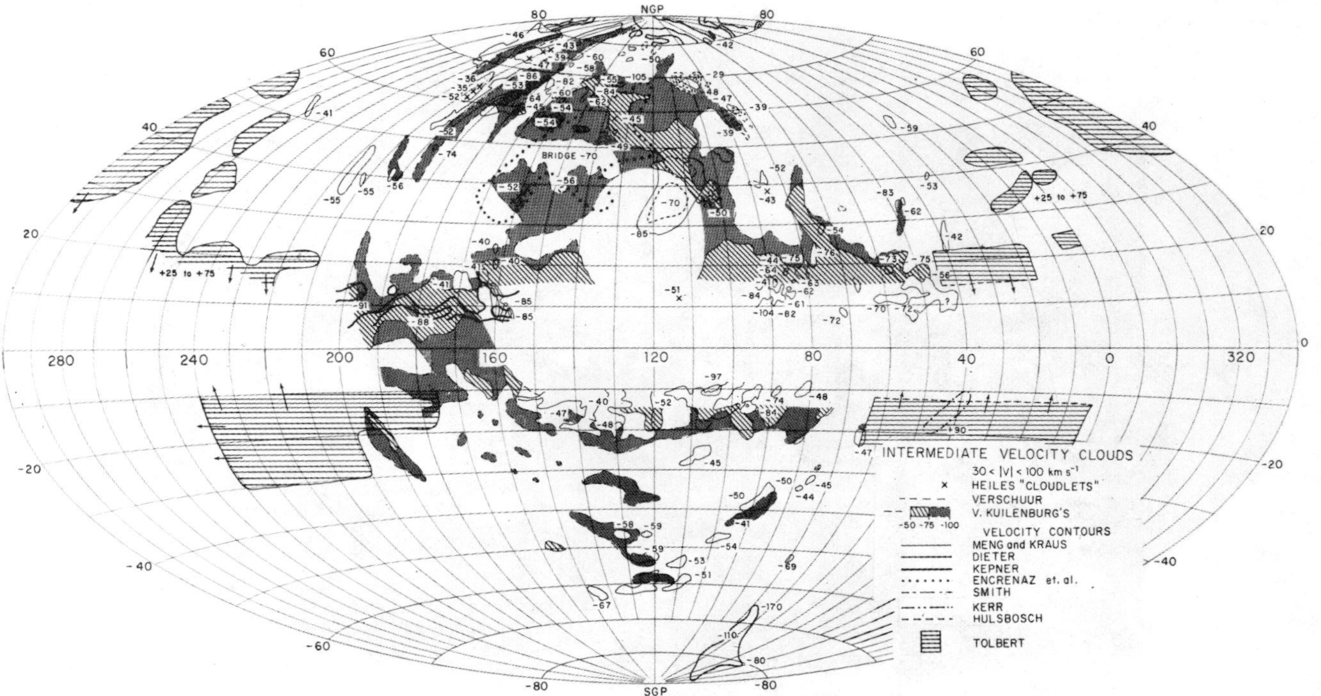

Figure 1 The distribution on the sky of the so-called intermediate-velocity clouds. Their velocities in km sec^{-1} with respect to the local standard of rest are indicated. For the source references see Verschuur (1973a).

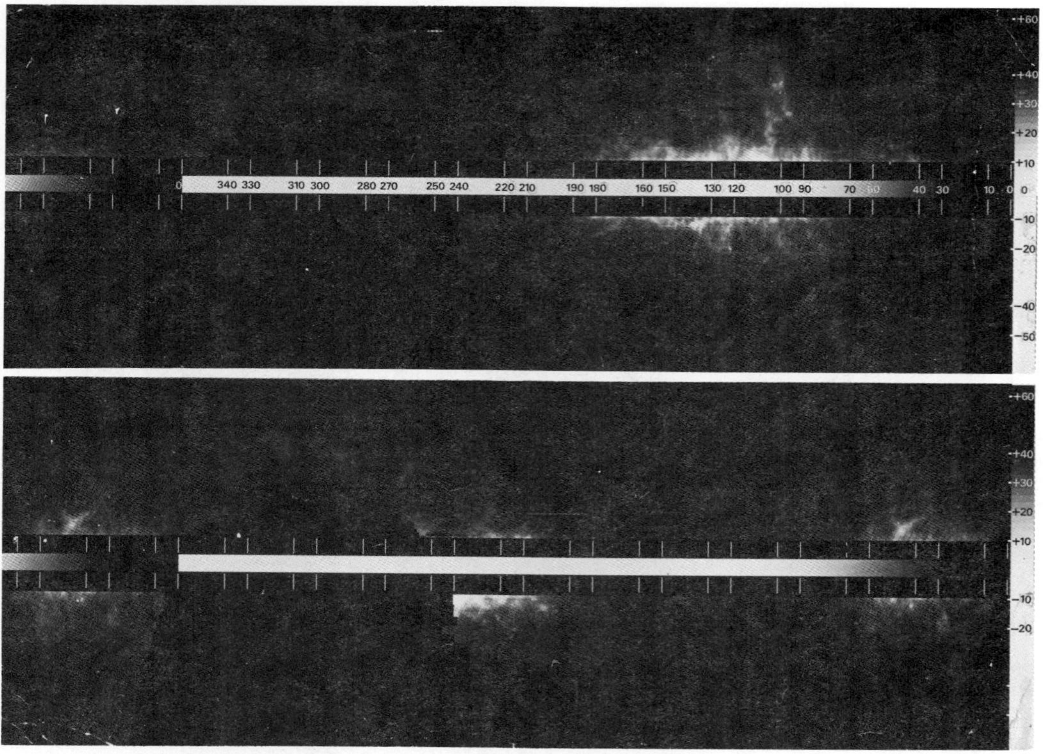

Figure 2 A photograph of the distribution on the northern sky of the intermediate-velocity hydrogen gas. *Upper*, the negative-velocity material between -20 and -90 km sec^{-1}; *lower*, the positive-velocity gas between $+20$ and $+70 \text{ km sec}^{-1}$. The strip of sky between latitudes $+10°$ and $-10°$ was not filled in with data in this survey. The dark area at the left-hand side of the map indicates that part of the sky inaccessible from the Hat Creek radiotelescope of the University of California. Galactic coordinates are indicated. The brightness of the emission is representative of the column density of the gas. Reproduced by kind permission of C. Heiles and E. Jenkins.

A much more dramatic way of presenting the data is in the form of the map produced by Heiles & Jenkins (1973). In Figure 2 we show their maps of the distribution of negative-velocity gas between -92 and -20 km sec^{-1} and the positive-velocity gas between $+20$ and $+70$ km sec^{-1} (at latitudes greater than 10°) in the form of a photograph. Basically, one can again see (more clearly in this case) that the intermediate negative-velocity hydrogen is concentrated between longitudes 80° and 190°, with filamentlike structures emerging from this main concentration. This simple distribution is basically expected for matter in some of the distant spiral arms, such as the Perseus arm and the arms beyond the Perseus arm, if the hydrogen is broadly distributed in z. This point is discussed in detail below.

In addition, the presence of a large amount of negative-velocity matter in Figures 1 and 2 can be seen at high latitudes (+70°) around $l = 100°$ to greater than 240°. It will be suggested later that these clouds are distinctly different from the lower-latitude, intermediate-velocity clouds. At longitude 100° there is also a filamentary projection up to latitude 50°. The rather incomplete data at positive velocities shown in Figure 1 indicate a basic distribution of this material similar to the main concentration of negative-velocity matter, except that it is located at those longitudes where one expects positive velocities in the closest spiral arms around $l = 270°$.

The distribution of the highest-velocity clouds is shown in Figure 3, also taken from Verschuur (1973a). Missing in this figure is the location of the south pole cloud now known as the Magellanic Stream. This is shown in a subsequent diagram (Figure 4), although a part of this cloud is plotted in Figure 1. In Figure 3 the positive- and negative-velocity clouds are all plotted, and again we notice that in general they are distributed more or less as expected for distant matter suffering the effects of differential galactic rotation, that is, negative velocities in the first and second quadrant of longitude and positive velocities in the third and fourth quadrants. At the same time, anticenter clouds, which on a model of uniform galactic rotation would have no line of sight velocity aside from some small random component, are seen to possess the greatest anomalous velocities considering their longitude.

There is now more data on the existence of high-velocity matter near the plane which extends those data plotted in Figure 3. However, this leads us into the area of confusion as far as the definition of high-velocity clouds is concerned. The definition presupposes that they are unusual features that need explaining. This is true for the clouds well away from the plane, but once one starts to recognize that many of the clouds form parts of distant spiral arms, and in particular those clouds which lie closest to the plane, then it is a moot point whether one will define the low-latitude, high-velocity components as high-velocity clouds or spiral-arm segments. At the same time we should bear in mind that the higher-latitude features can just as well be referred to as spiral-arm features on some of the models to be discussed. This is particularly true for the lower-latitude, intermediate-velocity hydrogen which, in Figure 2, so clearly appears to belong to other spiral arms. The label of intermediate-velocity cloud may then be confusing. Indeed, it is

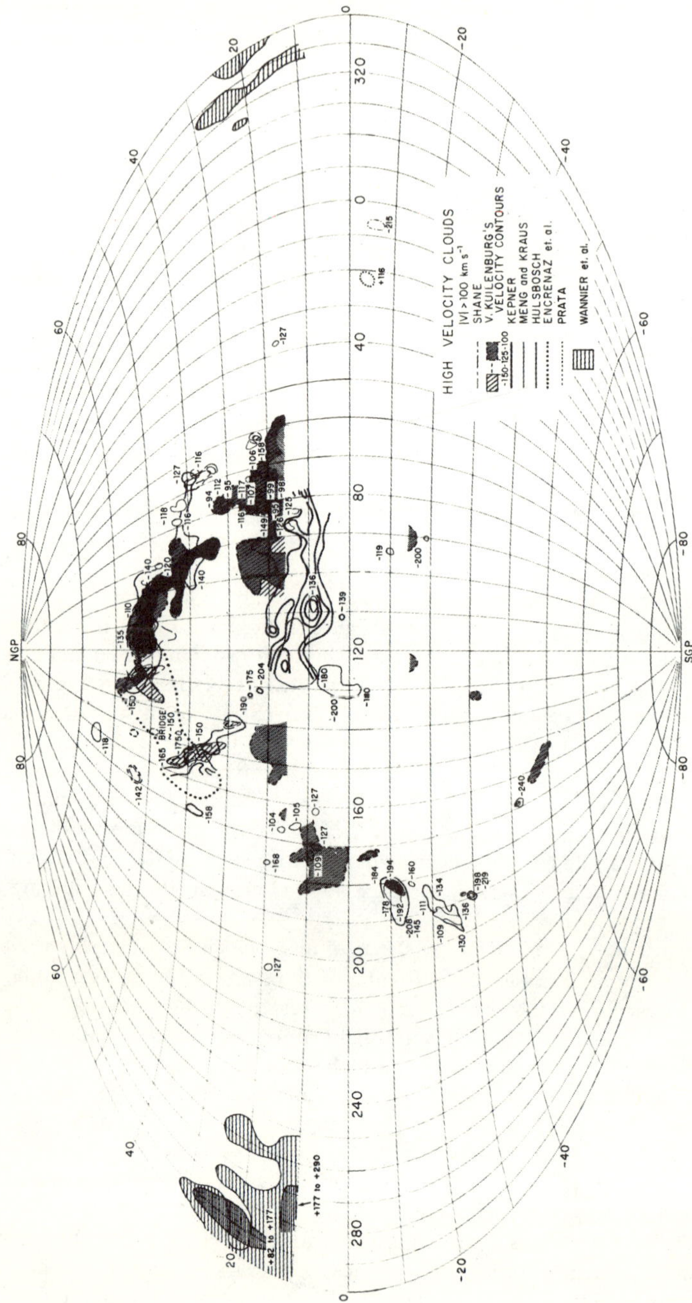

Figure 3 The distribution on the sky of the high-velocity clouds with velocities indicated in km sec^{-1} with respect to the local standard of rest. For the source references see Verschuur (1973a).

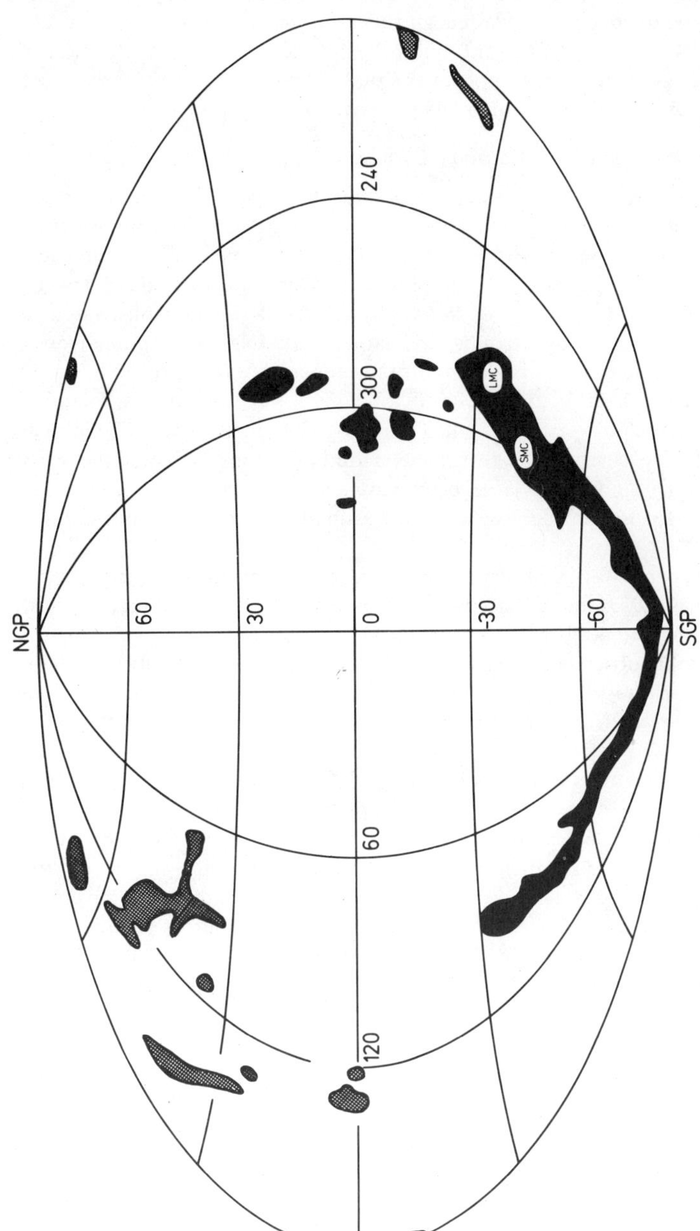

Figure 4 A schematic representation of the Magellanic Stream from Mathewson et al (1974). The positive-velocity, high-velocity clouds around $l = 290°$ are outlined as are the main negative-velocity, high-velocity clouds at positive latitudes.

shown below that the overall label of HVC or intermediate-velocity cloud (IVC) should possibly be dropped once we recognize where the clouds are. There would only be a few IVCs which would continue to bear that name as an indication of their strangeness. After all, we do not label clouds lying in the galactic disk and belonging to distant spiral arms with a label referring to their velocities.

Properties of the Anomalous-Velocity Clouds

Comparison of several of the properties of various types of clouds is made difficult by the fact that many of the observations have been made by different observers using different beamwidths and bandwidths. The considerable differences obtained on even the same cloud by different observers was illustrated by Verschuur et al (1972). They observed two high-velocity clouds previously studied by Hulsbosch (1968). For two clouds, the larger radiotelescope and narrower bandwidths led to differences in peak brightness temperature (a factor of 4–5), and differences in linewidth (a factor of 3) and total masses (as much as a factor of 4–5). These systematic differences were also found by Giovanelli et al (1973) in their more comprehensive survey of high-velocity clouds. It appeared that the early Dutch observations made with large beamwidths, incomplete survey grids, and large bandwidths led to considerably incorrect estimates of the properties of high-velocity clouds.

In the case of comparative observations of many intermediate-velocity clouds, Wesselius & Fejes (1973) found, based on their reasonably complete survey of the northern sky using the 40′ beamwidth of the Dwingeloo radiotelescope, that the agreement with other observers using larger radiotelescopes is reasonably good. This is not true for all clouds observed: examination of Verschuur's (1971a) data finds several having linewidths one third as great as those in the sample of Wesselius and Fejes.

Comparisons of the mass estimates for most of the clouds from one observer to the next are somewhat more impressive. Perhaps this is because integrations are performed over a large extent of the cloud and differences in linewidths from point to point in the cloud are then not so important as to produce very significant differences in the total mass. Data from various sources on cloud masses have been combined to produce the following set of histograms. These sources include Meng & Kraus (1970), Wesselius & Fejes (1973), Giovanelli et al (1973), Heiles (1968), Verschuur (1971a), and Rickard (1971). Giovanelli et al (1973) have made the most detailed survey of individual clouds and an example of one of their maps is given in Figure 5.

In Figure 6*a*, a histogram of cloud masses for all intermediate-velocity clouds, which are apparently part of the cloud complexes at $b = +70°$ (most are at $b > 45°$) is shown. The Heiles (1968) cloudlets are not included in this sample because the definition of a cloudlet was not taken into consideration in the other data used. For example, the data of Verschuur (1971a) showed that there were small-scale structures present in an intermediate-velocity cloud, but usually several of these were taken together in the integration to get the cloud mass. It could be said that

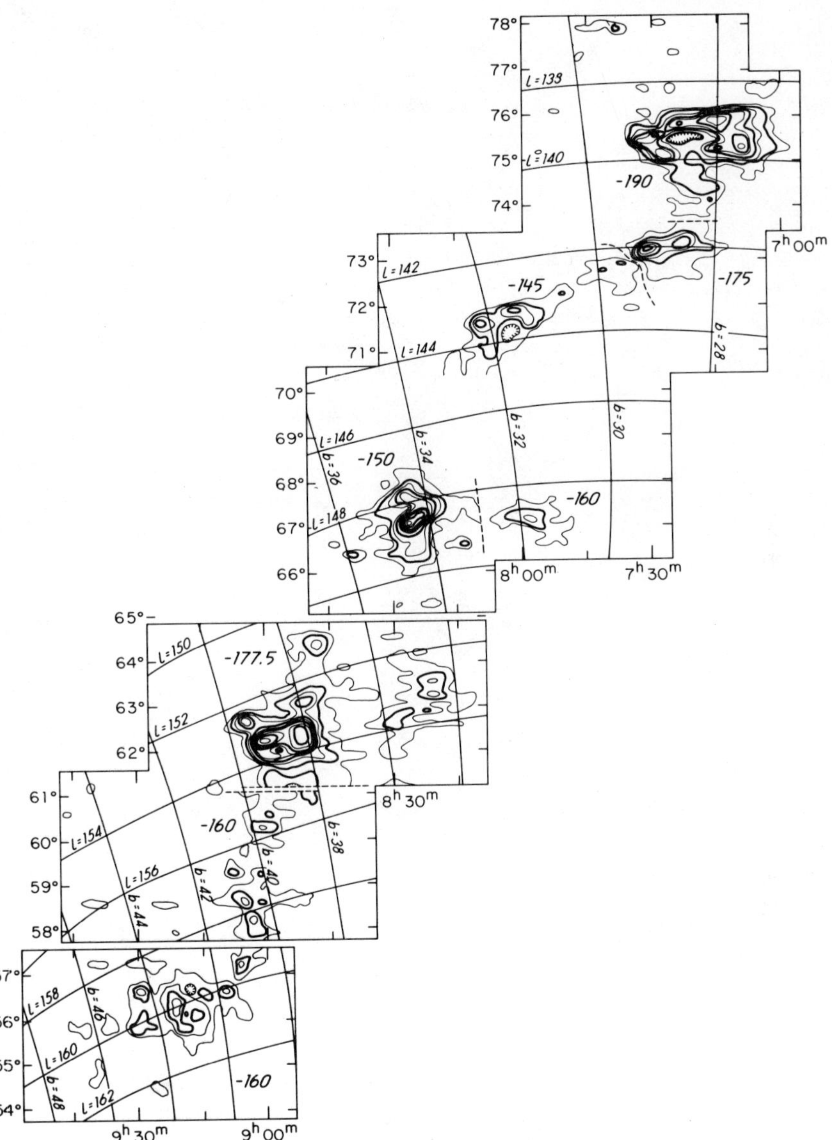

Figure 5 High-velocity cloud *A* as mapped by Giovanelli et al (1973). The coordinates are right ascension and declination and a representative contour map of each element of cloud *A* at the negative velocity indicated is shown. This cloud clearly breaks up into many discrete elements when studied with a resolution of 10′.

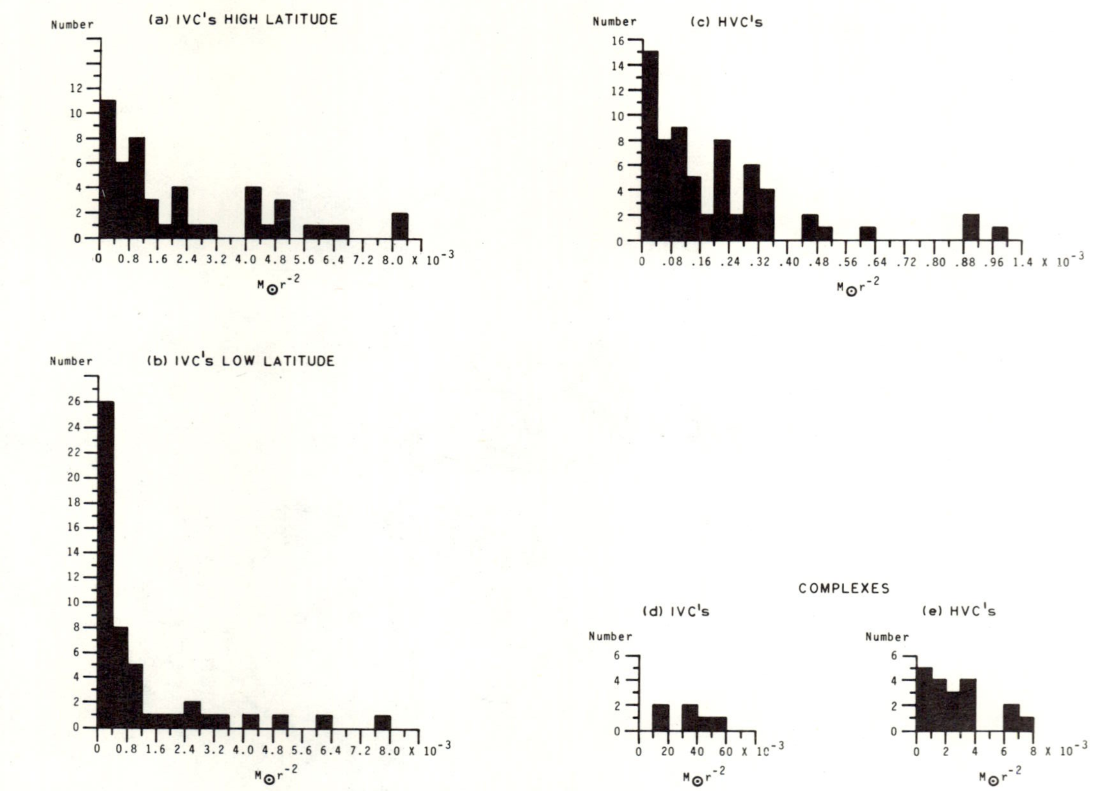

Figure 6 Histograms of cloud masses in units of solar masses times parsecs^{-2}: (*a*) intermediate-velocity clouds at $b > 45°$; (*b*) intermediate-velocity clouds at $b < 45°$; (*c*) high-velocity clouds; (*d*) intermediate-velocity cloud complexes; and (*e*) high-velocity cloud complexes.

the typical intermediate-velocity cloud consists of two to four cloudlets, but Heiles considered individual cloudlets.

In Figure 6*b* the histogram for the masses of the remaining intermediate-velocity clouds below latitude $+45°$ is shown. There appears to be a very distinct difference in the mass distribution for these two populations of clouds. There are many more higher-mass clouds in the former sample; the latter sample contains many small-mass objects. Inclusion of Heiles's data adds equal numbers (about 8) to the lowest mass bins in each category.

The mean mass of 51 clouds at the highest latitudes is $2.3 \times 10^{-3} M_{\odot} \mathrm{pc}^{-2}$; the mean mass for 52 clouds at the lower latitudes is $1.5 \times 10^{-3} M_{\odot} \mathrm{pc}^{-2}$. The mean mass for several complexes of intermediate-velocity clouds is $31 \times 10^{-4} M_{\odot} \mathrm{pc}^{-2}$ (Figure 6*d*).

Figure 6*c* is a histogram of cloud masses for high-velocity clouds and is very clearly different from those for the intermediate-velocity clouds. The masses are an order of magnitude different. We define *clouds* here as small-scale structures that are independent of one another, although they may occur in large complexes which make up some of the originally defined high-velocity clouds. We now know that incomplete surveying grids prevented the recognition of many of the individual clouds or concentrations earlier. A histogram of the masses of the high-velocity complexes is also shown (Figure 6*e*).

A very striking comparison between the intermediate- and high-velocity clouds can now be made. The mean masses of the various classes of clouds is summarized in Table 2. The last column gives the ratio of distances if the clouds are basically the same phenomenon, but are located at different distances. Here we picture the lower-latitude intermediate-velocity clouds as belonging to the Perseus arm and perhaps the arm immediately beyond that (about 3–6 kpc from the sun), but the high-velocity clouds as belonging to (or forming) the most distant spiral arms in the northern skies. Both types of clouds seems to show a hierarchy of structure with relative masses similar for the two classes of objects. The high-velocity clouds then appear to be about three times further away than the intermediate-velocity clouds. This therefore places them about 10–18 kpc from the sun, which is where both Verschuur (1973a–d) and Davies (1972a,b) place them.

Table 2 Summary of cloud masses

HVC Clouds	IVC Clouds	Distance ratio
$0.2 \times 10^{-3} M_{\odot} \mathrm{pc}^{-2}$	$1.5 \times 10^{-3} M_{\odot} \mathrm{pc}^{-2}$	2.7
HVC Complexes	IVC Complexes	Distance ratio
$2.5 \times 10^{-3} M_{\odot} \mathrm{pc}^{-2}$	$31 \times 10^{-3} M_{\odot} \mathrm{pc}^{-2}$	3.5
	Highest Latitude IVCs above 45°	
	$2.3 \times 10^{-3} M_{\odot} \mathrm{pc}^{-2}$	

In Table 2 the mean mass for the highest-latitude clouds thought to be associated with the supernova shell (see below) or a large mass of infalling hydrogen is given. Their mean mass has little significance for a comparison with the other clouds, as the former are thought to be clouds that are local and that have been accelerated to their presently observed velocities, as opposed to the other clouds or cloud complexes which are at rest with respect to the spiral arms in which they find themselves (at least to first order).

Both the high- and intermediate-velocity clouds appear to show the complexes consisting of many individual concentrations, and their relative masses indicate that the two classes of objects are either intrinsically very different, or similar and located at different distances from the sun.

Very high-resolution aperture-synthesis data have revealed that structures of the order of 50″ exist within one high-velocity cloud known as HVC 132+24−210 (Westerbork 1973). This cloud had been studied before by Verschuur et al (1972) and the discovery of such small-scale structure is consistent with this cloud being at a great distance and not very different from more local clouds in which much structure is seen. The linewidths of the high-velocity clouds have been studied by Giovanelli & Brown (1973) and indicate that there might be at least three temperatures in the interstellar medium at which hydrogen can exist in a stable phase. Hughes et al (1971) have set upper limits on the optical depth in two intermediate-velocity clouds of 0.06 and 0.02.

The Distance of High-Velocity Hydrogen

Oort (1969) has noted that some of the intermediate-velocity clouds lie in the direction of stars which show calcium absorption lines at the same velocities as the hydrogen emission lines. There are in fact three stars listed by him containing some five lines common to the calcium and hydrogen spectra. These three stars are all more than 1 kpc from the sun. Three stars at high latitudes closer than 1 kpc from the sun do not show calcium lines, although there are intermediate-velocity clouds in their direction.

One of the more distant stars shows an absorption line at $+20$ km sec^{-1} and for its longitude of 320° we expect matter belonging to a spiral arm about 1500 pc away to occur at this velocity. In the case of the four calcium lines seen toward the other two stars, we find that the velocities are indeed anomalous for any simple, uniform model of the galaxy.

Rickard (1972) has also observed calcium absorption lines in the direction of a number of high-latitude stars. He finds that the calcium K-line profiles are well correlated with 21-cm line observations. The radial velocities are consistent with a quiescent interstellar medium showing no peculiar motions out to distances of about 1500 pc. For all the stars with a z distance greater than 600 pc, intermediate-velocity calcium components are also seen. No simple dependence of velocity on z distance is evident.

These data are taken together to suggest that the intermediate-velocity hydrogen is galactic and located at least 1 kpc from the sun. At the same time, no upper limit for their distance has yet been established.

Concerning the distance of any of the other high-velocity hydrogen clouds, we have even less information to go on. Only indirect arguments invoking the virial theorem give any clue about distance, and these are usually ignored. For the highest-velocity clouds to be gravitationally stable we find that they are well and truly extragalactic, whereas some of the intermediate-velocity clouds might be of the order of kiloparsecs away. Only for the Magellanic Stream, which can be fairly clearly linked to the Magellanic Clouds, do we have a distance.

Several globular clusters are known to have high-velocity interstellar Ca II K-line components in their directions; because the distances to the clusters are known, Kerr & Knapp (1972) searched for high-velocity neutral hydrogen features in those directions. Of the six clusters, only one (M10) was found to have a hydrogen feature at nearly the same velocity as the Ca line. The hydrogen cloud is at least 20′ in size. The location of the cluster M10 is $l = 15°$ and $b = +23°$, and the velocity of the feature is about -80 km sec^{-1}.

Because the cluster distance is 6.2 kpc, it is suggested that there is high-velocity gas at distances of this order or smaller in this direction. The intermediate-velocity cloud found by Kerr and Knapp is not in any previous catalog of intermediate- or high-velocity matter and it is far from certain what can be derived from this observation about the more general high-velocity cloud phenomenon.

In the direction of M92 they did not find any association of neutral hydrogen emission with the Ca line, but 1° away from the cluster position they found 21-cm emission at the same velocity as the Ca line (-100 km sec^{-1}). The location ($l = 68°$, $b = +34°$) and distance of the cluster (8 kpc) would allow the gas to be located only as far as the solar circle and not in distant spiral arms outside the solar circle. However, we can ask what the likelihood is of chance coincidence. In this direction high-velocity gas is common. The cluster lies just beyond the edge of a large high-velocity cloud complex (see Figure 3). If the high-velocity Ca line is due to a process related to the cluster itself, then we should calculate the probability that a neutral hydrogen high-velocity cloud, which is part of a large-scale phenomenon over the northern sky, would lie in this direction. The velocities are not quite coincident, differing by 7 km sec^{-1}.

Clearly, if large numbers of associations or near associations could be found between Ca lines in the directions of objects lying within the boundaries of known high-velocity clouds, then a strong argument could be made for the distances of the hydrogen gas.

The South Pole Cloud or Magellanic Stream

Figure 1, showing the distribution of the intermediate-velocity hydrogen on the sky, includes a cloud near the South Galactic Pole which has a velocity that ranges from -50 to -170 km sec^{-1}, that is, it spans both the velocity ranges considered above. Since the data for Figure 1 were collected, Wannier & Wrixon (1972) found that their high-sensitivity surveys showed that this cloud extended much further in position and velocity toward the galactic plane more or less along a line of constant longitude around 90°. The velocity increased systematically to -400 km sec^{-1} and the change of velocity with position was consistent with solid-body rotation (at least to first order). Several workers in the field noted, usually in private, that when extrapolated along its length in the other direction, the cloud reached the Magellanic Clouds in position and even velocity. Observations have now been made which show that this is in fact the case.

Mathewson et al (1974) reported their observations as follows:

> A southern sky survey of HI in the velocity range -340 km $\sec^{-1}$ to $+380$ km $\sec^{-1}$ has shown that a long filament of HI extends from the region between the Magellanic Clouds down to the South Galactic Pole and connects with the long HI filament discovered recently by Wannier & Wrixon (1972) and van Kuilenburg (1972). There is also some evidence that this continues on the other side of the Magellanic Clouds and crosses the galactic plane at $l^{\text{II}} = 306°$. This filament, which follows very closely a great circle over its entire 180° arc across the sky, is given the name "The Magellanic Stream." It may have been produced by gravitational interaction between the Small Magellanic Cloud and the Galaxy during a close passage (20 kpc) of the SMC some 5×10^8 years ago, although it is impossible to account for the observed radial velocities along the Stream unless some force other than gravity is invoked to act on the Stream as well. There are also reasons for doubting whether the Magellanic Clouds have ever come closer than about 40 kpc to the galactic center. These arise from the smooth manner in which the radial velocity contours of the Inter-Cloud region connect with the radial velocity contour of HI in the main bodies of the Large and Small Magellanic Clouds and from the whole Magellanic System which indicate that a bridge of gas exists between the two galaxies and that they form a bound system. This implies that the Clouds did not make a close approach (20 kpc) to the galactic center. If this is correct, some other explanation must be sought to explain the Magellanic Stream (and the warp of the galactic plane). Maybe the HI in the Inter-Cloud region is foetal gas remaining after the formation of the LMC and SMC, and perhaps collisions with the intergalactic wind or explosions may have expelled gas from this region to form the Magellanic Stream.

In Figure 4 the outlines of this high-velocity cloud now known as the Magellanic Stream are shown. Also indicated are the outlines of some of the high-velocity clouds around $l = 120°$ and the positive-velocity, high-velocity clouds around $l = 270$–$290°$.

Wannier et al (1972), in their paper presenting the results of their survey of positive-velocity hydrogen between $l = 252°$ and $l = 322°$, found that positive-velocity material was visible at latitudes of over 20°, and they basically recognized two features, one with a velocity of ~ 150 km $\sec^{-1}$ and the other with a velocity of ~ 232 km $\sec^{-1}$. They suggested that since the former could readily be accounted for by invoking the existence of high-z extensions of a spiral arm known to exist at lower latitudes at these longitudes and velocities, perhaps the highest-velocity feature was also a distant arm. In that case, its z distance had to be about 13 kpc if one interpreted its velocity as a distance according to the Schmidt rotation law for our Galaxy. This feature would then be at a distance comparable to that of the Magellanic Clouds.

In view of the discovery of the Magellanic Stream, this now seems a quite reasonable place to position these high positive-velocity features. They would simply be very distant spiral features that sweep up out of the plane and then reach to the

Magellanic Clouds. They would in fact be part of an intergalactic bridge, albeit a very patchy one, between the Clouds and the Galaxy.

Picturing this configuration in Figure 4 we would expect that the distant arm (or arms) would sweep upward from the plane at around $l = 240°$, up to latitude about 20°, and then curve back down, crossing the plane at $l = 300°$ to reach the Magellanic Clouds.

Returning to the Magellanic Stream, Mathewson et al (1974) express concern about the stability of the Stream, considering the velocity between different parts of individual clouds within it and across the Stream itself (as much as 43 km sec^{-1}). We should bear in mind that this same problem, one of stability, exists throughout astronomy in many different contexts, and that this is just another area in which it is of concern. Self-gravitation is apparently too small to hold anything together, such as clusters of galaxies, and any of the high-velocity clouds themselves.

As far as theories to account for the present characteristics of the Stream are concerned, J. H. Oort and A. N. M. Hulsbosch (personal communication) have considered some of them. It is beyond the scope of the present review to critically consider Oort's points, some of which are dealt with by Mathewson et al (1974) already, but one comment is in order. Oort states that the most plausible explanation for the Stream is that it has been *drawn out of* the Large Magellanic Cloud. Perhaps we should consider whether or not such streams (or tails) could exist which were never in any galaxy, but which form structures in what would normally be called intergalactic space. These may be part of the formation processes of the galaxies themselves. They would be like superspiral arms made up of matter that was not used in the main body of the galaxies that formed from the protogalactic matter. Their present velocities and shapes should therefore be considered on the larger scale of galaxy formation, so that these features might perhaps have their own particular "rotation curve" with respect to a pair, or a group, of galaxies. After all, we are perhaps suffering from galaxy chauvinism here, by assuming that all extragalactic structures originate in galaxies and have been drawn out from them.

Despite these speculations, we can say with considerable certainty that at least some of the high-velocity clouds are extragalactic in nature and form the Magellanic Stream. No doubt further observations from the southern hemisphere will help clarify the picture.

High-Velocity Hydrogen Near Other Galaxies

The existence of high-velocity hydrogen, that is, hydrogen with velocities unexpected for its location or direction in space, has been reported in M31 (Whitehurst & Roberts 1972) and near M33 (Wright 1973). In the case of the anomalous-velocity gas in the direction of M31, the reason for its existence is not clear and several alternatives, including considerable z motions, have been noted. In the case of the gas near M33, it is the location of it that is peculiar, being two Holmberg radii from M33. However, the emission velocity of the cloud reported by Wright is quite possibly high-velocity gas in our Galaxy. The only thing that supports the extragalactic nature of this cloud is its apparently small angular extent.

More recently, Wright (1974) has found a very large and very high-velocity cloud, some 5° by 3° in angular extent with a velocity of -400 km sec^{-1} and located toward $l = 128°$ and $b = -33°$. This is about 30° beyond the presently mapped boundary of the Magellanic Stream. Wright favors this cloud as an intergalactic object and it seems possible to the present author that Wright's cloud forms a continuation of the Stream. The area between this new cloud and the Stream should be mapped carefully in a search for further emission. Wright seems to use the existence of this cloud and the Stream as an argument against the models of Davies and Verschuur, but in the section above, the present author has noted that although we might place most of the high-velocity clouds in the outer spiral structure of our Galaxy, we must stop thinking that our Galaxy is bounded and ordered. An extension of the model in which our Galaxy interacts with the Magellanic Clouds should allow us to locate elongated H I features out to enormous distances from our Galaxy, and the Magellanic Clouds. Such features would simply be part of a more comprehensive model which is far from two-dimensional.

A systematic search for hydrogen outside nearly edge-on spiral galaxies has been reported by Gerard (1973). He found no halos in any of the galaxies studied with upper limits which suggest that there is no phenomenon similar to the high-velocity clouds in our Galaxy in any of these other galaxies, provided that the models of the types suggested by Verschuur or Davies were valid. Of course, in order to have spiral structure distorted to large-z distances one needs interactions with other galaxies and Gerard says nothing about this in his paper. Also, his observations would have revealed none of the high-velocity clouds in our Galaxy had he been looking at a similar galaxy, because the smoothed H I column density would be very far below his detection limits. We are not dealing with a halo in our models, but with a very patchy spiral-structure phenomenon. A similar experiment should obviously be done with a synthesis-type antenna at some time in the future.

Most recently, M. S. Roberts and R. N. Whitehurst (personal communication) have found that the outer parts of M31 are warped and that this warp is up to 5 kpc above the plane of M31.

MODELS FOR THE GENERAL HIGH-VELOCITY HYDROGEN PHENOMENON

An Overview

Oort (1966, 1969, 1970) was the first to extensively review possible interpretations of the high-velocity clouds. In reexamining his model, which invokes the infall of matter toward the galactic plane, it should be borne in mind that all the earliest clouds discovered showed negative velocities, although we now know that this was probably a selection effect. In his paper, Oort listed several hypotheses that might be considered. The clouds could be:

1. Parts of nearby supernova shells
2. Condensations formed in a gaseous corona of high temperature
3. Ejecta from the galactic nucleus falling back into the plane
4. Ejecta from the galactic disk which were the principal constituents of the corona

5. Intergalactic gas accreted by the Galaxy
6. Satellites of the Galaxy or independent members of the Local Group of galaxies

For various reasons Oort rejected all of these hypotheses except number 5, that is, the "infall" theory. Since then, however, other workers have considered hypotheses 1 and 6 more seriously, and it now seems reasonably certain that the supernova hypothesis might well account for some of the intermediate-velocity clouds at the highest latitudes. The extragalactic nature of the clouds was considered by various workers (Kerr & Sullivan 1969, Verschuur 1969), and it is now known that at least one of the highest-velocity clouds is certainly outside our Galaxy at the distance of the Magellanic Clouds (Mathewson et al 1974).

A seventh possibility to account for the clouds is that they are parts of distant spiral arms. Oort did mention the work of Habing (1966) in his paper, but did not seriously consider this model. It is this model that has since gained considerable support.

The Infall Hypothesis and its Shortcomings

The main proponent of the hypothesis that the high-velocity clouds are manifestations of matter falling in toward the Galaxy has been Oort (1969, 1970). In his model he suggests that explosions in the galactic disk initially drive galactic matter out into the halo of the Galaxy where they are met by an intergalactic wind which decelerates them and then drives them back into the disk of the Galaxy. This model purports to explain the large negative velocities seen in high-velocity clouds. These velocities are greater than any free-fall velocity would be if the clouds were expelled and then fell back into the Galaxy without the effects of an intergalactic wind. We now examine the many assumptions that are made by Oort—some explicit, some implicit.

The first assumption has been that negative velocities mean motions toward the sun, or even toward the plane of the Galaxy. This is not necessarily so, as has been pointed out by many other workers in the field (F. J. Kerr, many conferences; Kerr & Sullivan 1969, Verschuur 1969), because a negative velocity implies only a *component* of a motion toward the observer. For example, matter in distant parts of the Galaxy in the first and second quadrants of galactic longitude appear to have negative velocities as a result of differential galactic rotation effects. Of course, for matter very near the galactic poles, one might not expect to see any component of motion in the line of sight due to such differential rotation effects. It has recently been found that the only truly high-velocity cloud near the pole (the south pole feature, now referred to as the Magellanic Stream), is a cloud at a very great distance from the Galaxy, and, in this case, we are certainly seeing only one component of its true space motion. Because the observed negative velocities of many of the originally discovered clouds, several other workers were led to propose alternative models which took into account the fact that we were seeing only one component of the actual space motion of the clouds. In the absence of any further data it seems highly improbable that negative velocities really indicate a true motion toward us or toward the galactic plane.

The second assumption concerns the apparent dominance of negative velocities for the clouds. The original clouds mapped by the Dutch radioastronomers were all in that part of the sky where one expects to find negative velocities for matter in distant parts of the Galaxy, or in intergalactic space. The exception is the anticenter clouds. In the anticenter we expect no line-of-sight component of motion for galactic clouds, assuming that all the hydrogen in the Galaxy is in purely circular orbits about the galactic center.

The Dutch radioastronomers were unable to see much of the southern skies; since the original surveys were made, however, we know that there are a large number of positive-velocity clouds in those parts of the southern skies where we expect such emission from distant matter. In the case of the positive-velocity, high-velocity matter we also find the emission up to very large latitudes. The implicit statement that high-velocity clouds have mostly a negative velocity does not appear reasonable at this time.

To proceed with an examination of Oort's model we need to accept two assumptions: 1. that negative velocities indicate motions toward the observer, and 2. that most, if not all, of the clouds show negative velocities.

We now have to figure out where the clouds are located, bearing in mind that an apparent division exists between high- and intermediate-velocity clouds. It has been noted that the division was a geographical one, as Leiden observers studied the emission at higher velocities ($|v| > 70$ km sec^{-1}) and Gröningen observers studied the emission from clouds between velocities ± 70 km sec^{-1}.

Using the data outlined above to prove the galactic nature of IVCs, Oort states that because there are no differences between the HVCs and the IVCs, the HVCs must also be galactic. Without wishing to question whether the HVCs are galactic or not, we should examine whether the HVCs and the IVCs are really similar. There are now several known systematic differences between them.

First, the spatial distribution of the two classes of objects on the sky is different in detail. The intermediate-velocity matter extends up to near the poles; the highest-velocity matter is found only up to latitude 60°. The highest-velocity clouds appear more cloudlike if one studies Figure 3, whereas the intermediate-velocity matter is more extended. The IVCs are found over a larger area of sky than are the HVCs. The HVCs, as presently defined, are found near the plane as well, with clouds at -200 km sec^{-1} seen toward $l = 130°$, $b = 2°$. Such clouds have no place in the infall theory. Of course, the IVCs are also seen in the plane, but we do not define the hydrogen near the plane at intermediate velocities as IVCs (everyone would accept the fact that distant spiral arms are indeed seen toward $l = 130°$ and that the gas in those arms shows velocities of -30 to -100 km sec^{-1} in the line of sight).

As far as the typical masses of the clouds are concerned, we saw above that the IVCs have a much higher typical mass, for those that have been mapped, as compared with the HVCs. The extrapolation made by Oort, from the observed correlation of a few calcium lines with IVCs to generalizing about the HVCs, seems to require further data.

Oort has also stated that the IVCs show a very outspoken flow toward the galactic plane. In the light of the data for other parts of the sky where the intermediate-velocity hydrogen shows a positive velocity, this is not the case. For example, from $l = 240°$ through the galactic center to $l = 20°$ and up to latitudes of as much as 58°, positive-velocity matter is found at intermediate velocities.

If we continue to allow the now somewhat doubtful assertion that the HVCs are indicative of infalling matter in the solar neighborhood, we can examine the estimates of the amount of matter involved. Oort used the data of Hulsbosch (1968) to estimate that the typical average surface density of the HVCs is 2×10^{19} atoms cm^{-2}. This agrees more or less with a rough value one can determine from the data of Meng & Kraus (1970) and Giovanelli et al (1973) where the derived value may be a factor of 2 higher (4×10^{19} atoms cm^{-2}). The next step in the argument is then to calculate the total amount falling into the Galaxy assuming that the clouds are only a tracer of this infall. Oort (1970) states that the estimate rests on the assumption that the extragalactic wind is homogeneous and also falls in the spaces between the clouds. It would seem that in the absence of further data such an assumption is of doubtful validity.

It is also mentioned that to understand the phenomenon of infall we need to have some knowledge about the way the gaseous halo of the Galaxy is maintained. Oort resorts to the data of Kepner (1970) and states that Kepner found distant "halo" clouds clustering about the velocities of distant spiral arms in the plane. This is not too surprising if we bear in mind that distant spiral arms are found to increase their z extent in a way that is consistent with theory (Verschuur 1973b). The definition of halo clouds, therefore, becomes a moot point. Of course, if one accepts a picture in which the Galaxy is 200 pc thick everywhere then perhaps the high-z extensions discussed by Kepner (1970) and others (Habing 1966, Verschuur 1973b) should be called halo clouds, but the fact that the outer parts of the Galaxy are just as thin as the inner parts has never been established. In fact, the opposite is the case (Verschuur 1973b).

Kepner herself stated that the high-z extensions were possibly just that, extensions of the spiral arms to higher z. Kellman (1972) has provided a theoretical picture of why this should be expected. Clouds in distant arms, where the gravitational attraction toward the plane is less due to the existence of less matter in the arms, would be expected to move to higher-z distances as a matter of course. We do not need to invoke explosions to account for their existence at great-z distances as Oort has found necessary.

Lastly, because the highest-velocity clouds have velocities considerably greater than that expected for free fall of gas into the Galaxy, Oort has to invoke the existence of an extragalactic wind to accelerate them to the large velocities. Such a wind would hardly blow clouds toward us from latitudes $+60°$ as well as zero and negative latitudes around the anticenter (down to latitude $-30°$), at velocities of up to -200 km sec^{-1}. Also, the existence of negative-velocity clouds above and below the plane in a given quadrant of longitude would rule out an intergalactic wind unless the wind were blowing in from all directions at once. In any event, one would then need to explain the existence of the positive-velocity clouds in some other way.

It therefore appears that unless more data that would support an infall hypothesis is forthcoming, we should turn to another model to explain the high- and intermediate-velocity clouds of hydrogen.

It should also be noted that the apparent excess of small negative velocities near the poles (here we mean velocities less than 20 km sec^{-1}) may indicate infall toward the galactic plane from both sides of the plane. Such gas is not within the realm of the present discussion, however, and such low velocities do not require an intergalactic wind for their existence.

Extragalactic Models

MEMBERS OF THE LOCAL GROUP Oort had considered this model, but rejected it on rather incomplete arguments, partly based on the incomplete knowledge at the time of the nature and velocities of the clouds. Burke (1967) later pointed out that for the clouds to be virially stable they would have to be located at distances of 500 kpc or more. Verschuur (1969) took up this point further and showed that, when the velocities of the highest-velocity clouds were considered with respect to the galactic center, they had the same distribution as the Local Group galaxies. Also, the Local Group galaxies seemed to lie more or less in the middle of that area of sky in which the high-velocity clouds were found. The HVCs would have to be at distances comparable to that of M31 for them to be stable, and their masses would then be similar to that of the Magellanic Clouds. The suggestion was that the clouds might well be the remains of intergalactic matter that had not been used in the formation of the Local Group galaxies, or that they were protogalaxies.

SATELLITES OF THE GALAXY Kerr & Sullivan (1969) showed that the high-velocity clouds had a velocity distribution as a function of longitude which was quite consistent with their being in orbit about the Galaxy at a distance comparable to that of the Magellanic Clouds, that is, about 50 kpc. They suggested that the clouds might be fragments of gas torn from the Galaxy by a close passage of the Magellanic Clouds in a close encounter between the Clouds and the Galaxy. The main purpose of the publication of these two models was to show that there were other ways of interpreting the data which could explain the high-velocity clouds as well as the infall model of Oort.

The High-Velocity Clouds and the Galactic Spurs

It was noted early by G. W. Rougoor (personal communication) that the high-velocity clouds lay nearly on the various galactic spurs of continuum emission, with the expectation that the North Polar Spur had no clouds associated with it. Meaburn (1965) followed up this point in a paper and put forward the idea that the clouds might be part of the supernova shells thought to be the origin of the spurs. Of course, because the clouds were apparently associated with the actual continuum ridges of the spurs, which were thought to be the edges of the supernovae shells, one would not expect that the highest-velocity clouds would be seen in those parts of the shell. Heiles (1974) has more recently noted that the high-velocity gas, certainly at intermediate velocities, still appears to be correlated with the minor spurs. Perhaps such a correlation is significant for the intermediate-velocity cloud patterns emerging from the plane around $l = 90°$, which would suggest that loop II is in another spiral arm! This location for loop II has recently been suggested by Sofue & Tosa (1974). As Heiles points out, the statistical significance of such correlations is highly questionable in the absence of more complete data for the southern hemisphere. Also, it seems that for two such large-scale phenomena as the loops and the highest-velocity clouds, which both cover large areas of the sky, the chances of accidental correlation is certainly finite.

Verschuur (1970) noted that the apparent elongations seen in the high- and intermediate-velocity clouds seemed to correlate well with the local magnetic field lines derived from optical polarization data (Mathewson 1968). It is now apparent that if clouds in distant spiral arms were elongated parallel to the axes of those arms, then they, by pure chance, would appear to be elongated parallel to the nearby helical magnetic-field line patterns. The agreement is, therefore, purely fortuitous.

Intermediate-Velocity Clouds

Rickard (1971) invoked a super-supernova to account for the motions of gas in the Perseus arms and, as a part of his model, he noted that perhaps the high-velocity clouds were thrown up out of the Perseus arm as part of this super explosion. His model would not account for the majority of the high-velocity clouds and certainly not the positive-velocity clouds since discovered.

Verschuur (1971a) showed that a group of intermediate-velocity clouds around $l = 180–200°$ and $b = +60–70°$ could be explained by a nearby supernova which occurred 10^6 years ago at a distance of about 130 pc. This supernova would also account for the absence of low-velocity matter in the same part of the sky. The clouds associated with this shell are then a separate phenomenon from the rest of the intermediate-velocity cloud distribution and are in no way related to the high-velocity clouds.

Wesselius & Fejes (1973) studied the overall distribution of the intermediate-velocity gas at latitudes greater than 15° and concluded that most of this gas at $b > 30°$ occurs in a few large complexes and showed a systematic velocity pattern. The largest complex coincides with the hole found in the low-velocity gas around zero velocity. They reconsidered the supernova model as the cause of this hole and as the accelerating source which would move the gas to intermediate velocities. They also considered a model which involved an interaction of a large gas complex with the galactic layer. They favored the latter model, which is a very specific version of the infall theory.

It seems established quite clearly that a "hole" in the gas, centered at around $l = 160°$, $b = +70°$, would be nearly filled in if the intermediate-velocity gas is taken into account in any integration of the total hydrogen content over the sky. The conclusion, that the presently observed intermediate- (negative-) velocity matter was once at rest, is therefore reasonable and this leads the authors to seek the cause of the acceleration of this gas. They reexamine the model proposed by Verschuur (1971a) more quantitatively. They assume that a shell of expanding material was produced by a supernova that occurred some 110 pc away, about 2×10^5 years ago. Although the energy requirements to move the observed mass of gas are met, as Verschuur (1971a) found, Wesselius and Fejes feel that the observed velocity pattern in the intermediate-velocity gas complexes is not explained, nor is the existence of some of the other higher-velocity clouds at intermediate latitudes. First, they are assuming that the supernova is exploding in a nicely uniform medium, in which case they might expect a particular velocity pattern in the subsequent motion of the shell, but this seems an oversimplification. The interstellar medium is just not that nicely organized, especially on the scale of 100 pc, which is the diameter of the present shell. The detailed velocity patterns that one is likely to observe will depend critically on just what the structure of the medium was before the explosion. As far as the existence of some of the higher-velocity clouds (in roughly the same direction of the sky) is concerned, it seems that they need not be thought of as being part of the same phenomenon. Wesselius and Fejes themselves make a strong case for the high-velocity clouds being different in many respects from the intermediate velocity clouds. They also suggest that the velocity cutoff between the two types of clouds be made at -90 km sec^{-1}.

They then consider the possibility that a big gas complex has collided with the galactic layer and study the consequent interaction. They find that such a model explains many of the properties of the observed hydrogen distribution quite well except the detailed structure of the disturbed area. This is because they are unable to estimate the nature of the incoming gas cloud to start with. The important point is that they are explicitly discussing a distinct clump of gas rather than any general intergalactic wind.

For both models the distance to the phenomenon is about 70 pc and in the case of the stream of gas it comes from $l \approx 120°$, $b \approx +40°$. They admit that they prefer the stream model but the supernova model remains a serious alternative. The interested reader is referred to the very detailed paper of Wesselius & Fejes (1973) as well as the papers by Fejes & Wesselius (1973) and Wesselius (1973) to obtain a good overall view of the detailed nature of the higher-latitude intermediate-velocity and low-velocity gas.

Bending of the Galactic Disk and High-z Extensions to Spiral Arms

It has been known for many years that our Galaxy shows a tilt in its outer regions, a tilt going to positive latitudes in the first and second quadrants of galactic longitudes and to negative latitudes in the third and fourth quadrants. The magnitude of the tilt was thought to be small based on the early observations (Oort et al 1958, Henderson 1967)—only of the order of up to 800 pc in the southern hemisphere.

However, Habing (1966) studied the distribution of 21-cm emission between $l = 42°$ and 142° and $b = +10°$ to $+15°$, and found that there is an H I component with a large negative radial velocity ($-130 < v < -70$ km sec^{-1}), with a low peak temperature ($\sim 4°K$), and broad emission linewidths. He surmised that this emission could be due to one of three phenomena: 1. The component originated in the Outer Arm of the Galaxy, in which case it extended to some 3–4 kpc from the galactic plane; 2. The component was a cloud close to the sun at a distance of about 300 pc, in which case it appeared similar to the high-velocity clouds which were thought to be at this distance by the Dutch school at the time (1966); or 3. The component was due to a large number of clouds located some 1–2 kpc from the sun, a sort of "cloud-bank" model.

Habing favored the first interpretation, although his suggestion did not get wide recognition at the time. In retrospect we now know that the peak temperatures seen in this high negative-velocity matter at these low latitudes is much greater than the 4°K he found.

A few years later, Kepner (1970) published data on a more comprehensive survey of the sky at intermediate latitudes between $l = 48°$ and $228°$. She also found that there was emission which could best be understood if it arose in distant arms extending to more than 3 kpc above the galactic plane. She studied several of the components found by Habing in some detail and concluded that they were parts of either distant spiral arms extending to high-z distance, or the high-velocity cloud phenomenon at an unknown distance. She found that several of the features were discrete, albeit large, clouds and that they did not, therefore, relate to galactic structure. Rather, they were thought to be high-velocity cloud complexes. Again, in retrospect, it should be noted that there is no a priori reason why parts of distant spiral arms should not be patchy. After all, many photographs of distant galaxies show just such an extreme patchiness in the outer parts of galaxies.

Kepner was led to conclude that the objects she was studying were high-velocity cloud complexes resulting from collisions with extragalactic gas. They were a sort of "rain" of clouds falling down toward the plane. I should note here that there is a strong influence from the Dutch school in this paper (as in Habing's paper) as can be noted by referring to the end of her paper.

A variation of the model in which distant arms are considerably extended in the z direction is found in the paper by Dieter (1971). She proposed that the Galaxy is surrounded by a shell extending to 3.5 kpc above and below the plane which had a rotational velocity appropriate to its distance from the galactic center with an added motion of some 135 km sec^{-1} toward the galactic center. The location of the shell was consistent with it being part of the outermost arms in the galaxy.

Several criticisms have been leveled at this model. In particular, it has been noted that the model depends heavily on the interpretation of the lowest levels in contour maps showing the intensity of H I emission as a function of velocity and longitude. Deiter found several jetlike features on her maps which have not been confirmed by other workers. In particular, P. L. Baker (personal communication) and G. L. Verschuur (unpublished data) checked the existence of many of the most striking of these jets and did not find them.

Burton (1971) noted that Dieter's model fit is controlled by varying the size of the Galaxy in different directions. This is more or less the same as saying that distant arms have varying distances as a function of longitude. It appears that what Dieter is suggesting is that, at least in the first and second quadrant of galactic longitude, the highest-velocity matter is part of distant spiral arms and one in fact need not invoke the additional inward motion if one places the matter a little further from the sun (and the galactic center) than she does. Burton also noted that the wings in her profiles were also visible at negative velocities at longitudes greater than 180°. This should not be the case for her model. Distant matter in the third quadrant of galactic longitude shows a more positive velocity as distance increases, and an addition of an infall motion of the type she suggests on top of that positive motion would not cause the emission to appear at the negative velocities in which the wings are still seen in her data.

Verschuur's Model

When the present author studied a high-resolution scan of 21-cm emission at latitude 5° covering longitudes 20–200°, it occurred to him that there was hydrogen emission at this latitude at velocities very similar to some of the high-velocity clouds at similar longitudes. If the highest-velocity clouds were simply parts of very distant spiral arms, which were at the same time very patchy, then one could explain the existence of the clouds by using a model in which the Galaxy was far from being a flat, uniformly rotating disk. The suggestion that the high-velocity clouds were part of the so-called normal spiral pattern was, therefore, outlined (Verschuur 1971b, 1972, 1973a). Observations of the way the Galaxy bent in its outer reaches were also used to extrapolate the work done earlier by others on just this point. By showing how much the Galaxy was bent in its outer parts in the first quadrant of longitude, and also how rapidly the thickness of the arms increased with galactocentric distance, it became possible to show that the intermediate- and high-velocity clouds could be thought of as being part of distant spiral arms.

We briefly illustrate the line of argument with a set of diagrams. Figure 7 shows how the z thickness and the center of gravity of distant arms depend on galactocentric distance (Verschuur 1973b). These data fit well with the theoretical estimates for these terms, except that the additional influence of the Magellanic Clouds in producing the z deflection is not worked out as yet. If one now pictures what a distant spiral structure looks like to an observer located on the solar circle, but on the far side of the Galaxy (in which direction the data for Figure 7 were collected), one finds that the observer will also see hydrogen emission at nonzero velocities well away from the galactic disk. That observer would label the clouds high velocity and intermediate velocity, whereas the clouds merely form a part of spiral arms near to him. Such a distant observer would no doubt also have a better picture of the way our anticenter clouds fit into the scheme of things. It is interesting to imagine what an observer located diametrically opposite us on the far side of the Galaxy would think of the anticenter clouds, since these would appear to radiate from the galactic nucleus as viewed by him! The clouds would have large negative velocities. No doubt such an observer is having a terrible time invoking some sort of dramatic explosive phenomenon at the galactic center based on observations of clouds that we know are at least 10 kpc beyond the center! The moral of this story is that not everything that appears in the direction of the galactic center is necessarily located anywhere near it!

Therefore, based on examination of the bending of the galactic plane and the velocity longitude maps at several latitudes, it is suggested that the data are consistent with the concept that there are many more spiral features in the Galaxy than has heretofore been realized. Combining the facts that these arms are patchy, the Galaxy is not flat, and the thickness of the matter in the disk increases with distance from the galactic center, we are able to account for the observed distribution of the high- and intermediate-velocity clouds. The exception is the location of the highest-latitude intermediate-velocity clouds mentioned before, which can well be a supernova phenomenon, and of course the South Pole Cloud.

The anticenter clouds are another apparent exception. The problem is that their velocities are so large in a region where zero velocities are expected on simple pictures of galactic rotation. The existence of these clouds, combined with the observed fact that the high-velocity clouds do not fit any simple galactic rotation curve, proves that noncircular motions exist in the outer parts of our Galaxy. Indeed, it is found that the Schmidt rotation law becomes less and less valid as the longitude increases, when one considers the possible location of distant high-velocity clouds as parts of distant spiral structure (Verschuur 1973c). This might be what is expected if the degree of noncircularity of the motions increased with distance from the galactic

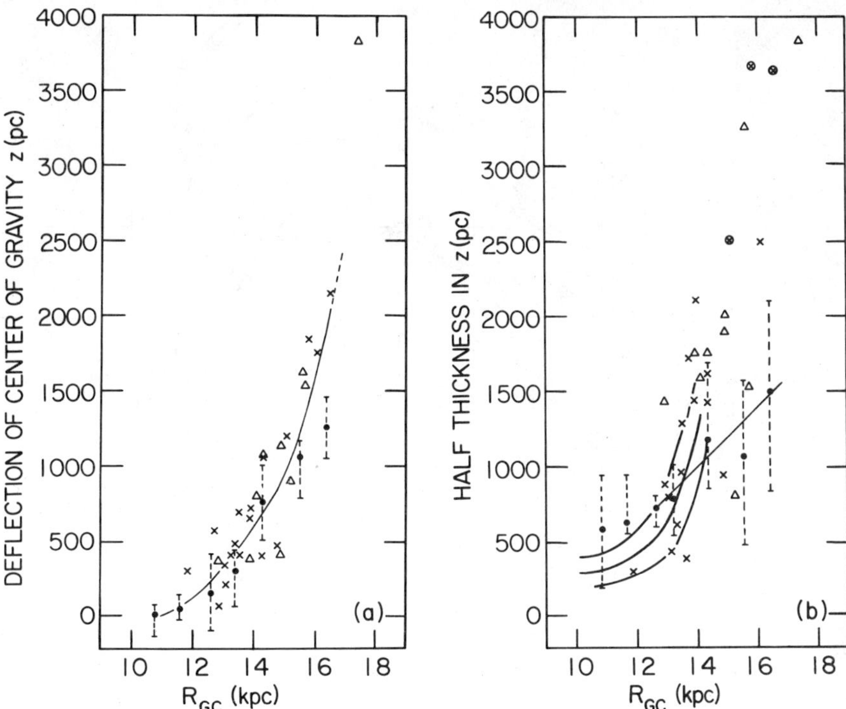

Figure 7 (*a*) The deviation of the center of gravity of the galactic disk as a function of galactocentric distance. (*b*) The variation of the half-intensity width of the galactic disk as a function of galactocentric distance. Both *a* and *b* are from Verschuur (1973b).

center. (Because of the trailing nature of the spiral arms in our Galaxy, a given spiral feature increases its distance from the galactic center as longitude increases.)

A possible way to account for this phenomenon, that is, the existence of non-circular motions, is to note that a particle located very far from the galactic center, say 25 kpc, will effectively see the Galaxy as a point mass and its orbit may be elliptical, thus generating a considerable line-of-sight component of its motion as seen by us. Detailed calculations on this likely effect, combined with the effect of a pull and subsequent release by the Magellanic Clouds in a close encounter, need to be made.

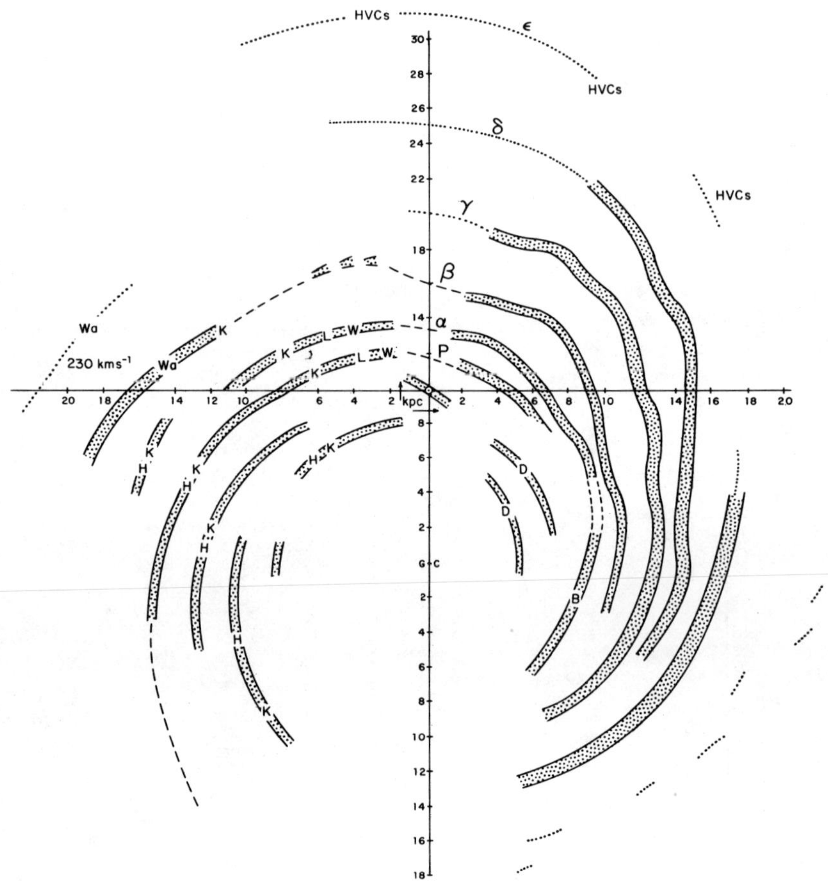

Figure 8 A schematic representation of the spiral structure of the outer parts of the Galaxy with the possible locations of the high-velocity cloud arms drawn in. See Verschuur (1973a) for references indicating the source of the information upon which this map was based.

Strong support for the existence of noncircular motions in the Galaxy comes from the work of Innanen (1969). He showed that a perturbation in the z direction in the outer parts of the Galaxy could be translated into considerable motion in the radial direction. Also Keenan & Innanen (1974) have found that the open cluster NGC 2420 ($l = 198°$, $b = 19.°6$) has a large radial velocity which suggests a very eccentric orbit about the galactic center, perhaps as a result of the passage of the Magellanic Clouds. The velocity found is a large positive velocity (~ 100 km sec^{-1}) in a region of the sky where the hydrogen gas shows a negative velocity.

Returning to the model in which the high-velocity clouds are simply part of "normal" galactic structure, it was proposed that intermediate-velocity clouds are in the nearer spiral arms, such as the Perseus arm and the α arm (Verschuur 1973d). The high-velocity clouds are in, or form the bulk of, the more distant arms. The author wishes to make one further comment. In a quite unprecedented way, the paper detailing this model (Verschuur 1973a) was accompanied by a note by Hulsbosch & Oort (1973) in the same issue of the journal in which the main paper appeared. They conceded that the distant-arm hypothesis was probably valid below latitudes of 25°, but above that they still favored the infall hypothesis. In another paper, Verschuur (1973b) countered those arguments by further work on the observed bending of the galactic disk. The latter paper, together with the work on the supernova or very localized infall region, described by Verschuur (1971a) and Wesselius & Fejes (1973), accounts for all the high-velocity matter at northern latitudes.

If the high-velocity clouds form part of distant spiral structure, then we should be able to use the data to make a new map of spiral structure, but this is very nearly impossible because of the existence of noncircular motions in these outer parts of the Galaxy. It has been argued that it would be best simply not to try making such a map. However, such a map should be made to make comparisons with the results of other workers, rather than to expect such maps to have any absolute validity. In Figure 8 a map of the spiral structure based on the work of Verschuur is shown. The Schmidt rotation law has been used where possible, that is, until it gave apparently meaningless distances for the hydrogen gas being observed. This map may be compared with others that have been produced, and it is also interesting to speculate how wrong such maps might really be, because we may never be able to know the answer unless we really could get outside our Galaxy.

Figure 9 summarizes some of the points made here. It is a latitude-velocity map made at longitude 75° which shows many of the phenomena discussed here, such as the bending of the galactic disk, high-z extensions, and the high-velocity clouds.

Davies's Model

At the time that Verschuur (1971b, 1972) was reporting on early versions of his model, Davies (1972a,b) was making observations of a different nature which led him to conclude that the high-velocity clouds were also far from the sun in the outermost spiral arms of the Galaxy. Davies made scans in latitude at constant longitude at many longitudes between 20° and 260°. He also found evidence for the existence of a number of new spiral features whose densities he estimated to be about 1% of the well-known spiral arms. The high-velocity clouds appeared to lie in bands, shown in Figure 10, which fan upward from these outermost arms in a trailing sense. Davies also suggested that the tilt of the galactic plane and the asymmetrical distribution of the high-velocity clouds were both results of the close passage of the Large Magellanic Cloud, which also introduced the noncircular component into the motions of some of the clouds.

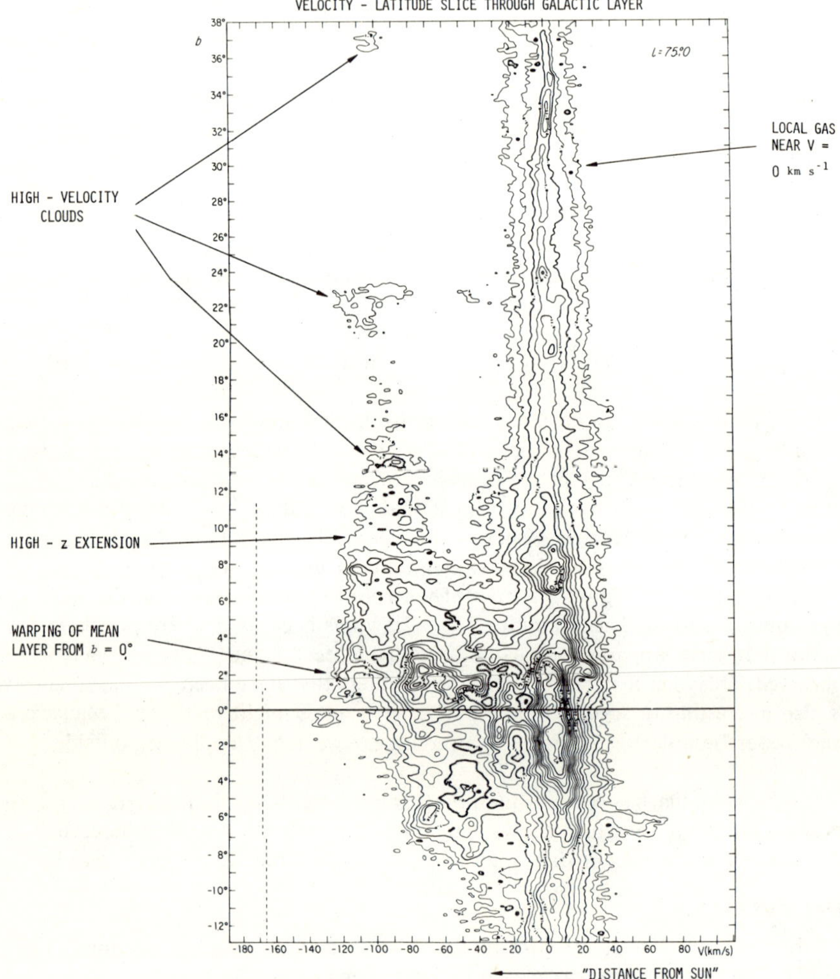

Figure 9 A velocity-latitude slice through the galactic neutral-hydrogen layer at $l = 75°$. The various phenomena discussed in this review are labeled in this figure. The high-velocity clouds shown are in the list prepared by Meng & Kraus (1970).

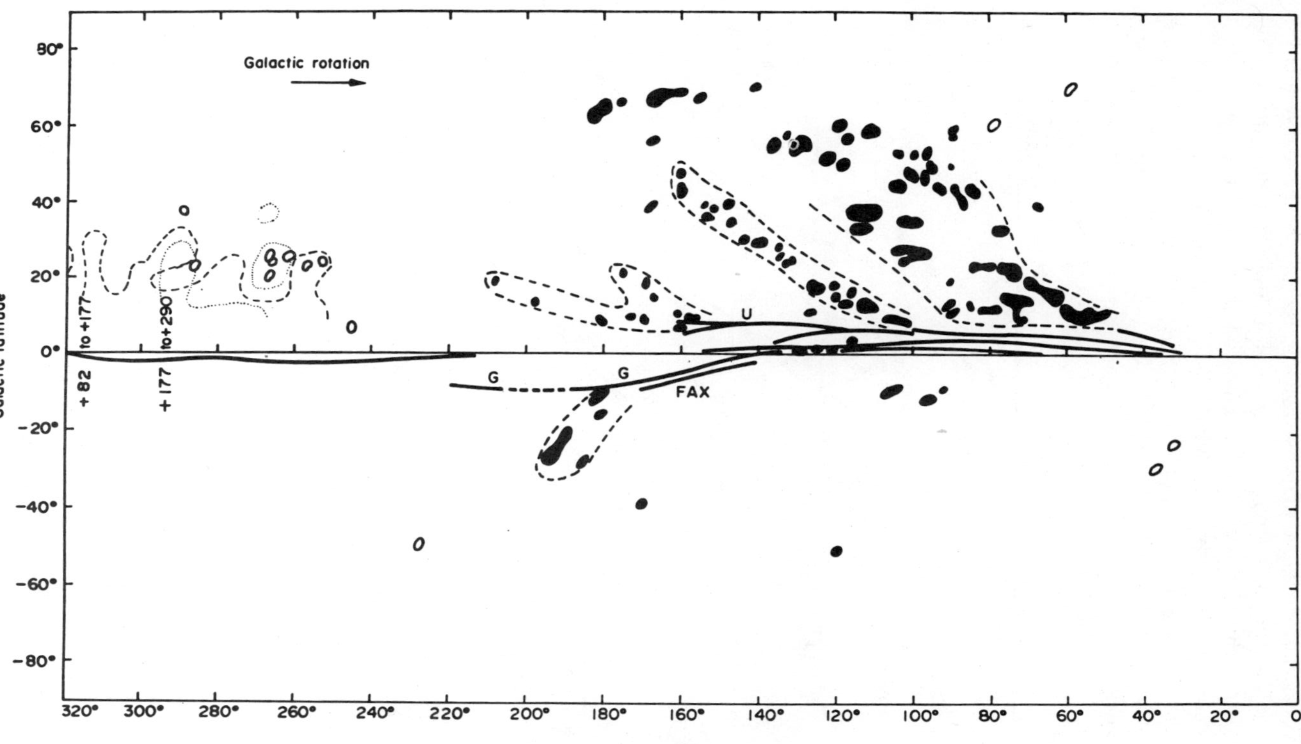

Figure 10 The plot of the high-velocity clouds on the sky prepared by Davies (1972b) to indicate the existence of several bands of clouds. Negative-velocity clouds are shaded in.

The model is basically similar to that of Verschuur, except that the high-velocity clouds are in bands which reach out from a spiral arm at lower latitudes, whereas in Verschuur's model the high-velocity clouds are indeed in bands, but these bands are independent spiral features simply reaching to very high latitudes. Davies also produced a map of his estimate of spiral structure. It compares very closely with the map shown in Figure 8 and is shown here as Figure 11. The scale of distance to the high-velocity clouds is virtually the same for the two maps, about 20 kpc around $l = 120°$, although Davies shows three arms beyond the Perseus arm and

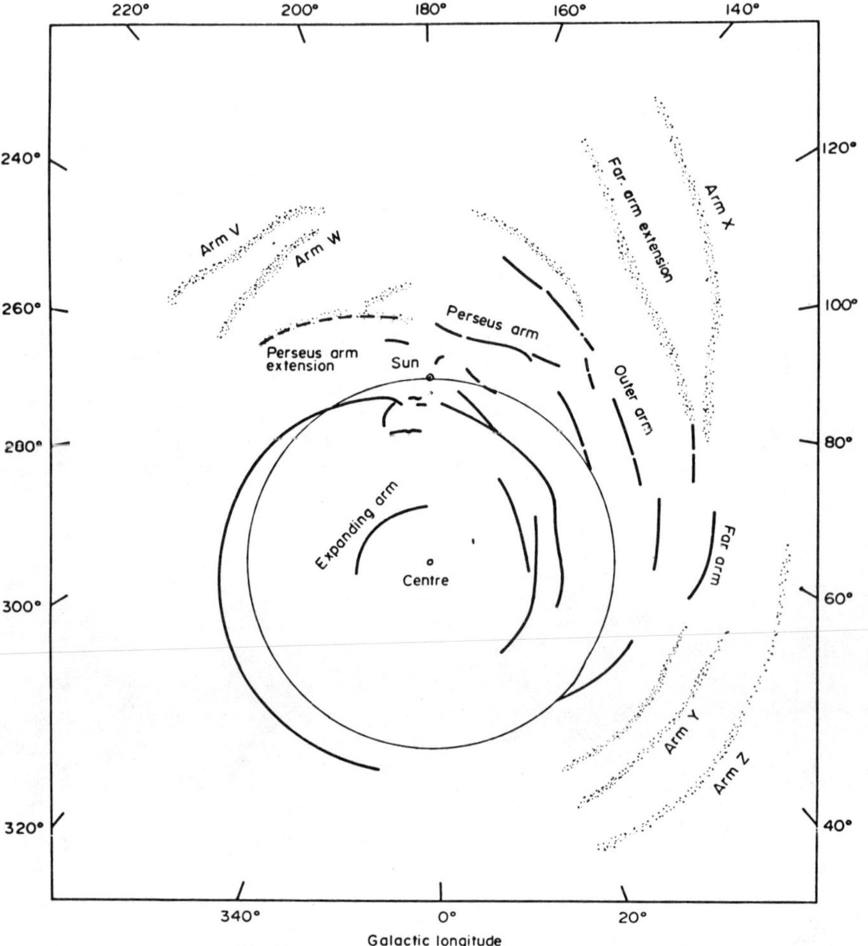

Figure 11 A schematic representation of the spiral structure of the Galaxy as devised by Davies (1972b), based on his model for the high-velocity clouds. Cf Figure 8, which was prepared independently.

Verschuur shows five. Actually, this discrepancy is only one arm because it has been pointed out that the hydrogen usually associated with the Perseus arm might well be a very strong arm which has been called the α arm by Verschuur (1973d). The outer arm in Figure 11 should be compared with the β arm in Figure 8. In view of the different approaches used by Verschuur and Davies, the similarities between the two derived maps of spiral structure are quite remarkable. The paper by Sofue & Tosa (1974) gives an interesting new view of the relationship between the Perseus and α arms.

The only real question is the absolute distance to the high-velocity clouds, or the outer spiral features of the galaxy, because the two are synonomous in these two models. Because no detailed models for the effect of the Magellanic Clouds on the Galaxy are yet available which might give a clue, we cannot be sure. It is quite possible that the outermost spiral features are much further away than either Verschuur or Davies estimated. If one starts examining the velocities and derives a distance, assuming the Schmidt rotation-law model, one finds reasonable distances for the outer arms at the lower longitudes, but then the distance estimates become meaningless as the longitude increases. This happens at different longitudes for the different spiral features shown in Figure 8. After one discovers that this has happened, one is reminded that merely because the assumption of circular rotation gave reasonable estimates of distance at some longitudes does not mean that non-circular motions were not present there too (one has no estimate of the correctness of the spiral structure maps). Resorting to presently available theory, such as density-wave models, does not help in these outer parts of the Galaxy.

The Magellanic Stream and the Anticenter Clouds as a Galactic Tail

Both Davies and Verschuur consider the anticenter clouds as part of the outermost structure of the Galaxy, with these clouds originally drawn out by the passage of the Magellanic Cloud(s) and now falling back in toward the Galaxy, probably in some sort of elliptical orbit.

It is the author's opinion at this time that the anticenter clouds might well be a very long tail to our Galaxy, something like a counterstream, so that a distant view of our Galaxy and the Magellanic Clouds would show the Magellanic Stream stretching out from the Clouds and away from both galaxies and the counterstream stretching out in nearly the opposite direction. (This would contain our anticenter clouds.) We see such groups of interacting galaxies elsewhere in the sky [see e.g. some of the work by Toomre & Toomre (1972)] and there is no reason at all to continue to hope that our Galaxy is a simple, uniformly rotating object with ordered spiral structure.

In picturing this distant view of our Galaxy and the Magellanic Clouds we also note again that the Magellanic Stream reaches nearly to M31. Wright (1974) has recently announced the discovery of a very high-velocity cloud near that galaxy and at a velocity that is nearly the same as the end of the Stream. Perhaps the Local Group is threaded by streams of hydrogen gas, continually dragged out after close interactions or streams that never entered any galaxy and are merely remnants of galaxy formation.

NOTE ADDED IN PROOF Mathewson & Cleary (1975) have recently reported the discovery of intergalactic H I in the Sculptor group of galaxies and suggest that the discovery gives support to the suggestions of Verschuur (1969) and Kerr & Sullivan (1969) that the HVCs may be intergalactic matter in the Local Group. This should be compared with the discussions above which place the HVCs at large distances from the Sun and tied to the outermost spiral structure of the Galaxy, bearing in mind the existence of the Magellanic Stream and a possible Counterstream in the anticenter.

Literature Cited

Blaauw, A., Tolbert, C. R. 1966. *Bull. Astron. Inst. Neth.* 18:405–12

Burke, B. F. 1967. *Proc. IAU Symp. No. 31. Radio Astronomy and the Galactic System. Noordwijk,* ed. H. van Woerden. New York: Academic. 299 pp.

Burton, W. B. 1971. *NRAO High Velocity Cloud Meeting.* Discussions

Davies, R. D. 1972a. *Nature* 237:88–91

Davies, R. D. 1972b. *MNRAS* 160:381–406

Dieter, N. H. 1965. *Astron. J.* 70:522–58

Dieter, N. H. 1971. *Astron. Ap.* 12:59–74

Dieter, N. H. 1972a. *Astron. Ap. Suppl.* 5: 21–80

Dieter, N. H. 1972b. *Astron. Ap. Suppl.* 5: 313–68

Fejes, I., Wesselius, P. R. 1973. *Astron. Ap.* 24:1–13

Gerard, F. 1973. *Astron. Ap.* 28:95–98

Giovanelli, R., Verschuur, G. L., Cram, T. R. 1973. *Astron. Ap.* 12:209–62

Giovanelli, R., Brown, R. L. 1973. *Ap. J.* 182: 755–66

Habing, H. J. 1966. *Bull. Astron. Inst. Neth.* 18:323–52

Heiles, C. 1968. *Ap. Lett.* 2:31–35

Heiles, C. 1974. *Proc. IAU Symp. No. 60,* ed. C. Simonson. Reidel. p. 13

Heiles, C., Habing, H. 1974. *Astron. Ap. Suppl.* 14:1–555

Heiles, C., Jenkins, E. 1973. *Sky Telesc.* 45: 214

Henderson, A. P. 1967. PhD thesis. Univ. Maryland, College Park, Md.

Hughes, M. P., Thompson, A. R., Colvin, R. S. 1971. *Ap. J. Suppl.* 23:323–70

Hulsbosch, A. N. M. 1968. *Bull. Astron. Inst. Neth.* 29:33–39

Hulsbosch, A. N. M., Oort, J. H. 1973. *Astron. Ap.* 22:153–54

Hulsbosch, A. N. M., Raimond, E. 1966. *Bull. Astron. Inst. Neth.* 18:413–20

Innanen, K. A. 1969. *J. Roy. Astron. Soc. Can.* 63:260–63

Keenan, D. W., Innanen, K. A. 1974. *Ap. J.* 189:205–7

Kellman, S. A. 1972. *Ap. J.* 175:353–62

Kepner, M. 1970. *Astron. Ap.* 5:444–69

Kerr, F. J., Knapp, G. R. 1972. *Astron. J.* 77:354–59

Kerr, F. J., Sullivan, W. J. 1969. *Ap. J.* 158:115–22

Mathewson, D. S. 1968. *Ap. J. Lett.* 153: L47–L54

Mathewson, D. S., Cleary, M. N. 1975. *Ap. J. Lett.* 195:L97–100

Mathewson, D. S., Cleary, M. N., Murray, J. D. 1974. *Ap. J.* 190:291–96

Meaburn, J. 1965. *Nature* 207:179–80

Meng, S. Y., Kraus, J. D. 1970. *Astron. J.* 75:535–62

Muller, C. A., Raimond, E., Schwarz, W. J., Tolbert, C. R. 1966. *Bull. Astron. Inst. Neth. Suppl.* 1:213–44

Oort, J. H. 1966. *Bull. Astron. Inst. Neth. Suppl.* 18:421–38

Oort, J. H. 1969. *Nature* 224:1158–63

Oort, J. H. 1970. *Astron. Ap.* 7:381–404

Oort, J. H., Kerr, F. J., Westerhout, G. 1958. *MNRAS* 118:379–89

Rickard, J. J. 1971. *Astron. Ap.* 11:270–78

Rickard, J. J. 1972. *Astron. Ap.* 17:425–31

Sofue, Y., Tosa, M. 1974. *Astron. Ap.* 32: 461–64

Tolbert, C. R. 1971. *Astron. Ap. Suppl.* 3: 349–454

Toomre, A., Toomre, J. 1972. *Ap. J.* 178: 623–66

van Kuilenburg, J. 1972. *Astron. Ap.* 16: 276–81

Verschuur, G. L. 1969. *Ap. J.* 156:771–77

Verschuur, G. L. 1970. *Proc. IAU Symp. No. 34,* ed. H. Habing, 150–67

Verschuur, G. L. 1971a. *Astron. J.* 76:317–21

Verschuur, G. L. 1971b. *Bull. Am. Astron. Soc.* 3:369

Verschuur, G. L. 1972. *Rep. Spiral Struct. Meet. Tuscon, Arizona,* ed. B. J. Bok, C. Cordwell, R. Humphreys

Verschuur, G. L. 1973a. *Astron. Ap.* 22: 139–51

Verschuur, G. L. 1973b. *Astron. Ap.* 24:

407–11
Verschuur, G. L. 1973c. *Astron. Ap.* 27: 73–76
Verschuur, G. L. 1973d. *Astron. Ap.* 24: 193–200
Verschuur, G. L., Cram, T. R., Giovanelli, R. 1972. *Ap. J.* 11:57–61
Wannier, P., Wrixon, G. T. 1972. *Ap. J. Lett.* 173:L119–23
Wannier, P., Wrixon, G. T., Wilson, R. W. 1972. *Astron. Ap.* 18:224–31
Wesselius, P. R. 1973. *Astron. Ap.* 24:35–39
Wesselius, P. R., Fejes, I. 1973. *Astron. Ap.* 24:15–34
Westerbork Observatory. 1973. *Ann. Rep.—Netherlands Foundation Radio Astron.*
Whitehurst, R. N., Roberts, M. S. 1972. *Ap. J.* 175:347–52
Wright, M. C. H. 1973. *Ap. J.* 179:253–60
Wright, M. C. H. 1974. *Astron. Ap.* 31: 317–22

UNSEEN ASTROMETRIC COMPANIONS OF STARS

×2083

Peter van de Kamp
Astronomical Institute, University of Amsterdam, The Netherlands; Sproul Observatory, Swarthmore College, Swarthmore, Pennsylvania 19081

1 INTRODUCTION

In this survey I attempt to gather the available information on unseen companions of stars, revealed either as unresolved astrometric binaries that have been found and studied from perturbations in the proper motion of a star previously considered single, or as an irregularity in the orbital motion of a visual double star, which thus proves to have a third component. The astrometric method is particularly suited for the discovery and subsequent study of perturbations with long periods and large amplitudes. A list is given of well-established as well as of provisional and uncertain results (section 7).

The companion revealed by a perturbation may be unseen for a number of reasons. It may be an object of low luminosity, a faint dwarf, close to and lost in the image of the primary star or too faint to be recorded photographically, even at a large angular separation from the primary. It may be an object emitting no light whatsoever, but nevertheless having sufficient mass to cause a measurable perturbation of its primary. In this case the unseen companion could be a neutron star or a black hole, or one or more planetary companions of sufficiently large mass to cause a measurable perturbation. Parallax, orbital elements, magnitude difference, observational coverage, and obtainable accuracy all play a role in determining the chances of discovery.

The discovery of unseen companions has been carried out by spectroscopic methods (Doppler shift) for a long time. The unseen objects thus found are always of comparatively large mass, and stellar in nature. The spectroscopic approach is particularly suited for the discovery of short-period binaries, which show a large range in radial velocity. The astrometric method is particularly suited for the discovery and subsequent study of long-period orbits of the order of several years or decades, with corresponding large amplitudes of the perturbation.

While the astrometric discovery of perturbations increases our knowledge of double stars, generally with appreciable magnitude differences, it also offers the possibility of finding objects less massive than the visible stars of known smallest mass (0.06 $\mathfrak{M}_{\odot}$). We also should be prepared for the discovery of massive planets or planetlike objects, at least for nearby stars. In other words, we may obtain information about objects in the gap between 0.06 $\mathfrak{M}_{\odot}$ and the largest known planetary mass, Jupiter (0.001 $\mathfrak{M}_{\odot}$).

2 HISTORICAL

Before going into some detail about current methods of studying unresolved astrometric binaries, we give, for the sake of perspective, a brief survey of some of the high points in the history of astrometry.

Astrometry, the oldest branch of astronomy, deals with the space-time relationships of celestial objects. These have been studied for more than two millennia, during which the various branches of astrometry were sequentially revealed in order of the ease of discovery, that is, the angular size of the phenomena. Because of its large angular value—50″ yearly—the *precession* of the equinoxes, which causes the secular changes in the equatorial coordinates of celestial objects, was discovered by Hipparchus as early as 125 B.C.

The next important step is the *heliocentric* viewpoint, established in 1543 by Nicholas Copernicus, stressing the Sun as the preferred origin to which to refer the motions of the planets; these motions were still accepted as circular.

In 1609 Johannes Kepler introduced the concept of (1) elliptical motion subject to the (2) law of equal areas. These two, the first and second law of Kepler, are referred to as *Keplerian motion.* In 1619 Kepler stated his famous third or *harmonic law,* which relates the orbits of the different planets, known at the time, by the simple relation: cube of the semimajor axis is proportional to the square of the period of revolution.

The inherent angular changes, or *proper motions* of individual stars, were not recognized until 1718, when Edmond Halley detected that some bright stars showed relative displacements about the size of the Moon's diameter from the positions recorded by Ptolemy in the second century A.D. Subsequent studies revealed the constancy in amount and direction of the proper motions of thousands of stars, the most beautiful demonstration of Newton's law of inertia. Exceptions to this fact gradually appeared, however, and proved to be of particular interest as we will see later.

Stellar aberration due to the finite velocity of light (which had been discovered by Olaus Römer in 1675) and the Earth's annual orbital motion around the Sun was discovered (its total amplitude is nearly 41″) and explained in 1726 by James Bradley who, a few years later, found the nutation, a periodic precessional term with a total amplitude of 23″ and period of 18.6 yr.

Solar motion, the Sun's motion relative to the stars, was first deduced by William Herschel in 1783 from the secular parallactic effect in stellar proper motions. In 1803 Herschel also noted relative *orbital motion* for the two components of Castor and other binary stars. Further studies of numerous binaries showed that the two components describe similar orbits around their common center of mass, which in turn has a uniform and rectilinear motion, as single stars have. The two orbits differ 180° in phase whereas their scales are inversely proportional to the masses of the respective components. Hence the components of a binary moving in nonrectilinear paths are the result of the uniform rectilinear motion of the center of mass and the orbital motion of the individual components.

An accurate determination of annual *stellar parallax* was made in 1838 by Friedrich Wilhelm Bessel for the double star 61 Cygni, at that time the star of largest known proper motion (5.″22 yearly) and in some contemporary books referred to as *Bessel's star*.

Subsequent astrometric discoveries included star streaming or *preferential* motion by Jacobus Cornelis Kapteyn in 1905; radial velocity measures contributed to the discovery of *asymmetry* in 1924 by Gustav Strömberg. Both phenomena were successfully explained by the theory of *galactic rotation* developed by Bertil Lindblad in 1926. This theory in turn led to Jan Hendrik Oort's simple mathematical formulation for differential galactic rotation in 1927; earlier (1921) Oort had made a study of high-velocity stars, which also exhibit the abovementioned asymmetry.

In 1844 Bessel announced the presence of a *perturbation* in the proper motions of Sirius and Procyon. His observations were based on visual observations with transit instruments over an interval of one century, and their positional accuracy was about one-tenth of 1″. The perturbations had amplitudes of several seconds of arc and thus were well established. Bessel explained both perturbations by the presence of faint secondary components. The unseen companions of Sirius and Procyon were seen with large telescopes in 1862 and 1896 respectively.

In 1846 an unseen planet in our own solar system—later named Neptune—was found from perturbations in the orbital motion of Uranus, and subsequently seen with the telescope. In 1930 the distant planet Pluto was discovered in a similar fashion. In the second part of the nineteenth century perturbations were found in the orbital paths of the binaries Zeta Cancri C in 1888 (Seeliger 1914) and Xi Ursae Majoris by Nörlund (1905). The companions in these systems still remain unseen; the companion of Zeta Cancri C appears to be a white dwarf (section 7).

The basic technique and methods of the current photographic study of unresolved astrometric binaries are described elsewhere (van de Kamp 1943, 1944, 1956, 1962, 1966, 1967, 1972); they are briefly reviewed in sections 3–6.

3 LONG-FOCUS PHOTOGRAPHIC ASTROMETRY

3.1 *Accuracy*

The study of unresolved astrometric binaries received new impetus during the first half of the twentieth century. Considerable improvement in positional accuracy became possible through the combination of long-focus refractor and photographic plate. The first successful precision work in long-focus photographic astrometry was done by Frank Schlesinger and by Ejnar Hertzsprung. Schlesinger developed both technique and methods as applied to the problems of annual stellar parallax; Hertzsprung did the same for photographically resolved ("wide") double stars, for which he obtained accurate relative positions of the components, using multiple exposures and objective gratings. We use an extension of Schlesinger's approach in the study of unresolved astrometric binaries.

The high positional accuracy is the result of several factors: long focal length, that is, large-scale portrayal, stability of photographic emulsions, and precision-measuring engines. The attainable accuracy is limited primarily by atmospheric and optical effects, which are kept small by maintaining stability of the optical parts, reducing the spectral bandwidth, and observing one and the same star field always close to the same hour angle, preferably on the meridian.

The refractor has proven well suited for precise photographic measures. As an illustration, the Sproul visual refractor (aperture 61 cm, focal length 10.93 m, scale 1 mm = 18.″87) is used in conjunction with 5 × 7 inch Eastman Kodak 103aG plates and a "minus blue," at present Schott OG-515, filter. A bandwidth of about 600 Å is obtained around wavelength $\lambda5607$ for minimum focal length. The effective wavelength ranges from about $\lambda5480$ for spectral type A0 to $\lambda5525$ for spectral type M0. The diameter of a well-defined sharp, well-blackened star image ranges from about 1 to 2″. Early parallax work was done mostly by refractors; reflectors were considered unsuitable because of limited field and changes in figure. The situation has now changed, however, witness the successful operation and results obtained with the quartz astrometric reflector of 155-cm aperture at the Flagstaff station of the U.S. Naval Observatory (USNO) (Strand 1971). Other astrometric reflectors (Fan Mountain, Pino Torinese) have been built and are ready for operation.

Positional measurements are made on a background of several faint stars, about magnitude 10 or fainter. Ideally all reference stars should have the same brightness and spectrum (color) as the central star; in practice this can only be approximated. The brightness of the central star, if necessary, is reduced by a rotating sector to obtain approximate magnitude compensation between central and reference stars; grating techniques may also be used. Several plates may be taken on any one night, each containing from one to five exposures. Reduction of film errors is obtained by the use of "double plates," effected by turning the same photographic plate 180° in its own plane between two successive sets of exposures.

The measured positions of any one series of plates are reduced by a linear transformation to the scale and orientation of a standard frame defined by the reference stars. Three reference stars are a minimum requirement; a larger number leads to a small increase in accuracy, which is primarily determined and limited by the positional accuracy of the central star. Generally there is little reason to use more than six reference stars unless one wants to introduce nonlinear terms or terms involving magnitude and color.

The positional accuracy obtained from one exposure of a star relative to the reference stars is approximately 2 μm probable error (pe) or smaller. By taking several exposures and more than one plate, a pe of 1 μm (~0.″02) or less may be obtained for the position determined on any one night. By combining the observed positions over several nights, a high accuracy may be reached with internal probable errors of 0.″005 or less.

As frequently happens, the law of diminishing returns enters the picture. There is evidence of a probable "year error" of about ±0.″002 in the Sproul astrometric positions, an amount that appears to set a limit to the accuracy obtainable in any one year; the same amount is found for normal points based on six or more multiple-exposure photographs of relative positions of the components of resolved double stars. Although high accuracy may thus be reached in terms of accidental errors, there remains the serious possibility of systematic errors, particularly for series of observations extending over long time intervals and therefore of particular concern in studies of perturbations.

Let us first repeat the classical statement expressed by Kapteyn (1922):

> I know of no more depressing thing in the whole domain of Astronomy, than to pass from the consideration of the accidental errors of our star-places to that of their systematic errors.

This concern, referring to visual meridian circle observations, holds equally for the current long-focus photographic technique; changes in the optical characteristics of objective, emulsion, and filter are undoubtedly the principal cause. It is still too early to pass judgment on the long-range performance of reflectors now active in long-focus photographic astrometry. As to the refractors, some have been in use now for over six decades, and there is evidence of systematic changes, corresponding with events in the history of the objective.

Precise, quantitative information on systematic errors is slowly becoming available. We give a brief indication of what appears to be known for the long-range series of photographs obtained at the Sproul Observatory. The systematic accumulation of plates begun in 1937 (there are now well over one hundred thousand plates) has gradually revealed the following effects on positions of a central star referred to a limited number (three–six) of reference stars of approximately the same magnitude:

1. A systematic difference or "equation" coinciding with lens adjustment in 1941.82. This break appears to be dependent on color; most stars in the Sproul program are nearby red dwarfs, but there are a limited number of earlier-type stars also
2. A systematic difference coinciding with the introduction of the new cast-iron cell, to replace the old aluminum cell, in 1949.21. At the same time the emulsion was changed from Eastman Kodak C to G. This break may not be color related; it may well be dependent on declination.

   Although both breaks appear more pronounced in right ascension measurements, they are not absent in declination
3. Subsequent lens adjustments in 1957 and 1966 appear to have had little or no systematic effects

It is no easy matter to determine the amounts of the breaks mentioned under 1 and 2, particularly their relationship to magnitude, color, and declination. Even with the comparative intensive observing coverage currently aimed for at Sproul Observatory, an understanding of these effects is being gathered in a painfully slow way. Only by continued efforts, at other observatories also, can this perennial problem of systematic errors be better understood, if not completely solved. The amounts of these breaks are of the order of 1 μm and never more than 3 μm or about $0\rlap{.}''05$ for the Sproul 24-inch refractor.

Fortunately there are several well-established perturbations. As of this moment there are 15 perturbations (listed in Table 1, see section 7), which have total amplitudes well over 0".05 and there need be no doubt about their reality. Two other stars, Ross 614 and VW Cephei, had perturbations well over 0".1 and did not remain unresolved for very long; they are now resolved astrometric binaries. There are, however, several other cases of perturbations of a provisional, suspected, or uncertain nature. In most of these cases the total amplitude is small, hovering near or below the critical value of 0".05 mentioned above. Hence the possible presence of systematic errors becomes a most serious matter. One way to check the reality of a small perturbation is to confront it with the behavior of residuals derived from parallax (and proper-motion) determinations of a number of stars with no suspected perturbations. This may be done by averaging the residuals for several stars for different portions of the sky, such as different declination zones (Lippincott 1971). Another approach is the thorough analysis of the path for any one star, using a variety of possible parameters (Hershey 1973a). In any case, and above all, it is desirable to follow the observed paths of as many stars as possible over as long an interval as possible, taking as many plates as possible. Obviously, at any one observatory this cannot be done for many stars. It is also desirable that the paths of some of the same stars should be followed at several observatories.

It is hoped that the concern with systematic errors, a first sign of developing maturity in our current studies, will be intensively, indefinitely, and effectively pursued at several observatories. Needless, if not trivial, to say, it is important to keep the optical performance of the telescope, as well as the observing techniques, as constant as possible.

Before dealing with the discoveries and analyses of perturbations, we briefly review the analyses and methods used in the study of parallax for single stars (section 3.2), and parallax and orbital motion for resolved binaries (section 3.3) and unresolved binaries (section 3.4).

3.2 *Parallax*

Parallax calculations make use of the following equations of condition for the observed path:

$$X = c_x + \mu_x t + q_x t^2 + \pi P_\alpha$$

and 3.1

$$Y = c_y + \mu_y t + q_y t^2 + \pi P_\delta,$$

where X and Y represent the position of the central star in right ascension (reduced to great circle) and declination measured and reduced to the standard reference frame. The first three terms on the right-hand side of the equations represent the heliocentric path, the fourth the parallactic displacement. The parallax factors P_α and P_δ represent the projected fractional contribution of the relative parallax π in the respective coordinates. The quadratic (acceleration) term may be required for series on nearby stars covering several decades.

Greatest parallactic shifts are obtained shortly after dusk and shortly before dawn. Whereas most parallax information is obtained from measurements in the right-ascension coordinate, it is customary to measure the declination coordinate also. On the average, the weight of parallax determinations in declination is only about 15% of that of determinations from right-ascension measures.

A conventional parallax determination based on some 20–30 plates, each with two or three exposures taken at extreme parallactic shifts extending over a few years, yields a probable error of 0″.01 or less for the relative parallax. By increasing the observational material, and from multiple determinations at different observatories, higher accuracy may be reached with errors probably down to 0″.002 or less.

The systematic errors described in section 3.1 would hardly influence parallax determinations. The latter might be affected, however, by systematic errors with the not unlikely period of one year. It may be considered a source of some reassurance that, on the average, for Sproul parallax determinations, there is a difference of only $+0''.001 \pm 0''.001$ between parallax determinations from the two coordinates in the sense right ascension *minus* declination. At the same time the Sproul determinations average only $+0''.003 \pm 0''.001$ more than the values given in the *General Catalogue of Trigonometric Parallaxes* and its supplement (Lippincott 1971).

The measured relative parallax requires a small reduction to absolute parallax, ranging from $\sim 0''.002$ to $0''.007$ and averaging $\sim 0''.003$.

The absolute magnitude M is obtained from the absolute parallax p and the apparent magnitude m by the relation

$$M = m + 5 + 5 \log p. \qquad 3.2$$

3.3 *Resolved Binaries*

A decade or so after the first determinations of stellar parallaxes by long-focus photographic astrometry, it became evident that by extending the observational material the same technique and methods could be used to determine the ratio of masses in binary systems. In this case another term has to be added to the equations of condition (3.1) of the stellar path.

Although the visual study of binaries permits the resolution of pairs down to a few tenths of 1″ or even less, the photographic method generally does not yield clearly separated images below about 2″. It is true that under excellent seeing conditions and for nearly equal magnitudes of the components, apparently well-separated images may be obtained down to ~1″. On the other hand, for large magnitude differences, clear separation may not be obtained below 3 or 4″. In all these cases of close proximity of images, or partial overlap, one must reckon with the possibility of appreciable systematic errors in the measurement of the separation, and the results are to be regarded with caution, if not suspicion. Even for apparently well-separated images it is not at all certain that systematic "proximity" effects may not exist up to separation as far as 5 or 6″, as indicated in a recent study by John L. Hershey at the Sproul Observatory for the "wide"-binary Xi Bootis.

Generally, therefore, results for orbital motion measured photographically must be evaluated carefully, with particular attention given to the separation of the components, which of course may vary widely over the interval studied. The simplest ideal case is that of material dealing with "wide" separation of the components and/or large magnitude difference. In the latter case exposure and magnitude compensation with the reference stars may be adjusted to give a measurable image of the primary while the light contribution of the secondary remains nil.

For the case of the resolved astrometric binary, the path of the primary is represented by the equations of condition

$$X = c_x + \mu_x t + q_x t^2 + \pi P_\alpha - \mathrm{B}\Delta X$$

and 3.3

$$Y = c_y + \mu_y t + q_y t^2 + \pi P_\delta - \mathrm{B}\Delta Y,$$

where B is the fractional mass $\mathfrak{M}_B/(\mathfrak{M}_A + \mathfrak{M}_B)$ of the secondary in terms of the combined mass of primary and secondary. The scale of the orbit of the primary is B times that of the relative orbit of the two components; the difference in phase (position angle) is 180°. By definition B is the fainter (but not necessarily the least massive) component; ΔX and ΔY represent the separation, that is, the relative position of the secondary B referred to the primary A. Similar equations hold for the secondary, replacing $-\mathrm{B}$ by $(1-\mathrm{B})$. If no outside information is available, that is, if ΔX and ΔY are not known from another source, such as a well-established relative orbit, one set of equations obviously is used. In case of outside information both sets of equations may be used, solving for B and $1-\mathrm{B}$.

It is often useful, and more explicit and elegant, to use the following equations (for the primary):

$$X = c_x + \mu_x t + q_x t^2 + \pi P_\alpha + \alpha Q_\alpha$$

and 3.4

$$Y = c_y + \mu_y t + q_y t^2 + \pi P_\delta + \alpha Q_\delta.$$

Here $\alpha = \mathrm{B}a$ is the semimajor axis of the primary, relative to the barycenter; a is the semimajor axis of the relative orbit of B around A. The quantities Q_α and Q_δ represent the projected fractional contribution of α in the measured coordinate; they are calculated from the orbital elements. These "orbital factors" are analogous to the parallax factors; the latter refer to the star's parallactic orbit, the former to the star's own apparent orbit. Similar equations may be used for the secondary component.

Observational material extending over an interval sufficiently large to separate the proper motion and, if need be, the quadratic time effect from the orbital motion will yield values of B or α, depending on which formulas we use.

The fractional mass B is found either directly, using equation 3.3 or from the relation

$$\mathrm{B} = \frac{\alpha}{a} \qquad 3.5$$

if we use equations 3.4.

3.4 *Unresolved Binaries*

In the case of photographic blending the formulas in section 3.3 require adjustment. This is the case for known binaries resolved visually but unresolved photographically, as well as unseen companions revealed as hitherto unknown unresolved astrometric binaries.

What position do we measure for an unresolved astrometric binary? The smallest photographic star images are rarely below 1″ diam. Blended exposures of components separated by 1″, sometimes more, generally present circular images. A close luminous companion, too faint to be detected visually or spectroscopically, will draw the center of light toward the center of mass. Hence the blended image of an unresolved binary may still appear circular, but the orbit of the measured center of the image will be smaller than the actual orbit described by the primary.

We introduce the concept of *photocenter,* ideally the weighted center of light intensity of the two components. In this case the fractional distance β of primary to photocenter in terms of the separation between the two components is given by

$$\beta = \frac{l_B}{l_A + l_B}, \tag{3.6}$$

where l_A and l_B are the luminosities of the components, or, in terms of magnitude difference Δm, companion minus primary, we have

$$\beta = (1 + 10^{0.4\Delta m})^{-1}. \tag{3.7}$$

The scale of the photocentric orbit is therefore $(\mathrm{B} - \beta)$ times that of the relative orbit, and formulas 3.3 are changed to

$$\begin{aligned} X &= c_x + \mu_x t + q_x t^2 + \pi P_\alpha - (\mathrm{B} - \beta)\Delta X \\ Y &= c_y + \mu_y t + q_y t^2 + \pi P_\delta - (\mathrm{B} - \beta)\Delta Y. \end{aligned} \tag{3.8}$$

Various investigators have found deviations of the observed values of β from the simple theoretical relation 3.7 (Hall 1951, Feierman 1971). The discrepancies depend on telescope, emulsion, separation, and other factors. For the Sproul refractor the observed value of β agrees with the theoretical value up to values of $\Delta m = 1$. For larger magnitude differences and for separations generally less than 1″ (0.05 mm), the observed values of β are smaller than the theoretical values by approximately 0.05, or even by larger amounts for separations over 1″. In other words, for values of $\Delta m > 1$, the influence of the companion on the location of the photocenter is less than what follows from the theoretical relation 3.7. Ideally, as suggested by Feierman, if β could be evaluated as a function of the changing separation, the observed positions could be corrected and the following formulas could be used for the determination of B:

$$X - \beta\Delta X = c_x + \mu_x t + q_x t^2 + \pi P_\alpha - \mathrm{B}\Delta X$$

and (3.9)

$$Y - \beta\Delta Y = c_y + \mu_y t + q_y t^2 + \pi P_\delta - \mathrm{B}\Delta Y.$$

Equation 3.7 remains of interest for the case that β may be assumed constant throughout the series of observations in 3.4:

$$X = c_x + \mu_x t + q_x t^2 + \pi P_\alpha + \alpha Q_\alpha$$

and

$$Y = c_y + \mu_y t + q_y t^2 + \pi P_\delta + \alpha Q_\delta,$$

where $\alpha = (\mathrm{B} - \beta)a$ is now the semimajor axis of the *photocentric orbit.*

The sign of α is determined by the sign of $(\mathrm{B} - \beta)$. For visual binaries on the main sequence (Feierman 1971) B ranges from 0.32 to 0.49, with values of β ranging from 0 to 0.116, that is, $\mathrm{B} - \beta$ is always positive. For a main-sequence star accompanied by a white dwarf, or a similar object of appreciable mass but little or no luminosity, β is virtually zero and again $\mathrm{B} - \beta$ is positive. $\mathrm{B} - \beta$ can be negative, that is, β can exceed B in absolute value, only for a comparatively small magnitude difference and a very small fractional mass for the fainter component, a condition contrary to established knowledge of mass and luminosities, but in principle not excluded.

For the case of constant β, the fractional mass is thus found to be

$$\mathrm{B} = (\alpha/a) + \beta. \qquad 3.10$$

This relation explicitly reveals the accuracy of B as it depends on errors in α, a, and β; the latter is often the principal source of error. This equation is of particular significance in the analysis of unresolved objects not previously known or recognized as binaries (section 6).

Individual masses may be obtained if the combined mass can be found with adequate accuracy. The space-time dimensions of the orbit yield the following relation for the combined mass

$$\mathfrak{M}_A + \mathfrak{M}_B = 4\pi^2/G \cdot a^3/P^2 \qquad 3.11$$

in cgs units. Here a, the semimajor axis of the B relative to the A component, is expressed in linear measure. Choosing as units the Sun's mass, astronomical unit, and sidereal year, $G = 4\pi^2$ and we have the "harmonic" relation

$$\mathfrak{M}_A + \mathfrak{M}_B = a^3/P^2, \qquad 3.12$$

or, since the parallax is required to obtain the linear value of a, we write

$$\mathfrak{M}_A + \mathfrak{M}_B = \frac{a^3}{P^2}\frac{1}{p^3}, \qquad 3.13$$

where a is now the (unprojected) semimajor axis of the relative orbit expressed in seconds of arc.

At present the accuracy of the combined mass is primarily limited by the attainable accuracy in the parallax p, and for close binaries, of the value of a. Frequently the mass ratios are much better known than the masses because of the triple effect of errors in p and a. Accurate masses have been determined for a few dozen binaries. Astrometric results have thus contributed to our knowledge of the mass-luminosity relation in our neighborhood for stars with masses less than about twice the Sun's mass. The relation closely resembles, in fact is, the main-sequence relation, with the striking exception of the three white dwarfs: Sirius B, Procyon B, and 40 Eridani B. The past decades have witnessed continued advances in our knowledge of the fainter portion of the mass-luminosity relation. Whereas before 1955 the smallest known mass of a visible star was that of Krüger 60 B (0.16 $\mathfrak{M}_\odot$), we have since found Wolf 424 A and B (0.065 ± 0.013 $\mathfrak{M}_\odot$ each) (Heintz 1972), Ross 614 A (0.114 ± 0.016 $\mathfrak{M}_\odot$), and Ross 614 B (0.062 ± 0.009 $\mathfrak{M}_\odot$) (Lippincott & Hershey 1972).

The three stars of known smallest mass are slightly below though very likely within the errors, both observational and theoretical, of the lower mass limit calculated for the main sequence (Grossman 1970, Straka 1971).

4 DISCOVERY OF PERTURBATIONS

Long-focus photographic astrometry has proven suitable and useful for the discovery and study of perturbations, both revealed either as "variable" proper motion of a star previously considered single, or as an irregularity in the Keplerian motion of a visual binary, which thus proves to have an additional component.

The classical outstanding example of a perturbation found photographically was that of Ross 614. The story of this object, which was later resolved visually, has been told before (Lippincott 1955a, b) and is discussed in more detail in section 7. The discovery of this perturbation as well as that of several others was not the result of a planned program. Systematic searches were begun in the early 1930s, however, at the Leander McCormick Observatory and have been carried out more intensively since 1937 at the Sproul Observatory, where all stars within 10 pc, sometimes more, and within reach of the 61-cm refractor are photographed on a regular basis, preferably at each observing season and, for the very nearest stars, on several nights each year. Obviously the majority of stars in this program are nearby late-type dwarf stars; because of their relatively small masses the chances of finding a perturbation are much more favorable than they are for, say, a solar-type star. Several other observatories have contributed to this field of research, notably Allegheny.

Thousands of parallax determinations of "single" stars have been made from material spread over a few years. Perturbations of "short" periods, of the order of a year or less, generally have small amplitudes and are not easily found. With few exceptions (G 24-16, section 7) it has usually been possible to represent a short-time path satisfactorily by uniform proper motion and a parallactic orbit; moreover, no increase in the probable error of one plate has been found for stars of large parallax. Perturbations of "long" periods, of the order of several decades, do not reveal any measurable perturbation in a few years; continued observations are required. Orbital motion may be partly absorbed in proper and parallactic motion, or may escape detection because of temporal gaps in the series of observations. Narrow hour-angle requirements cause annual gaps of six to seven months, which may result in spurious periods; positions in successive cycles of a short-period orbit may be interpreted by a multiple of the true period, or vice versa. As long as the binary is not resolved, the analysis lacks the control of the harmonic relation (equation 3.12), which for resolved binaries may serve as a guide for the period. Interpreting a limited material by a fortuitous orbit may be avoided through additional observations.

The multiple-exposure technique mentioned earlier is ideally suited for the discovery and study of perturbations in well-resolved binaries. For close binaries, visual techniques, micrometer or others, remain a worthwhile and desirable observational approach.

A satisfactory perturbation orbit generally is not obtained until all phases of the orbit have been covered. Correct dynamical interpretation is aided by the fact that, as a rule, Keplerian motion is observed in two coordinates. We must also reckon with the possibility that the observed perturbation is the result of two (or more) perturbations caused by two (or more) companions.

5 ANALYSIS OF PERTURBATIONS

5.1 *Dynamical Elements*

The remainders R from a solution for proper motion and parallax reveal the perturbation over the interval covered by series of observations. If no prior orbital elements are known, an attempt is made to determine the dynamical elements P, T, and e. If the interval exceeds the period of the perturbation (assuming for the moment one companion only), the period P will be revealed by a repetition of the pattern of the remainders after the first period. There are a few cases on record where the series of observations have covered more than two periods. If the remainders clearly show that one full period has not yet been covered, but if the amplitude of the observed portion of the perturbation is sufficiently large, estimates of the period may be made and provisional results obtained for the other elements.

Periastron passage T and eccentricity e may be derived from the (apparent) photocentric orbit, though this may be awkward, since it involves locating the invisible barycenter. A different method is much preferred, namely, using the time-displacement curves R/t in any coordinate, ideally and preferably right ascension and declination. This procedure is useful for "linear" orbits as represented by visual orbits seen nearly edgewise, and has long been in use for spectroscopic binaries. The method is also applicable to any "open" visual or photocentric orbit. The success of the method, as of any other, depends on adequate knowledge of the apparent, in our case, photocentric orbit; it depends therefore on the precision with which the proper motion of the barycenter is known, and can be allowed for. This generally implies coverage of the orbital motion over more than one period. If a substantial portion of the period has been covered and the observations are of sufficient precision, however, a provisional value of the proper motion of the barycenter may be obtained and allowed for and the method for determining T and e may be applied. In this respect there is no difference from the general procedure used for resolved binaries, where provisional orbits are always determined when the time seems ripe. In any case a series of attempts may be made and successive approximations used. The provisional dynamical information and the ephemeris to test future observations are the obvious justifications for determining provisional elements.

Having plotted the remainders after correcting any proper motion of the barycenter, time-displacement curves are drawn that satisfy the remainders. Next, periastron and apastron are located by the fact that their epochs differ by half a period (or we may say that their mean anomalies differ by 180°), and their ordinates are equal and of opposite signs when referred to the center of the orbit. Periastron and apastron are conveniently located by making a copy of a displacement curve, reversing it along the central line representing the center of the orbit, and shifting the reversed curve half a period along the time axis. Generally two pairs of intersections result: the single one on the shorter, steeper branch, and the middle intersection on the longer branch represent periastron and apastron respectively. The slopes of the displacement curve represent projected velocities dR/dt. The ratio of the true (unprojected) velocity vectors at periastron and apastron is $-(1+e)/(1-e)$, their directions being opposite. This ratio does not alter in projection; hence the ratio of the slopes $(dR/dt)_P$ and $(dR/dt)_A$ at periastron and apastron respectively amounts to $-(1+e)(1-e)$. We thus find the eccentricity of the orbit, independently of the barycenter, through the relation

$$e = (dR/dt)_P + (dR/dt)_A/(dR/dt)_P - (dR/dt)_A. \tag{5.1}$$

For the sake of completeness we give the corresponding relation for analysis of the radial velocity pattern V/t of the spectroscopic binary:

$$e = \left(\frac{d^2R}{dt^2}\right)_P^{1/2} + \left(\frac{d^2R}{dt^2}\right)_A^{1/2} \Bigg/ \left(\frac{d^2R}{dt^2}\right)_P^{1/2} - \left(\frac{d^2R}{dt^2}\right)_A^{1/2}. \tag{5.2}$$

This method becomes unreliable when periastron and apastron are close to the extreme displacements, in which case a combination of graphical and analytical methods may be used (van de Kamp 1947a).

5.2 *Geometric Elements*

The geometric elements (B), (A), (G), and (F) refer to the photocentric orbit of the unresolved system and are given in parentheses to distinguish them from the geometric elements B, A, G, and F, which refer to the orbit of the companion relative to the primary. The latter elements are also called *natural elements* or *Thiele-Innes constants*. They appear in the apparent orbit as the projected rectangular equatorial coordinates B, A of periastron and the coordinates G, F of the point $E = 90°$ on Kepler's auxiliary circle. In the photocentric orbit, (B), (A) appear as the projected rectangular coordinates of periastron and (G), (F) appear as the point $E = 90°$ on the auxiliary circle of the photocentric orbit. The latter is located in the orbital plane tangent to the orbit in periastron and apastron; the radius of Kepler's circle equals the unprojected semimajor axis α of the photocentric orbit.

The geometric elements of the photocentric orbit are related to the Thiele-Innes constants as follows:

$$\begin{aligned} (B) &= -(\alpha/a)B \qquad (G) = -(\alpha/a)G \\ (A) &= -(\alpha/a)A \qquad (F) = -(\alpha/a)A, \end{aligned} \tag{5.3}$$

where α and a are the semimajor axis of the photocentric and relative orbit respectively (van de Kamp 1967).

The following formulas are used for the orbital displacements in the two equatorial coordinates:

$$(B)x+(G)y \quad \text{(in right ascension)},$$
$$(A)x+(F)y \quad \text{(in declination)}. \qquad 5.4$$

Here x and y are the elliptical rectangular coordinates in the (dimensionless) unit orbit; they are the well-known functions of the dynamical elements P, T, and e:

$$x = \cos E - e$$

and 5.5

$$y = \sin E(1-e^2)^{1/2}.$$

The eccentric anomaly E is related to P, T, and e as follows:

$$E - e\sin E = \frac{2\pi}{P}(t-T) = M \text{ (mean anomaly)}, \qquad 5.6$$

where t is the epoch of observation. Tables exist for x and y, as functions of e and M, so that E need not even be computed (Franz & Mintz 1964). Current computer programs conveniently perform these computations as the need occurs.

5.3 *General Formulas*

The general formulas for analyzing the path of photocenter for proper, parallactic, and orbital motion are therefore

$$X = c_x + \mu_x t + q_x t^2 + \pi P_\alpha + (B)x + (G)y$$

and 5.7

$$Y = c_y + \mu_y t + q_y t^2 + \pi P_\delta + (A)x + (F)y.$$

Generally the observed path is first analyzed for proper motion and parallax (and quadratic time effect if need be).

Hence, after P, T, and e have been determined, they may be used via the elliptical rectangular coordinates in unit orbit, x and y, to calculate the geometric elements (B), (G), (A), and (F) and corrections to c, μ, q, and π, if desired.

From the natural elements the conventional elements α, i, ω, and Ω may be derived through the following relations

$$\alpha^2 = j+k$$

where

$$\left.\begin{aligned} k &= \{(A)^2+(B)^2+(F)^2+(G)^2\}^{1/2} \\ m &= (A)\cdot(G)-(B)\cdot(F) \\ j &= (k^2-m^2)^{1/2} \\ i &= \text{arc cos } m\alpha^{-2} \end{aligned}\right\} \qquad 5.8$$

and

$$\left.\begin{array}{l} \tan(\omega+\Omega) = \dfrac{(B)-(F)}{(A)+(G)}, \\ \text{where } \sin(\omega+\Omega) \text{ has the same sign as } (B)-(F) \\ \tan(\omega-\Omega) = \dfrac{-(B)-(F)}{(A)-(G)}, \end{array}\right\} \quad 5.9$$

where $\sin(\omega-\Omega)$ has the same sign as $-(B)-(F)$. The quadrants of $\omega+\Omega$ and $\omega-\Omega$ are unambiguously determined. However, we may add 360° to $\omega+\Omega$ (or to $\omega-\Omega$) and thus obtain two solutions

$$\omega, \Omega \quad \text{and} \quad \omega \pm 180°, \Omega \pm 180°.$$

Unless a distinction is possible, the solution is adopted corresponding to $\Omega < 180°$. [For further definitions see *IAU Trans.* V:332 (1935); see also *IAU Trans.* VIII B:146 (1967).]

6 DYNAMICAL INTERPRETATION: MASS FUNCTION

The analysis of the photocentric orbit of an unresolved astrometric binary yields the two data of primary interest: the period P (in years) and the semimajor axis α, in astronomical units, assuming the parallax to be known. In case of a companion which is nonluminous or sufficiently separated from the primary, the following mass function is found:

$$\frac{\alpha^3}{P^2} = \mathrm{B}^3(\mathfrak{M}_A+\mathfrak{M}_B). \quad 6.1$$

The situation is analogous to the mass function for a spectroscopic binary with one component visible,

$$\frac{(\alpha \sin i)^3}{P^2} = \mathrm{B}^3 \sin^3 i(\mathfrak{M}_A+\mathfrak{M}_B), \quad 6.2$$

except that for the astrometric binary the inclination may be determined. We may write equation 6.1 as follows:

$$\mathfrak{M}_\mathrm{B} = \alpha P^{-2/3}(\mathfrak{M}_A+\mathfrak{M}_B)^{2/3}, \quad 6.3$$

or

$$\mathfrak{M}_\mathrm{B} = \mathrm{B}(\mathfrak{M}_A+\mathfrak{M}_B) \quad 6.4$$

where

$$\mathrm{B} = \alpha/a. \quad 3.5$$

Adopting a value for the sum of the masses, which implies equation 3.12, an adopted value for the semimajor axis a of the relative orbit and the mass of the companion is uniquely determined, as seen from equations 6.3 or 6.4 and 3.5.

Generally, however, we do not know whether the observed orbit refers to the pure image of the primary or to the photocenter of primary and companion. In the latter case $\alpha = (\mathrm{B}-\beta)a$ and the following mass function is found:

$$\frac{\alpha^3}{P^2} = (\mathrm{B}-\beta)^3(\mathfrak{M}_A+\mathfrak{M}_B), \qquad 6.5$$

which may be written as

$$\mathfrak{M}_B - \beta(\mathfrak{M}_A+\mathfrak{M}_B) = \alpha P^{-2/3}(\mathfrak{M}_A+\mathfrak{M}_B)^{2/3}. \qquad 6.6$$

This expression gives a lower limit for the mass of the companion for an adopted value of the combined mass.

In the present case

$$\mathrm{B} = \frac{\alpha}{a} + \beta. \qquad 3.10$$

Because β is not known, but generally can be limited to a limited range of values, a corresponding range of values for B, and hence of $\mathfrak{M}_B$, is found. Values of the mass of the companion (and of the primary) may be derived for an adopted range in the sum of the masses ($\mathfrak{M}_A+\mathfrak{M}_B$) and of the fractional luminosity β (or Δm). The adopted values for the sum of the masses together with the period P yield values for the semimajor axis a of the relative orbit of primary and companion. The values of a, as well as the corresponding values of β or Δm, generally will limit the adopted ranges; a larger value of a and/or a small value Δm would possibly have led to earlier visual detection of the companion. On the other hand, such values, together with knowledge of greatest separation, will aid to predict the next greatest elongation favorable for possible detection of the companion.

The various constraints generally lead to rather narrow limits for the mass of the companion and an upper limit for the luminosity.

7 INDIVIDUAL OBJECTS

We now review individual results obtained for unseen companions of stars by the technique and methods of long-focus photographic astrometry. The observatories which thus far have contributed to our knowledge of perturbation are primarily Allegheny, Leander McCormick, Sproul, U.S. Naval, Van Vleck, and Yale. The relative accuracy of the results depends on several factors other than adequate observational coverage, such as size of parallax and observed range of displacement. With some exceptions most results refer to stars within 10 pc. Whereas the data for the perturbation normally depend on observations made at one observatory only, the adopted value for the absolute parallax generally is based on determinations made at several observatories.

Table 1 lists well-established perturbations. The sequence is in order of right ascension. The salient data are Period P, semimajor axis α of photocentric orbit, and the adopted absolute parallax p; probable errors are used. For the sake of completeness periastron passage T and eccentricity e are included. The last two columns give the absolute visual magnitude and mass for primary and companion, based on the considerations and calculations discussed in section 6. Generally the mass of the primary is an adopted value, which leads to a value of the mass of the companion within narrow limits, while the luminosity of the companion is indicated by a likely upper limit.

For the majority of stars in Table 1 the perturbation is larger than the parallax, a not unexpected selection effect. Striking exceptions are 1. Epsilon Eridani and BD+43° 4305, for which the periods are over two decades but the inferred masses are below 0.05 $\mathfrak{M}_\odot$; 2. Xi Ursae Majoris and G 24-16 for which the periods are less than two years but the inferred masses are above 0.07 $\mathfrak{M}_\odot$; and 3. Barnard's star for which the small perturbation is attributed to two companions with masses comparable to Jupiter.

Table 2 lists objects with perturbations of a provisional, suspected, or uncertain nature. Gradually some of the objects in Table 2 may be transferred to Table 1. These perturbations, as well as several others not listed, are not ready for analysis at this time. There are several reasons for excluding these objects: insufficient accuracy, and changing perturbation pattern over the years, or rather decades, which may be of instrumental origin or due to the stars having more than one companion. There are puzzling contradictions in certain cases between results obtained at different observatories.

The two objects in Table 3 were first studied as unresolved binaries from the observed perturbations. More observations of these difficult resolved objects are needed to obtain better knowledge of the semimajor axes of the relative orbits and hence the masses of components.

Objects not included in the present survey are long-period spectroscopic binaries which reveal a measurable astrometric perturbation. Although such objects were discovered and studied spectroscopically, their astrometric studies furnish a most desirable supplement. Examples are Algol, Epsilon Aurigae, and VV Cephei.

Four perturbations are illustrated in Figures 1–3. Ross 614 is the classical example of a very accurately determined perturbation for a single star which led to the visual discovery of the companion. BD+66° 34 A is a fine example of a perturbation in the component of a visual binary. BD+6°398 is an example of a perturbation with a long period but for which good provisional results may be obtained. BD+67°552 is a perturbation whose accuracy is typical and indicative of results obtained at present for several objects.

All these perturbations were obtained from plates taken with the Sproul 61-cm refractor. The observations in the diagrams are normal points for intervals of one year (Figures 1 and 2) or sometimes longer (Figure 3). In all cases proper motion and parallax have been removed.

Brief comments on the objects listed in Tables 1 and 3 are given below.

BD+66°34 A

This is the brighter component of the visual binary Milburn 377 (Vyssotsky 2, ADS 433), with absolute visual magnitudes 10.51 and 12.4, and spectra dM 2.5e and dM 4.5. The perturbation due to the invisible companion *a* was discovered by

Table 1 Unresolved astrometric binaries. Semimajor axes, parallaxes, periods and eccentricities, and absolute visual magnitudes and masses

Name	Right Ascension Declination (1950)	Visual Magnitude / Spectrum	α / p	P (yr) / e	Absolute Visual Magnitude of Primary Companion	Mass of Primary Companion ($\mathfrak{M}_\odot$)
BD + 66°34 A	$0^h29^m.3$	10.51	0″.125 ± 0″.002	15.92 ± 0.22	10.5	0.4
	+66°58′	M2.5e	0″.100 ± 0″.002	0.05 ± 0.07	> 14	0.13
μ Cas	$1^h4^m.9$	5.18	~ 0″.2	~ 22	5.7	0.8
	+54°40′	G5VI	0″.127 ± 0″.003	—	> 7.7	0.2
BD + 6°398	$2^h33^m.3$	5.82	0″.221 ± 0″.008	50	6.5	0.7
	+6°39′	K3V	0″.138 ± 0″.004	0.6	> 11.5	0.10
ε Eri	$3^h30^m.6$	3.73	0″.019 ± 0″.002	25	6.1	0.74
	−9°38′	K2eV	0″.305 ± 0″.003	0.5	> 9.1	0.05–0.006
γ Gem	$6^h34^m.8$	1.93	0″.065 ± 0″.011	12.6	−0.6	3.0
	+16°27′	A0IV	0″.031 ± 0″.004	0.75	≧ 0.4	~ 1
ζ Cnc C	$8^h9^m.3$	6.7	0″.191 ± 0″.001	17.5 ± 0.1	5.1	0.90
	+17°48′	G2	0″.047 ± 0″.005	0.11 ± 0.03		0.90
BD + 67°552	$8^h31^m.9$	9.3	0″.158 ± 0″.004	23.0	8.5	0.6
	+67°28′	dM1	0″.070 ± 0″.004	0.73	10.5–11.5	0.36–0.27
ξ UMa A	$11^h15^m.5$	4.32	0″.055	1.832	4.9	0.83
	+31°49′	G0V	0″.130	0.56	10:	0.31

α Oph	$17^h32^m.6$	2.1	0.″065	8.5	1.0	3.0
	+12°36′	A5III	0.″060 ± 0.″003	0.4	3–5	1.3–0.6
BD + 68°946	$17^h36^m.7$	9.15	0.″102	24.5	10.7	0.274
	+68°23′	M3.5V	0.″209 ± 0.″004	0.9	> 15.7	0.026
Barnard's star	$17^h55^m.4$	9.54	{0.″0093 {0.″0053	{11.5 {20–25	13.25	0.15
	+4°33′	M5 V	0.″552 ± 0.″001	0.0	— —	{0.001 {0.0004
χ Dra	$18^h22^m.0$	3.58	0.″048 ± 0.″005	0.77	4.2	1.5
	+72°43′	F7V	0.″129 ± 0.″005	0.45	6.5	1.1
δ Aql	$19^h23^m.0$	3.36	0.″058	3.42	2.6	1.2
	+3°1′	F0IV	0.″069 ± 0.″003	0.40	> 5.6	0.5
G 24-16	$20^h27^m.4$	13.05	0.″029	1.5	13.3	0.14–0.20
	+9°31′	M6 est	0.″113 ± 0.″002	0.5	> 16.3	0.07–0.11
BD + 27°4120	$21^h35^m.8$	9.8	0.″181	40	9.0	0.46
	+27°30′	M0e	0.″069 ± 0.″005	0.43	≧ 13.0	0.17
ζ Aqr *B*	$22^h26^m.3$	4.59	0.″097	25.5	2.8	0.85
	−0°17′	dF1s	0.″043	0.20	> 6.8	0.28
BD + 43°4305	$22^h44^m.7$	10.2	0.″039 ± 0.″002	28.9	11.7	0.25
	+44°5′	M4.5e	0.″200 ± 0.″003	0.44	> 14.7	0.023–0.009

Table 2 Stars with perturbations of provisional, suspected, or uncertain nature

Name	Right Ascension / Declination (1950)	Visual Magnitude / Spectrum	Parallax	Absolute Visual Magnitude	Reference
van Maanen's star	$0^h46^m.5$	12.37			van de Kamp 1971a
	$+5°9'$	DG	$0''.239 \pm 0''.004$	14.3	Gatewood 1974a
L 725-32	$1^h9^m.9$	11.6			
	$-17°16'$	M5e	$0''.261 \pm 0''.005$	13.6	Heintz 1973
BD+63°238	$1^h44^m.1$	5.63			
	$+63°36'$	K0V	$0''.098 \pm 0''.004$	5.6	Lippincott 1974
BD+18°863	$4^h40^m.0$	9.94			
	$+18°53'$	M2.5eV	$0''.097 \pm 0''.004$	9.9	Lippincott 1974
Ross 986	$7^h6^m.6$	11.48			
	$+38°38'$	dM5e	$0''.165 \pm 0''.005$	12.6	Appelbaum 1972
BD+5°1668	$7^h24^m.7$	9.82			
	$+5°23'$	M5	$0''.270 \pm 0''.003$	11.9	van de Kamp 1971b
Ross 434	$9^h41^m.7$	10.7			
	$+76°17'$	dM2.5	$0''.065 \pm 0''.006$	9.8	Alden 1951
BD+48°1829	$9^h59^m.3$	10.07			
	$+48°21'$	dM2	$0''.067 \pm 0''.005$	9.2	Lippincott 1974
BD+20°2465	$10^h16^m.9$	9.43			
	$+20°7'$	M4.5eV	$0''.204 \pm 0''.004$	11.0	Lippincott 1969
Lal 21185	$11^h0^m.6$	7.5			Lippincott 1960
	$+36°18'$	M2eV	$0''.397 \pm 0''.004$	10.5	Gatewood 1974b
CC 986	$16^h22^m.6$	10.27			
	$+48°28'$	dM3	$0''.138 \pm 0''.006$	11.0	Lippincott & Yang 1967
61 Cyg A	$21^h4^m.7$	5.22			van de Kamp 1971b
	$+38°30'$	K5eV	$0''.296 \pm 0''.004$	7.6	van de Kamp 1973a
Wolf 922	$21^h28^m.6$	11.95			
	$-10°1'$	dM4.5e	$0''.134 \pm 0''.007$	12.6	Osvalds & Osvalds 1959
BD+56°2966	$23^h10^m.9$	5.57			
	$+56°54'$	K3V	$0''.146 \pm 0''.004$	6.4	Lippincott 1974

Table 3 Astrometric orbits with limited resolution data

Name	Right Ascension	Visual Magnitude	α	P (yr)	Absolute Visual Magnitude of	Mass of	Reference
	Declination (1950)	Spectrum	p	e	Primary Companion	Primary Companion ($\mathfrak{M}_{\odot}$)	
Ross 614	$6^h26^m.8$	11.07	$0''.312 \pm 0''.002$	16.60	13.0	0.114 ± 0.016	
	$-2°46'$	dM4.5e	$0''.248 \pm 0''.003$	0.38	16.5:	0.062 ± 0.009	Lippincott & Hershey 1972
VW Cep	$20^h38^m.0$	7.1	$0''.130 \pm 0''.002$	30.4	5.2	1.47 ± 0.5	
	$+75°25'$	G5	$0''.041 \pm 0''.002$	0.60	8.1	0.56 0.5	Hershey 1974

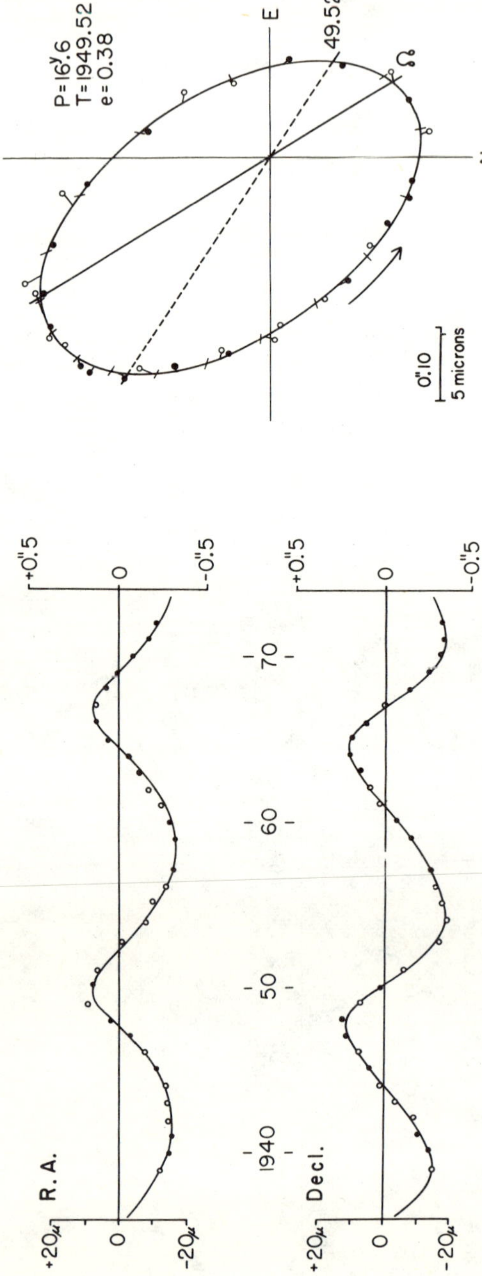

Figure 1 Ross 614. *Left*, observed positions and displacement curves showing the orbital effect of the perturbation in right ascension and declination; *right*, photocentric orbit.

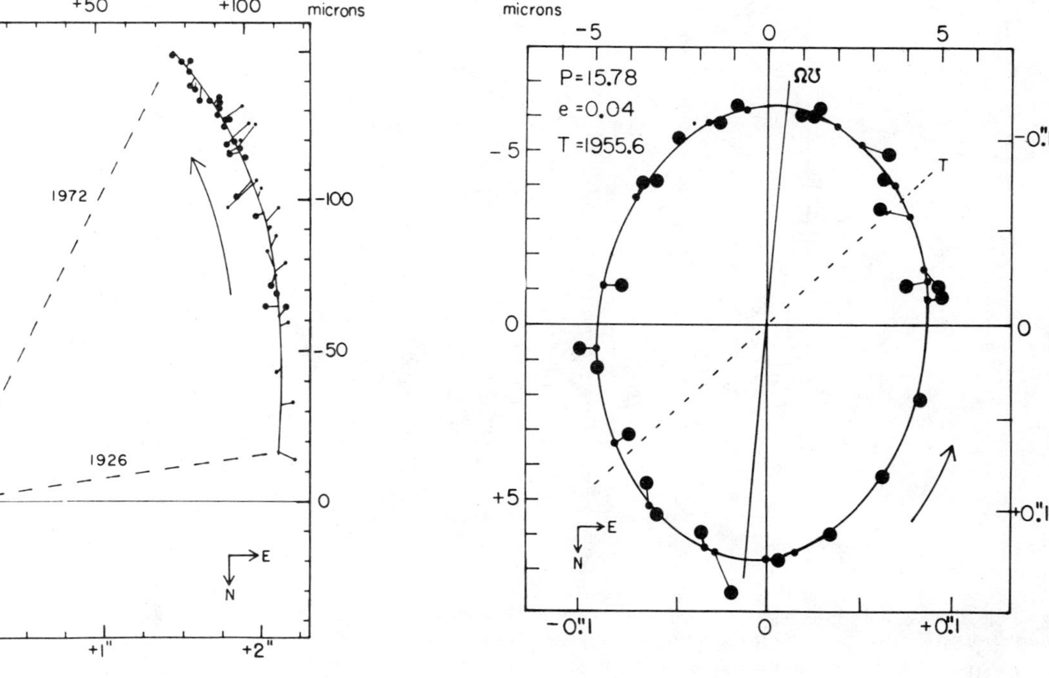

Figure 2 BD+66°34. *Left*, observed positions of *A* and *B* from 1926 to 1972 referred to the center of mass of the *Aa*, *B* orbit. The helix is the photocentric path of *A*. The broken line in the helix is the barycentric path of *A*, *a*; *right*, photocentric orbit of BD+66°34 *A*.

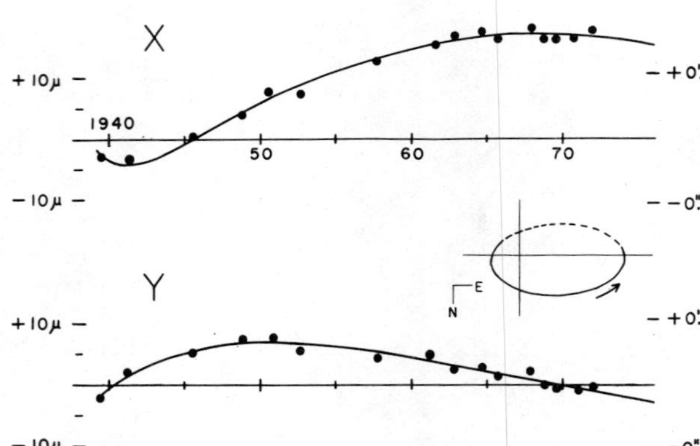

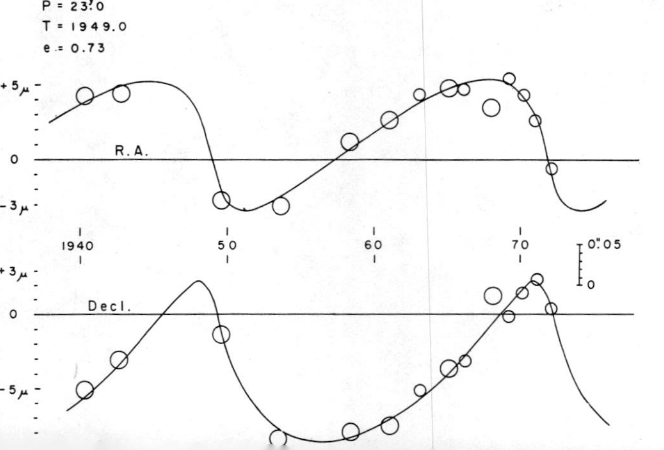

Figure 3 BD+6°298 and BD+67°552. *Above left* (BD+6°398), observed positions and displacement curves showing perturbation in right ascension and declination. Insert shows photocentric orbit. Solid curve indicates part covered by observations; *below left* (BD+67°552), observed position and displacement curves showing perturbations in right ascension and declination; *right*, photocentric orbit.

H. L. Alden at the Leander McCormick Observatory as a "by-product" of a first attempt to determine the mass ratio of the visible components A and B.

A study was made by Hershey (1973b) based on 575 Sproul Observatory plates taken on 163 nights over the interval 1938–1972. Eight reference stars were used; the plates were measured on the two-coordinate Grant machine. The orbital results and the adopted absolute parallax are: $P = 15.9 \pm 0.2$ yr (pe); $T = 1956.0 \pm 2.8$; $e = 0.05 \pm 0.07$; $\alpha = 0''.125 \pm 0''.002$; and $p = 0''.100 \pm 0''.002$. This is one of the most accurately known perturbations, which is partly due to the coverage of the observational material over more than two cycles. The absolute visual magnitude M_v is 10.51 for A and 12.4 for B. The mass ratio determined for the large orbit Aa, B (period 320 yr) yields $\mathfrak{M}_{Aa} = 0.53\ \mathfrak{M}_\odot$, $\mathfrak{M}_B = 0.12\ \mathfrak{M}_\odot$. Without blending effect, that is, $\beta = 0$, the small orbit yields $\mathfrak{M}_a = 0.13\ \mathfrak{M}_\odot$; small blending effects up to $\Delta m = 3.5$ do not change the second decimal of this figure. The same value is found if we assume the value $\mathfrak{M}_A = 0.4\ \mathfrak{M}_\odot$ from known mass-luminosity data. The error in the mass value due to various sources is probably less than $0.02\ \mathfrak{M}_\odot$.

Failure to see the companion at predicted maximum separation $0''.5$ with the large USNO 155-cm reflector indicate a visual Δm of 3.5 or more, hence an $M_v > 14$ for the unseen component a.

The unseen component is most likely a late M dwarf, like the B component.

Mu Cassiopeiae

This is a G5 VI star, of visual magnitude 5.18. The perturbation was discovered by N. E. Wagman as a by-product of two parallax series at the Allegheny Observatory.

A recent study was made by S. L. Lippincott, based on Sproul plates taken over the interval 1937–1973. Four reference stars were used; the plates were measured on the two-coordinate Grant machine.

A period of about 22 yr and a semimajor axis of about $0''.2$ appear to better represent the more recent material than an earlier orbit (Lippincott & Wyckoff 1964). The interpretation remains the same as before. Since the companion has not yet been seen (maximum separation over $1''$) it is assumed that $\Delta m = 3$ or more, which for a normal mass of about $0.8\ \mathfrak{M}_\odot$ for the primary yields a mass of $0.2\ \mathfrak{M}_\odot$ for the secondary. The absolute visual magnitudes are 5.7 for the primary and 7.7 or fainter for the companion.

The unseen companion appears to be an M dwarf.

BD+6°398

This K3V star (PGC 588) has a visual magnitude 5.82 with a proper motion companion (11.65 dM4) at a distance of $165''$. The perturbation was established from Sproul plates in 1968; three more years of observations were included in the present study by Lippincott (1973), based on 434 plates taken on 117 nights over the interval 1937.86–1971.94. Three reference stars were used. The plate measurements were divided over the one-screw Ridell and the two-coordinate Grant machine; common measurements showed no systematic difference. Although a complete cycle has not yet been covered, the observed amplitude is sufficiently large (6 μm or $0''.11$ in x, 9 μm or $0''.17$ in y) so that a reliable provisional orbit may be determined. A wide range of periods, 40–60 yr, satisfies the observations; the following elements were adopted: $P = 50$ yr; $T = 1940.0$; $e = 0.6$; $\alpha = 0''.221 \pm 0''.008$; and $p = 0''.138 \pm 0''.004$. The absolute visual amplitude of BD+6°398 is 6.55, the expected mass 0.7 $\mathfrak{M}_\odot$. Though the predicted maximum separation (1974–1978) is about 2''.5, the companion has not been detected, suggesting a value of Δm of 5 or more, hence no blending effect. The corresponding value for the mass of the companion is 0.10 $\mathfrak{M}_\odot$, and its abolute magnitude is probably 11.5 or less. (The absolute visual magnitude of the distant proper motion components is 12.4; the estimated mass, from the spectrum, is 0.10 $\mathfrak{M}_\odot$.) A recent study from 105 plates taken with the Leander McCormick refractor over the interval 1915–1974 yields the following elements: $P = 60.0$ yr; $T = 1937$; $e = 0.45$; and $\alpha = 0''.257 \pm 0''.008$; and a corresponding value of 0.12 $\mathfrak{M}_\odot$ for the mass of the companion (Martin & Ianna 1975).

Epsilon Eridani

This is one of the nearest single stars, visual magnitude 3.73, spectrum K2V ($p = 0''.305$), which, together with Tau Ceti, visual magnitude 1.92, spectrum G8 ($p = 0''.273$), has been at times considered as a candidate for a sunlike star with possibly a life-supporting companion or companions. Three reference stars were used; the measurements were divided over the one-screw St. Clair-Kasten and the two-coordinate Grant machine; no systematic differences between the results from the two machines were found.

A recent study (van de Kamp 1973b) based on 900 Sproul plates taken on 238 nights over the interval 1938–1972 indicated a perturbation with a total amplitude of over 1 μm (0''.02) in right ascension and over 1.5 μm (0''.03) in declination. The deviations from a solution for parallax and proper motion may be represented by the following elements: $P = 25$ yr; $T = 1944$; $e = 0.5$; and $\alpha = 0''.0191 \pm 0''.0018$. The adopted absolute parallax is 0''.305. The predicted separation is over 2'' during more than half the period, but no companion has been detected. We may assume Δm to be three or more, that is, there is no significant blending effect. The absolute visual magnitude of Epsilon Eridani is 6.15, the corresponding value of the mass is 0.74 $\mathfrak{M}_\odot$. Values for the corresponding mass of the companion range from 0.05 $\mathfrak{M}_\odot$ for $\Delta m = 3$ to 0.006 $\mathfrak{M}_\odot$ for $\Delta m = \infty$.

The inferred value of the small mass of the companion appears to be very vulnerable to the adopted value of Δm. Or, we may express it the other way around: a range of less than 10 in the calculated values of the mass corresponds to an infinite range in the corresponding luminosities.

Ross 614

This visual binary, first resolved visually (and photographically) in 1955 by Walter Baade with the 5-m Hale reflector, was first discovered as an unresolved astrometric binary from the observed perturbation in a series of 25 parallax plates at the Leander McCormick Observatory (Reuyl 1936), extending over nine years. The latest "definitive" photocentric orbit is derived from 868 Sproul plates taken on 252 nights covering the interval 1938–1972 (Lippincott & Hershey 1972). Four reference stars were used; the measurements were divided over the St. Clair-Kasten, the Grant, and the USNO machines. The results were: $P = 16.60$ yr; $T = 1949.52$;

$e = 0.38$; $\alpha = 0''.312 \pm 0''.002$; and $p = 0''.248 \pm 0''.003$. Strictly speaking this object does not belong in the present listing; nevertheless it has been listed for historical, sentimental, and pedagogical reasons.

Observations of the resolved pair, made by W. Baade, G. Van Biesbroeck, and C. E. Worley, yield a ratio of 0.345 for the scale of the photocentric compared to the relative orbit of the two components. Making allowances for blending effect resulting from the estimated visual magnitude difference $\Delta m = 3.5$, the values for the masses are found to be $\mathfrak{M}_A = 0.114\ \mathfrak{M}_\odot$, $\mathfrak{M}_B = 0.062\ \mathfrak{M}_\odot$ and the absolute visual magnitudes are $M_A = 13.0$ and $M_B = 16.5$. Ross 614B represents the lowest value for the well-determined mass of a visible star. Similar values have been found for the components of the close, semiresolved, binary Wolf 424 (see section 3).

Gamma Geminorum

Gamma Geminorum is an A0 IV star of visual magnitude 1.93. Recognized as a long-period spectroscopic binary, its period, first found from astrometric measurements, later was more precisely determined from spectroscopic data.

A study was made by Kamper (1971) from 130 plates taken over the interval 1932–1958 with the 50-cm refractor of the Van Vleck Observatory. Nine reference stars were used; the plates were measured on the USNO automatic machine. The results were: $P = 12.6$ yr; $T = 1941.30$; $e = 0.75$; $\alpha = 0''.065 \pm 0''.011$; and $p = 0''.031 \pm 0''.004$. Because there is no spectroscopic evidence of the companion, Δm is assumed to be >1.0, possibly >1.5. Assuming $\mathfrak{M}_A = 3.0\ \mathfrak{M}_\odot$, it is likely that $\mathfrak{M}_B$ is close to one solar mass. The absolute visual magnitudes of primary and companion are -0.6 and $+0.4$ or fainter.

Zeta Cancri C

This binary is the distant companion of the well-known binary Zeta Cancri AB. In 1880 Seeliger discovered a perturbation in the system *AB-C*, which has been confirmed since and proves to belong to the component *C* (Seeliger 1914). The very fact that micrometer observations established a perturbation for a star as far away as 20 pc shows that the unseen companion *D* must have an appreciable mass, and most likely is a white dwarf (van de Kamp 1947b).

The visual magnitudes of *A*, *B*, and *C* are 6.2, 6.5, and 6.7; the combined spectrum of *AB* equals F7; the spectrum of *C* equals G2. The most recent comprehensive study is by Gasteyer (1954), who used material from several observatories, multiple-exposure photographic observations from the Potsdam, Union, Dearborn, Sproul, and Yerkes observatories, and visual observations over the interval 1831–1951. A photographic series taken at Yerkes Observatory was used to determine parallax and mass ratio. Three reference stars were used, and the measurements were made with the long-screw Gaertner machine of the Yerkes Observatory. The following results were found: for orbit *A*, *B*, $P = 59.7 \pm 0.1$ yr; $T = 1930.0 \pm 0.3$; $e = 0.32 \pm 0.2$; and $a = 0''.884 \pm 0''.009$. For orbit of center of mass of *C* and *D*, around center of mass of *A* and *B*, $P = 1150 \pm 34$ yr; $T = 1960 \pm 135$; $e = 0.26 \pm 0.03$; and $a = 7''.96 \pm 0.01$. The results for the photocentric orbit of the perturbation in *C* were $P = 17.5 \pm 0.2$ (me); $T = 1944.4 \pm 0.7$; $e = 0.11 \pm 0.03$; and $\alpha = 0''.191 \pm 0''.001$, a very well-established perturbation indeed.

The absolute parallax of the Zeta Cancri system is $0''.047 \pm 0''.005$, the corresponding absolute visual magnitudes for A, B, and C are 4.6, 4.9, and 5.1. Assuming no light for D, which has never been seen, though the separation of C to D is about $0''.4$, the following mass values are found: $\mathfrak{M}_A = 0.99\,\mathfrak{M}_\odot$; $\mathfrak{M}_B = 0.88\,\mathfrak{M}_\odot$; $\mathfrak{M}_C = 0.90\,\mathfrak{M}_\odot$; and $\mathfrak{M}_D = 0.90\,\mathfrak{M}_\odot$. We conclude that the fourth unseen component in this interesting system is a white dwarf.

BD+67°552

This dM1 star (Ci 20 475) has a visual magnitude 9.29. A study by Lippincott (1973) was based on 328 plates, taken with the Sproul refractor on 96 nights over the interval 1939–1972. Five reference stars were used, and measurements were divided over the one-screw Ridell and the two-coordinate Grant machine; measurements in common showed only a minute systematic difference, well below 1 μm. The results were: $P = 23.0$ yr; $T = 1949.0$; $e = 0.73$; $\alpha = 0''.158 \pm 0''.004$; and $p = 0''.070 \pm 0''.004$. If we make different assumptions for the sum of the masses and for Δm, the most likely results for the masses of the components are found to be $\mathfrak{M}_A = 0.6\,\mathfrak{M}_\odot$, in agreement with the mass-luminosity relation and a range from 0.36 to 0.27 $\mathfrak{M}_\odot$ for the unseen companion, corresponding to a Δm of 2–3. Maximum separation of $0''.6$ around 1978 should be utilized for possible detection of the companion with a large telescope.

The values for the absolute visual magnitudes are 8.5 for A and >10.5 for the companion.

Xi Ursae Majoris A

This is the brighter component of the well-known binary. Its visual magnitudes are 4.32, 4.80, its spectra are G0V.G0V, with a period of 59.84 yr and semimajor axis of $2''.53$. From visual observations Nörlund (1905) found a perturbation which had been noted spectroscopically by Wright in the brighter component A, and was confirmed astrometrically through multiple-exposure photographs by Hertzsprung. The fainter component B is a spectroscopic binary, thus making this interesting object a quadruple system.

The latest comprehensive study by Heintz (1966) yielded the following elements for the perturbation of ξ UMa A caused by the unseen companion a: $P = 1.832$ yr; $T = 1935.410$; $e = 0.56$; $\alpha = 0''.055$; and $p = 0''.130$; from which is found:

absolute visual magnitudes $M_A = +4.9$, $M_a = +10$ (est.)

masses $\mathfrak{M}_A = 0.83\,\mathfrak{M}_\odot$, $\mathfrak{M}_a = 0.31\,\mathfrak{M}_\odot$.

Heintz assumed a negligible luminosity for the unseen companion a, which most likely will turn out to be a main-sequence M dwarf.

Alpha Ophiuchi

This is an A5III star of visual magnitude 2.1. The perturbation was discovered by Wagman from two parallax series of plates taken with the Allegheny 76-cm refractor; a provisional period of nine years was found.

A later study was based on Allegheny plates taken on 183 nights over the interval 1918–1963 and 152 Sproul plates taken on 43 nights over the interval 1946–1964 (Lippincott & Wagman 1966). Three reference stars were used for the measurements of the Sproul plates which were divided over the one-screw Gaertner and St. Clair-Kasten machines. The results of combining the Allegheny and Sproul data were: $P = 8.5$ yr; $T = 1944.2$; $e = 0.4$; $\alpha = 0''.065$; and $p = 0''.060 \pm 0''.003$.

The absolute visual magnitude of A is $+1.0$, corresponding to a likely mass of 3.0 $\mathfrak{M}_\odot$. If we assume $\Delta m = 2$, the corresponding value of the mass of the unseen companion would be 1.3 $\mathfrak{M}_\odot$; $\Delta m = 4$ would yield 0.6 $\mathfrak{M}_\odot$. The corresponding absolute visual magnitudes would be $+3$ for $\Delta m = 2$, $+5$ for $\Delta m = 4$.

For a range of 2–6 $\mathfrak{M}_\odot$ of the total mass of the system, the maximum separation of the stellar companion ranges from $0''.4$ to $0''.6$, to be reached in 1974. If the companion could be seen and the separation measured, a first direct determination of the mass of an AIII-type star would result. The importance of visual scrutiny of this star for detection of duplicity is obvious.

BD + 68°946

Cin 18,2354 is an M3.5V star of visual magnitude 9.15. A perturbation in declination was found from plates taken with the Sproul refractor over the interval 1938–1948. The Lippincott (1967) study was based on 702 plates taken on 183 nights over the interval 1938–1966. Three reference stars were used; the measurements were made on the St. Clair-Kasten machine. The results were: $P = 24.5$ yr; $T = 1961.9$; $e = 0.90$; $\alpha = 0''.102$; and $p = 0''.209$. High eccentricity was indicated; values of 0.7–0.97 were considered. The scale α of the photocentric orbit was very sensitive to the choice of eccentricity, ranging from $0''.06$ for $e = 0.7$ to $0''.13$ for $e = 0.95$. For the Lippincott study $e = 0.90$ was adopted; a range of values of Δm was considered together with values for the combined mass. Again, for the present study $\Delta m \geqq 5$ was assumed; smaller values of Δm yielded abnormally high values for the mass of the companion, judging by current knowledge of the fainter portion of the mass-luminosity relation.

Maximum separation of about 1″ occurred in 1967. As long as the companion has not been seen there is some support for a high value of Δm. The possibility of more than one unseen companion must not be excluded, and may be a likely explanation of the possibly spurious high eccentricity.

On the one-companion hypothesis we adopt the following:

absolute visual magnitudes $M_A = 10.7$, $M_B > 15.7$

masses $\mathfrak{M}_A = 0.274\mathfrak{M}_\odot$, $\mathfrak{M}_B = 0.026\mathfrak{M}_\odot$.

Barnard's Star

The nearest star (except for the Sun) in the northern equatorial hemisphere is Barnard's star. Its spectrum is M5V; its visual magnitude is 9.54.

The first signs of a perturbation with a period exceeding two decades were found from Sproul material in 1956 and reported in 1963. Subsequent studies gave a choice of one perturbation with a highly elliptical orbit or two perturbations with circular orbits. In either case the unseen companions had masses of the order of Jupiter. The latest current analysis was based on some four thousand plates taken over one thousand nights over the interval 1938–1974, plus 20 plates taken over the interval 1916–1919. Three reference stars were used; all plates were measured on the two-coordinate Grant machine. After allowing for parallax, proper motion, and a quadratic time effect, and taking into account systematic errors, the remainders can be represented by a perturbation resulting from two unseen companions (van de Kamp 1974). The reality of a companion with a period of 11.5 yr and total amplitude of 0″.02 appears to be well established and is confirmed by observations and measurements made elsewhere; the companion has a mass close to that of Jupiter. The orbital characteristics of a second companion are less certain at this time: the period appears to be between 20 and 25 yr with a total amplitude of only 0″.01, corresponding to a mass of nearly half that of Jupiter. Observations are continuing at Sproul, and I trust at other observatories, to test and improve the results for this interesting object.

Comments on this and other perturbations are given by Martin (1974).

Chi Draconis

Chi Draconis is an F7V star of visual magnitude 3.57; it has the shortest known period of any unresolved astrometric binary. A known spectroscopic binary, its apparent orbit was first studied by H. L. Alden in 1936. A spectroscopic study was made by Vinter Hansen in 1942. A recent study (Breakiron & Gatewood 1974) is based on 86 plates taken with the Allegheny refractor over the interval 1924–1972. Eleven reference stars were used, and the plates were measured on the USNO measuring machine. Adopted values from the spectroscopic analysis by Vinter Hansen are $P = 0.77$ yr, $T = 1920.31$, and $e = 0.4515$. The resulting value for the semimajor axis α_{pg} of the photocentric orbit is $0''.0485 \pm 0''.0047$ (me).

The absolute parallax is $0''.129 \pm 0''.005$ (pe). Making use of the spectroscopic mass-ratio determination by M. Spite, the following values are found:

absolute visual magnitudes $M_A = +4.2, M_B = +6.5$

masses $\mathfrak{M}_A = 1.5\,\mathfrak{M}_\odot, \mathfrak{M}_B = 1.1\,\mathfrak{M}_\odot$.

Delta Aquilae

Delta Aquilae is an F0IV star of visual magnitude 3.36. Its perturbation was discovered by H. L. Alden in 1929. The Osvalds (1958) study was based on 131 plates taken at the Yale southern station over the interval 1936–1943, and on 170 plates taken with the Leander McCormick refractor over the interval 1914–1954. The results were: $P = 3.42$ yr; $T = 1934.16$; $e = 0.40$; $\alpha = 0''.058$; and $p = 0''.069 \pm 0''.003$. The companion, at the close separation of about 0″.2, has not yet been seen and is assumed to be several, say at least three, magnitudes fainter than the primary. A mass of $1.2\,\mathfrak{M}_\odot$ may be assumed for the primary, with a corresponding mass of $0.5\,\mathfrak{M}_\odot$ for the companion. The absolute visual magnitude of the primary is 2.6, that of the secondary >5.6.

GV 24-16

This is a main-sequence red dwarf with visual magnitude 13.05, $B-V=+1.64$, $U-B=+1.30$. An anlysis by R. K. Riddle in 1970 for parallax based on photographs taken with the 155-cm astrometric reflector of the USNO at Flagstaff, Arizona, over the interval 1964.4–1969.8 revealed a perturbation. The present analysis (Harrington 1971) was based on 88 plates taken over the interval 1965.7–1970.7. The results were: $P=1.5$ yr; $T=1967.7$; $e=0.5$; $\alpha=0''.029$; and $p=0''.113\pm0''.002$, corresponding to an absolute visual magnitude of 13.3.

The predicted separation is about $0''.1$; Δm_v is estimated to be >3. Assuming the companion to be truly unseen, the following ranges in values for the masses are found: $\mathfrak{M}_A=0.14$–$0.20\ \mathfrak{M}_\odot$; and $\mathfrak{M}_B=0.07$–$0.11\ \mathfrak{M}_\odot$. The companion appears to be a subluminous dwarf (possibly degenerate).

The striking message from the star is that a short-period perturbation, in this case only 1.5 yr, can be discovered in a few years, provided adequate observational coverage and precision exist.

VW Cephei

VW Cephei is a W UMa-type eclipsing variable. Its visual magnitude is 7.1; its spectrum is G8-K1, with period changes over several decades. The possibility of a light-time effect due to an unseen companion was raised as early as 1941. An earlier study by Villamediana based on Sproul material revealed a perturbation in fair agreement with an orbital interpretation of the light-time data, but indicated that the third star should be visible. An astrometric analysis (Hershey 1974, 1975) based on 610 plates taken on 164 nights with the Sproul refractor over the interval 1942–1973 confirmed a large amplitude ($\sim0''.2$) in both right ascension and declination, which leads to the following elements: $P=30.4$ yr; $T=1966.5$; $e=0.60$; $\alpha=0''.130\pm0''.002$; and $p=0''.041\pm0''.002$. The companion *C*, probably a late K dwarf, has subsequently been seen, close to the predicted location, at an observed separation of $0''.64$ and with an estimated Δm of 2.9 (Heintz 1975). The masses are found to be $1.47\pm0.5\ \mathfrak{M}_\odot$ for *AB* and $0.56\pm0.2\ \mathfrak{M}_\odot$ for *C*. The absolute visual magnitudes are 5.2 and 8.1 for *AB* and *C* respectively.

Ironically, the eclipse light-time residuals, which prompted long-term astrometric surveillance, have turned out to be unrelated to a third object, being caused by a combination of sudden and gradual changes in the period. The astrometric efforts nevertheless resulted in the discovery of an unseen companion at a distance of over 20 pc! How many more similar situations may exist?

BD+27°4120

BD+27°4120 is a spectral-type M0e with a visual magnitude of 9.8. The perturbation was discovered and analyzed using 172 Sproul plates taken on 47 nights over the interval 1939–1963 (Bieger 1964). Three reference stars were used, and all plates were measured on the one-screw Ridell machine. Various periods were tried ranging from 24 to 60 yr. Because the star has not been resolved visually, the longer periods were excluded. A period of 40 yr was adopted for Bieger's study. Though the period was tentative, the amplitude of the perturbation over the interval covered by the observations was about 0".2 in each coordinate; the amplitude was therefore substantial and the provisional analysis was worthwhile.

The resulting data were: $P = 40$ yr; $T = 1960.0$; $e = 0.43$; $\alpha = 0''.181$; $p = 0''.069 \pm 0''.005$. Again, since the companion has not been seen (estimated separation about 0".7) it is assumed that $\Delta m \geqq 4.0$, with the following results:

absolute visual magnitudes $M_A = 9.0$, $M_B \geqq 13.0$

masses $\mathfrak{M}_A = 0.46\, \mathfrak{M}_\odot$, $\mathfrak{M}_B = 0.17\, \mathfrak{M}_\odot$.

Zeta Aquarii B

Zeta Aquarii is a well-known double star whose visual magnitudes are 4.42 and 4.59; its spectra are dF2, dF1s, with a long period of 600 yr. It has an unseen companion *C* discovered in 1942 by K. Strand, later found to be attached to the fainter (*B*) of the two visual components.

The most recent analysis (Franz 1958) made use of visual observations since 1780 up to and including 1955, and from 55 multiple-exposure plates taken at the Potsdam, Johannesburg, Lick, Sproul, Yerkes, Bosscha, and Dearborn observatories, covering the interval 1914–1956. In addition 17 parallax plates taken with the Yerkes refractor were measured with the long-screw Gaertner machine at Dearborn Observatory; three reference stars were used. The perturbation was found to belong to the fainter component. The results of this investigation were: for visual orbit *A*, *B*, $P = 600$ yr; $T = 1972.75$; $e = 0.45$; and $a = 4''.013$; for photocentric orbit *B*, *C*, $P = 25.5$ yr; $T = 1954.5$; $e = 0.20$; and $\alpha = 0''.097$. An absolute parallax of 0".043 was adopted. The separation *B*, *C* is about 0".4; *C* has never been seen, however, and is assumed to be several, perhaps four magnitudes fainter than *B*. The following interpretation was found:

absolute visual magnitudes $M_A = 2.6$, $M_B = 2.8$, $M_C > 6.8$

masses $\mathfrak{M}_A = 1.13\, \mathfrak{M}_\odot$, $\mathfrak{M}_B = 0.85\, \mathfrak{M}_\odot$, $\mathfrak{M}_C = 0.28\, \mathfrak{M}_\odot$.

An improved parallax value for this interesting object is very much desired.

BD+43°4305

This star (EV Lacertae), whose visual magnitude is 10.2 and spectrum is M 4.5e, showed sign of a perturbation in an analysis of Sproul plates covering the interval 1937–1963. The latest analysis (van de Kamp & Worth 1972) is based on 717 plates taken on 210 nights, covering the interval 1937–1971.

The results were: $P = 28.9$ yr; $T = 1970.1$; $e = 0.44$; $\alpha = 0''.039 \pm 0''.002$; and $p = 0''.200 \pm 0''.003$. The estimated separation is well over 1" over the interval 1950–1955, but no elongation of the images is evident; hence we estimate that $\Delta m \geqq 3$. When a mass of 0.25 $\mathfrak{M}_\odot$ was adopted for the primary, the following results were found for the unseen companion:

$\Delta m = 3$	4	∞
$M = 14.7$	15.7	∞
$\mathfrak{M}/\mathfrak{M}_\odot = 0.023$	0.015	0.009

The absolute visual magnitude of the primary is 11.7.

8 GENERAL SURVEY

We now make a survey of the best-established perturbations, as listed in Table 1; the objects listed in Table 3 are also included.

Among the stars nearer than 5.2 pc, 32 stars appear single, 11 double, and 2 triple; these figures refer to visible stars and do not include unseen companions. Of the 20 established unseen companions, 14 are attached to single primaries, whereas as many as 6 are found in known binary systems (van de Kamp 1971b).

Density Function

The density function, that is, the space density or frequency at different distances, is presented in the following summary, which includes the companions of Ross 614 and VW Cephei, once unseen but no longer so. The number of unseen companions within spheres up to different distances, is given in one but the last column of Table 4.

The apparent decrease in space density with increasing distance is very striking. The overwhelming preponderance of unseen companions found for the very nearest stars is striking though not unexpected. If we assume that the space density beyond 5.8 pc is the same as that within the inner core, values are predicted, as given in the last column of Table 4 and illustrated in Figure 4. These "relative" estimated values do not take into account the obvious incompleteness due to uneven observational coverage of the sky. But they do suggest a high degree of incompleteness with increasing distance. There appears to be a most promising field to discover large numbers of unseen companions.

Table 4 Number of unseen companions

			Number of Unseen Companions	
Parallax	Distance (pc)	Relative Volume	Observed	"Estimated"
$\geqq 0''.171$	<5.8	1	6	6
$\geqq 0''.100$	<10	5	12	30
$\geqq 0''.063$	<16	20	15	120
$\geqq 0''.040$	<25	78	19	468
$<0''.040$	>25	—	20	—

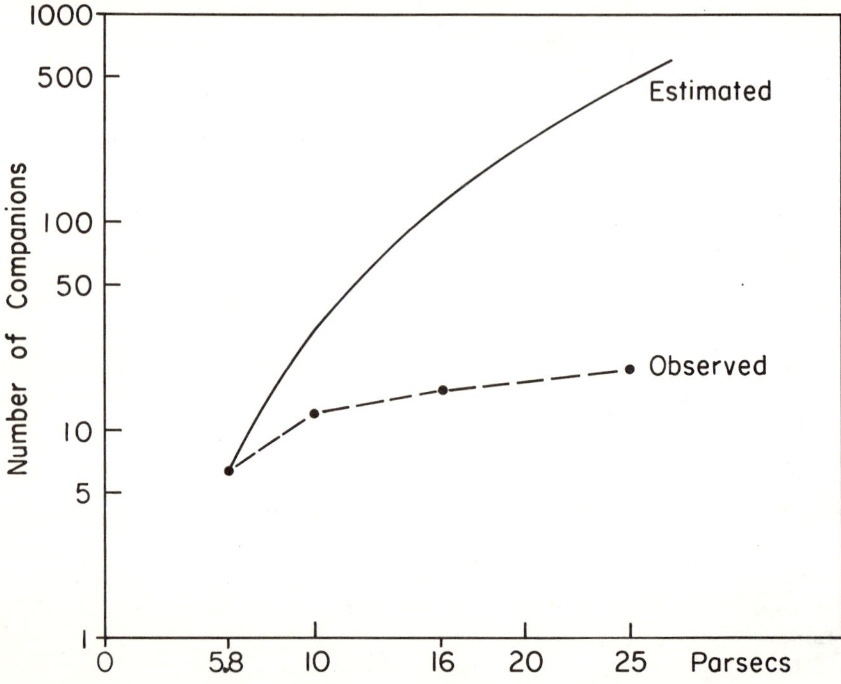

Figure 4 Number of unseen companions up to different distances. Because the density up to 5.8 pc must be a minimum value, the extrapolated values beyond 5.8 pc likewise are minimum values, assuming the density function to remain constant with increasing distance.

Luminosity Function

Although it is not possible for the unseen companions in Table 1 to derive a luminosity function, that is, frequency of different luminosities, limiting information may be derived, that is, frequency of upper limits for absolute visual magnitude.

Table 5 gives the frequency distribution of the upper limits of successive ranges of absolute visual magnitudes of the unseen companions in Table 1 (and Table 3); for comparison the luminosity function of absolute visual magnitudes is given for the (seen) stars within 5.2 pc. A comparison of the two sets of values is not particularly significant, except for the not unexpected increase of frequency with increasing absolute magnitude.

Mass-Luminosity Relation

Similarly we may make an attempt to study the relation between *masses* and *luminosities* of the unseen companions listed in Tables 1 and 3. The masses are rather well determined, within narrow limits. The upper limits of the luminosities are fairly well known; therefore, we plot a mass-luminosity diagram, indicating the limitations (Figure 5). With few exceptions all of the unseen companions discovered

Table 5 Luminosity function

Absolute Visual Magnitude (Upper Limit)	Number of Unseen Companions	Number of Visible Stars within 5.2 pc
0 to +2.4	1	2
2.5 to +4.9	1	3
5.0 to +7.4	3	6
7.5 to +9.9	3	6
10.0 to +12.4	3	19
12.5 to +14.9	3	18
15.0 to +17.4	3	6
Uncertain (ζ Cnc D)	1	—
Planetary	2	—

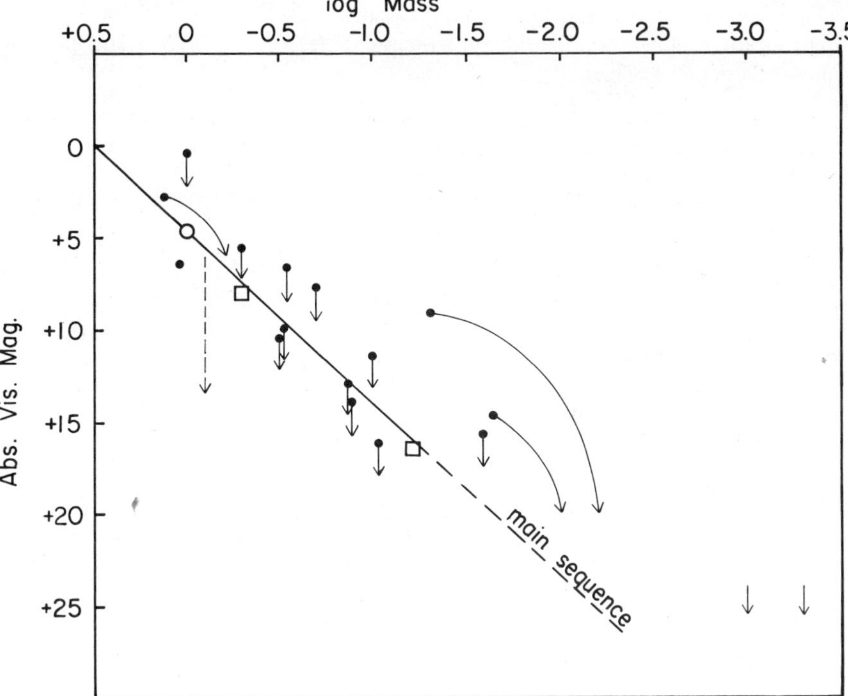

Figure 5 Mass-luminosity relation for unseen companions. Except for the two companions of Barnard's star (*lower right*), all unseen companions are represented by upper limits of luminosities. The lengths of the vertical arrows are not significant. For Alpha Ophiuchi, Epsilon Eridani, and BD+43°4305, upper limits of both mass and luminosity are given. Zeta Cancri appears on the left just to the right of the Sun (*open circle*), which has been plotted as a useful reference point. The two squares represent the former unseen companions of VW Cephei and Ross 614.

astrometrically appear to be bona fide stars, that is, they are objects with masses greater than 0.06 times the Sun's mass and may well be on or close to the main sequence (Harris, Strand & Worley 1963). One striking exception is Zeta Cancri *D*, the unseen companion of Zeta Cancri *C*. Other exceptions are the companions of Epsilon Eridani, of BD+68°946, of BD+43°4305, and of course the planetlike companions of Barnard's star. It remains uncertain and a debatable question whether these objects may be situated on what might be an extension of the main sequence toward very low luminosities and very low masses.

9 CONCLUSIONS

The present survey is based on a minute and selective sample of the total population of unseen companions. We recall the vast number of spectroscopically discovered unseen companions—the spectroscopic binaries with only one spectrum visible—which led W. W. Campbell and H. D. Curtis, as long ago as 1905, to be impressed by and draw attention to the abundance of unseen companions. We quote (Campbell & Curtis 1905).

> In fact, the star which seems not to be attended by dark companions may be the rare exception. There is the further possibility that the stars attended by massive companions, rather than by small planets, are in a decided majority; suggesting, at least, that our solar system may prove to be an extreme type of system, rather than a common or average type....

The astrometrically discovered unseen companions hold for an entirely different portion of the stellar population, namely comparatively nearby objects, that is, primarily main-sequence primaries, whose comparatively small masses render these objects ideal to reveal measurable perturbations.

As in so many branches of research, some of the finest and most interesting results have been found not as part of a planned investigation, but more or less accidentally, as a *by-product* of other research. Outstanding examples are the initial discovery of the perturbation of Ross 614, due to procrastination from 2 to 10 yr in obtaining the material for a first parallax determination (van de Kamp 1961); the perturbations in the visual binaries Zeta Cancri and Zeta Aquarii; the perturbation in BD+66°34, found by Alden in a premature attempt to determine the mass-ratio of this visual binary; and the recent study by Hershey of the perturbation in VW Cephei, an object, as it turned out, put on the observing program for the wrong reason.

It behoves us to be on the alert for unplanned discoveries. Nevertheless planned studies should be continued, and started, at several observatories. While the systematic studies of the Sproul Observatory are primarily concerned with stars within 10 pc and, to a lesser extent, with a limited number of stars beyond, it is of interest to note the several perturbations already found for stars beyond 10 pc. The question arises whether one should start systematic programs for more distant stars, up to 25 pc or more. The chances for finding perturbations of short periods, say, less than 10 yr, diminish with increasing distance. Perturbations with longer periods, say, 25 yr or more (or even much more) will be and, for more distant stars, can only be discovered by series of observations spread over decades, if not centuries.

It is clear that discoveries of companions of very small mass—say, less than the generally adopted value of 0.06 $\mathfrak{M}_{\odot}$ for the lowest mass of a normal star—and of planetary companions, are most promising for the very nearest stars. There is reason therefore to reach for the highest possible positional accuracy attainable for a limited number of the very nearest stars. Barnard's star is an example; would that an intensive and extensive series of material be attained on Alpha Centauri *C*!

The discovery of unseen objects of very low mass remains a ticklish problem. In our present study, we find only a few such objects, namely, the unseen companions of Epsilon Eridani, of BD+43°4305 and of BD+68°946, besides the planetary companions of Barnard's star. Note that all these objects are within 5 pc.

Accurate positions of stars have been determined since the middle of the eighteenth century. The principal outcome of these continued observations, carried out by meridian instruments whose precision has been gradually increasing, is knowledge of the positions and proper motions of tens of thousands of stars. Meridian observations were sufficiently accurate to reveal perturbations of large amplitude caused by unseen companions of Sirius and Procyon as early as the first half of the nineteenth century. The end of this work is not in sight; an essential aspect is the continued study of systematic errors so that the accuracy of the positions and proper motions of stars may gradually be improved.

The history of meridian work provides useful perspective on the long-focus photographic work begun in the beginning of this century. The planned search for perturbations in the proper motions of a limited number of nearby stars was not begun until less than four decades ago. This time interval is small compared with the more than two centuries covered in the aforementioned field of absolute positions and proper motions. What conclusions may we draw? The potential accuracy of the photographic work is much higher than that reached by meridian observations and therefore should reveal the presence of perturbations of comparatively short periods and correspondingly small amplitudes. The perturbations discovered thus far often have periods of the order of the time interval spanned by the observations. This is a natural selection, an example of the observer creating the universe in his own image, in this case limitations in accuracy and above all in time. As times goes on, perturbations with long periods, and correspondingly larger amplitudes will gradually be discovered. We are dealing with phenomena which often cover long intervals of time. The situation does not differ from that of resolved binaries where several decades, even centuries, may be required before even a provisional orbit may be determined. Moreover, for the intriguing case of companions of planetlike or planetary nature, the perturbations are so small as to hover around the threshold value of attainable accuracy.

Several perturbations discovered thus far are not well established, because of limitation in accuracy and the specter of systematic errors. Doubt exists about the reality of a number of results. This is no reason for discouragement, however.

History repeats itself. Conflict and controversy arising for perturbations of small amplitude were nicely discussed at the end of the nineteenth century in an article with the quaint title "Invisible Double Stars" by Burnham (1891). The classical controversy was that of the perturbation of Zeta Cancri *C*, discovered by Seeliger but questioned by Burnham. By now there is no question of the reality of the perturbation, the unseen companion of large mass causing an effect of appreciable angular dimensions for this object with comparatively small parallax (section 7).

Many nearby stars observed intensively for over three decades show no indication of a perturbation. This may mean that there is no measurable perturbation or that period and amplitude of the perturbation are so small as to lie below the threshold of detection. However, it may also mean that the perturbation has such a long period that the orbital effect over the interval is essentially linear and therefore cannot be detected. At the Sproul Observatory it is our policy in cases of this kind to stop observing the star every year, but make observations at intervals of 2, 5, or even 10 years, depending on a number of factors. We must face the fact that there are celestial phenomena taking place on a large time scale. Continued observation year after year, decade after decade, remains an obvious requirement, if we are to do justice to this field of study.

Long-focus photographic programs have just begun; it would be shortsighted to think otherwise. There are short-term astrometric problems; a precise parallax can always be determined in a few years. But when it comes to orbital motions, either of resolved or unresolved astrometric binaries, no time limitations can be set. It is imperative, therefore, that the search for unseen companions by the techniques and methods of long-focus photographic astrometry be continued indefinitely.

ACKNOWLEDGMENTS

I am grateful to the National Science Foundation without whose support only a fraction of the studies referred to in this summary could have been done. My thanks go to the Netherlands Foundation for Zuiver Wetenschappelyk Onderzoek (ZWO), to the Fulbright-Hays Commission, and to the Astronomical Institute of the University of Amsterdam for providing me with the opportunity to clarify my thoughts and prepare this review article during my temporary leave of absence from the Sproul Observatory. Acknowledgments are due to my colleagues at the Sproul Observatory and at the Astronomical Institute for their assistance and cooperation.

I should like to quote some of the requirements that have played and continue to play a role in the present study. From the last sentence of my first Swarthmore article, "The Invisible Universe," (van de Kamp 1938): "persistence, curiosity and faith"; from the coat of arms of the city of Amsterdam: "heldhaftig, vastberaden, barmhartig," freely translated as "courage, resolution, charity." In all these ideals and aims I was continually supported by my wife Olga to whose memory I dedicate this article.

Literature Cited

Alden, H. L. 1951. *Astron. J.* 56:34
Appelbaum, L. T. 1972. *Astron. J.* 77:518
Bieger, G. S. 1964. *Astron. J.* 69:812
Breakiron, L. A., Gatewood, G. 1974. *Publ. Astron. Soc. Pac.* 86:448
Burnham, S. W. 1891. *Mon. Notic.* 51:388

Campbell, W. W., Curtis, H. D. 1905. *Lick. Obs. Bull.* 3:145

Feierman, B. H. 1971. *Astron. J.* 76:89

Franz, O. G. 1958. *Astron. J.* 63:329

Franz, O. G., Mintz, B. F. 1964. *Astron. Pap. Am. Ephemeris Nautical Alm.*

Gasteyer, C. 1954. *Astron. J.* 59:243

Gatewood, G. 1974a. *Astron. J.* 79:52

Gatewood, G. 1974b. *Astron. J.* 79:815

Grossman, A. S. 1970. *Ap. J.* 161:619

Hall, R. G. Jr. 1951. *Astron. J.* 55:215

Harris, D. L. III, Strand, K. Aa., Worley, C. E. 1963. *Stars and Stellar Systems,* 3:265, 285. Chicago, Ill.: Univ. Chicago Press

Harrington, R. S. 1971. *Astron. J.* 76:930

Heintz, W. D. 1966. *Veroeff. Sternwarte Muenchen* 7:35

Heintz, W. D. 1972. *Astron. J.* 77:160

Heintz, W. D. 1973. *Astron. J.* 78:780

Heintz, W. D. 1975. *IAU Inform. Circ. Visual Double Stars,* No. 64

Hershey, J. L. 1973a. *Astron. J.* 78:421

Hershey, J. L. 1973b. *Astron. J.* 78:935

Hershey, J. L. 1974. *Bull. Am. Astron. Soc.* 6(3): Pt. 1, 305

Hershey, J. L. 1975. *Astron. J.* 80:662

Kamper, K. W. 1971. *Astron. J.* 77:85

Kapteyn, J. C. 1922. *Bull. Astron. Inst. Neth.* No. 14:71

Lippincott, S. L. 1955a. *Astron. J.* 60:379

Lippincott, S. L. 1955b. *Sky Telesc.* 14:364

Lippincott, S. L. 1960. *Astron. J.* 65:445

Lippincott, S. L. 1967. *Astron. J.* 72:1349

Lippincott, S. L. 1969. *Astron. J.* 74:224

Lippincott, S. L. 1971. *Publ. McCormick Obs.* 16:59

Lippincott, S. L. 1973. *Astron. J.* 78:303

Lippincott, S. L. 1974. *Astron. J.* 79:974

Lippincott, S. L., Hershey, J. L. 1972. *Astron. J.* 77:679

Lippincott, S. L., Wagman, N. E. 1966. *Astron. J.* 71:122

Lippincott, S. L., Wyckoff, S. 1964. *Astron. J.* 69:471

Lippincott, S. L., Yang, C. Y. 1967. *Astron. J.* 72:840

Martin, A. R. 1974. *J. Brit. Interplanet. Soc.* 26:643, 881

Martin, G. E., Ianna, P. A. 1975. *Astron. J.* 80:321

Nörlund, N. E. 1905. *Astron. Nachr.* 170:9

Osvalds, V. 1958. *Astron. J.* 63:222

Osvalds, Z., Osvalds, V. 1959. *Astron. J.* 64:265

Reuyl, D. 1936. *Astron. J.* 45:133

Seeliger, H. 1914. *Astron. Nachr.* 199:273

Straka, W. C. 1971. *Ap. J.* 165:109

Strand, K. Aa. 1971. *Publ. U.S. Naval Obs.* 20: Pt. I, 9

van de Kamp, P. 1938. *Sigma Xi Quart.* 26:103

van de Kamp, P. 1943. *Publ. Astron. Soc. Pac.* 55:263

van de Kamp, P. 1944. *Astron. J.* 51:7

van de Kamp, P. 1947a. *Astron. J.* 52:185

van de Kamp, P. 1947b. *Astron. J.* 53:207

van de Kamp, P. 1956. *Vistas Astron.* 8:1040

van de Kamp, P. 1961. *Publ. Astron. Soc. Pac.* 73:389

van de Kamp, P. 1962. *Stars and Stellar Systems* 2:487. Chicago, Ill.: Univ. Chicago Press

van de Kamp, P. 1966. *Vistas Astron.* 8:215

van de Kamp, P. 1967. *Principles of Astrometry.* San Francisco: Freeman. 227 pp.

van de Kamp, P. 1971a. *IAU Symp.* No. 42:32

van de Kamp, P. 1971b. *Ann. Rev. Astron. Ap.* 9:103

van de Kamp, P. 1972. *Advan. Space Sci. Technol.* 11:437

van de Kamp, P. 1973a. *Astron. J.* 78:1099

van de Kamp, P. 1973b. *Astron. J.* 79:491

van de Kamp, P. 1974. *Bull. Am. Astron. Soc.* 6(3): Pt. 1, 306; 1975. *Astron. J.* 80:658

van de Kamp, P., Worth, M. D. 1972. *Astron. J.* 77:762

EQUATION OF STATE AT ULTRAHIGH DENSITIES

PART 2

V. Canuto

NORDITA, Copenhagen and Institute for Space Studies, Goddard Space Flight Center, NASA, New York, NY 10025

Ah, but a man's reach should exceed his grasp, or what's a heaven for?

Browning

1 INTRODUCTION

In Part 1 of this review (Canuto 1974) we analyzed the present knowledge about the behavior of matter at densities up to a few times nuclear density. We concluded by presenting in Tables 5 and 10 the best relations $P = P(\varepsilon)$ for a system of pure neutrons in the region $10^4 \leqq \rho \leqq 2 \times 10^{14}$ g cm^{-3}. In this second part we review the work done so far in the high density region $\rho \geqq 2 \times 10^{14}$ g cm^{-3}.

First, we review the work concerning the appearance of hyperons at densities higher than nuclear density and their influence on the equation of state. Second, we review the work concerning the possible existence of a solid neutron core. The hypothesis that a liquid of neutrons could solidify at high enough densities has recently been the subject of several microscopic detailed computations; the present state of the art is discussed. In Section 4 we discuss several (so far) unrelated attempts to describe the behavior of matter in the relativistic region, that is, at densities greater than 10^{16} g cm^{-3}. All the studies of this region published so far deal with a relativistic liquid of neutrons; should the neutrons indeed solidify at lower densities, the previous relativistic treatment ought to be modified.

One of the most disturbing features of the superhigh-density region is the disagreement in the predicted value for the velocity of sound c_s (in units of c) in the relation $P = c_s^2\varepsilon$. Some theories predict $c_s^2 \to 1$, whereas others predict $c_s^2 \to \frac{1}{3}$ (free particles) or even lower values.

The available data regarding neutron stars, in particular the moment of inertia, are shown in Section 5 to be insufficient to pin down a specific value of the velocity of sound at superhigh densities. A possible answer is found, however, by analyzing the multiparticle production in the high-energy proton-proton (p-p)

scattering. The Landau hydrodynamic model is reviewed and it is found that the value $c_s^2 \rightarrow 1$ seems favored, thus providing a way of continuing the equation of state into the superhigh-density region.

2 THE HYPERONIC LIQUID

2.1 *Generalities*

After the clusters of neutrons and protons that form the crust of a neutron star have dissolved at a density around 10^{14} g cm^{-3}, the system is left in a state composed predominantly of neutrons. If one neglects the possible existence of other particles, like hyperons, the treatment of a pure neutron fluid has already been presented in Section 5 of Part 1, with the corresponding equation of state given in Table 10 (1). If the treatment of the region $\rho > 2 \times 10^{14}$ g cm^{-3} as a fluid of pure neutrons is a good approximation, it is nevertheless true that physically the situation is much more complex, due to the energetically favorable appearance of hyperons. Before going into any detail, it is important to summarize the results by saying that all the computations of the hyperonic liquid published so far have reached the same conclusion, that is, the equation of state is not severely altered from the one corresponding to a pure neutron gas. This statement has to be taken with extreme care. The fact that almost all the computations reach the same conclusion does not make that conclusion necessarily right. In fact, all the computations share the same defect: the hyperonic potentials are taken to be almost identical to the nucleon-nucleon (*NN*) case. Should future studies of the hyperonic forces reveal unexpected features, the equation of state could be drastically changed. From the historical point of view the first to recognize that hyperons could be present were Cameron (1959) and Ambartsumyan & Saakyan (1960). Salpeter (1960) concluded on general grounds that Σ^- should appear first, a conclusion confirmed by all the detailed computations published since.

2.2 *Formulation of the Problem*

Consider a system composed of k different species, characterized by a concentration n_k. We want to find n_k subject to the condition that the total number of baryons is constant and that the total charge of the system is zero. The problem is solved by evaluating a simultaneous set of chemical potential equations obtained by minimizing the Gibbs free energy. For instance, for a system of e^-, n, p, Λ, Σ, these equations read

$$\begin{aligned} &\mu_n = \mu_{e^-} + \mu_p, \qquad \mu_{\Sigma^-} = \mu_n + \mu_{e^-}, \\ &\mu_\Lambda = \mu_n, \qquad \mu_{\Sigma^0} = \mu_n, \qquad \mu_{\Sigma^+} = \mu_n - \mu_{e^-}. \end{aligned} \tag{2.1}$$

We can easily find some interesting results even without knowing the exact form of the μ's. For example, at a density of 2.39×10^{14} g cm^{-3}, $\mu_e = 103.57$ MeV, $\mu_n = 20.37$ MeV$+m_n c^2$, and $\mu_p = -81.92$ MeV$+m_p c^2$ [Table 2 (Part 1)]. The first relation (2.1) is satisfied. If there is no interaction among the particles, the minimum value of μ is just the rest mass. At 2.39×10^{14} g cm^{-3}, Λ cannot exist as yet, because $\mu_\Lambda \approx m_\Lambda c^2 = 1115$ MeV; even Σ^- cannot be present either;

in fact, $\mu_n + \mu_e \approx 124$ MeV$+ m_n c^2$ is smaller than $m_\Sigma \sim 1190$ MeV; even worse is the situation for Σ^0 and Σ^+. Admittedly, these conclusions can be significantly altered if account is taken of the potentials acting among the particles.

2.3 *The Multicomponent Many-Body Theory*

If we want to treat the particles in a realistic model, their attraction and repulsion must be included in the computation. Two difficult problems immediately arise:

1. The two-body potential between hadrons is considered known as long as we deal with neutrons and protons. In fact, one has at hand many good potentials, among which one of the favorites nowadays is the one constructed by Reid
2. The situation is very different for any two hyperons or even for the hyperon-nucleon (Y-N) case. The experimental information is still insufficient to determine the hyperonic potentials with the same degree of reliability as the nucleon-nucleon case. No high-energy data exist on YY or YN scattering to allow one to fix the parameters of the potentials in the high angular momentum waves. No experimental information exists, for instance, about $\Sigma\Sigma$ or $\Lambda\Lambda$ for the P waves, which, however, are of primary importance at high densities

The simplest possible choice for the hyperonic potentials consists in assuming that they are the same as the NN case, if one excludes the one pion contribution.

Reasonable though it could sound, this procedure is by no means entirely satisfactory, because it is known (Brown, Downs & Iddings 1970) from ΛN cross-section and singlet (triplet) scattering lengths that much better results are obtained if the ΛN scattering is considered to go through an intermediate state ΣN: pions can then be exchanged. Besides, between two Σ's, new mesons can be exchanged that are not contained in Reid's potential, for example, the K meson with $T = \frac{1}{2}$. An explicit example of how misleading this procedure could be will be given in 2.4.4.

To improve upon this method, one has to employ a full meson-exchange hyperonic potential, which, in principle, can be derived by the exchange of mesons. Evidently the coupling constants entering into the final form of V_{12} cannot be determined with the same accuracy as in the NN case. This point is discussed in more detail below. For the time being, we proceed with the presentation of the many-body formalism as if the potentials were indeed known. When the particles interact with one another the general expression for μ is

$$\mu = (c^2 p^2 + m^2 c^4)^{1/2} + U(p), \tag{2.2}$$

where $U(p)$ is the one-body potential felt by the particle with Fermi momentum p. From Brueckner's many-body theory (Brueckner et al 1968), we know that the one-body potential felt by the particle i is given by

$$(2\pi)^2 U(p_i) = \int_0^{p_i} p_j^2 \, dp_j [K(p_i/p_j) - \text{exch}] + \sum_{l \neq i}^{n_k} \int_0^{p_l} p_j^2 \, dp_j \, K(p_i/p_j). \tag{2.3}$$

In the first integral we integrate over the particle j, which we take to be of the same kind as i. The second integral refers to the interaction of i with a second particle j, when j is of a different nature. Because the particles are then distinguishable, the Pauli principle does not apply and therefore there is no exchange. Each integral over p_j runs up to p_l, where p_l is the Fermi momentum of the species l. Evidently, the various Fermi momenta are the unknown of the problem, since they are related to the concentrations.

The K-matrix is proportional to the matrix element of the two-body potential, taken between the perturbed and the unperturbed wave functions. The perturbed wave function is in turn the solution of an integral equation (Schrödinger equation with the appropriate boundary conditions), whose Green function is again given

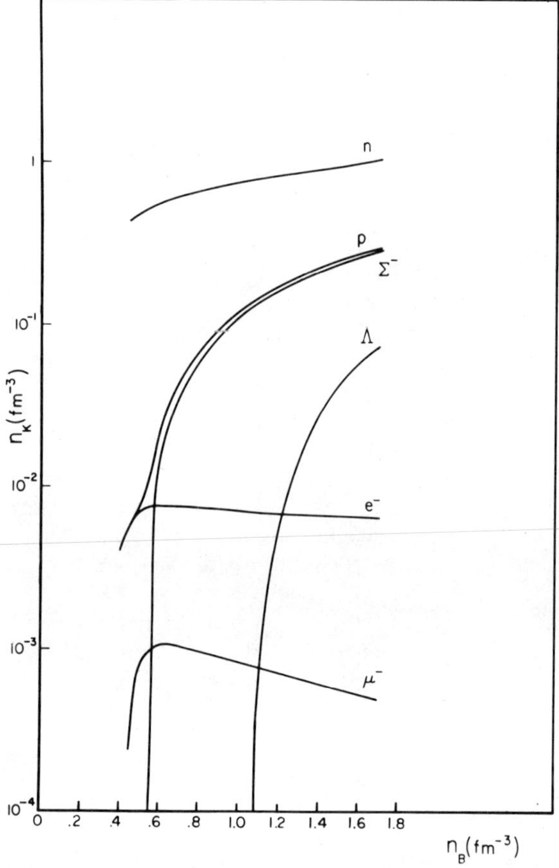

Figure 1 Concentrations of a free hyperonic liquid vs baryonic density, as from the work of Ambartsumyan & Saakyan (1960).

in terms of the one-body potential $U(p)$. This self-consistency is the most important feature of the Brueckner many-body theory.

If it is difficult to handle such a feature in the nuclear matter case where there are only neutrons and protons, it is clear that it becomes a tremendously involved problem when we want to deal with say 8 or 10 different species. Unfortunately, none of the results for the hyperonic liquid published so far has treated this essential feature in a satisfactory way.

2.4 *Review of the Results*

2.4.1 THE WORK OF AMBARTSUMYAN & SAAKYAN (1960) When the particles are noninteracting, the system 2.1 can be solved almost analytically. Ambartsumyan & Saakyan (1960) solved such a system by considering n_n as the independent variable, instead of the customary baryonic number density, n_B. We have repeated the computation by solving the system of equation 2.1 as a function of n_B, in order to make the future presentation of other results easier. We have considered eight variables: e^-, μ^-, n, p, Λ, Σ^0, Σ^-, and Σ^+. In the absence of interaction the chemical potentials are given by 2.2, with $U(p) = 0$.

The system 2.1 must be solved simultaneously with the constraint of charge and baryonic number conservation

$$e^- + \mu^- + \Sigma^- = p + \Sigma^+, \qquad n + p + \Lambda + \Sigma = n_B, \tag{2.4}$$

where the symbol of a particle stands for its numerical density. The results of Ambartsumyan and Saakyan are shown in Figure 1.

2.4.2 THE WORK OF LANGER & ROSEN (1970) The Levinger-Simmons NN nonlocal potential, equation 3.9 (1), was adopted for any two hadrons. We do not know at present if this assumption is correct. The hyperon-nucleon data at low energy indicate that the YN potentials are rather different from the NN potentials, as indicated in Figure 3. The authors considered the following particles: n, p, e^-, μ^-, Σ^-, Σ^0, Σ^+, Δ^-, and Δ^0.

The treatment was based on a Hartree-Fock definition of the one-body potential $U(p)$; no self-consistency requirement was employed nor was a multicomponent many-body theory as outlined before. The results are shown in Figure 2. The first hyperon to enter is indeed Σ^-, and its appearance is accompanied by a decrease in the number of electrons. This is clearly due to charge conservation, because one can use Σ^- to counterbalance the proton charge. Σ^- are more advantageous than e^-, which are rather expensive due to their small mass: their energy goes up rather fast with density, because they lack any attractive interaction to lower the energy.

The Levinger-Simmons potential is given in two different versions, the so-called V_α and V_β potentials, that differ slightly in their analytic behavior. The results presented here correspond to the choice of V_α, the alternative choice giving rather similar results.

As we said before, these results can substantially change if the potentials are treated more realistically and the many-body effects are included.

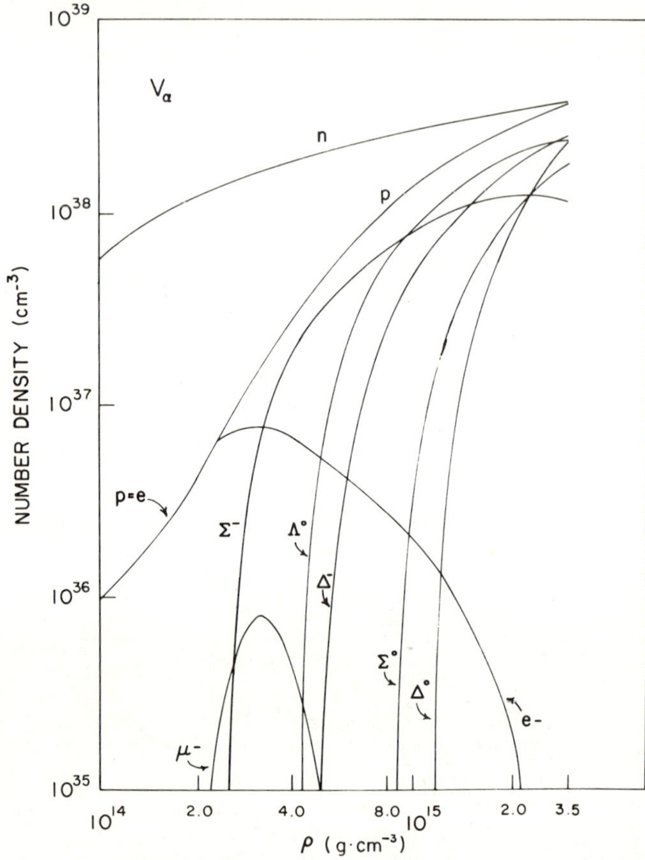

Figure 2 Hyperonic concentrations vs matter density as from the work of Langer & Rosen (1970).

2.4.3 THE WORK OF BUCHLER & INGBER (1971) Were it not for the limited number of particles employed, the work of Buchler & Ingber (1971) would be a considerable improvement upon the previous work, in that it is based on a self-consistent many-body calculation with realistic potentials. The two-body potential used was the one previously derived by Ingber & Potenza (1970), based upon a phenomenological meson-nucleon Lagrangian. It is nonlocal, that is, it contains momentum-dependent terms and has a vanishing tensor force at short distances. The potential was checked against nucleon-nucleon phase shifts, the quadrupole moment of the deuteron and binding energy of nuclear matter. The authors included the following particles: n, p, e^-, μ^-, and π^-, and performed the calculation in the density region $2.5 \times 10^{37} \leqq n_B \leqq 3.55 \times 10^{38}$ cm^{-3}. Their results were presented in Figure 12 of

Part 1 because they constitute a continuation of the structure after the coalescence of the nuclei of the crust.

The flattening of the concentration of μ^- at $n_B = 2.66 \times 10^{38}$ cm^{-3} is entirely due to the presence of pions. However, the pions are not treated correctly because their strong interaction with nucleons, repulsive in S-state and attractive in P-state, has been ignored.

The real progress with respect to the previous work lies in the self-consistency requirement; the concentration of each species depends on each of the other species and not only on the average matter density.

2.4.4 THE WORK OF PANDHARIPANDE & GARDE (1971, 1972) In a first paper, Pandharipande (1971) studied the composition of a baryonic liquid using a many-body theory based upon Jastrow's variational approach: the two-body wave function was taken as satisfying a simplified form of the Bethe-Goldstone equation, in which the complicated inhomogeneous term, representing the Pauli principle, was omitted but simulated by imposing a healing condition. Instead of a state-dependent correlation function, an average was used and only its spin dependence (singlet or triplet) was introduced.

The one-body potential $U(p)$ was assumed to depend only upon the total density. Because it has been stressed that the density dependence of $U(p)$ is a complicated function of the concentration of each species, the approximation used in this work is far from exhaustive.

Three models were studied with different choices for the constituents and the potentials. In particular:

$A \quad n, p, \Lambda, \Sigma, \Delta, e^-, \mu^-$

$B \quad n, p, \Lambda, \Sigma, e^-, \mu^-$

$C \quad n, p, \Lambda, \Sigma, \Delta, e^-, \mu^-.$

In A, the same interaction was used for any two baryons. In B, the Δ hyperons were excluded, but otherwise the potential was the same as in A. The potentials were assumed to be purely central and, in particular, between two hyperons or a hyperon and nucleon they were taken to be ($x = 0.7r$).

$$
\begin{aligned}
l = 0 \quad & V(r) = -1650.6\frac{e^{-4x}}{x} + 6484.2\frac{e^{-7x}}{x} \\
l = \text{odd} \quad & V(r) = -933.48\frac{e^{-4x}}{x} + 4151.1\frac{e^{-6x}}{x} \\
l = \text{even}, \neq 0 \quad & V(r) = -12.322\frac{e^{-2x}}{x} - 1112.6\frac{e^{-4x}}{x} + 6484.2\frac{e^{-7x}}{x}.
\end{aligned}
\tag{2.5}
$$

To check the validity of this assumption we have used equation 2.5 ($l = 0$) to construct the ΛN cross section for center of mass energies up to 20 MeV and the results are shown in Figure 3. The curve A is obtained by using equation 2.5, whereas curve B is obtained by employing a meson theoretical potential. The potential 2.5 gives a rather poor fit to the experimental data.

As far as the $l \neq 0$ waves are concerned, one cannot make the same statement or for that matter any statement because we do not possess the high-energy data to determine the strength of the P and D waves. However, if these potentials are compared with the ones obtained by exchange of heavy mesons after using a phenomenological hadron-meson Lagrangian, the result is found that the former potentials are on the average too repulsive.

In case C, it was recognized that the ΛN interaction (as well as the $\Lambda\Lambda$) cannot be as strong as the NN case, because the ΛN system does not form a bound state (Brown et al 1970). To this end, an arbitrary factor of $\frac{9}{10}$ was introduced to decrease the attraction when dealing with hyperons. As pointed out by the author himself, this reduction of 10% has no theoretical basis: it is simply a change of an order of magnitude to study the sensitivity of the concentrations.

Even though a multiplicative factor is quite arbitrary, a more justifiable number would actually be either $\frac{2}{3}$ or $\frac{4}{9}$ as explained below. Model C is at any rate more credible than either A or B.

It is important to notice that the energy per particle has a rather strange behavior in models A and B, because it decreases with density, indicating a negative pressure. Such a strange behavior is not present in model C, which, as we said, is slightly more realistic. The sensitivity of the concentrations upon the form of the potential was checked in another paper by Pandharipande & Garde (1972), who introduced tensor forces into the problem. The resulting composition is shown in Figure 4.

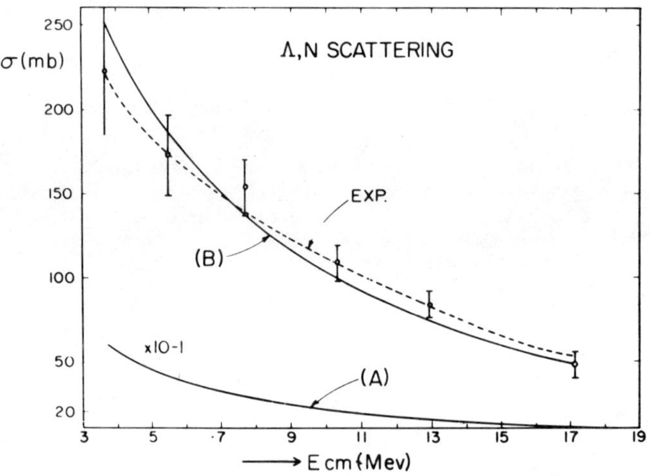

Figure 3 The ΛN scattering cross section as computed by the author (unpublished). Curve A is obtained using $V(r)$ as given by 2.5. The fit is rather poor. Curve B is obtained by using Canuto & Datta (1974) hyperonic potential.

The energy per baryon in C is slightly smaller than in the pure neutron case, but the pressure vs energy-density relation is not sensibly altered. This is the first indication that the hyperonic liquid might not have a great effect in determining the mass and the radius, even though it can be important for the dynamics of the star. As an example, we can quote the work of Langer & Cameron (1969), who showed that the vibrational energy of neutron stars with hyperons is damped very rapidly (on an astronomical scale). For example, a neutron star with $\rho_c > 2.6 \times 10^{14}$ g cm^{-3} will not sustain vibrations.

2.4.5 THE WORK OF MOSZKOWSKI (1974) The problem of self-consistency was not solved in this paper either, but at least its importance was fully appreciated and several physical hypotheses were introduced to account for its presence. It was explicitly recognized that the potential felt by any particle should depend not only upon a second particle, but also upon a third one and so on and even more so at high density, where the Pauli principle loses its importance and the short-range repulsion becomes of primary importance through the dispersion effect. A many-

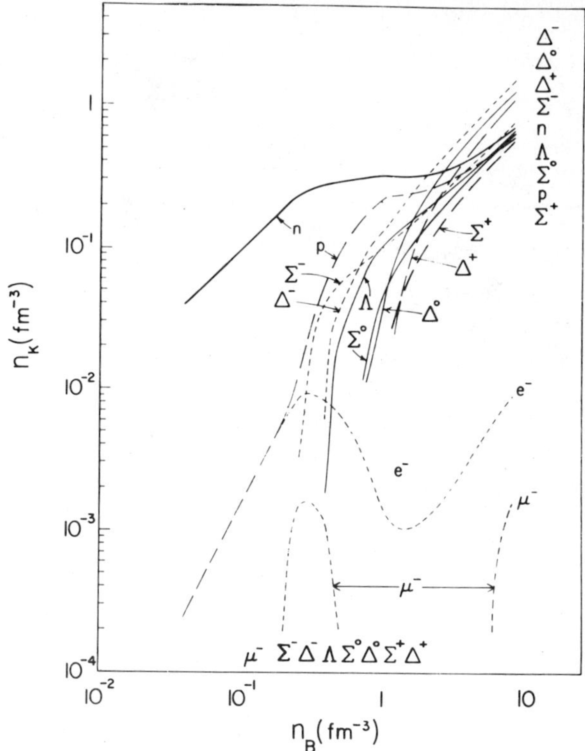

Figure 4 Hyperonic concentrations vs baryonic density as from Pandharipande & Garde (1972).

body treatment of the type explained before was adopted, but the problem of full self-consistency was avoided by choosing a specific form for the G-matrix between any two baryons. This was done in the following way: when the system is composed of only neutrons and protons, the G_{ik}'s ($i, j = n, p$) are taken to be dependent upon $n_n + n_p$. However, when hyperons are present, the extension is not trivial and the author pursued the calculations using two different parameterizations for the density dependence of the G-matrix. In the first one, he chose (a)

$$G_{pp}(n, p, \Sigma, \text{etc}) \equiv G_{pp}[n+p, p/(n+p)]$$
$$G_{\Sigma\Sigma}(n, p, \Sigma, \text{etc}) \equiv G_{pp}[n+\Sigma, \Sigma/(n+\Sigma)]$$

for identical baryons, and

$$G_{ij} \equiv G_{np}\left[n_i + n_j; \frac{\min(n_i + n_j)}{n_i + n_j}\right]$$

for unlike particles.

A second parameterization was chosen by virtue of which the G-matrix depends upon the total baryonic number, that is, (b)

$$G_{nn}(n, p, \Sigma, \text{etc}) \equiv G_{nn}(n_B, n/n_B)$$
$$G_{pp}(n, p, \Sigma, \text{etc}) \equiv G_{pp}(n_B, n_p/n_B)$$
$$G_{\Sigma\Sigma}(n, p, \Sigma, \text{etc}) \equiv G_{pp}(n_B, n_\Sigma/n_B)$$

for identical particles, and

$$G_{ni} \equiv G_{np}(n_B, n_i/n_B)$$
$$G_{ij} \equiv G_{np}(n_B, \{n_i + n_j\}/n_B) \qquad (i, j \neq n)$$

for unlike particles.

Choices a and b reduce the knowledge of any $G_{\alpha\beta}$ to the knowledge of G_{nn}, G_{pp}, and G_{np}, the neutron-neutron, proton-proton, and neutron-proton G-matrices. These were taken from unpublished work of T. Sawada and C. W. Wong, who extended their original nuclear-matter program to the case of unequal numbers of neutrons and protons.

Four tables are given by the author for the quantities G_{nn}, G_{pp}, $G_{np}(T=0)$, and $G_{np}(T=1)$, for four values of y defined as

$$y(n_n + n_p) = n_p \qquad y = 0, \tfrac{1}{6}, \tfrac{1}{3}, \tfrac{1}{2}$$

and six values of Q_f defined as

$$2Q_f^3 = 3\pi^2(n_n + n_p) \qquad Q = 0.5, 1, 1.5, 2, 2.5, 3.$$

In the Sawada-Wong extension of the $n \neq p$ case, the self-consistency was not fully accounted for in that the single-particle energies entering in the Green's functions were parameterized in a form dependent upon three coefficients fitted to the nuclear matter case ($y = \frac{1}{2}$) and pure neutron case ($y = 0$).

As far as one can see, the self-consistency problem was treated, if not exactly, at least extensively enough to give one the feeling that the sensitivity of the problem to this aspect has been brought to light. Evidently a final answer can be obtained only after a fully self-consistent approach has been made.

The second serious difficulty encountered by Moszkowski concerns the potentials involving hyperons. Here too, the author decided that, lacking any fundamental approach to the problem, the most sensible way to proceed was to test the results under different assumptions.

The most obvious one is to suppose that the interaction between any two like (or unlike) baryons is the same as the one between $p-p$ (or $n-p$). This will be referred to as choice 1. We have already discussed the poor fit given by such a hypothesis to the ΛN cross section. It suffices to mention that such a postulate (combined with either choice a or b of the G-matrix) gave rather strange results, for instance, negative energy or negative pressure.

These results are a clear indication of something that we already knew: hyperonic forces are essentially different from nucleon-nucleon forces. Faced with this problem, Moszkowski altered the Reid potential in a clever way to incorporate the different nature of the mechanism of exchange of mesons occurring when we deal with hyperons (choice 2). It is known that the intermediate part of the NN potential is dominated by the exchange of a two-pion system. Traditionally this effect is simulated by the exchange of a σ-meson. According to the quark model a π^0 is made of

$$2^{-1/2}(n\bar{n}-p\bar{p}),$$

where n is a neutron quark, and $\bar{n}$ is an antineutron quark. Given a vertex containing the same hyperon (in and out), the coupling constant for Λ and Σ is then $\frac{2}{3}$

Table 1 Models $2a$ and b of Moszkowski (1974)[a]

n_B	n	p	$\Sigma^- + \Delta^-$	Λ	E/A (MeV)
Model $2a$					
0.068	0.065	0.003	0	0	7.67
0.131	0.124	0.008	0	0	11.6
0.228	0.164	0.035	0.029	0	13.4
0.362	0.219	0.074	0.069	0	14.0
0.541	0.279	0.132	0.129	0	22.4
0.770	0.365	0.203	0.202	0	48.2
1.056	0.450	0.303	0.303	0	99.7
1.825	0.757	0.534	0.534	0	335
Model $2b$					
0.068	0.065	0.003	0	0	7.67
0.228	0.196	0.021	0.010	0	17.4
0.541	0.404	0.075	0.062	0	50.1
1.056	0.824	0.131	0.101	0	174
1.824	1.625	0.166	0.033	0	457

[a] The baryonic density and concentrations are in particles fm^{-3}.

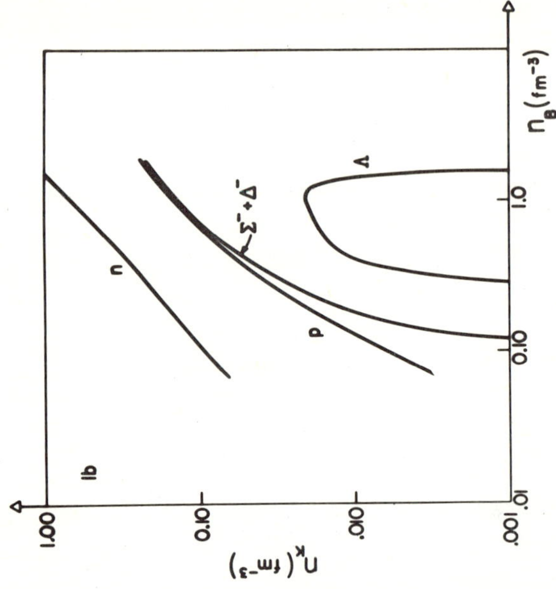

Ib
n
p
Λ
$\Sigma^- + \Delta^-$
1.00
0.10
0.010
0.001
0.10
1.0
n_B (fm^{-3})
n_K (fm^{-3})

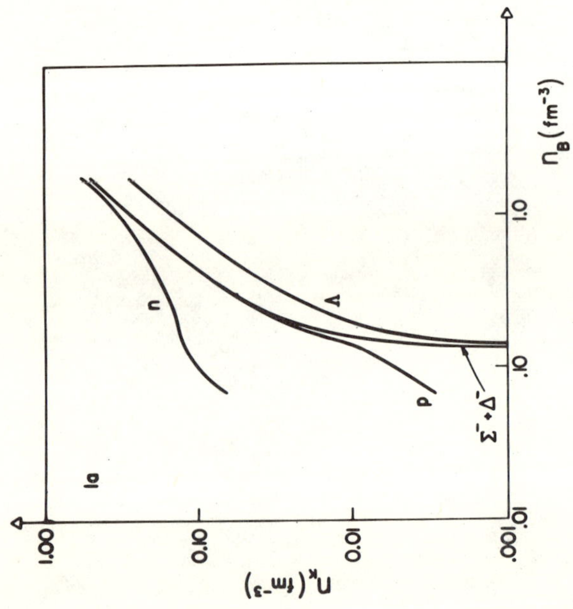

Ia
n
p
Λ
$\Sigma^- + \Delta^-$
1.00
0.10
0.01
.001
.01
.10
1.0
n_B (fm^{-3})
n_K (fm^{-3})

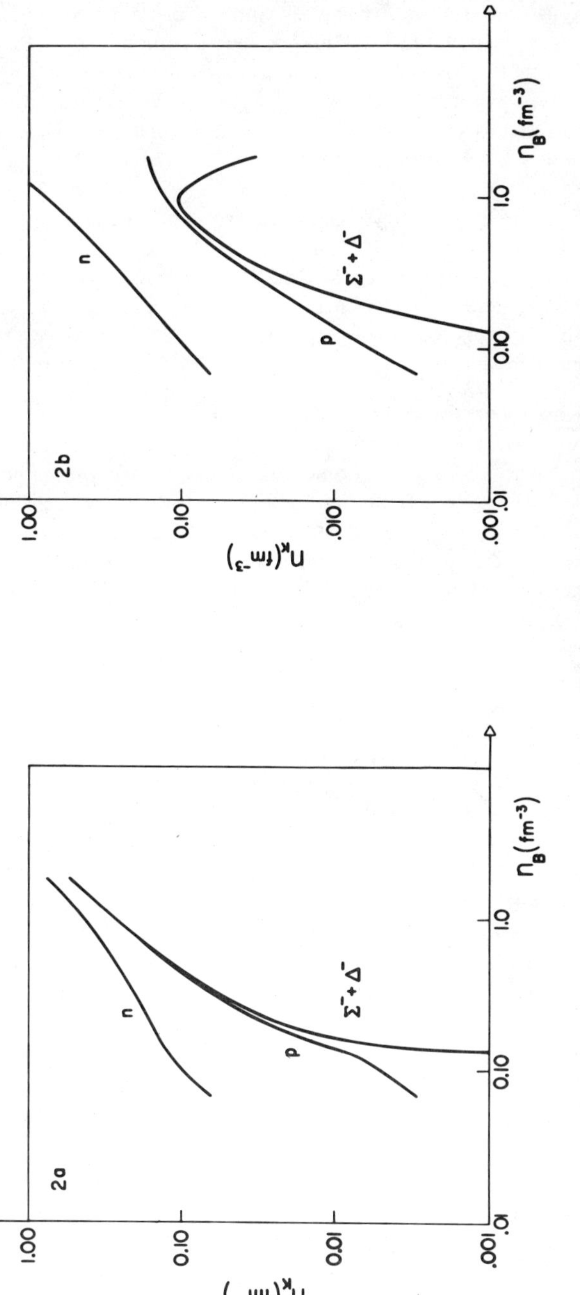

Figure 5 Hyperonic concentrations vs baryonic density following the models 1*a,b* and 2*a,b* of Moszkowski (1974).

as large with respect to N or Δ because Λ and Σ contain two, whereas N and Δ contain three, nonstrange quarks. Thus the reaction matrix for ΛP scattering is $\frac{2}{3}$ of NP and the one for $\Lambda\Lambda$ is $\frac{4}{9}$ of the one for NP. This factor is sensibly different from $\frac{9}{10}$ employed before. As an example, Reid's ${}^1S_0$ potential has a depth of ~ 90 MeV, whereas for ΛN derived from meson exchange it is only ~ 60 MeV, a factor of about $\frac{2}{3}$. The numerical coincidence indicates that this line of reasoning contains some truth.

The resulting energy per baryon and relative concentration of each species vs the baryonic number are presented in Table 1, for the choices $2a$ and b. For the reasons given before, the choices $1a$ and b seem less credible than the others. Between $2a$ and b there is a difference in the density dependence of the G-matrix and at the time of this writing it is difficult to decide with certainty in favor of any of them. The relative concentrations are presented in Figure 5.

2.4.6 THE WORK OF BETHE & JOHNSON (1974) As we have repeatedly stressed, the most important ingredient in any many-body computation is the NN potential. This is no place to discuss this topic; it suffices to say that among the most simple forms so far proposed, the one published by Reid has been most successful in fitting the experimental phase shifts. Reid's potential is a superposition of Yukawas of different strengths and ranges, and even though the analytic structure is highly suggestive of those derived theoretically from meson theories, still its purely phenomenological nature is evident in the fact that one cannot write it in the general form

$$V_{LSJ}(r) = V_c(r) + V_\sigma \boldsymbol{\sigma}_1 \cdot \boldsymbol{\sigma}_2 + V_T S_{12} + V_{LS} \mathbf{L} \cdot \mathbf{S} + V_Q (\mathbf{L} \cdot \mathbf{S})^2, \tag{2.6}$$

but one must be content with apparently unrelated forms for each partial wave. On the basis of the meson theory of nuclear forces, the ω-vector meson should dominate the behavior of $V_{LSJ}(r)$ at short distances. Moreover, because ω is (vector) isoscalar, the core should be the same in all states, the range of it should correspond to that of the ω's, and finally the strength of it should be given by the ω-coupling constant. None of these features is present in Reid's potential. Because at high densities the behavior of the core is of critical importance, Bethe and Johnson undertook the task of rebuilding an NN potential that would satisfy the previous requirements. In Table 1 of their paper, five sets of coefficients C_n and n are given, entering in the definition of the Bethe-Johnson potential ($x = 0.7r$):

$$V_{LSJ}(r) = \sum_n C_n(LSJ) \frac{e^{-nx}}{x} + V_T(r). \tag{2.7}$$

The coefficient C_1 and the tensor V_T are taken from the one-pion-exchange model.

The new potentials reproduce experimental phase shifts, nuclear matter binding energy, and deuteron quadrupole moment as well as Reid's. The work of Bethe and Johnson consists of two distinct parts: 1. determination of the ground-state energy and therefore equation of state for a pure neutron gas; and 2. derivation of the same properties when hyperons are present. Let us first discuss the results for a pure neutron liquid.

The many-body technique employed is the lowest-order constraint variation (LOCV) as developed by Pandharipande (1971). It is stated by the authors that such a method is a good approximation of the more exact treatment developed by Pandharipande & Bethe (1973) and, moreover, that LOCV is entirely satisfactory for numerical calculations. We shall see, however, how recent computations have shown that this is indeed not the case and that such a method gives too large an energy when compared with an essentially exact Monte Carlo computation. Let us first present the results and then discuss this point.

Five different models have been studied and we cannot but give the summary of their results. For Model I an average process was used, whereby only even and odd potentials were used for any two baryons. The resulting equation of state reads (n_B in fm^{-3}, E_B in MeV, P_{35} in units of 10^{35} dynes cm^{-2}; $x = n_B$, $a = 1.54$)

$$E_B = 236x^a, \qquad P_{35} = 5.831x^{a+1}, \qquad c_s^2 = x^a(1.01+0.648x^a)^{-1}. \tag{2.8}$$

When compared with the values for a pure neutron gas, given in Table 10 (Part 1) obtained by using Reid's potential, equation 2.8 gives values consistently larger. For example, at $n_B = 1$, we read from Table 10 (Part 1) that $E_B = 157.47$ and $P_{35} = 4.029$, whereas Model I predicts $E_B = 236$ and $P_{35} = 5.831$. The present equation of state is much stiffer. Moreover, the velocity of sound c_s exceeds the velocity of light at $n_B = 1.98$ fm^{-3}. In Model II, the average over the potentials was not performed and a more correct evaluation was carried out of the ^3P_J waves. The resulting energies were much lower than in Model I, but still higher than those of Table 10 (Part 1).

Model III yields results very similar to Model II, the main difference being the treatment of the $^1P_1$ wave, that is here less repulsive. Again, E_B is higher than that given in Table 10 (Part 1).

Models IV and V are interesting in that the range of the repulsion corresponds exactly to $m_\omega c/\hbar = 3.85$. Model V has a less repulsive core in even states and a more repulsive core in odd states with respect to Model IV. The resulting energies for particles do not differ significantly as shown in Table 2, where the results of all the models are collected. The same many-body technique was employed by the authors to study a hyperonic liquid composed of p, Λ, Σ, and Δ. The hyperonic

Table 2 Energy per particle (MeV) for neutron (N) and hyperon (Y) liquids following different models of H. A. Bethe and M. B. Johnson (1974)

n_B	II		III		IV		V		Reid
(fm^{-3})	N	Y	N	Y	N	Y	N	Y	N
0.5	65	65	63	63	61	61	63.5	63.5	57
1.0	189	189	185	189	181	191	191	185	156
2.0	610	690	590	560	560	575	570	480	490
4.0	1890	1920	1810	1610	1580	1450	1650	1200	1450
10.0	7500	7300	7200	6200	6600	5900	6000	4000	—

potential was constructed in the following way: the authors define an even- and an odd-state potential such that the even one is the same as the $^1D_2$ potential for the NN case and the odd one is a spin-isospin average of the ^3P_J states NN potential (less OPEP).

The authors do not give a numerical comparison of the results that such potentials would yield for experimental hyperonic data such as the one presented in Figure 3 for the ΛN cross section and we are therefore in no position to judge the reliability of such a hyperonic potential. Granting, without justification, that one can use the NN potential in some way, the even state has been chosen here more correctly than in Pandharipande (1971), because $^1D_2$ is less attractive than $^1S_0$ and it therefore resembles more closely the true ΛN potential as derived from meson exchange.

We feel less confident about the procedure used for the odd waves. In fact, it is known that $^3P_1$ for NN is repulsive at any distance, whereas the $^3P_1$ theoretically

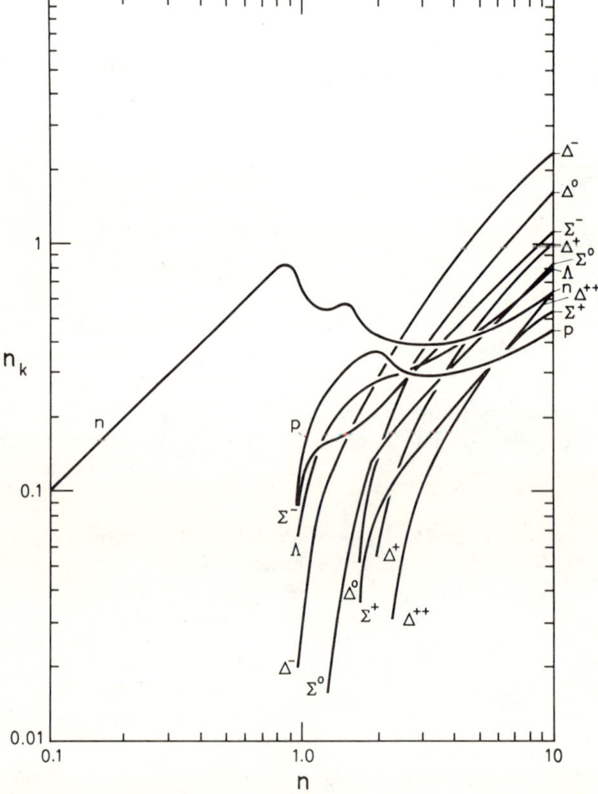

Figure 6 Hyperonic concentrations (in particles fm^{-3}) vs baryonic density (same units) as from the work of Bethe & Johnson (1974).

derived (OBEP) for ΛN is attractive and repulsive. It therefore seems that the odd potential used by Bethe and Johnson is too repulsive. The result for the ground-state energy when hyperons are included is also given in Table 2 and the relative concentrations (for Model V) are presented in Figure 6.

Before turning to the equation of state, we ought to comment on the many-body theory employed. A hypothetical but indicative calculation for a pure neutron liquid with a simplified potential of the form ($x = 0.7r$),

$$V(x) = 6484.2 \frac{e^{-7x}}{x}, \tag{2.9}$$

has been recently solved by several groups, with the aim of revealing the approximations used in the different many-body techniques so far adopted. If we take the Monte Carlo results of Cochran & Chester (1973) as a gauge to judge the other methods, the results are as follows (Figure 7):

1. The energies obtained by using LOCV of Pandharipande (1973) are higher than the Monte Carlo energies up to 200 MeV at 3.2×10^{15} g cm^{-3} (curve 1). The LOCV method has been employed by Bethe & Johnson (1974)
2. The many-body technique employed by Shen & Woo (1974) yields results that never exceed the Monte Carlo energies by more than 25 MeV (curve 2)
3. The full variational method (and not just the LOCV) as developed by

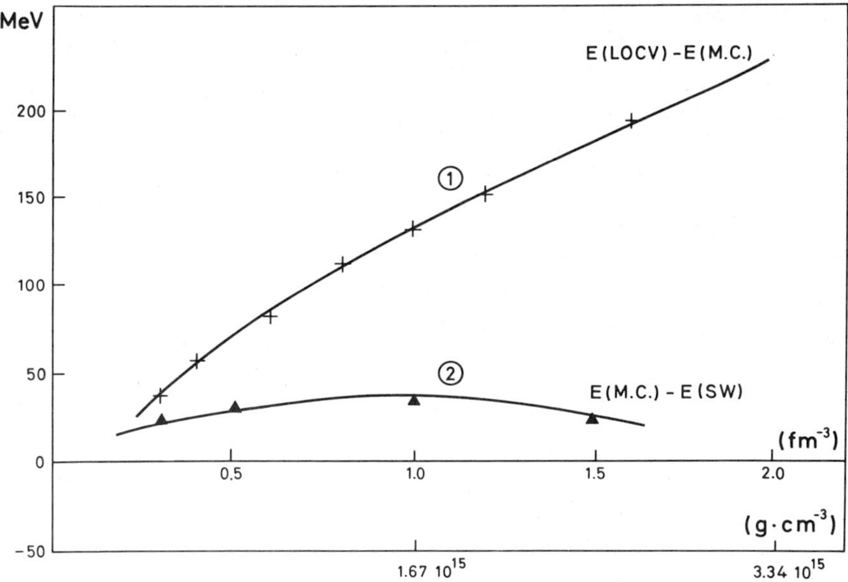

Figure 7 Energy per baryon of a neutron liquid as from Pandharipande (1973) (LOCV) and Shen & Woo (1974). Both results are judged against the almost exact Monte Carlo (MC) results of Cochran & Chester (1973). The LOCV yields too high energy.

Pandharipande & Bethe (1973) gives energies in agreement with Monte Carlo results (H. A. Bethe 1974, personal communication)

In view of the much higher reliability of the methods employed in 2 and 3, it would be highly desirable to repeat the Bethe & Johnson computations using the new potentials but employing the more exact many-body formulation either in the form 2 or 3. An overall concern about the Bethe-Johnson potential is that the ω-meson coupling constant $g_{\omega}^2/4\pi\hbar c$ is 22.1, 29.6, 47, 47, and 47 for models V, IV, III, II, and I, respectively. This has to be compared with the "experimental value" of 10 ± 2.

2.5 *Concluding Remarks on the Hyperonic Liquid*

Despite the considerable amount of work that has gone into the study of the hyperonic liquid, none of the previous works can claim the same degree of reliability

Table 3 ΛN potential as derived by meson exchanges[a]

	$\eta(548, T=0, Ps)$	$k(495, T=1/2, Ps)$
	$\eta \equiv 2.78\, f^{-1}$	$k \equiv 2.35\, f^{-1}$
V_σ	$-19.39(e^{-\eta r}/r)$	$+(-)^{L+S}57.7(e^{-kr}/r)$
V_T	$-58.18 f(r)(e^{-\eta r}/r)$	$+(-)^{L+S}173.3 f(r)(e^{-kr}/r)$
	$f(r) = \frac{1}{3} + 0.36/r + 0.13/r^2$	$f(r) = \frac{1}{3} + 0.43/r + 0.18/r^2$
	$V_c = V_{LS} = V_Q = 0$	
	$\sigma(490, T=0, S)$	$\omega(888, T=0, V)$
	$\sigma \equiv 2.48\, f^{-1}$	$\omega \equiv 4.51\, f^{-1}$
V_c	$1131.08(e^{-\sigma r}/r)$	$+6815.03(e^{-\omega r}/r)$
V_σ	$+0.98 f(r)(e^{-\sigma r}/r)$	$+[616.78 - 43.61 f(r)](e^{-\omega r}/r)$
	$f(r) = 0.402/r + 0.32/r^2 + 0.13/r^3$	$f(r) = 0.22/r + 0.09/r^2 + 0.02/r^3$
V_T	0	$-925.18 f(r)(e^{-\omega r}/r)$
		$f(r) = \frac{1}{3} + 0.22/r + 0.05/r^2$
V_{LS}	$-135.90 f(r)(e^{-\sigma r}/r)$	$-5754.49 f(r)(e^{-\omega r}/r)$
	$f(r) = 0.40/r + 0.16/r^2$	$f(r) = 0.22/r + 0.05/r^2$
V_Q	$-1.98 f(r)(e^{-\sigma r}/r)$	$+87.23 f(r)(e^{-\omega r}/r)$
	$f(r) = 0.16/r^2 + 0.13/r^3$	$f(r) = 0.05/r^2 + 0.02/r^3$
	$k^*(890, T=1/2, V)$	
	$k^* \equiv 4.42\, f^{-1}$	
V_c	$(-)^{L+S}552.41(e^{-k^* r}/r)$	
V_σ	$(-)^{L+S}[353.10 - 147.82 f(r)](e^{-k^* r}/r)$	$f(r) = 0.22/r + 0.10/r^2 + 0.02/r^3$
V_T	$-(-)^{L+S}529.65 f(r)(e^{-k^* r}/r)$	$f(r) = \frac{1}{3} + 0.22/r + 0.05/r^2$
V_{LS}	$-(-)^{L+S}1495.69 f(r)(e^{-k^* r}/r)$	$f(r) = 0.22/r + 0.05/r^2$
V_Q	$(-)^{L+S}295.65 f(r)(e^{-k^* r}/r)$	$f(r) = 0.05/r^2 + 0.02/r^3$

[a] Λ, N potential: $T = 1/2$; $V = V_c + V_\sigma \boldsymbol{\sigma}_1 \cdot \boldsymbol{\sigma}_2 + V_T S_{12} + V_{LS} \mathbf{L} \cdot \mathbf{S} + V_Q W_{12}$.

that characterizes that of the pure neutron liquid. One of the main reasons is the inadequate treatment of the hyperonic forces. A full OBEP hyperonic potential has recently been determined by Canuto & Datta (1974) on the basis of data from Brown, Downs & Iddings (1970). The general potential is written in the form 2.6 and the contributions from the several bosons are listed separately. As an example, we present in Table 3 the ΛN potential that was used in the construction of curve *B*, Figure 3. The equilibrium calculations with such a potential, however, have not yet been performed. As far as the many-body techniques are concerned, Figure 7 clearly indicates that the LOCV should be replaced with either method 2 or 3 just mentioned.

A final criticism of all the previous computations has been raised by Sawyer (1972). He has stressed the point that in deciding whether a given hyperon will appear or not, a parameter equally as important as the potential, and perhaps even more, is the mass shift that every hyperon acquires because of the dense surrounding medium. In a model calculation for Δ^-, Sawyer has shown that the mass shift is of such a magnitude as to quench the appearance of Δ^- until densities higher than 10^{16} g cm^{-3} (see Figures 2 and 4). None of the previous computations has included any mass shift and only when such an effect is thoroughly investigated and embodied in the calculations can the previous results be totally trustworthy.

2.6 *The Equation of State with Hyperons*

We have already presented in Tables 11 and 12 of Part 1 two $P = P(\varepsilon)$ relations including hyperons. To these we should now add the one computed by Bethe and Johnson. This is given in Tables 4*a* and *b*.

Moszkowski's results [Table 12 (Part 1)] can be compared with either Tables 11 (Part 1) or 4 only numerically, because both the many-body theory and the baryonic potentials are different. We have already commented previously about the many-body theory employed in Tables 11 (Part 1) and 4. The same type of comment cannot be made concerning Moszkowski's theory. However, we feel that as far as the hyperonic potential is concerned, his work is more realistic.

In view of the difficulties still plaguing all the hyperonic equations of state published so far, it is our feeling that it is safer not to use any of them as yet and consider the region $>2 \times 10^{14}$ g cm^{-3} as a pure neutron liquid, thus using Tables 10 (Part 1) or 4*a* of this paper up to a point where the relativistic effects become important (see Section 4).

3 THE SOLID CORE

3.1 *Generalities*

The uncertainties in the results presented in the previous section, though important, are of a rather technical nature, regarding both the similarity (or lack of similarity) of the full hyperonic potential to the NN case and the so far little investigated mass-shift effect. In both cases the physical idea is clear. Time and an extra good deal of work will almost certainly clarify the situation. On the other hand, the region we are about to discuss has been characterized by a much deeper, more fundamentally

Table 4a Equation of state for a pure neutron gas after Bethe & Johnson (1974)

n_B (fm^{-3})	ρ ($g\,cm^{-3}$)	$\mathscr{E}$ (MeV)	P (dynes cm^{-2})	$P/\rho c^2$
0.1	1.70×10^{14}	12.6	1.19×10^{33}	—
0.15	2.55×10^{14}	16.6	2.93×10^{33}	0.013
0.2	3.42×10^{14}	21.2	6.00×10^{33}	0.019
0.25	4.31×10^{14}	26.0	1.09×10^{34}	0.028
0.3	5.20×10^{14}	32.2	1.83×10^{34}	0.039
0.4	7.04×10^{14}	46.9	4.09×10^{34}	0.087
0.5	8.95×10^{14}	64.4	7.61×10^{34}	0.095
0.6	1.09×10^{15}	83.7	1.26×10^{35}	0.128
0.7	1.31×10^{15}	109.0	1.99×10^{35}	0.169
0.8	1.53×10^{15}	135.0	2.85×10^{35}	0.207
0.9	1.76×10^{15}	160.0	3.71×10^{35}	0.234
1.0	2.01×10^{15}	190.0	4.92×10^{35}	0.272
1.1	2.28×10^{15}	224.0	6.23×10^{35}	0.304
1.25	2.70×10^{15}	274.0	8.58×10^{35}	0.353
1.4	3.16×10^{15}	327.0	1.14×10^{36}	0.401
1.5	3.48×10^{15}	360.0	1.34×10^{36}	0.428
1.7	4.19×10^{15}	442.0	1.85×10^{36}	0.493
2.0	5.35×10^{15}	560.0	2.76×10^{36}	0.574
2.5	7.70×10^{15}	788.0	4.83×10^{36}	0.698
3.0	1.06×10^{16}	1040.0	7.62×10^{36}	0.800

physical question. Do neutrons solidify at high enough density? We have so far treated the neutrons and for that matter the hyperons as a liquid even though it is perfectly legitimate to ask oneself whether the NN potential is strong enough to force nucleons into a regularly arranged structure.

The old hard-core potential almost by definition gives rise to a solid structure at densities of the order of $n_B = (\frac{4}{3}\pi r_c^3)^{-1}$. In fact, because at these densities each particle feels an infinitely strong repulsion everywhere around it, it is caged, and such a localization leads to a crystalline structure. Such a potential is therefore of no interest, in addition to having been superseded by more modern versions that call for the existence of a soft core due to the exchange of vector mesons. With a soft but infinite potential, like the ones constructed by Reid and by Bethe and Johnson, it is not clear if the system will eventually crystallize, because the potential can be too soft, can let the particles slip through it, and can never produce the localization necessary for a crystal structure to set in, unless perhaps when the distance is exceedingly small and the repulsion exceedingly high. If that is the case, it could well be that a crystallization density exists but is so high that it is of no interest even for the superdense interior of neutron stars. A third possibility is that the repulsive potential is finite at the origin, that is, a gaussian type of potential of the form proposed by the Japanese school (Otsuki et al 1964). If the gaussian form is such that it provides enough repulsion to localize the particles, it will

Table 4b Equation of state for hyperonic matter after Bethe & Johnson (1974)

n_B (fm^{-3})	ρ ($g\ cm^{-3}$)	$\mathscr{E}$ (MeV)	P ($dynes\ cm^{-2}$)
0.1	1.70×10^{14}	12.6	1.19×10^{33}
0.15	2.55×10^{14}	16.6	2.93×10^{33}
0.2	3.42×10^{14}	21.2	6.00×10^{33}
0.25	4.31×10^{14}	26.0	1.09×10^{34}
0.3	5.20×10^{14}	32.2	1.83×10^{34}
0.4	7.04×10^{14}	46.9	4.09×10^{34}
0.5	8.95×10^{14}	64.4	7.61×10^{34}
0.6	1.09×10^{15}	83.7	1.26×10^{35}
0.7	1.31×10^{15}	109	1.99×10^{35}
0.8	1.53×10^{15}	134	2.84×10^{35}
0.9	1.76×10^{15}	160	3.36×10^{35}
1.0	2.01×10^{15}	189	4.02×10^{35}
1.1	2.26×10^{15}	215	5.02×10^{35}
1.25	2.66×10^{15}	254	6.76×10^{35}
1.4	3.08×10^{15}	296	8.81×10^{35}
1.5	3.38×10^{15}	324	1.03×10^{36}
1.7	4.01×10^{15}	382	1.38×10^{36}
2.0	5.04×10^{15}	475	2.02×10^{36}
2.5	7.04×10^{15}	640	3.40×10^{36}
3.0	9.38×10^{15}	815	5.20×10^{36}

cease to do so when the energy of localization, $\varepsilon = \hbar c n_B^{1/3}$ is greater than V_0, the value of the potential at $r = 0$. For $V_0 = 1$ GeV, $n_B \approx 10^3$ fm^{-3}, or $\rho \approx 10^{18}$ g cm^{-3}. It is not clear a priori whether a repulsive gaussian will produce a crystal, but if it does it will be for only a finite interval in density. For densities such that $\varepsilon \gg V_0$, the crystalline structure ceases to exist.

Because the mesonic theories of nuclear forces indicate that, near the origin, the potential provided by the ω-meson has a Yukawa shape, in what follows we consider exclusively such types of potentials.

The work on the solidification problem can be divided into two categories: the first is of an exploratory nature, whereas the second is of a microscopic nature, because it employs the best available nucleon-nucleon data and many-body techniques. Before reviewing the work, we should perhaps comment on the results: all but one of the computations so far performed indicate a solidification density well within the range of the values expected in the interior of neutron stars, the lowest being 5×10^{14} g cm^{-3} and the highest 3×10^{15} g cm^{-3}. We first review the work in some detail and then critically summarize the results in Section 3.3.

3.2 *Review of the Results*

3.2.1 THE WORK OF CAZZOLA, LUCARONI & SCARINCI (1966) As far as this author can recollect, this work represents the first published paper in which the idea of a solid structure was proposed and a model computation of the corresponding equation of state was carried out. The authors do not actually show that a crystal structure indeed occurs. They postulate that the nucleon-nucleon potential is repulsive enough to produce a localization, which is then accounted for by assuming that each nucleon is trapped in a finite region of space, characterized by a potential $U(r)$. The single-particle energies are found by solving a one-particle Dirac equation in which $U(r)$ is taken to be a square well. Once the eigenvalues are known, the pressure is easily evaluated with the result

$$P = 0.32 n_B^2 10^{39} \text{ dynes cm}^{-2}, \tag{3.1}$$

where n_B is the baryonic number density (particles per fm^3). This expression does not join in any way to the values given in Tables 10–12 (Part 1), giving $P = 6.27$ 10^{38} dynes cm^{-2} at $n_B = 1.4$, as compared with 1.189 10^{36} from Table 12 (Part 1). The merit of the present work is clearly not in the expression for the equation of state, but in its having pointed out a physically attractive alternative to the liquid structure.

3.2.2 THE WORK OF BANERJEE, CHITRE & GARDE (1970) These authors performed an exploratory computation of the equation of state for matter at high densities by localizing the neutrons in a bcc lattice and performing a classical lattice-dynamics analysis. The interaction potential between two opposite neutrons was taken to be an average of Reid's $^1S_0$ potential and the repulsive part of the same wave. The total energy is contributed by the rest energy, the static energy, and the vibrational energy. This last contribution required an elaborate lattice-dynamics analysis.

In the spirit of classical lattice dynamics, the solid structure was thought to set in when the oscillator frequency of each nucleon at its lattice site ω, proportional to the second derivative of the NN potential, becomes real. This occurred at 8×10^{14} g cm^{-3}, which was taken to be the solidification density.

If compared with the presently known E/N for the liquid state, the solid energies so obtained are far too large. The comparison is not meaningful, however, since the solid has been computed with zero-spread single-particle wave functions. A fully quantum-mechanical computation was shortly thereafter initiated by Canuto & Chitre (1973a,b), but before its completion, several other results appeared and we review them first.

3.2.3 THE LAW OF CORRESPONDING STATES The approach to the problem of solidification of neutron matter through the application of the so-called law of corresponding states was first suggested by Anderson & Palmer (1971) (see also Palmer & Anderson 1974). It is known that the potential between two rare gas atoms can be represented by a 6-12 function of the Lennard-Jones type (LJ)

$$V(r) = 4\varepsilon\left\{\left(\frac{\sigma}{r}\right)^{12} - \left(\frac{\sigma}{r}\right)^{6}\right\}. \tag{3.2}$$

Because only two parameters, an energy (ε) and a length (σ), are necessary to specify a given substance, de Boer (1948) was able to show that the dimensionless quantities

$$V^* = V/N\sigma^3, \qquad kT^* = kT/\varepsilon, \qquad P^* = P\sigma^3/\varepsilon \tag{3.3}$$

for several known rare gases fall in straight lines when plotted against the quantum parameter $\Lambda = h(m\varepsilon\sigma^2)^{-1/2}$, nowadays known as the de Boer parameter. If, in the same way, we were able to find the Λ corresponding to a system of neutrons, then the solidification pressure would be readily found by simply looking at the experimental (universal) curve P_s^* vs Λ. By so doing, Anderson & Palmer (1971) found the following result:

$$P_s = 0.4\varepsilon\sigma^{-3} = 4 \text{ MeV fm}^{-3} = 6.7 \times 10^{33} \text{ dynes cm}^{-2}; \Lambda \approx 3.4. \tag{3.4}$$

The similarity with He^3 ($\Lambda \sim 3.5$) is striking. The authors arrived at the result without fitting an *NN* potential to a LJ function. This was later attempted by Clark & Chao (1972), who constructed a density-dependent *NN* potential that is a superposition of ${}^1S_0$ and ${}^1D_2$ waves. When such a potential was fitted to a LJ form, the following results were found:

$$P_s = 4.7 \times 10^{33} \text{ dynes cm}^{-2}; \Lambda = 5.3. \tag{3.5}$$

The corresponding solidification densities can be found by continuing the $P = P(\varepsilon)$ relation given in Table 10 (*Part* 1) with the solid structure. The results are

$$\rho_s = 3.7 \times 10^{14} \text{ g cm}^{-3}; \ \rho_s = 3 \times 10^{14} \text{ g cm}^{-3} \tag{3.6}$$

for Anderson and Palmer and for Clark and Chao, respectively. We critically review these results in Section 3.3.

3.2.4 THE WORK OF COLDWELL (1972) In a paper submitted in April 1971, Coldwell performed an exploratory computation with the goal of investigating whether a simple Hartree-Fock computation would predict neutron solidification. The analysis is incomplete with respect to the work described in the following sections, because it neglects the distortion of the two-body wave function caused by the short-range repulsion and accounted for by the introduction of a correlation function of the type described in Section 5.2 of Part 1. An analogous computation was performed years back by Nosanow & Shaw (1962) for ${}^3He$. Without a correlation function, the single particle energies are bound to be overestimated by a considerable amount, as happened in solid ${}^3He$.

The importance of Coldwell's work lies in the fact that for the first time he performed a double computation, that is, he treated liquid and crystal with the same method, thus making the comparison meaningful. The novelty of this computation with respect to the ones performed before and even after is that the single-particle wave functions were taken to be eigenfunctions of a fictitious potential

$$V(c, x) = c \sin^2 (\pi x/a). \tag{3.7}$$

For $c \to 0$, the single-particle wave function becomes a plane wave (liquid), whereas for $c \neq 0$, $V(c, x) \sim x^2$, as $x \to 0$. The corresponding wave functions are then localized gaussians (solid). No analytic expression can be obtained for the ground-state energies (computed using Reid's potential) and the work has to be done entirely numerically. The results can be summarized as follows: Up to a density of

3.98×10^{14} g cm^{-3}, the minimum of the energy is provided by a ferromagnetic arrangement of the nucleons and not by a crystal. For densities higher than 3.98×10^{14} g cm^{-3}, however, the minimum of the energy is given by localizing the particles in a crystalline structure. The energies obtained by Coldwell are too high when compared with the ones given in Table 10 (Part 1) and Table 2. However, the absolute values for the energy of the liquid and solid are probably off by the same amount and consequently the solidification pressure and density can be considered to be correct.

3.2.5 THE WORK OF CANUTO & CHITRE (1973, 1974) The work presented so far has clearly indicated that a full quantum-mechanical computation is needed to study the solidification problem. This is a formidable task in that a microscopic computation can be considered conclusive only if performed simultaneously for liquid and solid with the same many-body theory and the same potential, that is, the ingredients must be identical. There are essentially two many-body techniques presently available. The variational method, extensively used for ^{3}He, has the unfortunate feature of not lending itself to a straightforward handling of spin and angular momentum, features that characterize the nuclear forces. On the other hand, it has the advantage that by imposing certain restrictions on the correlation function, one can handle the higher many-body contributions in such a way as to render their contribution unimportant. There is in fact one computation for ^{3}He (Hetherington et al 1967), where a restricted variational correlation function was chosen to make the three-body clusters reasonably small. Even though this is a brute-force way of doing things, at least one has an empirical way of making the cluster expression converge. We must notice, however, that if the three-body contribution is a serious problem for liquids it is perhaps of less importance for solids.

In systems that are translationally invariant (i.e. liquids) the unperturbed wave functions are usually taken to be plane waves. Brueckner's theory is a low-density expansion, and it is not obvious that such a method should work for solid hydrogen or solid helium. Solid hydrogen has recently been studied by Østgaard (1971, 1972) and the t-matrix approach (including up to two-body clusters) was found to work rather well. This is partly due to the fact that instead of plane waves, one employs gaussian wave functions, which already incorporate several correlations. One can therefore hope that the cluster expansion up to second order will be adequate for solids but not necessarily for liquids.

If the variational and t-matrix techniques are considered up to the second order, the advantages offered by the t-matrix are clearly superior because the state dependence of the NN potentials can be fully accounted for to any degree of accuracy. On the basis of this last feature and the probable unimportance of the three (and more)-body correlations for solids, Canuto and Chitre decided to adopt the t-matrix method. Evidently the whole method can be judged only after having tested it against a well-known quantum solid such as ^{3}He.

It is shown later that such a method actually produces the best E/N vs molar volume so far published, the deviation from the experimental data being less than 1°K.

Even though the author is not aware of any such computation, it is almost certainly true that the same method, when applied to a liquid, say, liquid ^{3}He, would probably produce rather poor results. This poor performance has nothing to do with the situation for the solid, however, where, as we have just said, it works better (or equally well) than any other method. Evidently, the choice of a method that works well for a solid and perhaps fails for a liquid violates the requirement of handling liquid and solid with the same method. Once the energies for the solid are obtained, the only way of determining if such a configuration is actually solid, in the commonly accepted definition, is by studying its elastic properties, that is, finding out if it has resistance against shearing stresses and if so, for which density interval.

The results indicate that the shear modulus C_{44} is positive for densities greater than (1–3) $\times$ 10^{15} g cm^{-3}, thus providing a lower limit for the stability of a neutron crystal.

Let us now consider a system of neutrons described by the Hamiltonian

$$H = -\frac{\hbar^2}{2m}\sum_i \nabla_i^2 + \tfrac{1}{2}\sum_{i<j} V_{ij}. \tag{3.8}$$

The Slater determinant for the system is built up of single-particle wave functions of the gaussian form

$$\phi(i) = \frac{\alpha^{3/2}}{\pi^{3/4}}\exp\frac{-\alpha^2}{2}|\mathbf{r}_i - \mathbf{R}_i|^2, \qquad \alpha^2 = \frac{m\omega}{\hbar}. \tag{3.9}$$

Here, $\mathbf{R}_i$ is the ith lattice site around which the particle performs an oscillatory motion under the influence of the remaining $(N-1)$ particles. The t-matrix expansion gives the following expression for the energy per particle up to and including two-body clusters:

$$\frac{E}{N} = \tfrac{3}{4}\hbar\omega + \sum_{ij}\frac{\int \psi_{ij}^* V_{ij}\phi_{ij}\, d^3r_i\, d^3r_j}{2N\int \psi_{ij}^*\phi_{ij}\, d^3r_i\, d^3r_j} = \tfrac{3}{4}\hbar\omega + \tfrac{1}{2}\sum_k n_k\,\varepsilon_k. \tag{3.10}$$

Here ϕ_{ij} is the uncorrelated two-body wave function $[=\phi(i)\ \phi(j)]$, whereas ψ_{ij} is the correlated two-body wave function to be determined by solving the homogeneous Bethe-Goldstone equation ($\mathbf{\Delta} = \mathbf{R}_1 - \mathbf{R}_2$)

$$\left[\frac{-\hbar^2}{m}\nabla_r^2 + \tfrac{1}{4}m\omega^2(\mathbf{r}-\mathbf{\Delta})^2 + V(r)\right]\psi(\mathbf{r}) = [-\tfrac{3}{2}\hbar\omega - 2U(0)]\psi(\mathbf{r}). \tag{3.11}$$

The most difficult part of the problem lies in the solution of equation 3.11. In fact the term $\mathbf{r}\cdot\mathbf{\Delta} = r\Delta\cos\theta$, much like the Stark effect, couples even with odd waves. If an angular momentum expansion is made of $\psi(\mathbf{r})$, then an infinite set of coupled differential equations results.

All previous work that dealt with such an equation invariably averaged over $\mathbf{r}\cdot\mathbf{\Delta}$, thus avoiding the angular-momentum problem.

To judge the t-matrix method and the handling of equation 3.11, the most appropriate test is solid ^{3}He. In Figure 8 we present several results. For the time being the significant comparison should be made between the results of Guyer (1969), who applied the method just described with the $\mathbf{r}\cdot\mathbf{\Delta} \approx r\Delta$ approximation, and the results of Canuto et al (1974a), who expanded ψ in its angular-momentum components and solved a resulting set of 25 coupled differential equations. The improvement on the previous result is significant. If the angular-momentum expansion is important for ^{3}He, whose two-body interaction is spherically symmetric, it is even more so when we deal with the NN case, where each partial wave has a different potential.

It is clear that, if we perform an average over the solid-state term $\mathbf{r}\cdot\mathbf{\Delta}$, then we have devoided the dynamic equation 3.11 of its more important features and any study of the importance of the angular momentum dependence of $V(NN)$ in the solidification problem has been irreparably undermined in its credibility. For a complete description of the neutron system we must also introduce the spin variables. Equation 3.11 will then split into three sets corresponding to $S = 0$, $M_s = 0$, and $S = 1$, $M_s = \pm 1, 0$. Canuto and Chitre solved the three sets of 7, 13, and 18 coupled differential equations by including up to $l = 6$. This was found to be large enough for the system to be stable. An fcc configuration was found to be more energetically favorable than a bcc one.

The detailed results are displayed in Table 5, where we list 1. the baryonic density, 2. the energy density, 3. the nearest neighbor distance Δ, 4. the spread of the wave function α^{-1}, and 5. the energy per baryon in MeV and then the elastic constants C_{11}, C_{12}, and C_{44} in units of 10^{-36} dynes cm^{-3}. The first

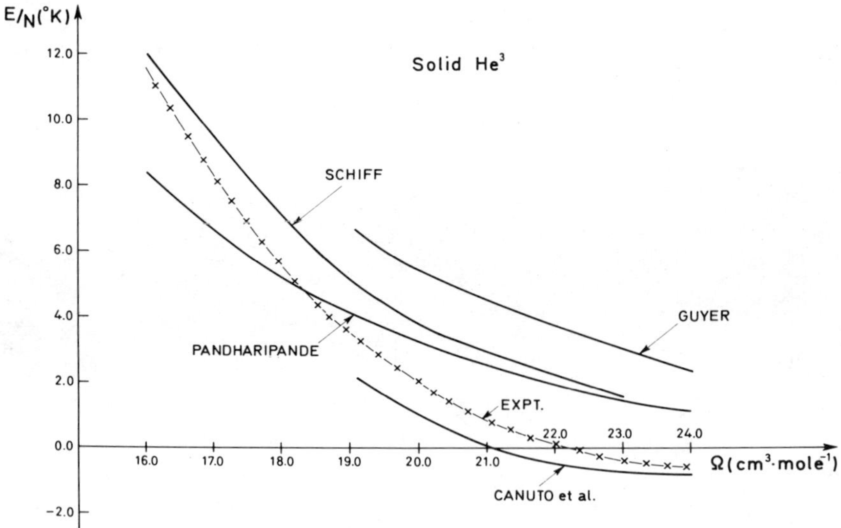

Figure 8 Ground-state energy vs molar volume for solid ^{3}He as obtained by Canuto et al (1974a). The results of Guyer (1969), Schiff (1973), and Pandharipande (1973) are reported for comparison.

Table 5 Results of the computation of V. Canuto and S. M. Chitre (1973, 1974)

n_B (fm^{-3})	$\rho \cdot 10^{15}$ ($g\ cm^{-3}$)	Δ (fm)	α^{-1} (fm)	$E(n_B)$ (MeV)	C_{11}	C_{12} (10^{-36} dynes cm^{-2})	C_{44}
0.84	1.4	—	—	—	0.16	0.15	−0.03
0.96	1.6	—	—	—	0.40	0.24	0.03
1.09	1.83	1.089	0.342	163	0.89	0.37	0.09
1.44	2.4	0.995	0.312	216	2.65	1.07	0.48
2.0	3.34	0.891	0.278	322	6.78	2.97	1.47
2.63	4.4	0.817	0.254	500	17.59	6.69	3.43
2.99	5.0	0.779	0.238	612	27.57	10.33	5.71

important observation is that the oscillation around a lattice site, α^{-1}, is only $\sim 30\%$ of the first neighbor distance, Δ. In view of empirical laws like the Lindenman melting rule, there is no probability of such an oscillation producing an instability and making the crystal disappear.

The stability against elastic deformation is studied by analyzing the behavior of the elastic constants vs density, as shown in Table 5. It can be seen that the shear modulus C_{44} is positive only for $\rho \approx 1.4\ 10^{15}$ g cm^{-3}, indicating that the fcc structure of neutrons is actually a solid, that is, it can withstand shearing stresses for densities higher than this value.

3.2.6 THE WORK OF SCHIFF (1973) Another way of studying the solidification problem was devised by Schiff, who applied an idea of Wu & Feenberg (1962). At relatively high density, the major feature of a system of nucleons is undoubtedly the strong repulsion and probably not the Fermi statistics. One can therefore treat the Pauli principle as a perturbation to a Bose system, for which it is hoped that one can compute the ground state energy to good accuracy. The hypothetical boson system is studied in two steps:

1. The repulsion is considered to be represented by hard spheres. Kalos, Levesque & Verlet (1974) numerically solved the Schrödinger equation for a system of 256 hard-sphere bosons. The results are in excellent agreement with the previous results obtained using a Jastrow-type of wave function (Hansen et al 1971)
2. The second step is to treat the attractive part of the two-body potentials by first-order perturbation theory. The hard-sphere ground-state energy is lowered by up to 20%. The most complicated part of the problem consists in taking into account the Pauli principle (PP), and this is done following the prescription of Wu and Feenberg

The solid phase is treated analogously, except that the corrections due to the statistics are neglected. The final result is that a solid structure does indeed occur at a density of $(2.9 \pm 0.5) \times 10^{15}$ g cm^{-3} and a pressure of $(4.7 \pm 1) 10^{30}$ atms.

If this method is applied to solid ^3He, one obtains the results presented in Figure 8. Even though they are not as good as those of Canuto et al (1974a), still they have the very enviable feature of getting better as the density increases. This is indeed very pleasing, even though it could be misleadingly interpreted as a mark of merit for the method when applied to a neutron system. Such is indeed not the case, because the background hard-core bosons that produce such good results for ^{3}He cannot be considered as an appropriate basis for neutrons, whose strong interaction is far from being a hard-core.

3.2.7 THE WORK OF NOSANOW & PARISH (1974) A different approach to the solidification problem was adopted by Nosanow and Parish, who employed a Monte Carlo technique to calculate the many-body effects of the short-range correlations. The problem is formulated within the framework of the variational approach, and we have on several occasions stressed the difficulties encountered in extending such a method to cope with the angular-momentum dependence of nuclear forces. The best one can do is to use a simplified force taken to be a superposition of a singlet ($^1S_0$) and triplet potential $V_c(^3P_2)$. As in the case of Coldwell's work, the best feature of the work of Nosanow and Parish is that they treat both liquid and solid with the same many-body technique. From the methodological point of view, this is a most welcome feature.

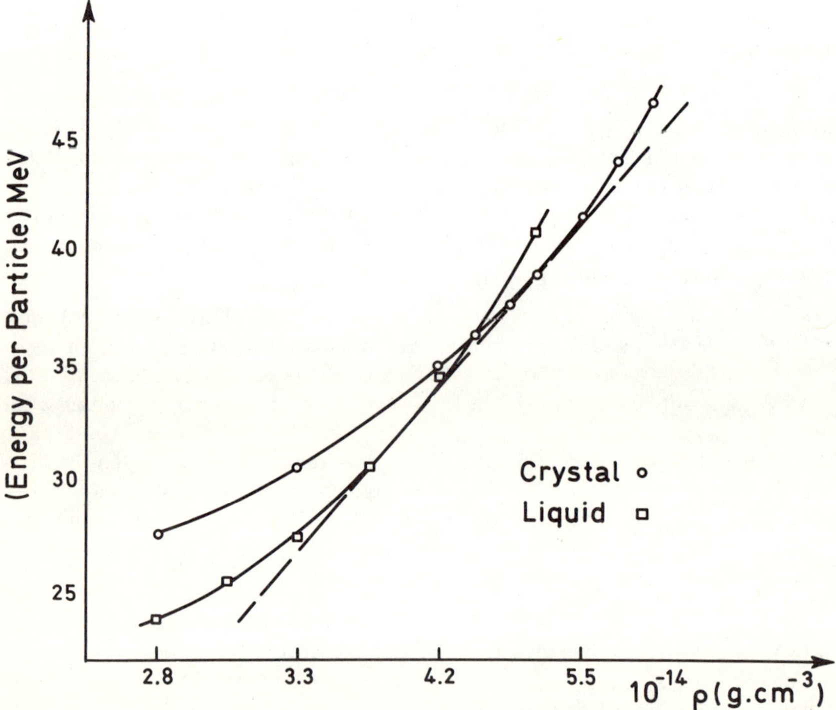

Figure 9 Liquid-solid transition in a neutron liquid as from Nosanow & Parish (1974).

The results of their computation are shown in Figure 9, where the liquid and solid energies are plotted against density. The liquid-solid transition occurs around 4.4×10^{14} g cm^{-3}; the solidification pressure is $\sim 10^{34}$ dynes cm^{-2}. These numbers are in good agreement with the ones obtained by Anderson and Palmer and by Clark and Chao but are much smaller than the ones obtained by Canuto and Chitre and by Schiff.

Nosanow and Parish do not give any indication of how sensitive the results are to the choice of potentials. Evidently there is more than one way to construct a singlet and triplet potential, and the sensitivity of the results to the amount of attraction could and should be checked. Such an analysis is still missing. More recently the same method failed to find solidification when the potential was chosen to be purely repulsive (Canuto et al 1974b). This has led us to believe that the treatment has not been fully investigated from the numerical point of view.

3.2.8 THE WORK OF PANDHARIPANDE (1973) A variational approach to the solidification problem was employed by Pandharipande (1973) who, following van Kampen (1961), expanded the ground-state energy E in clusters, truncated at the second order

$$E = \tfrac{3}{4}\hbar\omega + \tfrac{1}{2}\sum_{ij} C_2(ij). \qquad 3.12$$

The term C_2 contains three parts. The first two do not involve the potential and they almost cancel each other. The author requires that the third part, called I_3, be minimized. In this way one obtains a differential equation for ψ of the form

$$\left[-\frac{\hbar^2}{m}\nabla^2 + \tfrac{1}{4}m\omega^2(\mathbf{r}-\mathbf{\Delta})^2 + V(r)\right]\psi(\mathbf{r}) = \varepsilon\psi(\mathbf{r}). \qquad 3.13$$

This is precisely equation 3.11 in the Canuto and Chitre formalism. The basic idea of the computation consists in imposing reasonable restrictions on the correlation function, f, $\psi = \phi f$, so as to make the truncated expansion reasonable. Such a method is called LOCV. In the next step, 3.13 is changed into an equation for f, by filtering ϕ to the left.

We have already discussed how difficult it is to evaluate equation 3.11 because of the strong angular-momentum dependence contained in the $\mathbf{r}\cdot\mathbf{\Delta}$ term and in $V(r)$. The physical angular momentum is the one that characterizes ψ; by splitting ψ into ϕf, one introduces two unphysical angular momenta l_1, l_2:

$$\psi(\mathbf{r}) = \sum_l \psi_l(r) Y_l = \sum_{l_1} \phi_{l_1} Y_{l_1} \sum_{l_2} \phi_{l_2} Y_{l_2}.$$

Because both quantities $V(r)$ and $\mathbf{r}\cdot\mathbf{\Delta}$ are operators in the angular-momentum space that act on ψ_l, their action on either Y_{l_1} or Y_{l_2} has no meaning. This in turn implies that one cannot transport $\phi(\mathbf{r})$ to the left of 3.13, to get an equation for f. When the LOCV is applied to the test case of solid ^{3}He, Pandharipande obtains the results shown in Figure 8. The energies are much less satisfactory than those of Canuto et al (1974a). Schiff's results get better as the density increases, whereas Pandharipande's get worse. This is rather surprising in view of the fact that the basic merit of LOCV was precisely the ability to handle the high density regime.

Having learned from the ^{3}He case the importance of the solid state term $\mathbf{r} \cdot \mathbf{\Delta}$, we must doubt the validity of the subsequent application of LOCV to the investigation of the importance of the l-dependence of $V(NN)$ in the solidification problem. For this reason we believe that the meaningful part of Table 6, where Pandharipande explores the importance of the l-dependence of nuclear forces, is only in the first and second case, where there is one potential for all partial waves, either purely repulsive or slightly attractive. Case 3 cannot be considered definitive in that the energies are too close. Case 4 is clearly in favor of a liquid instead of a solid structure, however.

Even if the physical idea is correct, that is, that the presence of too much attraction will favor the liquid, it is our opinion, based on the previously presented arguments and experimental facts on ^{3}He, that Pandharipande's (1973) way of handling the dynamic equation is not convincing.

3.3 *Critical Review of the Work on the Solidification Problem*

The results of the calculations so far performed are presented in Figure 10. Even supposing that all of them are correct, the spread in the reported solidification densities is still too large. Several considerations have to be made. First, not all the computations have the same degree of reliability and just looking at the various results without having this in mind can be highly misleading. Second, none of the computations have yet been repeated by any other group to check the results. An independent check, especially of the more microscopic detailed computations, would be highly desirable.

The results of Nosanow and Parish are inscrutable in that they are almost totally numerical and, since the present author is far from being an expert on Monte Carlo methods, one cannot but hope that the authors would check time and again the sensitivity of their results to the possible numerical pitfalls. The results based on the law of corresponding states, after a pleasant and even convincing

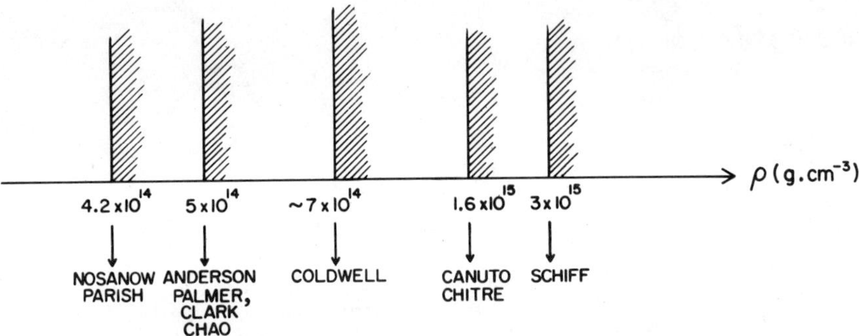

Figure 10 Present status of the results on the solidification problem.

start, face the serious problem of being too severely linked to the possibility of representing $V(NN)$ in a Lennard-Jones fashion. Even supposing that the core of $V(NN)$ is hard enough (which it is probably not) to be represented by a LJ form, still its fundamental feature of being state dependent cannot be accounted for by a simple analytic form. Despite several clever tricks devised by many people, the idea seems to have exhausted its fruitfulness.

Coldwell's results are more difficult to esteem. In fact his claim that the lack of a correlation function at short distances equally affects the solid and liquid, so as to make the results correct on a relative basis, is difficult to check.

Logically the next step is to perform a microscopic calculation. One way of dealing with that problem is due to Schiff. Here the impression is that a system of hard spheres is probably not a good representation and that in assuming such a system one actually introduces too much repulsion. One can always argue that the same amount of repulsion is also present in the liquid and therefore it should not matter very much. Even so, one cannot help feeling that if the system that Schiff has studied indeed solidifies, it has little to do with a real assembly of neutrons, whose repulsion is not, as far as we know, a hard-core. Here too, there is no way to improve upon the computation unless one has the corresponding solution of 256 Schrödinger equations with a potential less repulsive than a hard-core. This would render Schiff's treatment much more realistic.

This leaves us with only two microscopic computations by Canuto and Chitre on one hand and Pandharipande on the other. Canuto and Chitre based their reason for choosing the t-matrix on 1. the excellent results obtained for ^3He (Figure 8); 2. the smallness of the parameter κ, indicating the relative importance of high-order corrections; and 3. the possibility of dealing exactly with the state dependence of $V(NN)$. On the other hand, Pandharipande preferred the variational method (LOCV) to deal with points 1 and 2 because he believed he would have a better way of handling the high-density region. We have already stressed the poor performance of the LOCV method for liquids (Figure 7) and solids (Figure 8) in the high-density region.

In addition to the purely many-body aspect of the problem, one must add the series of approximations made in the case of the solid. In fact, had Pandharipande solved the dynamic equation correctly, the comparison of the results would have been very significant. As shown in Table 6, Pandharipande's contention is that by letting the potentials go from purely replusive to more realistic, the solid structure is not preferred over the liquid.

We have already stressed how this statement is not fully convincing; it is unfortunately based on an equation of motion that has, from the very beginning, been devoided of its angular-momentum features.

In the only two instances in which that equation of motion can be used with confidence, the solid actually exists (Table 6, first two cases).

On the other hand, in Canuto and Chitre's calculation, there is one unsatisfactory feature, whose implications have not been fully understood. The term $\mathbf{r} \cdot \mathbf{\Delta}$, being parity-violating, couples l with $l \pm 1$, that is, brings into the problem odd waves that should not be present. At high densities, the potentials of those waves are repulsive and therefore help solidification, so to speak. Such help should not be there, and only when a method is devised to eliminate such a spurious effect will the role of the attraction be ascertained.

In a still unpublished work, R. A. Guyer has succeeded in eliminating the unwanted waves. When he deals with a purely repulsive potential, equation 2.9, his energies are in good agreement with those of Canuto, Lodenquai & Chitre (1974b), as are those of Chakravarty et al (1974). Unfortunately, they all disagree with those of Pandharipande (column 1, Table 6), indicating that even for a unique state-independent potential, the LOCV method gives too high an energy for the solid. This is in line with what was shown in Figure 7, where the LOCV produced too high an energy for the liquid.

In conclusion, the whole problem seems to have to wait until the odd-waves contribution in Canuto and Chitre's treatment has been removed in order to check the trend of Table 6, that is, that the softness of Reid's potential prevents solidification. At the same time, within the variational framework it seems clear that the LOCV has shown its limitation in both liquid and solid phase and that a more reliable method has to be worked out.

3.4 *The Equation of State in the Solid Region*

Due to the still uncertain nature of the solid core and the rather small changes that its existence would imply in the relation $P = P(\varepsilon)$, we feel that one can safely treat this region as a liquid.

4 THE REGION $\rho > 8 \times 10^{15}$ g cm^{-3}

4.1 *Generalities*

From a judicious appraisal of the difficulties encountered in the two previous sections, we must derive a very clear if somewhat negative message. Once we exceed a density 10 times the nuclear density, the behavior of matter can no longer be described by the two conventional tools: 1. nonrelativistic many-body theories and 2. the concept of a potential.

The inclusion of hyperons offered the possibility of keeping the energy safely nonrelativistic and this prompted the extension of nonrelativistic many-body theories to the hyperonic region. The hope did not last long, because the lack of detailed knowledge of the hyperonic potentials seriously undermined the reliability of the results. Even assuming that hyperons can be used to keep the energy nonrelativistic, it is clear that even the most trustworthy many-body methods cannot be stretched beyond 10 times nuclear densities. What about the *NN* potential? Even before one reaches densities at which the concept of a potential itself breaks down, we are already facing another problem. Different potentials, equally reliable on the basis of the phase-shift fitting, give rise to unpleasantly different ground-state energies. This is a simple manifestation of the fact that the region around 10^{15} g cm^{-3} is extremely sensitive to the hard-core region that is left untouched by the phase-shifts criterion. In addition to this strictly technical question, the concept of static or quasistatic *NN* potential is bound to break down. As the density increases, the meson

Table 6 Energy per particle (MeV) vs density (fm^{-3}) for different types of NN potentials

n_B	Boltzman Statistics				Fermi Statistics			
(fm^{-3})	Solid	Liquid	Solid	Liquid	Solid	Liquid	Solid	Liquid
0.4	245.6	229	—	—	—	—	—	—
0.6	408.2	399.8	—	—	—	—	—	—
0.8	589.3	594.5	—	—	—	—	—	—
1.0	788	810.5	—	—	—	—	—	—
1.2	1002	1043	370.9	337.3	364.1	334.8	—	—
1.6	1461	1543	545.7	526.7	536.8	508.6	—	—
2.0	1961	2085	738.8	736.3	729.0	702.0	—	—
2.4	2494	2660	948.1	959.6	939.3	909.7	802.9	697.8
2.8	3054	3263	1172	1204	1165	1137	1010	893.4
3.2	3637	3890	1410	1456	1405	1373	1232	1105.2
3.6	4240	4527	1656	1720	1656	1621	1467	1332.8
	$V(r) = 6484.2\, e^{-7x}/x$ $x = 0.7r$		$V(r) = \frac{1}{4}[V(^1D_2)+3V(^3P_2)]$		$V(r) = V(^1D_2)$ $S = 0$ $V(r) = V_c(^3P_2)$ $S = 1$		$V(r) = V(^1S_0)$ $S = 0,\ l = 0$ $V(r) = V(^1D_2)$ $S = 0,\ l \geqq 2$ $V(r) = V_c(^3P_2)$ $S = 1$	

degrees of freedom cannot be eliminated in favor of a NN potential. They must be explicitly taken into account.

All these considerations call for a relativistic treatment of a many-particle system. The work that has been published in this area and that we are about to review is still of an exploratory nature. This is not a point of demerit. In a density region such as the one we are about to describe, the unknowns are so many and the pitfalls so frequent and unexpected that any type of incursion is useful, if for nothing else but to eliminate a few cases from the list of possible alternatives.

4.2 *Relativistic Hadronic Lagrangian*

Let us consider a system of nucleons interacting via scalar and vector mesons. We shall omit the pseudoscalar π, because at high density it cannot play a significant role. The scalar interaction is a simulation of a 2π system that dominates the NN interaction at intermediate densities. Even if unnecessary for most of the presentation to follow, we will think of the vector meson as the ω meson. This identification will become useful when we need the appropriate coupling constant.

The Lagrangian describing such a system is ($\hbar = c = 1$)

$$\begin{aligned}\mathscr{L} = &-\bar{\psi}(\gamma_\mu \partial_\mu + M)\psi - \tfrac{1}{2}(m_s^2\phi^2 + \phi_\mu^2) - \tfrac{1}{2}m_v^2 A_\mu^2 \\ &- \tfrac{1}{4}F_{\mu\nu}^2 + g_v \bar{\psi}\gamma_\mu \psi A_\mu + g_s \bar{\psi}\psi\phi.\end{aligned} \tag{4.1}$$

The first term is the Dirac Lagrangian for free fermions, the second the Lagrangian for a free scalar boson, the third and fourth the Lagrangians for a massive vector field (Proca field, $F_{\nu\mu} = A_{\mu\nu} - A_{\nu,\mu}$), whereas the fifth and the sixth terms are the interaction terms.

There seems to be little point at the moment in discussing how realistic equation 4.1

actually is. The derivation of a nucleon-nucleon potential is an art all by itself and it would be entirely out of place to get involved in any meaningful discussion of how many mesons ought to be included in a Lagrangian in order to be in a position to trust the resulting $V(NN)$. Needless to say, many more mesons than just a scalar and a vector are needed. If Lagrangian 4.1 does not contain all the necessary mesons, it surely does not leave out the most important ones, one attractive and one repulsive. We therefore consider 4.1 as a good reasonable starting point.

4.3 *Review of the Results*

4.3.1 THE MEAN-FIELD APPROXIMATION From equation 4.1 we can derive the equation of motion for the three fields ψ, ϕ, and A_μ as

$$(\gamma_\mu \partial_\mu + M - g_v \gamma_\mu A_\mu - g_s \phi)\psi = 0 \tag{4.2}$$

$$(\Box^2 - m_s^2)\phi = -g_s \bar{\psi}\psi \tag{4.3}$$

$$(\Box^2 - m_v^2)A_\mu = -g_v \bar{\psi}\gamma_\mu \psi + \partial_\nu \partial_\mu A_\nu. \tag{4.4}$$

In principle one can think of eliminating ϕ and A_μ in favor of ψ, via 4.3 and 4.4. The solution of 4.2 will then yield the nucleon spinor, ψ. If so, the Lagrangian 4.1 can be rewritten as a function of ψ only, thus providing the desired $P = P(\varepsilon)$ relation through the evaluation of the diagonal terms of the energy momentum tensor

$$T_{\mu\nu} = \mathscr{L}\,\delta_{\mu\nu} - \frac{\partial \mathscr{L}}{\partial \psi_\mu}\psi_\nu \qquad \psi_\mu \equiv \partial\psi/\partial_{x_\mu}. \tag{4.5}$$

Needless to say, such a program has no chance of being implemented unless one resorts to some kind of approximation. The method employed can be called a mean-field approximation or, alternatively, a relativistic Hartree approximation. It was first used by Marx (1956) and by Marx & Nemeth (1964) for a purely scalar field, and more recently rediscussed in the context of high-density matter by Kalman (1974). It was first employed for a pure vector interaction by Zeldovich (1962), who made an extensive study of the high-density limit of the relation $P = c_s^2\varepsilon$, both classically and quantum mechanically, reaching the well-known conclusion that $c_s^2 \to 1$, as $\varepsilon \to \infty$, that is, at superhigh density the velocity of sound approaches the velocity of light. From now on, the question of the asymptotic behavior of c_s^2 will frequently arise and we had better prepare the reader for a series of discrepant results. Maximum stiffness corresponds to $c_s^2 \to 1$, whereas maximum softness corresponds to $c_s^2 \to 0$. For a free gas, the well-known result is $c_s^2 \to \frac{1}{3}$. In the opinion of the author, the balance is presently tilted in favor of $c_s^2 \to 1$, even though it is very hard to find data, astrophysical or otherwise, through which to settle the question unequivocally. The more extensive study of 4.1 for both scalar and vector interaction has been performed by Walecka (1974); we review this work.

In the spirit of the mean-field approximation, we perform an average on both sides of 4.3 and 4.4, by substituting ϕ and A_μ with $\langle\phi\rangle$ and $\langle A_4\rangle$. The solution of 4.3 and 4.4 is then trivial. Upon substituting $\langle\phi\rangle$ and $\langle A_4\rangle$ in 4.2, we obtain a new Dirac equation in which both the mass and the energy get renormalized. Once

Dirac's equation is solved, ψ is substituted back into 4.3 and 4.4, thus providing the self-consistent solutions for $\langle \phi \rangle$ and $\langle A_4 \rangle$. The equation of state is then readily obtained. We will not write down the analytic expression, equations 3.51 and 3.56 of Walecka's paper. By inspecting the $P = P(\varepsilon)$ relation, it is easy to show that as $\varepsilon \to \infty$, the contribution of the scalar meson is negligible compared to the vector meson. In the same limit $\varepsilon \to \infty$, Walecka recovers the Zeldovich result:

$$P = c_s^2 \varepsilon, \qquad c_s^2 \to 1. \tag{4.6}$$

The numerical values for Walecka's $P = P(\varepsilon)$ are presented in Table 7.

Even though several points remain to be studied, for example, the stability of the ground state against fluctuations and the inclusions of more mesons, the idea of a

Table 7 Equation of state for a pure neutron gas after Walecka (1974)

n_B (fm^{-3})	$\rho = \varepsilon/c^2$ (g cm^{-3})	P (dynes cm^{-2})	$P/\rho c^2$
0.70	1.149×10^{15}	2.829×10^{35}	0.273
0.80	1.367×10^{15}	4.066×10^{35}	0.330
0.90	1.604×10^{15}	5.469×10^{35}	0.379
1.0	1.859×10^{15}	7.034×10^{35}	0.420
1.1	2.132×10^{15}	8.756×10^{35}	0.456
1.25	2.576×10^{15}	1.163×10^{36}	0.501
1.4	3.060×10^{15}	1.484×10^{36}	0.539
1.6	3.768×10^{15}	1.964×10^{36}	0.579
1.8	4.547×10^{15}	2.503×10^{36}	0.611
2.0	5.397×10^{15}	3.100×10^{36}	0.638
2.2	6.315×10^{15}	3.755×10^{36}	0.661
2.4	7.303×10^{15}	4.468×10^{36}	0.680
2.6	8.360×10^{15}	5.239×10^{36}	0.696
2.8	9.485×10^{15}	6.067×10^{36}	0.711
3.0	1.068×10^{16}	6.952×10^{36}	0.723
3.2	1.194×10^{16}	7.894×10^{36}	0.735
3.4	1.327×10^{16}	8.893×10^{36}	0.745
3.6	1.466×10^{16}	9.948×10^{36}	0.754
3.8	1.612×10^{16}	1.106×10^{37}	0.762
4.0	1.765×10^{16}	1.223×10^{37}	0.770
4.2	1.925×10^{16}	1.345×10^{37}	0.777
4.4	2.091×10^{16}	1.474×10^{37}	0.783
4.6	2.264×10^{16}	1.607×10^{37}	0.790
4.8	2.443×10^{16}	1.747×10^{37}	0.794
5.0	2.629×10^{16}	1.892×10^{37}	0.800
5.2	2.822×10^{16}	2.043×10^{37}	0.804
5.4	3.021×10^{16}	2.199×10^{37}	0.809
5.6	3.226×10^{16}	2.361×10^{37}	0.813
5.8	3.438×10^{16}	2.529×10^{37}	0.817
6.0	3.657×10^{16}	2.702×10^{37}	0.821

constant mesonic field at superhigh density seems a fruitful one. The advantage of this model, when compared with the ones based on the concept of a $V(NN)$ potential, is that it is not a perturbative approach with the coupling constant as a smallness parameter. In a restricted sense, the coupling constant is treated exactly to all orders of perturbation theory. From the many-body point of view, clearly only the lowest-order many-body diagrams have been included. For example, neither the nucleon self-energy nor the polarization tensor in the equation of motion for the mesons are included in this formalism.

4.3.2 THE PERTURBATIVE APPROACH An effort in this direction has been recently undertaken by Zimmerman and collaborators (Bowers & Zimmerman 1973a,b, Bowers, Campbell & Zimmerman 1973a–c). From the relativistic many-body theory it is known that the pressure and energy density can be computed if one knows the two-point Green's function, solution of the Dyson equation

$$(\gamma_\mu \partial_\mu + M + \Sigma)G = 1. \tag{4.7}$$

The general expression for Σ requires the knowledge of G itself and the exact meson Green's functions, say $\mathscr{D}$. In turn $\mathscr{D}$ is given as a function of the polarization tensor, Π, whose computation again requires the knowledge of G. This formidable chain of self-consistent equations is approximated by the authors at the first stage, that is, in the evaluation of Σ, by identifying $\mathscr{D}$ and G with the corresponding free-particle Green's functions, thus obtaining for Σ

$$\Sigma = ig^2 \int \frac{d^4k}{(2\pi)^4} \gamma G^{(0)} \gamma \mathscr{D}^{(0)}. \tag{4.8}$$

Given the exploratory nature of this analysis, this series of approximations seems natural; if the incentive is great enough, one could look for the changes brought in by the inclusion of more terms. However, this is probably not the most significant part. In fact, a more serious difficulty is represented by the treatment of the coupling constant only up to the second order. In this respect, the present treatment faces the same kind of criticism as those employing a NN potential. A convergence analysis would imply the evaluation of not just one, but several terms in powers of g^2, with the almost inevitable result that such a series does in fact not converge under the present circumstances. In our opinion this is the weakest and most difficult to amend feature of the whole treatment.

The explicit evaluation of 4.8 is carried out by taking $\gamma \equiv \gamma_5$, that is, by supposing that the nuclear force is mediated by pions. Unfortunately, this is not a very realistic choice for the high-density regime, where the pions will clearly play a negligible role and the vector mesons will undoubtedly dominate, as we have seen explicitly in Walecka's work. It turns out that in the high-density regime the present computations predict $c_s^2 \to \frac{1}{3}$. This has given rise to some misunderstandings about the behavior of the velocity of sound as $\varepsilon \to \infty$ (Ruffini 1973, Cameron & Canuto 1973). Although we have no quarrel with the relation obtained by Zimmerman and collaborators, we would like to stress that it cannot be taken as a proof that c_s^2 actually approaches $\frac{1}{3}$ in the real world. The exchange of pions is a good representation of the low-density regime and the extrapolation of the resulting $P = P(\varepsilon)$ relation to the high-density regime has only an academic interest. We do not know with any certainty the behavior of c_s^2 as the density increases. The whole point, however, is that the present calculation cannot be used in any meaningful way in that respect.

If, instead of a pion, the authors had exchanged a vector meson, the comparison with Walecka's results would have been more instructive. Due to the general nature of the Zeldovich argument, we feel that the relativistic treatment of Zimmerman and collaborators is bound to recover the relation $c_s^2 \to 1$, when the vector meson is introduced. It is indeed hard to believe that the introduction of the self-energy operator could change the density dependence in such a way as to upset the $c_s^2 \to 1$ relation. For the sake of completeness, we reproduce below the $P = P(\varepsilon)$ relation obtained by Zimmerman and collaborators, even though it can be used meaningfully only in the low-density regime.

$$P = P_0 x^4[1-24x^{-4}+4\,e^{-x}\,x^{-4}(6+6x+3x^2+x^3)]$$

$$\varepsilon = \varepsilon_0[9/4(8+x^4)-9\,e^{-x}(2+2x+x^2)-2x^3] \qquad 4.9$$

$$P_0 = \frac{729}{500}\frac{M^4c^5}{\pi^2\hbar^3} \qquad \varepsilon_0 = \tfrac{4}{3}P_0 \qquad x = \frac{5}{9}\frac{\hbar k_F}{Mc}.$$

4.3.3 THE HAGEDORN-TYPE FORMULATIONS The two previous investigations, though different in their conception and technical aspects, nevertheless have several features in common. The work we are about to review departs radically from any field theoretical consideration and the results are therefore more difficult to compare with anything we have already presented.

Even supposing that we could clear up all the problems that still plague the two previous treatments, there always remains the possibility that we should actually go beyond the basic baryonic octet and include the whole host of baryonic resonances that comprise about 354 baryons up to 2.2 GeV. We do not actually know if such excited states ought to be included or not, but it is clear that should the answer be positive, none of the previous formulations would be adequate for such a job.

We review in what follows a series of papers that have dealt with this problem by postulating that such a host of baryonic states should indeed be accounted for because it actually determines the behavior of the $P = P(\varepsilon)$ relation at superhigh densities. We also present a serious criticism that has caused some misgivings about the validity of the models.

The simplest way to evaluate the equation of state for a system of excited baryons is to suppose that the constituents are free, but with their masses adjusted to account for the interaction. Sawyer (1972) first noticed that the masses of the resonances that we read from the tables refer to free decaying resonances and may be significantly different from the ones we are dealing with in a dense medium. Since the mass shift is actually unknown, Leung & Wang (1973) postulated that the effective mass spectrum due to the mass shifts obeys a power law, that is, that the number of baryons below a certain mass m can be represented as

$$N(m) = Am^{2a}. \qquad 4.10$$

The Equation of state for a free ensemble of baryons with such a mass spectrum is easily found with the result

$$P = c_s^2 \varepsilon, \qquad c_s^2 = (3+2a)^{-1}. \tag{4.11}$$

If we postulate that the parameter a is the same as the one obtained by fitting equation 4.10 to the experimentally known masses, then $a = 2.9$ and $c_s^2 \rightarrow 0.11$, an extremely low value, if compared with the previous values of $\frac{1}{3}$ or 1.

Leung and Wang's work is a particular case of a more general line of approach due to Hagedorn (1970), who proposed a quasifree or asymptotically free model for baryonic matter by pointing out that, at very high densities, the interaction may be (partially?) accounted for by precisely including the whole spectrum of baryonic states. The point being made is that the existence of the baryonic states is per se a manifestation of the interaction and therefore instead of trying the impossible job of taking into account the interactions, one just counts the existing baryons, establishes a mass spectrum of the form 4.10, or more complicated (Hagedorn 1970), and then treats the system as free (Frautschi et al 1971, Wheeler 1971). The idea is very simple and very appealing. A general feature common to all the results obtained with the previous method, or modification of it, is that the resulting equation of state is very soft. We quote here an expression due to Wheeler (1971) and Hagedorn (unpublished):

$$\begin{aligned} P &= \rho c^2 [\ln(\rho/\rho_0)]^{-1} \\ \rho_0 &= 2.5\ 10^{12}\ \text{g cm}^{-3}. \end{aligned} \tag{4.12}$$

For $\rho \rightarrow \infty$, the velocity of sound goes to zero, contrary to $c_s^2 \rightarrow 1$, derived before. How can we assess the value of such theories? We discuss in the next section the possible asymptotic values of c_s^2 and what the experimental data seem to indicate. For the time being, we would like to present a theoretical criticism to the previous way of computing $P = P(\varepsilon)$, due to Sawyer (1972).

In one way or the other, the previous approaches rely on the existing data on excited baryonic states to fix the parameters of the supposed mass spectrum. This is clearly true in the polynomial fit of Leung and Wang, and even if it is not trivially clear in the Hagedorn type of approach, where the parameters of the more complicated spectrum

$$N(m) = m^{\alpha} \exp(\beta m) \tag{4.13}$$

are derived theoretically, it is a common practice to show that such behavior at least does not contradict the existing data. The main point is that the data in one way or another are actually used. Sawyer has made the valid point, however, that such masses refer to free decaying baryons, whereas the ones we are interested in concern particles imbedded in a dense medium. We said before that the Hagedorn formulation can be rephrased by saying that the chemical potential of one particle

$$\mu = (m^2 + p^2)^{1/2} + U(p) \tag{4.14}$$

can actually be rewritten as

$$\mu = (m^{*2} + p^2)^{1/2}, \tag{4.15}$$

that is, the two-body interaction can be absorbed in an effective mass. Sawyer's point is that the most important contribution to the effective mass is not actually originating from the two-body potentials, but from the mass shift that the dense surrounding medium exerts on that particle. Sawyer computed the mass shift by evaluating the self-energy operator Σ (see equation 4.7) for a Δ^- (1236 MeV) immersed in a neutron medium. The energy shift increases with density and goes from $+140$ to $+400$ MeV in the density region $0.5 \leqq n_B \leqq 5$ (fm^{-3}), even in the absence of a two-body interaction.

The effect is of a general nature and is therefore expected to hold for any other type of resonance. If so, two consequences are immediately obvious:

1. Because each resonance is actually heavier than we thought, it will require a higher density to have it come in and all the hyperonic configurations presented before can therefore be drastically changed. For instance, the appearance of Δ^- at 1.5×10^{15} g cm^{-3} (Figure 6) can actually be quenched. The same argument would apply to all the previous computations
2. The baryonic-level density formula, as used previously, seems now quite unlikely. In fact the energy shift can become so large that no bound state beyond the basic baryonic state is ever populated for any density. If so, the equation of state would be actually much harder than the one proposed by the Hagedorn type of formalism

4.4 *Concluding Remarks on the High-Density Region*

We can only concur with Sawyer that his computation has actually shown, if not the incorrectness of treating the baryonic resonances as a free system with the mass spectrum given by the experimental values, at least the strong need for a much more careful examination of the treatment. In view of this serious difficulty, we conservatively regard Walecka's computation as the most adequate and reliable way, presently known, of describing the region $\rho \geqq 8 \times 10^{15}$ g cm^{-3}.

5 THE RELATION $P = c_s^2 \varepsilon$ IN THE LIMIT $\varepsilon \rightarrow \infty$

5.1 *Generalities*

We concluded Section 2 on the hyperonic liquid suggesting that the most reliable $P = P(\varepsilon)$ relation in the interval $2 \times 10^{14} \leqq \rho \leqq 7.7 \times 10^{15}$ g cm^{-3} is the one provided by Bethe and Johnson. We concluded Section 4 indicating how Walecka's work can be taken to continue the previous equation of state from 8×10^{15} g cm^{-3} upwards. If so, we can consider concluded the problem of the high-density behavior of matter. The feeling remains, however, that in deciding among the various possibilities, we should have presented what the experimental data, astrophysical or otherwise, suggest, instead of using theoretical arguments only.

We have intentionally used theoretical arguments only because the experimental data have been, until very recently, of little or no use at all. The situation has now somewhat changed and we describe in what follows the way one can use the high-energy data on *p-p* collisions in a meaningful way.

5.2 *Neutron Stars*

Despite the fact that neutron stars are indeed the denser objects presently available in astrophysics, they are not dense enough to allow us to ascertain the behavior of matter at densities 10 times higher than nuclear density. Rhoades & Ruffini (1971) computed the mass and radius of a stable neutron star by using a very hard ($c_s^2 \to 1$) and a very soft ($c_s^2 \to 0$) equation of state starting at 10 times the nuclear density. The two possibilities are indicated in Figure 12.

The largely expected result was that neither quantity was sensitive to the high-density behavior of P vs ρ. Because neither mass nor radius can be measured, one often employs the moment of inertia. In the popular dipole model for the Crab nebula it is thought that the energy loss is optical and X-ray synchrotron radiation in the amount of (Baldwin 1971)

$$0.2 \times 10^{38} \left(\frac{\text{ergs}}{\text{sec}}\right) \frac{1}{(\text{kpc})^2}$$

must be replenished by the loss of rotational energy $I\Omega\dot{\Omega}$. Because the distance of the Crab pulsar is quite uncertain and it can be anywhere between 1.25 and 2.5 kpc (Trimble & Woltjer 1971, Börner 1973), it follows that I must at least be such as to satisfy the condition

$$0.288 \times 10^{38} < I\Omega\dot{\Omega}.$$

For the Crab pulsar $P = 2\pi/\Omega = 0.033$ sec, and $P\dot{P} = 1.4 \times 10^{14}$, so that I must be at least greater than 0.62×10^{44} g cm^2. If we assume an average distance of 2 kpc, then I must be greater than 1.72×10^{44} g cm^2. From the work of Rhoades and Ruffini, the moment of inertia for the two extreme cases can be computed and found to satisfy these lower bounds. This reiterates the well-known fact that observational properties of neutron stars cannot be used to study the behavior of matter at 10 times the nuclear density.

5.3 *High-Energy* p-p *Collisions: The Hydrodynamic Model*

Canuto & Lodenquai (1975) have recently proposed that some of the properties of high-energy *p-p* collisions could be the most useful data available to study the behavior of the speed of sound at superhigh density. Experimentally it is known that the collision of two energetic (10^2–10^4 GeV) protons is accompanied by the production of a host of other particles whose multiplicity increases with the energy of the incoming protons.

The two incoming protons are strongly Lorentz contracted in the direction of motion, the contraction factor being M/E_c. One possible model describing the *p-p* collision, the so-called statistical hydrodynamical model, calls for the formation, right after the collision, of a hot compressed disc with dimensions

$$V = \frac{4\pi}{3}\left(\frac{\hbar}{m_\pi c}\right)^3 \left(\frac{2Mc^2}{E_c}\right) \tag{5.1}$$

that subsequently expands under its own pressure. The energy density in the initial disc is easily computed to be

$$\rho = \varepsilon/c^2 = E_c/c^2 V = 1.5 \times 10^{14} E_L \text{ (g cm}^{-3}\text{)}, \tag{5.2}$$

where $2E_L = E_c^2$ is the lab energy in GeV. Typically, for $E_L \sim 10^3$ GeV, $\rho \sim 10^{17}$ g cm^{-3}. The hadronic matter inside the hot disc has therefore an energy density greater than the one encountered at the center of a neutron star. It is to be expected that the development of such a hadronic matter will strongly depend on the assumed equation of state. The original model for such a hot hadronic gas given by Fermi (1950), followed by a revision due to Pomeranchuck (1951), was systematically worked out in its full mathematical complexity and implications by Landau (1953). Continuous refinements and improvements on the original Landau model have been accomplished over the years, notably by Feinberg and his school [(1965); see (1972) for a full account of the model].

The central point of the Landau model is that the original hot disc evolves in time in a way that must be described by the relativistic Navier-Stokes Equation

$$[(P+\varepsilon)U_\mu U_\nu + P\delta_{\mu\nu}]_{,\nu} = 0. \tag{5.3}$$

This requires an equation of state. Landau, upon using $P = \frac{1}{3}\varepsilon$, arrived at a series of important relations that only recently, with the advent of new experimental data, have been restudied and reanalyzed. The program of many of the modern versions of the Landau model consists in solving 5.3 for a general c_s^2, whose value is then adjusted to fit the data. Needless to say, the job is difficult and several alternatives exist, but the important point is that, contrary to the neutron star case, there is a clear prospect of achieving some positive results.

Suhonen et al (1973) analyzed two types of data: the multiplicity N vs energy and the distribution of $N(\eta)$ vs η, the so-called rapidity distributions. The multiplicity was found to be

$$N \sim E^{(1-c_s^2)/(1+c_s^2)}. \tag{5.4}$$

If 5.4 is correct, we must conclude that c_s^2 cannot be 1, because we know that N increases with E. The authors' contention is that c_s^2 should be close to 0.28 or $\frac{1}{3}$ in order to fit the existing data. The argument is not correct, however. In the first place, formula 5.4 is given only within a log E term, because it is derived from statistical mechanics. This alone implies that $c_s^2 = 1$ is perfectly legitimate, since the multiplicity can easily be fitted by log E, if the fitting is started at, say, 10^2 GeV, as one should, for the statistical model is certainly not applicable at lower energies. In the second place, it is not at all clear that the measured multiplicity has anything to do with the one computed from 5.4. Many intermediate processes have intervened and the prehistoric age to which 5.4 refers has been obliterated to the point of being irrelevant.

The rapidity distribution curve specifically depends upon the Landau model and the authors claim to have solved 5.3 numerically for $P = c_s^2\varepsilon$. Two sets of experimental data from the Pisa-Stony Brook collaboration have been employed and the result is that c_s^2 can be either $\frac{1}{2}$ or $\frac{1}{3}$ but not 1. We cannot express any sound opinion on this second point because we do not know how accurate the numerical

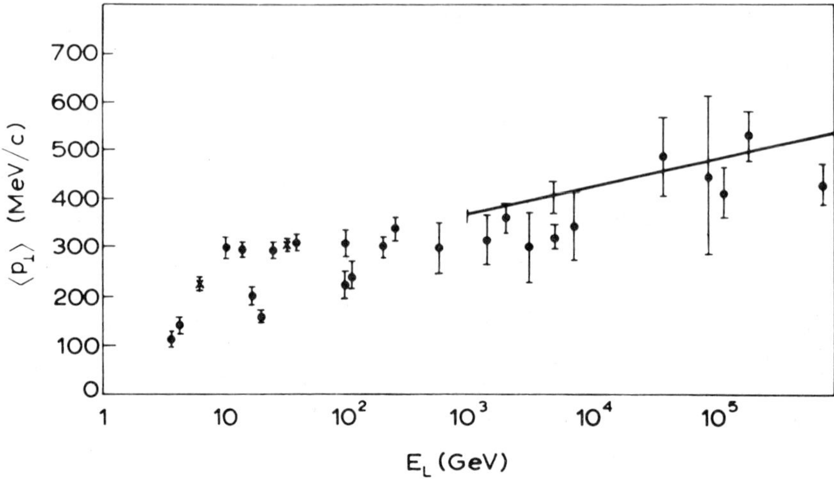

Figure 11 Transverse momentum vs E_L. The data are from Feinberg (1972). The solid line corresponds to $c_s^2 = \frac{1}{3}$.

solution actually is. The boundary conditions imposed by Landau are rather tricky and can lead to inconsistencies if not treated properly. We have to reserve judgement on this point.

In recent work by Satz and collaborators (Chaichian et al 1974) the point was made that the more reliable experimental quantity is probably the constancy of the transverse momentum with respect to the energy of the incoming protons. This fact, known for many years, prompted Cocconi (1959) to propose the existence of an ultimate temperature, an idea at the very basis of the Hagedorn model. Using the solution of the Landau model for a general c_s^2, Satz and collaborators derived the following expression for the transverse momentum

$$p_T = p_0 F^{\frac{c_s^2(1-c_s^2)}{1+c_s^2}}. \tag{5.5}$$

In Figure 11 we present the fit to several experimental points as from $c_s^2 = \frac{1}{3}$, that is, $p_T \sim E^{1/6} \sim E_L^{1/12}$. The fit is indeed excellent. Very satisfactory agreement can also be obtained by using $c_s^2 = 1$, since both 5.4 and 5.5 are given to within a

Table 8 Value of the velocity of sound (in units of c) from several pionic Lagrangians

Lagrangian	$L_0{}^{\mathrm{a}} - \lambda\phi^{2n}$	$L_0 - \lambda\phi^{2n} - v\phi^{2k}$	$[1 - l^4(\dot{\phi}_\mu^2 - m_\pi^2\phi^2)]^{1/2}$	$\frac{\phi_\mu^2}{(1+F^{-2}\phi^2)^2} + \frac{m_\pi^2\phi^2}{(1+F^{-2}\phi^2)}$
c_s^2	$\frac{n-1}{n+1}$	$\frac{n-k}{2kn+k-n}$	0	1
Constancy of p_T	yes if $n \gg 1$	yes if $n-k \sim kn$	?	yes

^{a} L_0 is the Free-pion Lagrangian.

ln E term. The value $c_s^2 = 0$ is excluded because with no pressure, there is no hydrodynamic expansion. As discussed by Canuto & Lodenquai (1975) the value $c_s^2 = 1$ is perhaps preferable since the ln E behavior so obtained is the same as the one predicted by the multiperipheral model. We therefore assume that c_s^2 indeed goes to one as the energy density goes to infinity. This being the case, one can not only have a handle on the high-density behavior of the equation of state, but also use the relation $c_s^2 \to 1$ ($\varepsilon \to \infty$) to decide among several pionic Lagrangians that have been proposed over the years. Canuto & Lodenquai (1975) analyzed four well-known cases with the results presented in Table 8.

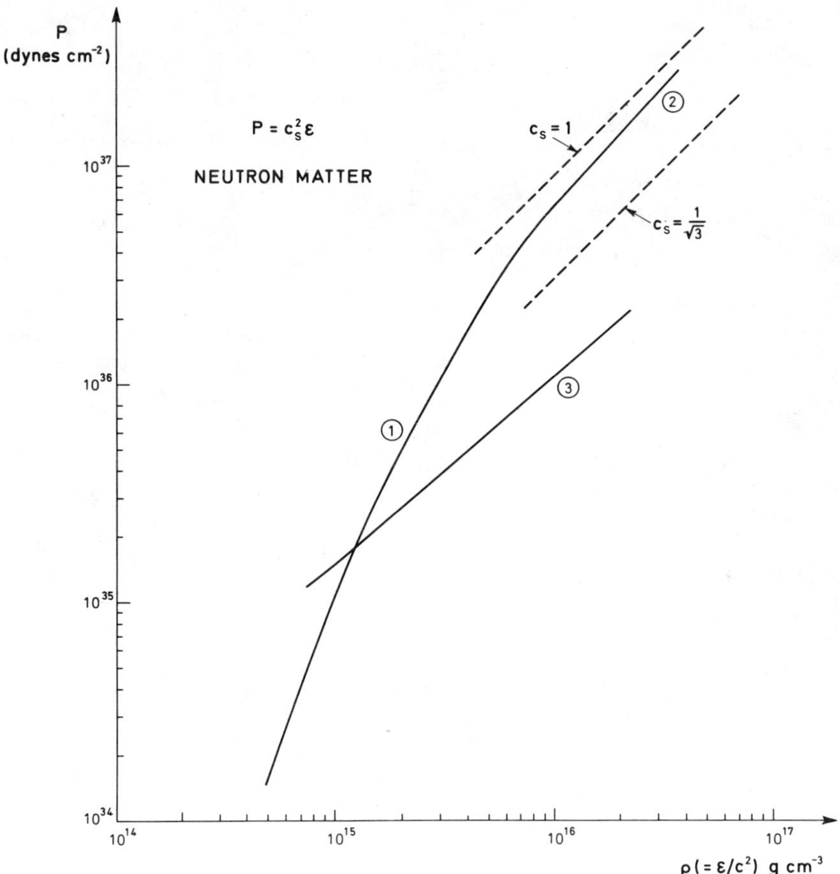

Figure 12 The "best" composite $P = P(\varepsilon)$ relation presently available. Up to 7.7×10^{15} g cm^{-3} (*curve 1*) we employ the Bethe-Johnson results. From 7.7×10^{15} g cm^{-3} (*curve 2*) we joined smoothly with the relativistic expression of $P = P(\varepsilon)$ given by Walecka (1974). Also shown are the causality limit $c_s = 1$ and the free-particle case $c_s = 1/(3)^{1/2}$. The Hagedorn-Wheeler equation of state (*curve 3*) is also shown for comparison.

The first two cases were already studied by Milekhin (1962). The third Lagrangian, originally proposed by Born & Infeld (1934), was used in this context by Heisenberg (1952). The last one has been recently proposed by Weinberg (1968). The first and second case can fit the constancy of p_T under appropriate restrictions on the indices n and k. The third one gives $c_s^2 \to 0$ and therefore no hydrodynamic expansion. The last Lagrangian satisfies the p_T test because it yields $c_s^2 \to 1$. It is the most modern and reliable Lagrangian and it is comforting that it is also in agreement with the behavior of matter at very high density.

5.4 *Conclusions*

We conclude this review by suggesting what, in our opinion, is the most reliable equation of state one can presently have covering the range of 10^4–10^{18} g cm^{-3}:

1. Up to a density of 2×10^{14} g cm^{-3}, any of the relations $P = P(\varepsilon)$ presented in Table 5(1) can be used
2. From 2×10^{14} up to 7.7×10^{15} g cm^{-3}, one can use any one of the $P = P(\varepsilon)$ relations given in Tables 10–12 of Part 1, or Table 4 of the present article. Because we do not actually know how reliable the computations that include hyperons are, it is perhaps more appropriate to regard this density regime as a neutron liquid. This will limit the choice to Tables 10 (Part 1) or 4*a*. We would suggest Table 4*a*
3. Starting at 7.7×10^{15} g cm^{-3}, we can adopt Table 7. Such a composite relation is shown in Figure 12, curves 1 and 2, together with the free-particle case ($c_s^2 = \frac{1}{3}$), the causality limit ($c_s^2 = 1$), and the Hagedorn-Wheeler equation 4.12, curve 3.

In conclusion, we would like to stress that, in spite of the impressive advances accomplished in the past few years, the problem of high-density matter cannot be considered satisfactorily understood unless further study and work is undertaken to reveal the missing facets of the hyperon liquid, the neutron solid, and the superhigh-density regime.

ACKNOWLEDGMENTS

The author would like to thank Professors B. Strömgren, A. Bohr, B. Mottelson, and J. Hamilton for their kind hospitality at NORDITA, Copenhagen, where this manuscript was written. He would also like to thank Drs. J. Lodenquai, C. G. Källman, and B. Datta for their thorough reading of the manuscript.

Literature Cited

Ambartsumyan, V. A., Saakyan, G. S. 1960. *Sov. Astron.* 4:187
Anderson, P. W., Palmer, R. G. 1971. *Nature Phys. Sci.* 231:145
Baldwin, J. E. 1971. *Proc. IAU Symp. No.* 46:22
Banerjee, B., Chitre, S. M., Garde, V. K. 1970. *Phys. Rev. Lett.* 25:1125
Bethe, H. A., Johnson, M. B. 1974. *Nucl. Phys. A* 230:1
Born, M., Infeld, L. 1934. *Proc. Roy. Soc. London Ser. A* 144:475
Börner, G. 1973. *Springer Tracts Mod. Phys.* 69:1
Bowers, R. L., Zimmerman, R. L. 1973a. *Ap. J.* 184:305
Bowers, R. L., Zimmerman, R. L. 1973b. *Phys. Rev. D* 7:296
Bowers, R. L., Campbell, J. A., Zimmerman, R. L. 1973a. *Phys. Rev. D* 7:2278
Bowers, R. L., Campbell, J. A., Zimmerman, R. L. 1973b. *Phys. Rev. D* 7:2289
Bowers, R. L., Campbell, J. A., Zimmerman, R. L. 1973c. *Phys. Rev. D* 8:1089
Brown, J. T., Downs, B. W., Iddings, C. K. 1970. *Ann. Phys.* 60:148
Brueckner, K. A., Coon, S. A., Dabrowsky, J. 1968. *Phys. Rev.* 168:1184
Buchler, J. R., Ingber, L. 1971. *Nucl. Phys. A* 170:1
Cameron, A. G. W. 1959. *Ap. J.* 130:884
Cameron, A. G. W., Canuto, V. 1973. *Neutron Stars: General Review. Int. Solvay Conf., 16th, Bruxelles*
Canuto, V. 1974. *Ann. Rev. Astron. Ap.* 12:167
Canuto, V., Chitre, S. M. 1973a. *Phys. Rev. Lett.* 30:999
Canuto, V., Chitre, S. M. 1973b. *Nature Phys. Sci.* 243:63
Canuto, V., Chitre, S. M. 1974. *Phys. Rev. D* 9:1587
Canuto, V., Datta, B. 1974. Preprint
Canuto, V., Lodenquai, J., Parish, L., Chitre, S. M. 1974a. *J. Low. Temp. Phys.* 17:179
Canuto, V., Lodenquai, J., Chitre, S. M. 1974b. *Nucl. Phys. A* 233:521
Canuto, V., Lodenquai, J. 1975. *Phys. Rev. D* 11:233
Cazzola, P., Lucaroni, L., Scarinci, C. 1966. *Nuovo Cimento B* 43:250
Chaichian, M., Satz, H., Suhonen, E. 1974. *Phys. Lett. B* 50:362
Chakravarty, S., Miller, M. D., Woo, C. W. 1974. *Nucl. Phys. A* 220:233
Clark, J. W., Chao, N. C. 1972. *Nature Phys. Sci.* 236:37
Cocconi, G. 1959. *Nuovo Cimento* 33:643
Cochran, S., Chester, G. V. 1973. Preprint
Coldwell, R. L. 1972. *Phys. Rev. D* 5:1273
de Boer, J. 1948. *Physica* 14:139
Feinberg, E. L. 1965. *Tr. Fiz. Inst. Akad. Nauk. SSSR* 29:155. Reprinted 1967 in *Quantum Field Theory and Hydrodynamics,* ed. D. V. Skobel'tsyn, 151. New York: Consultant Bureau
Feinberg, E. L. 1972. *Phys. Rep. C* 5:237
Fermi, E. 1950. *Progr. Theor. Phys.* 5:570
Frautschi, S., Bahcall, J. N., Steigman, G., Wheeler, J. C. 1971. *Comments Ap. Space Sci.* 3:121
Guyer, R. A. 1969. *Solid State Commun.* 7:315
Hagedorn, R. 1970. *Astron. Ap.* 5:184
Hansen, J. P., Levesque, D., Schiff, D. 1971. *Phys. Rev. A* 3:776
Heisenberg, W. 1952. *Z. Phys.* 133:65
Hetherington, J. H., Mullin, W. J., Nosanow, L. H. 1967. *Phys. Rev.* 154:175
Ingber, L., Potenza, R. M. 1970. *Phys. Rev. C* 1:112
Kalman, G. 1974. *Phys. Rev.* 9:1656
Kalos, M., Levesque, D., Verlet, L. 1974. To be published
Landau, L. D. 1953. *Izv. Akad. Nauk. SSSR* 17:31. Engl. Transl. 1965 in *Collected Papers of L. D. Landau,* ed. D. ter Haar. NY: Gordon & Breach
Langer, W. D., Cameron, A. G. W. 1969. *Ap. Space Sci.* 5:213
Langer, W. D., Rosen, L. 1970. *Ap. Space Sci.* 6:217
Leung, Y. C., Wang, C. G. 1973. *Ap. J.* 181:895
Marx, G. 1956. *Nucl. Phys.* 1:660
Marx, G., Nemeth, J. 1964. *Acta Phys. Acad. Sci. Hung.* 8:77
Milekhin, G. A. 1962. *Izv. Akad. Nauk. SSSR Ser. Fiz.* 26:635
Moszkowski, S. 1974. *Phys. Rev. D* 9:1613
Nosanow, L. H., Shaw, G. L. 1962. *Phys. Rev.* 119:968
Nosanow, L. H., Parish, L. J. 1974. VI *Texas Symp. Relativistic Ap., 6th,* 224:226. New York: NY Acad. Sci.
Østgaard, E. 1971. *J. Low Temp. Phys.* 5:237
Østgaard, E. 1972. *J. Low Temp. Phys.* 8:479
Otsuki, S., Tamagaki, R., Wada, M. 1964. *Progr. Theor. Phys.* 32:220
Palmer, R. G., Anderson, P. W. 1974. *Phys. Rev.* 9:3281
Pandharipande, V. R. 1971. *Nucl. Phys. A* 174:641
Pandharipande, V. R., Garde, V. K. 1972. *Phys. Lett. B* 39:608
Pandharipande, V. R., Bethe, H. A. 1973.

Phys. Rev. C 7:1312
Pandharipande, V. R. 1973. *Nucl. Phys. A* 217:1
Pomeranchuk, I. Ya. 1951. *Dkl. Acad. Nauk. SSSR* 78:889
Rhoades, C. E., Ruffini, R. 1971. *Ap. J. Lett.* 163:83
Ruffini, R. 1973. Discussion after a talk by A. G. W. Cameron and V. Canuto at XVI *Int. Solvay Conf., 16th* (see Cameron & Canuto 1973)
Salpeter, E. E. 1960. *Ann. Phys.* 11:393
Sawyer, R. F. 1972. *Ap. J.* 176:205
Schiff, D. 1973. *Nature Phys. Sci.* 243:130
Shen, L., Woo, C. W. 1974. *Phys. Rev.* 10:371
Suhonen, E., Enkenberg, J. Lassila, K. E., Sohlo, S. 1973. *Phys. Rev. Lett.* 31:1567
Trimble, V., Woltjer, L. 1971. *Ap. J. Lett.* 163:97
van Kampen, N. G. 1961. *Physica* 27:783
Walecka, J. D. 1974. *Ann. Phys.* 83:491
Weinberg, S. 1968. *Phys. Rev.* 166:1568
Wheeler, J. C. 1971. *Ap. J.* 169:105
Wu, F. Y., Feenberg, E. 1962. *Phys. Rev.* 128:943
Zel'dovich, Ya. B. 1962. *Sov. Phys. JETP* 14:1143

ASTROPHYSICAL PROCESSES NEAR BLACK HOLES[1]

Douglas M. Eardley[2]
Department of Physics, Yale University, New Haven, Connecticut 06520

William H. Press[2]
Department of Physics, Princeton University, Princeton, New Jersey 08540

1 INTRODUCTION AND OVERVIEW

Astronomy is restricted to studying phenomena that occur in our Universe. As regards black holes, this limitation is important, because the surface of a black hole is a boundary of our Universe, separating it from an interior region that is observationally unknowable—causally disconnected from us. Although theory predicts remarkably singular behavior for matter inside a black hole (and its predictions might therefore be suspect), these predictions are of no relevance to the external astrophysicist.

It is remarkable that the Universe can be equipped with boundaries of causality (called *event horizons*) that are compact and localized; even more remarkable when we note that these boundaries gravitate externally and behave—viewed from afar—like compact massive objects. They carry mass, angular momentum, and (in principle) charge; they can be tidally perturbed by an external mass distribution; they can accrete matter and become larger, or they can merge with each other. Details of these specifics are calculated using Einstein's general relativity [the Brans-Dicke (1961) theory of gravity gives almost identical results]. The physical necessity for black holes in most (but not all) relativistic theories of gravity is something like this: 1. causal structure is determined by the paths of photons (and other zero-rest-mass null particles) because no information-carrying particle can escape its own future light cone; 2. gravity exerts an attractive force on all particles, for example, a body redshifts photons moving outward and deflects inward those moving transversely; 3. the strength of gravity increases as a body becomes more compact; so, typically, at some point, light and thus all information becomes trapped.

[1] Supported in part by the National Science Foundation (GP-36687X, GP-40682) at the California Institute of Technology, Pasadena, California.

[2] Previous address during preparation of this review: Kellogg Radiation Laboratory, California Institute of Technology, Pasadena, California 91109.

For Newtonian gravity, and for a particulate model of light, this was first calculated by Laplace (1798).

The formation of a black hole by the collapse of a star or galactic nucleus is a violent event which couples normal astrophysical complexities (equations of state, hydrodynamics, nuclear energy release) with the relativistic complexities of strong, dynamical gravity. A large number of isolated facts are known and from them a whole fabric of what is "generally supposed" has been woven. In Section 2 we survey the situation: spherical collapse is fairly well understood; when an object of mass M or "mean" density ρ approaches its Schwarzschild radius $2M_*$ (units: length)

$$2M_* \equiv 2\frac{G}{c^2}M = \left(\frac{3\pi}{8}\frac{c^2}{G\rho}\right)^{1/2} = 3.0\left(\frac{M}{M_\odot}\right)\text{km} = \left(\frac{\rho}{1\text{ g/cm}^3}\right)^{-1/2} 1.3 \times 10^{14}\text{ cm} \quad 1.$$

a "trapped surface" of photons attempting escape—and therefore an event horizon—forms. A "stiff" equation of state can stop the formation of low-mass ($\sim 1M_\odot$) holes by making degeneracy-supported white dwarfs or neutron stars larger than their Schwarzschild radius. For nonrotating degenerate dwarfs, equilibrium configurations are possible up to the Chandrasekhar mass,

$$M_C \approx 1.2M_\odot\left(\frac{\mu}{2}\right)^{-2} \quad 2.$$

(where μ is the mean number of nucleons per electron, 2 for He^4). Rapidly, differentially rotating degenerate dwarfs can have somewhat greater masses (Ostriker 1971); the upper-mass limit for stable configurations is poorly known, but is probably $\sim 3M_\odot$ (see also Section 2.3). For neutron stars the corresponding upper-mass limit, the Oppenheimer-Volkoff (1939) mass M_{OV}, is not so accurately known because it depends on the equation of state of high density nuclear matter. It is probably in the range $0.7M_\odot < M_{OV} < 2.5M_\odot$ (for a review see Börner 1973), and on very general grounds it is almost certainly less than about $3.5M_\odot$ (Nauenberg & Chapline 1973, Rhoades & Ruffini 1974, Sabbadini & Hartle 1973).

Spherical objects of a larger mass cannot achieve a permanent configuration supported by the pressure of degenerate fermions. They must either eventually collapse to black holes or find some other way of blowing themselves apart, sending their matter to infinity. Relativity theory allows black holes to have any mass, but the mass ranges that will actually be populated with holes are determined by the vagaries of stellar evolution (for black holes in the stellar-mass range) and by the largely unknown nature of super-massive star formation ($\sim 10^2$–$\sim 10^6 M_\odot$) and the collapse of protogalactic nucleii ($\sim 10^7$–$\sim 10^{12} M_\odot$). In Section 2 we discuss some of these issues.

From the spherical picture we go to what is "generally supposed" (a mixture of rigorous mathematical reasoning, physical argument, extrapolation, and blind faith): when an object of any shape becomes as small as equation 1 (perhaps with a factor of order unity) in *all* dimensions, then an event horizon forms. Once a horizon is present, its interior state is irrelevant to the further development of the system. The exterior gravitational field adjusts itself by radiating gravitational (and sometimes electromagnetic) waves, seeking a stationary equilibrium final state. If the perturbing effects of external matter are small, the unique stationary final state is the Kerr (1963) black hole.

Black holes mature quickly. Within a few hundred light-travel times around the hole, $\sim 10^{-2}\ (M/M_{\odot})$ sec, the dynamical behavior of space-time has settled down. With gravity now virtually static, the complexities of the hole's astrophysical interaction are of a much less exotic nature: one has to model only the hydrodynamical behavior of accreting matter, the transport of the radiation that it generates, and the point-mass dynamical effect of a hole in a cluster of other bodies. In Section 3 we discuss this passive interaction of the hole with its environment. Because the hole is so dense compared with ordinary matter, it is a good approximation to neglect the perturbing influence of external matter on it, even, for example, if it collides with an ordinary star. This approximation of a fixed background metric probably fails in only two astrophysical situations: 1. a hole orbited by a massive self-gravitating disk (cf Bardeen & Wagoner 1971, Salpeter 1971), which might be present around a massive hole (10^6–$10^9 M_{\odot}$) in the nucleus of a protogalaxy; 2. a collision with a neutron star or another black hole that returns a stellar-mass hole to a dynamical state (cf Lattimer & Schramm 1974).

In Section 4 we discuss one particular situation in greater detail: the interaction of a black hole in a binary system with its companion star. This, one is quickly coming to realize, is the arena where the predictions of the theory are first being tested. The X-ray data on Cyg X-1 (and similar sources) potentially set the first strong-field observational test of general relativity, and there is at present a concentration of effort in connecting theory to observation for this situation.

The more exotic possibilities implicit in the theory are summarized in Section 5: small black holes ($\ll 1 M_{\odot}$; formed in the big bang) and their "quantum evaporation," holes inside stars, white holes, wormholes, and—necessarily—caveats.

Table 1 shows the main astrophysical processes relevant to black holes of various possible masses and indicates sections of this review where the processes are discussed. For other reviews on black holes we recommend the collection edited by DeWitt & DeWitt (1973) (especially Novikov & Thorne 1973), Bardeen (1974), Chandrasekhar (1974), Peebles (1972a,b), Rees (1974), Wheeler (1974), and the textbook by Misner, Thorne & Wheeler (1973).

2 GRAVITATIONAL COLLAPSE

2.1 *Nonrotating Stars*

The endpoint of a total gravitational collapse—formation of a black hole—occurs when a star's gravitational binding energy has become comparable to its total mass energy

$$\frac{GM^2}{R} \sim Mc^2 \Rightarrow R \sim \frac{GM}{c^2} \approx 1.5 \frac{M}{M_{\odot}} \text{km}. \qquad 3.$$

The *onset* of collapse is determined by a different criterion. Roughly speaking, when the gravitational binding energy becomes comparable to the total mass of

Table 1 Black-hole masses and astrophysical interactions

Mass of hole	How can it form?	What are its principal interactions?
10^{-5} g	only primordially, in big bang (§5.3)[a]	dominated by quantum gravity effects; may decay in 10^{-43} sec (§5.4)
10^{-5}–10^{15} g	only primordially, in big bang (§5.3)[a]	quantum effects; may have decayed to photons, neutrinos, over 10^{10} yr (§5.4)
10^{15}–10^{33} g	only primordially, in big bang (§5.3)[a]	might be cosmological "missing" mass, but difficulties with this (§5.3) if inside a star, can probably grow to swallow star in $< 10^{10}$ yr (§5.5)
1–$10^2 M_\odot$	collapse of normal stars (§§2.1, 2.3, 2.5, 2.6)	violent collapse generates strong gravitational, possibly electromagnetic waves; possible expulsion of matter (§2.6) unseen companion in a binary system (§§3.1, 4.6) accretion of interstellar matter produces IR through UV radiation; time variable; may produce γ's weakly (§3.2) accretion in close binary system forms an accretion disk, produces X rays; time variable; effects of hole's rotation (§§3.3, 4–4.6) exotic and unlikely: explosive energy-extraction effects; effects due to a substantial charge (§§5.1, 5.2)
10^2–$10^5 M_\odot$	collapse of supermassive stars, but unlikely (§2.2)	
10^5–$> 10^{12} M_\odot$	collapse of supermassive stars, galactic nuclei, or superdense star clusters (§§2.2, 2.4, 2.5, 2.6)	accretion in galactic center or in cluster of galaxies produces IR, optical radiation (§3.3) dynamical effects on surrounding stars (§3.4) if cosmological "missing" mass, lens effect on background QSOs (§5.3)

[a] § numbers refer to sections of this review.

particles *that are supplying the pressure,* then the effective adiabatic index approaches the critical value of 4/3 and dynamical collapse takes over. If ε is the fractional mass of the total that is responsible for the pressure, then

$$\frac{GM^2}{R} \sim \varepsilon Mc^2 \Rightarrow R \sim \frac{1}{\varepsilon}\frac{GM}{c^2}. \qquad 4.$$

For ordinary matter (or for electron-degenerate matter), where the pressure is supplied by electrons, $1/\varepsilon \sim m_p/m_e \approx 1836$; when radiation pressure becomes important $1/\varepsilon \to \infty$, reflecting the tendency toward instability in radiation-dominated stars. Neutron stars with $1/\varepsilon \sim 1$ can be near their gravitational radii without collapsing dynamically. Otherwise, independent of whether the body is temperature supported or degeneracy supported, collapse will begin at $\gtrsim 10^3$ gravitational radii.

The question that must be answered by detailed stellar evolution calculations is this: What range of initial stellar masses will lead to evolved stellar cores which at some point (*a*) come inside this collapse danger zone, and (*b*) have masses larger than the upper-mass limit of a neutron star? At present there is no agreement on this: on one hand, some model calculations [see e.g. Barkat et al (1974), and earlier references cited in Novikov & Thorne (1973)] find a range of initial models that are massive enough to have passed quasistatically through the stages of nuclear burning up to an iron core, but not so massive as to have had their evolution interrupted by instabilities such as electron-positron pair formation. The massive cores eventually collapse to black holes. Using the solar-neighborhood stellar birth-rate function, one then tentatively concludes that $\sim 0.1\%$ of the mass of the Galaxy could be in the form of black holes today (Novikov & Thorne 1973). On the other hand, some other recent models (Arnett 1973, cf Arnett & Schramm 1973) find that stars up to 32 or perhaps $64 M_\odot$ are able to eject their loosely bound envelope at one stage or another, leaving a core of ~ 1.4–$2.2 M_\odot$, which will preferentially become a degenerate dwarf or neutron star. More massive stars in these models leave no remnant at all.

One does not know, therefore, how optimistic to be about the formation of black holes in the normal stellar evolution of isolated, nonrotating stars. Even in the most pessimistic picture, however, it is likely that some black holes *will* be formed in binary systems, where a white dwarf or neutron star remnant can at a later stage accrete matter from its companion. Also, stellar rotation may aid the formation of black holes in isolated stars. We return to these issues below.

2.2 *Supermassive Stars and Galactic Nuclei*

In the range of masses $\sim 50 M_\odot < M < \sim 10^5 M_\odot$ the formation of black holes by ordinary star formation is probably inhibited. For the metal-rich abundances of population I stars, condensing protostars of $\gtrsim 50 M_\odot$ develop H II regions; their temperature and pressure increase and effectively halt further collapse [see Larson (1973) and references cited therein for much of this and the following material]. For population II metal-poor abundances, protostars in the above mass range can probably condense and ignite their hydrogen (although there are uncertainties regarding this). The trouble is that nuclear energy release turns on too rapidly and tends to bounce the system out to infinity or [unless rotation is important, cf Stothers (1974)] endow it with nonlinear pulsation modes which can lead to its eventual disruption. At a mass of $\sim 4 \times 10^5 M_\odot$ the supermassive protostars become able to collapse gravitationally in spite of their energy release (Fricke 1973), but this is just about the mass where condensation of the interstellar medium is inhibited for population II abundances as it was at much smaller mass for population I abundances, so the effective birthrate may be negligibly small.

Another possibility is that supermassive objects are formed in galactic nuclei by the evolution of dense clusters of normal stars, rather than by condensation of a gaseous medium. Spitzer & Thuan (1972), Spitzer & Hart (1971), and others have considered the evolution of such clusters. The basic time scale is the relaxation time (Spitzer 1956 or Chandrasekhar 1960).

$$t_{\text{relax}} \approx 8 \times 10^5 \frac{N^{1/2}}{\ln N}\left(\frac{R}{1\text{ pc}}\right)^{3/2}\left(\frac{m_{\text{star}}}{M_\odot}\right)^{-1/2} \text{yr}, \qquad 5.$$

where N is the number of stars, m_{star} is their typical mass, and R is the cluster radius. In a time of order 20 t_{relax} the clusters tend to develop density singularities in their center; at some point before this time, stellar collisions become evolutionarily important and—perhaps—lead to the formation of supermassive objects: a model by Sanders (1970) finds a $10^7 M_\odot$ cluster of 0.1 pc radius evolving to a supermassive body in 10^7–10^8 yr. Collapse to a black hole is presumed to follow.

More radical have been the suggestions of Gold, Axford & Ray (1965), Lynden-Bell & Wood (1968), and others that there might be collective-mode effects in the evolution of clusters which proceed much faster than the relaxation time scale and lead directly to a relativistically dense cluster and subsequent gravitational collapse [see Fackerell et al (1969) for references].

Another speculative scenario for the existence of black holes in galactic nuclei (Ryan 1972) is to imagine that they are the primordial $\sim 10^7 M_\odot$ "seeds" around which galaxies condense in the first place. (The formation of primordial black holes is discussed in Section 5.3.)

No mechanism above is outstandingly compelling. The main theoretical motivation for positing supermassive black holes is to make them available for use in models for quasars or for energetic processes in galactic nuclei (e.g. Lynden-Bell 1969, Lynden-Bell & Rees 1971, Wolf & G. R. Burbidge 1970).

2.3 *Rotating Stars*

As a star of mass M collapses, the supporting effect of its angular momentum J becomes more and more pronounced, until at a characteristic radius

$$R \sim \frac{J^2}{GM^3} \sim 50\left(\frac{J}{10^3 J_\odot}\right)^2\left(\frac{M}{10 M_\odot}\right)^{-3} \text{km} \qquad 6.$$

centrifugal forces become comparable to the force of gravity. [Here $J_\odot$ is the apparent solar angular momentum, not accounting for the possibility of a fast-rotating core (Dicke 1964). Early-type stars with $M \sim 10 M_\odot$ can typically have $J \sim 10^3 J_\odot$.] The upper-mass limits for normal stars, degenerate dwarfs, or neutron stars are not formally applicable to these rotation-supported configurations, and one is thus entitled to wonder whether massive rotating configurations can avoid entirely the rate of black-hole collapse. There are good indications that they cannot.

If a rotating star is to exceed the applicable spherical mass limit by any substantial factor, then rotational support must dominate pressure support. In this situation, the star is highly flattened. Such configurations are known to be unstable against bar-mode and fragmentation instabilities—they come apart into a pancake swarm of orbiting lumps (Goldreich & Schubert 1967, Ostriker & Bodenheimer 1973). The instabilities are dynamic rather than secular, so the fragmentation time is comparable to the system's dynamical time

$$\tau_{\rm dyn} \sim \frac{J^3}{G^2 M^5} \sim 4 \times 10^{-4} \left(\frac{J}{10^3 J_\odot} \right)^3 \left(\frac{M}{10 M_\odot} \right)^{-5} \text{ sec.} \qquad 7.$$

The mechanism of further evolution now depends on whether the radius of the pancake (equation 6) is of order of the gravitational radius GM/c^2. If it is, then

$$J \lesssim GM^2/c \qquad \text{or} \qquad \frac{J}{J_\odot} \lesssim 5.2 \left(\frac{M}{M_\odot} \right)^2. \qquad 8.$$

If equation 8 is true, then (*a*) the system's angular momentum is small enough (or nearly so) that it *can* become a Kerr black hole (which *always* must satisfy equation 8), and (*b*) the system will lose angular momentum by gravitational radiation and spiral in to smaller radii in a time comparable to the dynamical orbit time, so that in fact it *will* easily collapse to a Kerr hole, with attendant chaotic splashes of gravitational radiation. The characteristic time for radiation spiraling is

$$\tau_{\rm rad} \sim \tau_{\rm dyn} \left(\frac{Rc^2}{GM} \right)^n \sim \tau_{\rm dyn} \left(\frac{cJ}{GM^2} \right)^{2n}, \qquad 9.$$

where the exponent n is 5/2 for bar-mode instabilities and larger (comparable to the total number of fragments) for smaller-scale fragmentation without a bar mode. This is the "collapse, pursuit, and plunge" scenario of Ruffini & Wheeler (1971).

If, on the other hand, $J \gg GM^2/c$, then the pancake swarm is much larger than its gravitational radius. Gravitational radiation still acts, but by equation 9 it may now be very slow. The work of Ostriker & Bodenheimer (1973), Ostriker & Peebles (1973), and others, however, strongly indicates that collective instabilities in the pancake will continue until the random kinetic energy of the matter has increased to a fraction about unity of the system's rotational energy. For stellar-mass objects where the dissipative mechanism is violent bulk turbulent collision of matter, this internal energy buildup will heat the matter to a temperature of order

$$kT \sim m_p c^2 \left(\frac{GM}{Rc^2} \right) \sim m_p c^2 \left(\frac{GM^2}{cJ} \right)^2 \qquad [m_p c^2/k \approx 10^{13}\,{}^\circ\text{K}] \qquad 10.$$

in a time comparable to the dynamical time (equation 7). The hot pancake thickens to an oblateness only of order unity, and it begins to cool by radiation.

There are at present no model calculations to guide us beyond this point, but on grounds of energy and angular-momentum conservation (and keeping in mind that if the pancake cools and becomes thin again, new instabilities will arise) the direction of its subsequent evolution is probably this: a hot central core loses energy and transfers its angular momentum (via instability-induced turbulence) to outlying matter; the center condenses, eventually becomes of order of its gravitational radius, and collapses to a Kerr hole, probably one with nearly maximal angular momentum. [Work by Bardeen & Wagoner (1971) on the quasistatic evolution of relativistic disks lends some support to this picture.] The outlying matter, which may contain only a small fraction of the original mass, but which must carry most of the original angular momentum, may cool to a thin disk, now gravitationally stabilized by the central black-hole mass. Viscous processes will gradually cause it to transfer angular momentum (along with some mass) outward to infinity, but much of the matter will eventually accrete onto the central hole. The process here is substantially the same as the accretion of matter by a hole in a binary system, which is discussed in Section 4.

To summarize, it seems likely that the effects of stellar rotation can delay the formation of a black hole only for a time comparable to the gravitational-radiation orbital decay time (equation 9), or comparable to the radiative cooling time of the hot, puffed-up pancake, whichever is *shorter*. In some circumstances, moreover, rotation may in fact be conducive to the formation of black holes (Wheeler & Rosenwald 1973). In situations where the spherical collapse of a low-mass stellar core to a neutron star is violent enough to blow-off the massive envelope, the effect of rotation may be to let the core settle down gradually, with the envelope subsequently settling down and joining the core in a standing shock. In this way some additional range of stars in the interval, say 8–60$M_\odot$, may develop massive cores and lead to black holes by a natural evolutionary path.

2.4 *Rotating Supermassive Stars and Rotating Galactic Nuclei*

If star formation has occurred before or during the collapse of a supermassive object, then gas-dynamical dissipation is no longer present. The internal heating of the resulting thin pancake by collective instabilities now yields random motions of the point-mass stars only in the plane of the pancake. The time scale for the pancake to thicken up (for the "temperature" to isotropize) may now be much larger than a dynamical time; the relevant scale is the gravitational relaxation time which is of order $N/\ln N$ times as long (equation 5), where N is the number of stars. After the thickening, the formation of the core and halo, and the energy loss of the core, cannot be due to radiation but must occur via stellar evaporation or collective effects. It is not reliably known how fast these processes proceed.

2.5 *Description of Spherical Collapse*

Having sketched some scenarios leading to collapse, let us now review what happens when a collapsing body does approach its gravitational radius. Outside of any spherically symmetric body, even a highly dynamic one, space-time is not only spherically symmetric but also *static*; there are no spherical gravitational waves in general relativity. When the surface of the body passes through the event horizon at $r = 2M_* = 2GM/c^2$ (here, r is the radial coordinate that makes $2\pi r$ the proper circumference of the body), it can no longer causally influence our Universe and we need consider it no further. This passage occurs after a perfectly finite time as measured by a clock on the surface of the body, roughly the Newtonian free-fall time to the gravitational radius. Light rays from the surface find it harder and harder to escape to infinity, however, and as measured by the clock of a distant observer they take longer and longer to do so. The distant observer never sees the body's surface pass the event horizon. He sees the infalling particle's clock become "frozen" at that particular value of time at which the event horizon was passed. The distant observer sees all physical processes on the collapsing surface slow down. In particular, outgoing radiation becomes more and more redshifted. At late times, the asymptotic redshift is

$$\nu_{\text{observed}}/\nu_{\text{emitted}} \approx \exp(-c^3 t/4GM) \sim \exp\left(\frac{-t}{2 \times 10^{-5}\,\text{sec}}\frac{M_\odot}{M}\right). \qquad 11.$$

The decay of total luminosity of the star is slower,

$$L \sim \exp(-c^3 t/3^{3/2} GM), \qquad 12.$$

because much of the luminosity is stored in spiral photon orbits that take a long time to climb away from the black hole (see Podurets 1964, Ames & Thorne 1968). Because a finite number of photons were emitted from the stellar surface before it crossed the horizon, there comes a time when the last of these has escaped; then the hole is completely black. For further review of spherical collapse see Misner, Thorne & Wheeler (1973).

2.6 *Nonspherical Collapse*

Without the assumption of spherical symmetry it has not been practical to calculate even simple models of gravitational collapse to a black hole. This bottleneck is unfortunate for two reasons: 1. collapse of a rotating body is intrinsically nonspherical, and 2. even for a nonrotating body, highly nonspherical collapse is probably the generic case, because departures from spherical symmetry tend to grow as $1/r$ in a dynamical collapse, which may start at $\lesssim 10^3$ gravitational radii.

Nevertheless, the key elements of spherical collapse, that is, the formation of an event horizon, the asymptotic disappearance of the star by infinite redshift, and the rapid decay to a "black" stationary state, are believed to occur in general nonspherical collapse (reviewed by Thorne 1972). The main additional element present is the decay of distortions in the geometry of the event horizon and of the external space-time: this decay takes the form of emission of gravitational radiation. If electromagnetic fields are present in the collapsing star, for example, if it has a magnetic dipole moment, then electromagnetic radiation also will be emitted.

The basis for all these beliefs is the theorem that "a black hole has no hair," that is, the final-state configurations of a gravitational collapse are severely limited. The only possible stationary vacuum final state is the Kerr (1963) black hole, whose structure is completely determined by its mass M and angular momentum J. All the non-Kerr higher-multipole asphericity in the gravitational field that may have developed during the collapse, and all electromagnetic moments [except a net electric charge, which is implausible in an astrophysical context (see Section 5.2)] must, loosely speaking, detach themselves and blow off to infinity as radiation. It is beyond the scope of this review to summarize the strong theoretical evidence for the "no-hair" theorem. Among the key mathematical results have been those of

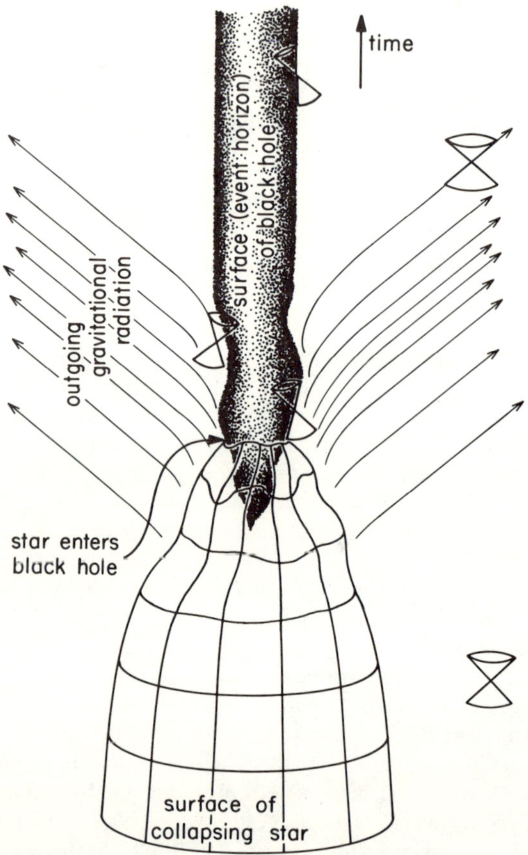

Figure 1 Nonspherical collapse of a star and formation of a black hole. An equatorial cross section is shown at successive instants of time (plotted upward), in Eddington-Finkelstein-type coordinates. As the star collapses asymmetries grow as $1/r$. Gravitational radiation is produced with increasing efficiency as the system approaches its event horizon (the surface of the black hole). The horizon is created in a nonspherically dynamical state, but it exhibits a strong "rigidity" and, damped by radiation, evolves to a state uniquely determined by the mass and angular momentum that have entered the hole. The outgoing edges of light cones originating on the horizon are tangent to it. Cones outside the hole admit outward-moving particles; inside the black hole (not shown) the future light cones are tilted inevitably inward.

Hawking (1972), Israel (1968), Carter (1973), Robinson (1975); calculated examples of Price (1972a,b). Cohen & Wald (1971), Hartle (1971), Teitelboim (1972), and Bekenstein (1972) have illuminated the phenomenon.

Knowing that the final state is a stationary black hole, one can describe in more detail the dynamical phase of collapse leading up to it. The first local sign that the formation of a black hole is occurring is the appearance within the collapsing matter of a *trapped surface,* a closed spatial surface such that outgoing null rays perpendicular to it are *converging,* focused by the strong gravitational field, instead of diverging normally to infinity. (A null ray or null geodesic is a trajectory through space-time of a photon, neutrino, etc.) These converging null rays will never reach infinity; the trapped surface is therefore *inside* an event horizon or boundary of causal influence, the surface of a black hole. When a trapped surface forms, a black hole has already been formed, although there is no locally measurable way to decide exactly where its horizon is while the dynamical collapse is still going on. The only certain method is to follow many different null rays and see which ones eventually make it out to infinity. The event horizon is, at each instant of time, the two-dimensional surface that is traced out by the "generating" null rays which are just marginally unable to escape to infinity, yet just marginally able to avoid convergence and collapse (Figure 1). In a stationary situation (long after the collapse) the proper surface area A of the event horizon is obviously constant. Hawking (1972) proved the remarkable theorem that in *any* situation, even a highly dynamical one, the area of the event horizon must be nondecreasing

$$dA/dt \geqq 0. \tag{13}$$

This law (called the *second law of black-hole dynamics* by analogy with the second law of thermodynamics) is equally valid if two black holes, say of areas A_1 and A_2, coalesce to form one black hole of area A_3:

$$A_3 > A_1 + A_2. \tag{14}$$

Thus, once a black hole forms, it can never disappear. It can be shown also that black holes cannot bifurcate.

Although the event horizon is an ideal surface, not a locally measurable "thing," it is often useful to think of it as having a certain physical reality and as being physically endowed with the mass M and angular momentum J of the black hole. It exhibits a strong rigidity, a strong tendency, if perturbed, to return quickly to a stationary state. If as a result of the dynamical asymmetric collapse the event horizon first appears in a distorted state, it radiates away its distortion as gravitational radiation in a few characteristic time periods of $GM/c^3 \sim 10^{-5}\,(M/M_\odot)$ sec.

For a typical, messy, asymmetric collapse, what fraction of the total mass energy Mc^2 of a collapsing body is radiated away in the form of gravitational waves? There are at present no reliable calculations. Scaling things up from some limiting cases that *are* known [e.g. the radial infall of a small particle into an already-formed black hole (Davis et al 1971)] one expects at least

$$E_{\text{radiation}} \gtrsim 0.01\,Mc^2. \tag{15}$$

For a rotating body with $J \gtrsim GM^2/c$ the "collapse, pursuit, and plunge" scenario above might give as much as $\lesssim 0.4\ Mc^2$ over the time of radiation of equation 9. [This latter fraction, more accurately 0.42, is the largest that can be released when a small mass gradually spirals into a maximally rotating black hole; cf Bardeen (1970).] Even without substantial rotation, collapsing masses much larger than the mass limit of a neutron star, M_{OV}, may develop a number of "independently collapsing" degrees of freedom, say M/M_{OV} of them. Simple arguments (Press 1975) then lead to a guess

$$E_{\text{radiated}} \sim (0.01 \text{ to } 0.1)\ Mc^2 \ln (M/M_{OV}) \qquad 16.$$

in a time of order a few GM/c^3; here, efficiency increases logarithmically with mass.

Because chaotic collapse radiates by a mixture of different multipoles, the radiation can carry off some net linear momentum, leaving the hole with a corresponding opposite velocity. Making some assumptions about the relative strengths of quadrupole and octupole radiation, Bekenstein (1973a) has estimated that velocities of $c/300 \approx 1000$ km/sec might occur, independent of the black hole's mass. The astrophysical consequence would be that black holes forming in binary systems would frequently escape the system. There are also more conventional disruption mechanisms (see Section 4.2).

Rees (1973) has suggested that strong waves produced by chaotic collapse could have another effect: matter immediately outside the collapse might be accelerated outward by its interaction with the wave. A dimensional calculation (W. H. Press and S. A. Teukolsky, unpublished) gives this result: if the characteristic velocities of the inner regions of collapse are of order c, and if some fraction of order unity of the total mass goes to gravitational radiation, then the net changes in velocity imparted to nearby matter by the interaction of the wave with its chaotic turbulent dissipation is *also* of order c. This suggests that matter ejection is not beyond possibility; the geometry of the gravitational wave could make the ejection somewhat directional.

We can estimate the electromagnetic radiation emitted in the collapse of a star with a magnetic field by "turning off" in a time $\tau \sim GM/c^3$ the multipole moments that have built up in the collapsing body and using the standard formulas for radiation from time-varying multipoles. Equivalently, one can say that the total magnetic energy (built up by adiabatic compression during the collapse) is radiated in a time τ and at frequencies of order $1/\tau$ when the star approaches its horizon:

$$E_{\text{radiated}} \sim \left(\frac{B_{\text{initial}}^2}{8\pi}\right)\left(\frac{4\pi}{3} R_{\text{initial}}^3\right)\left(\frac{R_{\text{initial}}}{GM/c^2}\right) \qquad 17.$$

or, for parameters approximating a degenerate magnetized white-dwarf core,

$$\sim 10^{46}\left(\frac{B}{10^8 \text{ gauss}}\right)^2\left(\frac{R_{\text{initial}}}{10^4 \text{ km}}\right)^4\left(\frac{M}{M_\odot}\right)^{-1} \text{ ergs.} \qquad 18.$$

This energy is small compared to the expected energy release from other sources.

2.7 *The Kerr Black Hole*

After the collapse, when all residual matter has either blown away, fallen into the black hole, or settled into stable orbit (typically with subsequent gradual accretion onto the hole), and after all the geometrical distortions of the event horizon have been emitted away as gravitational radiation, the black hole settles into a stationary state that is uniquely determined by its mass M and angular momentum J. If the mass of close-orbiting matter is small compared with the hole, so that the hole is not strongly distorted by tidal effects, then the space-time geometry is that of the Kerr (1963) metric,

$$ds^2 = -(1-2M_* r/\Sigma)c^2 dt^2 - (4M_* ar\sin^2\theta/\Sigma)c\,dt\,d\phi + (\Sigma/\Delta)dr^2 + \Sigma\,d\theta^2 + (r^2+a^2+2M_* a^2 r\sin^2\theta/\Sigma)\sin^2\theta\,d\phi^2, \qquad 19.$$

where $M_* \equiv GM/c^2$, $\Delta \equiv r^2 - 2M_* r + a^2$, $\Sigma \equiv r^2 + a^2\cos^2\theta$, and a is the angular momentum per unit mass in geometrized units, $a \equiv J/Mc$. The maximum allowed value of a, for an extreme-rotating hole, is $a = GM/c^2 = M_*$ (the limit of equation 8). In the metric (19), t is the time coordinate, and $r\theta\phi$ are spatial coordinates that tend to oblate spheroidal coordinates far from the hole. In the nonrotating limit of $a = 0$, the Kerr metric becomes the Schwarzschild metric.

The event horizon is the surface defined by $\Delta = 0$, or

$$r = r_+ \equiv M_* + (M_*^2 - a^2)^{1/2}. \qquad 20.$$

The Kerr black hole for $a \neq 0$ is rotating, not only in the sense that it contains the angular momentum of the collapsed object that formed it, but also in the direct physical and observational sense that infalling particles just crossing the event horizon will be seen by a distant observer to be "dragged" around with the rotation of the hole at a fixed angular velocity

$$\Omega = \frac{J}{2M}\frac{1}{M_*^2 + M_*(M_*^2 - a^2)^{1/2}}. \qquad 21.$$

The surface of a Kerr black hole [in fact, even an arbitrarily distorted stationary black hole (Carter 1973)] rotates uniformly: Ω does not depend on the colatitude θ.

Outside the horizon, the dragging effect becomes weaker with increasing radius. Nevertheless, out to a radius $r = r_0 \equiv M_* + (M_*^2 - a^2\cos^2\theta)^{1/2}$ called the *static limit*, the dragging effect is so strong that no finite acceleration can counteract it. All physical matter, regardless of the forces acting on it, will be dragged around the hole. Outside the static limit there is still a dragging, but it is possible to counteract it with a finite acceleration and remain fixed with respect to the distant universe (see Figure 2). The region between the static limit and the horizon is called the *ergosphere* for reasons explained in Section 5.1.

Radiation from an infalling particle or the infalling surface of a collapsing star is infinitely redshifted in qualitatively the same fashion as for a nonrotating Schwarzschild black hole, except that the rotation of the hole leads to a periodic

Doppler effect. This is strongest for observers in the equatorial plane, and vanishes for an observer on the polar axis. If this Doppler effect is factored out, the decay of frequency is governed by the law (generalization of equation 11)

$$\nu_{\text{observed}}/\nu_{\text{emitted}} \sim \exp(-t\kappa/c), \qquad 22.$$

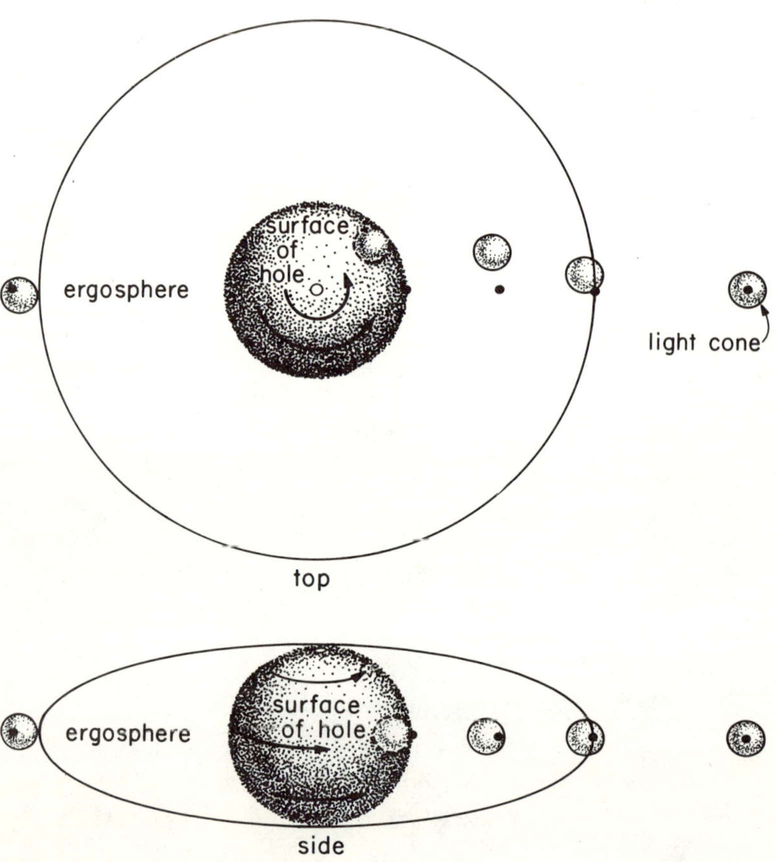

Figure 2 Spatial views of the exterior of a rotating Kerr black hole. Light cones are now depicted as points (sources of light) and spheres (expanding spherical wavefronts an instant later). Two effects are evident close to the hole: the hole's mass pulls the light cone inward, and the hole's angular momentum drags the light cone around in the direction of the hole's rotation. The azimuthal dragging becomes irresistible at the "static limit," shown as an oblate spheroidal boundary; in the "ergosphere" within, the light-cone sphere is detached from the point of emission, indicating that all observers are dragged around, regardless of their acceleration. They can, however, still spiral outward in radius and escape. Only inside the horizon, the actual surface of the hole, is the inward pull insurmountable.

where

$$\kappa = \frac{c^2}{2M_*}\frac{(M_*^2 - a^2)^{1/2}}{M_* + (M_*^2 - a^2)^{1/2}} \qquad 23.$$

is called the *surface gravity* of the black hole. This surface gravity is merely the acceleration that the distant observer would ascribe naively to the particle after observing its increasing redshift for a short time, if he did not know that it was falling into a hole but imagined that it was accelerating away from him. It is a remarkable fact that κ, like Ω, is constant over the surface of a stationary black hole (event horizon), even a strongly distorted one (Carter 1973, Bardeen, Carter & Hawking 1973).

3 INTERACTIONS OF THE HOLE WITH ITS ENVIRONMENT

3.1 *Orbits Around a Black Hole*

Many gravitational radii GM/c^2 from a black hole, a Newtonian physicist can describe its gravitational effect on matter in terms of a Newtonian gravitational acceleration

$$\mathbf{a} = \frac{GM}{r^2}\hat{\mathbf{r}}\left[1 + \mathcal{O}\left(\frac{GM}{rc^2}\right)\right]. \qquad 24.$$

Close to the hole, when the relativistic correction term in square brackets becomes important, a Newtonian description is no longer valid. The orbits of test bodies must now be computed as geodesics of the space-time metric representing the hole's exterior.

It turns out that there are relatively few contexts where the relativistic orbital equations need to be used. The reason is that black holes are so dense that test bodies orbiting in a region where relativistic effects would be important are typically disrupted by tidal forces, so that "clean" orbital problems become "messy" hydrodynamic ones. Suppose that we ask for relativistic effects to exceed 0.1, that is

$$r \lesssim 10\,\frac{GM_{\text{hole}}}{c^2}. \qquad 25.$$

The Roche criterion for the avoidance of tidal breakup of a body of density ρ orbiting a mass M_{hole} is in order of magnitude (cf Chandrasekhar 1969)

$$\rho \gtrsim \frac{M_{\text{hole}}}{r^3}. \qquad 26.$$

Combining equations 25 and 26 gives

$$\rho \gtrsim 3 \times 10^{14}\left(\frac{M_{\text{hole}}}{M_\odot}\right)^{-2} \text{g cm}^{-3}. \qquad 27.$$

Thus, for holes $\lesssim 10^4 M_\odot$ the only "test" bodies that do not come apart hydrodynamically are neutron stars; and only for $M_{\text{hole}} \gtrsim 10^8 M_\odot$ can we speak of a normal star in a relativistically close orbit around the hole. [The relativistic Roche problem near a Kerr hole has been considered in detail by Fishbone (1973); for a Newtonian treatment of the problem of *dynamical* tidal distortion, e.g. of a body on an elliptical orbit, see Nduka (1971); the work of Lattimer & Schramm (1974) and Mashhoon (1972) is also relevant.]

There are, nevertheless, some interesting points of principle that arise in considering the orbits around a Kerr hole. The axial symmetry of the hole guarantees only one component of conserved angular momentum. There is no reason that *total* angular momentum should be conserved in Kerr orbits (a Newtonian analog is the orbit around an irregular-shaped mass: tidal torques exerted on a test particle can change its angular momentum). Nevertheless, the Kerr metric does miraculously conserve a quantity somewhat analogous to total angular momentum. This was found by Carter (1968), who noticed that the Hamilton-Jacobi eikonal equation in the Kerr metric separates completely into angular eigenfunctions. In such a separation, the separation constants are conserved constants of particle motion.

With four conserved constants (energy, ϕ-angular momentum, Carter's constant, and rest mass of orbiting particle) one can classify the possible orbit trajectories in detail [see Stewart & Walker (1973) and references cited therein]. The special case of equatorial motion has been investigated by de Felice (1968) and summarized by Bardeen et al (1972) and Misner, Thorne & Wheeler (1973). A general description of these results is as follows: In comparison with Newtonian orbital equations, the radial force of gravity becomes stronger at small distances (a few Schwarzschild radii) from the hole. This rounds off the Newtonian centrifugal barrier to a "hill" of finite height. Orbits whose energies take them to "high on the hill" spend relatively long times near pericenter. Since the relativistic equation for ϕ-motion is not too different from the Newtonian case, these orbits can spiral around the hole many times before returning to large radii (this is a spectacular limiting case of the perihelion advance of Mercury around the sun). Orbits whose energy takes them "over the potential hill" fall into the hole.

When the hole is rotating, another effect is superposed: the hole drags inertial frames (and the trajectories of orbits) around in its direction of rotation. The angular frequency of dragging varies with the mass and angular momentum of the hole, and the radius r as

$$\omega = \frac{2craM_*}{(r^2+a^2)^2 - a^2 \sin^2\theta (r^2 - 2M_* r + a^2)} \qquad 28.$$

and goes to the finite constant Ω (equation 21) on the event horizon. The effect on equatorial orbits is to increase the centrifugal barrier for direct orbits and decrease it for retrograde ones. This enables direct orbits to have much closer stable (against falling-in) pericenters than retrograde orbits can have. For example, the smallest stable orbit of a maximally rotating Kerr hole has a radius (in the coordinates of equation 19) of GM/c^2, whereas the smallest stable retrograde orbit is nine times larger.

For nonequatorial orbits the frame dragging adds to the change in a trajectory's longitude during its latitude excursion; for example, the analog of an inclined circular orbit is a trajectory that is dragged around on a sphere as it travels between equal positive and negative latitudes.

In the regime where relativistic orbital effects are important, there are also strong relativistic effects in the propagation of light from the orbiting body out to infinity. The appearance of a point source of radiation that radiates isotropically in its own rest frame as it moves around a Kerr hole in an equatorial orbit has been calculated by Cunningham & Bardeen (1973) (see also de Felice et al 1974, Polnarev 1972). There are several images of the source that are due to different paths available to photons skirting the hole. Gravitational focusing effects and Doppler shifts cause the intensity of the images to vary sharply with time over an orbital period, especially as seen by an observer situated close to the orbital plane; here focusing effects are the strongest. Although the Roche limit mentioned above restricts the direct applicability of this work, it may well be relevant to radiation from "hot spots" in an orbiting disk of matter (see Section 4.4).

3.2 *Accretion of Matter onto an Isolated Black Hole*

We turn now to a hole's hydrodynamic interaction with a surrounding medium. An idealized problem, which gives intuition about more realistic cases, is to consider the spherical accretion of a perfectly adiabatic gas ($P = K\rho^{\gamma}$) onto a nonrotating black hole of mass M. The solution to this problem (for a Newtonian central mass) is due originally to Bondi (1952). Its application to black holes was considered by Salpeter (1964) and more recently by Zel'dovich & Novikov (1971), Shvartsman (1971), Novikov & Thorne (1973), and Pringle et al (1973). Far from the hole, the gas has some temperature T_∞ or equivalently some sound velocity a_∞. The attraction of the hole is negligible outside an "accretion radius" r_i, where thermal velocities and escape velocities become comparable,

$$r_i \sim \frac{2GM}{a_\infty^2} = 1 \times 10^{14}\left(\frac{M}{M_\odot}\right)\left(\frac{T_\infty}{10^4\,{}^\circ\mathrm{K}}\right)^{-1}\ \mathrm{cm}. \qquad 29.$$

Inside the accretion radius, r_i, the gas is practically in free fall. Mass flows into the hole at a rate of order

$$\frac{dM}{dt} \sim \frac{4\pi G^2 M^2 \rho_\infty}{a_\infty^3} = 5 \times 10^{10}\left(\frac{M}{M_\odot}\right)^2\left(\frac{\rho_\infty}{10^{-24}\,\mathrm{g/cm^3}}\right)\left(\frac{T_\infty}{10^4\,{}^\circ\mathrm{K}}\right)^{-3/2}\ \mathrm{g/sec}. \qquad 30.$$

Its density increases as $\rho \propto r^{-3/2}$, and its temperature increases as $T \propto r^{-3(\gamma-1)/2}$. The infall becomes supersonic at a radius

$$r_s = \left(\frac{5-3\gamma}{4}\right)\frac{GM}{a_\infty^2}. \qquad 31.$$

(For $\gamma = 5/3$ the sonic radius is not really zero, but rather is of order $GM/a_\infty c$ instead of GM/a_∞^2.)

Notice that (neglecting cooling processes) the temperature at the event horizon comes out to be of order

$$kT_{(r=2GM/c^2)} \sim m_p c^2\left[\frac{m_p c^2}{kT_\infty}\right]^{(3\gamma-5)/2} \qquad 32.$$

The dimensionless factor in brackets is very large, about $3 \times 10^9(T_\infty/3000°K)$, and this leads to an important result: accretion temperatures near the hole are relativistically high only for gases with a "stiff" adiabatic index $\approx 5/3$. When the adiabatic index is "soft," temperatures can remain relatively low all the way into the hole.

We can now generalize this picture to include cooling (e.g. from free-free emission, and thermal synchrotron radiation from trapped, infalling magnetic fields): a cooling process lowers the "effective" γ of the infalling gas because compressional energy is lost to cooling (analogous to being lost to internal molecular degrees of freedom). This, it turns out, usually makes spherical accretion rather inefficient as a means of turning rest-mass energy into observable radiation. If the infalling gas can radiate, then its tendency is to stay cold!

Pringle et al (1973) describe the process in this way: Because radiative losses increase with gas density, there may be some critical radius $R_{\rm crit}$ at which adiabatic heating and radiative cooling rates are comparable. From infinity to this radius a gas will heat up, roughly adiabatically; inside this radius a gas will cool as it falls. The total energy radiated per unit mass is comparable to the *internal* energy per unit mass at this critical radius. This is in turn comparable to the rest-mass energy of infalling matter only when both $R_{\rm crit} \sim GM/c^2$ and $\gamma \approx 5/3$, a situation that typically does not occur for astrophysically plausible parameters. For accretion onto large ($\gtrsim 10^4 M_\odot$) holes, $R_{\rm crit}$ is usually so large that the gas cools as it falls.

For stellar-mass black holes, calculations by Shvartsman (1971) and Shapiro (1973) show that cooling is inefficient all the way in ($R_{\rm crit}$ is never reached). The total luminosity is of order

$$L \sim (10^{26}\text{–}10^{29})\left(\frac{T_\infty}{10^4 °K}\right)^{-3}\left(\frac{M}{M_\odot}\right)^3\left(\frac{n_\infty}{1\ \text{cm}^{-3}}\right)^2 \text{ erg/sec.} \qquad 33.$$

The radiation is principally from thermal synchrotron radiation. As the interstellar magnetic field is dragged in with the accreting matter it is compressed, sheared, and amplified. These model calculations have assumed that reconnection of field lines and turbulence of the magnetohydrodynamic flow bring about an equipartition of energy between magnetic energy, turbulent energy, internal gas energy, and—most importantly, but also with the least theoretical foundation—infall kinetic energy. This last assumption is equivalent to the most optimistic assumption of an effective adiabatic index of 5/3. Because magnetic field densities enter crucially into the radiation efficiency, there is a large model dependence in the calculations of luminosity, as indicated in equation 33. In Shapiro's model, the flux is predominantly in the IR, while Shvartsman's spectrum extends through the visible to the UV (resembling the spectrum of a DC white dwarf). A characteristic of this turbulent accretion may be rapid variations in the flux on times of order 10^{-3}–$10^{-4}(M/M_\odot)$ sec.

The low luminosities and uncertain spectra of these models seem to make the observational detection of an isolated, spherically accreting hole quite difficult. An observational program by V. F. Shvartsman (1974) inspecting DC white dwarf candidates for rapid time variations, has thus far yielded no identifications.

An interesting possibility has been pointed out by Dahlbacka et al (1974), however: for accretion on stellar-mass holes, infall times near the hole are much less than the electron-ion coupling time for the gas. This may cause the ions (now with adiabatic index 5/3) to ultimately become relativistically hot, and a significant luminosity of pions might be produced. Decay of the neutral pions would produce a "universal" (independent of mass of hole or density at infinity) background spectrum of black-hole-generated gamma rays, characteristically (but very broadly) peaked at ~ 20 MeV; detection of such an anomalous feature, or upper limits for it, can put limits on the total number of accreting holes in our galaxy or in the universe. The γ luminosity per hole is $\sim 10^{22}$ ergs/sec with the same dependence on parameters as equation 33.

When the black hole is moving with velocity v through a homogeneous medium, the picture is not too different from the spherical case. The accretion radius r_i is now located where the hole's gravity dominates both thermal motion and the relative velocity

$$r_i \sim \frac{2GM}{a_\infty^2 + v^2}. \tag{34.}$$

Inside this radius the flow becomes essentially spherical in the hole's frame. The accretion rate (compare with equation 30) is of order

$$\frac{dM}{dt} \sim \frac{4\pi G^2 M^2 \rho_\infty}{(a_\infty^2 + v^2)^{3/2}}. \tag{35.}$$

When the velocity of the hole through the medium is supersonic, $v > a_\infty$, then there will be a bow shock located roughly a distance r_i in front of the hole. The energy dissipated in this shock is, however, only of order

$$\frac{dE}{dt} \sim v^2 \frac{dM}{dt} \sim 10^{24} \left(\frac{M}{M_\odot}\right)^2 \left(\frac{\rho_\infty}{10^{-24}\ \mathrm{g\ cm^{-3}}}\right)\left(\frac{v}{10\ \mathrm{km/sec}}\right)^{-1/2} \mathrm{ergs/sec}. \tag{36.}$$

For details and references, see Novikov & Thorne (1973).

3.3 *Accretion Disks*

The behavior of accreting matter is quite different if, on the average, it has some angular momentum with respect to the hole. In this case it cannot fall directly into the hole, but goes into an orbit at the characteristic radius where its angular momentum per mass balances gravity. In terms of the mass M of the hole and the transverse velocity $v_\perp$, which matter initially has at some radius R, the orbital radius r is

$$r \approx \frac{v_\perp^2 R^2}{GM}. \tag{37.}$$

This will be larger than the gravitational radius of the hole when

$$\left(\frac{v_\perp}{c}\right)\left(\frac{Rc^2}{GM}\right) \gtrsim 1. \tag{38.}$$

Suppose that the angular momentum arises from turbulence in the interstellar medium; then substituting the accretion radius r_i (equation 29) for R gives the condition

$$v_{\perp} > \left(\frac{a_{\infty}}{c}\right)^2 c \approx 10^2 \left(\frac{T_{\infty}}{10^{4\circ}\mathrm{K}}\right) \mathrm{cm/sec}. \qquad 39.$$

Observations of interstellar gas put its turbulent velocity at 10^6 cm/sec on scales of 100 pc. The velocities decrease with decreasing scale l as l^q. The exponent q is 1/3 (the Kolmogorov value) on large scales, steepens to $\geqq 1/2$ if magnetic viscosity is important, and to 3 if conventional viscosity ever comes into play (Kaplan & Pikel'ner 1970). Because r_i is typically a factor of $\sim 10^6$ smaller than 100 pc for stellar mass holes (equation 29), the uncertainty in q prevents a decisive answer to the question of whether or not disk accretion is the generic case for an isolated hole. The formation of a disk is, however, much more certain in the case of systems where angular momentum is dynamically important, for example, gas which settles to the center of a rotating galaxy, or mass flow onto a compact object in a binary star system.

D. Lynden-Bell (1960, unpublished) and Prendergast & G. R. Burbidge (1968) emphasized the importance of angular momentum in accretion; disk models go back to as far as Lüst (1952) (see Section 4.4 for further references). When matter approaches the orbital radius appropriate to its angular momentum its orbit becomes quickly circularized, and it relaxes to a common plane (that of its average angular momentum) as a result of dissipative interaction among the various matter streams. If, as is typically but not always the case (Section 4.4), the matter is able to cool so that its internal thermal energy becomes much less than the orbital binding energy, then the disk is thin, scale thickness h is much less than radius r, and the matter goes into essentially a Keplerian orbit.

The matter in a radial band of width $\Delta r \sim h$ forms a single dynamical entity on times comparable to the Kepler time scale $\tau_{\mathrm{dyn}} = (r^3/GM)^{1/2}$. Matter in the band Δr shares angular momentum and energy rapidly. On the other hand, separate bands Δr a distance much greater than h apart are nearly dynamically uncoupled. The long-term evolution of the disk is governed by the viscous stress driven by differential rotation between successive bands Δr, that is, driven by the shear of the Kepler flow. The viscous stress transports angular momentum outward and matter is therefore able to spiral inward. Viscosity has two additional effects: 1. it transports energy along with angular momentum, and 2. it heats the disk by viscous dissipation, causing thermal radiation. The viscosity is presumably magnetic or turbulent in origin, but only the most rudimentary calculations of these effects are available, and the nature and properties of this postulated viscosity are the greatest uncertainties in the model.

The inner edge of the disk around a black hole lies at the innermost stable circular orbit; after passing this radius, matter spirals into the hole on a dynamic (not a viscous) time scale, with negligible further dissipation. The total thermal energy liberated by matter in traversing the disk is just the binding energy in this innermost stable orbit. For a nonrotating hole, the luminosity is therefore $L =$

$0.057(dM/dt)c^2$; for an extreme-rotating hole, $a = M_*$, it is $L = 0.42(dM/dt)c^2$; for $a = 0.998M_*$ (see Section 4.3) it is $L = 0.32\ (dM/dt)c^2$. In contrast to the case of spherical accretion, disk accretion will almost always liberate these very large efficiencies. The reason is that the gradual spiraling in of matter transports energy outward and gives it time to radiate, whereas for radial infall most energy goes down the hole in the form of internal heat or radial kinetic motion.

Because of the existence of X-ray, optical, and radio data on X-ray sources associated with evolved binary systems, the structure of the accretion disk is currently the most important point of contact of black-hole theory and astronomical observation. For this reason, we return to the quantitative discussion of binary star accretion disks in Section 4.4. Accretion disks around large black holes in galactic nuclei or in the center of clusters of galaxies have been considered by Lynden-Bell (1969), Lynden-Bell & Rees (1971), Pringle et al (1973), and others. These may radiate typically in the IR and optical, and have a luminosity on the order of

$$L \sim 0.3\frac{dM}{dt}c^2 \sim 3\times 10^{46}\left(\frac{M}{10^{11}M_\odot}\right)^2\left(\frac{n_\infty}{10^{-3}\ \text{cm}^{-3}}\right)\left(\frac{T}{10^{6\circ}\text{K}}\right)^{-3/2} \text{ergs/sec.} \qquad 40.$$

Because there are numerous instabilities to which accretion disks *may* be prone (Section 4.3), the luminosity *might* vary on time scales as rapid as $\sim 10^6$ sec$(M/10^{11}M_\odot)$. This feature has suggested the application of accretion disks to quasar models.

3.4 *A Black Hole's Effect on a Surrounding Star Cluster or Galaxy*

The possibility that massive black holes can form in the centers of dense star clusters by core-relaxation effects was mentioned in Section 2. Even in systems like elliptical galaxies, whose relaxation time is very long, arguments have been advanced that gas shed by stellar evolution might settle in the center of the gravitational potential well and form massive condensed objects. Here we ask, What are the (potentially observable) consequences of a central black hole on the stellar density distribution of the cluster or galaxy? On scales much larger than its size, of course, the black hole acts like a Newtonian point mass. What follows is from the clear discussion of this issue by Peebles (1972a).

Well away from the black hole, but still at the center of the cluster, suppose that the root mean square velocity dispersion of stars is $\bar{v}$. Then the hole of mass M will exert a significant influence out to a radius (cf equation 29)

$$r_i \sim \frac{2GM}{\bar{v}^2} \sim 8\left(\frac{M}{10^5M_\odot}\right)\left(\frac{\bar{v}}{10\ \text{km/sec}}\right)^{-2} \text{pc.} \qquad 41.$$

The density distribution within this radius depends on the relative size of some different time scales: τ_{dyn}, the crossing or dynamical time of the cluster; $\tau_{\text{relax}} \sim \tau_{\text{dyn}} N/\ln N$, where N is the number of stars (see equation 5); τ_{growth}, the length of the period during which the black hole grew; and τ_{age}, how long ago the formation took place.

If τ_{age} is bigger than any of the other times, then the cluster has had time to relax to a stationary state around the hole. This will be true for black holes in the center of globular clusters, whose relaxation times are of the order of 10^7–10^9 years. One's first inclination is to suppose that the density would relax to the Boltzmann form, say

$$\rho \propto \exp(-3GM/r\bar{v}^2) \qquad 42.$$

near the hole; but as Peebles points out, this equilibrium does not satisfy the boundary condition that stars are swallowed by the central hole. Instead, a stationary *non*equilibrium state is reached, where stars inside r_i drift into the hole on an appropriate relaxation time scale, and they have a density which probably varies as

$$\rho \sim \rho_0(r_i/r)^{9/4}, \qquad r < r_i, \qquad 43.$$

where ρ_0 is the central density away from the hole (Peebles 1972b).

For galactic nuclei (such as the center of M31) and for elliptical galaxies, τ_{relax} is much greater than the age of the universe. We can also assume $\tau_{growth} \gg \tau_{dyn}$ and $\tau_{age} \gg \tau_{dyn}$ because the mass that makes up the central condensation can hardly have appeared there in less than a crossing time. Star orbits then have been drawn to the hole adiabatically, with a resultant density

$$\rho \sim \rho_0(r_i/r)^{3/2}, \qquad r < r_i. \qquad 44.$$

This strong dependence will show up as a cusp in surface brightness $\sigma(r) \propto r^{-1/2}$, which can in principle be detected photographically.

Present observational evidence [see Peebles (1972a,b) and Tremaine et al (1975) for discussion and references] puts upper limits of about $10^7 M_\odot$ for a collapsed object in the nucleus of M31, and about $10^{10} M_\odot$ for the center of NGC 4472, the brightest elliptical in the Virgo cluster.

4 BLACK HOLES IN BINARY SYSTEMS

The UHURU X-ray satellite discovered a class of galactic, close-binary X-ray sources, generally thought to be powered by accretion onto a compact object [Gursky 1973, Giacconi 1973, see Blumenthal & Tucker (1974) for a recent review]. These binary systems each consist of a "normal" star (mass $M_1 = 1$–$50 M_\odot$) and a less massive, compact companion ($M_2 = 0.25$–$15 M_\odot$, $M_2/M_1 \lesssim 0.5$): a degenerate dwarf, neutron star, or black hole. Typical orbital periods are $P = 0^d.2$ to 10^d, implying a distance of centers

$$a = 4.2 R_\odot [(M_1 + M_2)/M_\odot]^{1/3} (P/1d)^{2/3} \approx 1\text{–}100 R_\odot. \qquad 45.$$

Because a star's radius far exceeds this size during post-main-sequence evolution, mass exchange or mass loss has almost certainly taken place during the history of these systems, explaining why the presently more highly evolved, compact object is *less* massive, and also possibly providing a source of accretion flow from the normal star.

4.1 *The Roche Lobes*

Let stars M_1 and M_2 be in circular orbit with angular velocity $\mathbf{\Omega}$; idealize them as mass points. In corotating coordinates $\mathbf{r}$, the gravitational-plus-inertial acceleration on a test particle with velocity $\mathbf{v}$ is

$$\frac{d\mathbf{v}}{dt} = -2\mathbf{\Omega} \times \mathbf{v} - \nabla\Phi. \qquad 46.$$

Here $-2\mathbf{\Omega} \times \mathbf{v}$ is the Coriolis force, and the effective potential Φ is the sum of gravitational and centrifugal terms,

$$\Phi = -GM_1/|\mathbf{r}-\mathbf{r}_1| - GM_2/|\mathbf{r}-\mathbf{r}_2| - \tfrac{1}{2}|\mathbf{\Omega} \times \mathbf{r}|^2. \qquad 47.$$

The equipotential surfaces of Φ are the surfaces of hydrostatic equilibrium for corotating matter. Very near to M_1 (or M_2) these equipotentials are nearly spherical, forming a deep well. Farther out they become increasingly distorted, until one equipotential is a figure-of-eight, surrounding both stars (see Figure 3). The interior of this equipotential surrounding each star is its Roche lobe, and the point between them is the Lagrange point L_1. Plavec (1968) gives an approximate formula for the mean radius R_{1c} of the Roche lobe around star M_1,

$$R_{1c} = a(0.38 + 0.2 \log_{10} M_1/M_2), \qquad 48.$$

valid for $0.3 \lesssim M_1/M_2 \lesssim 20$.

During post-main-sequence evolution, the radius R_1 of star M_1 tends to increase. An attempt to expand to a radius $R_1 \geqq R_{1c}$ (*if* the envelope corotates; see below) results in gas spilling through the saddle point L_1, streaming down into the potential well of star M_2, and accreting onto it. Alternatively, if the gas stream gains some energy hydrodynamically, it can escape to infinity. An evolving star may lose up to $\sim 2/3$ of its mass in this way.

Coriolis forces affect the motion of matter that is not at rest in the corotating frame (equation 46). If a star does not corotate, there will still be a limiting surface for shedding of the envelope, but the surface will not coincide with the Roche lobe. For stellar evolution the existence of the limiting surface is the crucial fact, and the exact size does not matter very much. One often assumes, however, that an observed star in a close binary is "at its Roche lobe" to derive its radius; the size of the true limiting surface may be somewhat different.

4.2 *Stellar Evolution in Close Binaries*

For reviews, see Plavec (1968), Paczyński (1971), and Martynov (1972). Systems that are low enough mass that the end state of the originally more massive star is a degenerate dwarf ($M_{1(\text{final})} \lesssim 1.2 M_\odot$) are fairly well understood. Later accretion onto the dwarf seems to power novae and cataclysmic variables. The history of systems massive enough that one star ends as a neutron star or black hole is less understood, not only because the evolution of single stars to these endpoints is poorly known, but also because the formation of a neutron star or black hole necessarily involves a catastrophic collapse of the stellar core. This almost certainly blows off the stellar envelope, and possibly disrupts the binary system leaving two single objects.

There are several possible mechanisms for such disruption: the sudden loss of envelope mass ΔM to infinity decreases the gravitational binding energy of a binary (Blaauw 1961). The system is disrupted if $\Delta M > (M_1+M_2)/2$ (for a circular orbit). But for the close binaries of interest here, the exploding star will typically be the *less* massive because of previous mass exchange or loss, and this process rarely disrupts the system: $\Delta M < M_1 < (M_1+M_2)/2$ (Paczyński 1971, Sutantyo 1973).

If the explosion is asymmetric, or the central collapse forms an initially asymmetric black hole that emits a gravitational wave pulse carrying linear momentum (see Section 2.6), or the explosion ablates matter from the companion star, the recoil can disrupt the binary system. Little is known about these processes [see Bekenstein & Bowers (1974) for review and references]. Failing disruption, any of these processes can impart a recoil velocity of the order of 10^2 km/sec to the whole bound system.

Observationally, there is some evidence for disruption: 1. Only one pulsar (out of ~ 80) is binary, even though many ($\gtrsim 30\%$) of their progenitor massive stars are. 2. There exists a class of "runaway" OB stars with high ($\sim 10^2$ km/sec) peculiar velocities which could be escapees of disruption events (Blaauw 1961). Alternatively, although there is no direct evidence for this, some of these stars may be recoiling binary systems with unseen collapsed companions (Paczyński 1971, Bekenstein & Bowers 1974). The X-ray binaries probably do contain neutron stars or black holes, and this fact is presently the best evidence that disruption does not always occur.

4.3 *Accretion-Powered X-Ray Sources in Close Binaries*

After a black hole of mass M_1 forms in a binary system, it remains quiet until the companion star M_2 evolves off the main sequence (van den Heuvel & De Loore 1973), probably as an OB supergiant. The latter then loses mass, either as a stellar wind driven by the star's radiation pressure, or as Roche-lobe overflow driven by expansion of the envelope. For OB supergiants stellar winds may range up to $\dot{M} \approx 10^{-5} M_\odot$/yr (Pottasch 1970); the envelope probably expands on the thermal time scale of $\gtrsim 10^4$ yr, generating $\dot{M} \lesssim 10^{-3} M_\odot$/yr (van den Heuvel & De Loore 1973). In either case, there is plenty of mass flow to power X-ray source by accretion into the black hole; a luminosity $L \approx 10^{37}$ erg/sec requires only $\dot{M} \approx 10^{-9} M_\odot$/yr, assuming high efficiency, $L/\dot{M} \approx 0.1\, c^2$.

For a wind, the gas flow is supersonic and accretion is probably roughly spherical near the hole. For overflow thorough the Lagrange point, however, the gas stream picks up enough angular momentum via Coriolis forces to orbit the hole at a distance $\sim a_2/4$, where a_2 is the hole's distance from the center of mass [cf Prendergast & Taam (1974) for the case that M_2 is a normal star and Flannery (1975) for the case of a compact white dwarf]. An accretion disk forms inside this radius (Figure 3).

The luminosity function of observed compact X-ray sources has a sharp peak at $L \approx 10^{37-38}$ erg/sec (Margon & Ostriker 1973), with a sharp cutoff at higher luminosities. Apparently the hole manages to regulate its accretion to $\dot{M} \approx 10^{-9} M_{\odot}$/yr, only 10^{-4}–10^{-6} of the gas flow through the Lagrange point. It is not known how this is effected, although there are several possibilities. If $\dot{M}$ into the hole were to exceed $10^{-8}(M_{\rm BH}/M_{\odot})$, the luminosity would exceed the Eddington (radiation pressure) limit, $L \approx 10^{38}$ erg/sec$(M_{\rm BH}/M_{\odot})$. For accretion from a stellar wind, this is a strong regulatory mechanism (Shakura & Sunyaev 1973, Davidson & Ostriker 1973). For accretion fed by Roche-lobe overflow, however, things are more complicated: the inner portion of the disk is disrupted; Shakura & Sunyaev (1973) suggest that when $\dot{M}$ in the disk exceeds the Eddington-limiting value $\dot{M}_{\rm Ed} \approx 10^{-8} M_{\rm BH}$/yr, only about $\dot{M}_{\rm Ed}$ will manage to get down the hole, and the rest will be expelled more or less along the axis. In this case the X-ray source will be partially or entirely blanketed by the expelled matter, and the luminosity $L \approx 10^{38}$ erg/sec $(M_{\rm BH}/M_{\odot})$ will be processed down to the UV or visible. This does not

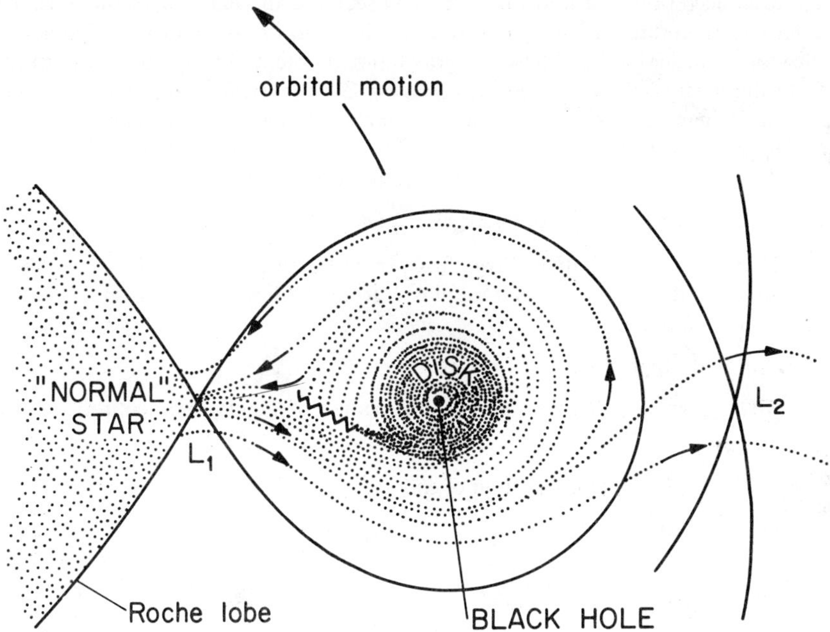

Figure 3 Mass flow onto a black hole in a binary system (figure adapted from Novikov & Thorne 1973). The primary star, late in its stellar evolution, swells to fill its Roche lobe and spills mass through the Lagrange point L_1. The flow of this gas is governed by hydrodynamic, gravitational, and coriolis forces. Some of the gas is deposited into a disk orbiting the black hole; it gradually spirals inward under the influence of viscous forces. The rest of the gas (most of it) either escapes to infinity or is deposited back onto the primary. Details are probably more complicated than shown here. Solid lines in this figure are equipotentials of the gravitational plus centrifugal potential. Dotted lines indicate matter flow, and the jagged line represents a shock or region of supersonic turbulence.

seem to be taking place in observed X-ray sources. Therefore, for a disk fed by Roche-lobe overflow, the regulation may be hydrodynamic and unrelated to the Eddington limit: the kinetic-energy turnover of the whole mass flow is $\sim \dot{M} v_{\text{orbit}}^2 \sim 10^{35-37}$ erg/sec (where v_{orbit} is the system's orbital velocity). If a few percent of the X-ray luminosity impinges on the gas streams, it will begin to influence strongly the flow at $L \approx 10^{37-38}$ erg/sec; thermal-pressure forces in the flow may then be sufficient to reject the unwanted matter (to infinity or back onto the companion star) and regulate the accretion rate.

4.4 *The Accretion Disk*

RADIAL DISK STRUCTURE As presently known, the theory of the structure of accretion disks splits into independent discussions of the radial and vertical disk structure (Shakura 1972, Novikov & Thorne 1973). Assuming a stationary state, at least in some time-averaging sense, the radial structure is determined entirely by the total accretion rate into the hole $\dot{M}$ (g/sec) and the mass M of the hole. The external gravitational field is due only to the hole (mass of disk negligible); in Newtonian order it follows from the gravitational potential GM/r, or relativistically from the form of the Kerr metric, equation 19. The explicit formulas below use the Newtonian approximation, with the angular velocity of Keplerian orbits given by $\Omega_{\text{Kepler}} = (GM/r^3)^{1/2}$.

Let the total viscous stress per unit circumference be W,

$$W \equiv \int (\phi - r \text{ component of viscous stress})\, d(\text{thickness of disk}).$$

Let $2F$ be the total radiative energy flux per unit area emitted by the disk because of viscous heating (F from the top, plus F from the bottom). Then the equations of radial structure are

$$d\dot{M}/dr = 0 \quad \text{(conservation of mass)},$$

$$\dot{M} d(\Omega_{\text{Kepler}} r^2)/dr = d(2\pi r^2 W)/dr \quad \text{(conservation of angular momentum)},$$

and

$$F = 3\Omega_{\text{Kepler}} W/4 \quad \text{(conservation of energy)}. \qquad 49.$$

Solving for F gives

$$F(r) = \frac{3\Omega_{\text{Kepler}}}{4}\left(\frac{\dot{M}\Omega_{\text{Kepler}}}{2\pi} - \frac{C}{r^2}\right), \qquad 50.$$

where C is a constant of integration that may be given as an inner boundary condition. If, as an inner boundary condition, it is prescribed that the matter is gently removed from orbit with insignificant torque, then set $F(r_{\min}) = 0$ to fix C. For the fully relativistic, black-hole case, this is the correct physical boundary condition at the innermost stable circular orbit, and the relativistic form of F tails smoothly off to zero there. At large $r, F(r)$ approaches the Newtonian-limiting form

$$F(r) = \frac{3}{8\pi}\frac{GM\dot{M}}{r^3}, \qquad 51.$$

which is three times the value one computes from loss of potential plus kinetic energy alone. The difference is made up by energy transported outward from somewhat smaller radii by the viscous stresses. The greatest importance of this latter effect is in moving the radius of peak flux outward so that half the energy is radiated at radii $\gtrsim 5GM/c$ from the black hole. The fully relativistic case is discussed by Page & Thorne (1974).

VERTICAL STRUCTURE The local energy flux $F(r)$ is, we see, uniquely fixed by conservation laws, but to determine the local mass density, the spectral shape, and possible time variation of the radiation, and to check the self-consistency of the thin disk approximation (matter loses energy to radiation as fast as it gains energy from viscous dissipation), fairly careful models of the local vertical structure of the disk are appropriate. Only crude models have been calculated to date (Shakura 1972, Shakura & Sunyaev 1973, Pringle & Rees 1972, Pringle et al 1973, Novikov & Thorne 1973, Lightman 1974a,b). Tentative conclusions of such calculations for a reasonable black-hole case ($M = 3M_\odot$, $\dot{M} = 10^{17}$ g/sec, $L \approx 10^{37}$ erg/sec) are: 1. The local spectrum will be black body only for local temperatures smaller than 0.1 keV or so, occurring in outlying parts of the disk, $r \gtrsim 10^3 GM/c^2$ where little total energy is developed. Compton-scattering opacity, leading to a high albedo and hence low emissivity, causes elevated effective emission temperatures, $T \lesssim 10T_{bb}$, where T_{bb} is the equivalent black-body temperature for essentially all of the observable X-ray flux. The spectrum of radiation received at infinity (an average over the disk) is peaked at the order of 10 keV, with substantial flux up to 50 or 100 keV. The observed spectrum is the sum of thermal spectra of different temperatures, and looks nonthermal. 2. If the disk is ever optically thin (i.e. translucent rather than opaque) it is prey to a thermal instability that might disrupt it entirely. Present models are marginally in this regime for a narrow radial band, $2GM/c^2 \lesssim r \lesssim 6GM/c^2$. 3. The viscosity law adopted in most of this work is the innocent ansatz

$$(\phi - r \text{ component of viscous stress}) = \alpha \times (\text{total hydrodynamical pressure}),$$

where α is a dimensionless parameter $10^{-3} \lesssim \alpha \lesssim 1$; this is roughly equivalent to postulating an effective Reynolds number of α^{-1}. But this disk model is then secularly unstable whenever radiation pressure dominates, as it does for $r \lesssim 40GM/c^2$, and it tends to break up violently into clumps or rings (Lightman & Eardley 1974, Lightman 1974b). Therefore, the existence of a stable thin-disk model depends crucially on the details of the viscosity, which are at present unknown.

FLUCTUATIONS Because the disk is turbulent, probably contains regions of magnetic field reconnection resembling solar flares, and may be locally unstable as mentioned above, one expects strong local time fluctuations in the flux at any particular point, subject to the necessary condition that the time-average flux is $F(r)$. From a particular radius r, the frequency of fluctuation should be roughly the orbital period $2\pi/\Omega_{\text{Kepler}}$ for several reasons: 1. The vertical hydrodynamical time scale is of this order. 2. The epicyclic frequency Ω_{ep} characteristic of inplane disturbances is equal to Ω_{Kepler} in the Newtonian limit, although $\Omega_{\text{ep}}/\Omega_{\text{Kepler}}$ goes to zero at the innermost circular orbit around a black hole. 3. The intensity of a point source orbiting a black hole ($r \lesssim 20GM/c^2$) as perceived by a distant observer near the equatorial plane ($i \gtrsim 60°$) varies strongly at a period of exactly $2\pi/\Omega_{\text{Kepler}}$, because of Doppler shift and focusing of light by the gravitational lens effect of the hole as the source passes behind it (Section 3.1). All these effects may lead to strong fluctuations in the X-ray signal at frequencies up to the frequency of the innermost stable circular orbit. This minimal period is about $70(M/M_\odot)$ μsec for an extreme-rotating hole, and about $500(M/M_\odot)$ μsec for a nonrotating hole. The detection of nearly periodic fluctuations on these short time scales would not only provide strong evidence for the presence of a black hole, but thus also provide a chance to measure the hole's rotation (Sunyaev 1973).

EVOLUTION OF THE HOLE The flow of matter from the disk into the hole changes both its mass $M_* \equiv GM/c^2$ and its specific angular momentum $a = J/Mc$, and always tends to increase toward the limiting value $a/M_* = 1$ (Bardeen 1970). But for values of a/M_* very close to 1, an appreciable amount of radiation from the disk is intercepted by the hole and a/M_* thereby tends to decrease. These two contrary effects balance stably at $a/M_* \approx 0.998$ (only slightly dependent on particular assumptions about the angular distribution of emitted radiation), as found by Thorne (1974). Any hole that has grown in mass by a factor exceeding ~ 2.5 via disk accretion will have assumed this canonical value (but the time scale for such growth is $\gtrsim 10^8$ yr); the collapse of a rotating star may well lead to $a/M_* \gtrsim 0.9$ in any case.

CASE OF ORBITAL PLANE AND HOLE'S ROTATION AXIS NOT ALIGNED So far, the discussion has implicitly assumed that the rotation axis of a rotating black hole is parallel to the average angular momentum of the accreting matter. But a chaotic collapse event could well leave even a nearly maximally rotating hole in a random orientation. The case of an accretion disk around a rotating hole with some finite, acute angle between the two axes has been studied by Bardeen & Petterson (1975) (see also Bardeen 1974).

The angular-momentum vector of each ring of matter spiraling in through the disk is initially parallel with that of the binary system. The relativistic frame-dragging induced by the hole causes each ring to precess around the hole's axis; because the ring spirals in slowly, the integrated precession angle becomes important rather far from the hole: for typical parameters at $r \approx 300GM/c^2$. The ring then enters a complicated "transition" region where successive rings have slight differential precessions with respect to one another, leading to viscous torques on each ring, the average effect of which is to damp the component of each ring's angular momentum perpendicular to the hole's axis. This damping is complete by about $r = 100GM/c^2$, and inside this radius the disk lies in the equatorial plane of the hole. In swallowing matter, therefore, the direction of the hole's angular momentum does not change—the "offensive" component of angular momentum has been carried out to infinity by viscous forces between the radial rings. Note that the disk configuration, although warped and complicated, is stationary in the transition region. The main observational consequence of this effect would be that the transition region intercepts a good deal of X-ray flux from the inner region, presumably reradiating it as softer X rays or UV. The details have not been worked out.

A further effect of a rotating misaligned hole would be that the hole itself would precess; this would be "precession of the equinoxes" for an inhabitant of the hole! For known systems the time scale would be unobservably long, $\gtrsim$ (orbital period) $\times$ (orbital velocity/c)$^2 \gg 10^3$ yr.

4.5 *Cygnus X-1 (Cyg X-1) and Other X-ray Sources that May Be Black Holes*

The X-ray source Cyg X-1 has attracted a great deal of attention as a possible black hole. Here observation directly confronts theory with the following questions: 1. Must a strong ($L \approx 10^{37}$ erg/sec), compact (time variations over $\lesssim 0.1$ sec, therefore size $\lesssim 3 \times 10^9$ cm) X-ray source be a compact stellar object (degenerate dwarf, neutron star, or black hole), or are other models feasible? 2. What is the maximum mass of a degenerate dwarf or neutron star? 3. How well can one estimate the mass of an unseen secondary in a close binary system? For Cyg X-1, the answers that indicate a black hole are: 1. An accreting, compact object is the most likely model for the X-ray properties. 2. A compact object of mass $\gtrsim 3M_\odot$ must be a black hole. 3. The optically identified, binary system has an unseen secondary of mass $M > 4M_\odot$ (and probably $M \approx 10M_\odot$), which is the X-ray source, which therefore must be a black hole.

The weakest link in the chain of reasoning is in answer 3, that is, that the unseen secondary is the X-ray source. The secondary could be a normal main-sequence star, and it would then be so faint as to be indeed unseen. The X-rays would then have to be produced elsewhere in the system, perhaps by powerful flares in strong magnetic fields linking the two stars, wound up by differential rotation (Bahcall, Kulsrud & Rosenbluth 1973), or in a third $1M_\odot$ neutron star orbiting the secondary (Bahcall et al 1974, Fabian, Pringle & Whelan 1974). Such bizarre models seem motivated by a healthy but extreme caution in accepting the reality of black holes, and it will be important to rule them out. Here, however, let us assume that the secondary is the X-ray source. Cyg X-1 is a strong X-ray source with an apparently nonthermal, hard-energy spectrum from $\lesssim 1$ keV to $\lesssim 100$ keV [Schreier et al (1971) and references cited there]. Its intensity is chaotically variable, by $\sim 25\%$ down to time scales of ~ 50 msec (Oda et al 1971). There is some evidence for changes as short as 1 msec (Rothschild et al 1974). A weak, very precisely located ($\lesssim 0.5''$) radio source lies within the rather large X-ray error box (Braes & Miley 1971, Wade & Hjellming 1972); in March 1971 this radio source suddenly appeared simultaneously with a sudden and permanent spectral change in the X-ray (Tananbaum et al 1972); therefore the two sources are presumably identical. In turn, at the radio position lies the visible star HDE 226868, an OB supergiant, which is a single-line spectroscopic binary ($P = 5\overset{d}{.}60$) with out-of-phase emission lines indicating gas streaming (Webster & Murdin 1972).

There is no eclipse in the X ray. However, the optical identification of Cyg X-1 has been confirmed by observations of changes in the X-ray spectrum correlated with the optical orbital period, specifically a recurrent, low-energy dip near the superior conjunction of the secondary (Sanford et al 1974, Li & Clark 1974, Mason et al 1974). These observations strongly suggest that the X-ray source is coincident with or very near to the secondary.

The primary star HDE 226868 is spectroscopically a normal 09.7Iab supergiant (Walborn 1973); therefore it is probably a helium core-burning and/or hydrogen shell-burning star of mass $M_1 \approx 30M_\odot$ (*if* it has lost no mass) and absolute visual magnitude $m_V \approx -6.5$. From its apparent magnitude $m_V = 8.9$ (corrected for interstellar reddening, $A_V \approx 3.3$), its distance should be $D \approx$ 2–2.5 kpc, which agrees well with the best estimate $D = 2.5$ kpc and firm lower limit $D \geqq 2$ kpc derived from a careful study of interstellar absorption in the field (Margon et al 1973, Bregman et al 1973). This distance measurement rules out the alternate possibility that the star is a closer ($D \lesssim 1$ kpc), less luminous, and much less massive B star (Trimble, Rose & Weber 1973).

The main uncertainties in the mass determination are now 1. The angle of inclination i between our line of sight and the system's axis is unknown. 2. The star HDE 226868 could have a significantly smaller mass and still exhibit the same spectrum and luminosity. It probably has a $\sim 10M_\odot$ helium core, and even if it has lost almost all of its hydrogen envelope (all but $\sim 0.5M_\odot$) because of its membership in a close binary system, the remainder of the envelope would swell up to present much the same appearance (van den Heuvel & Ostriker 1973).

Luckily, one can put a lower bound on the mass M_2 of the secondary, without any assumptions about M_1 (an argument due to Paczyński, 1974): From the spectral classification 09.7Iab, the effective temperature is $\log T_{\rm eff} = 4.4$, and the bolometric correction is $\Delta m_{\rm bol} - -2.2$. Using the measured distance $D = 2.5$ kpc, the photometry implies $\log(L/L_\odot) = 5.36$. From $L = 4\pi\sigma R_1^2 T_{\rm eff}^4$, one solves for the radius R_1 of the star,

$$R_1 = 26R_\odot. \qquad 52.$$

This value is in agreement with that inferred from the spectral classification. Now enforce the geometrical constraint that the system not eclipse in the X ray, $R_1 \leqq a \cos i$. The observed orbital parameters are (Brucato & Zappala 1974, Bolton 1972): $a_1 \sin i = 7.9R_\odot$, $M_2^3 \sin^3 i/(M_1+M_2)^2 \approx 0.21M_\odot$. Upon varying i there follows an absolute lower bound on the mass M_2 of the secondary,

$$M_2 > 5.8M_\odot\left(\frac{D}{2.5\ \text{kpc}}\right)^2, \qquad 53.$$

where the dependence on D has been reinserted.

If one further assumes that the primary is corotating (consistent with but not forced by the spectroscopy), the additional geometric constraint that it not overfill its Roche lobe, $R_1 \leqq R_{1c}$ (equation 48), gives

$$\begin{aligned} M_2 &> 9.5M_\odot \\ M_1 &> 42M_\odot. \end{aligned} \qquad 54.$$

(Assuming generous errors on all measured quantities, one can beat M_2 down to $\sim 2.5M_\odot$, but only if $M_1 \approx 1.2M_\odot$, which is absurd for the spectral class and luminosity of HDE 226868. If one assumes these generous errors, but takes $M_1 \geqq 10M_\odot$, the absolute bound above always holds.)

Likely values of M_2 are only poorly determined. If there is a near-eclipse, then $M_2 = 6\text{–}7M_\odot$ and $i = 40\text{–}65°$ for assumed values $M_1 = 10\text{–}30M_\odot$. On the other hand, the photometric light curve is double-humped, indicating gravitational distortion of the primary; the size of this effect suggests $i \approx 30°$ (Hutchings et al 1973, Lyutyi et al 1973), whence $M_2 = 8\text{–}15M_\odot$ for assumed values $M_1 = 10\text{–}30M_\odot$. As a likely range, one might adopt

$$6M_\odot < M_2 < 15M_\odot. \tag{55}$$

Equations 53 and 55 strongly exclude the possibility that the compact object is a degenerate dwarf or neutron star, even a rapidly rotating one (see discussion in Sections 1 and 2). The weakest bound of equation 53 could only barely allow a neutron star with a most optimistic equation of state. The conclusion of choice is that the compact object is a black hole.

The main remaining uncertainties in this conclusion are 1. the assumption that the X-ray source is a compact object and 2. the identification of that compact object with the secondary. These points will best be clarified by further extensive X-ray observations by earth satellite, with good coverage over the $5^d\!.60$ orbital period. X-ray observations with very high time resolution ($\lesssim 1$ msec) and good counting statistics (cf Press & Schechter 1974) will be valuable in that one may well see the quasiperiodic effects arising near the hole, as discussed in Section 4.4.

Binary X-ray sources rather similar to Cyg X-1 are 3U0900-40 (≡Vel X-1) (Wickramasinghe et al 1974), 3U1700-37 (Wolff & Morrison 1974), 3U0115-73 (≡SMC X-1) (Osmer & Hiltner 1974), and 3U1516-56 (≡Cir X-1) (Jones et al 1974), all but the last optically identified with blue supergiants with unseen companions. These companions seem to be significantly less massive than Cyg X-1, $0.5M_\odot \lesssim M_2 \lesssim 4M_\odot$, and they could well be neutron stars, black holes, or degenerate dwarfs. All these sources eclipse. The future identification of an eclipsing source of *large* mass would provide a more strenuous test of the black-hole hypothesis. For recent work on accretion-disk models of Cyg X-1, see Thorne & Price (1975) and Eardley et al (1975).

4.6 *Optical Binaries that May Contain Black Holes*

If in a single-line spectroscopic binary the unseen secondary has large mass, $M_2 \gtrsim 2M_\odot$, it might conceivably be a black hole. Unfortunately, it is difficult to rule out the simpler explanation that the secondary is less luminous than the primary. This difficulty is compounded by the possibility that evolution and mass exchange may have left the primary overluminous for its mass, or the secondary underluminous; these effects are indeed observed.

Lists of single-line spectroscopic binaries with large mass function were given by Zel'dovich & Guseynov (1966) and Trimble & Thorne (1969). If a binary system now containing a black hole or neutron star survived a stellar collapse without disruption, it may have been left with a high eccentricity or a high recoil velocity (Section 4.2): Gibbons & Hawking (1971) discuss several single-line spectroscopic binaries with high eccentricities [for a comment see Batten & Olowin (1971)]; Gott (1971, 1972) discusses binaries with high peculiar velocity. Bekenstein & Bowers (1974) list 55 "runaway" OB stars, and suggest that some may be binaries with collapsed companions, but none are known to be binaries.

Recent evolutionary work [see, e.g. van den Heuvel & De Loore (1973) and references cited therein] indicates that in close binaries containing a neutron-star or black-hole secondary, $M_2 \ll M_1$ because of previous exchange and loss. Suntantyo (1973) critically examines these earlier studies in this light, and suggests that systems with *small* mass functions are the best candidates. Indeed the binary X-ray sources, by far the best candidates presently, generally have small mass functions. As a whole, these studies have produced no convincing candidates.

Wilson (1973) summarizes careful studies of individual binary systems with black-hole candidates. Again, the case is not proven.

Van den Heuvel (1973) estimates on evolutionary grounds that there may be $\gtrsim 10^3$ binaries with a quiet neutron star or black hole in the galaxy, as well as $\gtrsim 10$ binary X-ray sources (as observed). But the evidence for the quiet systems is unconvincing because of its essentially negative character. If one of these binaries should exhibit additional positive evidence, such as weak X-rays or characteristic optical emission from accretion, the case would be greatly strengthened.

5 EXOTIC POSSIBILITIES

5.1 *Energy Extraction from Rotating Black Holes*

A clever rocket pilot can increase the efficiency of his rocket by coasting in an orbit that passes very near a star, and expending his fuel mass M_{fuel} near periastron. This strategy leaves the exhaust deep in the star's potential well, and so by conservation of energy gives greater ultimate kinetic energy to the rocket. If the star is a black hole, and if the exhaust velocity is relativistic, this extra energy is of order $M_{\text{fuel}}c^2$. Penrose (1969) first noticed that if the hole is *rotating,* the extra energy can *exceed* $M_{\text{fuel}}c^2$. This occurs when the rocket flies through the "ergosphere" (between the static limit and the horizon) and ejects the exhaust into an orbit of negative total energy, one for which binding energy exceeds rest mass. The exhaust falls into the hole and decreases the hole's angular momentum, so the extra energy is being extracted from the hole's rotational energy. Details have been studied by Christodoulou (1970), Bardeen et al (1972), and Wald (1974a). Wheeler (1970) has suggested that the tidal disruption of a star in the ergosphere of a massive, galactic-core hole might leave some of the ejecta on orbits of negative total energy, allowing the rest to escape to infinity with more energy that $M_{\text{star}}c^2$. But a study of the feasibility of this process by Mashhoon (1972, 1973) reaches no firm conclusion. It presently seems unlikely that the Penrose process happens naturally.

A rotating black hole also amplifies the energy of certain electromagnetic and gravitational wave modes upon scattering (Misner 1972, Zel'dovich 1971, 1972). For a wave with t and ϕ dependence, $\exp(-i\omega t + im\phi)$, the condition for this "superradiant scattering" is $0 < \omega/m < \Omega_{BH}$, where Ω_{BH} is the hole's angular velocity of rigid rotation (equation 21). Significant amplification occurs only for wavelength comparable to size of hole. Calculations were made possible by Teukolsky's (1972, 1973) discovery of a separable wave equation, and they show a maximum energy amplification of 138% for gravitational waves and 4.4% for electromagnetic waves (Starobinsky & Churilov 1973, Teukolsky & Press 1974). *Infinite* amplification would be equivalent to spontaneous instability of a rotating hole, but numerical work indicates that such instabilities do not occur (Press & Teukolsky 1973, Stewart 1974).

The astrophysical relevance of superradiant scattering depends on possible feedback mechanisms to repeatedly amplify an electromagnetic wave. A mirror around a rotating black hole would in principle explode because of buildup of reflected and amplified radio waves (Press & Teukolsky 1972), but the natural formation of such a reflecting cavity strong enough to permit the extraction of a significant amount of energy seems unlikely. Possible similar effects of other sorts of nearby conductors (e.g. a plasma accretion disk) remain to be explored.

A particle orbiting a rotating black hole and radiating gravitational waves might superradiantly gain as much wave energy from the hole as it loses to infinity (Misner 1972, Press & Teukolsky 1972); it would be in a nondecaying "floating orbit," acting as a catalyst to extract rotational energy from the hole. It is not yet known whether this effect is actually possible.

5.2 *Charge on Black Holes*

In principle a black hole can carry an electric charge Q (units: esu); the hole is completely specified by the parameters M, J, and Q, which must obey the constraint $M^2 \geqq (cJ/GM)^2 + Q^2/G$. The maximum theoretical charge is thus $Q_{max} = G^{1/2}M$. The hole's magnetic moment is fixed by M, J, and Q, $\mu = QJ/Mc$; the gyromagnetic ratio is 2, just as for a Dirac-theory electron. Newman et al (1965) first obtained an explicit solution of the Einstein-Maxwell equations that describes this "Kerr-Newman" black hole; the nonrotating, charged solution is called the Reissner-Nordstrom black hole. Many of the beautiful, abstract results on black holes generalize to the charged case, and there emerge new effects, such as strong mixing of electromagnetic and gravitational radiation near a hole with $Q \approx Q_{max}$ (Johnston, Ruffini & Zerilli 1973, Gerlach 1974, Olson & Unruh 1974).

However, black holes in the real universe must always be nearly uncharged, because of the very large ratio of electromagnetic to gravitational forces upon electrons and protons, expressed by the ratio $e/G^{1/2}m_p \sim 10^{18}$. If a hole has a charge exceeding $10^{-18}Q_{max}$, electrical forces on a distant plasma overwhelm gravitational forces, causing charge separation, and drawing particles of opposite charge to the hole, neutralizing it. The mass of plasma necessary to neutralize even a maximally charged hole in this way is

$$M_{plasma} \sim 10^{-18}M \sim 10^{15}(M/M_{\odot})\text{ g}, \tag{56.}$$

which is easily available in a binary system, and is available at a radius of $\sim 10^{13}$ cm in the interstellar medium. So $Q \lesssim 10^{-18}\ Q_{max}$ in the astrophysical environment;

in particular the magnetic field generated by the hole is limited,

$$B \lesssim m_p c^2/(GM/c^2)e \sim 10(M_\odot/M) \text{ gauss}. \quad 57.$$

Even in strict vacuum, intense electric fields would cause copious, spontaneous quantum-mechanical creation of electron-positron pairs (Schwinger 1951) for a black hole with charge

$$Q \gtrsim G^2M^2m_e^2/ehc \approx 10^{-5}(M/M_\odot)Q_{\text{max}}, \quad 58.$$

quickly reducing the charge below this limit (Gibbons 1974).

5.3 *Primordial Black Holes*

In the present universe, black holes can form only for mass above the Chandrasekhar or Oppenheimer-Volkoff limits, $M \gtrsim 1\text{–}3M_\odot$. However, if the early universe was chaotic (Misner 1968), with strong fluctuations on all length scales, "primordial" black holes of any mass could have been produced (Hawking 1971). A typical hole would have been formed by a lump of moderate overdensity, $\delta\rho/\rho \sim 1$, as measured at the time when light had just had time to traverse the lump since the big bang, that is, when it was first within its particle horizon. Holes that formed when the universe had average density ρ would have a characteristic mass

$$M \sim (c^6/\rho G^3)^{1/2} \sim 10M_\odot(\rho/10^{15} \text{ g cm}^{-3})^{-1/2}. \quad 59.$$

Because densities were arbitrarily high sufficiently early, black holes could have formed with arbitrarily small masses, down to the limit of the Planck (1899) mass $M_p = (hc/G)^{1/2} = 2 \times 10^{-5}$ g, where quantum effects dominate and one cannot properly speak of classical black holes. At the other extreme, a hole just forming today would have a mass on the order of the mass of the observable universe (and would create a huge cosmological anisotropy which is not seen).

We can immediately deduce (cf Novikov & Thorne 1973) that holes produced in this way must have been very rare in the early universe, *if* the universe was radiation dominated. The main constraint is just that the total present mass density of holes cannot much exceed the critical density to close the universe (because it is still expanding), $\rho_{\text{holes}} \lesssim \rho_{\text{crit}} \approx 10^{-29}$ g cm^{-3}. As the radiation-dominated early universe evolves, its mass energy redshifts away adiabatically as $1+z$. Any mass in black holes, however, does not redshift away. Thus the fraction of total mass in black holes increases as $(1+z)^{-1}$ as time goes on. Presently, $\rho_{\text{radiation}}/\rho_{\text{crit}} \approx 10^{-4}$; so for $z \gtrsim 10^4$

$$\rho_{\text{holes}}/\rho_{\text{total}} \lesssim 10^4/(1+z). \quad 60.$$

This is a stringent limit at $z \approx 10^{21}$, when 10^{15} g holes formed! The implication is that the early universe could not have been both "maximally chaotic" and radiation dominated; Rees (1972) has proposed a maximally chaotic, cold big bang on other grounds.

Granting the formation of a primordial hole, one can ask what the hole's subsequent interactions would be. It would grow by accretion, but only increasing its mass asymptotically by a factor of order unity [Carr & Hawking (1974); this careful calculation disproves the previous estimate of Zel'dovich & Novikov (1966) which showed a catastrophic growth during the radiation era.] The hole's effects on the present universe would depend on its mass M. For $M \lesssim 10^{15}$ g, holes might have "evaporated" by now (Section 5.4), releasing a possibly observable burst of hard radiation. For $10^{19}\ \text{g} \lesssim M \lesssim 10^{33}$ g, a primordial hole inside a star would grow and swallow it (Section 5.5). For $M \gtrsim 1M_\odot$, the primordial hole is of course indistinguishable from holes formed by gravitational collapse at later epochs.

For all masses $10^{15}\ \text{g} \lesssim M \lesssim 10^{17}M_\odot$, primordial holes might form the cosmological "missing mass" that would close the universe. For $10^5\ \text{g} \lesssim M \lesssim 10^{20}$ g, holes would strike the earth at a rate of $\sim 10^{-5}/\text{yr}\ (M/10^{15}\ \text{g})^{-1}$ if holes are bound in our galaxy's halo, or of $\sim 10^{-8}/\text{yr}\ (M/10^{15}\ \text{g})^{-1}$ if cosmologically uniform. These holes would interact with normal matter mainly through their Newtonian gravitational field, as their size is very small, $M_* \approx 10^{-13}\ \text{cm}\ (M/10^{15}\ \text{g})$. Such a hole would pierce right through the earth. Jackson & Ryan (1973) ingeniously suggested that a 10^{20-22} g hole could have caused the Tunguska meteor event of 1908; this possibility seems remote in view of the low flux, even neglecting the simpler explanation of a meteor or comet (Wick & Isaacs 1974). For much larger masses, $M \gtrsim 10^5 M_\odot$, a cosmological density of holes would cause observable gravitational-lens effects in images of quasars at $z \gtrsim 1$ [Press & Gunn (1973) and references cited therein]. Double images would be common, with a separation $\sim 2.6''(M/10^{12}M_\odot)^{1/2}$. For $M \gtrsim 10^{17}M_\odot$, the nearest hole would cause a grotesque optical distortion of angular extend $\gtrsim 1°$ on the sky behind. This hole's gravity would also distort the local velocity field of galaxies at a noticeable level, $\gtrsim 600$ km/sec.

5.4 *Quantum Processes of Microscopic Black Holes*

Very recently attention has focused on the spontaneous quantum interaction of a microscopic hole. Here *quantum* does not mean quantum gravity (which is crucial only for $M \lesssim 10^{-5}$ g), but rather the coupling of the strong gravitational field near the hole to other particle fields: photons, neutrinos, pions, etc. The subject is too new to assess adequately, but we summarize some reported results:

Starobinsky (1973) first noted that corresponding to the classical "superradiant" wave amplification of a rotating hole (above) there should be a spontaneous quantum emission of zero-rest-mass particles (or massive particles whose Compton wavelength is bigger than the hole). The energy of particles emitted is of order

$$\hbar\omega \sim \frac{c^3 h}{GM} \sim 100\ \text{MeV}\left(\frac{10^{15}\ \text{g}}{M}\right) \qquad 61.$$

and the rate of energy emission is

$$\frac{d\varepsilon}{dt} \sim \hbar(c^3/GM)^2 \sim 10^{20}\ \text{ergs/sec}\ (10^{15}\ \text{g}/M)^2. \qquad 62.$$

Notice that small holes are *more* active (for stellar-mass holes, the effect is totally negligible).

Hawking (1974a,b) has reported the even more striking finding that even *non*-rotating black holes emit quanta roughly according to equations 61 and 62. He considers the collapse that forms the hole: coupling to the strong gravitational field essentially fills all available phase space with quanta of wavelength comparable to the radius of curvature of space (size of hole). Because quanta can remain trapped near the hole for very long times before leaking off to infinity, the phase space available includes a nearly constant future flux (equation 62). Moreover, the spectrum of the flux, determined by this phase space, is that of a blackbody, and Hawking identifies an effective temperature for the hole which is related to its surface gravity (equation 23),

$$kT = \hbar\kappa/2\pi c = \hbar c^3/8\pi GM \approx 10\ \text{MeV}(10^{15}\ \text{g}/M) \tag{63}$$

(for a nonrotating hole).

With this flux, a black hole of mass M will completely evaporate in a time

$$\tau \sim 10^{10}\ \text{yr}\left(\frac{M}{10^{15}\ \text{g}}\right)^3. \tag{64}$$

Thus, if correct, this process would have evaporated all $\lesssim 10^{15}$ g holes by the present epoch. The hole gets hotter as it shrinks (equation 63), so the evaporation accelerates catastrophically. The last $\gtrsim 10^{30}$ erg is released explosively in $\lesssim 0.1$ sec (see equation 64). Hawking speculates that the large entropy per baryon of the present universe may possibly be due to the radiative evaporation of small holes that formed early on.

Hawking's identification of an effective temperature allows black holes to be included within existing thermodynamic laws [something first attempted by Bekenstein (1973b); see also Bardeen, Carter & Hawking (1973)]. A black hole contains entropy in proportion to the surface area of its horizon. The non-decreasing character of the horizon's area in a classical context is thus considered a special case of the second law of thermodynamics, whereas a decrease in area from quantum effects must be balanced by an increase in entropy outside the hole.

5.5 *Black Holes Inside Stars*

Either a stellar-mass hole, $M \gtrsim M_\odot$, or a small primordial hole, $M \ll M_\odot$, may find itself at the core of a star. The hole would affect the structure of the star by its gravitational field, and it would grow by accretion and contribute to the star's luminosity by the energy released. Unfortunately, no calculations of these latter processes are available. The basic question is whether the accretion energy is efficiently converted into outward photon flux, thereby choking the accretion flow by radiation pressure ("Eddington-limited accretion") and enforcing a long time scale for growth ($\gtrsim 10^8$ yr); or whether the accretion energy is swallowed along with the inflowing matter, allowing very rapid growth, limited only by a sonic throat.

For Eddington-limited accretion, radiation pressure balances gravity. This requires

$$\eta \dot{M} c^2 \approx L_{\rm Ed} = 4\pi G M m_p c/\sigma_T \approx 1.3 \times 10^{38}(M/M_\odot)\ \text{erg/sec}, \qquad 65.$$

where $\eta < 1$ is an efficiency, and it is assumed that opacity is dominated by Thomson scattering (as is usually the case). The hole grows on the time scale $t_{\rm Ed} = 4 \times 10^8$ yr,

$$\dot{M}/M \approx 1/(\eta t_{\rm Ed}) \Rightarrow M \propto \exp(t/\eta t_{\rm Ed}), \qquad 66.$$

which is independent of M and of conditions in the surrounding matter (as long as sufficient $\dot{M}$ is available). It is alternatively possible that an efficiently accreting hole would create a very hot, low-density region near it, and reduce $\dot{M}$ *below* the Eddington limit by means of a sonic throat there. Then the growth time scale would be even longer.

If, in contrast to the Eddington-limited case, the accretion-produced energy is unable to get out, then the accretion rate $\dot{M}$ is controlled only by a sonic throat, appropriate to the central temperature of the star (equation 29). The growth time scale is short,

$$M/M \approx (M/M_{\rm star})/t_{\rm star}, \qquad 67.$$

where $t_{\rm star} \approx 10^2(\rho_{\rm central}/100\ \text{g cm}^{-3})$ sec is the central dynamical time scale of the star. If the initial mass of the black hole is M_i, the entire star is swallowed in a time $t \approx t_{\rm star}(M_{\rm star}/M_i)$. In particular, if the sun had initially contained a black hole, $M_i \gtrsim 10^{19}$ g, the sun would be gone by now. If $M_i \approx M_{\rm star}$, the star disappears in a dynamical time; the "bottom drops out."

5.6 *White Holes and Wormholes*

A white hole is the hypothetical time-reverse of a black hole; matter or information can flow outward but not inward. Since a black hole forms in a collapse and exists forever after, a white hole must have existed forever, that is, since the big bang. Quiescent initially, the white hole explodes in a burst of outgoing matter at an unpredictable time. Because matter emerges from the initial singularity within the white hole at this late time, the white hole is a "lagging core" of the big bang. Novikov (1964) and Ne'eman (1965) proposed that QSOs are such lagging cores. However, current indications are that white holes cannot exist in the present universe; any that failed to explode within $\sim 10^6(M/10^9 M_\odot)$ sec after the big bang would have been unstable against turning into black holes (Eardley 1974).

A wormhole (Wheeler 1963) is a tunnel connecting two holes, either far apart in one universe, or in two otherwise disconnected universes. In the Kerr and Reissner-Nordstrom black-hole solutions there exists a timelike wormhole that can be traversed by an observer who falls into the black hole, so that he emerges through a white hole into another universe. However, the evidence is that such a wormhole cannot actually form and be stable in real gravitational collapse. Wormholes share the same instability as do white holes; further there exists within them a "Cauchy horizon" [a limit of predictability; see Hawking & Ellis (1973)], which is violently unstable (Simpson & Penrose 1973). This internal instability probably renders the wormhole impassible.

5.7 *Caveats and Naked Singularities*

The "naked singularity," a space-time singularity that is not hidden inside an event horizon but rather is visible from infinity, is a skeleton in the closet of black-hole theory. A naked singularity can emit matter and information in an unpredictable way, and is therefore disastrous for the physicist who desires to predict the behavior of systems. If the gravitational collapse of a system produces a naked singularity, either instead of or in addition to an event horizon, then most of the theoretical results concerning black holes are simply inapplicable, and we have very little idea of the nature of the final state of gravitational collapse (Penrose 1973). It is piously hoped that naked singularities cannot occur in general relativity: this is Penrose's "cosmic censorship hypothesis." The hypothesis is unproved, but there are various pieces of evidence supporting it. Perturbation analysis of spherical collapse indicates that naked singularities do not form in nearly spherical collapse (Doroshkevich et al 1965, Price 1972a,b), except for "shell-crossing"-type singularities directly ascribable to unrealistic equations of state (Yodzis et al 1973). Valiant attempts to formulate examples of nonspherical collapse that violate the principle of nondecreasing horizon, thus indicating the existence of a naked singularity, have failed (Gibbons 1972, Penrose 1973, Hawking 1973). If the angular momentum of a rotating black hole could be forced above the limit $a/M_* = 1$ by dropping in matter carrying angular momentum, then a naked singularity would appear. But it seems impossible to exceed this limit, either naturally (Thorne 1974) or ideally (Wald 1974b). Nevertheless, the difficult question of ruling out naked singularities is the most fundamental outstanding problem in pure black-hole theory. Progress can come from two directions: theoretical calculations of realistic, asymmetric collapse will, one hopes, show clearly the formation of an event horizon that surrounds any developing singularities; most importantly, the rapidly increasing observational evidence for black holes itself lends credence to the entire theory.

ACKNOWLEDGMENTS

We thank James Conwell for assistance in surveying the published literature. We are grateful to a number of colleagues for discussions and correspondence, among them J. M. Bardeen, M. Demianski, A. P. Lightman, D. N. Page, J. Roberts, D. N. Schramm, S. A. Teukolsky, and K. S. Thorne.

Literature Cited

Ames, W. L., Thorne, K. S. 1968. *Ap. J.* 151:659
Arnett, W. D. 1973. In *Explosive Nucleosynthesis,* ed. D. N. Schramm, W. D. Arnett. Austin: Univ. Texas Press
Arnett, W. D., Schramm, D. N. 1973. *Ap. J. Lett.* 184:L47
Bahcall, J. N., Dyson, F. J., Katz, J. I., Paczyński, B. 1974. *Ap. J. Lett.* 189:L17
Bahcall, J. N., Kulsrud, R. M., Rosenbluth, M. N. 1973. *Nature Phys. Sci.* 243:27
Bardeen, J. M. 1970. *Nature* 226:64
Bardeen, J. M. 1974. In *Gravitational Radiation and Gravitational Collapse,* ed. C. DeWitt-Morette. Dordrecht: Reidel
Bardeen, J. M., Carter, B., Hawking, S. W.

1973. *Commun. Math. Phys.* 31:161
Bardeen, J. M., Petterson, J. A. 1975. *Ap. J. Lett.* 195:L61
Bardeen, J. M., Press, W. H., Teukolsky, S. A. 1972. *Ap. J.* 178:347
Bardeen, J. M., Wagoner, R. V. 1971. *Ap. J.* 167:359
Barkat, Z. K., Wheeler, J. C., Buchler, J.-R., Rakavy, G. 1974. *Ap. Space Sci.* 29:267
Batten, A. H., Olowin, R. P. 1971. *Nature* 234:341
Bekenstein, J. D. 1972. *Phys. Rev. D* 5: 2403
Bekenstein, J. D. 1973a. *Ap. J.* 183:657
Bekenstein, J. D. 1973b. *Phys. Rev. D* 7: 2333
Bekenstein, J. D., Bowers, R. L. 1974. *Ap. J.* 190:653
Blaauw, A. 1961. *Bull. Astron. Inst. Neth.* 15:265
Blumenthal, G. R., Tucker, W. H. 1974. *Ann. Rev. Astron. Ap.* 12:23
Bolton, C. T. 1972. *Nature Phys. Sci.* 240:124
Bondi, H. 1952. *MNRAS* 112:195
Börner, G. 1973. In *Springer Tracts in Modern Physics,* ed. G. Höhler, Vol. 69
Braes, L. L. E., Miley, G. K. 1971. *Nature* 232:246
Brans, C., Dicke, R. H. 1961. *Phys. Rev.* 124:925
Bregman, J. et al 1973. *Ap. J. Lett.* 185: L117
Brucato, R. J., Zappala, R. R. 1974. *Ap. J. Lett.* 189:L71
Carr, B. J., Hawking, S. W. 1974. *MNRAS* 168:399
Carter, B. 1968. *Commun. Math. Phys.* 10: 280
Carter, B. 1973. In DeWitt & DeWitt (1973)
Chandrasekhar, S. 1960. *Principles of Stellar Dynamics.* New York: Dover
Chandrasekhar, S. 1969. *Ellipsoidal Figures of Equilibrium.* New Haven: Yale Univ. Press
Chandrasekhar, S. 1974. *Contemp. Phys.* 15:1
Christodoulou, D. 1970. *Phys. Rev. Lett.* 25:1596
Cohen, J. M., Wald, R. M. 1971. *J. Math. Phys.* 12:1845
Cunningham, C. T., Bardeen, J. M. 1973. *Ap. J.* 183:237
Dahlbacka, G. H., Chapline, G. F., Weaver, T. A. 1974. *Nature* 250:36
Davidson, K., Ostriker, J. P. 1973. *Ap. J.* 179:585
Davis, M., Ruffini, R., Press, W. H., Price, R. H. 1971. *Phys. Rev. Lett.* 27:1466
de Felice, F. 1968. *Nuovo Cimento B* 57: 351
de Felice, F., Nobili, L., Calvani, M. 1974. *Astron. Ap.* 30:111
DeWitt, C., DeWitt, B. S., eds. 1973. *Black Holes—Les Astres Occlus.* New York: Gordon & Breach
Dicke, R. H. 1964. *Nature* 202:432
Doroshkevich, A. G., Zel'dovich, Ya. B., Novikov, I. D. 1965. *Zh. Eksp. Teor. Fiz.* 49:170 [trans. *Sov. Phys. JETP* (1966) 22:122]
Eardley, D. M. 1974. *Phys. Rev. Lett.* 33: 442
Eardley, D. M., Lightman, A. P., Shapiro, S. L. 1975. *Ap. J. Lett.* In press
Fabian, A. C., Pringle, J. E., Whelan, J. A. J. 1974. *Nature* 247:351
Fackerell, E. D., Ipser, J. R., Thorne, K. S. 1969. *Comments Ap. Space Phys.* 1:134
Fishbone, L. G. 1973. *Ap. J.* 185:43
Flannery, B. 1975. *Ap. J.* In press
Fricke, K. J. 1973. *Ap. J.* 183:941
Gerlach, U. H. 1974. *Phys. Rev. Lett.* 32: 1023
Giacconi, R. 1973. *Ann. NY Acad. Sci.* 224: 149
Gibbons, G. W. 1972. *Commun. Math. Phys.* 27:87
Gibbons, G. W. 1974. "Vacuum Polarization and the Spontaneous Loss of Charge by Black Holes." Preprint. Univ. Cambridge, England
Gibbons, G. W., Hawking, S. W. 1971. *Nature* 232:465
Gold, T., Axford, W. I., Ray, E. C. 1965. In *Quasi-Stellar Sources and Gravitational Collapse,* ed. I. Robinson, A. Schild, E. L. Schucking, 93–98. Chicago: Univ. Chicago Press
Goldreich, P., Schubert, G. 1967. *Ap. J.* 150:571
Gott, J. R. 1971. *Nature* 234:342
Gott, J. R. 1972. *Ap. J.* 173:227
Gursky, H. 1973. In DeWitt & DeWitt (1973)
Hartle, J. B. 1971. *Phys. Rev. D* 3:2938
Hawking, S. W. 1971. *MNRAS* 152:75
Hawking, S. W. 1972. *Commun. Math. Phys.* 25:152
Hawking, S. W. 1973. *Commun. Math. Phys.* 33:323
Hawking, S. W. 1974a. *Nature* 248:30
Hawking, S. W. 1974b. "Particle Creation by Black Holes." Preprint. Univ. Cambridge, England
Hawking, S. W., Ellis, G. F. R. 1973. *The Large Scale Structure of Space-Time.* Cambridge, Engl.: Cambridge Univ. Press
Hutchings, J. B., Crampton, D., Glaspey, J., Walker, G. A. H. 1973. *Ap. J.* 182:549
Israel, W. 1968. *Commun. Math. Phys.* 8: 245
Jackson, A. A., Ryan, M. P. 1973. *Nature*

245:88
Johnston, M., Ruffini, R., Zerilli, F. 1973. *Phys. Rev. Lett.* 31:1317
Jones, C., Giacconi, R., Foreman, W., Tananbaum, H. 1974. *Ap. J. Lett.* 191: L71
Kaplan, S. A., Pikel'ner, S. B. 1970. *The Interstellar Medium.* Cambridge, Mass.: Harvard Univ. Press
Kerr, R. P. 1963. *Phys. Rev. Lett.* 11:237
Laplace, P. S. 1798. Trans. in Hawking & Ellis (1973)
Larson, R. B. 1973. *Ann. Rev. Astron. Ap.* 11:219
Lattimer, J. M., Schramm, D. N. 1974. *Ap. J. Lett.* 192:L145
Li, F. K., Clark, G. W. 1974. *Ap. J. Lett.* 191:L27
Lightman, A. P. 1974a. *Ap. J.* 194:419
Lightman, A. P. 1974b. *Ap. J.* 194:429
Lightman, A. P., Eardley, D. M. 1974. *Ap. J. Lett.* 187:L1
Lüst, R. 1952. *Z. Naturforsch. A* 7:87
Lynden-Bell, D. 1969. *Nature* 223:690
Lynden-Bell, D., Rees, M. 1971. *MNRAS* 152:461
Lynden-Bell, D., Wood, R. 1968. *MNRAS* 138:495
Lyutyi, V. M., Sunyaev, R. A., Cherepashchuk, A. M. 1973. *Sov. Astron.* 17:1
Margon, B., Bowyer, S., Stone, R. P. S. 1973. *Ap. J. Lett.* 185:L113
Margon, B., Ostriker, J. P. 1973. *Ap. J.* 186:91
Martynov, D. Ya. 1972. *Usp. Fiz. Nauk* 108:701 [*Sov. Phys. Usp.* (1973) 15: 786]
Mashhoon, B. 1972. PhD thesis. Princeton Univ., Princeton, NJ
Mashhoon, B. 1973. *Ap. J. Lett.* 181:L65
Mason, K. O., Hawkins, F. J., Sanford, P. W., Murdin, P., Savage, A. 1974. *Ap. J. Lett.* 192:L65
Misner, C. W. 1968. In *Batelle Recontres in Mathematics and Physics,* ed. C. DeWitt, B. DeWitt, 117. New York: Benjamin
Misner, C. W. 1972. *Phys. Rev. Lett.* 28: 994
Misner, C. W., Thorne, K. S., Wheeler, J. A. 1973. *Gravitation.* San Francisco: Freeman
Nauenberg, M., Chapline, G. 1973. *Ap. J.* 179:277
Nduka, A. 1971. *Ap. J.* 170:131
Ne'eman, Y. 1965. *Ap. J.* 141:1303
Newman, E. T. et al 1965. *J. Math. Phys.* 6:918
Novikov, I. D. 1964. *Astron. Zh.* 41:1075 (trans. in *Sov. Astron. AJ* 8:857)
Novikov, I. D., Thorne, K. S. 1973. In DeWitt & DeWitt (1973)
Oda, M. et al 1971. *Ap. J. Lett.* 166:L1
Olson, D. W., Unruh, W. G. 1974. *Phys. Rev. Lett.* 33:1116
Oppenheimer, J. R., Volkoff, G. 1939. *Phys. Rev.* 55:374
Osmer, P. S., Hiltner, W. A. 1974. *Ap. J. Lett.* 188:L5
Ostriker, J. P. 1971. *Ann. Rev. Astron. Ap.* 9:353
Ostriker, J. P., Bodenheimer, P. 1973. *Ap. J.* 180:171
Ostriker, J. P., Peebles, P. J. E. 1973. *Ap. J.* 186:467
Paczyński, B. 1971. *Ann. Rev. Astron. Ap.* 9:183
Paczyński, B. 1974. *Astron. Ap.* 34:161
Page, D. N., Thorne, K. S. 1974. *Ap. J.* 191: 499
Peebles, P. J. E. 1972a. *Gen. Relativity and Gravitation* 3:63
Peebles, P. J. E. 1972b. *Ap. J.* 178:371
Penrose, R. 1969. *Riv. Nuovo Cimento Ser. I,* Vol. 1, Numero Speciale:252
Penrose, R. 1973. *Ann. NY Acad. Sci.* 224: 125
Planck, M. 1899. *Sitzungsber. Deut. Akad. Wiss. Berlin, Kl. Math.-Phys. Tech.,* p. 440
Plavec, M. 1968. *Advan. Astron. Ap.* 6:201
Podurets, M. A. 1964. *Astron. Zh.* 41:1090 [trans. in *Sov. Astron. AJ.* (1965) 8:868]
Polnarev, A. G. 1972. *Astrofizika* 8:461 [trans. in *Astrophysics* (1974) 8:273]
Pottasch, S. R. 1970. *Interstellar Gas Dynamics, IAU Symp. No. 39,* ed. H. Habing., 272. Dordrecht: Reidel
Prendergast, K. H., Burbidge, G. R. 1968. *Ap. J. Lett.* 151:L83
Prendergast, K. H., Taam, R. E. 1974. *Ap. J.* 189:125
Press, W. H. 1975. In *Gravitation and Relativity Proc. Int. Conf. Gen. 7th.* New York: Wiley
Press, W. H., Gunn, J. E. 1973. *Ap. J.* 185:397
Press, W. H., Schechter, P. 1974. *Ap. J.* 193:437
Press, W. H., Teukolsky, S. A. 1972. *Nature* 238:211
Press, W. H., Teukolsky, S. A. 1973. *Ap. J.* 185:649
Price, R. H. 1972a. *Phys. Rev. D* 5:2419
Price, R. H. 1972b. *Phys. Rev. D* 5:2439
Pringle, J. E., Rees, M. J. 1972. *Astron. Ap.* 21:1
Pringle, J. E., Rees, M. J., Pacholczyk, A. G. 1973. *Astron. Ap.* 29:179
Rees, M. J. 1972. *Phys. Rev. Lett.* 28:1669
Rees, M. J. 1973. *Ann. NY Acad. Sci.* 224: 118
Rees, M. J. 1974. *Observatory* 94:168

Rhoades, C. E., Ruffini, R. 1974. *Phys. Rev. Lett.* 32:324
Robinson, D. C. 1975. *Phys. Rev. Lett.* 34:905
Rothschild, R. E., Boldt, E. A., Holt, S. S., Serlemitsos, P. J. 1974. *Ap. J. Lett.* 189:L13
Ruffini, R., Wheeler, J. A. 1971. *Relativistic Cosmology and Space Platforms, ESRO Reprint SP-52,* Paris. Reprinted 1974 in *Black Holes, Gravitational Waves and Cosmology,* ed. M. Rees, R. Ruffini, J. A. Wheeler. New York: Gordon & Breach
Ryan, M. P. 1972. *Ap. J. Lett.* 177:L79
Sabbadini, A. G., Hartle, J. B. 1973. *Ap. Space Sci.* 25:117
Salpeter, E. E. 1964. *Ap. J.* 140:796
Salpeter, E. E. 1971. *Nature Phys. Sci.* 233:5
Sanders, R. H. 1970. *Ap. J.* 162:791
Sanford, P. W., Mason, K. O., Hawkins, F. J., Murdin, P., Savage, A. 1974. *Ap. J. Lett.* 190:L55
Schreier, E., Gursky, H., Kellogg, E., Tananbaum, H., Giacconi, R. 1971. *Ap. J. Lett.* 170:L21
Schwinger, J. S. 1951. *Phys. Rev.* 82:664
Shakura, N. I. 1972. *Astron. Zh.* 49:921 (trans. in *Sov. Astron. AJ* 16:756)
Shakura, N. I., Sunyaev, R. A. 1973. *Astron. Ap.* 24:337
Shapiro, S. L. 1973. *Ap. J.* 180:531; 185:69; 189:343
Shvartsman, V. F. 1971. *Soviet Astron. AJ* 15:377
Shvartsman, V. F. 1974. Results presented at *IAU Symp. No. 67,* Moscow
Simpson, M., Penrose, R. 1973. *Int. J. Theor. Phys.* 7:183
Spitzer, L. Jr. 1956. *Physics of Fully Ionized Gases.* New York: Wiley
Spitzer, L. Jr., Hart, M. H. 1971. *Ap. J.* 164:399
Spitzer, L. Jr., Thuan, T. X. 1972. *Ap. J.* 175:31
Starobinsky, A. A. 1973. *Zh. Eksp. Teor. Fiz.* 64:48 (trans. in *Sov. Phys. JETP* 37:28)
Starobinsky, A. A., Churilov, S. M. 1973. *Zh. Eksp. Teor. Fiz.* 65:3 (trans. in *Sov. Phys. JETP* 38:1)
Stewart, J. 1974. Preprint
Stewart, J., Walker, M. 1973. In *Springer Tracts in Modern Physics,* ed. G. Höhler, Vol. 69
Stothers, R. 1974. *Ap. J.* 192:145
Sunyaev, R. A. 1973. *Sov. Astron. AJ* 16:941
Suntantyo, W. 1973. *Astron. Ap.* 29:103
Tananbaum, H., Gursky, H., Kellogg, E., Giacconi, R., Jones, C. 1972. *Ap. J. Lett.* 177:L5
Teitelboim, C. 1972. *Phys. Rev. D* 5:2941
Teukolsky, S. A. 1972. *Phys. Rev. Lett.* 29:1114
Teukolsky, S. A. 1973. *Ap. J.* 185:635
Teukolsky, S. A., Press, W. H. 1974. *Ap. J.* 193:443
Thorne, K. S. 1972. *Magic without Magic: John Archibald Wheeler,* ed. J. Klauder, 231. San Francisco: Freeman
Thorne, K. S. 1974. *Ap. J.* 191:507
Thorne, K. S., Price, R. M. 1975. *Ap. J. Lett.* 195:L101
Tremaine, S. O., Ostriker, J. P., Spitzer, L. Jr. 1975. *Ap. J.* 196:407
Trimble, V., Rose, W. K., Weber, J. 1973. *MNRAS* 162:1P
Trimble, V. L., Thorne, K. S. 1969. *Ap. J.* 156:1013
van den Heuvel, E. P. J. 1973. *Close Binaries: Lectures Summer Sch. Phys. Ap. Compact Objects,* Cambridge, England
van den Heuvel, E. P. J., De Loore, C. 1973. *Astron. Ap.* 25:387
van den Heuvel, E. P. J., Ostriker, J. P. 1973. *Nature Phys. Sci.* 245:99
Wade, C. M., Hjellming, R. M. 1972. *Nature* 235:271
Walborn, N. R. 1973. *Ap. J. Lett.* 179:L123
Wald, R. M. 1974a. *Ap. J.* 191:231
Wald, R. M. 1974b. *Ann. Phys.* 82:548
Webster, B. L., Murdin, P. 1972. *Nature* 235:37
Wheeler, J. A. 1963. *Relativity, Groups and Topology,* ed. C. DeWitt, B. DeWitt. New York: Gordon & Breach (1964)
Wheeler, J. A. 1970. *Nuclei of Galaxies,* ed. D. O'Connell. Amsterdam: North-Holland (1971)
Wheeler, J. A. 1974. In *Proc. XVIth Solvay Conf.* Brussels, Belgium: (Ed. Univ. Bruxelles)
Wheeler, J. C., Rosenwald, R. D. 1973. *Ap. Lett.* 15:75
Wick, G. L., Isaacs, J. D. 1974. *Nature* 247:139
Wickramasinghe, D. T., Vidal, N. V., Bessell, M. S., Peterson, B. A., Perry, M. E. 1974. *Ap. J.* 188:167
Wilson, R. E. 1973. *Ann. NY Acad. Sci.* 224:263
Wolf, A. M., Burbidge, G. R. 1970. *Ap. J.* 161:419
Wolff, S. C., Morrison, N. D. 1974. *Ap. J.* 187:69
Yodzis, P., Seifert, H. J., Müller zum Hagen, H. 1973. *Commun. Math. Phys.* 34:135; 37:29
Zel'dovich, Ya. B. 1971. *Pisma Zh. Eksp.*

Teor. Fiz. 14:270 (trans. in *Sov. Phys. JETP Lett.* 14:180)
Zel'dovich, Ya. B. 1972. *Zh. Eksp. Teor. Fiz.* 62:2076 (trans. in *Sov. Phys. JETP* 35:1085)
Zel'dovich, Ya. B., Guseynov, O. M. 1966. *Ap. J.* 144:840
Zel'dovich, Ya. B., Novikov, I. D. 1966. *Astron. Zh.* 43:758 (trans. in *Sov. Astron. AJ* 10:602)
Zel'dovich, Ya. B., Novikov, I. D. 1971. *Relativistic Astrophysics,* Vol. 1. Chicago: Univ. Chicago Press

INSTRUMENTAL TECHNIQUE IN X-RAY ASTRONOMY

Laurence E. Peterson
University of California, San Diego, La Jolla, California 92037 and Tata Institute for Fundamental Research, Bombay, India 400 005

INTRODUCTION

Although cosmic X-ray sources were first discovered in 1962 and have been studied since then using instruments on rockets, balloons, and small satellites, the full scope of X-ray astronomy became apparent only after the extended observations provided by the satellite UHURU. X-ray measurements now play a role in modern astrophysics comparable to that of radio, optical, and infrared astronomy. In addition to increasing the number of detected sources fourfold because of the increased sensitivity, the new observations indicate the importance of high-energy processes in many classes of galactic and extragalactic emitters. Long-term studies have also discovered time variations on scales inaccessible to rockets and balloons, providing new insights into the role of compact objects in stellar evolution.

The development of the instrumental concepts and their implementation on space vehicles has of course paralleled the advances of X-ray astronomy. Although scientific progress has been well documented by two major IAU Symposia devoted to the subject (Gratton 1970, Bradt & Giacconi 1973), there has been no major discussion of the instrumental development since the review by Giacconi and his associates (Giacconi et al 1968). Workers at the Goddard Space Flight Center (GSFC) and the University of Maryland have, however, produced a monograph on experimental technique in high-energy astrophysics (Ögelman & Wayland 1970), and Gursky & Schwartz (1974) have an extensive discussion in *X-Ray Astronomy* (Giacconi & Gursky 1974). IAU Symposium No. 41, *New Techniques in Space Astronomy* (Labuhn & Lust 1971), also contains recent material, as does a general discussion by Gursky (1973) in a review on galactic X-ray sources. Extensive papers on specific instrumental topics such as grazing-incidence devices (Giacconi et al 1969), linear and rotating modulation collimators (Bradt et al 1968, Schnopper et al 1968), high-resolution spectroscopy, and polarimetry (Novick 1973) have appeared. Instruments have also been discussed from the viewpoint of application to specific vehicles such as balloons (Peterson et al 1972), rockets (Giacconi 1971), and the High Energy Astronomical Observatory (HEAO) (Peterson 1972). Although many papers describing specific devices or techniques now appear in the open literature, much of the experimental lore resides in laboratory or government reports and proposals, or simply in the minds of the practitioners.

This review is confined to nonfocusing, high-sensitivity counter techniques used in the detection of cosmic photons in the 0.20–300 keV range. Therefore, there is no discussion of grazing incidence, or high-resolution spectroscopic techniques, because, as indicated, these have been reviewed recently (Giacconi et al 1969, Giacconi 1971). There is also little discussion of instruments in the 0.3–10 MeV gamma-ray regime, even though the principles and many techniques extend to these energies. Proportional and scintillation counters have been used in atomic and nuclear physics experiments for many decades. The use of these detectors in laboratory X-ray spectroscopy, however, in conjunction with accelerator experiments, or in ground-based low-level counting systems, poses constraints quite different from the measurement of weak cosmic fluxes from space platforms.

STATUS OF X-RAY ASTRONOMY

General

Because the instrumental development and requirements must be discussed in the context of scientific progress, we wish to review here the present status of X-ray astronomy. The discoveries made by the UHURU on galactic and extragalactic sources have been summarized by Tananbaum (1973) and Kellogg (1973), respectively; several recent, more general expositions (Friedman 1973, Peterson 1974) have appeared.

X-ray astronomy conveniently divides itself into energy ranges where instrumental technique, state of development, and physical processes all differ. In the 1–10 keV range (1.2–12 Å),[1] where the early discoveries were made and where the UHURU operates, observations are made with collimated proportional counters. At lower energies, 0.2–2.0 keV, proportional counters with very thin windows of organic materials have measured absorption effects and structure due to the interstellar medium (Friedman et al 1973, Hayakawa 1973) and soft X rays from relatively cool supernova remnants (Pounds 1973). One-dimensional reflecting telescopes have been used by Gorenstein et al (1971a,b) and Rappaport (Borken et al 1972) as "concentrators" in conjunction with very thin-window proportional counters to determine the soft X-ray structure of these remnants. Although a focusing device on the Copernicus satellite, described by Fabian et al (1973) and Bowles et al (1974), has also provided new information on these as well as other objects (Fabian et al 1974), a grazing-incidence X-ray telescope that can form images directly as in conventional astronomy has not yet been flown with enough area to permit definitive observations of cosmic sources. Focusing instruments on Skylab (Vaiana 1973) have, however, obtained high-resolution pictures ($\sim 2''$) of solar-active phenomena at wavelengths as short as 4 Å (Vaiana et al 1973).

Scintillation counters are usually employed above 20 keV, where observations

[1] A simple formula, $\lambda(\text{Å}) = 12.4/E$ (keV), relates photon energy to wavelength.

may also be made from high-altitude balloons (Peterson et al 1972, Peterson 1973). Here, the steep cosmic spectra, combined with the difficulty of implementing large-area low-background detectors, has restricted spectral measurements to perhaps only 40 of the strongest sources and then usually only to energies less than 100 keV. Long-term observations with a 7–550 keV telescope on OSO-7 (Peterson 1973) have complemented the UHURU coverage and provided extended spectra of the binary X-ray sources (Ulmer 1975) and cèrtain extragalactic objects (Baity et al 1975a).

The concept of an X-ray measurement with a collimated counter on UHURU (Giacconi et al 1971) or any other rotating and precessing vehicle such as OSO (Clark et al 1973) is shown in Figure 1. As the spacecraft scans, a source passes through the aperture, resulting in an increase in the counting rate. The upper panel of Figure 2 shows a typical result as the instrument, spinning at $\sim 1/10$ rpm,

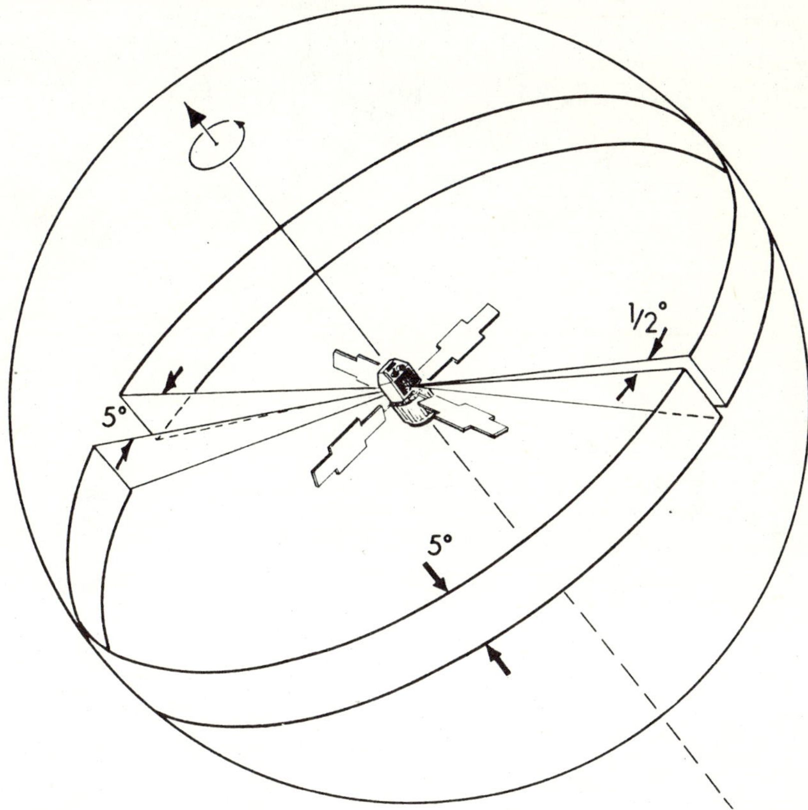

Figure 1 The concept of an X-ray astronomy observation as exemplified by UHURU. Proportional counters collimated to a field of view a few degrees wide scan across a source as the vehicle rotates. (From Giacconi et al 1971.)

passed over the strong source Cen X-3 (Schreier et al 1972). The envelope of the response is due to the aperture; the periodic variations during the transit are due to 4.8^+-sec pulsations of the source. The background rate, due to diffuse cosmic X rays and charged cosmic rays and their secondaries, was obtained before and after the scan.

UHURU, launched 12 December 1970, has a sensitive area of 840 cm^2, and detected sources as weak as $\sim 2 \times 10^{-3}$ counts cm^{-2} sec^{-1}, or about one-five-

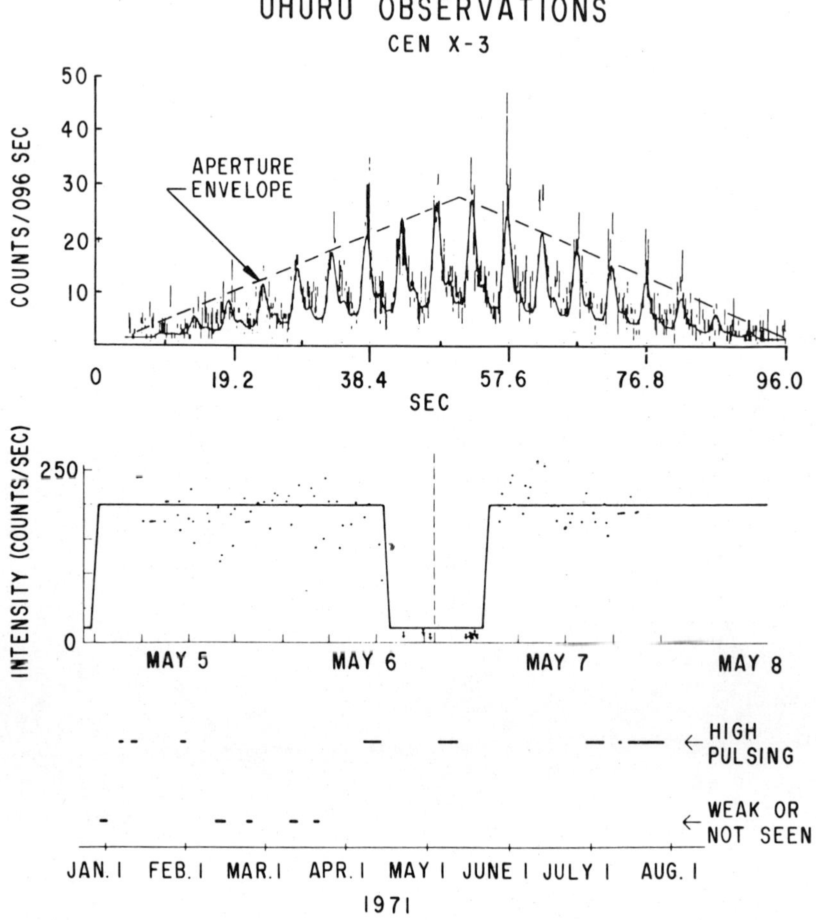

Figure 2 Time variations observed on three different scales from the binary source Cen X-3. *Upper curve,* the 4.8^+-sec pulses during a single UHURU scan; *middle panel,* the 2.067-day variation associated with the binary eclipse; *lower panel,* the long-term quasi-random "on" and "off" states. (Adapted from Schreier et al 1972.)

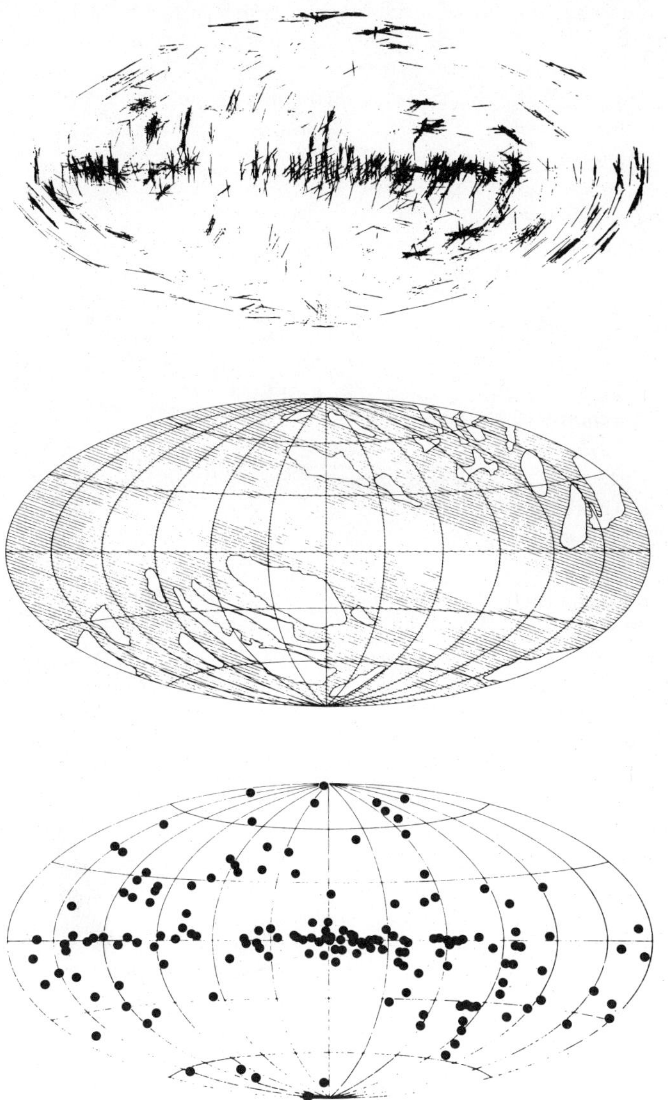

Figure 3 The X-ray sky map in galactic coordinates obtained by UHURU from approximately 125 days of operation. *Upper panel,* the 1878 "lines of position" from which the 161 source locations in the bottom panel were obtained; *central panel* (*shaded*) areas of the sky covered to a sensitivity better than 10 counts sec^{-1}, or about 10^{-2} that of the Crab nebula. (From Giacconi et al 1974.)

hundredth the flux from a strong source such as the Crab nebula. This corresponds to an energy input in the range 10^{-8}–10^{-11} ergs cm^{-2} sec^{-1}, or for sources at a few kiloparsecs, to X-ray luminosities, L_x, of 10^{35}–10^{38} ergs sec^{-1}. An extragalactic source at 10 Mpc must radiate $L_x \gtrsim 10^{41}$ ergs sec^{-1} in 1–10 keV X rays for detection. The X-ray sky map in galactic coordinates as obtained by UHURU is shown in Figure 3 (Giacconi et al 1974). Also shown are the 1878 "lines of position" from which 161 source locations are obtained and the regions of the sky surveyed to a sensitivity of 10 UHURU counts sec^{-1} or better. The sources are located to positional areas typically a few tenths of a square degree in size, depending considerably on source strength, number of observations, and background conditions.

Only about 40 X-ray sources have been identified with known radio or optical objects. Although about 100 of the 161 sources are concentrated in the galactic plane, their locations do not coincide in detail with known galactic features such as OB-associations, H II regions, novae, nonthermal radio sources, etc. The 64 sources at $b^{II} \geqq 20°$ are generally of a weaker nature. Although some are associated with known extragalactic objects, and a few are galactic, the majority are simply unidentified (Kellogg 1973). If these are indeed extragalactic, some must be X-ray galaxies (Giacconi et al 1971).

The photon spectrum, which is a signature of the emission process, can be obtained from pulse-height analysis of the proportional or scintillation-counter events to an ultimate resolution no better than 5–10%. Emission from an optically thin hot gas at 10^7–10^8 °K produces an exponential spectrum, whereas synchrotron emission and Compton scattering from power-law electron distributions also produce power-law spectra. Blackbody emission in X rays may also occur from an optically thick region.

Discrete *K*- or *L*-shell X rays, expected from 10^7 °K thermal plasmas with normal cosmic abundances, have been searched for extensively. Although several observations have tentatively indicated iron-line emission for Sco X-1 (Novick 1973), the most conclusive detection seems to be oxygen lines from the Cygnus loop (Stevens & Garmire 1973) and the 5.8 keV iron line from Cyg X-1 (Serlemitsos et al 1974). Gamma rays, expected in the 20 keV–1 MeV range due to nuclear processes following catastrophic events, have also been searched for. Although only upper limits at about 10^{-3} photons cm^{-2} sec^{-1} exist for lines from the most likely object, the Crab nebula (Peterson & Jacobson 1970), there is a recent indication of line structure near 480 keV from the galactic center (Johnson & Haymes 1973). Novick and his colleagues (Novick 1973) have measured polarization in 2–6 keV X rays from the Crab at $15.4 \pm 5.2\%$. Obtaining high resolution X-ray and gamma-ray spectroscopic and polarimetric data is a major objective of the next generation of X-ray astronomy instrumentation.

Galactic Sources

BINARY X-RAY SOURCES The sources known to be associated with binary systems fall into two classes: those that pulsate with eclipsing, and those that simply eclipse or have other variations. About eight such systems have been identified, all of which have tentative optical identication. The observational status of binary X-ray sources as of 1974 has been reviewed by Gursky & Schreier (1974).

Figure 2 also shows counting rates observed for Cen X-3 on a number of time scales by the UHURU workers (Gursky 1973, Schreier et al 1972). The rates during a single 90-sec scan are not only modulated by the aperture response of the detector but are pulsed at a 4.8-sec rate, due to the source. The average rate over many such scans (*center panel*) shows a characteristic on/off feature, which repeats with a 2.067-day period. Analysis shows the 4.8^{+}-sec period is frequency modulated in phase with the same 2.067-day period.

These effects are understood under the concept that the pulsating X rays are produced by a compact object revolving about a larger one, which occults the small one during a portion of the orbit. The change in pulse arrival time, interpreted in terms of a Roemer shift, gives the projected orbit size as 1.19×10^{12} cm. The fast pulsation requires a source region of radius less than 10^9 cm. Cen X-3 also shows quasi-random "off" periods, which last for weeks or months, as shown in Figure 2 (*lower panel*). Because of its location in a highly obscured region, Cen X-3 has only recently been tentatively identified with a reddened early OB star by Krzeminsky (1974) and Vidal et al (1974).

The second detected but more extensively studied pulsating binary system is Her X-1, which has a 1.24^{+}-sec period, a 1.7-day eclipse, and a quasi-regular 11-day "on," 24-day "off" periodicity. Her X-1 has now been identified with the variable stellar object HZ Her (Bahcall et al 1974). Although HZ Her has the general appearance of a late A or early F star, its intensity is modulated $\Delta m_V \cong 1$ at the 1.7^{+}-day period.

X-ray emission of compact objects in binary systems is usually explained in terms of accretion (e.g. Ostriker & Davidson 1973), where gas flowing out of the visible star accumulates in an orbiting disc around the compact object. If the latter is a rotating neutron star with an intense magnetic field oblique to the rotation axis, matter from the disc may fall or "accrete" onto the magnetic polar surfaces, become gravitationally heated, and emit X rays. The pulsations are due to modulation at the rotation rate by the geometry of the emission region. Simple consideration shows that an accretion rate of $10^{-9} M_{\odot}$ yr^{-1} onto a 10 km radius, $1.0 M_{\odot}$ neutron star will easily result in a kinetic temperature of 10^8 °K, and an emission of 10^{37} erg sec^{-1} from a polar region less than 1 km across. The most difficult feature of Her X-1 to explain is the modulation at the $\sim$34.85-day period.

Six other sources exhibit an X-ray periodicity between 4.8 h (Cyg X-3) and 8.95 day (Vela X-1). Although most of these show considerable variability during noneclipsed phases, none has shown fast periodic pulsations like Cen X-3 or Her X-1. Cyg X-3, which has been associated with a remarkable series of radio outbursts discussed by Hjellming (1973), has now been positively identified with a variable infrared object (Becklin et al 1973). The X-ray variations are nearly sinusoidal, as are the in-phase IR periodicities. It has been suggested that Cyg X-3 may be a binary system, where the compact object is a white dwarf. The 4.7-day SMC X-1 is in the Small Magellanic Cloud, and has an intrinsic X-ray luminosity $L_x \sim 3 \times 10^{38}$ ergs sec^{-1}, considerably above the typical galactic binary of $\sim 5 \times 10^{37}$ ergs sec^{-1}.

Another well-studied object is Cyg X-1, the "black-hole" X-ray source whose extreme and complex variability on all time scales has been seen from many rocket (Rothschild et al 1974), balloon (Overbeck & Tananbaum 1968), and satellite (Oda et al 1972) observations. The source is now positively identified with the 5.6-day spectroscopic binary system HD 226868. Because 5.6-day X-ray periodicities are not seen, the assumption is that the X-ray emission occurs from an unseen, uneclipsed compact companion accreting matter onto its surface. From the velocity and period, assuming the visible object is a $\geqq 20 M_{\odot}$ BOIb supergiant, the companion has a mass $\geqq 3 M_{\odot}$. This object, if collapsed, has a radius inside the Schwarzchild limit and therefore has the properties of a black hole. The intense,

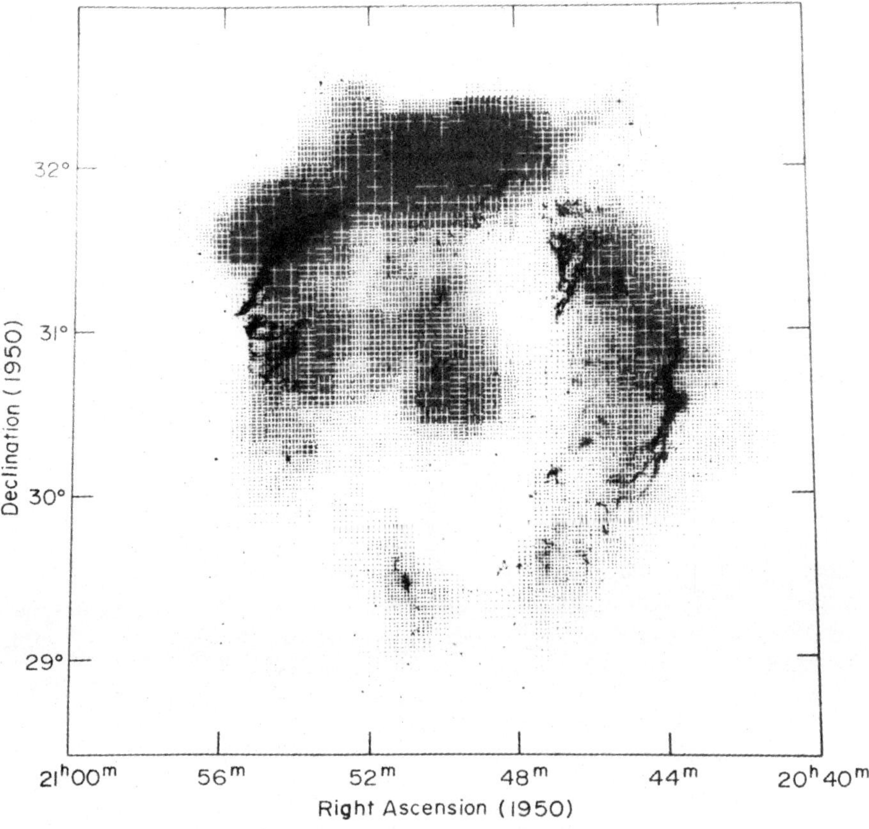

Figure 4 X-ray surface brightness map at 0.15–0.85 keV superimposed on an optical photograph of the Cygnus loop. The density of dots is linearly proportional to the X-ray intensity. Where this intensity is less than 5% of the maximum, zero intensity is displayed. (From Rappaport et al 1974; photograph copyright National Geographic Society—Palomar Sky Survey.)

highly variable X-ray emission, requiring extreme compactness, and the unseen nature of the companion are all consistent with the "black-hole" hypotheses. Other sources, such as Vela X-1, Cir X-1, and 3U1700-37 may be in the same category. Although the "black hole" hypothesis for Cyg X-1 is currently most discussed, alternative explanations, such as ternary systems or stars with usual spectral properties have also been discussed.

SUPERNOVA REMNANTS These form a distinct class. In addition to the Crab nebula, a unique source discussed next, X rays have now been observed from Cas A (Fabian et al 1973), SN 1572, Tycho's SN, the Cygnus loop, Pup A, Vela X, and possibly others. As discussed by Pounds (1973) and Gursky (1973), these latter objects have steep spectra characteristic of a hot gas at a few million degrees, although the Cal Tech group (Stevens & Garmire 1973, Stevens et al 1973) and the Wisconsin group (Coleman et al 1973) have observed complex, multicomponent spectra from several sources. The Cygnus loop is not even seen in 2–6 keV counters because of its soft spectrum, but "maps" of the emission region have been obtained at 0.2–0.5 keV and compared with optical and radio structure. Figure 4 shows a recent measurement of the X-ray-emitting structure by Rappaport et al (1974).

Because the distances to supernova remnants are usually well known, luminosities may be determined, and it is found that these fall in a class 10^{35}–10^{36} ergs sec^{-1}, which is less than the typical galactic emitter of 10^{37}–10^{38} ergs sec^{-1}. The X-ray emission is thought to result from an expanding gaseous shell which may go through an extreme heating phase during its evolution. The relation of this gas to the energetic electron populations producing nonthermal radio emission usually observed in these remnants is unclear.

CRAB NEBULA AND NP 0532 The mass of data now available on this object and some interpretations of the data have been reviewed recently by Apparao (1973). Early observations, reviewed by Peterson & Jacobson (1970), indicated that the X-ray emission was distributed over a diameter of about 100″ (Bradt et al 1968), approximately entered on the optical synchrotron-emitting region, and that the spectrum over the 1–500 keV range had a power-law shape with a negative photon number index of 2.1. About 15% of the X-ray emission is pulsed at the $\sim$33 msec rate of the radio and optical pulsar NP 0532 in phase with the optical. Figure 5 shows the hard X-ray spectrum of both the nebula and NP 0532, as determined by a recent University of California, San Diego (UCSD) balloon investigation (Laros et al 1973). The pulse shape, obtained by folding the entire 5000 sec of balloon data modulo the 33^{+} $msec^{-1}$ period, is shown in the inset.

It is now generally accepted that NP 0532 is a rotating neutron star and the nebula is powered by kinetic energy converted from its slowdown. The details of the particle acceleration in the neutron star magnetosphere and the transfer of energy to the nebula are not understood, however. If the emission of the extended region is indeed due to synchrotron radiation of electrons injected from NP 0532, then their lifetime in the $\sim 4 \times 10^{-4}$ gauss nebular field is insufficient for them to have propagated to the observed distances, and an additional acceleration process must be operative. Crucial to this question is the size of the region at higher energies, which may be determined by balloon observations during recent lunar occultations (Ricker et al 1974, 1975) or by using a wire grid "modulating collimator" in front of proportional or scintillation counters (Bradt et al 1968).

Source spectra often indicate absorption in the 0.2–2.0 keV range, which may be

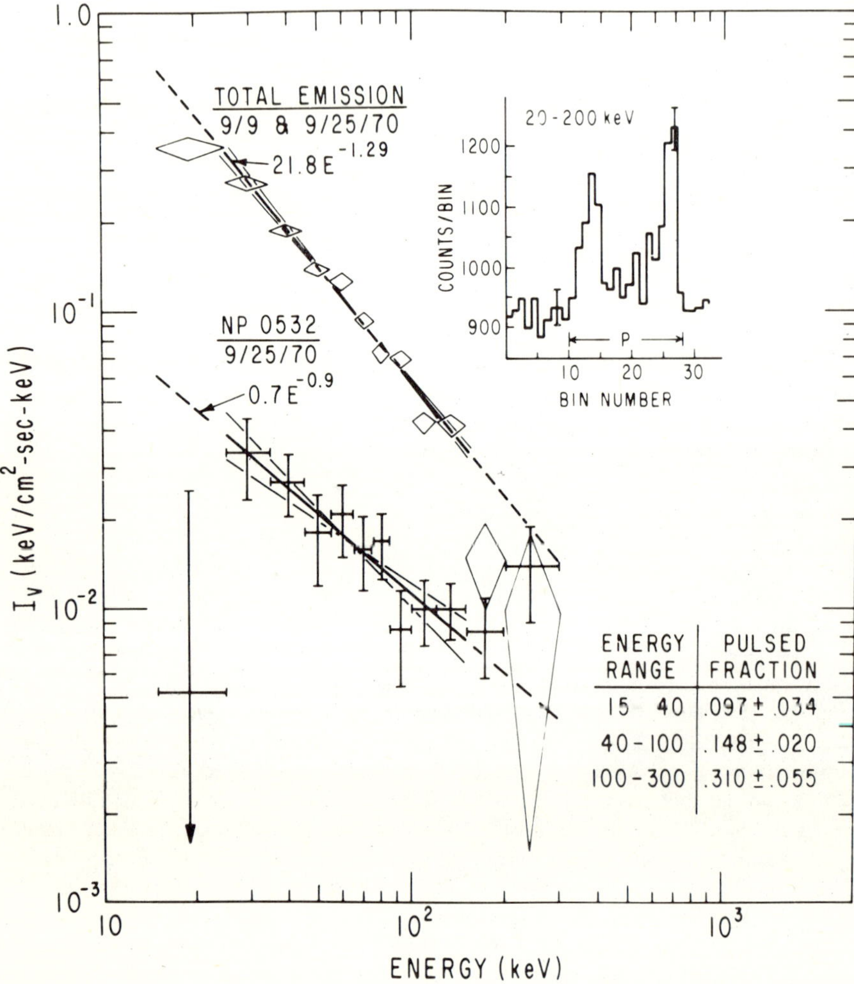

Figure 5 A balloon observation of the hard X-ray spectrum of the Crab nebula and NP 0532. The power-law spectrum over the 1–200 keV range is believed due to synchrotron radiation from energetic electrons in the $\sim 10^{-4}$ gauss nebular field. The mechanism of the pulsed component and its relation to the nebular emission is unclear at present. (From Laros et al 1973.)

of either interstellar or local origin. Although the origin may be uncertain in compact sources, the observed absorption from the Crab nebula is clearly of interstellar character. Because the X-ray absorption is independent of the state of the un-ionized matter, that is, atomic, molecular, or granular, X-ray observations indicate the total hydrogen column density. Although there is some disagreement among measurements, rocket observations indicate that the absorption from the direction of the Crab indicates a column density of 3.5×10^{21} H cm^{-2}, or about twice that of the 21 cm radio observations (Iyengar et al 1975, Margon 1974).

TRANSIENT X-RAY SOURCES Since 1967, about five strong sources have been observed to abruptly appear and then decay into the background level after a few weeks or months, as discussed by Gursky (1973) and Gursky & Schreier (1974). These sources show a spectrum which softens with time and decays in intensity similar to that of galactic novae, hence the term *X-ray novae.* For example, Cen X-4 has been observed with the Vela satellites (Evans et al 1970) and Cep X-4 with OSO-7 (Ulmer et al 1973). None of these has been seen during the rising phase, and none is positively identified with known novae or with any optical or radio object. There is increasing evidence that transient X-ray phenomena are occurring at more than the generally accepted rate of two per year and that such objects will be seen more often as sky coverage and instrument sensitivity increase. The recently discovered transient gamma-ray sources, also observed with the Vela satellites (Strong et al 1974), are apparently of a different nature, having very hard spectra and time scales of only tens of seconds.

OTHERS This includes the majority of the galactic sources. Some, such as Sco X-1 and Cyg X-2, have been observed extensively since the early days of X-ray astronomy and have well-known optical counterparts. Sco X-1 (Loh & Garmire 1971, Pelling 1973) may be explained in terms of a 50×10^6 °K gas of 10^{16} atoms cm^{-3} having a radius of 10^8 cm. Such a configuration becomes optically thick in the near IR, consistent with observations. Some, such as Cir X-1, observed by UHURU (Jones et al 1974) and OSO-7 (Baity et al 1975b), have extreme variability on all time scales.

The energy input to these sources is a major problem. Therefore it has been suggested that *all* galactic X-ray emitters, with the exception of the Crab and other supernova remnants, are of compact binary nature, in which the binary characteristics have been suppressed because of the configuration or absorption by thick envelopes (Tananbaum 1973). It is clear that strong X-ray sources are a rare galactic phenomenon, with perhaps only a few hundred in our galaxy, all emitting in the 10^{37}–10^{38} ergs sec^{-1} range and, with the exception of the supernova remnants, there is apparently no large population in the 10^{35}–10^{36} ergs sec^{-1} class.

Extragalactic Sources

Much less is known about extragalactic X-ray emitters because their fluxes are inherently weaker and their spectra are usually steeper than galactic sources. Their X-ray emission is important in terms of galactic structure and evolution, and may contribute importantly to cosmological effects. Perhaps 20 of the approximately 60 UHURU sources with $b^{II} > 20°$ have been identified with known extragalactic objects, of both extended and compact nature.

EXTENDED OBJECTS X-ray emission has been observed over extended regions, $\gtrsim 0.7°$ diam from the galactic clusters Perseus, Virgo, and Coma by UHURU (Gursky et al 1972) and Copernicus (Griffiths et al 1974, Fabian et al 1974). The UHURU flux is typically < 50 counts $\sec^{-1}$, implying $L_x \sim 3 \times 10^{44}$ ergs $\sec^{-1}$ for Perseus and Coma. It has been suggested that the emission is due to a hot intercluster gas, $\sim 10^8$ °K, sufficient to provide the "missing mass" required by the virial

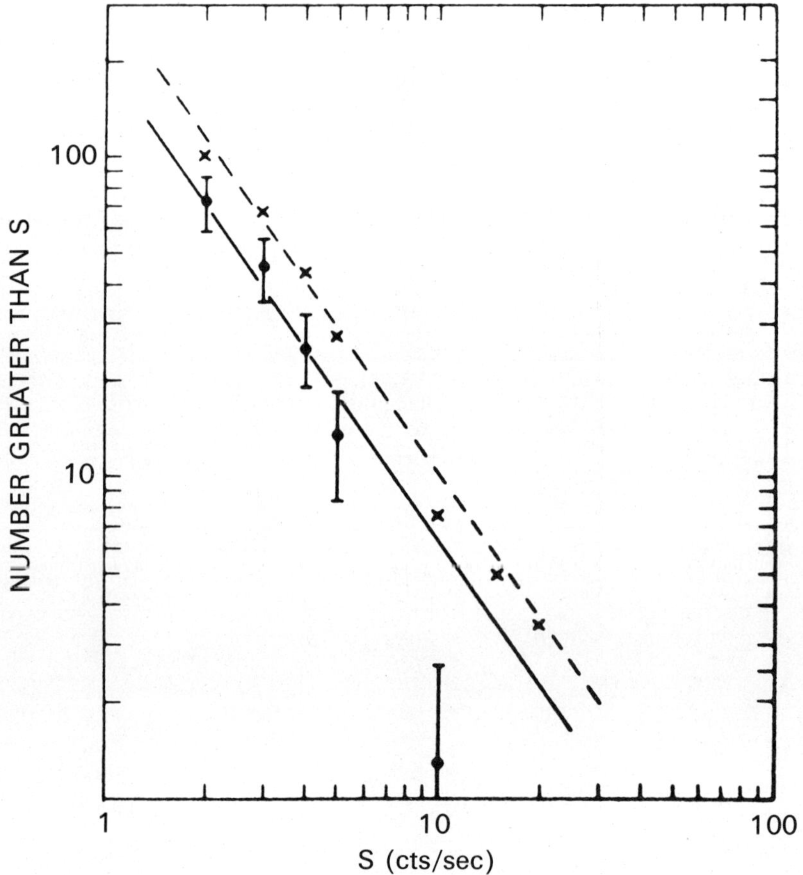

Figure 6 The integral intensity distribution plot ($\ln N - \ln S$) for the high-latitude sources indicates the -1.5 slope (*dashed* and *solid lines*) expected from an isotropic distribution. The crosses are for all sources $b^{II} \geqq 20°$, and the bars are for those not identified with known objects. (From Holt et al 1974.)

theorem for dynamic stability. The X-ray-emitting mass required, however, depends critically on the "clumpiness" and other parameters, and the whole problem is controversial at present. Alternatively, the X-rays may originate from scattering on the 3°K radiation by an intracluster population of cosmic-ray electrons leaking from the galaxies.

Because these strong sources are associated with "rich" clusters, the luminosity may be related to the richness class. About 12 additional UHURU sources in the 2–7 counts sec^{-1} range have been tentatively associated with Abell clusters of varying richness (Kellogg et al 1973). At these low fluxes, however, it is presently impossible to determine the extent of the sources or their spectra.

COMPACT OBJECTS These include the brightest QSO, 3C273, the Seyfert galaxies NGC 4151 and NGC 1275, and the radio sources Cen A and Cyg A, all of which are detected at ~ 5 counts sec^{-1} by UHURU (Kellogg 1974). 3C273 has a luminosity $\sim 4 \times 10^{45}$ ergs sec^{-1} at its Hubble-law distance, which is about one-twentieth of the IR power. Other sources are in the range 10^{41}–10^{43} ergs sec^{-1}. The mechanism is unclear, however, in the strong radio sources. Compton scattering seems likely, and in fact Cen A (NGC 5128) is observed to have a power-law spectrum measured to ~ 100 keV, whose number index is in the range 1.5–2.0 (Peterson 1973). Based on the UHURU data, most of the emission from Virgo is from a diffuse source, although the Lockheed group (Catura et al 1974) has shown that the possibility of a component from M87 cannot be excluded. A recent detection of 7–110 keV X rays from NGC 4151 (Baity et al 1975a) has shown this Seyfert to have the hardest X-ray spectrum of any cosmic source and a luminosity $L_x = 3 \times 10^{43}$ ergs sec^{-1}.

UNIDENTIFIED EMITTERS The majority of the high-latitude sources are simply unidentified and may in fact be a new class of extragalactic objects (Kellogg 1973, Schwartz & Gursky 1974), although a population of local galactic sources with $L_x \sim 10^{34}$ ergs sec^{-1} cannot be presently excluded (Holt et al 1974a). Figure 6 shows the integral number intensity plot (ln N–ln S) for high-latitude sources, both total and unidentified. The high-latitude sources show the -1.5 law, and if extragalactic, L_x must be about 5×10^{43} ergs sec^{-1}. Normal galaxies, such as our own, and M31, are in the range 0.2–1.0×10^{40} erg sec^{-1}, consistent with the nondetection from other such galaxies.

The Diffuse Component

Related to the problem of the extragalactic sources is the diffuse component of cosmic X rays, whose 2–7 keV intensity is $\sim 10^{-8}$ ergs cm^{-2} sec^{-1}-sr^{-1}, and which has been measured to be isotropic and uniform to a few percent on both large and small scales by Schwartz (1970) on OSO-III. Although the spectrum has now been extended to over 100 MeV, and can be grossly described in terms of $dN/dE \sim E^{-2.2}$ according to Pal (1973), the details indicate a complex, multi-component origin (Schwartz & Gursky 1973). Figure 7 shows the spectrum, as determined from selected data (Schwartz & Gursky 1973) at low energies. The

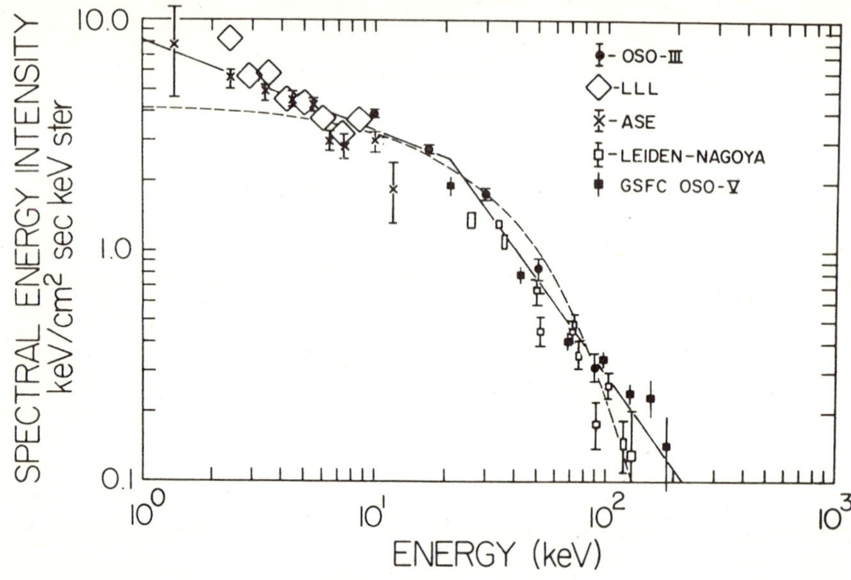

Figure 7 Measurements of the diffuse X-ray spectrum. Experiments selected for this figure either spanned a wide range of geomagnetic conditions or directly detected and rejected charged-particle effects. A single power law does not appear to fit the data between a few keV and 100 keV. (From Schwartz & Gursky 1974.)

Apollo 15 (Trombka et al 1973) has extended the spectrum to over 10 MeV, and SAS-2 (Fichtel et al 1973) to over 100 MeV.

Three possibilities for the origin have been reviewed by Felten (1973) and by Silk (1973). Compton scattering of intergalactic electrons on the 3°K radiation produces a spectrum which is naturally a power law, with changes in slope due to diffusion and lifetime of the electrons. An $\sim 7 \times 10^7$°K intergalactic medium of sufficient density to close the universe ($\sim 10^{-29}$ g cm^{-3}) produces more than the observed X-ray intensity, depending somewhat on the big-bang or steady-state model. A reduction in the Hubble constant to <40 km sec^{-1} Mpc^{-1}, among other effects, could remove this problem. One would still have to involve an added component, however, to explain the radiation at energies >50 keV. The diffuse flux may be due to the superposition of many discrete sources. The lack of "lumpiness" in the background, however, implies a very large number of sources ($\geqq 10^7$). The volume emissivity of the known discrete extragalactic objects fails to account for the observed diffuse intensity by a factor of 6 (Schwartz & Gursky 1973). The unidentified sources, when extended to a Hubble distance, can only account for about one-half the background in the 2–7 keV range.

Soft X Rays

Recent rocket experiments have made important discoveries on diffuse cosmic X rays in the 0.2–1.0 keV range. Here the galaxy is no longer transparent, so fluxes measured represent a competition between photons from the diffuse component, or possibly other extragalactic sources, and absorption and production within the galaxy (Hayakawa 1973). Observations by Kraushaar and his colleagues (McCammon et al 1971) and the U.S. Naval Research Laboratory (NRL) (Friedman et al 1973, Friedman 1973) show an X-ray sky markedly different in the 40–60 Å range (~0.35 keV) from that shown by the UHURU. The NRL "map" shown in Figure 8 indicates that the most intense emission is observed at high galactic latitudes where the hydrogen column density is a minimum.

Explaining the measured dependence on galactic latitude in terms of an origin characterized by a simple scale height above the galactic plane does not explain

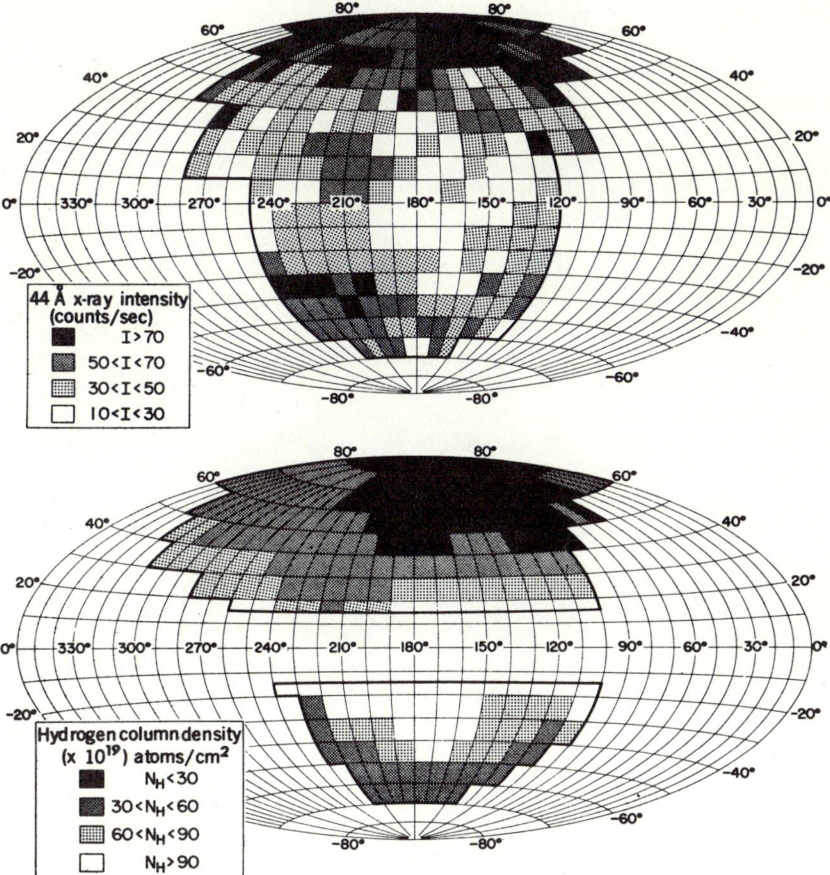

Figure 8 A map in new galactic coordinates of the soft X-ray sky (*top panel*), centered on the galactic anticenter, at 44–60 Å (~0.28 keV). *Bottom panel,* the hydrogen column density with lowest concentrations darkest to show the relationship to the low-energy X-rays. *Heavy lines* indicate regions of the sky for which data are available. (From Friedman et al 1973.)

the observations at all wavelengths. Adding an isotropic extragalactic flux permits construction of a model that reconciles the data (Friedman et al 1973). The extragalactic flux, however, should appear absorbed by nearby galaxies. This effect is not observed for the Small Magellanic Clouds (McCammon et al 1971) and M31 (Margon et al 1974) so the entire question of the origin of the soft X-ray background is unclear (cf Silk 1973). It is certain, however, that an extragalactic flux much in excess of that predicted by extrapolation of the diffuse component as a power law to lower energies is not required.

In addition to the general dependence on galactic latitude, the soft X-ray flux shows special features which may be associated with North Polar Spur (Bunner et al 1972), a galactic radio feature, or the Vela supernova remnant. As discussed above, soft X rays have been studied from the Cygnus loop and Puppis remnant, and possibly others. The galactic soft X-ray flux requires a superposition of sources, which may be stellar or nebular. Additional supernova remnants, radio spurs, accretion onto main-sequence stars, and heating in active, convective stars such as T-Tauri type have been suggested (Hayakawa 1973).

PHOTON INTERACTIONS

The detection of cosmic photons depends on their interaction with matter. In the energy range discussed here, the photon wavelength is comparable to or less than the interatomic spacing of solid materials; hence the principal interaction is with individual atoms and their constituents, rather than processes such as reflection and refraction, which involve large-scale or coherent effects. Grazing-incidence reflection at low energies ($\lambda \gtrsim 3$ Å) depends on the wavelength encompassing many lattice planes when projected tangentially onto a solid.

Three principal noncoherent interactions occur to the photons considered here: 1. photoelectric absorption by the entire atom; 2. Compton scattering on the individual bound electrons of each atom; and 3. positron-electron pair production in the Coulomb field of the nucleus. It is beyond the scope of this work to discuss these processes from first principles; these are discussed in standard textbooks (Rossi 1952, Evans 1955, Jackson 1962). Comments from a phenomenological viewpoint are appropriate, however. Blumenthal & Gould (1970) and Greisen (1971) have also reviewed recently the interactions from first principles, with particular reference to astrophysical applications. Approximations involved in the various calculations and the experimental checks are discussed in the National Bureau of Standards (NBS) tabulations by Hubbell (1969) where attenuation coefficients in the 10 keV–10 MeV range are given for many materials. This work also includes a discussion of earlier compilations and is an updated version of an earlier NBS publication by White-Grodstein (1957).

A beam of penetrating photons at energy E entering a medium is attenuated by catastrophic interactions. The statistical probability of a flux F surviving from an initial beam strength of F_0 after traversing a distance l in a material of density ρ is

$$F = F_0 \exp(-\mu\rho l), \qquad 3.1.$$

where μ is the mass attenuation coefficient, in $cm^2\ g^{-1}$. The path in the medium may be expressed more conveniently as the mass column density $x = \rho l\ g\ cm^{-2}$. Because the photon may not lose all of its energy in an interaction, the energy absorption coefficient differs from the mass attenuation coefficient. Occasionally

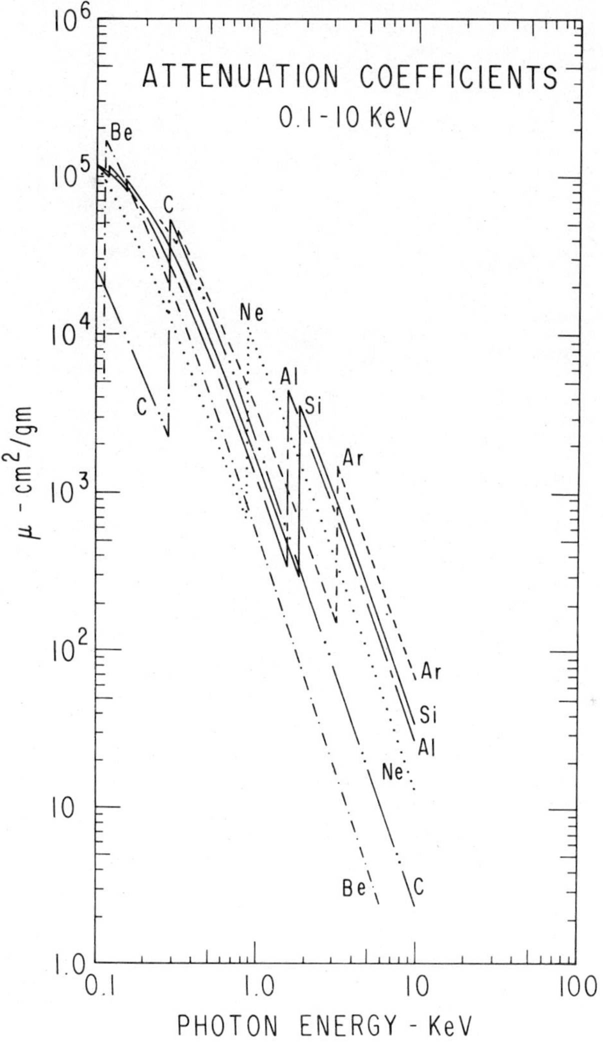

Figure 9 Photoelectric attenuation coefficients in the energy range 0.1–10 keV for various materials used in proportional counters. The sharp discontinuities are at the *K*- or *L*-shell electron binding energies. (From the tabulation of Henke & Elgin 1970.)

the term *attenuation length* or *photon range* ($\lambda = 1/\mu$ g cm^{-2}) is used to describe the depth for a $1/e$ reduction in intensity.

Photoelectric absorption is the dominant process at lower energies, dependent considerably on the atomic species being targeted. The cross section decreases very sharply above the wavelength corresponding to the binding energy of K or L electrons, and at longer wavelengths shows structure corresponding to the various atomic levels. At an energy E above an absorption edge, E_k, the cross section σ_{ph} varies $\sim(E/E_k)^{-2.8}$. The photoelectric attenuation coefficient is related to the cross section by

$$\mu_{ph} = \sigma_{ph}(N_0/A) \text{ cm}^2 \text{ g}^{-1}. \quad 3.2.$$

Figure 9 shows photoelectric attenuation coefficients for selected materials in the range below 10 keV, taken from the work of Henke & Elgin (1970); Figure 10 shows the photon range extended to higher energies, where other processes dominate, as compiled by Fillius (1963).

The Compton-scattering cross section per electron at the lowest energies reduces to the Thomson cross section

$$\sigma_t = 8\pi^3/3(r_e^2) = 6.67 \times 10^{-26} \text{ cm}^{-2} \quad 3.3.$$

for free electrons, and is modified slightly by binding effects in high Z atoms. The Compton attenuation coefficient

$$\mu_c = \sigma_c(Z/A)N_0 \text{ cm}^2 \text{ g}^{-1} \quad 3.4.$$

varies slowly with increasing atomic number through the term Z/A. The total cross-section σ_c, which is the integral of the Klein-Nishina formula for the differential cross-section $d\sigma/d\Omega$, decreases at higher energy, being 0.5 σ_t at 400 keV and 0.1 σ_t at 7 MeV. Binding effects reduce the cross section slightly at low energies and for forward scattering. Coherent or Rayleigh scattering, important for high Z materials at long wavelength, tends to increase the cross section slightly, and is an $\sim$6% effect at 10 keV for NaI.

Not all the energy in a Compton interaction is transferred to the electron; some is associated with the longer wavelength-scattered photon, which may, however, be rescattered, degraded in energy, and finally absorbed in the medium, depending on the thickness. Forward-scattered photons transfer very little energy to the electron, whereas backward-scattered photons transfer a maximum value,

$$\varepsilon_{max} = 2E/mc^2/(1+2E/mc^2), \quad 3.5.$$

because here the electron is ejected forward. This results, for example, in the phenomenon of the "Compton edge" in scintillation-counter spectra, and a "back-scatter peak" for certain detector and gamma-ray source geometries. The average energy transferred to the electron in Compton scattering is only about 2% at 10 keV, 14% at 100 keV, and rises to 44% at 1 MeV for free electrons. A complete discussion of the Compton process, with cross sections, energy-angle relations, and average energetics, is given in an earlier NBS publication by Nelms (1953).

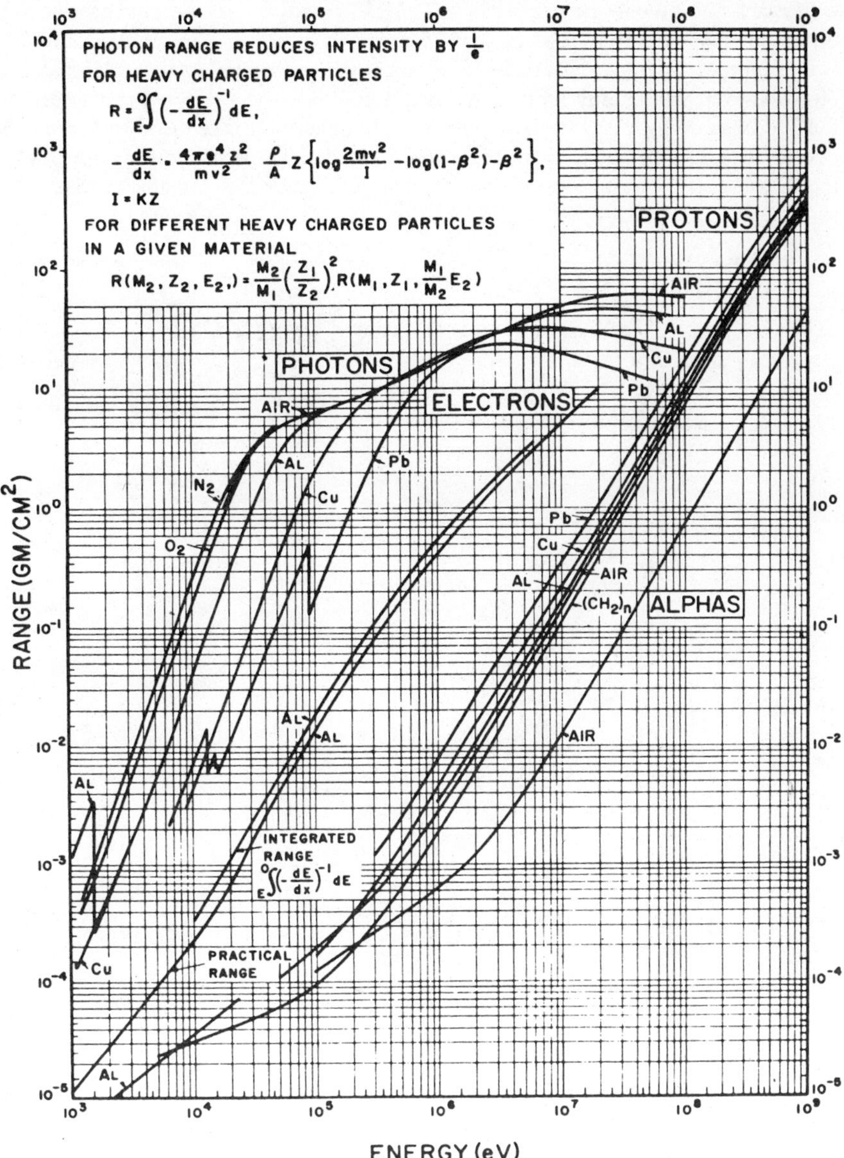

Figure 10 Range-energy curves for penetrating radiations in the ~ 1–10^6 keV range. At increasing energies the dominant processes for photons are Compton scattering and ultimately pair production. (From Fillius 1963.)

Above a photon energy of $2\ mc^2 = 1.02$ MeV, positron-electron pair production in the intense Coulomb field of high-Z nuclei is possible. The cross-section σ_p varies approximately as Z^2; it rises monotonically from threshold and levels off above 50 MeV to a high-energy limiting value. At 10 MeV the corrected cross section for hydrogen is 2.1×10^{-3} b (b = barns). With higher Z it is reduced by screening effects to 1.83×10^{-3} b/Z^2 in the case of lead. The asymptotic values with increasing energy are about 6.4 mb for H and 5.2b for Pb. On the average, the

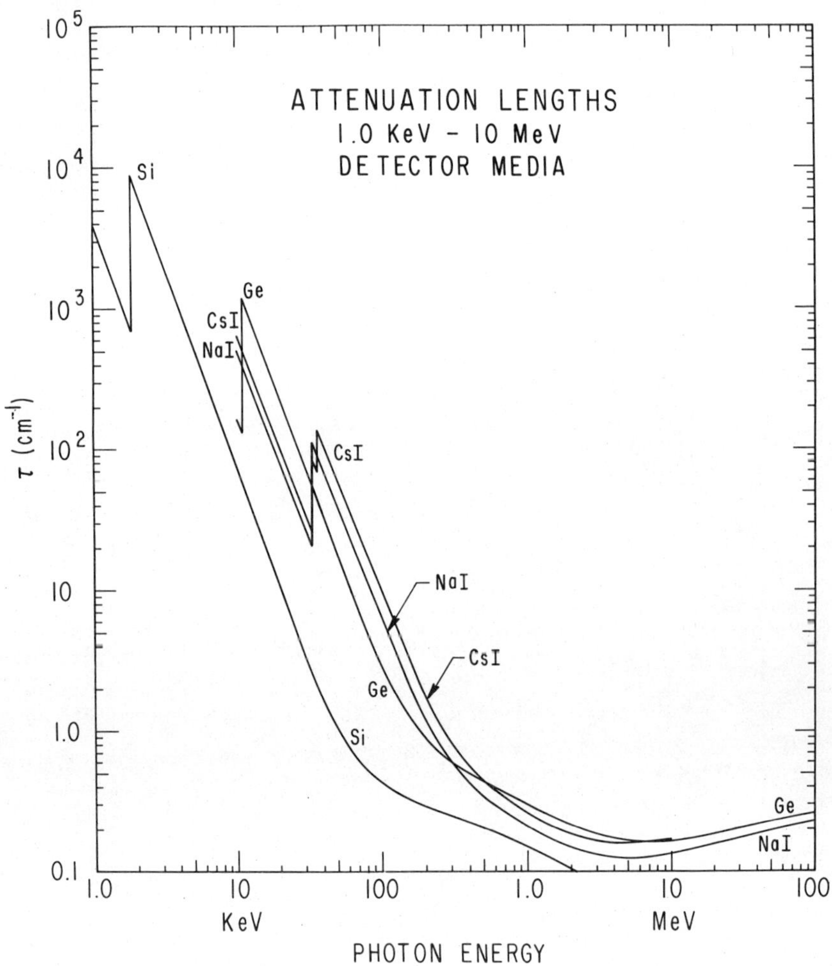

Figure 11 Attenuation lengths for 1 keV–10 MeV photons in common scintillation and solid-state detector materials. (From materials in Hubbell 1969 and elsewhere.)

pair shares the energy equally, and the positron annihilates upon stopping, producing two 0.51 MeV gamma rays.

In addition to the attenuation coefficient for photons, the penetrating power or range of charged particles is of considerable interest in detector design and for understanding effects of the radiation environment. The range of electrons and positrons has also been discussed and compiled in an NBS publication by Nelms (1956, 1958). Figure 10, due to Fillius (1963), compares the range of electrons, protons, and heavier particles with that of photons in several elements used in the construction of counters as detection, collimation, shielding, or support materials. Electron range varies only slowly for low-Z materials, hence the aluminum curve is typical. Note that in most materials over the energy interval of interest here the range of an ejected photoelectron, Compton, or pair electron will be only 10^{-2}–10^{-3} that of a photon at the same energy.

Figure 11 shows attenuation lengths (cm^{-1}) of various processes in CsI and NaI, the commonly used scintillation materials, and in Si and Ge, the detection elements of solid-state counters.

As a final note on ranges, Figure 12 shows the attenuation length in the atmosphere as well as the altitude, based on the 1972 COSPAR International Reference Atmosphere (ICSU 1972). The operating altitudes required of balloons and rockets for X-ray astronomy observations are also indicated.

The probability of a photon interaction indicates only that the photon at energy E is removed from the beam, and does not imply total energy absorption. In the case of each interaction, the detectable energy loss depends on ionization by an electron resulting from the interaction process. In the photoelectric case, partial losses may occur also because of K X-ray escape (Liden & Starfelt 1954, Wapstra et al 1959). The Compton process gives a fraction of the energy to the electron; the rest is carried away by the scattered photon, as indicated above. The positron-electron pairs may undergo bremsstrahlung, or the annihilation photons may escape.

In all cases, the range of the electrons may be greater than the residual material thickness and hence partial energy losses occur because of geometrical considerations. As shown in Figure 10, this is particularly true above several MeV, where the electron range becomes comparable to the photon range. Here evaluation of the energy lost may involve a cascade calculation, which can be done only on a statistical basis and is a most important consideration for detection of photons at energies above a few MeV.

DETECTOR PRINCIPLES AND APPLICATIONS

Photon detection is an indirect process: incoming photons must interact with matter in the detector, producing electrons that cause ionization, leading to effects that may ultimately be measured electronically. Here we discuss the three principal devices used in X-ray astronomy: proportional, scintillation, and solid-state counters. Other devices, such as X-ray conversion and imaging devices, crystal spectrometers with open photomultipliers, etc also may use similar physical effects but differ considerably in the technique of producing the electronic signal and in the information available from a photon interaction event, or a statistically added ensemble thereof (Giacconi 1971).

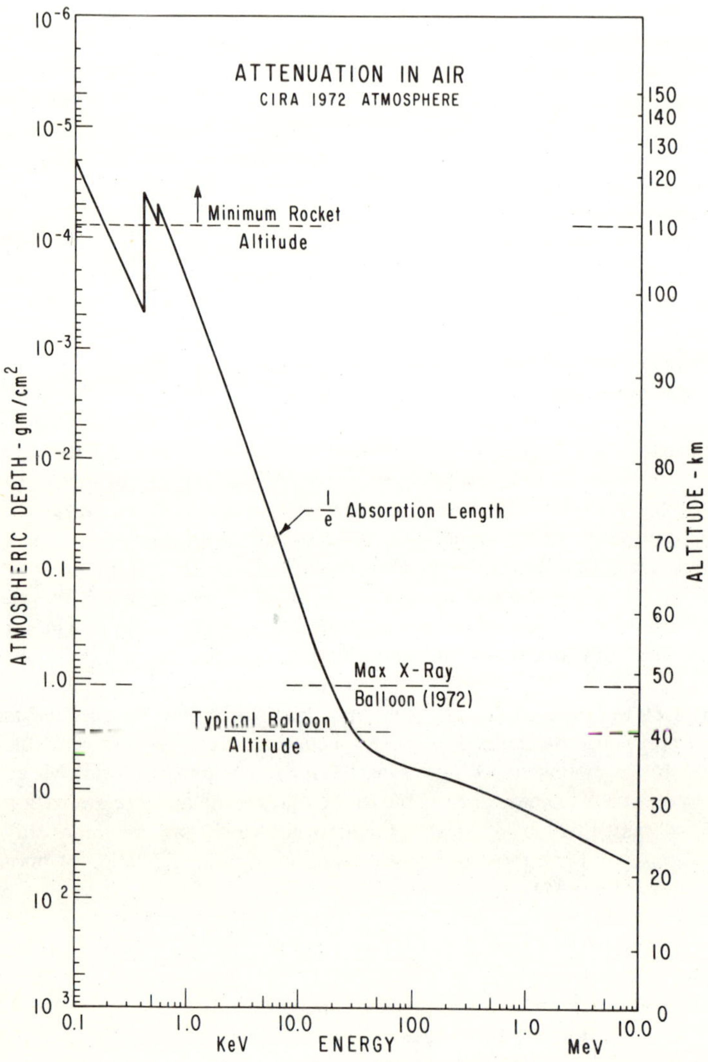

Figure 12 Attenuation of 0.1 keV–10 MeV photons in the 1972 COSPAR International Reference Atmosphere, with the 1/*e* absorption length plotted as a function of energy and altitude or atmospheric depth. Typical operating altitudes are indicated. (From ICSU 1972.)

Proportional Counters

Proportional counters have been discussed in standard references (Rossi & Staub 1949, Sharpe 1954), and their application to X-ray astronomy has been discussed with the subject (cf Giacconi et al 1968). In no other application has the large-area, thin-window, and low-background combination been required. A proportional counter consists of a central anode wire of ~ 0.01 mm diam at a potential of up to several thousand volts in a coaxial chamber filled with a gas mixture. A photon entering the gas may interact photoelectrically. The resultant energetic photoelectrons and Auger electrons from deexcitation of the positive ion ionize the gas, producing electrons and positive ions. Electrons diffuse toward the central wire and the positive ions toward the outer cylinder. If the potential is low, and no recombination occurs within the gas, a charge $q(t)$ proportional to the energy loss is collected at the anode and one has an ionization chamber. The charge produces a change in potential across an anode resistor and its distributed capacity, which may be amplified externally or integrated to measure the total dosage of ionizing radiation over some time interval.

As the voltage on the central wire is increased, the field gradient becomes large and the electrons, after diffusing toward the central wire, may gain more than the ionization potential of the gas in one collision mean-free path, and therefore produce additional ion pairs. These electrons may further ionize, producing a "gas multiplication" which may be on the order of 10^3–10^5 and a much larger charge pulse. For example, a 1 keV photoelectron in argon will produce about 37 ion pairs, and therefore a charge of 6×10^{-18} coulombs, undetectable above the noise level associated with ordinary wide-band amplifiers, whereas the pulse after a gas multiplication may be about 10^{-14} coulombs, which is easily detected and amplified.

Because of the fast diffusion rate, the electron pulse is collected at the anode in a few times 0.1 μsec, whereas the positive ions drift toward the cathode only after 10–100 μsec. This slower pulse, which results in a dead time for the counter, is usually not measured because of use of a fast anode time constant. At higher potentials, photons from deexcitation of ions near the central wire produce photoelectrons in the gas or the walls of the counter, resulting in a continuous discharge, as in a Geiger counter.

Noble gases such as argon or xenon operated near atmospheric pressure are generally used because the lack of low-level excitation states tends to reduce non-ionizing collision losses and spurious counts from free electrons. About 10% of an organic quench gas such as CH_4 or CO_2 will absorb UV photons, prevent the formation of secondary electrons near the walls, and cause the ions to deexcite collisionally. Furthermore, this quench gas tends to stabilize counter action because very small quantities of impurities in a nearly pure noble gas can determine the multiplication and discharge properties. Counters are often filled with a 90% argon, 10% methane gas mixure, known as *P-10*.

Simple proportional counters, as used in X-ray astronomy, have been reviewed by Giacconi et al (1968, 1971). The volume is generally rectangular with sides, rear, and ends forming part of the structure and housing. The front is covered by a thin metal or organic window, such as beryllium or aluminized mylar, often supported against blowout by the collimator assembly. Ground-plane wire grids may be inserted to section a common gas volume into many elements, each with its own anode, as shown in Figure 13 (Holt 1970). This has the advantage of making "wall-less" counters, and the possibility of using the outer elements for anti-coincidence. Such devices, known as *proportional chambers,* have been extensively developed at the Goddard Space Flight Center (GSFC) (Serlemitsos 1971, Bleach et al 1972).

The efficiency, $\varepsilon(E)$, is determined by the transmission of the window and by the absorption of the gas, and is therefore energy dependent:

$$\varepsilon(E) = \exp(-\mu_w x_w)[1-\exp(-\mu_g x_g)], \qquad 4.1.$$

where μ_w and μ_g are the absorption coefficients of the window and gas, respectively, and x_w and x_g are the thicknesses expressed in g cm^{-2}. The response vs energy

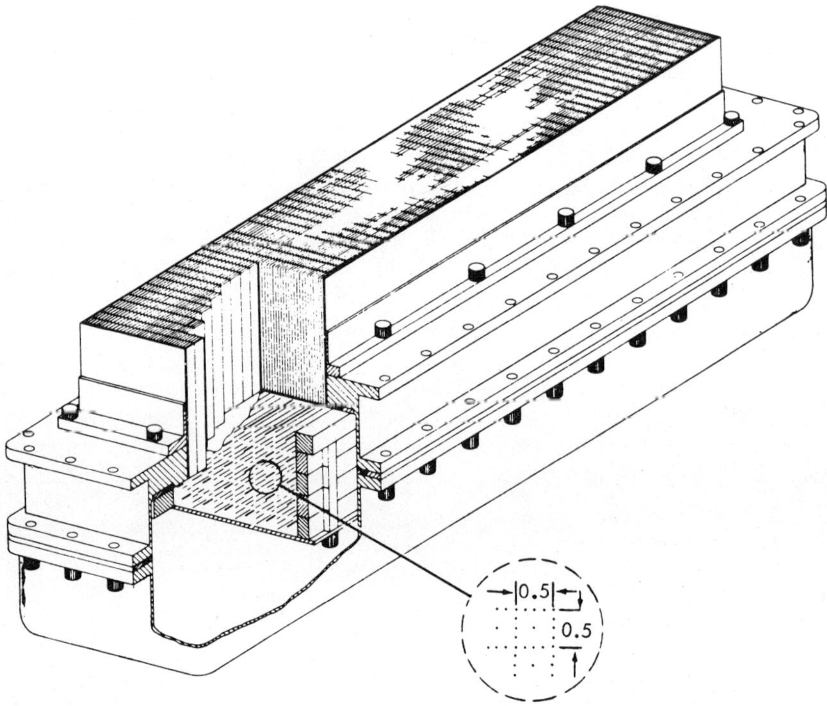

Figure 13 Proportional chamber used in rocket X-ray observations. The collimator (*top*) is formed of rectangular tubing and supports the thin mylar window against the 1 atm pressure differential. The central wire in each cell is the anode and cells on the edge may be used in anticoincidence, thus creating a "wall-less" counter. (From Holt 1970.)

of typical counters used in recent rocket flights to measure very soft X rays is shown in Figure 14 (Friedman et al 1973). These counters have very thin organic windows which result in transmission below the carbon *K*-edge at 284 eV.

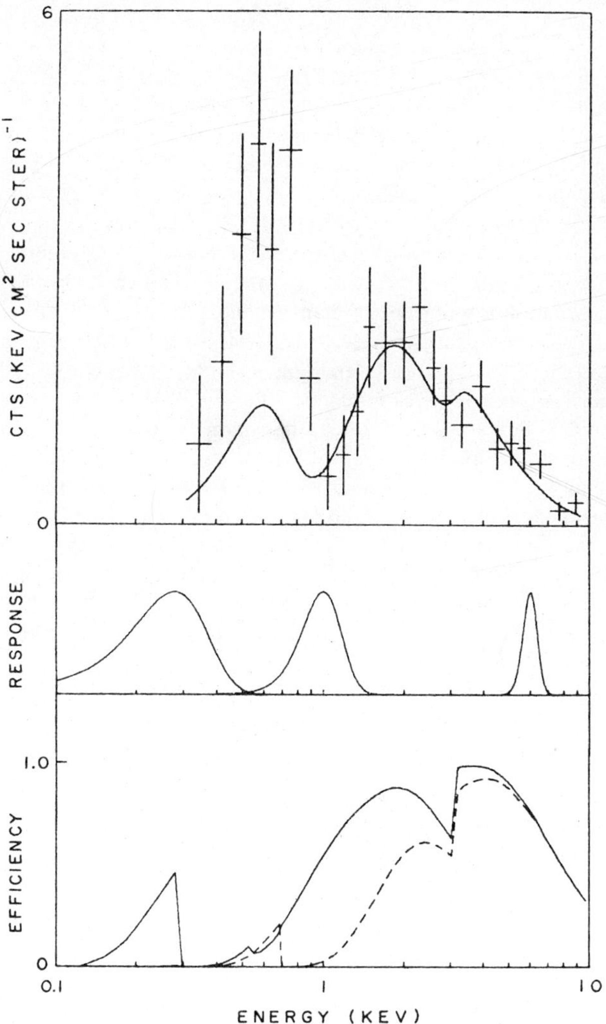

Figure 14 Spectral resolution of a gas-filled proportional counter. *Bottom,* the efficiency of an argon-filled counter with a 2 μm polycarbonate (*solid curve*) and 3.2 μm teflon (*dashed curve*) window. *Center,* the response to line emission at 0.28, 1.0, and 6.0 keV. *Top,* data from near the galactic pole, together with the theoretical response to a simple power-law spectrum (*solid curve*) which fits the data above 1 keV; the low energy excess is clearly apparent. (From Friedman et al 1973.)

The energy resolution $\Delta E/E$ depends primarily on random statistics associated with the initial production of ions, modified somewhat by fluctuations in the multiplication process, and by geometrical effects associated with uniformity of the anode wire and field configuration near its ends. The mean number of ion pairs (or photoelectrons) is $\bar{N} = \alpha E$, where α has the dimensions of ion pairs keV^{-1}. The fluctuations about this mean are approximated by Gaussian statistics, with a standard deviation $\alpha = \bar{N}^{-1/2}$. This yields an energy resolution $\Delta E/E\ E^{-1/2}$. Figure 15 shows both the Gaussian probability curve $P(N)\,dN$ and its integral. The full-width at half-maximum (FWHM) experimental resolution parameter is 2.35σ.

Because the average ionization potential in argon is about 27 eV, one expects about 37 independent ion pairs per keV, which gives a 7.1% FWHM at the 5.9 keV Fe *K* X-ray line. Based on a study of the resolution of proportional counters in the $0.3 < E < 10$ keV range, Charles & Cooke (1968) found an empirical relation, resolution $(\mathrm{FWHM}) = 0.14/(E)^{1/2}$, with E in keV. This gives a 5.8% FWHM resolution at 5.9 keV, slightly better than usually obtained in flight instruments. Figure 14 shows the response of a thin-window argon-methane proportional counter to monochromatic X rays of various energies.

The counter response is also modified by fluorescence, as some ionized atoms may deexcite by producing a *K* X ray, rather than by emitting an Auger electron. This X ray may escape from the gas, resulting in an ionization loss less than that of the incoming photon. Giacconi et al (1968) have tabulated absorption edges and fluorescent yields for several gases commonly used in proportional counters.

Recovery of source spectra from the proportional-counter data often requires careful analysis. Because the data are often limited in statistical precision and spatial and energy resolution, various corrections must be applied. Accordingly, model spectra, based on a preconception of the astrophysical processes, are usually

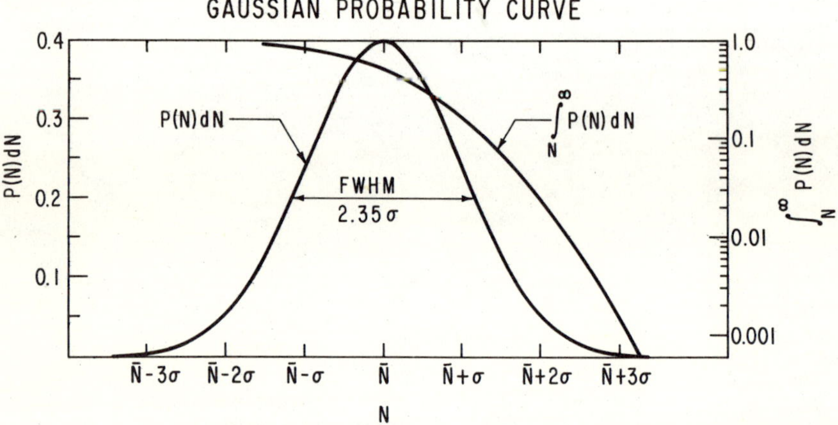

Figure 15 Gaussian probability curve, showing the relative probability of observations of value N given the true population mean $\bar{N}$ and standard error σ. The curve is normalized for an integral of unity. The integral probability for an observation $N' \geqq N$ is also shown.

fit to the data using statistical tests to estimate the validity of the model and the uncertainties of the fitted model parameters. This procedure, which has been described in detail by Gorenstein et al (1968) and by Dolan (1972), requires accurate knowledge of detector response.

Figure 14 also shows a result of this procedure for data obtained from the galactic pole regions with thin organic window counters. The solid curve indicates the expected response to a power-law spectral form for the diffuse flux entering the counter aperture, fitted to the data above 1 keV. The excess counting rate at low energies is taken to indicate an additional flux, which may be of galactic or extragalactic origin (Friedman et al 1973).

There are many properties of proportional counters that may be used to advantage. Charles & Cooke (1968) have reviewed the factors affecting the resolution, lifetime, and stability of counters used in X-ray astronomy. The shape of an anode pulse, which depends on the type of ionizing particle, and the distribution of ionization in the counter volume may be used to distinguish fast background electrons (either ejected from the counter walls or entering the window) from short-range photoelectrons that have the same total energy loss (Gorenstein & Mickiewicz 1968). The charge due to a photoelectron is produced quite locally, because of the short electron range, and flows in two directions along the anode wire in amounts inversely proportional to the distance from the collection point to the end of the wire in a coaxial, or nearly coaxial, counter (Charpak et al 1968, Lacy & Lindsey 1973). Measurement of the ratio of charge collected at each end therefore permits location of the photon interaction along the wire, hence the term *position-sensitive* counter. Multiwire counters, with position-sensitive detection, have been used to read out the event location in two dimensions. Such a device at the focus of grazing-incidence collector or mirror can be used for X-ray imaging (R. Novick, personal communication). Counters filled with liquid xenon, which has the absorption properties of NaI, permit localization of gamma-ray interactions, particularly when combined with position-sensing techniques. A liquid xenon counter has been developed by Alvarez and his associates (Alvarez et al 1973) for use as a Compton-scattering telescope at energies ~ 1 MeV.

Scintillation Counters

The theory and practice of scintillation counting are treated in standard monographs (Birks 1964, Sharpe 1954). Energetic electrons following a photon interaction in a solid medium excite elevated atomic, lattice, or molecular binding states by collision. These may return to the ground states by radiative deexcitation, releasing optical photons. If the medium is transparent to its own radiation, the light may be collected on a multiplier cathode, thereby releasing photoelectrons that upon amplification produce an electrical pulse proportional to the energy loss. There are at least two types of scintillators: 1. inorganic, nonconducting alkali-halide crystals; and 2. organic molecules, in the form of crystals, plastics or liquids.

NaI or CsI, doped with 10^{-2}–10^{-3} mole per fraction Tl or Na impurity atoms produce luminescent centers intermediate between the filled valence electron band and the ordinarily empty conduction band. Collisions of an energetic electron with the lattice create hole-electron pairs that relax to thermal energies at the bottom of the conduction band in $\sim 10^{-13}$ sec, giving their energy to lattice vibrations. Most of these pairs also deexcite in this mode. A few, however, become attached to the luminescent center, where they have lifetimes of hundreds of nanoseconds, and produce upon recombination a photon whose wavelength is typically 4200 Å. Electron-hole pairs may also recombine in a radiationless process at the luminescent center, or at electron-hole traps due to other impurities or lattice defects. The scintillation efficiency and the decay time constant are temperature dependent; at 20°C the fractional energy loss converted into useful luminescence radiation in NaI(Tl) is about 12% (Birks 1964).

Luminescence or scintillation in organic materials is an inherent molecular property arising from the electronic structure of certain molecules and is independent of phase state. This process may be understood in terms of the orbital configurations of the carbon atom; the π-orbital of aromatic hydrocarbons and its excited states are of particular interest. The π-electron, if excited by photon absorption or by collisions into a higher singlet state, is rapidly (10^{-12} sec) degraded into the lowest vibration level of the first excited singlet state (S_{10}). The coupling is due to the overlap of the many excited states by the vibrational level substructure. The S_{10} state has a typical lifetime of 10^{-8} sec, and decays to the various vibrational sublevels S_{00}, S_{01}, etc of the ground state, producing the fluorescent spectrum. In many molecules, the first excited triplet state lies below the S_{10} state and is metastable to radiative deexcitation to the S_0 levels. This radiation, known as phosphorescence, is at a longer wavelength and is characterized by a much longer decay time ($>10^{-4}$ sec).

Table 1 indicates properties of scintillation materials commonly used in X-ray astronomy. Many additional properties, response characteristics, and practical considerations are listed in various manufacturers' catalogs (cf Harshaw Chemical Co. 1969). Various organic scintillating materials have been investigated and used in the form of crystalline solids, binary or ternary liquid or plastic solutions, and gases (Birks 1964). Because of their low stopping power for photons, plastic organic scintillators are used mainly for charged-particle detection and anticoincidence. A 1 cm thickness of NE-102 will produce about a 2 MeV energy loss for a relativistic $Z = 1$ particle, and will completely stop a 33 MeV proton, yet is essentially transparent to photons >30 keV. The high Z of the inorganic scintillators and higher luminescent efficiency makes them much preferred for X-ray and gamma-ray detectors. Although the manufacture of plastic scintillators in large sizes and arbitrary shapes has never been a problem, large-sized inorganic crystals are a fairly recent introduction. Because of its hygroscopic nature and brittleness, NaI(Tl) is rather unsatisfactory for the large sections required for anticoincidence absorbing shields in space applications. The development of CsI(Na) (Menefee et al 1967) has been particularly helpful for the latter application because it is nonhygroscopic, relatively easily machined, and lacks longer-lived fluorescent states that caused difficulties in early applications of large-volume CsI(Tl) shields (Frost et al 1966). Considerable information is now available on the mechanical and radiation properties of CsI (BBRC 1971, Crannell et al 1974).

Table 1 Properties of commonly used scintillation and solid-state counter materials

Material	$\overline{Z}$	$\overline{A}/\overline{Z}$	K-edge (keV)	Index of Refraction	Density (gm cm^{-3})	Scintillation Efficiency (Relative)	Decay Time (ns)	Band Gap (eV)	Approx. λ_{max} (A)	Energy, p-n pairs (eV)	Comments
Inorganics											
NaI(Tl)	32	2.34	33.1	1.78	3.67	210	240	5.38	4100	—	Very hygroscope, easily fractured
CsI(Tl)	54	2.40	36.0	1.84	4.51	55	450	5.67	4200–5700	—	{ Easily machined.
CsI(Na)	54	2.40	36.0	1.84	4.51	178	1100	5.67	4200	—	CsI(Na) surface effects due to water vapor
Organics											
NE-102	3.2	1.83	0.284	1.58	1.03	65	3	—	4250	—	{ Plastics used for
Pilot B	3.2	1.83	0.284	1.58	1.06	52	≤ 3	—	4100	—	anticoincidence
Anthracene	3.9	1.89	0.284	1.62	1.25	100	35	—	4400	—	Reference only
Solid State											
Si(Li)	14	2.00	1.84	—	2.35	—	—	1.21	—	3.6	Cooled to LN_2 during operation.
Ge(Li)	32	2.27	11.1	—	5.36	—	—	0.785	—	2.9	Must be maintained at LN_2, storage and operation

Following a photon interaction, light is reflected or diffused to the cathode of a carefully selected photomultiplier, often placed in direct optical contact with the scintillation medium. Photoelectrons produced at the cathode are amplified a factor of 10^5 or 10^6 by the secondary electron-multiplying action of the phototube and the much enhanced charge pulse collected at the anode. Serving the same function as the gas multiplication counters, this amplification increases the almost undetectable charge produced by the direct physical effect following ionization loss to a level well above the equivalent input noise of ordinary, wide-band electronic amplifiers. Although the selection of large diameter phototubes has always been somewhat limited, especially in ruggedized versions, several recent types have rather excellent properties for space application (RCA 1970).

Assuming a 12% conversion of energy loss to light for NaI(Tl), about 40 luminescence photons are produced in situ at the interaction site for a 1 keV energy loss. Because the light collection to the photocathode is rarely better than 50%, and the photo-efficiency of an *S*-11 cathode is only about 10%, about two photoelectrons/keV are produced for amplification by the electron multiplier and external electronics. Single electron thermal emission at the cathode or first multiplier dynode (the dark-current pulse spectrum) constitutes the essential limitation for detection at low energies (Jerde et al 1967). Only with small crystals in an ideal geometry and with selected low-noise, high-quantum efficiency phototubes has it been possible to resolve X rays at a few keV from the noise at low counting-rate levels (Harshaw 1969). The energy resolution is determined by the photoelectron statistics, assuming uniform light collection and photocathode. For the 5.9 keV Fe *K*-line, one could expect about 12 photoelectrons, or a 66% FWHM resolution, a number only recently achieved with spaceflight detectors. The response and resolution of scintillation crystals is nonlinear at low energies, as has been investigated by Aitken et al (1967).

The practical application of simple scintillation-counter technique to a balloon-borne instrument used for measurement of the diffuse X-ray spectrum over the 20–200 keV range by the Leiden group (Bleeker & Deerenberg 1970, Bleeker 1971) is shown in Figure 16. Here a thin NaI(Tl) crystal is coupled through a Lucite light pipe to a 5-inch photomultiplier. Shielding and collimation for X rays outside the aperture are provided by a combination of "graded" absorbers, which have the property that valleys in the absorption just below the *K*-edges are covered by the strong absorption above the *K*-edge of the adjacent material. The plastic anti-coincidence shield vetos direct effects of cosmic rays and X-ray production in the shield associated with passage of cosmic rays or their charged secondaries. Devices using some of these techniques have been used by MIT in early balloon work (Clark et al 1968), by Lawrence Radiation Laboratory (LRL) in rocket observations (Harri et al 1969), by the Physical Research Laboratory (PRL) (Kasturirangan 1971) and the Tata Institute for Fundamental Research (TIFR) (Manchanda et al 1972) in balloon observations at low geomagnetic latitude. The Leiden device also has a "shutter" to separate fluxes entering the aperture from background inherent to the detector. More advanced anticoincidence systems use CsI or NaI as the shield and

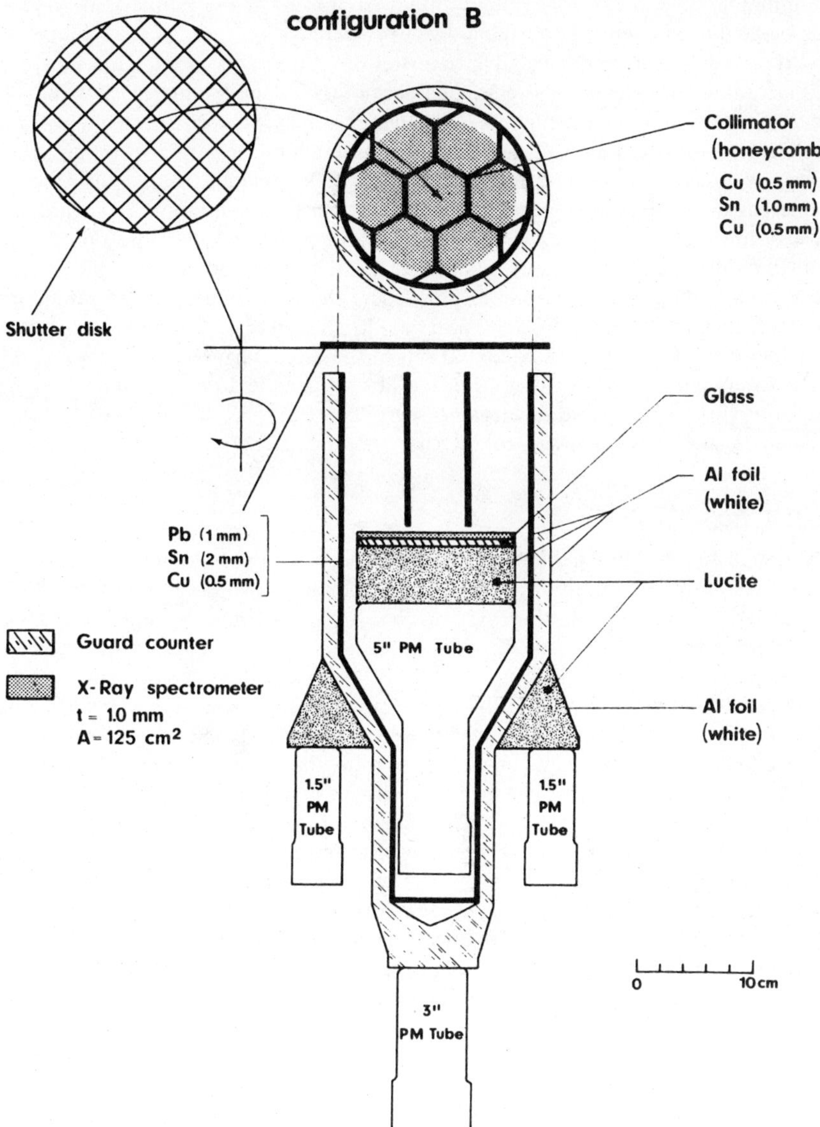

Figure 16 NaI(Tl) scintillator 20–80 keV telescope with 117.5 cm^2 effective area, 18° FWHM field of view and plastic guard counter, used for balloon observations by the Leiden group. Special features include a graded inactive shield, a collimator and shutter disk, and a thick light guide that helped achieve the 5% uniformity. (From Bleeker 1971.)

collimator, or a *phoswich* configuration to remove the phototube mass and its associated production from inside the active volume.

Many particular (and peculiar) properties of scintillation counters may be used to advantage in detection systems. The decay time of the luminescence in CsI(Tl) depends on the specific ionization and therefore may be used as an indicator of electron ionization following a gamma-ray interaction or a slow-stopping heavy particle from a neutron recoil. In the phoswich technique two phosphors of different decay time are optically coupled to the same photomultiplier and to a decay-time discriminator used to distinguish the pulses and therefore the medium in which the energy loss occurs. Omnidirectional NaI(Tl) crystals with plastic anti-coincidence jackets have been used to measure the total diffuse cosmic gamma-ray spectrum from Ranger 3, ERS-18, and Apollo 15 and 16 (cf Trombka et al 1973). NaI(Tl)/CsI(Na) cylindrical combinations have been configured so that one crystal acts as an active anticoincidence photon-absorbing shield for the other (Peterson et al 1972). Plastic or liquid organic scintillators may be "loaded" with high-Z elements such as tin or lead to increase the effectiveness as an anticoincidence absorber.

Solid-State Counters

Although counters based on the properties of electrons and holes in semiconducting materials have been used in particle physics for over two decades (Sharpe 1954), their use in photon-counting systems has occurred only recently due to the development of large-volume drifted detectors and exceptionally low-noise charge pulse preamplifiers.

A solid-state detector consists of a large volume of intrinsic Si or Ge sandwiched between and in electrical contact with thin layers of n-type and p-type material, hence the term *p-i-n diode*. As in gas and scintillation counters, photons interacting in the detection medium (the intrinsic volume) produce energetic electrons that lose their energy by ionization and collision. In solid-state counters, however, electrons from the valence band are raised into the conduction band, forming hole-electron pairs that diffuse under the applied electric field and are collected as a charge pulse at the electrode. Because few competing energy-loss or recombination processes occur in sufficiently pure intrinsic material, a large and definite fraction of the energy loss goes into hole-electron pair production. The average energy required to produce a charge-carrier pair is therefore not much more than the band gap. Table 1 shows the properties of Si and Ge that are most useful in detector design. Manufacturer literature contains much useful information on the application of solid-state detectors (ORTEC 1970, Princeton Gamma-Tech. Inc. 1970).

In the usual construction, lithium atoms (n-type) are diffused (drifted) under controlled field and temperature conditions through p-type silicon or germanium to compensate exactly the p-centers over a large volume, thus rendering the material conductorless or "intrinsic." Excess Li forms the n-layer. Because the Li atoms will rediffuse at room temperature, Ge(Li) detectors must be continually kept below about $-50°C$ with LN_2 or an equivalent cryogen. In the case of either Ge(Li) or Si(Li), operation at very low temperatures is desirable to reduce thermal hole-electron pair creation, which would give large noise currents and cause recombination of photon-produced conduction electrons. Recently, progress in producing large volumes of exceedingly pure Ge (10^9–10^{10} impurity atoms cm^{-3}) indicates that truly intrinsic (rather than drifted) detectors may soon be available. These Ge(Li) counters will have the considerable practical advantage that low temperatures will be required only during actual operation and not during storage.

The collection time depends on the electric field, the mobility of the carriers, and the geometrical path from the interaction location to the electrodes. A field of about 1000 V cm^{-1} saturates the drift velocity at a maximum of nearly 10 cm μsec^{-1}, hence several tenths of a μsec may be required to collect the charge in a large-volume Ge(Li) detector. The charge shape therefore varies from event to event.

Because $\varepsilon = 2.9$ eV are required to produce a hole-electron pair in Ge, the resolution at 10 keV would in principle be 4.0 %. In practice, however, the resolution is determined by other factors. Unlike proportional or scintillation counters, where only a small fraction of the total energy loss goes into the physical effect (the ion pairs or photoelectrons) detected, a major fraction here goes into the creation of electron-hole pairs. Poisson statistics are not obeyed because the total energy in the created ion pairs nearly equals the energy loss; with no statistical fluctuation in the latter, the resolution is better than expected from random statistics. This improvement is expressed in terms of a Fano (1947) factor

$$F = \frac{\sigma^2}{E/\varepsilon}, \tag{4.1.}$$

which has the limiting value $F = 0$ in the case of no statistical fluctuations, and the limit $F = 1$ if the probability of hole-electron pair creation is small. In Ge, typically, $F = 0.13$, so the expected peak width at 10 keV,

$$\Delta E = \text{FWHM} = 2.35(EF/\varepsilon)^{1/2} = 47 \text{ eV},$$

or 0.47%, a small value indeed.

Because of the small charge, 5×10^{-16} coulomb for a 10 keV energy loss, the resolution is in fact determined by the amplifier noise, and therefore by the distributed capacity across the input of a field-effect transistor (FET) charge-sensitive preamplifier. Goulding (1972) and Radeka (1972) have recently discussed the critical problem of pulse-shaping in low-noise preamplifiers. Because of these limitations, an energy-independent resolution of ~3.5 keV FWHM was obtained in the ~10 cc-volume Ge(Li) detectors flown on balloons in exploratory work in the 1966 era (Jacobson 1968, Womack 1969, Macklin 1969). Advances in amplifiers and techniques now permit resolutions of 1.2 keV FWHM in laboratory Ge(Li) instruments (ORTEC 1970). A resolution of 2.0 keV at the 123 keV Co^{57} lines has recently been obtained by an X-ray telescope containing four 40 cc Ge(Li) detectors flown on a balloon to search for cosmic gamma-rays in the 50–1200 keV range (A. S. Jacobson, personal communcation).

At energies below about 200 keV, the efficiency is essentially unity, as preparation of detectors greater than $1/\mu(E)$ thickness presents no problem. The cutoff at lower energies is determined by the windows needed for the cryogenic system and, depending on the geometry, the dead layer due to the Li drifting operation. Typically the latter can be reduced to 300 μm in Ge, corresponding to 20% attenuation at 15 keV.

Discussion of solid-state counters has been directed toward the upper end of the X-ray regime where exploratory work has occurred. Here nuclear gamma-ray lines from such astrophysical processes as radioactive decay of *r*-process elements or

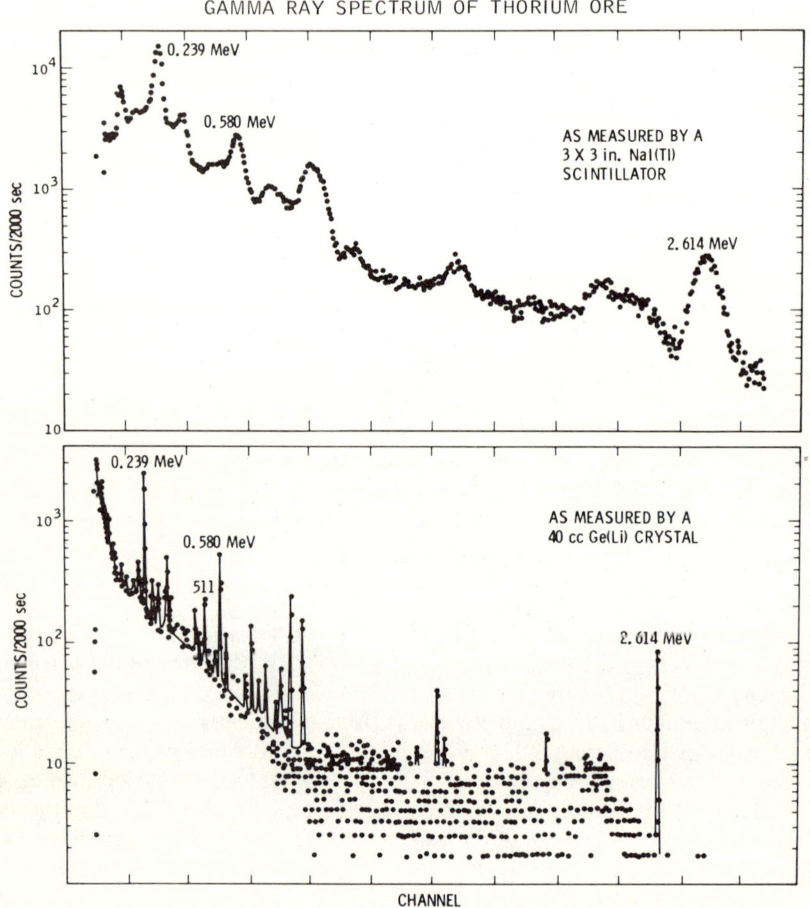

Figure 17 Comparison of spectra measured by a NaI(Tl) crystal scintillator (*top*) and a Ge(Li) solid-state detector. The resolution available with the solid-state device is clearly far superior. The sensitivity of the (typically larger) scintillator is greater above ~1 MeV. (Courtesy Dr. A. E. Metzger.)

decay chains following Si burning, supernova explosions, positron annihilation in dilute gases, etc may be expected. Because of the weak fluxes and possibly complex spectra, the extreme resolution obtainable with Ge(Li) detectors is required. Figure 17 compares the resolution and counting rate of a large-volume (40 cm^3) Ge(Li) system, with that of a 7.6×7.6 cm right-cylindrical NaI(Tl) scintillation counter, often used as a standard in nuclear spectroscopy. The advantages of the solid-state counter at lower energies are obvious.

Figure 18 shows the device used by Jacobson (1968) to obtain upper limits of $\sim 10^{-3}$ photons cm^{-2} sec^{-1} for gamma-ray line emission from the Crab nebula. The detector is in a vacuum cryostat mounted on a cold finger continually immersed in an LN_2 dewar. The balloon gondola and supporting system have been described in various publications by the UCSD group (Peterson et al 1967, 1972). The system used by Womack (1969) was similar to that of Jacobson, and used a similar collimation and shielding technique. A 20 cm^3 Ge(Li) counter telescope with a wide aperture has now been flown on a spacecraft in a polar orbit (Nakano et al 1973a) to study terrestrial emission, background effects (Imhof et al 1974), and the diffuse cosmic component in the 20–1500 keV range (Nakano et al 1973b). Although Figure 18 illustrates a trapezoidal geometry, various other drifting and electrode configurations are available (cf ORTEC 1970).

Figure 18 also illustrates the use of the active anticoincidence shield and collimator often used at higher energies. This concept, which Frost introduced (cf Frost et al 1966) and used for detectors flown on OSO-2 and OSO-5 (Dennis et al 1973), has been developed extensively by the UCSD group for balloon and satellite studies of the $\sim$10–200 keV spectra of cosmic X-ray sources (Peterson et al 1972, Peterson 1970, 1973), as well as by other workers. This technique has resulted in the lowest specific background of any shielding and collimation method thus far, and has been used in building collimated detectors operating over the 0.3–10 MeV range (Haymes et al 1969, Gruber 1974, Ling 1974).

Electronic and Data Handling Technique

All rocket, balloon, and spacecraft X-ray observations are supported by electronic systems which translate the information contained in the charge pulses of the detector(s) into a code suitable for telemetry or storage. The pulses may contain information on photon energy (pulse height), absolute or relative timing (coincidence), or event type (pulse shape). The analog pulse processing requires a trade-off based on the rate at which events occur in the system. Good energy resolution usually requires relatively "slow" electronics to achieve the lowest system noise and to obtain the maximum information about the event type before analysis. Fast electronics are required for the best anticoincidence systems and to minimize the spectral distortions at high rates due to pulse pile-up effects (Datlowe 1975).

The data system may be either of the "accumulating" or "event-by-event" type. In the accumulating technique, pulses within a given energy width (channel) are counted for a fixed time increment and transmitted as a total number of events; this technique is capable of handling high counting rates, but may be limited in energy or time resolution. This method is most adaptable to large area devices on rotating

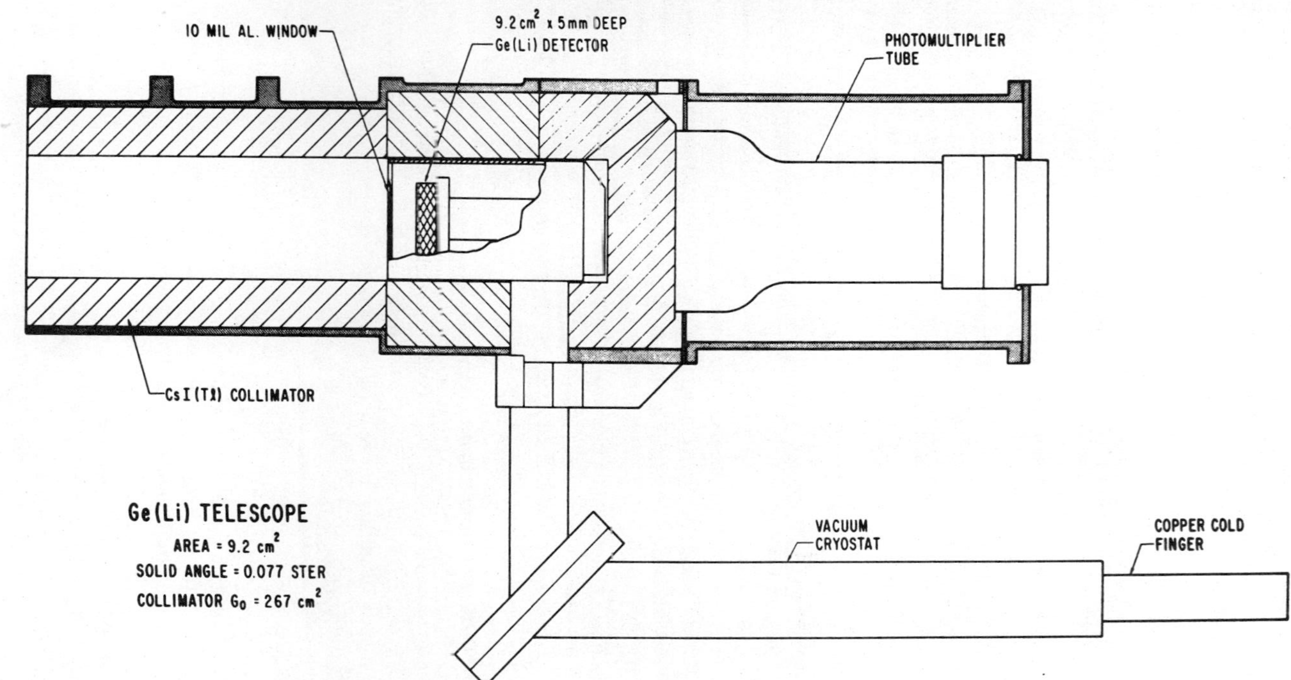

Figure 18 Ge(Li) telescope flown on a balloon in 1967 by Jacobson. The detector was maintained at low temperature by a copper heat conductor or "cold finger" immersed in liquid nitrogen. Although the intrinsic resolution for a 50 keV X-ray line is ~ 0.3 keV, the actual resolution, determined by amplifier noise, was ~ 3.5 keV. (Jacobson 1968.)

platforms where the accumulation and/or readout time is often determined by the scan rate and the instrument angular resolution, or for studying intense, variable sources such as the Sun. Jagada et al (1975) have described the accumulator-type electronics of the UHURU satellite. In the event-by-event technique, all of the information associated with each interaction is encoded (usually digitally) and transmitted or stored. This method provides the maximum information on each event, but often at a very limited rate, and is therefore often used with small-area devices, where counting rates are low, and in applications where the desired signal is comparable to the noise, so that events must be carefully selected. Commandable or automatic modes may permit a variable format and bit rate to match specific observations, or to compensate partly for unexpected operational conditions. The use of on-board minicomputers is now providing increased flexibility for such complex format manipulations. Determination of the vehicle and instrument aspect history (so the measured fluxes can be accurately related to the celestial sphere) may also pose formidable problems, as discussed by Schwartz & Gursky (1974).

Although it is beyond the scope of this work to enter into details of the electronic systems required to support the detectors under discussion, an example of an "event-by-event" system is given. Figure 19 shows the electronic "block" diagram of the UCSD X-ray detector on the OSO-7 (Baity et al 1970, Peterson 1973, MacKay et al 1975). The central detector consists of a 64 cm^2 effective area and 1.0 cm thick NaI(Tl) crystal viewed through a 1-inch Pyrex light pipe by a 3-inch RCA Cl02A photomultiplier. The active anticoincidence collimator and a geometry factor of 610 cm^2 for an isotropic flux. The aperture, formed by 10 2.54 cm holes drilled in the Cs(Na) collimator, has a 6.4° FWHM response. The detector and collimator are each covered with 0.05 cm Be windows, resulting in a 1/e cutoff at 5 keV.

Charges at the phototube anodes are amplified and shaped into pulses several microseconds long which trigger threshold level discriminators, generating logic signals of about 7 μsec duration. Logic operations on these pulses qualify the event for pulse-height (energy) analysis and transmission through the telemetry. The event is formatted and transmitted as a binary signature, which in this case also contains a time readout accurate to 0.625 msec. In addition, counting rates of the level discriminators, various logic functions, and other housekeeping data are transmitted for monitoring and diagnostic purposes.

SENSITIVITY CONSIDERATIONS

Signals from cosmic X-ray sources are received on a photon-by-photon basis by detectors. Observable properties of sources such as intensity, location, angular extent, or spectral shape are therefore statistical quantities to be estimated from the observational data to within a statistically determined error. Many of the more important observations occur at the limit of an instrument's sensitivity. This threshold may be optimized with an appropriate aperture or collimator structure.

Analysis of the data stream from an observation seldom reduces to the simple case of an on-source, off-source comparison, because the motions of balloon and

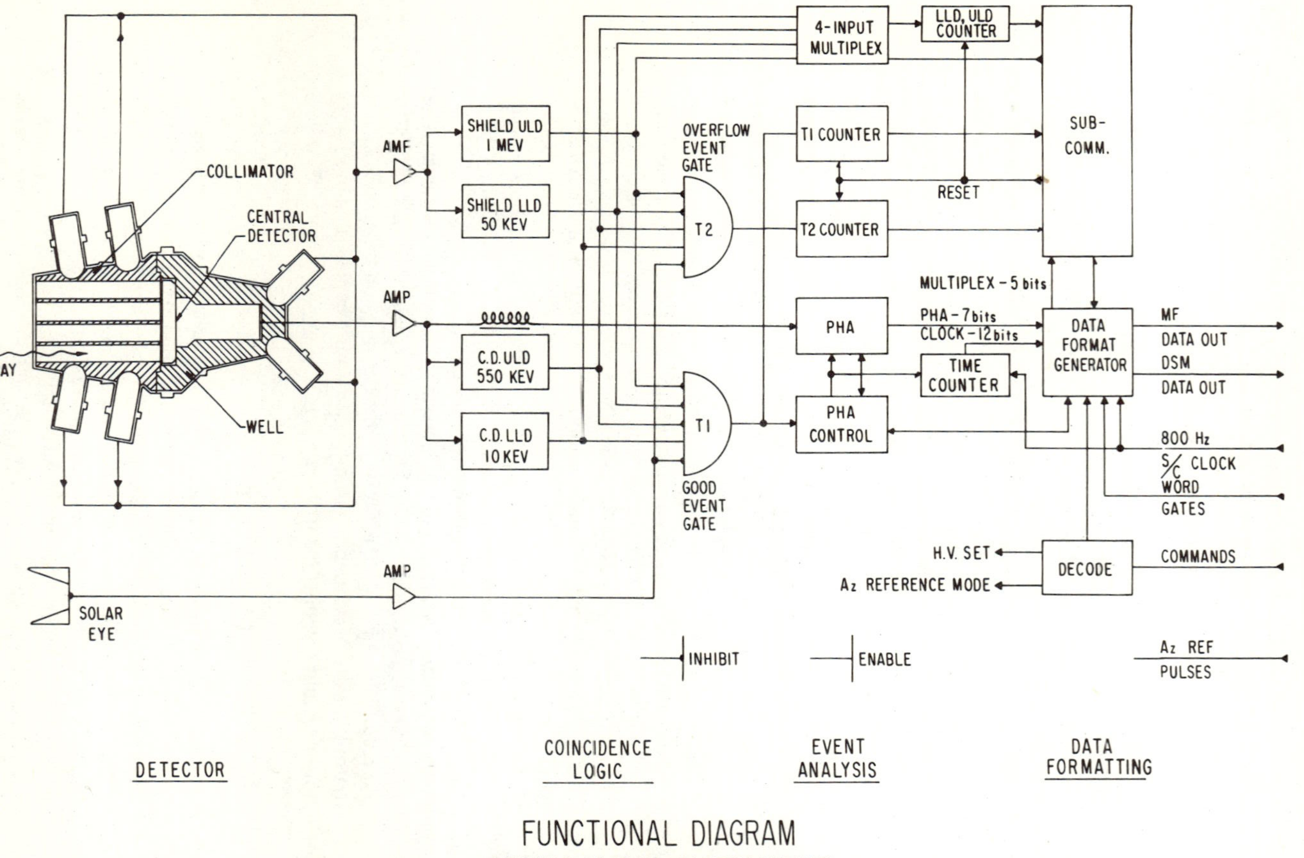

Figure 19 Functional diagram for the UCSD OSO-7 cosmic telescope, sensitive in the 7–500 keV range, with 64 cm^2 effective area and a 6.5° FWHM field of view. Sensitivity of this device was largely limited by the spacecraft telemetry allocation. (From Baity et al 1970.)

space platforms are often complex and sometimes uncontrollable. A model involving a history (aperture function) of the detector exposure to the source and a parameterized background is generally fit according to standard statistical procedures to a time series of counting-rate averages. Obtaining maximum information about a source requires careful procedures.

Statistical Limits

The statistics of nuclear counting as applied to X-ray astronomy have been most recently treated by Gursky & Schwartz (1974); the earliest published work consistent with the definitions and viewpoint used here appears to be that of Fisher & Meyerott (1966).

FUNDAMENTAL RELATIONS In general, measurement of a localized source must be made by subtracting a background rate B. Here we express B as the differential rate in counts cm^{-2} sec^{-1} keV^{-1}; the flux from a localized source, F, in photons cm^{-2} sec^{-1} keV; and the flux from a diffuse source, j, in photons cm^{2} sec^{-1} keV^{-1} sr^{-1}. If the observed quantities are stationary during the period of measurement, then the signal-to-noise ratio S is determined by Poisson counting statistics. The number of counts N_s from a source of strength F observed by a counter of effective area A (cm^2), efficiency ε, over a time interval τ (sec), and an energy interval ΔE (keV) is

$$N_s = \varepsilon A \tau \Delta E F. \qquad 5.1.$$

For a strong source, $\varepsilon F \gg B$, the fluctuations are due to statistics of N_s, and therefore the standard error or noise figure for $N_s \gg 1$ is

$$\delta N_s = (\varepsilon A \tau \Delta E F)^{1/2}, \qquad 5.2.$$

and likewise the signal-to-noise ratio

$$S = \frac{N_s}{\delta N_s} = (\varepsilon A \tau \Delta E F)^{1/2}. \qquad 5.3.$$

This is of course just the standard deviation of the measurement. For equation 5.3 to hold, it is only necessary that any changes ΔB in the background rate during the observation period τ be small compared to statistical fluctuations, or

$$\Delta B \ll \left(\frac{\varepsilon F}{A \tau \Delta E}\right)^{1/2} \text{ (Strong source)}. \qquad 5.4.$$

For a weak source, $\varepsilon F \lesssim B$, fluctuations in the background counts limit detection. The minimum detectable flux, F_{min}, at a chosen signal-to-noise ratio S, is determined in terms of the number of standard deviations $\sigma = N_s/\delta N_b$ above B, or

$$F_{min} = \frac{S}{\varepsilon}\left(\frac{\varepsilon B}{A \tau \Delta E}\right)^{1/2} \text{ photons cm}^{-2}\text{ sec}^{-1}\text{ keV (Fundamental Relation)}. \qquad 5.5.$$

Here we have assumed that B is well known, and that the only uncertainty is due to statistical fluctuations of B during the time τ the source is under observation. As shown in Figure 15, choosing S is equivalent to choosing a confidence level for the measurement, that is, the a priori probability that a sample σ standard deviations from the background mean is not a statistical fluctuation. A frequent choice, $S = 3$, corresponds to a 99.8% confidence level. Note that F, B, ε, and possibly even A may be energy dependent. Equation 5.5 shows that improvement in sensitivity varies only as the square root of most of the controllable parameters. A typical pre-UHURU rocket observation with a thin-window proportional counter of area 10^3 cm^2, a 5 keV bandwidth, and a background level $\sim 10^{-2}$ counts cm^{-2} sec^{-1} keV^{-1}, in a 10 sec observation could detect sources of strength about 1.3×10^{-3} photons cm^{-2} sec^{-1}, or about one-fiftieth of the Crab nebula flux.

In the case of X- or gamma-ray line emissions, an average over the linewidth ΔE must be used, and therefore

$$F_{\min} = \frac{S}{\varepsilon}\left(\frac{B\Delta E}{A\tau}\right)^{1/2} \text{ photons cm}^{-2}\text{ sec}^{-1} \text{ (Discrete X rays).} \qquad 5.5a.$$

The advantage of a high-resolution, high-efficiency counter is obvious.

BACKGROUND TERMS The background B may consist of a number of components, as discussed in detail in the following section. Here we distinguish between fluxes entering the aperture, such as the diffuse cosmic background B_d and, in the case of a balloon-borne counter, the background B_a resulting from fluxes produced in the residual atmosphere, from the inherent background rate B_i due to detector effects. The latter is often called the "true detector background" (Peterson et al 1972). For a detector whose aperture is Ω, the total background rate is

$$B = B_i + B_a + B_d, \qquad 5.6.$$

where $B_a = \varepsilon\Omega j_a$ and $B_d = \varepsilon\Omega j_d$.

The diffuse flux j_d and the atmospheric flux j_a (both photons cm^{-2} sec^{-1} keV^{-1} sr^{-1}) may be estimated from previous measurements of the diffuse and atmospheric fluxes. For a rocket or satellite experiment looking away from the earth $j_a = 0$, and in equation 5.6 we have implicitly assumed that no X-ray-producing masses are located within the aperture. Determination of B_i may not be so simple, because this results from the interaction of the detector with the total radiation environment. This B_i term frequently dominates. In such a case the use of focusing optics (possible below a few keV) can increase the area A_c for collection of source photons above the effective area A_b for background, thus improving the sensitivity expressions 5.4 and 5.5 by a factor $(A_c/A_b)^{1/2}$.

In addition, confusion caused by the presence of more than one source in the aperture is a source of background noise (cf Pelling 1973) and may have been the limiting factor for UHURU (D. Schwartz, personal communication).

DIFFUSE FLUXES Measurements of isotropic cosmic X-ray background j_d requires either $B_i \ll \varepsilon\Omega j_d$ or an independent measurement of B_i to an accuracy consistent with the desired precision of j_d. For measurements of sources that are diffuse in the sense that the angular size is large compared to the aperture, such as the

Cygnus loop (shown in Figure 4), equation 5.6 applies, and the minimum detectable surface brightness, $j_{\min}$, is

$$j_{\min} = \frac{S}{\varepsilon\Delta\Omega}\left(\frac{B}{A\tau\Delta E}\right)^{1/2} \text{ photons cm}^{-2}\text{ sec}^{-1}\text{ keV}^{-1}\text{ sr}^{-1}$$

(Surface Brightness for Extended Object), 5.7.

where τ is the observing time on each spatial resolution element $\Delta\Omega$.

The cosmic X-ray background B_d is usually the important contributor to the diffuse term in the background. However, for measurements on such objects as NP 0532, buried in the extended Crab nebula (Laros et al 1973), and NGC 1275 in the Perseus cluster (Fabian et al 1974), the surface brightness of the nebular region contributes an additional background term that may in fact dominate B_d.

SCANNING EXPERIMENTS For observation, many instruments make use of the motion of uniformly scanning vehicles, such as the UHURU or OSO-7 spacecraft, the spinning and precessing of rockets, or the earth's rotation for balloon detectors. Sources are observed for a time depending on the aperture FWHM, $\theta_{1/2}$, in the scan direction. We assume that the aspect is well enough known so n-independent scans can be added to increase the precision of the observations according to $(n)^{1/2}$. In a total observing time T, the time τ on the source is just

$$\tau = \frac{\theta_{1/2}}{2\pi} f T, \tag{5.8.}$$

where we now explicitly consider the fractional live time $f \leq 1$ due to anticoincidence shields, telemetry limitations, earth occultation, or operational effects.

The solid angle for an aperture $\theta_{1/2}$ FWHM in the scan direction, and $\phi_{1/2}$ FWHM in the perpendicular direction, is just $\Omega = \phi_{1/2}\theta_{1/2}$, for $\Omega \ll 4\pi sr$. Therefore,

$$F_{\min} = \frac{S}{2\varepsilon}\left(\frac{B_i + \varepsilon\theta_{1/2}\phi_{1/2}(j_d + j_a)}{A \cdot (\theta_{1/2}/2\pi) f T \cdot \Delta E}\right)^{1/2} \text{ (Including all Background Terms).} \tag{5.9.}$$

Small background In a detector whose aperture is sufficiently large so that the diffuse fluxes dominate the internal background, $B_i \ll \sigma\theta_{1/2}\phi_{1/2}(j_d + j_a)$, and

$$F_{\min} = \frac{S}{2}\left(\frac{2\pi\phi_{1/2}(j_d + j_a)}{{}^{\varepsilon}Af T\Delta E}\right)^{1/2} \text{ (Diffuse Limited Sensitivity).} \tag{5.10.}$$

Here we explicitly assume the modulation caused by a triangular aperture over the time T, resulting in a factor of 1/2. The sensitivity is independent of the aperture in the scan direction, and the efficiency ε, assumed to be the same for both the diffuse and source fluxes, now appears as the root.

If the rocket or spacecraft scans at the angular rate ω_s, and precesses at a rate ω_p, the total observing time on a source is also proportional to $\phi_{1/2}$, and is

$$\tau = \frac{\theta_{1/2}}{2\pi}\frac{\phi_{1/2}}{\omega_p} f.$$

Under these conditions,

$$F_{\min} \sim S \left[\frac{2\pi\omega_p(j_d)}{\varepsilon A f \Delta E} \right]^{1/2} \text{(Sensitivity for Precessing Vehicles)}, \tag{5.11.}$$

independent of the aperture. Such observations are normally made outside the atmosphere, hence $j_a = 0$. A slow precession increases the sensitivity.

The aperture size on precessing vehicles may also be chosen by the need to reduce source confusion, by the observational program, and by vehicular and constructional constraints. On nonprecessing platforms, such as the inertial wheel-stabilized Small Astronomy Satellite, or balloon detectors fixed in azimuth, $\phi_{1/2}$ is determined by a compromise between reducing background due to diffuse fluxes and the opportunity for scanning a larger region.

Finite background In the case where B_i is not negligible, decreasing the aperture much beyond

$$\theta_{1/2}\phi_{1/2} \lesssim \frac{B_i}{\varepsilon(j_d + j_a)} \tag{5.12.}$$

does not improve $F_{\min}$, but only increases requirements on aspect and pointing. For scanning instruments with a circular aperture ($\phi_{1/2} = \theta_{1/2}$) the sensitivity is

$$F_{\min} = \frac{S}{\varepsilon} \left[\frac{B_i + \pi/3(\theta_{1/2})^2 \varepsilon(j_d + j_a)}{A f T \theta_{1/2}/2\pi} \right]^{1/2} \text{(Circular Aperture)}, \tag{5.13.}$$

and there is an angular aperture $\theta^*_{1/2}$, which maximizes the sensitivity. This is given by

$$\theta^*_{1/2} = \frac{2B_i}{\pi\varepsilon(j_d + j_a)}. \tag{5.14.}$$

The collimator used by Jacobson (Figure 18) was designed with optimum aperture, as the inherent background could be estimated from previous experiments, and there are no other strong sources in the vicinity of the Crab nebula. Relations for other special cases can be obtained also.

SYSTEMATIC EFFECTS The expressions in the previous section assume that systematic variations are small, and that sensitivity can be made arbitrarily large by increasing the area A or observing time T. In practice, however, it is necessary to consider the signal-to-background ratio,

$$\delta = \frac{\varepsilon F}{B}, \tag{5.15.}$$

at the threshold flux $F_{\min}$. To realize the theoretical sensitivity, variations of systematic parameters (especially background rates) over the observing period, including the interval used for determining B, must be negligible or correctable to within an error $\ll \delta$. Expressed in terms of the uncertainty ΔB in B,

$$\frac{\Delta B}{B} \ll \delta_{\min} = \frac{\varepsilon F_{\min}}{B} = \sigma \left(\frac{1}{BA\tau\Delta E} \right)^{1/2}. \tag{5.16.}$$

The use of long observing and integration times to achieve high sensitivity requires not only an exceptionally stable or well-monitored background, but also that the area, efficiency, and energy-conversion gain must remain constant, to better than δ_{min}.

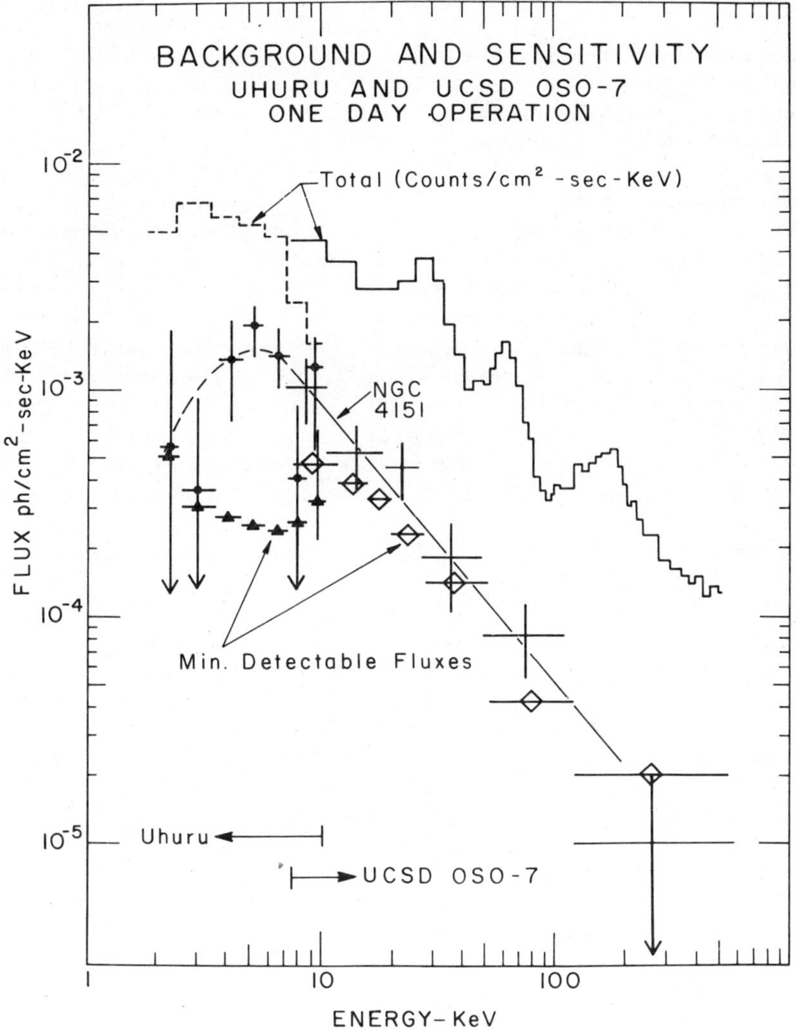

Figure 20 Background level and theoretical minimum detectable flux for a 1-day observation by the UCSD OSO-7 cosmic telescope and by the 5° × 5° UHURU detector. Also shown is the measured flux from a weak source, the Seyfert galaxy NGC 4151. The error in the flux scale assigned to the UHURU NGC 4151 results is estimated at not more than 30%. (UHURU background spectrum courtesy of D. Schwartz. NGC 4151 results from Baity et al 1975a and Kellogg 1973.)

If one has an a priori estimate of the systematic uncertainties inherent in the measurement, one can determine for a level σ a maximum observing time τ_{max} after which there is no usable increase in sensitivity. This is given by

$$\tau_{max} = \frac{\sigma^2}{BA\Delta E}\left(\frac{B}{\Delta B}\right)^2. \qquad 5.17.$$

For example, in a recent detection of 1–10 MeV gamma-rays from the Crab nebula by Gruber (1974) during a 4 h balloon observation using a large shielded scintillation counter, the statistical uncertainties over channels of width $E_2/E_1 \cong 3$ were $\sim 1.0\%$. To determine the flux at a level of $\delta \cong 3\%$ required an investigation of many possible systematic effects, an exceptionally accurate background model, and sophisticated statistical tests to verify the parameters of that model.

Figure 20 shows background and detection levels for 1970 generation instruments, the $5° \times 5°$ FWHM proportional counter on the UHURU (Giacconi et al 1971) and the 6.4° FWHM $\sim$ 7–500 keV scintillation-counter telescope on the OSO-7 (Peterson 1972). The measured background when looking at the sky is shown. Minimum detectable fluxes over the energy intervals shown are based on UHURU $A = 840$ cm^2 and OSO-7 $A = 64$ cm^2, a one-day (10^5 sec) operation and $f = 0.6$, which approximately allows for earth occultation and dead time due to trapped particle effects. The minimum detectable fluxes indicated are actually somewhat higher if allowance is made for source confusion noise (Condon 1974). NGC 4151, whose spectrum is shown, is one of the weakest sources to be detected by either instrument. Apparently, observing times, systematic effects, and data-reduction procedures are such that sources near the theoretical F_{min} are in fact detected.

Design of Aperture and Collimators

Collimation techniques in the proportional-counter energy regime are relatively simple (Giacconi et al 1968, Gursky & Schwartz 1974). Here a mechanical, egg-crate, or honeycomb structure of aluminum or beryllium suffices. Because photon ranges are short, rather thin sections may be used. Figure 13 shows a typical collimator configuration for soft X rays. Collimation in the 0.2–2 keV range does not present any particular difficulty, although some care in material selection and geometry is needed to avoid reflections. Simple designs give a triangular response whose FWHM can be tailored for maximum signal-to-noise to suit the observing program or the requirements for specific source studies. At energies above $\sim$20 keV, however, the photon range increases and thick sections are required for attenuation. Furthermore, photons at large angles from the detector axis may leak through housings and shields and interact in the detector medium, resulting in a high inherent background. The design of the collimator and shield therefore is largely concerned with reduction of this background component.

SHIELD CRITERION Assuming that an efficient anticoincidence system removes all prompt background effects due to the passage of cosmic rays through the detector, a simple criterion for determining the average shielding thickness is that the inherent background, here ascribed entirely to leakage of photons through the sides and rear of the collimator, be equal to or less than the diffuse fluxes entering the aperture. A detector above or near the top of the atmosphere is bombarded by a nearly isotropic flux of unwanted photons. The diffuse cosmic flux, atmospheric gamma rays, and production in vehicles or spacecraft usually conspire to make this assumption remarkably correct, at least for photons above ~50 keV and below 1 MeV. For such a case the aperture condition

$$\frac{\Omega}{4\pi} \geqq \exp\left[-\mu(E)\bar{t}\right] \qquad 5.18.$$

must be maintained. Here $\bar{t}$ (g cm^{-2}) is the average thickness of the shield. Due to the steep dependence of μ on energy, this criterion is easily met at lower energies, and is in fact rarely considered. Above a few hundred keV very massive shielding components become necessary, and no detector with small apertures has yet been flown meeting this criterion above ~1 MeV. Here other methods of providing directional information or collimation, such as Compton-coincidence telescopes, Fourier-transform telescopes, or track-indicating devices, such as spark chambers, are needed (cf Ögelman & Wayland 1970).

ACTIVE COLLIMATORS The requirements for thick-sectioned collimators of low total mass and low background production have led to the introduction of the active anticoincidence drilled collimator (Peterson et al 1967, Lampton et al 1972) of CsI. This concept as used on the OSO-7 is shown in Figure 19. The balloon-borne detector (Peterson 1970, Peterson et al 1972), used by Matteson (1971) in his studies of Cyg X-1 time variations and by Laros et al (1973) for determination of the spectrum of the Crab nebula and NP 0532 shown in Figure 5, also used this configuration. The latter telescope also uses a NaI(Tl)/CsI(Tl) phoswich as a central detector, which effectively removes all nonactive mass from inside the anticoincidence. The shield thickness was chosen to meet the criteria (equation 5.18) up to an energy of ~150 keV; over much of this range, $B_i \approx B_a \approx B_d$ at 3.0 g cm^{-2}, $\Lambda = 42°$ (Peterson et al 1972). Total B was ~2×10^{-3} c cm^{-2} sec^{-1} keV^{-1} at 50 keV. Similar devices have also been used by other workers. In principle, the optimum aperture for lowest F_{min} depends on the radiation environment; therefore, different geometries of the same general design may be optimal at different locations.

The geometrical theory for optimizing the areal efficiency and the response of drilled collimators has been developed by Aitken et al (1967). For the device in Figure 20, this was 65%. It is important that the ratio of effective area to volume for the detection element be as large as possible, consistent with reasonable efficiency at the highest desired operating energy, because many mechanisms responsible for the inherent background, such as Compton scattering of higher-energy "leakage" photons, neutron interactions, and spallation radioactivity, are volume effects.

As a compromise between the totally active and totally passive collimators, several groups have installed passive graded assemblies, such as that shown in Figure 16, inside an annulus of CsI or NaI. Lewin et al (1971) have used a detector of this type for a number of observations in the 15–150 keV range from extremely high altitude balloons. This scheme has the advantage of providing high areal efficiency, but is of limited effectiveness at higher energies. Above ~300 keV, the required thickness is so large that production by neutrals results in a background higher than that contributed by diffuse fluxes from wide angles.

MODULATION COLLIMATOR Determination of locations to accuracies on the order of minutes of arc is essential for source identification, especially in crowded fields. Obtaining source structure is important for understanding the source-emission mechanisms. For simple collimators these can be obtained only to within the order of the aperture size. For a source of strength F whose diameter is much less than $\theta_{1/2}$, location can be determined to an accuracy on the order

$$\delta\theta \approx \theta_{1/2}\left(\frac{F_{\min}}{F}\right)^{1/2}, \qquad 5.20.$$

assuming the aspect is determined to better than $\delta\theta$. Even for strong sources, systematic effects probably limit the accuracy to $\sim 0.1\ \theta_{1/2}$. Use of mechanical collimators with $\theta_{1/2}$ in the minutes-of-arc range presents formidable problems in construction, alignment, and vehicle pointing (Giacconi et al 1968).

To overcome these limitations, Oda has devised the modulation collimator technique. This has been reviewed extensively in the literature (Bradt et al 1968, Oda 1971), and is mentioned here only briefly. An arrangement consisting of aligned planes of grid wires of diameter d and center-to-center spacing usually $2d$, the planes perpendicular to the view axis of a detector system, gives a series of narrow transmission bands over a wide field of view. For two grid planes spaced by a distance L, a point source will give a triangular modulation of FWHM $\theta_m = d/L$, and this response pattern will be repeated over the entire $\theta_{1/2}$ of the primary aperture. The limit of geometrical optics applies if the grids are of sufficient thickness to be opaque. Additional grid planes of identical spacing and thickness inserted at distances $L/2$, $L/4$, $L/8$, etc from the detector plane result in the removal of one out of two, three out of four, seven out of eight, etc individual response θ_m functions. An n-grid collimator thus consists of triangular response functions of θ_m half width repeating every $2^{n-1}\theta_m$. For a nearly uniform scan this gives unambiguous one-dimensional angular structure to a resolution of θ_m for sources whose size is less than $2^{n-1}\theta_m$. The observing time per resolution element is reduced, however, by the factor $1/2^{n-2}$. The rocket-borne instrument used by Oda and his colleagues (Bradt et al 1968) to determine the location of Sco X-1 and the structure of the Crab nebula had four grid planes of gold wires, of $d = 0.12$ mm diam, in a collimator whose total length L was 67 cm, giving $\theta_m = 40''$.

Data Analysis

FLUX AND POSITION FITTING The field of view of a detector is commonly swept across a source by the rotation of its rocket or satellite, or of the earth. With the usual circular or fan-shaped collimator this uniform motion gives a simple or rounded triangular functional dependence on time for the exposed area. The centroid of the triangle, or the time t_0 of maximum exposure, is related to the source direction through the aspect solution. A more complicated aperture function $A(t-t_0)$ may result for a different choice of collimator, or if multiple scans are superposed. If the source strength can be assumed to hold constant over the period of observation, analysis for source strength simply amounts to the measurement of a term proportional to $A(t-t_0)$, usually by the least-squares method, in the time series of count-rate averages. The duration of the time bins for averaging is usually selected for convenience, but an optimum binning follows from a Nyquist criterion, namely half the shortest time over which the aperture function changes significantly, or alternatively, the binning frequency is twice the upper limiting frequency of the power spectrum of $A(t)$. Finer binning serves no purpose.

Although the background rate is normally assumed to be constant during an observation, and great ingenuity is sometimes expended to ensure such constancy, an analysis for source strength can perfectly well be conducted in the presence of a time-varying background $B(t)$, if its functional dependence is well known, is sufficiently different from $A(t)$, and can be suitably parameterized. Such an analysis for a balloon observation is reported by Gruber (1974).

If the source position is to be determined, then t_0 is also treated as an adjustable parameter. Linear regression analysis normally cannot be used, but simple and fast computer routines for nonlinear least-squares fitting are readily available.

For an observation in which the total number of counts is small or the measured flux is barely above the threshold, a more reliable confidence level for the result may be obtained with the method of maximum likelihood (O'Mongain 1973). The Poisson probability for the observed data is calculated given the best-fit source and background models; a second Poisson probability is calculated based on the best-fit background-only model (null hypothesis). For both a ratio of 20 indicates a positive flux observation at the 95% confidence level, a ratio of 100 indicates the 99% level, etc.

SPECTRAL FITTING Simultaneous flux observations at a number of energies may be used to determine a source spectrum. If the detector response is simple and well known, then the observed rates may readily be corrected to spectral fluxes which, in turn, may be fit with model spectra related to the astrophysical processes considered possible for the source (Gorenstein et al 1968). In addition to a multiplicative intensity factor, these model spectra depend in a more complicated way on one or more adjustable parameters such as spectral index or kT value. Best fit values and error estimates for these require nonlinear regression analysis. Alternative spectral models are accepted or rejected on the basis of χ^2 probabilities. However, if the energy binning is much finer than the Nyquist criterion, the systematic variance can be negligible compared to the statistical, and the test is not optimally sensitive. The method of maximum likelihood may also be used here.

For observations of weak statistical significance there may be no simple estimate for the errors in the determination of spectral parameters, that is, the contours of $\chi^2 > \chi^2_{min}$ in parameter space are not simple ellipsoids. In this case, the results of spectral analysis are properly reported as plots in parameter space of limiting contours corresponding to some standard confidence level. Margon et al (1975) and Kellogg et al (1975) have discussed the proper determination of these contours; what constitutes a standard confidence level is subject to varying interpretations.

Only the Ge(Li) counter among those discussed here has a truly sharp spectral response. Both the proportional and scintillation counters have an indefinite relationship between the observed pulse height and the incident photon energy, partly because of counting statistics for ion pair or photoelectron production, and partly because of loss of a portion of the interaction energy from the counter volume. Comparison of data with model is usually accomplished for these data by first passing the model spectrum through the transformation properties of the detector (often done with a Monte Carlo simulation), followed by the usual regression analysis performed in pulse-height space.

Methods have been developed (Dolan 1972, Trombka 1970) to bypass this laborious and sometimes ambiguous procedure by converting the observed n-channel pulse-height spectrum into an n-channel photon-energy spectrum. The observed pulse-height spectrum and the incident photon spectrum are related by a series of transformations

$$\mathbf{S}_0(h) = R(h, E'')A(E'', E')T(E', E)\mathbf{S}_i(E),$$

where T represents the transmission of the atmosphere and detector window for which Compton scattering may cause a redistribution in energy, A represents the detector-effective area and energy loss from escape of secondary photons if possible, and R represents the resolution function. If considered a matrix problem, solution for the input photon vector $\mathbf{S}_i$ in terms of the observed $\mathbf{S}_0$ requires inversion of the matrix $M = RAT$. Inversion of the triangular matrices A and T is simple, but the R matrix is singular, thus there is no unique solution to the problem. The technique of Dolan gives an approximate solution which should be considered representative only, although in the limit of perfect resolution it must converge to the correct input spectrum. With data below 100 keV, characterized by poor resolution, transformation of model spectra into pulse-height space for statistical testing may still represent the safest procedure.

RADIATION ENVIRONMENT

The background rate of an X-ray astronomy instrument is determined largely by the high-energy radiations at the detector. Background components for space instruments include penetrating cosmic rays, their secondaries produced in nearby matter, the diffuse component of cosmic X rays, and direct effects of trapped or quasi-trapped electrons and protons in the earth's magnetic field (Van Allen belts). In addition, radioactivity may be induced because of cosmic rays and passage through the trapped proton belts, or included explicitly or inadvertently in detectors or spacecraft materials. For counters operating below ~1 keV, even UV skyglow can produce undesirable effects. If there were no energetic particles or photons in space other than those from localized X-ray sources, reduction of detector background would indeed be a simple matter and large-area detectors would achieve remarkable sensitivities. Nature has, however, arranged a complex local radiation environment to make X-ray astronomy challenging.

Cosmic Rays

The intensity of relativistic galactic cosmic rays, isotropically distributed in near-earth space, is on the order of 1–3 nuclei $\text{cm}^{-2}\ \text{sec}^{-1}$. Meyer (1969) has presented a review of recent observations. About 15% of the cosmic rays are helium nuclei, and a few percent are CNO and heavier nuclei. Generally the spectrum is relatively flat below ~ 500 MeV nucleon^{-1}, and decreases rapidly with energies above a few BeV nucleon^{-1}. The lower-energy (< 100 MeV nucleon^{-1}) particle fluxes are particularly susceptible to solar cycle modulation and furthermore may show considerable enhancement following production at the Sun during solar flares. Figure 21 shows the differential spectrum of protons and helium nuclei outside the magnetosphere measured during solar minimum. Other nuclei show similar spectra, but with variations in isotope abundance vs energy, which are important for theories of the origin, acceleration, propagation, and modulation of cosmic rays. These details are irrelevant, however, for background considerations in instruments. Suffice it to say that although only $\sim 20\%$ of the flux above 500 MeV nucleon^{-1} is in the form of $A > 1$ nuclei, these nuclei carry $\sim 50\%$ of the energy in the cosmic-ray beam.

The geomagnetic field prevents cosmic rays below a certain cutoff rigidity, $\mathcal{R}_c$, from reaching a given location on the earth. The rigidity of a particle $\mathcal{R}$, usually expressed in billion volts (GV) is given by

$$\mathcal{R} \equiv pc/Ze = (E^2 + 2EAm_0c^2)^{1/2}/Ze \qquad 6.11.$$

for particles of species A, Z, momentum p, and kinetic energy E. The rigidity scale for protons is shown in Figure 21. Although the motion of particles in the earth's field is a complex problem, and the cutoffs at a given location depend on azimuth and zenith angle, the cutoff for particles vertically incident at a geomagnetic latitude λ_m is given by

$$\mathcal{R}_c = 14.7 \cdot \left(\frac{R}{R_e}\right)^2 \cdot \cos^4 \lambda_m (\text{GV}), \qquad 6.2.$$

where R/R_e is the distance from the earth center in earth radii. This, when folded into the integral of the spectra of Figure 21, gives the flux of primary particles vertically incident on a detector system at magnetic latitude λ_m and height R. To first order, penumbral effects at midlatitudes and east-west effects at low latitudes cancel; therefore, for detection systems immediately above the atmosphere, the effective solid angle Ω may be taken as 2π. For a detection system in space, the shadowing effect of the earth can probably be accounted to first order simply by the ratio $\Omega_{\text{sky}}/4\pi$.

Near the earth the primary beam is accompanied by a large number of lower-energy secondaries, mostly electrons and mesons. Because production increases with cosmic-ray energy, no simple relation exists between latitude effects of the secondary and primary particles. It has been noted empirically, however, that at extreme balloon altitudes ($\sim$3 g cm^{-2}), the counting rates of anticoincidence shields due to penetrating particles are about a factor of 2 above the primary rate over Palestine, Texas ($\lambda_m = 40°$) and a factor of 3 over Hyderabad, India ($\lambda_m = 8°$).

For spacecraft in low-altitude, low-inclination orbits, the local cutoff rigidity can be determined from the magnetic shell parameter L of McIlwain (1966), which is

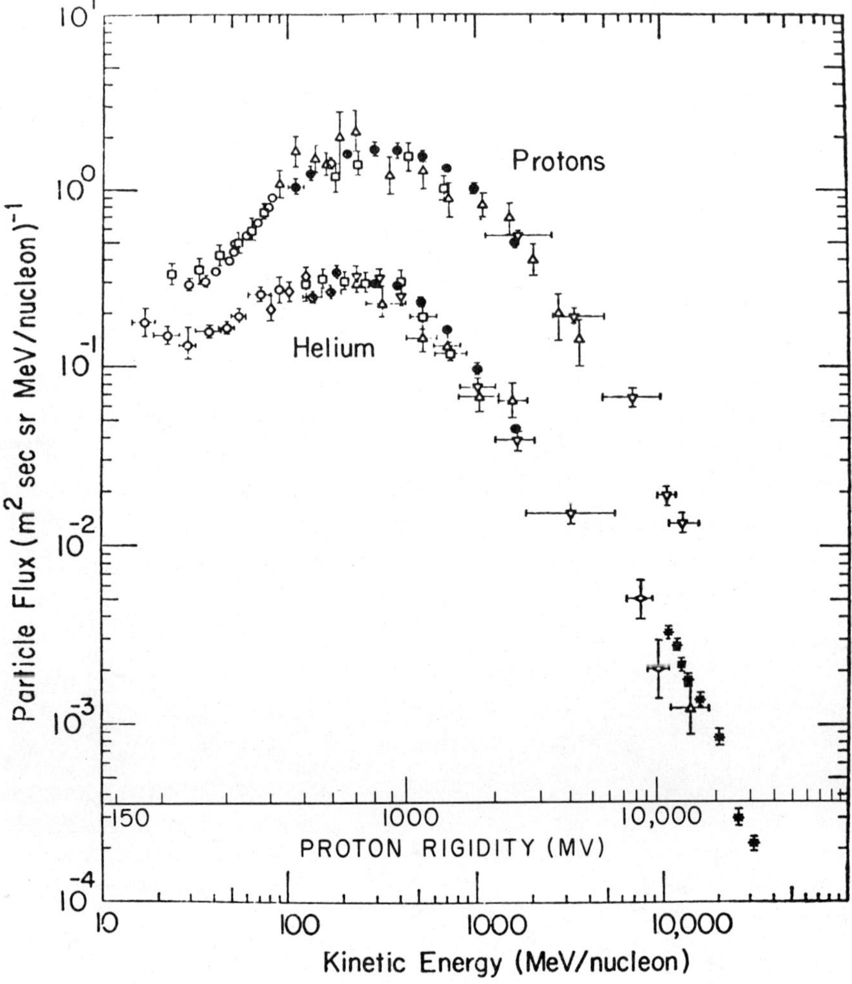

Figure 21 Energy spectrum of cosmic-ray protons and helium nuclei near earth. Measurements were made during solar minimum 1963–1965, when the spectra below 100 MeV nucleon^{-1}, which are strongly modulated by the solar cycle, were at a maximum. (Adapted from Meyer 1969.)

defined to account for nondipole terms in the earth's field, and inherently considers the orbital motions of trapped particles. In terms of L the invariant latitude is defined as

$$\Lambda \equiv \cos^{-1}\frac{1}{L^{1/2}}, \qquad 6.3.$$

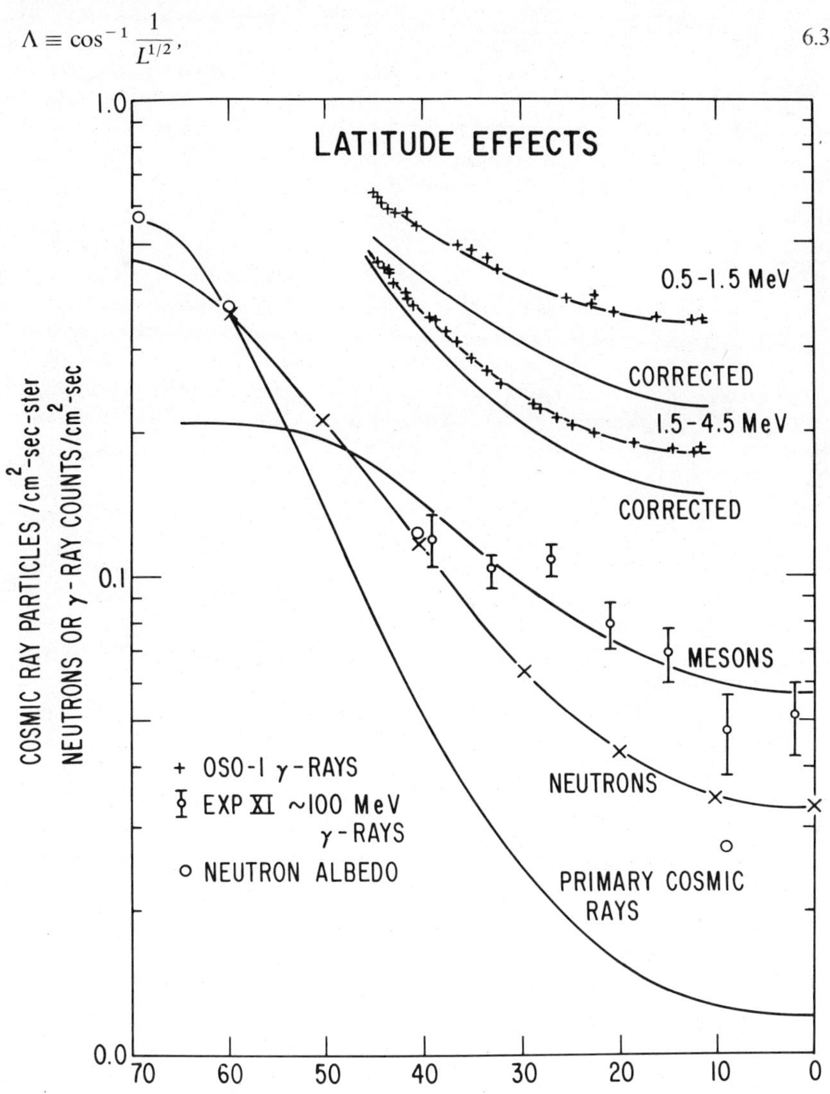

Figure 22 Variation with magnetic latitude for cosmic-ray primaries and various secondaries. The OSO-1 gamma rays are shown before and after correction for the contribution to counting rates of the omnidirectional counters by the cosmic diffuse component. (Peterson 1967.)

and may be used in place of the geomagnetic latitude λ_m. Hence the cosmic-ray cutoff rigidity at a spacecraft is $14.7/L^2$. It has been shown (Peterson 1967) that use of L and simple solid-angle considerations provide entirely adequate predictions of direct effects due to cosmic rays. Figure 22 summarizes latitude effects for cosmic-ray primary and secondary phenomena.

Atmospheric Gamma Rays

Atmospheric gamma rays produced by cosmic rays contribute in several important ways to the background observed on balloons and in low-altitude orbits. Gamma rays incident on the detector from outside the aperture, transmitted or scattered by the shield or collimator, may interact in the sensitive volume of a solid counter or produce a scattered electron ejected from the walls of a proportional counter. In addition, gamma rays produced by material within the aperture appear as incident fluxes, and therefore as background for observations of diffuse or point sources. The production of gamma rays in either the atmosphere, the spacecraft, or material immediately local to the sensitive volume may be regarded as equilibrium phenomena (Peterson 1967, Peterson et al 1973a) associated with the passage of cosmic rays through matter. Removing non-aperture photons is easy at low energies where the photon mean-free path is short; at higher energies this becomes increasingly difficult. For example, at ~ 3 MeV, the Compton-scattering length in air of ~ 40 g cm^{-2} is not much different from the interaction length of GeV protons of ~ 120 g cm^{-2} and the radiation length of 100 MeV electrons of ~ 80 g cm^{-2}. In a passive collimator at energies above ~ 300 keV there is more production than attenuation per unit length, when multiplicity is included.

The atmospheric gamma-ray flux has now been measured over a wide range of energies, altitudes, and latitudes, with a variety of directional and omnidirectional detectors (Peterson et al 1973a, Kasturirangan et al 1972, Chupp et al 1970, Haymes et al 1969). Figure 23 shows the counting-rate spectrum of atmospheric gamma rays measured over Palestine, Texas ($\Lambda = 42°$) at 3.5 g cm^{-2} with a 7.6×7.6 cm NaI(Tl) counter (Peterson et al 1973). The intensity over Hyderabad, India ($\Lambda = 8°$) is about a factor of 3 lower at the same altitude (Daniel et al 1972), whereas indirect evidence indicates the flux at high latitude $\lambda_m \gtrsim 55\%$ is about twice the $\Lambda = 42°$ value. Above a few MeV, the flux has been modeled theoretically (Daniel & Stephens 1974) and the description is rather complete. At lower energies, however, the work has been indeed piecemeal. Often the distinction between background originating from diffuse cosmic and atmospheric sources has not been made for directional detectors, but the total simply called "flux."

A gamma-ray source model, much more appropriate for evaluating fluxes incident on detector systems in and above the atmosphere, has been developed by Peterson et al (1973a) and Ling (1974, 1975). Here the production is expressed in terms of a source strength $S(E, x)$ photons g^{-1} sec^{-1} MeV^{-1}, derived empirically from omnidirectional counter data. The flux is obtained by integrating the source function over the volume of atmosphere seen by the detector. These calculations and the derivation of the source parameters are described elsewhere (Peterson et al 1973a, Ling 1975), but a few results important for background prediction are indicated here. Results applicable to 20–200 keV gamma rays have been described earlier (Peterson et al 1972).

The source strength over Palestine, Texas ($\lambda = 40°$) derived by Ling is shown in

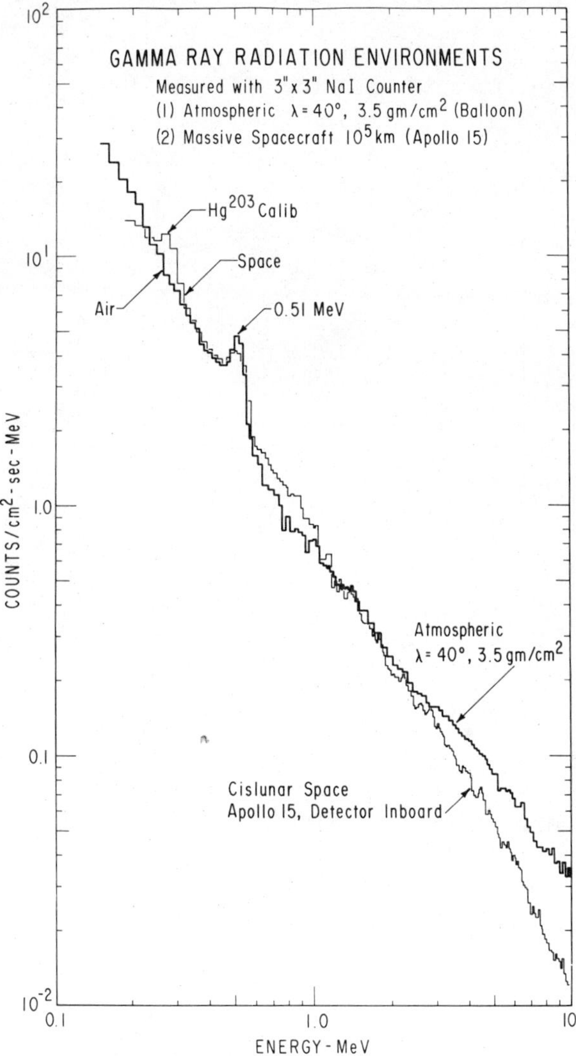

Figure 23 Spectra of environmental gamma radiation measured with $3'' \times 3''$ NaI(Tl) counters. The spectra obtained during a balloon flight at 3.5 g cm^{-2} over Palestine, Texas ($\Lambda = 42°$) and measured on the Apollo 15 spacecraft during transearth coast (*light line*) are quite similar. Cosmic-ray-produced gamma rays dominate both spectra.

Figure 24. At low energies there is very little dependence on depth. In general, the atmospheric flux incident on a detector at depth h is given by

$$F_a(E) = \int_{\Omega}\int_{r} \frac{S(E,x)}{4\pi r^2} \exp[-\mu(E)r] r^2\, dr\, d\Omega \text{ (photons cm}^{-2}\text{ sec}^{-1}\text{ MeV}^{-1}), \qquad 6.4.$$

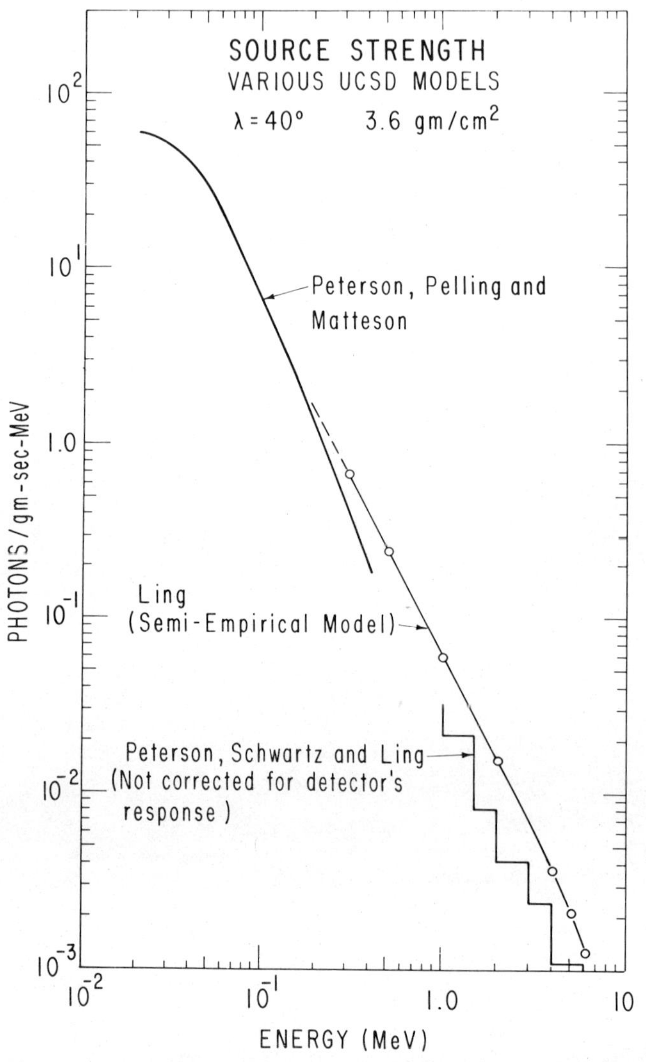

Figure 24 Source strengths for atmospheric gamma-ray production at a residual depth of 3.6 g cm^{-2} at $\Lambda = 42°$, based on the production models developed at UCSD. These models are derived empirically from observed counting rates. (Ling 1974.)

where the distance r to the source volume located at x is related to h by $r = |h-x|/\cos\theta$, $d\Omega$ is the differential solid angle, and $\mu(E)$ is the attenuation coefficient. Several important results are given here, under the assumption that $S(E, x)$ is depth independent over a photon mean-free path λ, that is, $\mu \gg 1/S(dS/dx)$. Practically, this means $E \lesssim 1$ MeV and $S_0(E)$ is used for $S(E, X)$.

1. The omnidirectional flux incident on a detector at depth h in the atmosphere is given by

$$F_a(E, h) = \frac{S_0}{2\mu}[2-\xi_2(\mu h)] \qquad 6.5.$$

where $\xi_2(\mu h)$ is the exponential integral of the second kind (Abramowitz & Stegun 1965). For large depths, that is, $\mu h \gg 1$, $F_a(E, h) = S_0/\mu$. $F_a(E, h)$ is the atmospheric gamma-ray radiation environment. The actual angular distribution of the radiation is expressed by the integrand in equation 6.4

2. For a directional detector at depth h, inclined at angle θ, the flux entering an aperture Ω (the aperture is assumed small, so variations over Ω are negligible) is given by

$$F_a(E, h) = \frac{S_0}{4\pi\mu}\Omega[1-\exp(-\mu h/\cos\theta)] \qquad 6.6.$$

3. For a detector looking down at the earth, the albedo flux entering the aperture is

$$F_a(E, h) = \frac{S}{\mu}\frac{\Omega}{4\pi}. \qquad 6.7.$$

Here we assume the earth fills the aperture, and that variations in S over the latitude seen by the detector are not large

4. The total earth albedo gamma-ray flux incident on a spacecraft at a distance R from the earth's center is given by

$$F_a(E, h) = \frac{S_0}{2\mu}[1-(1-R_e^2/R^2)^{1/2}]. \qquad 6.8.$$

This reduces to $S/4\mu(R_e/R)^2$ for large distances.

Although in principle the source function must be evaluated over the invariant latitude seen by the spacecraft, using the source functions derived at $\Lambda = 42°$ seems a practical procedure, as the latitude dependence of the source function has not been worked out in detail. To close order, this must simply scale as the flux measured at some intermediate energy, such as 1 MeV.

Spacecraft Production

Much less is known about gamma-ray production by cosmic rays in massive spacecraft. Here, because the spacecraft mass column density is usually less than or on the order of cosmic-ray interaction lengths, the production function in Figure 24, together with the latitude dependence in Figure 22, will certainly provide an upper limit, and in fact seems consistent with measurements thus far on OSO-1 (Peterson 1967), Ranger 3 (Metzger et al 1964, Peterson 1967), and Apollo 15 (Trombka et al 1973). That production in spacecraft is important was clearly demonstrated with Apollo 15 and 16, when a 7.6×7.6 cm NaI(Tl) counter placed on a retractable boom observed gamma-ray fluxes over the 0.2–27 MeV range. Figure 23 compares the spectrum measured with the detector inboard during cislunar transit with a balloon observation over Texas. The detectors were identical, and the production mass seen by the detectors in each case was slightly greater than 2π, so the comparison is relevant. The agreement in counting flux is quite remarkable: Apollo 15 was exposed to the full intensity of galactic cosmic rays in April 1971, and the production occurred in condensed matter whose average thickness was about 50 g cm^{-2}, whereas the atmospheric fluxes were produced in a dilute air mass of essentially infinite extent by a cosmic-ray spectrum cut off at rigidity 4.5 BV.

The source functions shown in Figure 24 thus appear to have considerable generality and may be used with considerable accuracy to predict the gamma-ray environment incident on detectors in spacecraft due to both local and atmospheric cosmic-ray production.

Trapped Radiation

The trapped radiation morphology has now been extensively investigated and its properties, at least for purposes of predicting effects in cosmic X-ray detectors, are now well known. These investigations have resulted in a series of models which describe the intensity of protons and electrons over a wide range of energies in B, L space (Teague & Vette 1974, Lavine & Vette 1970, Stassinopoulos 1970) and in a number of computer programs (J. I. Vette, personal communication) which give either instantaneous proton and electron intensities or integrated doses for orbits of arbitrary parameters.

The intensities are such as to render impossible low-background measurements over much of the near-earth environment, particularly for instruments on spacecraft in such desirable orbits as low-altitude polar, or earth synchronous at 6 R_e ($L \approx 6$). Most of the instruments flown on OSOs, SASs, and soon the HEAOs, are in low-inclination, $i \leqq 33°$, low-altitude, $400 \leqq h \leqq 600$ km, orbits. For these orbits, $0.20 \leqq B \leqq 0.47$ gauss and $1.1 \leqq L \leqq 2.2$. The desirability of low L and high B orbits is clear and the 500 km equatorial orbit, such as that used by the SAS, is most desirable. This orbit has, however, a nearly constant cosmic-ray intensity, and therefore advantage cannot be taken of latitude effects to separate out cosmic-ray coupled backgrounds, which is particularly useful for measurements of diffuse cosmic fluxes at energies $\gtrsim 300$ keV.

Figure 25 shows, as an example, isointensity contours for $\geqq 30$ MeV protons for a 500 km altitude taken from Stassinopoulos (1970). Also shown are the trajectories of a succession of 93^m OSO-7 orbits, which result in a history of radiation doses. The highest intensities occur in the region known as the South Atlantic Anomaly (SAA), where the earth's displaced and tipped dipole and local magnetic perturbations combine to give a particularly low B at a given altitude. The proton

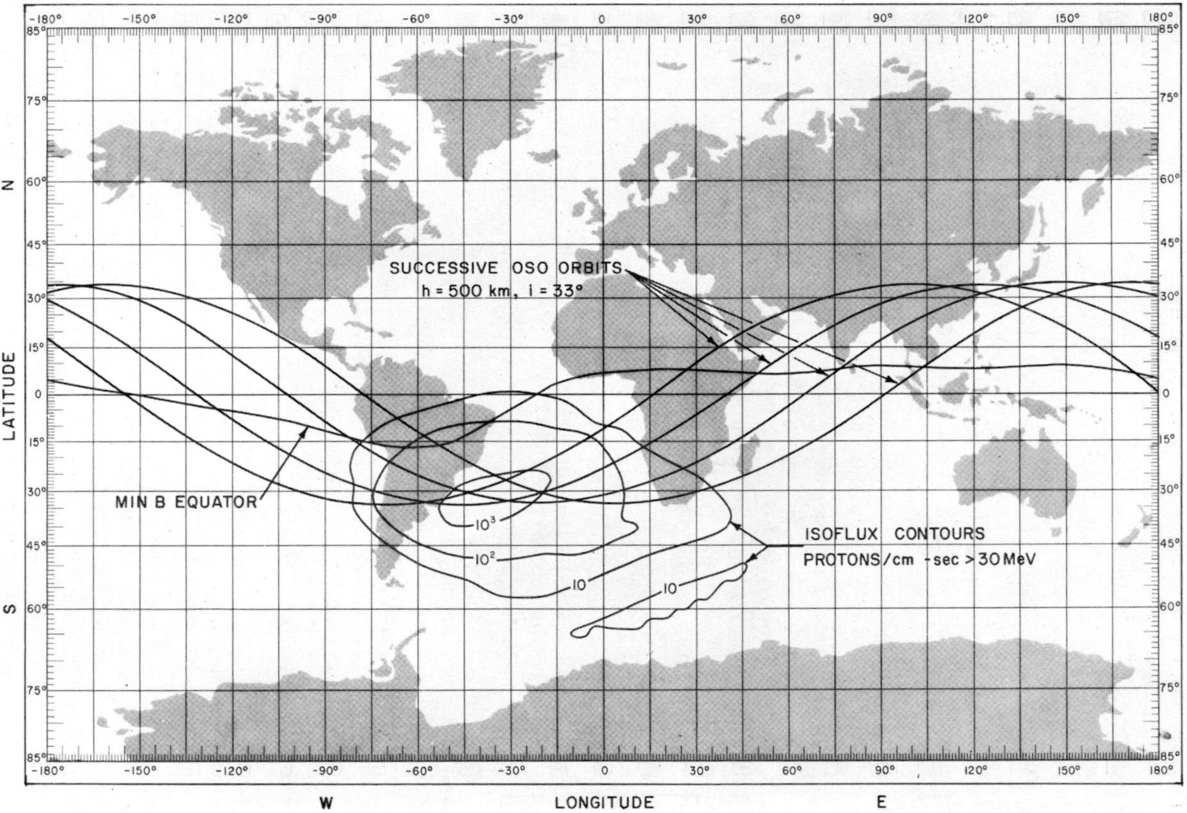

Figure 25 Orbits of the $i = 33°$, ~500 km OSO-7 intersecting isoflux contours of 30 MeV protons in the SAA. The intense radiation doses encountered during these transits cause activation of detector and spacecraft materials.

spectrum in this region may be represented as $\exp-(E_p/25\text{ MeV})$, $50 \leqq E_p \leqq 200$ MeV, and the electrons as $E_e^{-3.8}$, $200 \leqq E_e \leqq 1500$ keV. Proton intensities typically increase a factor of 2 each 30 km in the SAA region; electrons appear to be less predictable and more sporadic. There are also time variations associated with magnetic activity and the solar cycle.

Outside the instrument aperture, only the most energetic particles, $E_p \gtrsim 30$ MeV protons and $E_e \geqq 1$ MeV electrons, may reach the detector through the spacecraft and instrumentation structure. Nevertheless, the intensity of secondary gamma rays and neutrons produces sufficient background to cause very high counting rates in anticoincidence shields and detectors, thus preventing measurements of the weak cosmic fluxes. It has often proven desirable to turn off high-voltage supplies during passage through the SAA to prevent premature failure of proportional counters due to total count limitations or to prevent gain changes in scintillation-counter photomultipliers because of high anode currents. In this case, the SAA typically causes ~20–25% loss of observing time.

In addition, protons >1.3 MeV and electrons >50 keV may pass directly through ~1 mil Be windows used on proportional counters and protons >14 MeV and electrons >700 keV through a 40 mil Al cover on a >15 keV scintillation counter. Prediction of all detailed effects during passage through a radiation zone of high intensity, such as the SAA, is a difficult and probably unproductive endeavor.

Electrons of energy $E_e \geqq 3$ keV may penetrate the ~50 μg cm^{-2} organic windows used on soft X-ray astronomy and cause background counts at low energies not easily rejected with pulse-shape or "wall-less" counter techniques (Hill et al 1970). The quasitrapped or precipitating electrons associated with this effect show a very sporadic and unpredictable behavior, and have even been important in measurements on the diffuse components at higher energies (Schwartz & Peterson 1974). Although magnetic "brooms" may be used to deflect these electrons toward the collimator structure, the electrons may still scatter into the counter. Low inclination orbits show less of these particle contamination effects and seem to offer the most effective solution. A recent workshop (Holt 1974) has been entirely devoted to the subject of low-energy electron contamination.

Induced Radioactivity

Although the additional background caused by trapped electrons is present only during exposure, effects due to protons may linger because of induced radioactivity. These can be of two origins: 1. activation by capture of neutrons produced in reactions occurring in mass outside of the detector, or 2. spallation radioactivity due to breakup of target nuclei, particularly in high-Z scintillation or solid-state counters. Both effects have been observed, due not only to trapped protons in OSO and polar orbits (Peterson 1965, Dennis et al 1973, Imhof et al 1974), but also to cosmic rays in the atmosphere (Jacobson 1968) and in interplanetary space on Apollo 15 and 16 (Trombka et al 1973, Peterson et al 1973b). Fishman (1972) and Dyer & Morfill (1971) have attempted to calculate the induced radioactivity, and although their original calculations were qualitatively correct, there was considerable discrepancy in detail. There has been a considerable refinement in the calculations, however, using more sophisticated models of the trapped radiation, spacecraft geometry, and direct observations of the effects on Apollo 17 (Peterson et al 1973b). Judging from the effort devoted to the problem, we may expect a more complete understanding of induced and trapped radioactivity in the near future (cf Dyer 1973).

Diffuse Sky Background

The diffuse cosmic component, although of astrophysical importance in its own right, forms a background against which observations must be made. As already indicated, the diffuse flux entering the aperture gives a counting-rate contribution to the incident flux which competes with that of point sources. Above 300 keV, photons incident from wide angles on the spacecraft, rocket, or balloon instrument add another component to the gamma-ray background. Figure 7 shows the spectrum of the diffuse flux at lower energies, and Figure 26 is a compilation of recent measurements at higher energies. Although the enhanced flux in the 1–10 MeV region may be somewhat in doubt, the current spectrum is probably adequate for background calculations.

The total flux incident on a detector system is of course modified by attenuation in a spacecraft and by the solid angle occulted by the earth at low altitudes. At balloon altitudes, even at $\lambda = 0°$, the diffuse component contributes only a small fraction of the total flux compared to atmospheric gamma rays, particularly at higher energies. The counting rate caused by cosmic flux $j_d(E, 0)$ at energy E and zero depth entering a telescope at atmospheric depth h of aperture Ω is given by

$$j_d(E, h) = j_d(E, 0)\Omega \exp(-\mu h/\cos\theta) \text{ photons cm}^{-1} \text{ sec}^{-1} \text{ MeV}^{-1} \text{ sr}^{-1}. \qquad 6.9.$$

The total flux incident omnidirectionally on the detector system is

$$F_d(E, h) = \int j_d(E, h)\, d\Omega = 2\pi j_d(E, 0)\varepsilon_2(\mu h). \qquad 6.10.$$

Photons from the diffuse component may also scatter in the upper atmosphere and enter the aperture; however, these are explicitly accounted for in the source-function models (Ling 1975).

OSO-7 Background

Here the OSO-7 is used as an example of the effects of the radiation environment on a scintillation-counter instrument in a low-altitude, low-inclination orbit. The ~7–500 keV scintillation-counter telescope has been described by Peterson (1972) and is shown in Figure 19. The background effects in this instrument have been studied by MacKay et al (1975). The instrument was located in a wheel compartment of the 650 kg spacecraft, which was launched 29 September 1971 into a 330 × 575 km, $i = 33°$ orbit.

Figure 27 shows the history of counting rates for the various detector components vs time on 25 February 1972. Counting rates are modulated at twice orbital frequency due to cosmic rays and their secondaries at high latitudes, north and south. During a succession of passes through the trapped radiation over the SAA (Figure 25) starting at about 6^h, the spacecraft encountered a series of increasing doses. Shown are measured intensities of trapped electrons determined with on-board particle detectors. The induced radioactivity, which dominates the rate of the $\gtrsim 50$ keV discriminator, tends to decay more slowly than that measured by the $\gtrsim 2$ MeV discriminator. The *True Rate 1,* which measures central detector events

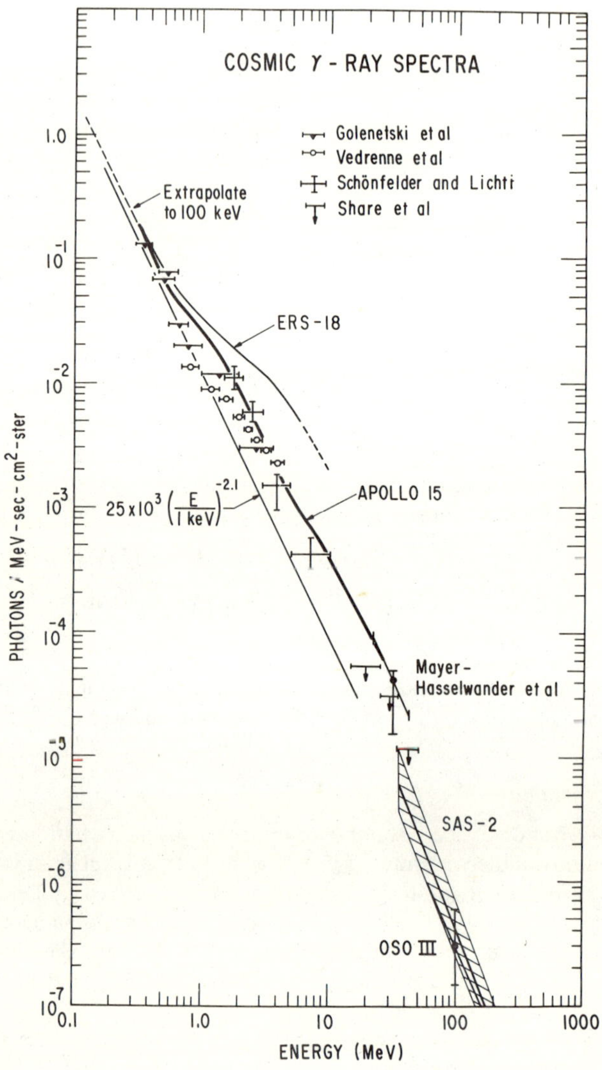

Figure 26 Spectrum above 0.3 MeV of the cosmic diffuse gamma-ray flux. Measurements in the 1–10 MeV range are several factors above a power-law extrapolation of the lower energy spectrum. (After Trombka et al 1973.)

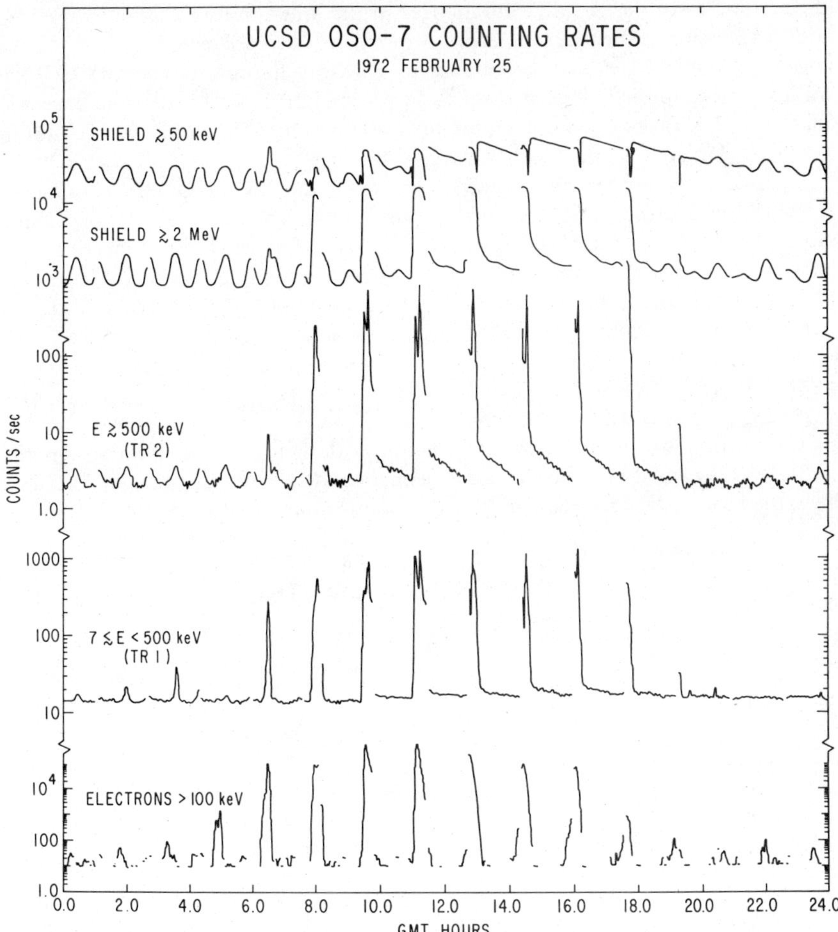

Figure 27 Counting rates of various detector functions for the UCSD OSO-7 experiment, shown in Figure 19, during orbital operation. The series of passes through the high fluxes of trapped particles causes activation of detector materials, resulting in elevated counting rates long after the region has been crossed. The counting rate of "good" detector events, True Rate 1, is, however, relatively constant.

$7 \lesssim E \leq 500$ keV, unaccompanied by an anticoincidence pulse, shows only small variations due to cosmic-ray effects and fast decays after dosage by trapped protons. The anticoincidence rejects about 97% of the particle-induced central detector events >7 keV.

The origin of the instrument background is indicated by the spectrum shown in Figure 28. The strong features at ~ 30, 60, and 180 keV show a complex dependence on the time since dosage, and on changes in the long-term irradiation caused by nodal precession of the eccentric orbit. Although it was anticipated there would be lines at 32 and 57 keV, corresponding to *K* X-ray fluorescence in the CsI(Na) collimator and to the first excited state of I^{127}, analysis indicates that the observed features are caused by more complex processes yielding 58^d Te^{125}, 60^d I^{125}, 100^d Te^{127}, 13^h I^{123}, etc as daughters.

Figure 28 also shows the expected spectrum due to the diffuse component entering the aperture, and the leakage due to the combined effects of diffuse, atmospheric, and spacecraft gamma rays. Clearly the induced effects increased the background at least 10 over that expected. The lower dosages encountered along the equatorial SAS orbit, or the 23°, 370 km orbit planned for HEAO, greatly improves the background picture for these missions.

CONTEMPORARY INSTRUMENTS

In this section we discuss typical instruments designed for observations in the 1970–1980 era. In addition to the nonfocusing, nondispersive instruments described here, the latter half of this period will see the introduction of Bragg spectrometers,

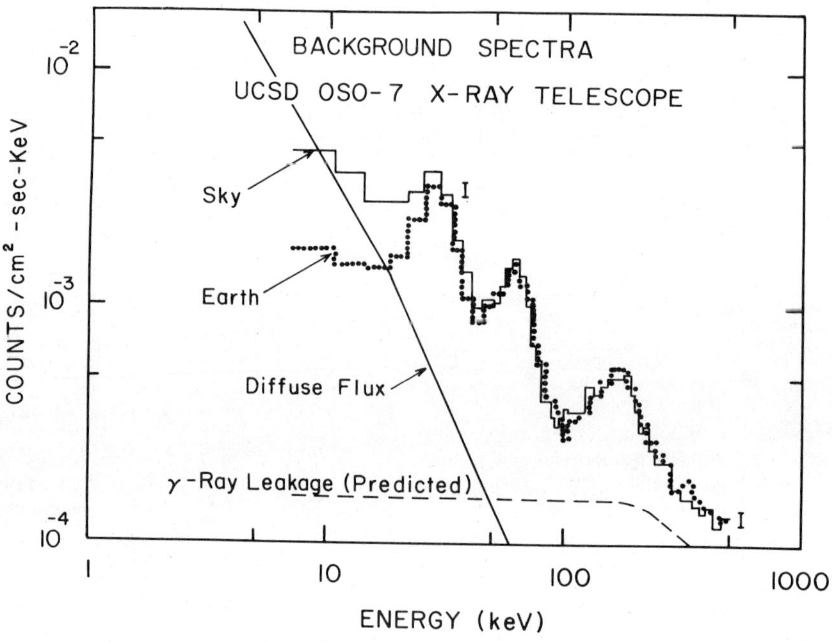

Figure 28 Background spectrum for the hard X-ray detector on OSO-7. When pointed at the sky, the diffuse cosmic flux entering the aperture dominates the background only below ~30 keV. The leakage component of the higher-energy background is dominated by induced radioactivity due to trapped protons.

polarimeters, and imaging systems at the focus of grazing-incidence mirrors or collectors on orbiting spacecraft. These instruments will have sufficient aperture and sensitivity to study relatively weak cosmic X-ray sources in a manner analogous to conventional astronomy.

Proportional Counter Techniques

Modern proportional-counter systems use a variety of techniques to simultaneously achieve large area and low background. Assuming that direct effects due to cosmic rays and charged secondaries passing through the counter can be eliminated with an anticoincidence scheme, and that there are few ambient energetic electrons, the background is due mainly to high-energy photons producing Compton electrons in the walls or possibly even in the gas. Because these electrons have energies of the order of several hundred keV, their range is comparable to the counter thickness, and therefore, unlike the low-energy photoelectrons, they produce ionization throughout the counter volume. The pulse-shape discrimination technique (Gorenstein & Mickiewicz 1968) requires no modification to counter geometry, but its use is limited at low energies by statistical fluctuations in the pulse shape, and at high energies by long electron ranges. Employment of this technique probably reduced the nonaperture background in the UHURU counters by about a factor of 10 to

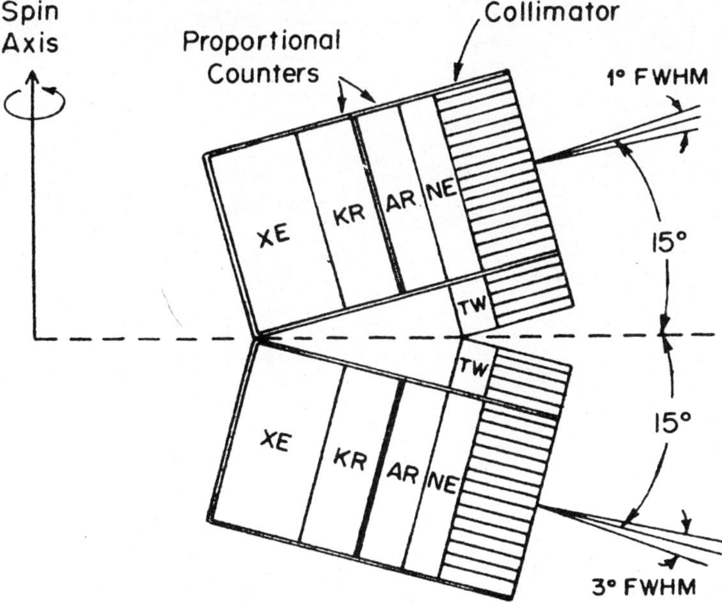

Figure 29 Schematic diagram of the MIT OSO-7 multilayer proportional counter for source variability and position determinations in the 1–60 keV range. Precession of the spin axis at a nominal 1°/day permits sources near the ecliptic plane to be observed periodically. (From Clark et al 1973.)

$B_i \sim 2 \times 10^{-3}$ counts cm^{-2} sec^{-1} keV^{-1} for $1.5 \leq E \leq 10$ keV. The diffuse flux is the dominant part of the UHURU background in the $5^\circ \times 5^\circ$ counters (see Figure 20).

The multiwire proportional chamber, developed at GSFC (Serlemitsos 1971, Bleach et al 1972) and shown in Figure 13, is a better and more versatile technique. Here the outer sections are used as anticoincidence elements, forming, as already discussed, "wall-less" counters. In fact, at higher energies the gas layer adjacent to the window also may be used in anticoincidence to reject quasitrapped electrons entering the aperture. Devices similar to that of Figure 13 with areas ~ 650 cm^2 have been flown on rockets (Holt et al 1974b) to observe Her X-1, among other sources, and achieved background fluxes apparently limited only by the diffuse component for apertures as narrow as $1^\circ \times 4^\circ$ ($\Omega = 0.0012$ sr). Such counters are scheduled for the OSO-I in 1975 and for the HEAO-A in 1977.

The simple one-window, one-gas counter has a rather limited energy range. To overcome this limitation, Clark and his colleagues at MIT (Clark et al 1973) have flown two multilayered proportional counters on the OSO-7 which observe in five spectral bands over the 1.0–60 keV. Figure 29 shows schematically the arrangement of counter layers. The window thicknesses, materials, and gas fillings indicated in Table 2 give the response characteristics shown in Figure 30. The two counters, each with slightly different fields of view, were placed to observe $+15^\circ$ and -15° from the OSO-7 spin plane. Because the OSO-7 spin axis motion was about

Table 2 MIT OSO-7 multilayer proportional counter

	Window		Gas		Unobstructed Areas (cm^2)	
Symbol	Material	Thickness (mg cm^{-2})	Material	Thickness (mg cm^{-2})	1° FWHM	3° FWHM
Four-Layer Counter						
NE	Be	4.6			66.1	75.4
			Ne	1.46		
			CO_2	0.40		
AR	Be	46.0			66.1	75.4
			Ar	5.41		
			CO_2	0.70		
	Al	68.5				
KR	Al	68.5			66.1	75.4
			Kr	23.9		
			CO_2	0.78		
XE	Al	411.0			59.2	67.5
			Xe	61.0		
			N_2	1.49		
Thin-Window Counter						
TW	Al	2.3			22.2	25.4
			Ne	1.28		
			CO_2	0.30		

1°/day, this resulted in a nominal 1–3 day observation of a source several times a year. Many results on source locations, spectra, and time variations have been reported from this instrument (cf McClintock 1974). The multilayer technique has already been incorporated into the proportional chamber of Bleach et al (1972).

The development of large-area, low-background proportional counters and their extension to low energies was pioneered by Friedman and his colleagues at NRL. Recently Kraushaar and co-workers (cf Bunner et al 1971) have made important contributions to technique. Soft X-ray counter construction has also been described by the Leiden-Nagoya group (de Korte et al 1974). Windows of materials such as

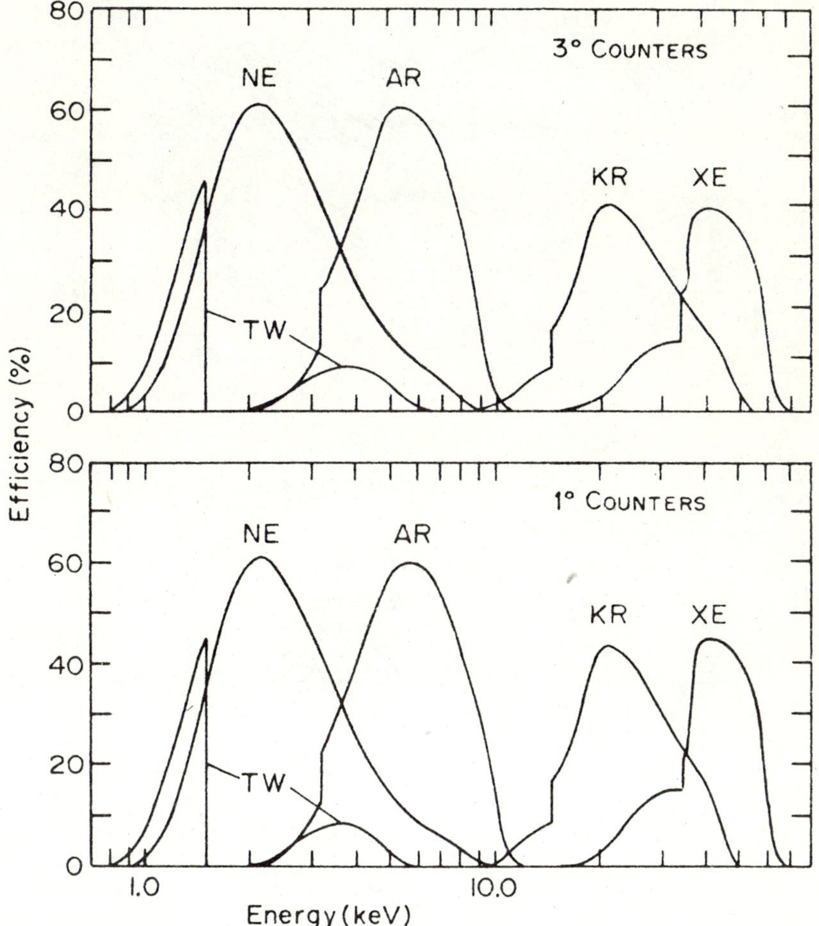

Figure 30 Theoretical detection efficiencies of the multilayer proportional counter shown in Figure 29 for X rays occurring at normal incidence on unobstructed portions of the front window. (From Clark et al 1973.)

Formvar and Kimfol (polypropylene) in thickness down to 0.5 μm are used, colloidal carbon is deposited on the windows to absorb UV photons, and the counters are carefully designed for low-noise and uniform collection, since only about eight pairs are produced by a 250 eV photon in P-10 gas. Although a multiwire-counter construction such as in Figure 13 is effective in reducing background electrons from the walls, the volume near the ends of the anode wires is not in anticoincidence. Kraushaar has overcome this limitation and improved collection uniformity in the

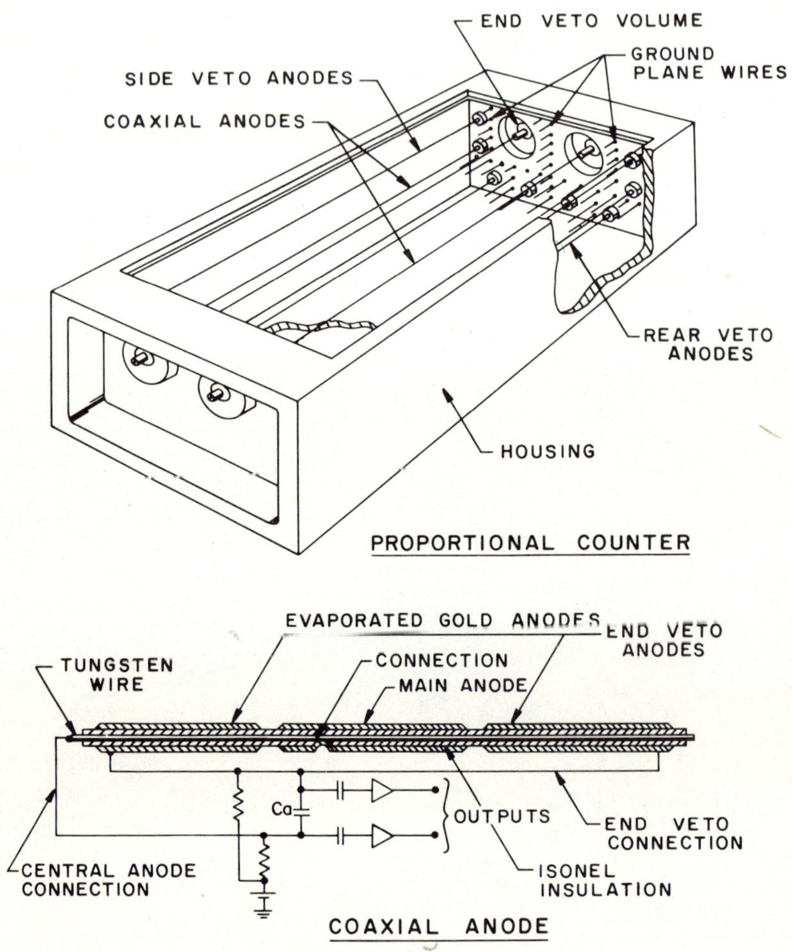

Figure 31 The coaxial anode as used in proportional counters designed for OSO-I by the University of Wisconsin. The end veto anodes detect background events from the counter walls, and result in a more uniform response for the central anode. (Courtesy of W. L. Kraushaar.)

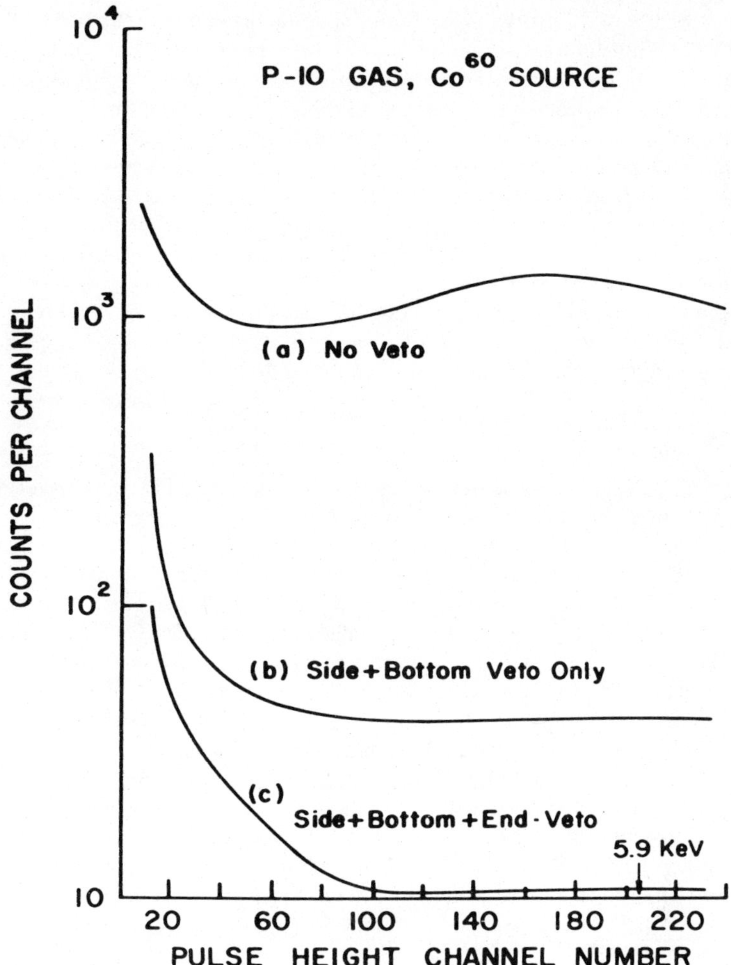

Figure 32 Reduction of counting rates in the coaxial-anode proportional counter under various veto conditions. Co^{60} gamma rays simulate the principal background-producing component at orbital altitudes. (After Bunner et al 1973.)

end region through development of a coaxial anode[2] (Bunner et al 1973). This construction applied to counters designed for the OSO-I, shown in Figure 31. Thus the sensitive volume can be completely surrounded by an active anticoincidence gas. Figure 32 shows the relative background rejection obtained using Co^{60} (1.17 and 1.33 MeV) gamma rays to simulate the high-energy photon radiation environment.

[2] U.S. Patent 3,812,358.

The fabrication of thin organic windows of large area has been described by Grader et al (1971). Gas diffusion through these windows presents no particular problem for short rocket flights, since refill and seal can be accomplished just prior to flight, but counters intended for long-term operation on spacecraft must have a gas replenishment system or be of a flow variety. In the former case, differential diffusion may preclude use of a two-component gas such as P-10 and require, for example, pure methane (Bunner et al 1973). M. Oda (personal communication, 1973) has described a gas and gain-control servosystem sensed by the counting rate of an Fe^{55} source in a small-volume auxiliary counter with high voltage and gas common to the large primary counter.

Extremely large area proportional counters on the order of 10^4 cm^2 have been flown by Seward on a Thor rocket (Seward et al 1971) and are also being used by NRL for a high-sensitivity survey over the 0.15–20 keV range on the HEAO-A (Peterson 1972).

Scintillation-Counter Telescopes

Because of the expense and weight required to provide completely absorbing active anticoincidence shields and collimators operating in the region above ~ 10 keV, detectors having both a large area ($A \gtrsim 200$ cm^2), and a low background ($B_i \lesssim 3 \times 10^{-4}$ photons cm^{-2} sec^{-1} keV^{-1}) have not been implemented. The background of the UCSD balloon-borne 35 cm^2 NaI(Tl) detector with a CsI(Tl) phoswich and a CsI(Na) honeycomb-collimated anticoincidence shield was of this order (Peterson et al 1972). Current efforts to obtain high sensitivities at $E \gtrsim 20$ keV therefore tend toward either very large areas with simple collimation or rather modest areas with sophisticated shielding.

K. J. Frost (personal communication) has developed a telescope for OSO-I with two 54 cm^2 effective area NaI(Tl) detectors operating in the 20 keV–5 MeV region. This has a 4π CsI(Na) anticoincidence shield with a drilled honeycomb collimator of circular aperture (FWHM 5°) for one of the two detectors but not the other, thus permitting an independent simultaneous measurement of certain components of the inherent background B_i. The total weight of the shield and collimator is about 45 kg. The detector field of view is offset by about 5° from the spin axis of the spacecraft which precesses at a 1° day^{-1} rate. This scheme, also used by the soft X-ray instruments on the same spacecraft, allows a relatively long integration period over a limited portion of the sky.

Very large area devices using banks of thin NaI(Tl) crystals, each having an area of roughly 100 cm^2, with simple passive shielding are being used at the TIFR (B. V. Sreekantan, personal communication) and at Nagoya University in Japan (Y. Tanaka, personal communication). These instruments are being flown from Hyderabad, India, where the low geomagnetic latitude ($\lambda_m = 8°$) and resultant low secondary cosmic-ray-produced background ($\sim 2.5 \times 10^{-3}$ counts cm^{-2} sec^{-1} keV^{-1}) give high sensitivities for total collecting areas on the order of 10^3 cm^2. As an extreme example of this approach, Bristol University, in collaboration with Leicester University, has constructed a detector system consisting of approximately 10^2 thin NaI crystals mounted on 12.5 cm photomultipliers for a total area of about

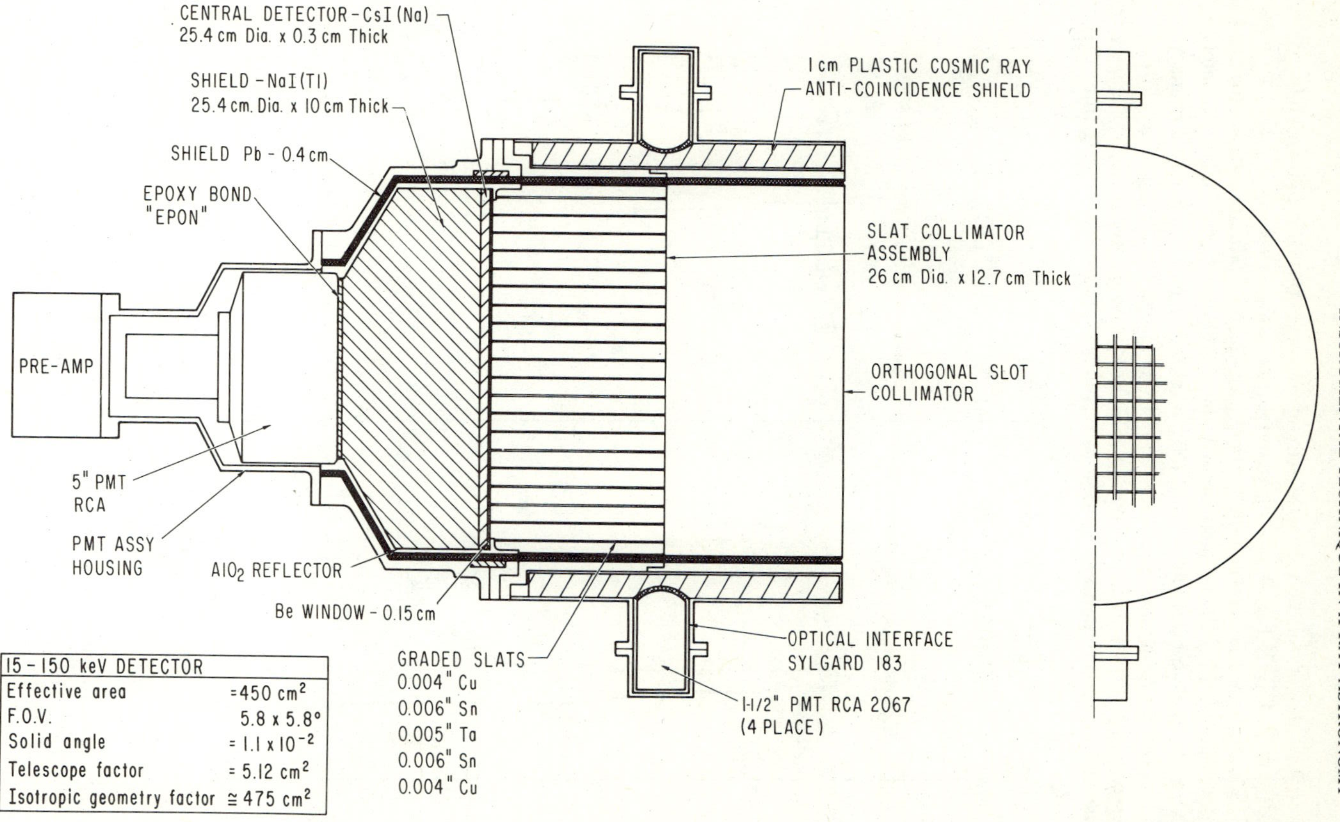

Figure 33 A large-area scintillation counter telescope for balloon observations of cosmic X-ray sources uses a CsI(Na)/NaI(Tl) phoswich with a very thick rear shield, and a combination of graded passive and active collimator elements to achieve low background.

10^4 cm^2. This device will be used primarily for observations of rapid variability of sources such as Cyg X-1 (R. Redfern, personal communication).

At UCSD, an approach providing low background and moderately large area has been developed (Pelling 1974). This system, shown in Figure 33, is modular, ultimately to consist of three modules each having 450 cm^2 effective area. An inverted primary/secondary crystal configuration with a thin CsI(Na) disc as the primary and NaI(Tl) as the secondary crystal is used. The NaI(Tl) shield crystal gives a high efficiency for interaction, detection, and rejection of secondary photons scattering from the

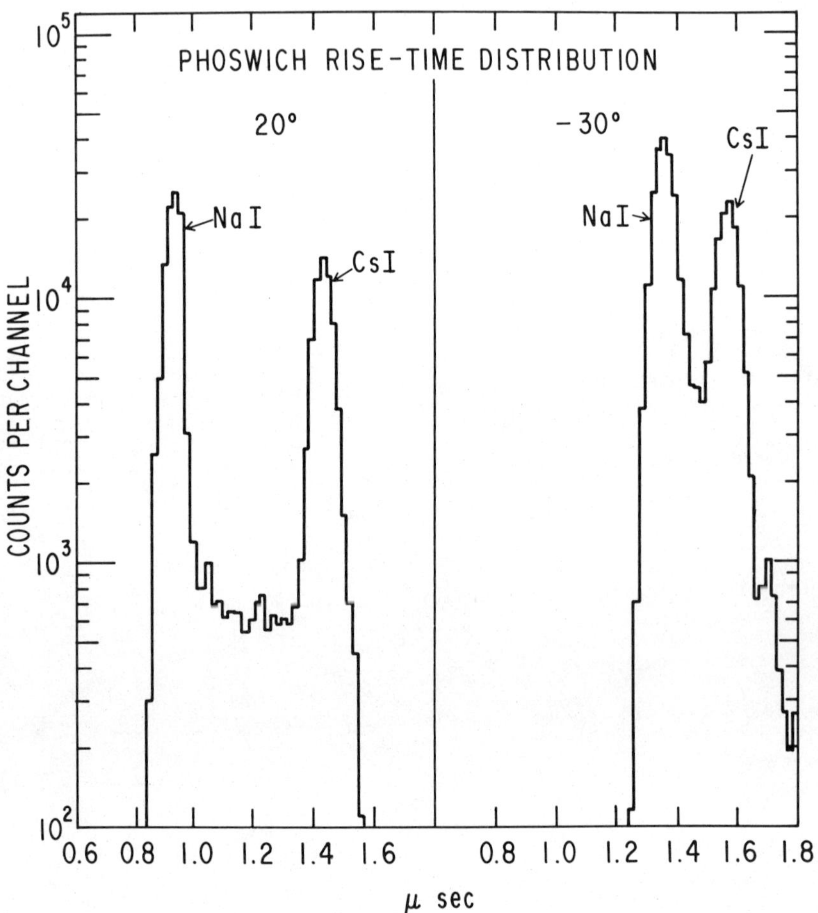

Figure 34 The distribution of zero crossing times for output pulses which have been singly integrated and doubly differentiated from a phoswich detector similar to that in Figure 33. The luminescence decay time of NaI(Tl) increases at lower temperature, resulting in less separation between event classes.

primary crystal. The plastic anticoincidence shield gives a further background reduction of about 30% by detecting cosmic-ray secondary particles entering the passive collimator. A similar device has been developed at MIT (Ricker et al 1975) with eight modules each having 100 cm^2 effective area. This system, although having somewhat less total area than the UCSD device, has been optimized to give state-of-the-art energy resolution for large-area scintillators of better than 20% at 60 keV.

Use of the phoswich technique requires effective discrimination between the 0.25 μsec pulse decay characteristic of NaI(Tl) and the 0.6 μsec or 1.1 μsec decay time of CsI(Na) and CsI(Tl), respectively. Figure 34 shows the distribution of zero-crossing times for shaped output pulses from the 12.0 cm diam. thick NaI(Tl)/CsI(Tl)

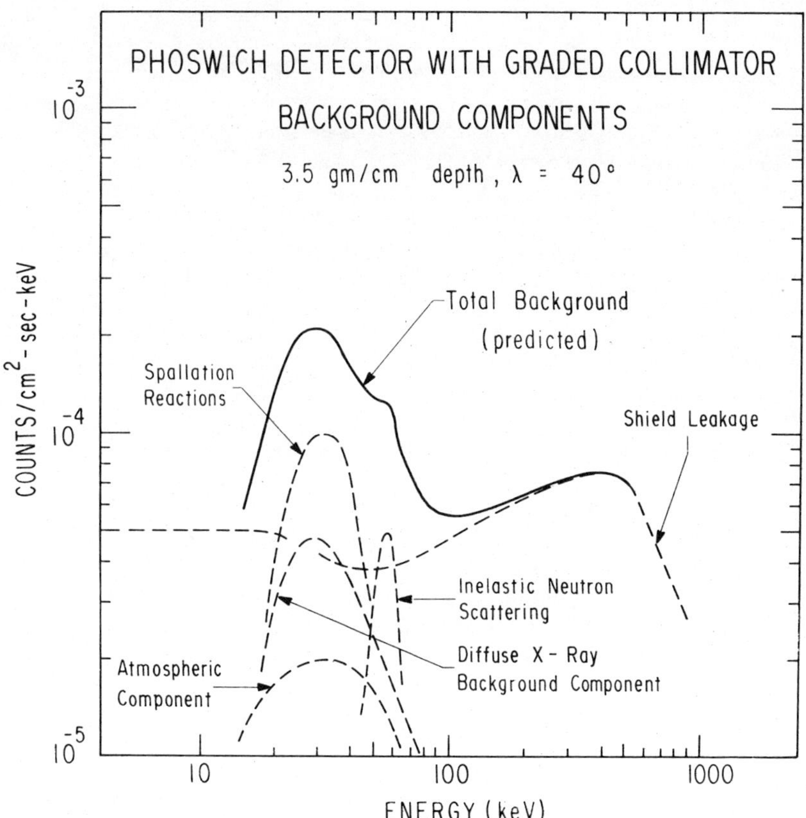

Figure 35 Background components predicted for the detector of Figure 33 in the radiation environment over Palestine, Texas, $\Lambda = 42°$, $h = 40$ km. Diffuse atmospheric and cosmic fluxes entering the aperture are small compared to leakage of higher-energy photons and spallation radioactivity due to cosmic-ray bombardment.

phoswich used by Gruber (1974) and Ling (1974) in studies of the Crab nebula and atmospheric gamma rays. The pulse-shape spectrum shows two peaks corresponding to pure energy losses in the NaI and CsI crystals as well as a continuum of mixed events which result from Compton-scattering interactions. Figure 34 also shows that discrimination is less efficient at low temperatures as the principal fluorescent decay times of NaI(Tl) and CsI(Tl) tend to become similar. Because of this effect, stable temperatures and monitoring of the rise-time spectrum are required to verify phoswich operation.

Figure 35 shows the background components in the radiation environment of 40 km over Palestine, Texas, calculated for the 450 cm^2, $5.8° \times 5.8°$ FWHM detector shown in Figure 33. The shield leakage due to higher-energy gamma rays is computed from the input spectrum shown in Figure 23 using Monte Carlo computational methods applicable to cylindrical geometries developed by Matteson (Peterson et al 1970, Ling 1974). Atmospheric and diffuse fluxes entering the aperture are calculated using the techniques described in a previous chapter. Spallation radioactivity and inelastic scattering from the first excited state of I^{127} at 57 keV are estimated from previous experiments and from the calculations of Fishman (1972) and Dyer & Morfill (1971). The estimated total background is less than 2×10^{-4} counts cm^{-2} sec^{-1} keV^{-1} over the nominal range $15 < E < 150$ keV. Preliminary results indicate the low predicted background is indeed obtained.

To meet the aperture criteria $\Omega/4\pi \leqq \exp(-\mu\bar{t})$ (equation 5.20) above ~300 keV for solid angles <0.3 sr, very massive shields are required. Figure 36 indicates the detection system for UCSD/MIT hard X-ray and gamma-ray instrument on HEAO-A. This combines seven detector modules of three different types with common active anticoincidence shields, all of which can operate in many commandable anticoincidence and data transmission configurations. Figure 37 shows the background components over the nominal energy range of the various detector modes predicted for the 23° inclination, 370 km circular orbit planned for the HEAO-A. The background due to spallation is an equilibrium value, and will vary slightly depending on the radiation history.

Recent advances in space flight photomultipliers and scintillation crystal-growing and packaging techniques have markedly increased the resolution and threshold at low energies. Figure 38 shows the response of the 12.5 cm diam. diffuse-mode phoswich for the UCSD/MIT HEAO-A instrument shown in Figure 36, to broad-beam illuminations of low-energy X rays. The resolution is 58% at 5.85 keV and 33% at 22 keV. Even better performance can be expected from optimum crystal geometry and phototube matching.

Modulation Collimators

Until large-aperture focusing X-ray telescopes with image readout referenced to the celestial sphere are available, the modulation collimator will provide source locations from scanning platforms, and in fact will always be useful in survey work. Accurate location of weak sources, however, requires large areas and a close-spaced grid structure. Figure 39 shows schematically a combined scanning and rotating modulation collimator designed to operate over the 1–15 keV range and scheduled

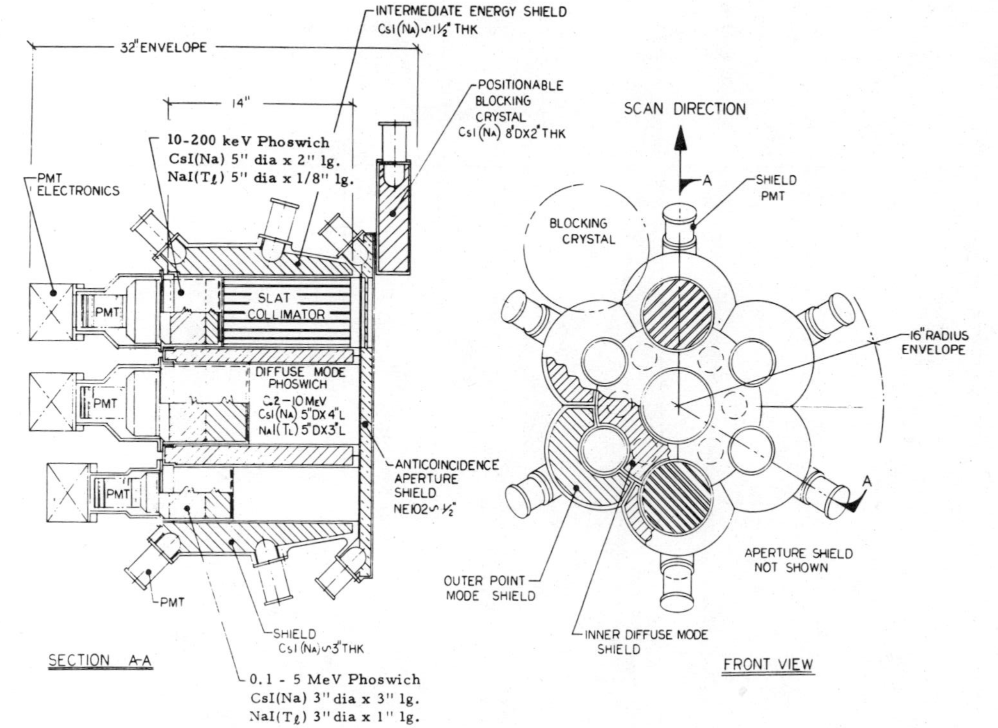

Figure 36 The UCSD/MIT scintillation-counter telescope (hard X-ray and low-energy gamma-ray experiment) to be flown on the HEAO-A in 1977. Three detector configurations, with different apertures and energy ranges, use common CsI(Na) anticoincidence shields. A movable anticoincidence blocking counter permits separation of background components.

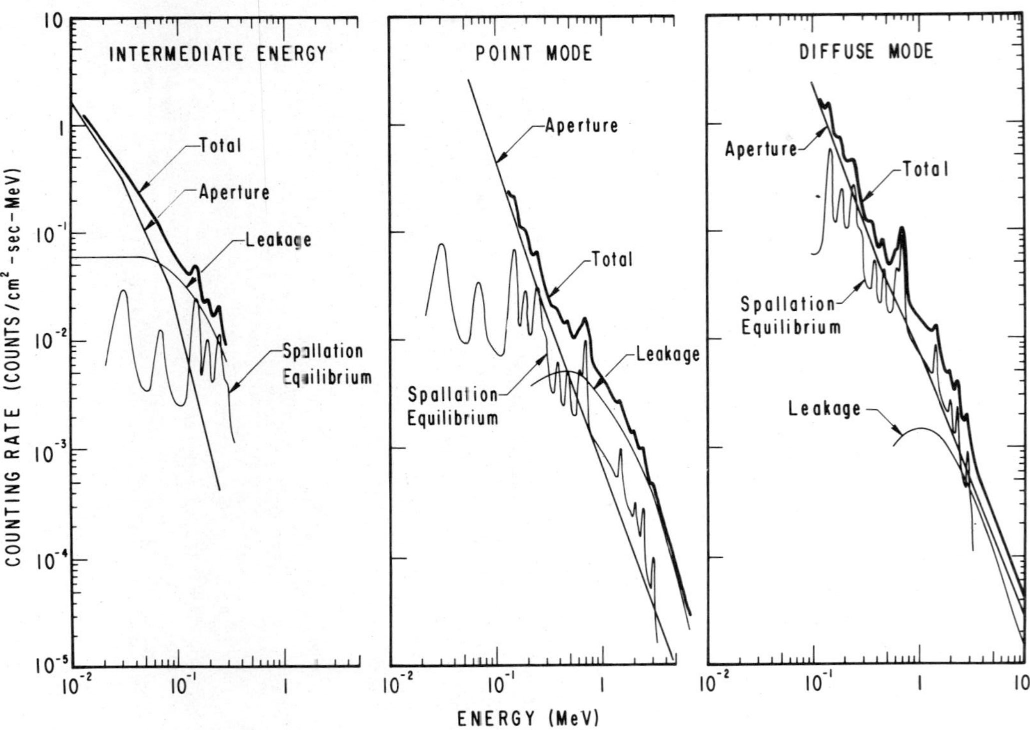

Figure 37 Background components predicted for the UCSD/MIT HEAO-A instrument in the $i = 33°$, $h = 370$ km orbit. The aperture flux is the diffuse cosmic component. The induced radioactivity will vary depending on the radiation history.

for the original HEAO-A mission (Peterson 1972). Although the instrument has been slightly reduced in area, and the rotating feature deleted, this device will still provide the definitive measurements of source sizes and positions in the $\sim 5''$ regime until the late 1970s.

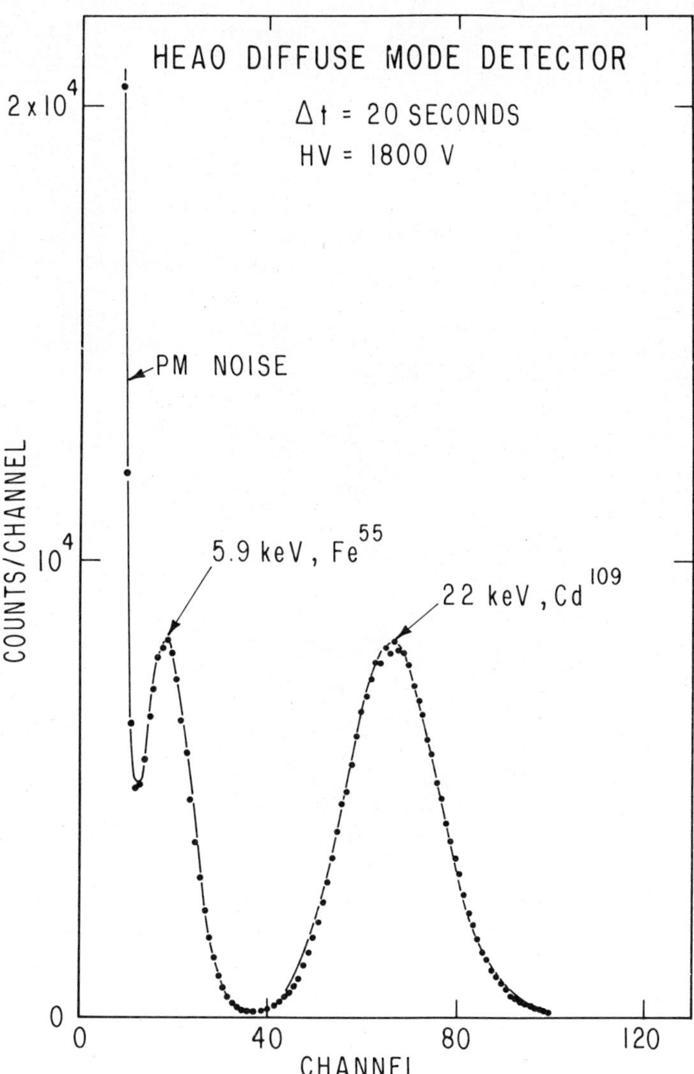

Figure 38 Resolution and noise obtained at low energies with modern photomultipliers and spaceflight packaging techniques for the 12 cm diam, 17.5 cm thick CsI(Na)/NaI(Tl) phoswich configuration in Figure 37.

Oda and his colleagues have constructed a collimator which operates in the 10–60 keV range to be used in conjunction with the detector module of Figure 33. This is intended to study the angular structure of the Crab nebula from a balloon, and consists of four planes each of layered etched Invar plates to form square grids 0.014 mm across, spaced 0.016 mm. The grids are mounted in an Invar structure 1.5 m long, resulting in response bands of 20″ FWHM at a spacing of 1.6′. A similar device has been proposed for an upcoming solar mission to study the angular sizes of the hard X-ray emitting regions in solar flares.

Spaceflight Missions

Rockets and balloons will continue to provide specific studies of the stronger sources during the 1975–1985 decade and inexpensive vehicles for test and development of instrumental concepts. Significant advances, however, will occur only from

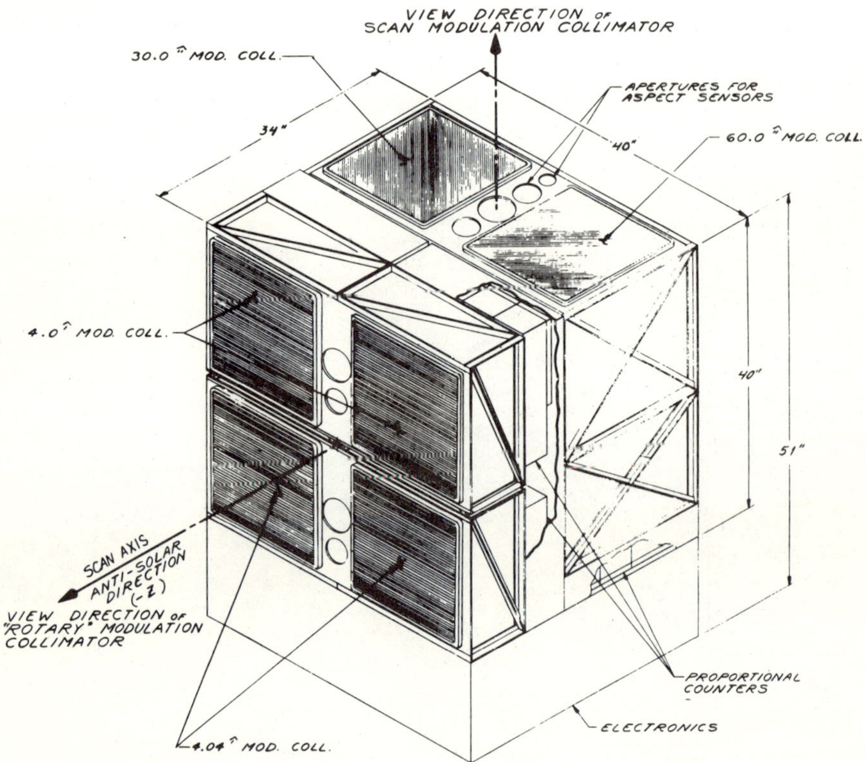

Figure 39 The large-area modulation collimator originally designed for HEAO had both rotating and scanning configurations. The HEAO-A scanning version can locate strong sources to ~5″, has an effective area of ~2200 cm², and operates over the 1–15 keV range. (From Peterson 1972.)

long-term observations that realize the ultimate sensitivity from detection systems, allow observation of many sources, and permit continuous studies of source-intensity fluctuations. Although this opportunity now exists only from orbiting vehicles, the development of large-volume superpressure balloons may ultimately provide an alternative for observations above 20 keV.

SMALL SCIENTIFIC SATELLITES A number of small scientific satellites launched or scheduled during 1974–1975 are substantially devoted to X-ray astronomy. Although the instruments are necessarily of modest area, they are often designated for specific studies, such as mapping of the galaxy in the 0.25–1.5 keV range, long-term observations of certain sources, wide-angle surveys for transient phenomena, fast time resolution, precise location, extension of spectra to higher energies, and high angular or spectroscopic resolution. These instruments often show ingenious use of counter and collimation techniques, vehicle properties, and data-handling methods to attain the objectives. Table 3 summarizes a number of these missions and their instrumentation. Several of the detectors have been discussed here, and many are described elsewhere in the literature.

HIGH ENERGY ASTRONOMICAL OBSERVATORY The next major advance in sensitivity, however, will occur with the High Energy Astronomical Observatory (HEAO). HEAO-A, scheduled for launch in Spring 1977, will scan the sky in the 0.15 keV–10 MeV range with instruments permitting significant improvements over earlier experiments in sensitivity and angular resolution. HEAO-B, scheduled one year later as a pointing mission, will carry grazing-incidence X-ray mirrors and collectors, with Bragg crystal and solid-state spectrometers, imaging devices, and polarimeters at the focus (Giacconi & Gursky 1974). HEAO-C, to be launched two years after HEAO-A, will search for gamma-ray lines in the 50 keV–10 MeV range and study elemental and isotopic abundances of cosmic rays. The present observatories are somewhat reduced from the original program plan (Peterson 1972).

The HEAO-A configuration is shown in Figure 40 and the nonfocusing X- and gamma-ray instrumental complement on HEAO-A and C is described in Table 4. HEAO-A will weigh approximately 3200 kg at launch and will be injected into a 370 km circular orbit with $i = 23°$, and thus will have a relatively low exposure to the trapped particles in the SAA. This HEAO will primarily scan at a 30-min rate, although its ability to point at selected targets for several hours will provide the unique capability of detailed measurements of discrete source phenomena at high sensitivity and with fast time resolution over the entire 0.15 keV–10 MeV range. HEAO-C will be similar to HEAO-A, except that the orbital inclination will be 45° to accommodate the cosmic-ray instruments.

Figure 41 shows a single 2200 cm^2 module of the NRL large-area proportional-counter array. One of these modules is coaligned with a modulation collimator similar to that shown in Figure 39, an array designed for diffuse source studies, and with the scintillation counter complex shown in Figure 36. The sensitivity for weak sources will be on the order of 10^{-6} Sco X-1 near 3 keV, while angular sizes and positions can be determined to $\sim 10''$, and the hard X-ray flux from power-law

Table 3 Small scientific spacecraft with cosmic X-ray instruments, 1974–1975

SPACECRAFT								
Designation	Country and Institution	Launch or Scheduled Launch	Orbit	Weight (kg)	Spin Rate	Objectives	Instrument	Institution
ANS	Netherlands	13 August 1974	Polar 1180 × 270 km	—	—	Stabilized UV X ray	UV telescope, soft X ray, hard X ray, Bragg Xtal spectrometer	HCO/SAO
UK-5	U.K.	15 October 1974	3° inc. 500 km	—	—	X-ray astronomy	Bragg spectrometer, polarimeter	Univ. of Leicester
							Rotating modulation collimator	MSSL
							Source monitor	MSSL
							Hard X-ray telescope	Imperial College
							Sky Survey	LU
							Pinhole camera	MSFC
SAS-C	USA, MIT	Spring 1975	3° inc. 500 km	—	Variable (inertial wheel stabilization)	X-ray astronomy 0.2–50 keV	Rotating modulation collimator	MIT and SAO
							Soft X-ray, concentrator	MIT
							All sky monitors	MIT
							Source monitors, with comparison FOV	MIL
							Sco X-1monitor	MIL
OSO-I	USA	Spring 1975	33° inc. 550 km	1070	6 rpm	Solar physics	X-ray crystal spectrometer	Columbia
							Stellar X-ray polarimeter	Columbia
						Soft and hard X-ray survey	Cosmic X-ray spectrometer	GSFC
							Hard X-ray telescope	GSFC
							Mapping X-ray heliometer	Lockheed
							Soft X-ray telescope	Univ. of Wisconsin
CORSA	Japan	February 1976	31° inc. 600 × 300 km	85	12 rpm	X-ray sources, variability	Survey: soft and very soft	Nagoya
							Variability: hard, soft, and very soft	ISAS/Tokyo
ARYABHATA	India, ISRO	April 21, 1975	51° inc.	360	10–90 rpm	X-ray astronomy	Prop. counter	PRL/ISSP
							Scin. counter telescope	
						Ionosphere	Suprathermal elec. detector	PRL
							UV detector	
						Solar neutrons and gamma rays	Scin. counter telescope	TIFR

Table 3 (*continued*)

INSTRUMENTS					
Detector	Energy Range (keV)	Effective Area (cm^2)	Field of View FWHM	FOV wrt Spin Axis	COMMENTS
Photomultiplier	3.8–8.3 eV				Pointing spacecraft, operating successfully >6 months.
	0.2–4	60	0.25 × 3°		
Prop. counters	2–20			~Perpendicular to sun	Hard X ray limited by geomagnetic particles
	1.8 and 2.0	80[a]	3°		
LiF Prop. counters	2–3.5	90 ea.	7°	Parallel	—
Xe/A/CO_2	3.5–7.7				
Prop. counters	2–20	600 (~100 eff.)	17° FWHM 15° fringe		
Channel multiplier	<2	10			
Prop. counters	2–30	~100	5°	Parallel	
NaI scin. with graded plastic shield	20–1000	12	10°	Parallel	
Prop. counters	2–20	150 ea.	0.7° × 10°	2 at 25°	
4 Be/A/CO_2	1–5			2 at 115°	
2 Position-sensitive prop. counters	3–6	1	90° × 4°	ea. ±45° w.r.t. to spin plane	
Prop. counter with Be window	1.5–10	330 eff.	2.4′ fringe	Parallel	
Prop. counter with filter wheel	0.15~1	40		Perpendicular	
4 Prop. counters 3 Ar 1 Xe	 1.3–13 1.3–15	~75 ea.	1° × 70° and 0.5° × 70°	Perpendicular	
4 Prop. counters 2 1 mil Be window 1 Ti window 1 Xe		~75 ea.	2° cir.	Perpendicular	
1 Prop. counter	1–10	25	12° × 70°	60°	
3 Prop. counters	2–8	2170[a]	3° × 80°	90°	
2 Prop. counters	2–6	800[a]	4° × 4°	180° (antiparallel)	
3 Prop. counters	2–60	244	5° × 5°	0° (parallel)	
	2–60	271	5° × 5°	175°	
	2–20	76	3° × 3°	180°	
CsI(Na)	20–5000	25	5° × 5°	175°	
6 Prop. counters	2–30	185	2′ × 2′	90°	
3 Prop. counters	0.13–35	155	5° × 5°	0°	
	0.13–35	155	5° × 5°	175°	
4 Prop. counters	0.2–20	140 ea.	2–5° × 60°	2 Perpendicular	
	2.0–20		2–10° × 10°	2 Parallel	
1 NaI scin. counter	>20	~20			
Ar + CO_2 filled prop. counter	2.5–15	25	8°	Parallel to spin axis	
NaI	10–100	11	19°	Perpendicular	2 scintillation telescopes, one blocked for instrumental background
Retarded potential analyzer	~3–100 eV	25	75°	Perpendicular to spin axis	
Ne filled, LiF and MgF windows	L_α and 1304 Å oxygen line	0.75	45°	Perpendicular	
CsI(Tl) with plastic anti	0.2–20 MeV gamma rays 10–500 MeV neutrons	26	~π Sr.	Perpendicular	Uses rise-time discriminator to separate neutron recoils from gamma rays

[a] Area of analyzing crystal.

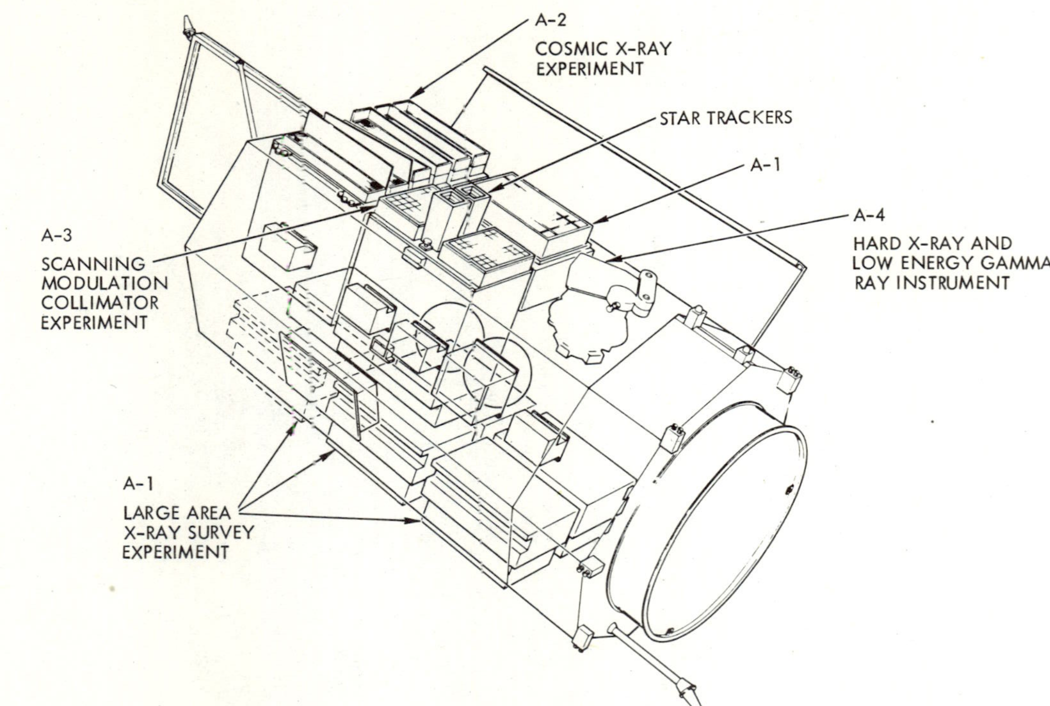

Figure 40 The HEAO-A spacecraft contains a number of instruments providing overlapping coverage in energy and angular resolution from 0.15 keV to 10 MeV. The spacecraft will primarily scan the sky at $\sim 1/30$ rpm, but will also have stopping and pointing capabilities. (From Peterson 1972.)

Table 4 High energy astronomical observatory (nonfocusing X- and gamma-ray instruments)

Mission	Experiment: Name	Experiment: Designation	Experiment: Principal Investigators	Energy Range (keV)	Field of View (°FWHM)	Area (cm^2)	Data Rate (kbps)	Weight (lb)	Power (W)	Detector Type	Primary Objective
HEAO-A	Large-area X-ray survey	A-1	Friedman	0.15–20	1° × 4 1° × ½ 8 × 2	8800 4400 2200	2.1	810	25	Prop. counters Xe/CO_2	Survey sky to limit of 10^{-4} Crab nebula. Measure spectrum of discrete sources
	Cosmic X-ray experiment	A-2	Boldt/Garmire	0.2–3.0	1½ × 3, 3 × 3, 6 × 3	2000	1.2	550	18	Prop. counter Propane	Measure spectrum and isotropy of diffuse X rays
				1.5–20	1½ × 3, 3 × 3	1000				P-10	
				2–60	1½ × 3, 3 × 3, 6 × 3	3000				Xenon/ methane	
	Scanning modulation collimator	A-3	Gursky/Bradt	1.5–15	4 × 4, ½ arc min modulation collimator	450	1.1	331	12.5	Prop. counter	Locate X-ray sources to 5″. Measure structure of extended sources on ½′–16′ scales
					4 × 4, 2 arc min modulation collimator	450				Ar/CO_2	
	Hard X-ray and low-energy gamma-ray experiment	A-4	Peterson/ Lewin	10–200 100–5000 200–10,000	1° × 20° 20° 40°	220 170 120	1.0	700	24.7	Scin. counters [NaI(Tl) detectors CsI(Na) shields]	Extend spectra of stronger point sources to ~1 MeV. Measure spectrum and isotropy of diffuse X rays and gamma rays
HEAO-C	High spectral-resolution gamma-ray spectrometer	C-1	Jacobson	50–10,000	30°	64	2.0	460	20	4 Ge(Li) detectors CsI(Na) shields	Search for gamma-ray lines in discrete sources and the diffuse component

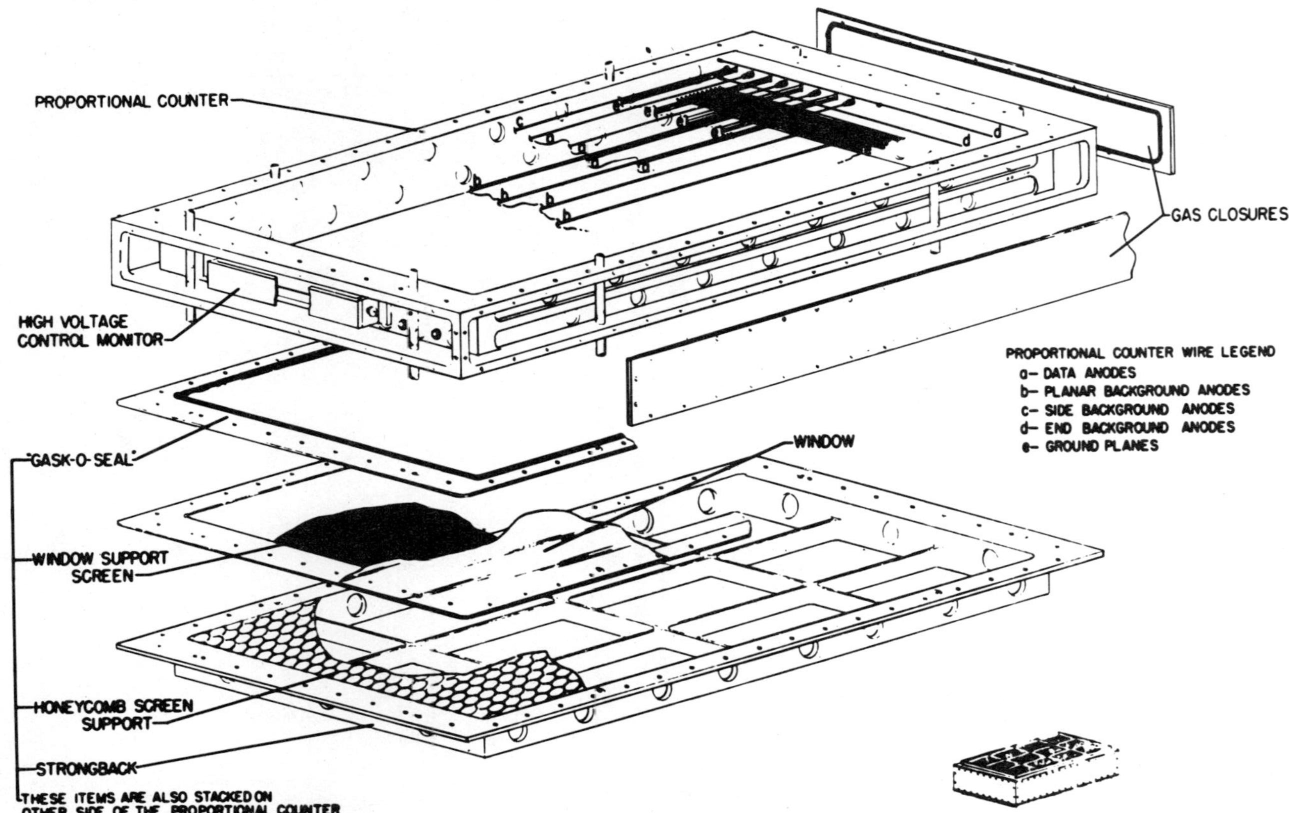

Figure 41 A single 2200 cm^2 module of the large-area proportional-counter array being developed by NRL for HEAO-A. Seven such modules, operating over the 0.15–20 keV range having various apertures and readout modes, comprise the array. (From Peterson 1972.)

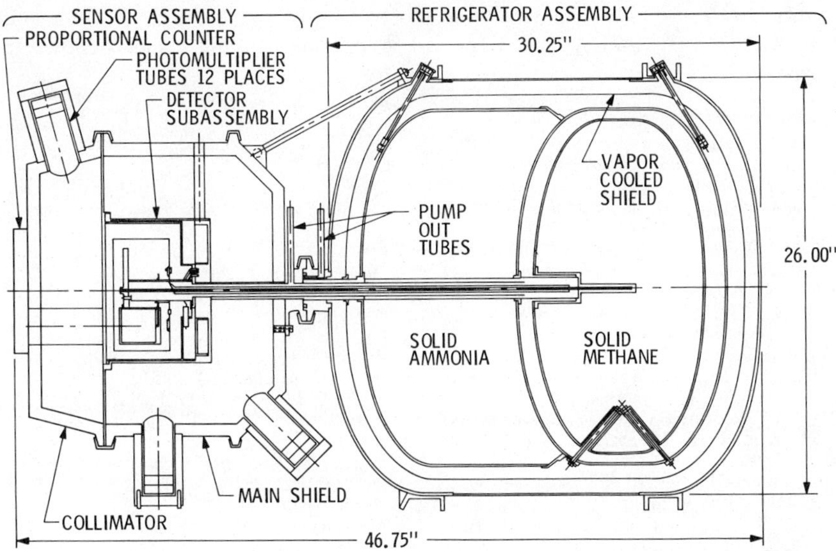

Figure 42 A high-resolution, high-sensitivity spectrometer to search for cosmic gamma-ray lines in the 50 keV–5 MeV range on the HEAO-C consists of four 60 cm^3 Ge(Li) detectors with an anticoincidence collimator and shield of CsI. This can detect fluxes as weak as $\sim 10^{-5}$ photons cm^{-2} sec^{-1}. The passive cryogenic system can maintain the detectors at $\sim -50°C$ for one year. (Courtesy Dr. A. S. Jacobson.)

sources such as the Crab can be determined to ~ 1 MeV, for sources as weak as $\sim 1/20$ Crab. The gamma-ray spectrometer on the HEAO-C, shown in Figure 42 (Hicks & Jacobson 1974), has four 60 cm^3 Ge(Li) detectors in a CsI(Tl) shield and at the design resolution of 2.5 keV near 1 MeV can detect discrete gamma-ray lines as weak as 10^{-5} photons cm^{-2} sec^{-1}. The two-stage refrigerator contains enough solid cryogen for a one-year lifetime.

The HEAO-B, with its complement of detectors at the focus of grazing-incidence mirrors and collectors, will complement the survey-type instruments on HEAO-A and C and provide high-resolution studies of specific sources in energy, time, and space in the region below ~ 3 keV. Gursky & Schwartz (1974) have described the HEAO-B instrumentation.

ACKNOWLEDGMENTS

The author is grateful to many colleagues at UCSD, TIFR, and elsewhere who have contributed materials or discussions for this review, and apologizes for particular detection schemes or ingenious techniques not included. Drs. W. A. Baity and D. E. Gruber performed a monumental task during the production, editing, and checking of the final draft and provided many calculations and ideas. Instrumental developments at UCSD have been accomplished under NASA contracts NAS-5-3177, NsG-318, NGL 05-005-003, NAS-5-11080 and 11081, and NAS-8-27974. The author was supported in part by a Guggenheim Fellowship during 1973–1974.

Literature Cited

Abramowitz, M., Stegun, I. A., Eds. 1965. *Handbook of Mathematical Functions*, 227–52. Dover: New York

Aitken, D. W., Beron, B. L., Yenicay, G., Zulliger, H. R. 1967. *IEEE Trans. Nucl. Sci. NS-14*, 468

Alvarez, L. W. et al 1973. *Berkeley Space Sci. Lab. Ser. 14, Issue 17*

Apparao, K. M. V. 1973. *Ap. Space Sci.* 25:3

Bahcall, J., Joss, P., Avni, Y. 1974. *Ap. J.* 191:211

Baity, W. A., Jones, T. W., Wheaton, W. A., Peterson, L. E. 1975a. *Ap. J. Lett.* 199: In press

Baity, W. A., Ulmer, M. P., Peterson, L. E. 1975b. *Ap. J.* In press

Baity, W. A., Peterson, L. E., Wheaton, W. A. 1970. *Univ. Calif., San Diego, UCSD SP-70-05*

Ball Brothers Research Corporation. 1971. *TR71-07*, ed. L. Wright. Boulder, Colo.: BBRC

Becklin, E. et al 1973. *Nature* 245:302

Birks, J. B. 1964. *The Theory and Practice of Scintillation Counting.* New York: Macmillan

Bleach, R. D., Boldt, E. A., Holt, S. S., Schwartz, D. A., Serlemitsos, P. J. 1972. *Ap. J.* 171:51

Bleeker, J. A. M. 1971. PhD thesis. Univ. Leiden

Bleeker, J. A. M., Deerenberg, A. J. 1970. *Ap. J.* 159:215

Blumenthal, G. R., Gould, R. J. 1970. *Rev. Mod. Phys.* 42:237

Borken, R., Doxsey, R., Rappaport, S. 1972. *Ap. J. Lett.* 178:115

Bowles, J. A., Patrick, T. J., Sheather, P. H., Eiband, A. M. 1974. *J. Phys. E., Sci. Instrum.* 7:183

Bradt, H., Garmire, G., Oda, M., Spada, G., Sreekantan, B., Gorenstein, P. 1968. *Space Sci. Rev.* 8:471

Bradt, H., Giaconni, R., Eds. 1973. *X- and Gamma-Ray Astronomy. IAU Symp. No. 55.* Dordrecht: Reidel

Bunner, A. N., Coleman, P. L., Kraushaar, W. L., McCammon, D. 1971. *Ap. J. Lett.* 167:L3

Bunner, A. N., Coleman, P. L., Kraushaar, W. L., McCammon, D. 1972. *Ap. J. Lett.* 172:L67

Bunner, A. N., Kraushaar, W. L., McCammon, D., Vanderhill, M., Williamson, F. 1973. *Rev. Sci. Instrum.* 44:418

Catura, R., Acton, L., Johnson, H., Zaumen, W. 1974. *Ap. J.* 190:521

Charles, M. W., Cooke, B. A. 1968. *Nucl. Instrum. Methods* 61:31

Charpak, G., Bouclier, R., Bressoni, T., Favier, J., Zupancic, C. 1968. *Nucl. Instrum. Methods* 62:235

Chupp, E. L., Forrest, D. J., Sarkady, A. A., Lavakare, P. J. 1970. *Planet. Space Sci.* 18:939

Clark, G. W. et al 1973. *Ap. J.* 179:263

Clark, G. W., Lewin, W. H. G., Smith, W. B. 1968. *Ap. J.* 151:21

Coleman, P. et al 1973. *Ap. J. Lett.* 185: L121

Condon, J. J. 1974. *Ap. J.* 188:279

Crannell, C. J., Kurz, R. J., Viehmann, W. 1974. *Nucl. Instrum. Methods* 115:253

Daniel, R. R., Joseph, G., Lavakare, P. J. 1972. *Ap. Space Sci.* 18:462–67

Daniel, R. R., Stephens, S. A. 1974. *Rev. Geophys. Space Phys.* 12:233

Datlowe, D. W. 1975. *Space Sci. Instrum.* Submitted for publication

de Korte, P. A. J., Bleeker, J. A. M., Deerenberg, A. J. M., Tanaka, Y., Yamashita, K. 1974. *Ap. J.* 190:L5

Dennis, B. R., Suri, A. N., Frost, K. J. 1973. *Ap. J.* 186:97

Dolan, J. F. 1972. *Ap. Space Sci.* 17:472

Dyer, C. 1973. *Gamma-Ray Astrophysics. NASA SP-339*, ed. F. W. Stecker, J. I. Trombka, 83

Dyer, C., Morfill, G. 1971. *Ap. Space Sci.* 14:243

Evans, R. D. 1955. *The Atomic Nucleus.* New York: McGraw-Hill

Evans, W., Belian, R., Conner, J. 1970. *Ap. J. Lett.* 159:L57

Fabian, A., Zarnecki, J., Culhane, J. 1973. *Nature Phys. Sci.* 242:18

Fabian, A. et al 1974. *Ap. J. Lett.* 189:L59

Fano, U. 1947. *Phys. Rev.* 72:26

Felten, J. 1973. In *X- and Gamma-Ray Astronomy. IAU Symp. No. 55*, 258

Fichtel, C., Kniffen, D., Hartman, R. 1973. *Ap. J. Lett.* 186:L99

Fillius, R. 1963. *Satellite instruments using solid state detectors. Res. Rep. SUI 63-26.* PhD thesis. State Univ. Iowa, Iowa City

Fisher, P. C., Meyerott, A. 1966. *IEEE Trans. Nucl. Sci. NS-13,* 580
Fishman, P. C. 1972. *Ap. J.* 171:163
Friedman, H. 1973. *Science* 181:395
Friedman, H., Fritz, G., Shulman, S. D., Henry, R. C. 1973. In *X- and Gamma-Ray Astronomy. IAU Symp. No. 55,* 215
Frost, K. J., Rothe, E. D., Peterson, L. E. 1966. *J. Geophys. Res.* 71:4079
Giacconi, R. 1971. In *New Techniques in Space Astronomy. IAU Symp. No. 41,* 104
Giacconi, R., Gursky, H., Eds. 1974. *X-Ray Astronomy.* Dordrecht: Reidel
Giacconi, R., Gursky, H., Van Speybroeck, L. P. 1968. *Ann. Rev. Astron. Ap.* 6:373
Giacconi, R., Kellogg, E., Gorenstein, P., Gursky, H., Tananbaum, H. 1971. *Ap. J. Lett.* 165:L27
Giacconi, R. et al 1974. *Ap. J. Suppl.* 237
Giacconi, R., Reidy, W. P., Vaiana, G. S., Van Speybroeck, L. P., Zehnpfennig, T. F. 1969. *Space Sci. Rev.* 9:3
Gorenstein, P., Gursky, H., Garmire, G. 1968. *Ap. J.* 153:885
Gorenstein, P. et al 1971a. *Science* 172:369
Gorenstein, P., Harris, B., Gursky, H., Giacconi, R. 1971b. In *IAU Symp. No. 41,* 183
Gorenstein, P., Mickiewicz, S. 1968. *Rev. Sci. Instrum.* 39:816
Goulding, F. S. 1972. *Nucl. Instrum. Methods* 100:493
Grader, R. J., Hill, R. W., McGoff, C. W., Salmi, D. S., Stoering, J. P. 1971. *Rev. Sci. Instrum.* 42:465
Gratton, L., Ed. 1970. *Non-Solar X- and Gamma-Ray Astronomy. IAU Symp. No. 37*
Greisen, K. 1971. *The Physics of Cosmic X-Ray, Gamma-Ray and Particle Sources.* New York: Gordon & Breach
Griffiths, R. E., Peacock, A., Davison, P. J. N., Rosenberg, F., Smart, N. C. 1974. *Nature* 250:471
Gruber, D. E. 1974. PhD thesis. Univ. Calif., San Diego, *UCSD SP-74-05*
Gursky, H. 1973. *Black Holes,* ed. C. Dewitt, B. S. Dewitt, 291. New York: Gordon & Breach
Gursky, H., Schreier, E. 1974. *SAO Preprint.* Presented at IAU Symp. No. 67, Moscow
Gursky, R., Schwartz, D. 1974. In *X-Ray Astronomy,* ed. R. Giacconi, H. Gursky, 25–98. Dordrecht: Reidel
Gursky, H. et al 1972. *Ap. J. Lett.* 173:L99
Harri, J., McGee, M., Toor, A. 1969. *Rev. Sci. Instrum.* 40:703
Harshaw Chemical Co. 1969. *Scintillation Phosphors.* Cleveland: Harshaw Chemical Co. 2d ed.
Hayakawa, S. 1973. In *IAU Symp. No. 55,* 235
Haymes, R. C., Glenn, S. W., Fishman, G. J., Harnden, F. R. 1969. *J. Geophys. Res.* 74:5792
Henke, B. L., Elgin, R. L. 1970. *X-Ray Absorption Tables for the 2-to-200 A Region, Advances in X-Ray Analysis, Vol. 13.* New York: Plenum
Hicks, D. B., Jacobson, A. S. 1974. *IEEE* NS-21:169
Hill, R. W., Grader, R., Seward, F. D. 1970. *J. Geophys. Res.* 75:7276
Hjellming, R. M. 1973. *Science* 182:1089
Holt, S. S. 1970. *Introduction to Experimental Techniques of High-Energy Astrophysics, NASA SP-243,* ed. H. Ogelman, J. R. Wayland, Chap. 2
Holt, S. S., ed. 1974. *Proc. GSFC Workshop Electron Contam. X-Ray Astron. Exp. GSFC X-661-74-130*
Holt, S. S. et al 1974a. *Ap. J. Lett.* 188: L97
Holt, S. S. et al 1974b. *Ap. J. Lett.* 190: L109
Hubbell, J. H. 1969. *Photon Cross Sections, Attenuation Coefficients, and Energy Absorption Coefficients from 10 keV to 100 GeV, National Bureau of Standards, NBS 29*
Imhof, W. L., Nakano, G. H., Johnson, R. G., Reagan, J. B. 1974. *J. Geophys. Res.* 79:565
International Council of Scientific Unions. 1972. *Cospar International Reference Atmosphere, 1972.* Berlin: Akademie-Verlag
Iyengar, V. S., Naranan, S., Sreekantan, B. V. 1975. *Ap. Space Sci.* 32:431
Jackson, J. D. 1962. *Classical Electrodynamics.* New York: Wiley
Jacobson, A. S. 1968. PhD thesis. Univ. Calif., San Diego, *UCSD SP 68-2*
Jagada, N., Austin, G., Mickiewicz, S., Goddard, R. 1975. *Ap. J.* In press
Jerde, R. L., Peterson, L. E., Stein, W. 1967. *Rev. Sci. Instrum.* 38:1387
Johnson, W. N., Haymes, R. C. 1973. *Ap. J.* 184:103
Jones, C., Giacconi, R., Forman, W., Tananbaum, H. 1974. *Ap. J. Lett.* 191:L71
Kasturirangan, K. 1971. *J. Geophys. Res.* 76: 3527
Kasturirangan, K., Rao, U. R., Bhavsar, P. D. 1972. *Planet. Space Sci.* 20:1961
Kellogg, E. 1973. In *X- and Gamma-Ray Astronomy. IAU Symp. No. 55,* 171
Kellogg, E. M. 1974. *X-Ray Astronomy,* ed. R. Giacconi, H. Gursky, 321–57. Dordrecht: Reidel
Kellogg, E., Baldwin, J. R., Koch, D. 1975.

Ap. J. In press
Kellogg, E., Murray, S., Giacconi, R., Rananbaum, H., Gursky, H. 1973. *Ap. J. Lett.* 185:L13
Krzeminski, W. 1974. *IAU Circ.*, 2612
Labuhn, F., Lust, R., Eds. 1971. *New Techniques in Astronomy. IAU Symp. No. 41.* Dordrecht: Reidel
Lacy, J. L., Lindsey, R. S. 1973. *High Resolution Readout of Multiwire Proportional Counters Using the Cathode Coupled Delay Line Technique.* 1973-778-208/489. Washington DC: GPO
Lampton, M., Margon, B., Bowyer, S., Mahoney, W., Anderson, K. 1972. *Ap. J. Lett.* 171:L45
Laros, J., Matteson, J., Pelling, R. 1973. *Nature Phys. Sci.* 246:109
Lavine, J. P., Vette, J. I. 1970. *NASA SP-3024*, Vol. 6
Lewin, W. H. G., Ricker, G. R., McClintock, J. E. 1971. *Ap. J.* 169:L17
Liden, K., Starfelt, N. 1954. *Ark. Fys.* 7:427
Ling, J. C. 1974. PhD thesis. Univ. Calif., San Diego, *UCSD SP-74-06*
Ling, J. C. 1975. *J. Geophys. Res.* In press
Loh, E. D., Garmire, G. P. 1971. *Ap. J.* 166:301
MacKay, H. B., Ulmer, M. P., Peterson, L. E. 1975. To be published
Macklin, R. L. 1969. *ORNL-TM-2675.* Oak Ridge, Tenn.: Oak Ridge Nat. Lab.
Manchanda, R. K. et al 1972. *Ap. Space Sci.* 15:272
Margon, B. 1974. *Nature* 249:24
Margon, B. et al 1974. *Ap. J. Lett.* 191:L117
Margon, B., Lampton, M., Bowyer, S., Cruddace, R. 1975. *Ap. J.* 197:25
Matteson, J. L. 1971. PhD thesis. Univ. Calif., San Diego, La Jolla
McCammon, D., Bunner, A., Coleman, P., Kraushaar, W. 1971. *Ap. J. Lett.* 168:33
McClintock, J. E. 1974. Paper presented at Int. Conf. on X Rays in Space, Univ. Calgary, Calgary, Canada
McIlwain, C. E. 1966. *Radiation Trapped in the Earth's Magnetic Field,* ed. B. M. McCormac, 45–61. Dordrecht: Reidel
Menefee, J., Cho, Y., Swinehart, C. 1967. *IEEE Trans. Nucl. Sci. NS-14*, 464
Meyer, P. 1969. *Ann. Rev. Astron. Ap.* 7:1
Metzger, A. E., Anderson, E. C., Van Dilla, M. A., Arnold, J. R. 1964. *Nature* 204:766
Nakano, G. H., Imhof, W. L., Johnson, R. G. 1973a. *IEEE Trans. Nucl. Sci. NS-21*, 159
Nakano, G. H., Imhof, W. L., Reagan, J., Johnson, R. G. 1973b. *Gamma-Ray Astrophysics, NASA SP-339*, ed. F. W. Stecker, J. I. Trombka, 71
Nelms, A. T. 1953. *NBS Circ.*, 542
Nelms, A. T. 1956. *Energy Loss and Range of Electrons and Positrons. NBS Circ.*, 577
Nelms, A. T. 1958. *NBS Circ.*, 577 (Suppl.)
Novick, R. 1973. In *X- and Gamma-Ray Astronomy. IAU Symp. No. 55*, 118
Oda, M. 1971. *IAU Symp. No. 41*, 89
Oda, M. et al 1972. *Ap. J. Lett.* 172:L13
Ögelman, H., Wayland, J. R., eds. 1970. *Introduction to Experimental Techniques of High-Energy Astrophysics, NASA SP-243*, Chaps. 2, 3
O'Mongain, E. 1973. *Nature* 241:376
ORTEC. 1970. *Instruments for Research Catalog*, 1002. Oak Ridge, Tenn.: ORTEC
Ostriker, J., Davidson, K. 1973. In *X- and Gamma-Ray Astronomy. IAU Symp. No. 55*, 143
Overbeck, J., Tananbaum, H. 1968. *Phys. Rev. Lett.* 20:24
Pal, Y. 1973. In *X- and Gamma-Ray Astronomy. IAU Symp. No. 55*, 279
Pelling, R. M. 1973. *Ap. J.* 185:327
Pelling, R. M. 1974. *Telescopy Systems for Balloon-Borne Research, NASA TM X-62-397*, 176
Peterson, L. E. 1965. *J. Geophys. Res.* 70:1762–65
Peterson, L. E. 1967. *Gamma-Ray Production by Cosmic Rays Observed on OSO-I. UCSD SP-68-01*
Peterson, L. E. 1970. In *IAU Symp. No. 37*, ed. L. Gratton. Dordrecht: Reidel
Peterson, L. E. 1972. *Sci. Technol.* 28:213
Peterson, L. E. 1973. In *X- and Gamma-Ray Astronomy. IAU Symp. No. 55*, 51
Peterson, L. E. 1974. *Bull. Astron. Soc. India* 2:51
Peterson, L. E., Gruber, D. E., Matteson, J. L., Vette, J. I. 1970. *ERS-18, Data Reduction and Analysis, UCSD SP-70-03*, Vol. 14
Peterson, L. E., Jacobson, A. S. 1970. *Publ. Astron. Soc. Pac.* 82:412
Peterson, L. E., Jerde, R. L., Jacobson, A. S. 1967. *AIAA J.* 5:1921–27
Peterson, L. E., Pelling, R. M., Matteson, J. L. 1972. *Space Sci. Rev.* 13:320
Peterson, L. E., Schwartz, D. A., Ling, J. C. 1973a. *J. Geophys. Res.* 78:7942
Peterson, L. E. et al 1973b. *Gamma-Ray Astrophysics, NASA SP-339*, 41
Pounds, K. 1973. In *X- and Gamma-Ray Astronomy. IAU Symp. No. 55*, 105
Princeton Gamma-Tech. Inc. 1970. *Ge(Li) Handbook*, Princeton: Princeton Gamma-Tech. 3d ed.
Radeka, V. 1972. *IEEE Trans. Nucl. Sci. NS-19*, 412
Rappaport, S., Doxsey, R., Solinger, A., Borken, R. 1974. *Ap. J.* 194:329

RCA Photomultiplier Manual. 1970. *Technical Series PT-61.* Harrison, NJ: RCA/Electronic Components
Ricker, G. et al 1974. *IAU Circ.,* 2704
Ricker, G. et al 1975. *Nature.* In press
Rossi, B. B. 1952. *High-Energy Particles,* Chap. 2. New York: Prentice-Hall
Rossi, B. B., Staub, H. H. 1949. *Ionization Chambers and Counters.* New York: McGraw-Hill
Rothschild, R. E., Boldt, E. A., Holt, S. S., Serlemitsos, P. J. 1974. *Ap. J. Lett.* 189: L13
Schnopper, H. W., Thompson, R. I., Watt, S. 1968. *Space Sci. Rev.* 8:534
Schreier, E. et al 1972. *Ap. J. Lett.* 172:L79
Schwartz, D. 1970. *Ap. J.* 162:439
Schwartz, D., Gursky, H. 1973. *Gamma-Ray Astrophysics, NASA SP-339,* 15
Schwartz, D., Gursky, H. 1974. *X-Ray Astronomy,* ed. R. Giacconi, H. Gursky, 359–88. Dordrecht: Reidel
Schwartz, D., Peterson, L. E. 1974. *Ap. J.* 190:297
Serlemitsos, P. 1971. *IAU Symp. No. 41,* 213
Serlemitsos, P., Boldt, E., Holt, S. S., Rothschild, R. E. 1974. Presented at Meet. AAS, 144th, Gainesville, Fla.
Seward, F. et al 1971. *The SXRE, a Large-Area, Thor-Carried Stellar X-Ray Experiment, UCID-15883*
Sharpe, J. 1954. *Nuclear Radiation Detectors.* London: Methuen. 2d ed.
Silk, J. 1973. *Ann. Rev. Astron. Ap.* 11:269
Stassinopoulos, E. G. 1970. *World Maps of Constant B.L. and Flux Contours, NASA SP-3054*
Stevens, J., Garmire, G. 1973. *Ap. J. Lett.* 180:L19
Stevens, J., Riegler, G., Garmire, G. 1973. *Ap. J.* 183:61
Strong, I., Klebesadel, R., Olson, R. 1974. *Ap. J. Lett.* 188:L1
Tananbaum, H. 1973. In *X- and Gamma-Ray Astronomy. IAU Symp. No. 55,* 9
Teague, M. J., Vette, J. I. 1974. *NASA Publ. NSSDC 74-03*
Trombka, J. I. 1970. *Nature* 226:827
Trombka, J. I. et al 1973. *Ap. J.* 181:737
Ulmer, M. P. 1975. *Ap. J.* In press
Ulmer, M., Baity, W., Wheaton, W., Peterson, L. E. 1973. *Ap. J. Lett.* 184: L117
Vaiana, G. S. 1973. *Solar Phys.* 32:81
Vaiana, G. et al 1973. *Ap. J. Lett.* 185:L47
Vidal, N. et al 1974. *Ap. J. Lett.* 191:L23
Wapstra, A. H., Nijgh, G. J., Van Lieshaut, R. 1959. *Nuclear Spectroscopy Tables.* Amsterdam: North-Holland
White-Grodstein, G. 1957. *NBS Circ.,* 583
Womack, E. A. Jr. 1969. PhD thesis. MIT, Cambridge, Mass.

ON THE PULSAR EMISSION MECHANISMS

×2087

V. L. Ginzburg
P. N. Lebedev Physical Institute, Academy of Sciences of the USSR, Moscow, USSR

V. V. Zheleznyakov
Radio-Physical Institute, Gorkii, USSR

INTRODUCTION

The theory of pulsars is primarily concerned with 1. the type and structure of the central body (star); 2. the models and dynamics of the star magnetosphere; and 3. the pulsar-emission mechanism.

Immediately after the discovery of pulsars (Hewish et al 1968) one could not foresee which of these problems would be the most complicated and which the first to be solved. The choice of the type of star and pulsation mechanism was in fact the first problem to be clarified most fully. Pulsar periods in the Crab and in the Vela (PSR 0532 and PSR 0833) are such that they can be associated only with a neutron star rotation period. Therefore the models of pulsating and rotating white dwarfs and pulsating neutron stars have been readily excluded. Logically, it is possible that pulsars corresponding to such models do exist; for lack of any evidence, however, all pulsars are currently regarded as rotating neutron stars. We proceed from this supposition.

The problem of the structure and dynamics of the rotating neutron star magnetosphere is far from solved, particularly when the axis of rotation does not coincide with that of magnetic symmetry (e.g. with the direction of the magnetic moment). Some models of the magnetosphere are analyzable, however; in these cases a rather definite picture of magnetosphere processes appears.

The foregoing cannot generally be said of pulsar-emission (and particularly radio-emission) mechanisms. Neither the nature of pulsar-emission beaming nor a concrete emission mechanism has yet been established. This situation is paradoxical: although we observe and study the pulsar emission, the *nature* of this emission is less known than the invisible structure of the neutron star itself.

Under these conditions we can only review what is known about pulsar-emission mechanisms and make some remarks. We do not touch upon X-ray pulsars, which are components of binary stellar systems (Cen X-3, Her X-1). For further exposition some information about the pulsar magnetosphere is presented in Section 1. [For more details see Ruderman (1972); for general reviews about pulsars see Hewish (1971), Ginzburg (1971), Smith (1972), ter Haar (1972).]

1 THE PULSAR MAGNETOSPHERE

Rotation of a massive star provides an observed high stability of pulse periods. At the same time a neutron star is the only equilibrium configuration of a stellar type that is stable when rotating with all observed pulsar periods (from 3×10^{-2} to 5 sec). In the absence of accretion, which is especially important in binary systems, a rotating star can be a pulsar only if it is magnetized. Polarization observations in particular give evidence for the presence of a magnetic field of pulsars; the expected field strength on the neutron star surface amounts to 10^{10}–10^{13} Oe (the most frequently adopted value is $H \sim 10^{12}$ Oe). Such an estimate readily follows from the considerations based on magnetic flux conservation for a typical star with a high electrical conductivity when contracted down to the dimensions of a neutron star (Ginzburg 1964, Woltjer 1964).

Because the radius r_0 of a neutron star is only about 10 km and its mass is comparable to the solar mass, acceleration due to gravity on the surface reaches values $g \sim 10^{14}$ cm sec^{-2}. Under these conditions the scale height of the atmosphere is $h = \kappa T/m_H g \sim 10^6\ T$. If, for example, the kinetic temperature of the neutron star surface $T \sim 10^6$ °K, $h \sim 1$ cm. Generation of the whole observable pulsar emission is unlikely to take place in a layer of such thickness. In fact, however, there is no difficulty here because the particle density near the star surface is determined not by the gravity and temperature but by electromagnetic forces. In particular, in creating the magnetosphere of a rotating neutron star with a strong magnetic

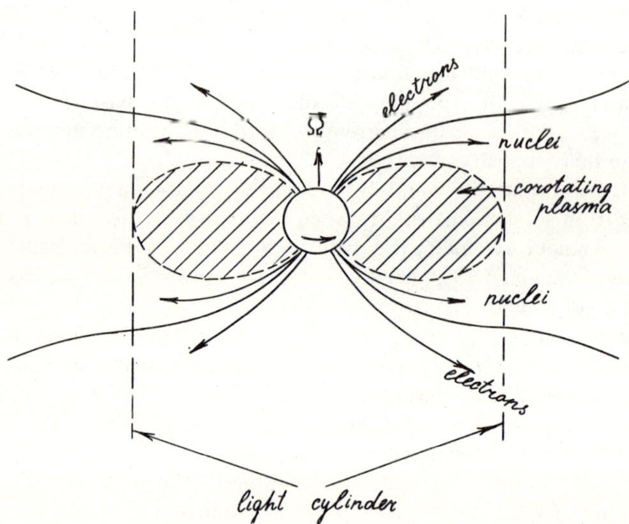

Figure 1 Model of a rotating neutron star magnetosphere.

field a decisive role can be played by an electric field of unipolar induction (Goldreich & Julian 1969).

Let the magnetic field outside a neutron star be the field of a magnetic dipole **m**, and assume that the highly conducting star rotates so that the vector **m** is parallel or antiparallel to $\mathbf{\Omega}$ (the angular velocity). Then in the vacuum around the star there arises a quadrupole electric field **E**. Near the star surface $E_0 \sim (\Omega r_0/c)H_0$, where H_0 is the magnetic field on the star surface and c is the velocity of light. For the pulsar in the Crab $\mathbf{\Omega} \approx 2 \times 10^2$ sec^{-1}, and setting $r_0 = 10^6$ cm, and $H_0 = 10^{12}$ Oe, we obtain a field $E_0 \sim 10^{10}$ CGSE. The force acting on an electron in such a field is 10^{13} times as large as that of gravity; the projection of the field **E** on **H** outside the star is nonzero. It is clear that under these conditions it is the field **E**, not gravity, that is responsible for the character of the particle motion around the star. If the vectors **m** and $\mathbf{\Omega}$ are parallel, the field **E** is directed toward the star surface in the neighborhood of the poles, and electrons will escape from this region. At lower latitudes protons and nuclei escape from the star. These electron and nuclear streams bear a charge capable of compensating considerably for the unipolar induction field **E** when the particle concentration in the streams acquires values of $N \sim \Omega H_0/4\pi ec \sim 10^{12}$ particle cm^{-3}. It follows that around a neutron star the appearance of a plasma envelope with $N \sim 10^{12}$ electrons cm^{-3} may be expected provided the outflow from the star surface is free. The latter is far from evident and depends on the concrete structure of the star surface. Thus Ginzburg & Usov (1972) have shown that under certain conditions there will be no emission of nuclei from the surface of a neutron star. Then the magnetosphere can arise from accretion of the interstellar plasma onto the star, that is, from external sources, or electron-positron pair creation (Sturrock 1971, Ruderman & Sutherland 1975).

Suppose, however, that in the vicinity of a star there exists an envelope (magnetosphere) with $N \sim 10^{12}$ particle cm^{-3}. A question arises about the density distribution of these particles and the magnetic field configuration in the magnetosphere. There is no definite answer to this question because a self-consistent solution of the problem of motion and distribution of plasma and fields in the vicinity of a rotating ball with a dipole magnetic moment has not yet been found.

The model of Goldreich & Julian (1969) (Figure 1) suggests that in the equatorial regions plasma is "carried along," that is, it will rotate with the star at a velocity $\mathbf{v} = [\mathbf{\Omega r}]$. It is possible for the plasma to rotate at the star's angular velocity Ω only of course up to distances $r < c\Omega$ from the rotation axis, that is, up to the boundary of the "light cylinder" whose points move at the velocity $v = c$. The region of perfect corotation will be limited by magnetic field lines "short circuited" by the star and touching the light cylinder. Outside this region the magnetic lines go beyond the cylinder together with particle streams.[1] It is clear that here particles

[1] Roberts & Sturrock (1972, 1973) consider that the magnetosphere with closed field lines is bounded by the region where the centrifugal force is comparable to the gravitational one. This region is much more close to the star than the light cylinder. Their considerations seem to be questionable, however, because the magnetosphere electric field that exceeds the centrifugal and gravitational forces by several orders of magnitude is ignored.

no longer move along the magnetic lines, for then their velocity would exceed that of light. Thus, in the model under consideration, plasma is carried along in the equatorial region and there is a system of streams (electronic and nuclear) leaving the star at higher latitudes.

It should be emphasized that this picture is for the most part intuitive; it is not yet clear to what extent it is realistic. Moreover, the model with $\mathbf{m} \| \mathbf{\Omega}$, and therefore symmetrical with respect to the rotation axis, is not yet a pulsar model: in this case there is no preferred direction in longitude where a local emission source could be placed sending (if its emission is directed) pulses to the Earth with period equal to that of the star's rotation. To approach a realistic pulsar model one must abandon axial symmetry and consider an oblique magnetic rotator, that is, the case where the magnetic dipole axis is inclined with respect to $\mathbf{\Omega}$. Under such conditions, however, it is even more difficult to solve the problem of the plasma distribution and the magnetic field configuration because of the absence of symmetry and the fact that a rotating magnetic dipole radiates low-frequency electromagnetic waves into the surrounding medium.

These complications have motivated extreme idealizations that permit simplified self-consistent solutions for the fields and charges in the magnetosphere of a rotating star. Michel (1973), for example, has considered a magnetosphere consisting of massless particles, because the gravitational forces are many orders of magnitude smaller than the electric forces. This supposition, however, is equivalent to looking for a configuration of fields and a space-charge distribution in the magnetosphere for which the forces acting on a charged particle are zero. Qualitative consideration of a possible magnetosphere structure on the assumption that each elementary magnetosphere volume contains only positive or negative particles (so that a neutral component is absent) led Holloway (1973) to the conclusion that surfaces dividing regions of opposite signs possess properties of *p-n* transition (a type of transition known in semiconductors). Ruderman & Sutherland (1975) supposed that a layer (gap) with a strong electric field up to 10^8–10^9 V cm^{-1} exists on the surface of a neutron star with $\mathbf{m}\downarrow\uparrow\mathbf{\Omega}$. A discharge followed by electron-positron pair formation takes place in this layer. Positrons accelerated in such a field up to

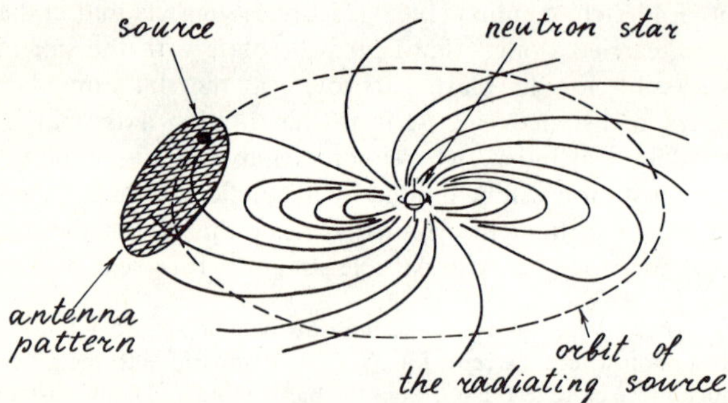

Figure 2 Schematic pulsar model.

energies $\mathscr{E} \sim 3 \times 10^6\ mc^2$ run into the neutron star magnetosphere. The magnetosphere plasma itself results from the creation of electron-positron pairs with energy $\mathscr{E} \sim 10^3\ mc^2$ by gamma-quanta emitted by energetic positrons arriving from the abovementioned gap near the star surface when they move in an inhomogeneous magnetic field (curvature radiation). Note that the idea about particle acceleration in an electric field parallel to **H** that resulted from the star rotation was put forward earlier by Sturrock (1970, 1971). He also investigated the process of pair production in the vicinity of a neutron star by gamma-quanta emitted by the curvature radiation of charged particles.

There is no definite opinion concerning either the position of the radiation source or particle acceleration mechanisms: Radhakrishnan & Cooke (1969), for example, suppose that the source is rather close to the star, far from the light cylinder; Gold (1969) considers the source to be near the light cylinder; and Lerche (1970) believes it to be far beyond the cylinder, in the low-frequency radiation field of a rotating magnetic dipole.

In this situation we can reasonably change our approach to the problem of pulsar radiation, by assuming the position of the source and seeing how our assumption accounts for the observed radiation characteristics. We can also try to ascertain which characteristics of the observable radiation depend on concrete radiation mechanism and which are connected with the chosen pulsar model.

Further we assume that a pulsar is a rotating neutron star with a strong dipole magnetic field, the magnetic moment **m** making an angle with the rotation axis $\mathbf{\Omega}$ (see Figure 2 with the version $\mathbf{m}\perp\mathbf{\Omega}$). The star is surrounded by a plasma magnetosphere that is partially corotating; the radiation source is localized somewhere in the magnetosphere, for example, above the magnetic pole. The position of the source depends on the concrete configuration of the magnetic field rotating with an angular velocity Ω, which is also the radiating region velocity. If the radiation is beamed, successive radiation pulses spaced by star periods will be observed on the Earth.

2 MECHANISMS OF ELECTROMAGNETIC RADIATION

Emission mechanisms acting in astrophysical conditions are usually divided into coherent and incoherent mechanisms. In a coherent mechanism the radiation intensity exceeds the total spontaneous radiation intensity of all individual particles in the source. By definition this is just the difference between coherent and incoherent mechanisms: for incoherent mechanisms the source radiation intensity cannot exceed the summed intensity of spontaneous radiation by individual particles. Let us consider the difference between coherent and incoherent radiation mechanisms in more detail (Ginzburg et al 1969, Ginzburg & Zheleznyakov 1970).

An example of a source with an incoherent mechanism is an object containing radiating particles with a thermodynamic equilibrium velocity distribution (radiation of such an object is called thermal). The brightness temperature of thermal radiation T_b never exceeds the kinetic temperature $T \approx \mathscr{E}/\kappa$ ($\mathscr{E}$ is the mean thermal particle energy, κ is the Boltzmann constant). The condition $T_b \lesssim \mathscr{E}/\kappa$ can be used also as a rough estimate of the particle energy for some other incoherent mechanisms unless the particle velocity distribution differs greatly from the equilibrium distribution (in the sense of a large enough distance from the instability threshold).

Specifically, for the pulsar in the Crab nebula, whose brightness temperature in the visible optics is about 10^{11}°K, the estimate $T_b \lesssim \mathscr{E}/\kappa$ entails values of $\mathscr{E} \gtrsim 10^7$ eV. Such comparatively small values of the radiating particle energy make it possible to attribute this optical radiation to incoherent mechanisms. Note that in the framework of incoherent mechanisms the elementary radiation process of each individual particle can be different, e.g. cyclotron and synchrotron radiation, inverse Compton effects, Cherenkov radiation.

For the radio emission of pulsars one finds values of $T_{\rm eff}$ up to 10^{30}°K; the above inequality for the mean radiating particle energy entails values of the order of 10^{26} eV. Particles of such fantastic energies have never been observed, however. They would have to have extremely high radiation losses and their acceleration even by pulsars seems improbable, to say nothing of the fact that under cosmic conditions such high-energy particles will not generally radiate primarily in the radio range. For pulsar radio emission a clear preference is given to coherent mechanisms: they can "raise" a radiation flux up to the observable level without assuming an extremely high radiating particle energy.

Thus, the great difference in emission brightness temperature indicates a difference between pulsar-emission mechanisms in the radio range and the optical and X-ray frequency ranges. This conclusion has some support in the fact that there are three types of pulsars: 1. pulsars emitting only in the radio range (includes most of the known pulsars); 2. pulsars emitting only at high frequencies, in the X-ray range (includes X-ray pulsars Cen X-3 and Her X-1); and 3. pulsars emitting both in the radio range and at high frequencies, in the optical and X-ray ranges (the pulsar in the Crab nebula).

In principle, two types of coherent mechanisms are possible. The first type is an antenna mechanism where coherence is provided by preliminary particle grouping. In this case the phases of the spontaneous radiation fields of individual particles are not random, so when the fields are summed, the radiation intensity exceeds the sum of the spontaneous radiation intensities. The second type is a maser mechanism. It works also in the absence of preliminary phasing or grouping. But in this case momentum and energy particle distributions that provide inverse population of energy states have to be created and then maintained. In such a system the reabsorption coefficient μ is negative, which is responsible for the intensification of spontaneous radiation appearing in the system. A coherent mechanism is thus again realized, that is, the radiation is phased to a certain extent. This phasing is now established automatically, however, under the action of the radiation itself.

The difference between antenna and maser mechanisms is the same as that between a tuned (LC) circuit and self-excited oscillator (e.g. a klystron); for the appearance of radiation (or oscillations) an instantaneously variable external electromotive force must be applied to the antenna (or tuned circuit). For the appearance of radiation (or oscillations) in the maser (or klystron) it is sufficient to switch on a constant electromotive source or maintain an energy distribution with a level inversion.

Of course at some stage the maser mechanism reduces, in some sense, to the antenna mechanism. In fact, in stationary amplification regime the electron motion in the system is equivalent to some complex set of microcurrents substantially varying in phase at distances of the order of a wavelength. A current system created and maintained artificially, that is, by applying electromotive forces, will give the same radiation as the maser. Such a system of currents, however, cannot be created practically without the maser effect, whereas in the maser system a current distribution necessary for obtaining powerful radiation occurs automatically under proper irradiation of the system.

The attempts to explain pulsar radiation on the basis of the antenna mechanism involve models with a more "regular" current distribution than that in the maser mechanism. It seems at first glance that such a distribution is easier to realize under cosmic conditions. Radiation of particle bunches (clusters), sets of bunches, current envelopes, expanding charge sheets, etc have been considered. One should bear in mind, however, that for such systems radiation coming to the point of observation from different particles has equal or nearly equal phases (so that fields, not intensities, are summed up) only if the linear dimensions of the bunches (clusters) are smaller than the wavelength. (It is usually assumed that the bunches themselves have a random spatial distribution.) Under cosmic conditions, and specifically in the pulsar magnetosphere, it is difficult to create and maintain for at least some fractions of a second such small plasma clusters. If the dimensions of a cluster exceed the wavelength, the radiation intensity decreases rapidly due to interference from different parts of the cluster (Ginzburg & Zheleznyakov 1970). These considerations, in our opinion, provide evidence against antenna mechanisms of pulsar radio emission (specifically, the mechanisms of Eastlund 1970, Lerche 1970, Tademaru 1971, and others).

Eastlund considered coherent magnetobremsstrahlung of arbitrarily chosen clusters of charged particles at low-gyrofrequency harmonics. In a number of articles Lerche discussed the antenna mechanism on the assumption that the observed pulsar radiation is "synchrotron" radiation of relativistic particles moving in variable low-frequency fields of a rotating magnetic dipole. Such radiation may also be treated as inverse Compton radiation resulting from collisions between low-energy photons and relativistic particles; sometimes it is called "synchro-Compton" radiation. To provide the cluster dimensions $l < \lambda$ (λ is a radiated wavelength) Lerche had to make l smaller and smaller. To explain the X-ray radiation he had to reduce the clusters to one particle and actually go over to the incoherent synchrotron radiation mechanism (this mechanism is described in Section 4). We discuss the antenna mechanism of Ruderman & Sutherland (1975) acting together with the plasma instability mechanism in Section 5, where the origin of pulsar radio emission is considered.

Coherent maser mechanisms in their turn may be divided into two groups. In the first group amplification takes place directly at radio waves. In the second group plasma waves are amplified that have no direct exist from the dense plasma into the interstellar medium; in this case radio emission is caused by conversion (transformation) of plasma waves into electromagnetic waves.

A coherent synchrotron mechanism acting in a system of relativistic electrons in a magnetic field is a mechanism of the first group. With an appropriate choice of the electron energy spectrum the reabsorption coefficient connected with synchrotron radiation can become negative and the waves radiated by individual relativistic electrons will be amplified within the system (Zheleznyakov 1966, Zheleznyakov & Suvorov 1972). The second group is represented by plasma wave amplification in a "beam-plasma" system (two-beam instability) and the conversion of the plasma waves into electromagnetic waves by scattering on plasma particles and on other plasma waves (Ginzburg et al 1969).

3 FORMATION OF THE RADIATION POLAR DIAGRAM (BEAMING MECHANISM)

For periodic pulses to appear from a rotating neutron star it is necessary that the star's radiation have a directed character. The polar (direction) diagram may be either fan shaped or pencil beam (Ginzburg et al 1969). Although this question has not been solved yet, only pencil-beam diagrams are under discussion in the literature now. Besides determining the character of the polar diagram, the causes of beamed radiation should be investigated. The observational data available do not give a definite answer to this question.

Among the possible causes of beamed radiation we mention the following: 1. maser effect; 2. radiation along magnetic field lines; and 3. relativistic motion of the source.

That the maser effect plays the main role in creating a sharp radio-emission diagram was supposed by Ginzburg & Zaitzev (1969). Let the amplification of electromagnetic waves take place in the source (due to plasma instability on these waves or to induced scattering of plasma waves when they transform into electromagnetic waves), that is, the reabsorption coefficient μ becomes negative. Then in a linear amplification regime the intensity of radiation leaving the generation region is given by

$$I = \frac{a}{|\mu|}[\exp(|\mu| L) - 1], \qquad 3.1.$$

and if the amplification is large enough ($|\mu| L \gg 1$), $I \approx a|\mu|^{-1} \exp(|\mu| L)$ (a is emissivity of a unit volume of the source with linear dimension L).

Due to the exponential dependence of I on μL even a weak change of the quantity μ, depending on the direction, will lead to the appearance of a sharp radiation beaming along the plane or the line on which μ is maximum. In this case a narrow fan-shaped or pencil-beam polar diagram is formed. The difficulties with such an interpretation in its simplest form arise from the necessity to suppose that the maser effect works for all kinds of pulsar radiation, from radio to X-ray and gamma-ray radiation; a weak-frequency dependence of the diagram width should be provided. Because $I \sim \exp(|\mu| L)$ and $\mu = \mu(\omega)$, the latter is not an easy task even for the entire radio range.

Another possible explanation of the beamed character of pulsar radiation was pointed out by Radhakrishnan & Cooke (1969). Later this hypothesis was widely discussed by Komesaroff (1970), Sturrock (1970), Manchester (1971), and others. Radhakrishnan and Cooke supposed that the radiation source is localized near the magnetic pole of a neutron star, and all the emission mechanisms form a narrow beam along the magnetic field. Due to the sharp directivity, in a given direction the part of the source that mainly radiates is that whose lines of force are tangent to this direction. Then the observed form of the pulsar pulse depends on the brightness distribution of the source. The source itself may be localized inside the light cylinder, in particular, rather close to the star surface. Of course, in such a model we observe only the pulsars whose magnetic axes lie near the pulsar-Earth line.

One difficulty with this model is that despite the obviously different character of the mechanisms generating radiation in different frequency ranges, all of them must provide the same orientation of the polar diagram along the magnetic field. This problem, however, seems much less serious than that in the pure maser model.

A third possible cause of directed radiation is rotation of the radiation source with relativistic velocity around the neutron star. Beaming and the pulsating character of pulsar radiation is associated in this case with relativistic transformation of the polar diagram of a quasi-isotropic source (when it is at rest) when it moves at a velocity comparable to that of light. A pencil-beam diagram may thus appear with the axis oriented along the tangent to the source orbit. An attractive feature of this hypothesis is that the beaming results from effects independent of concrete radiation mechanism.

This hypothesis was put forward by Smith (1969, 1970) and considered in detail by Zheleznyakov (1971) and Zheleznyakov & Shaposhnikov (1975). Some aspects of this interpretation were discussed by Ferguson (1971) and Smith (1973). Ferguson (1973), Cocke et al (1973), and Ferguson et al (1974) studied the possibility in order to explain in the framework of this hypothesis the polarization of an optical radiation of the pulsar in the Crab nebula. Zheleznyakov & Shaposhnikov (1972) used this hypothesis in the theory of synchrotron and Compton radiation of the same pulsar in the Crab.

The hypothesis about relativistic beaming is based on the supposition that the radiation source is localized near the light cylinder of a rotating neutron star. The position of the source in the star magnetosphere depends on the radiating particle distribution, which is finally fixed by a definite configuration of a magnetic field. For the majority of pulsars the pulse form averaged over a large number of periods is constant in time. This is apparently explained by both the stability of average characteristics of energetic particle sources and the fixed magnetic field configuration in the coordinate system rotating with the neutron star. At the same time the observed variability of the form of individual pulses indicates considerable variations in the distribution of radiating particles over times comparable with the star period. We may say that the "weather" on a neutron star is highly variable while the "climate" is very stable. There are exceptions to this rule: the pulsars PSR 0329 and PSR 1237 that have rapid "jumps" of the "climate" and whose pulse profiles undergo catastrophic reconstructions (Backer 1970b, Hesse et al 1973).

To conserve a stable magnetic field configuration it is necessary that the total particle energy in the source $\mathscr{W}_N \sim \mathscr{E} N L^3$ be less than the magnetic field energy $\mathscr{W}_H \sim (H^2/8\pi)L^3$ (N is the particle density, $\mathscr{E}$ their energy). It is more difficult to maintain this condition in a model with relativistic beaming (and thus with the source localized near the light cylinder) than it is in the model of Radhakrishnan & Cooke (1969), where the source is near the star, in the region with a stronger magnetic field. Nevertheless, the condition $\mathscr{W}_H > \mathscr{W}_N$ is satisfied in particular for a synchrotron mechanism of the optical and X-ray radiation in the Crab acting in a relativistically moving source (Zheleznyakov & Shaposhnikov 1972); the same condition $\mathscr{W}_H > \mathscr{W}_N$ must be valid for any realistic mechanism of radio-emission generation.

A consistent development of the notion of relativistic beaming has accounted for a number of important characteristics of the observable pulsar radiation [for more details see Zheleznyakov (1971)].

First, relativistic beaming accounts "automatically" for the pulsed character of the radiation by associating beaming immediately with the star's rotation and localization of the radiating region near the light cylinder. In this case a diagram is formed with the radiation pulse width (at half the peak intensity)

$$\Delta t = \frac{P}{2\pi}(1-\beta^2)^{3/2} b\left(1+\frac{b^2}{3}\right); \quad b = 2^{1/(3+\alpha)}-1. \qquad 3.2.$$

Here P is the pulsar period, $\beta = v/c$, and α is an index of the power-frequency radiation spectrum ($I \sim \omega^{-\alpha}$). Equation 3.2 is more exact than that presented by Zheleznyakov (1971); it is derived from the relations determining the change in the radiation flux when going over from the reference frame accompanying the source to the frame moving at a velocity v. Correct expressions for this transition are presented by Pacholzyk (1970), Zheleznyakov (1971), Ferguson (1971), and McCrea (1972); incorrect expressions are given by Smith (1970) and ter Haar (1972).

From equation 3.2 it follows, for example, that for pulsar PSR 0833 (period $P \approx 90$ msec, pulse duration is 1/45 of the period) the source velocity should be equal to 0.8 c. Note that with such a velocity the diagram width amounts to 40° and the distance from the source to the star is about 3000 km.

From the above example it is clear that the ratio of the pulse duration to the period is small even when the radiation polar diagram width is comparatively large. This happens due to the contraction of the observed pulse under forward radiation by a moving source (i.e. we are dealing essentially with an aspect of the Doppler effect). The wide diagram width enables one to receive radiation from a considerable fraction (up to one third) of existing pulsars despite the pencil-beam character of their polar diagrams.

From the Lorentz transformation for radiated power it follows that if in the A′ frame accompanying an isotropic source there is a power-law radiation spectrum, then in the A system connected with the observer the character of the spectrum is conserved. At the same time, as is seen from equation 3.2, the pulse duration for such a spectrum is frequency independent. Because a power-law radio-emission spectrum is generally typical of pulsars (see, e.g. Backer & Fisher 1974), the relativistic beaming effect thus accounts for the usual independence (or rather, weak dependence) of the pulsar pulse duration on the radio-emission frequency.

Let us now consider the reasons for the complicated character of the pulsar pulse profile. One reason is the presence of several local radiation sources in the magnetosphere. Pulsars PSR 0532, PSR 0950, and some others, whose interpulse is registered approximately halfway between the main pulses, may serve as examples. To explain these interpulses in the framework of the pulsar model with a pencil-beam radiation diagram (extended along the source motion velocity), it is necessary to assume the existence of a second source located almost exactly opposite the first one (e.g. over the opposite pole of the neutron star dipole magnetic field). Sometimes the observed displacement of interpulses halfway between the main pulses serves as an argument against the assumed location of the radiation source near the light cylinder (see e.g. Sturrock 1971). In a vacuum at large distances from the star the magnetic field will be rather close to the field produced by a rotating magnetic dipole even when the field near the star differs from the field produced by the dipole. In the framework of the above hypothesis a symmetrical location of the interpulse between the main pulses is established. However, the occurrence of charged particles with the energy density, being a marked part of the magnetic energy density (but not the same above both poles), may essentially change the magnetic field configuration at the light cylinder providing the necessary interpulse displacement. Different widths of the main pulse and interpulse may be connected with different velocities of the sources due to different distances from the rotation axis.

Figure 3 presents positions of optical and radio-emission sources of the pulsar in

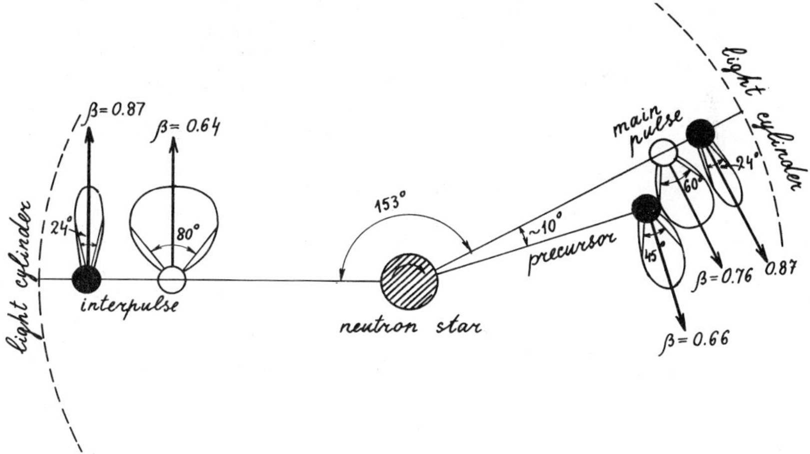

Figure 3 Location of the centers of optical (*light circles*) and radio emission (*dark circles*) of the pulsar in the Crab nebula.

the Crab nebula (projection on the equatorial plane of the star). The positions of the centers of the sources and corresponding polar diagrams of radiation providing the observed pulse duration are marked in the figure. The system of local sources of the pulsar in the Crab nebula is more complicated in the radio range than in the optical range: the main pulse has a precursor with a different pulse duration, polarization, and frequency spectrum. This precursor is apparently generated in a separate local source, also indicated in Figure 3. Note that the sources of optical and radio emission in the main pulse as well as in the interpulses are placed in the same meridional plane. This provides simultaneous radio and optical pulses on the Earth, proved by Conklin et al (1969) with an accuracy of at least 1 msec.

The relativistic character of the source's orbital motion can also explain some features typical of the second pulsar period found by Drake & Craft (1968). This second period characterizes subpulses arising inside a main pulse and gradually displacing in time against the main pulse profile ("running" subpulses). The simplest explanation of this phenomenon is periodic modulation of the source radiation intensity; if the modulation period is incommensurable with the pulsar period, the subpulses will displace against the pulse profile. The relativistic

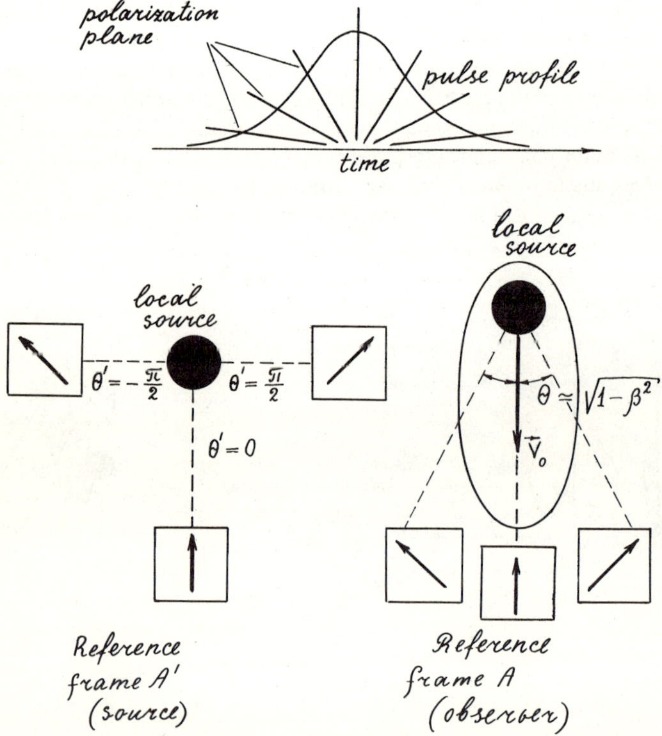

Figure 4 The optical polarization plane rotation in PSR 0532 in the Crab.

character of the source motion makes it possible to understand why the observed magnitude of the second period P_2 is different in the middle and at the ends of the pulse [at the ends of a pulse in pulsar PSR 1919, P_2 is approximately twice as large as in the middle part of the pulse profile (Backer 1970a)].

In fact, if in the A′ system, following the source, the radiation intensity modulation period is P_2^1, then the modulation period P_2 registered by an observer on the Earth (system A) will be different due to the Doppler effect; its value is connected with P_2^1 by the relation

$$P_2 = \frac{P_2^1(1-\beta\cos\theta)}{(1-\beta)^2}, \qquad 3.3.$$

where θ is the angle between the source velocity $\mathbf{v} = \boldsymbol{\beta}c$ and the direction to the Earth.

From equation 3.3 it is clear that when the source velocity is close to that of light (i.e. at $\beta \approx 1$) the subpulse duration at the ends of the pulse (where $\theta \approx \sqrt{(1-\beta)^2}$ becomes twice as large as in the middle ($\theta = 0$). This is in full agreement with the observational data of Backer (1970a) for pulsar PSR 1919.

Note finally that the aberration effect arising at a relativistic source motion also provides an insight into polarization features of optical radiation of the pulsar in the Crab and probably of pulsars' radio emission.

A characteristic polarization feature of optical radiation of the pulsar in the Crab is a systematic polarization plane swing during a pulse at a total angle of about 140° (Wampler et al 1969). The explanation of the rotation, essentially very simple, is illustrated in Figure 4, which shows radiation polarization states in the accompanied frame of reference A′ and from the viewpoint of the observer A.

Let us first stay in the frame of reference A′ accompanying the source. Suppose that the source in two opposite directions, $\theta' = \pm\pi/2$, orthogonal to the source velocity generates radiation of the following properties: it is polarized linearly, and the polarization plane is the same for both directions.[1a] This plane is at an acute angle σ to the source orbit plane and one can notice the following: when regarding the source from the two directions, we see this plane oriented at different angles. In one direction the plane bends to the right, in the other, to the left. Then, thanks to aberration at relativistic source motion, the two considered directions in the A′ system correspond to the ends of the polar diagram (i.e. to the beginning and the end of the pulse) in the A system connected with an observer. At different ends of the pulse the polarization plane will thus be different. Taking into account that an orbital source motion will gradually transform the polarization from one state to the other, we obtain the observed systematic swing of the polarization plane during the pulse. The magnitude of the turn during a pulse is $\mathscr{Y} = 2\sigma$ or $\mathscr{Y} = 180-2\sigma$, where σ is the inclination of the polarization plane to the source orbit at the boundary of the polar diagram. For the pulsar in the Crab the turn $\mathscr{Y}$ within the pulse is about 140°; then the angle $\sigma = 20$ or 70°. The choice between these values

[1a] The abovementioned feature of polarization is easily realized in the framework of the synchrotron mechanism of optical radiation (see Section 4).

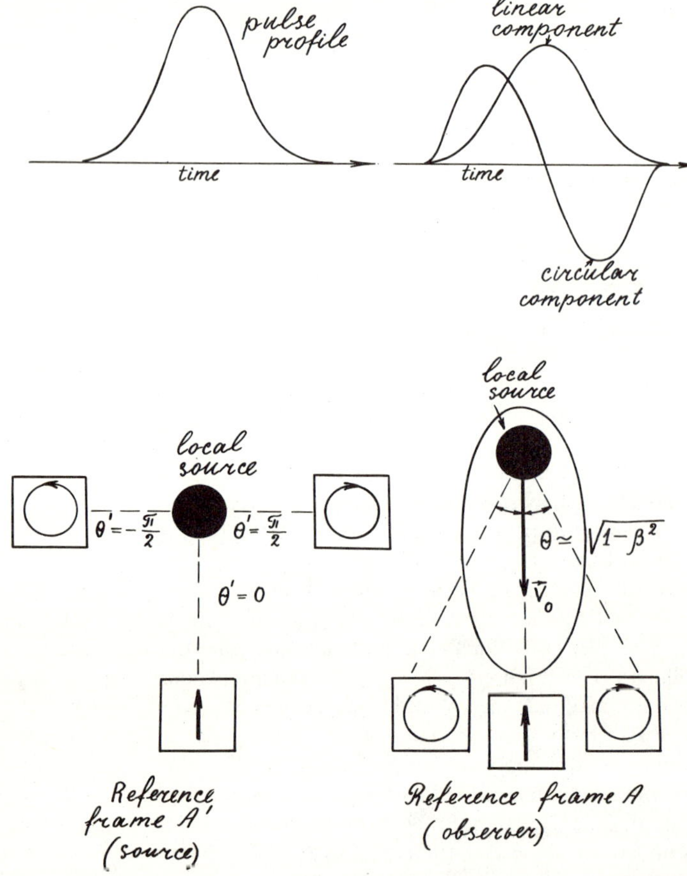

Figure 5 Radio-emission polarization features of PSR 0328.

of σ and the direction of the polarization-vector change during the pulse are determined by the magnetic field structure in the source.

The swing of the radio-emission polarization vector during one pulse, observed, for example, in PSR 0833, may be explained essentially in the same way as the polarization-plane swing in the optical emission of the pulsar in the Crab (provided, of course, that the polarization of radiation escaping the source does not change up to the exit into the interstellar medium). That the radio-emission polarization plane does not swing in some pulsars may be due to the fact that in the comoving system the polarization plane of the radio emission coincides with the orbital plane or is orthogonal to it for all emission directions. (It is assumed that an observer is in the pulsar equatorial plane.)

The same explanation may be given for the peculiar change in the polarization magnitude during the pulse in the radio emission of PSR 0328, that is, the sign of the circular polarization changes in the middle of the pulse when linear polarization is the highest (see Figure 5). Suppose that the character of polarization of ordinary and extraordinary waves in the source is circular along the magnetic field and linear across the field, the polarization ellipse gradually changing its shape with the change of the angle between the radiation direction and the magnetic field.[2] In magnetoactive plasma the polarization vector of an extraordinary wave is known to rotate in the same direction as an electron in the magnetic field. Therefore, if the source generates radiation corresponding to the waves of only one type (for example, extraordinary), the signs of the polarization vector rotation observed in the A′ system along and across the field will be different. If the magnetic field in the source is oriented across the motion velocity, then with the change of the angle θ' the character of the source polarization will change from circular (at $\theta' = -\pi/2$) to linear (at $\theta' = 0$); at $\theta' = 0$ the polarization changes in sign and again becomes circular at $\theta' = \pi/2$. This circumstance is seen in Figure 5 in the A′ system. Due to aberration, the change of polarization will repeat itself in the A system for the time equal to the pulse duration; observations on PSR 0328 confirm this conclusion.

Such an interpretation of the polarization of PSR 0328 imposes certain requirements on the character of ordinary and extraordinary waves in the source. This provides additional evidence for possible conditions in the source that can appear essential in the choice of the pulsar radio-emission mechanism.

Recently Manchester et al (1973) have mentioned the difficulties involved in the hypothesis of relativistic beaming mechanisms in application to some pulsars. Their main argument concerns equation 3.2, where the pulse duration of pulsars with power-law radiation spectra is ω-independent. Manchester et al emphasize that the bandwidth of the radio pulse of PSR 0328 stays constant, although the radiation spectrum deviates appreciably from a power law. The same argument also applies to the pulsar in the Crab since Becklin et al (1973) discovered that the shape of the pulse is the same in the optical and infrared ranges, although the corresponding spectral indices are probably different.[3]

However, it should be mentioned here that equation 3.2 is derived from the following assumptions: 1. the source is pointlike and its position in the pulsar magnetosphere does not depend on the radiation frequency; 2. the radiation polar diagram is isotropic at all frequencies in the corotating frame of reference A′; and 3. the radiation-frequency spectrum is a power law ($I \sim \omega^{-\alpha}$). Violation of even one of these conditions will lead to an ω-dependence of Δt. However, when conditions

[2] In nonrelativistic plasma such a normal wave polarization takes place at frequencies $\omega \sim \omega_H \gg \omega_L$ (ω_H is an electron gyrofrequency, ω_L is the plasma frequency).

[3] At present, however, it is possible that the spectral indices for PSR 0532 in the infrared and optical regions are of the same or close values $\alpha \approx -0.4$. The data available do not contradict this supposition, although it needs experimental verification, first of all in optics. [Measurements of Oke (1969) do not exclude that the spectrum in optics is a power law with a small index α, although it is better approximated by the Plank curve with $T = 10^4$ °K.]

1 and 3 are violated, a kind of compensation results. In a pulsar with a point source located "for radiation" at low frequencies farther from the star than for radiation at high frequencies, the pulse duration will increase with frequency. This may compensate the pulse duration decrease connected with the growth of the spectral index α to high frequencies at PSR 0328. This compensation is, of course, an exception.

Another way to account for the constancy of the pulse profile at different frequencies for a non-power-law radiation spectrum is to abandon the "pointlike" idealization of the source. If a distributed source occupies a magnetospheric region whose angular dimensions in longitude exceed the width of the polar diagram, only a part of the source will contribute radiation at a given moment: the pulse profile will then be defined by the distribution of emissivity over the source. The actual form of the relativistic diagram becomes inessential only if it is narrow enough. Dependence or independence of the pulse profile on frequency in this case is evidently connected with variability or invariability of the emissivity distribution over the source. The relativistic contraction and beaming effect in the diagram is then "hidden" and does not directly define the resulting radiation polar diagram. The form of the pulse may then be explained in the same way as in the model of Radhakrishnan & Cooke (1969); only the reason for the beaming of radiation from each element of the source is different. In this model the swing of the optical radiation-polarization plane during one pulse depends mainly on the change in the magnetic field orientation with respect to the source (see Lyne et al 1971, Smith 1971). It is quite possible also that the longitude dimensions of the source are comparable with the width of the polar diagram and that rotation of the polarization plane is due to both aberration and inhomogeneity of the field in the source (Zheleznyakov & Shaposhnikov 1972).

In summary, we may say that the question of the pulsar-radiation beaming mechanism cannot be considered solved as yet. Furthermore, the reasons for the formation of a source localized near the light cylinder (in the case of a relativistic beaming mechanism) or the appearance of a "polar cap" giving beamed radiation (in a model with radiation along the magnetic field) remain obscure; in fact, investigation of this problem has only started.

4 MECHANISM OF OPTICAL, X-RAY AND GAMMA-RAY EMISSION OF THE PULSAR IN THE CRAB

The brightness temperature of PSR 0532 in visible light ($T_b \sim 10^{11}$ °K) is rather low. We may thus use incoherent mechanisms to interpret the observed radiation, supposing the presence of radiating particles of energy $\mathscr{E} \sim 10^7$ eV in the source.

Figure 6 presents the results of numerous measurements on the radiation flux of PSR 0532 in the Crab at frequencies higher than 10^{14} Hz. The frequency spectrum of radiation apparently peaks in the optical (or ultraviolet) range. The detailed character of the spectrum in the transition interval between the optical to the X-ray range remains unknown. In the X-ray region, for photons of energy $\mathscr{E}$ exceeding 10 keV, the frequency spectrum corresponds to a power law: the flux $F_\nu =$

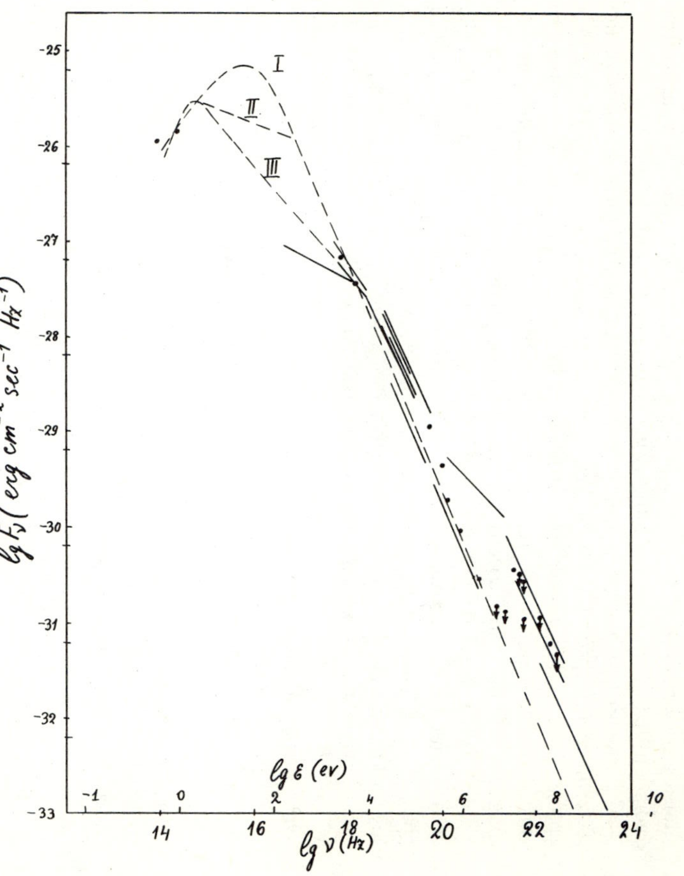

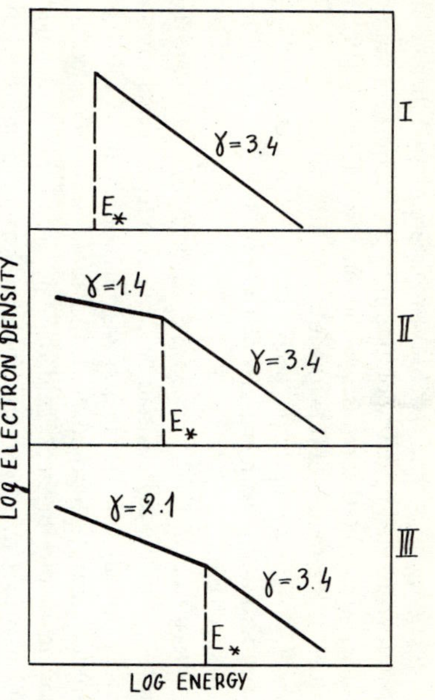

Figure 6 *Left*, the frequency spectrum of infrared, optical, X-ray, and gamma-ray radiation of PSR 0532 in the Crab nebula. Dark points and lines are the results of the measurements for the main pulse; dashed lines are theoretical frequency spectra. *Above*, three possible versions of relativistic electron-energy spectra.

$2\pi F_\omega \propto \omega^{-\alpha}$ with the index $\alpha \approx 1.2$. At an energy of about 10 keV there is a break; as for the values of the spectral index for $\mathscr{E} < 10$ keV there are disagreements. The data on the level of gamma radiation are contradictory. The flux of soft gamma emission with a photon energy greater than 0.6 MeV, registered by Hillier et al (1970), exceeds the flux values obtained by extrapolation into the gamma region of an X-ray spectrum with $\alpha = 1.2$. On the contrary, the data by Kurfess (1971) indicate that the flux values in the region of 0.1–10 MeV coincide with the extrapolated spectrum with $\alpha = 1.2$. At the same time, according to Albats et al (1972), the values of the gamma-radiation flux with energies of 10–100 MeV are above the continuation of the X-ray spectrum.

In a number of papers (Bertotti et al 1969, Ginzburg & Zheleznyakov 1969, 1970, Shklovsky 1970, Epstein 1973, Epstein & Petrosian 1973) an incoherent synchrotron mechanism of the infrared, optical, and X-ray pulsar emission in the Crab has been discussed. The authors have supposed that the radiation is created by relativistic electrons moving in the magnetic field of a neutron star. A detailed analysis of this mechanism was carried out by Zheleznyakov & Shaposhnikov (1972) on the assumption that relativistic beaming is responsible for the pulsed character of the radiation. They have considered three versions of the relativistic electron-energy spectrum presented in Figure 6 (*right*). The parameters of the electron spectra (indices of the power part of the spectrum γ, the "break" energy E_*, the number of particles), the magnitude of the magnetic field, and the source dimensions were varied to obtain the best agreement between the synchrotron radiation-frequency spectra and the available data on the radiation flux.

The synchrotron radiation spectra corresponding to the three versions of the electron-energy spectrum are marked by a dashed line in Figure 6. The various spectra differ mainly in the interval between the optical and the X-ray range. In this region, however, measurements are either absent or not quite reliable. To choose a definite energy spectrum of radiating electrons further measurements are needed. It can be said now only that the first version is less probable than the second and third. The point is that in the infrared region and in the optical spectrum F_ν fall toward the low frequencies (the index $\alpha = -1/3$); however, according to Oke (1969) the spectrum probably has a maximum in the optical.[4] The synchrotron radiation mechanism provides such a maximum for the spectra of types II and III. The "fall" of the spectrum in the infrared region is connected with synchrotron radiation reabsorption in the source.

The values of the source parameters for all three versions are rather close. We present, as an example, the parameters of the second version:

linear dimension	$L \sim 5 \times 10^7$	$\delta^{1/17}$ cm;
magnetic field	$H \sim 6 \times 10^4$	$\delta^{4/17}$ Oe;
"break" energy	$E_* \sim 3 \times 10^8$	$\delta^{-2/17}$ eV;
density of radiating electrons	$N \sim 6 \times 10^{11}$	$\delta^{-7/17}$ cm^{-3}.

[4] True, according to the latest data by Becklin et al (1973), the slope of the spectrum between the optical and the infrared is about $\alpha = -0.4$. This is close to the value of α obtained for energy spectrum I.

All the quantities above are expressed through δ, the ratio of the magnetic field energy in the source to the energy of all the electrons radiating at frequencies of the order or exceeding the infrared ones. In the framework of the synchrotron radiation this parameter remains unknown. One can only say that it must not become less than unity to maintain the stability of the system of charged particles.

The system parameters depend weakly on the value of δ. At the same time this quantity can be estimated from the observed gamma emission on the assumption that it arises from the inverse Compton effect (i.e. from scattering of optical and X-ray photons on relativistic electrons) in a radiating system. Such estimates give values of $\delta \sim 30 - 10^5$.

The assumption of the Compton mechanism for the pulsars' gamma radiation looks quite natural if the frequency spectrum of the gamma rays differs from that of the X rays. The ambiguous observational data allow another possibility: that the frequency spectra in the X- and gamma-ray regions are identical. In this case the assumption about different generation mechanisms in the two frequency ranges does not look natural. It would then be more reasonable to assume that gamma radiation is also for the most part synchrotron radiation and the inverse Compton effect is not essential.[5] This gives higher values of the parameter $\delta > 30 - 10^5$.

Knowing the magnetic field in the radiation source and the distance from the source to the star one can estimate the field on the surface of a neutron star. If the magnetic field in the magnetosphere does not differ very much from a dipole, that is, if the strength of the field H is inversely proportional to r^3, the magnitude of the magnetic field on the surface of the neutron star is about 10^{11} Oe. Rough as it is, this estimate generally agrees with that obtained for a neutron star from the requirement that the magnetic flux during the star's contraction must be constant.

For the synchrotron mechanism of optical and X-ray radiation, the pulsar luminosity in the optical, for a power-law spectrum and $H \sim r^{-3}$, is nearly proportional to $\Omega^{10.5}$, where Ω is the pulsar rotation frequency (Goldreich et al 1972). Such a strong dependence makes it understandable why optical radiation is found only in the Crab pulsar: this object has the highest rotation frequency Ω of all known pulsars.

Although the synchrotron mechanism seems rather probable as the source of optical and X radiation from the Crab pulsar, it is not the only possibility. For example, Kaplan & Tsytovich (1972a–c) believe that this radiation arises from inverse Compton effect on plasma waves. They think that in the radiation source, along with relativistic electrons, there is developed plasma turbulence; and plasma waves scattering on relativistic electrons produce optical and X-ray radiation. But it is necessary to explain the appearance in such a model of developed plasma turbulence with the appropriate parameters, which is not easy.

It should also be noted that many pulsar models, sometimes very complicated ones [e.g. the Lerche (1970) model], are in essence based on the synchrotron or Compton mechanisms of radiation in the optical and X ray regions.

[5] For the relative contribution of the synchrotron and Compton mechanisms in gamma radiation of the pulsar in the Crab see also Maraschi & Treves (1973).

5 MECHANISMS OF PULSARS' RADIO EMISSION

Section 2 presents arguments in favor of a coherent mechanism of pulsars' radio emission.

No definite conclusion can be drawn at present concerning a specific form of the coherent radio-emission mechanism in pulsars. None of the known models has been developed to the extent that one can make sure that it corresponds to the data on pulsar radio emission. We therefore do not think it reasonable to discuss in detail different views concerning the origin of radio emission; the reader is referred to the review by ter Haar (1972). We restrict ourselves to general remarks and mention some new works.

First of all, the proximity of even coincidence of the positions of the radio and optical sources in the magnetosphere of the Crab pulsar indicate that the orders of magnitude of the magnetic fields and probably the particle concentration are similar in both the sources. Therefore, radio-emission generation in pulsars (at least in the Crab) takes place under conditions essentially different from solar conditions. Whereas in the solar corona the radio-emission frequency is usually not lower than the plasma frequency $\omega_L = \sqrt{(4\pi e^2 N)/m}$ and gyrofrequency $\omega_H = eH/mc$, the radiation frequency in pulsars can be much lower (especially lower than ω_H). The problem of the radio-emission exit from the dense plasma in the source into the interstellar medium therefore becomes very important.

It has been suggested that not electrons but nuclei, radiating, for example, in the framework of the coherent synchrotron mechanism, are responsible for radiation at such low frequencies (Zheleznyakov 1973).

Another possibility is synchrotron radiation by an accelerated particle (electron, nucleus) moving along curved magnetic lines—curvature radiation.[6] For example, when electrons with $\mathscr{E}/mc^2 \sim 10^2$ move in the region near the light cylinder along magnetic lines with curvature radius $\rho \sim 10^8$ cm, the characteristic frequency of the radiation $\omega \sim (c/\rho)(\mathscr{E}/mc^2)^3 \sim 3 \times 10^8\ \text{sec}^{-1}$, that is, radio frequency (Goldreich et al 1972). The question of what kind of maser mechanism accounts for the high level of the observed radio emission is still unsolved here, however.

Soon after the discovery of pulsars, plasma mechanisms for their radio emission were discussed in some detail (Ginzburg et al 1969). It was assumed that streams of charged particles (possibly relativistic) pierced through the pulsar magnetosphere filled with a "cold" (nonrelativistic) plasma. In such a system intense plasma waves are excited (plasma turbulence arises) due to two-stream instability. Because of the high level of the plasma waves, conversion of these waves into electromagnetic radiation proceeds as induced scattering in the "cold" plasma.

If the electromagnetic radiation intensity I in the source is due only to scattering (spontaneous and induced), the change of I along the ray l is described by the transfer equation

[6] Suvorov, Chugunov & Eidman (1975) have pointed out that the motion of charged particles along curved magnetic lines is provided by the Lorentz force appearing when these particles drift perpendicularly to the plane tangent to the magnetic field lines.

$$\frac{dI}{dl} = \xi I_p + \chi I_p I. \qquad 5.1.$$

Here ξI_p is the emissivity due to spontaneous scattering of plasma waves with the intensity I_p; χI_p is the amplification coefficient of electromagnetic radiation due to induced scattering. In a homogeneous source (ξ, χ, and I_p are constant) the solution of equation 5.1 takes the form

$$I = \frac{\xi}{\chi} \exp(\chi^I p^l - 1). \qquad 5.2.$$

It is clear from equation 5.2 that, because of the effect of induced scattering, the radiation intensity increases along the ray according to the exponential law. Therefore the process of induced scattering provides a very effective conversion of plasma waves into electromagnetic ones. This circumstance in application to pulsars has been pointed out by Ginzburg et al (1969).

A somewhat different version of the plasma mechanism (Kaplan & Tsytovich 1969) is also possible in principle when the plasma frequency ω_L (and therefore the plasma-wave frequency) is much lower than the radio-emission frequency ω. High values of ω are obtained here due to induced scattering of plasma waves not in the "cold" plasma but on relativistic particles in the source.

Recently Ruderman & Sutherland (1975) have suggested a new type of plasma mechanism realizable in a relativistic electron-positron plasma moving in a neutron star magnetosphere with a beam velocity which corresponds to the energy of separate particles $\mathscr{E} \sim 10^3\ mc^2$. This plasma is pierced by a still more energetic positron beam with $\mathscr{E}/mc^2 \sim 3 \times 10^6$ (for the origin of particles with such energies see Section 2).

According to Ruderman and Sutherland such a system of an electron-positron plasma and a fast positron beam develops a two-stream instability. The latter causes plasma-wave excitation accompanied by bunching of charged particles. At the same time, when moving in the curved magnetic field of a neutron star, plasma electrons and positrons with $\mathscr{E}/mc^2 \sim 10^3$ give curvature radiation in the radio range; due to the bunching of these particles, their radiation is coherent, which accounts for the observed high level of pulsar radio emission.

In connection with such a generation mechanism the following should be noted. As is known, coherent radiation of an electromagnetic wave with the field $\mathbf{E} = \mathbf{a} \exp i(\mathbf{kr} - \omega t)$ arises only if

$$\int_t \int_V \mathbf{j}_{\text{ext}}(\mathbf{r}, t) \exp\{-i(\mathbf{kr} - \omega t)\}\, dt\, dV \neq 0. \qquad 5.3.$$

Here V is the volume occupied by a distribution of "external" macroscopic currents $\mathbf{j}_{\text{ext}}$ created by the charge-system motion. Equation 5.3 will evidently hold when in the expansion of $\mathbf{j}_{\text{ext}}$ into Fourier integral there is a component $\mathbf{j}_p \exp i(\mathbf{k}_p \mathbf{r} - \omega_p t)$ such that

$$\mathbf{j}_p \mathbf{a} \neq 0 \qquad 5.4.$$

and at the same time

$$\omega_p = \omega; \qquad \mathbf{k}_p = \mathbf{k}. \qquad 5.5.$$

Ruderman and Sutherland suppose that external current distribution $\mathbf{j}_{ext}(\mathbf{r}, t)$ appears as a result of the joint influence of the neutron star's inhomogeneous magnetic field and the plasma wave's electric field. Precisely, we deal with clusters or bunches, that is, concentration inhomogeneities moving along the lines of force and place at intervals $\Lambda_p = 2\pi/k_p$ (more precisely, this is the distance between the clusters of the same sign; the clusters of the opposite sign are displaced by $\Delta_p/2$; Δ_p is the plasma wavelength). We are interested in curvature radiation directed along the same magnetic lines. Under such conditions $\mathbf{j}_p \mathbf{a} \neq 0$ only when the curvature of the magnetic lines is taken into account. From equation 5.5 it is also clear that coherent radiation under the abovementioned conditions will appear only if the dispersion curves for electromagnetic waves $\omega = \omega(\mathbf{k})$ and for plasma waves $\omega_p = \omega_p(\mathbf{k}_p)$ coincide (or at least intersect, which provides the validity of equation 5.5 for one wave vector and one frequency).

It is possible, of course, that external current distribution has a form

$$\mathbf{j}_{ext} = \mathbf{j}_p(\mathbf{r}) \exp\{i(\mathbf{k}_p \mathbf{z} - \omega_p t)\}. \qquad 5.7.$$

Here $\mathbf{j}_p(\mathbf{r})$ is the plasma wave envelope with characteristic length $L \sim V^{1/3}$. The Fourier expansion will include harmonics with wave vectors within the phase volume $(\Delta k_p)^3 \sim 1/L^3$ near $\mathbf{k}_p$. In this case the coherent electromagnetic radiation appears under condition

$$\omega = \omega_p, \quad |k - k_p| \lesssim \Delta k_p \sim \frac{1}{L}. \qquad 5.8.$$

Criterion (5.8) depends on linear dimensions of generation region and dispersion characteristics of relativistic plasma; its validity for Ruderman and Sutherland's model should be analyzed seriously. Amplification of curvature radiation (as well as any other electromagnetic radiation) can be achieved by another process connected with the presence of plasma waves in the system. We mean induced scattering of plasma waves and their transition into electromagnetic waves. It has been noted above that this process will cause an exponential increase of the intensity of electromagnetic waves (appearing due to spontaneous scattering, curvature radiation, etc) within the source.

Hitherto, when mentioning plasma mechanisms in the pulsar magnetosphere, we have not discussed the very possibility of plasma-wave excitation. In a nonrelativistic plasma pierced by a particle beam this question was clarified long ago: two-stream instability leads to effective excitation of plasma waves and to the establishment of plasma turbulence. Many authors have leaned essentially upon this circumstance when discussing plasma mechanisms of radiation (Ginzburg et al 1969, Ichimaru 1970, Kaplan & Tsytovich 1972a–c, Buckee et al 1973). The situation, however, changes if plasma in the pulsar magnetosphere is relativistic, which is most probable. Recently Suvorov & Chugunov (1975) have shown that longitudinal waves with the phase velocity v_{ph} less than the velocity of light c cannot exist in a relativistic plasma having a one-dimensional particle velocity distribution in a strong magnetic field (with mean energy of particles $\mathscr{E} \sim mc^2$). Since $v_{\mathrm{ph}} > c$ for plasma waves in the abovementioned cases, plasma waves cannot be generated by two-stream instability because the particle velocity is always less than c. However, in other cases [in relativistic plasma with one-dimensional particle distribution of the type investigated by Kaplan & Tsytovich (1972a–c), for example] the phase velocity V_{ph} can be less than c and the beam instability of longitudinal waves is quite possible.

What has been said forces us to look carefully at beam instability in relativistic plasma; its realization for the concrete form of particle distribution requires special study. This remark seems to be essential when considering the paper by Ruderman & Sutherland (1975), although these authors treat electron-positron plasma as nonrelativistic in the comoving reference frame, where the mean plasma velocity is zero. The point is that in the case of a relativistic plasma with $\mathscr{E}/mc^2 \sim 10^3$ in the pulsar reference system the assumption of a practically complete absence of velocity spread seems to be unrealistic.

An adequate knowledge of the beaming mechanism could facilitate the choice of the radio-emission mechanism. However, if the correct model is that which connects beaming and a number of polarization characteristics of the radiation with relativistic source motion, the search for a concrete radio-emission mechanism is probably considerably hampered: the mechanism of radio emission in this case must account *only* for the high radiation intensity and the form of the frequency spectrum. These requirements do not sufficiently restrict a possible choice of the mechanisms. True, there also exists the radio-emission polarization which may be explained in two ways: 1. polarization can be associated completely with the conditions in the source and with the effect of relativistic source rotation around the star. This was just the way we explained the swing of the polarization plane and the change of the radio-emission polarization across the impulse in Section 4; and 2. these features can be connected with the conditions of the radio-emission exit from the dense magnetosphere into the interstellar medium [see Ginzburg et al (1969), Swarup et al (1969), Zheleznyakov (1970); for objections to the explanation of polarization effects due only to radio-emission exit conditions see Komesaroff et al (1971). By the way, the exit conditions may ensure strong nonregularity and variability of the polarization which takes place in radio emission from some pulsars. This circumstance has been pointed out by Rankin et al (1974)].

6 CONCLUSION

There is no doubt at present that pulsars (or at least "ordinary" pulsars) are rotating neutron stars with a strong magnetic field. One cannot yet be fully confident, however, about the charged-particle distribution and configuration of magnetic and electric fields in the magnetosphere of such a star. A number of pulsar radiation features can probably be explained by relativistic source motion around the star, but the validity of such a model has not yet been proved. The assumption of an incoherent synchrotron mechanism for infrared, optical, and X-ray radiation makes it possible to explain the pulsar frequency spectrum in the Crab, to establish parameters of the radiating region, and to give a reasonable independent estimate of the magnetic field strength on the neutron star surface. The question of the acceleration mechanism of radiating electrons, which must be rather effective, remains open. A specific pulsar radio-emission mechanism is unknown. We can only suppose that it must be a coherent maser mechanism acting (at least in the Crab) in a dense plasma with a strong magnetic field. An essential role here is possibly played by curvature radiation of relativistic electrons (or positrons) with such a velocity distribution that it could provide amplification (maser effect) of this emission in the radio-emitting region.

Further careful considerations should be given to the pulsar emission mechanisms, which are undoubtedly of great interest.

ACKNOWLEDGMENTS

This review was distributed as the P. N. Lebedev Physical Institute Preprint No. 130 (1974). We would like to thank all of our colleagues who submitted their critical remarks to us. We have tried to take into account all of these remarks, but it was too late and technically very difficult to include more than a few additional notes and references.

Literature Cited

Albats, R., Frue, G. M., Zych, A. D. 1972. *Nature* 240:221
Backer, D. C. 1970a. *Nature* 227:692
Backer, D. C. 1970b. *Nature* 228:1297
Backer, D. C., Fisher, J. R. 1974. *Ap. J.* 189:137
Becklin, E. E., Kristian, J., Matthews, H., Neugebauer, G. 1973. *Ap. J. Lett.* 86: L137
Bertotti, B., Cavaliere, A., Pacini, F. 1969. *Nature* 223:1351
Buckee, J. M., Grands, S., Miranda, L. C. M., ter Haar, D. 1973. Preprint
Cocke, W. J., Ferguson, D. C., Muncaster, G. W. 1973. *Ap. J.* 183:987
Conklin, E. K., Howard, H. T., Miller, J. S., Wampler, E. J. 1969. *Nature* 222:552
Drake, F. D., Craft, H. D. 1968. *Nature* 220:231
Eastlund, B. J. 1970. *Nature* 225:430
Epstein, R. I. 1973. *Ap. J.* 183:593
Epstein, R. I., Petrosian, V. 1973. *Ap. J.* 183:611
Ferguson, D. C. 1971. *Nature Phys. Sci.* 234:86
Ferguson, D. C. 1973. *Ap. J.* 183:977
Ferguson, D. C., Cocke, W. J., Genrels, T. 1974. *Ap. J.* 190:375
Ginzburg, V. L. 1964. *Dokl. Akad. Nauk SSSR* 156:43 (Engl. transl. *Sov. Phys.—Dokl.*)
Ginzburg, V. L. 1971. *Highlights of Astronomy*, 2:20. Dordrecht: Reidel
Ginzburg, V. L., Usov, V. V. 1972. *Sov. Phys.—JETP* 15:196
Ginzburg, V. L., Zaitzev, V. V. 1969. *Nature* 222:230
Ginzburg, V. L., Zheleznyakov, V. V. 1969. *UFN* 99:514, 524 (Engl. transl. *Sov. Phys.—Usp.*)
Ginzburg, V. L., Zheleznyakov, V. V. 1970. *Comments Ap. Space Phys.* 2:167, 197
Ginzburg, V. L., Zheleznyakov, V. V., Zaitzev, V. V. 1969. *Ap. Space Sci.* 4: 464
Gold, T. 1969. *Nature* 221:25
Goldreich, P., Julian, W. H. 1969. *Ap. J.* 157:869
Goldreich, P., Pacini, F., Rees, M. J. 1972. *Comments Ap. Space Phys.* 4:23
Hesse, K. N., Sieber, W., Wielebinski, R. 1973. *Nature* 245:57
Hewish, A. 1971. *Highlights of Astronomy*, 2:3. Dordrecht: Reidel
Hewish, A., Bell, S. J., Pilkington, J. D. H., Scott, P. F., Collins, R. A. 1968. *Nature* 217:709
Hillier, R. R., Jackson, W. R., Murray, A., Redfern, R. M., Sale, R. G. 1970. *Ap. J. Lett.* 162:L177
Holloway, N. J. 1973. *Nature Phys. Sci.* 246: 6

Ichimaru, S. 1970. *Nature* 226:731
Kaplan, S. A., Tsytovich, V. N. 1969. *Sov. Astron. J.* 46:192
Kaplan, S. A., Tsytovich, V. N. 1972a. *Astrophizika* 8:441
Kaplan, S. A., Tsytovich, V. N. 1972b. *Problemi Attuali di Scienza e di Cultura* 162:175
Kaplan, S. A., Tsytovich, V. N. 1972c. *Plasma Astrophysics*. Moscow: Science
Komesaroff, M. M. 1970. *Nature* 225:612
Komesaroff, M. M., Ables, J. G., Hamilton, P. A. 1971. *Ap. Lett.* 9:101
Kurfess, J. D. 1971. *Ap. J. Lett.* 168:L39
Lerche, I. 1970. *Ap. J.* 159:229; 160:1003; 162:153
Lyne, A. G., Smith, F. G., Graham, B. A. 1971. *MNRAS* 153:337
Manchester, R. N. 1971. *Ap. J. Suppl.* 23:283
Manchester, R. N., Tademaru, E., Taylor, J. H., Huguenin, G. R. 1973. *Ap. J.* 185:951
Maraschi, L., Treves, A. 1973. *Int. Cosmic Ray Conf., 13th, Denver,* No. 541
McCrea, W. H. 1972. *MNRAS* 157:35
Michel, F. C. 1973. *Ap. J.* 180:207
Oke, J. B. 1969. *Ap. J.* 156:491
Pacholczyk, A. G. 1970. *Radio Astrophysics*. San Francisco: Freeman
Radhakrishnan, V., Cooke, D. J. 1969. *Ap. J. Lett.* 3:L225
Rankin, J. M., Campbell, D. B., Backer, D. C. 1974. *Ap. J.* 188:609
Roberts, D. H., Sturrock, P. A. 1972. *Ap. J.* 173:33
Roberts, D. H., Sturrock, P. A. 1973. *Ap. J.* 181:161
Ruderman, M. 1972. *Ann. Rev. Astron. Ap.* 10:427
Ruderman, M. A., Sutherland, P. G. 1975. *Ap. J.* 196:51
Shklovsky, I. S. 1970. *Nature* 225:251; *Ap. J. Lett.* 159:L77
Smith, F. G. 1969. *Nature* 223:934
Smith, F. G. 1970. *MNRAS* 149:1
Smith, F. G. 1971. *Nature Phys. Sci.* 231:191
Smith, F. G. 1972. *Rep. Progr. Phys.* 35:399
Smith, F. G. 1973. *Nature* 243:207
Sturrock, P. A. 1970. *Nature* 227:465
Sturrock, P. A. 1971. *Ap. J.* 164:529
Suvorov, E. V., Chugunov, Yu. V. 1975. *Astrofizika*. In press
Suvorov, E. V., Chugunov, Yu. V., Eidman, V. Ya. 1975. *Ap. Space Sci.* 32:7
Swarup, G., Chitre, S. M., Sinha, R. P. 1969. *Rep. Pulsar Conf., Rome*
Tademaru, E. 1971. *Ap. Space Sci.* 12:193
ter Haar, D. 1972. *Phys. Rep.* 3c: No. 2
Wampler, E. J., Scargle, J. D., Miller, J. S. 1969. *Ap. J. Lett.* 157:L1
Woltjer, L. 1964. *Ap. J.* 140:1309
Zheleznyakov, V. V. 1966. *Zh. Eksp. Teor. Fiz.* 54:627
Zheleznyakov, V. V. 1970. *Radiofizika* 13:1842
Zheleznyakov, V. V. 1971. *Ap. Space Sci.* 13:74
Zheleznyakov, V. V. 1973. *Vestn. Akad. Nauk SSSR* 6:80
Zheleznyakov, V. V., Shaposhnikov, V. E. 1972. *Ap. Space Sci.* 18:141
Zheleznyakov, V. V., Shaposhnikov, V. E. 1975. *Ap. Space Sci.* 33:141
Zheleznyakov, V. V., Suvorov, E. V. 1972. *Ap. Space Sci.* 15:3

AUTHOR INDEX

A

Aannestad, P. A., 144
AARSETH, S. J., 1-21; 1-10, 12, 14-16
Ables, H. D., 235, 243
Ables, J. G., 533
Abramowitz, M., 477
Acton, L., 435
Adams, W. S., 133, 135
Agekyan, T. A., 7, 8
Ahmad, A., 2, 4
Aikens, R. S., 178, 180
Aitken, D. W., 452, 467
Ajzenberg-Selove, F., 70, 87, 88
Aksnes, K., 3
Albats, R., 528
Albrecht, P., 182
Alden, H. L., 314
Aldrovandi, S. M. V., 159
Aleshin, V. I., 7
Alladin, S. M., 16
Allen, C., 3, 8
Allen, D. A., 207
Aller, L. H., 113, 115, 119, 120, 123, 125, 233
Alloin, D., 122, 123
Alvarez, L. W., 449
Ambartsumyan, V. A., 191, 336, 338, 339
Ames, W. L., 389
Ammar, A., 139
Anderson, E. C., 478
Anderson, K., 467
Anderson, P. W., 356, 357
Andouze, J., 140
Andrew, B. H., 39
Andrews, P. J., 227
Andrillat, Y., 239
Apparao, K. M. V., 431
Appelbaum, L. T., 314
Appenzeller, I., 190
Ardeberg, A., 232
Arnett, W. D., 48, 55, 58, 60-62, 64, 129, 145, 150, 222, 223, 385
Arnold, J. R., 478
Austin, G., 459
Avni, Y., 429
Axford, W. I., 386

B

Baade, W., 16, 217, 235, 237, 246
Backer, D. C., 520, 521, 523, 533
Bagnuolo, W. G., 229
Bahcall, J., 429
Bahcall, J. N., 88, 151, 372, 409
Bahcall, N. A., 15, 46-51, 90, 91, 103
Bahner, K., 180, 182
Bahng, J. A. R., 182
Bailey, J. A., 30, 31
Baity, W. A., 425, 433, 435, 459, 460, 465
Baker, J. C., 249
Baker, P. L., 161
Baldwin, J. E., 25, 374
Baldwin, J. R., 198, 238, 470
Balick, B., 209
Ballabh, G. M., 16
Banerjee, B., 356
Baranov, A. S., 4
BARKAT, Z., 45-68; 47, 58, 60-62, 385
Barker, T., 114, 125
Barnes, C. A., 95, 107, 108, 110
Bates, D. R., 144
Batten, A. H., 412
Baum, W. A., 243
Beaudet, G., 47-49, 51, 56, 62
Becklin, E. E., 191, 208, 212, 213, 429, 525, 528
Bekenstein, J. D., 391, 392, 404, 412, 416
Belian, R., 433
Bell, R. A., 231
Bell, S. J., 511
Bennett, A. S., 25, 33
Benvenuti, P., 120-23
Beron, B. L., 452, 467
Bertaux, J. L., 139
Bertotti, B., 528
Bessell, M. S., 411
Bethe, H. A., 348-52, 354, 355
Bhavsar, P. D., 474
Bieger, G. S., 325
Biermann, P., 17
Bignami, G. F., 156
Birks, J. B., 449, 450
Blaauw, A., 260, 404
Black, J. H., 142-45, 156
Blamont, J. E., 139
Blanco, V. M., 218, 229
Bleach, R. D., 446, 486, 487
Bleeker, J. A. M., 452, 453, 487
Bless, R. C., 142, 149
Blumenthal, G. R., 402, 438
Bodenheimer, P., 387
Boegaard, A. M., 148
Boeshaar, G., 119, 123
Bohlin, R. C., 138
Bohuski, T. J., 123
Boldt, E. A., 409, 428, 430, 446, 486, 487
Bolton, C. T., 410
Bolton, J. G., 25, 29, 38, 40
Bond, H. E., 220
Bondi, H., 397
Boozer, A. H., 56
Borken, R., 154, 424
Born, M., 378
Börner, G., 374, 382
Bothwell, G. W., 178, 180
Bouclier, R., 449
Bouvier, P., 10, 18
Bowers, R. L., 370, 404, 412
Bowles, J. A., 424
Bowyer, S., 410, 467, 470
Boyce, P. B., 182
Bozis, G., 17
Bracewell, R. N., 183
Bradt, D. J., 180
Bradt, H., 423, 431, 432, 468
Braes, L. L. E., 409
Brandie, G. W., 38
Brans, C., 381
Breakiron, L. A., 324
Breger, M., 194, 197, 200
Bregman, J., 410
Bressoni, T., 449
Bridle, A. H., 35-37, 39
Brocklehurst, M., 115
Brookes, M., 178, 182
Broude, C., 166
Brown, J. T., 337, 342, 353
Brown, R. H., 140
Brown, R. L., 156, 158, 209, 272
Browne, J. C., 144
Brucato, R. J., 410
Brueckner, K. A., 337
Bruenn, S. W., 48, 61, 62
Buchler, J. R., 47, 60-62, 340, 385
Buckee, J. M., 532
Bunner, A. N., 154, 437, 438, 487, 489, 490
Burbidge, E. M., 96, 122, 123
Burbidge, G. R., 96, 122, 123, 386, 400
Burke, B. F., 258, 280
Burnham, S. W., 332

Burton, W. B., 161, 283
Busch, C. L., 88
Butterworth, E. M., 3

C

Calvini, M., 397
Cameron, A. G. W., 47, 48, 61, 64, 65, 150, 222, 224, 336, 343, 370
Campbell, D. B., 533
Campbell, J. A., 370
Campbell, J. W., 18
Campbell, W. W., 330
CANUTO, V., 335-80; 49, 64, 335, 342, 353, 356, 358, 360, 361, 363, 366, 370, 374, 377
Capps, R., 213
Capps, R. W., 212
Carr, B. J., 415
Carrasco, L., 190-92, 195, 201, 206, 211
Carter, B., 391, 393, 395, 396, 416
Caswell, J. L., 29, 30
Catura, R., 435
CAUGHLAN, G. R., 69-112; 69
Cavaliere, A., 528
Cazzola, P., 49, 355
Cesarsky, C. J., 140, 155
Cesarsky, D. A., 149
Chaichian, M., 376
Chakravarty, S., 366
Chandrasekhar, S., 3, 4, 245, 383, 386, 395
Chao, N. C., 357
Chapline, G., 382
Chapline, G. F., 150, 399
Charles, M. W., 448, 449
Charpak, G., 449
Chastian, J., 182
Chechetkin, V. M., 47
Chen, Y., 64
Cherepashchuk, A. M., 411
Chester, G. V., 351
Chevalier, R. A., 12, 154
Chin, C. W., 53, 55
Chincarini, G., 251
Chitre, S. M., 356, 358, 360, 361, 363, 366, 533
Chiu, H.-Y., 46, 48, 71, 88, 92
Chiuderi, C., 49
Cho, Y., 450
Chou, C. K., 49, 64
Christensen, C. G., 240
Christiansen, W. N., 181
Christodoulou, D., 412
Christy, R. F., 101
Chugunov, Yu. V., 530, 532
Chupp, E. L., 474
Churchwell, E. B., 114
Churilov, S. M., 413
Clark, G. W., 410, 425, 452, 485-87
Clark, J. W., 357
Clarke, R. W., 28
Clayton, D. D., 70, 92
Cleary, M. N., 262, 267, 274, 275, 277, 292
Cobern, M. E., 95
Cocconi, G., 376
Cochran, S., 351
Cochran, W. D., 135, 137, 140, 143, 144, 153, 157
Cocke, W. J., 519
Code, A. D., 138, 171
Cohen, E. R., 69
Cohen, J. M., 391
Cohen, L., 2, 4, 19
Cohen, M., 195, 197, 200
Coldwell, R. L., 357
Cole, D. J., 29
Coleman, P. L., 431, 437, 438, 487
Colgate, S. A., 48, 61-64, 89, 129, 150, 183
Colla, G., 23, 29-32
Collins, R. A., 29, 30, 511
Colvin, R. S., 161, 272
Comte, G., 121, 122
Condon, J. J., 27, 31-33, 37, 39, 41, 466
Conklin, E. K., 522
Conner, J., 433
Contopoulos, G., 17
Cooke, B. A., 448, 449
Cooke, D. J., 40, 41, 515, 519, 520, 526
Coon, S. A., 337
Cooper, M. S., 70
Costero, R., 116, 118, 119
Costrell, L., 178
Couch, R. G., 48, 61
Cox, D. P., 116, 154, 160
Craft, H. D., 522
Craine, E., 125
Cram, T. R., 261, 262, 268, 269, 272, 279
Crampton, D., 411
Crannell, C. J., 450
Crawford, D. F., 24-26, 28, 29
Crawford, D. L., 181
Critchfield, C. L., 96
Cruddace, R., 470
Cruz-Gonzalez, C., 3, 10
Culhane, J. L., 424, 431
Cunningham, C. T., 397
Cuperman, S., 19
Curtis, H. D., 330
Czyzak, S. J., 115, 119, 123, 125

D

Dabrowsky, J., 337
Dahlbacka, G. H., 399
Dalgarno, A., 142-46, 155, 156, 158, 160
Daltabuit, E., 116
Daniel, R. R., 474
Danziger, I. J., 114, 125, 198
Datlowe, D. W., 457
Datta, B., 342, 353
Davidson, K., 405, 429
Davies, I. M., 23, 29, 31, 32, 34, 42
Davies, J. G., 181
Davies, R. D., 259, 271, 287, 289, 290
Davis, J. H., 192, 210
Davis, M., 391
Davis, M. M., 23, 35-39, 41, 42
Davison, P. J. N., 434
Day, G. A., 29
Dearborn, D. S., 16
de Boer, J., 356
de Boer, K. S., 133, 137, 148, 149, 157
Deerenberg, A. J., 452
Deerenberg, A. J. M., 487
de Felice, F., 396, 397
de Groot, M., 232
de Korte, P. A. J., 487
De Loore, C., 404, 412
Dennis, B. R., 457, 480
Dennison, E. W., 173, 179, 181
Dent, W. A., 38
Deutsch, A. J., 242
De Vaucouleurs, G., 243
DeWitt, B. S., 383
DeWitt, C., 383
DeWitt, H. E., 70
De Zotti, G., 49
Dicke, R. H., 381, 386
Dickel, J. R., 30
Dickens, R. J., 221
Dickinson, D. F., 213
Dicus, D. A., 49, 51
Dieter, N. H., 261, 262, 283
Dixon, R. S., 35
D'Odorico, S., 120-23
Dolan, J. F., 449, 470
Doroshkevich, A. G., 418
Douglas, A. E., 145
Downs, B. W., 337, 342, 353
Doxsey, R., 424, 430, 431
Drake, F. D., 522
Drake, J. F., 133, 134, 139, 140, 143
Dressler, K., 133, 134, 140
Duboshin, G. N., 9
Duck, I., 101
Dufour, R. J., 115, 116, 119, 123, 127
Dunn, R. B., 183
Durdin, J. M., 23
Dwarakanath, M. R., 95, 109

Dyck, H. M., 196
Dyer, C., 480, 481, 494
Dyer, P., 95, 107
Dyson, F. J., 409

E

EARDLEY, D. M., 381-422; 407, 411, 417
Eastlund, B. J., 517
Ehman, J. R., 35
Eiband, A. M., 424
Eidman, V. Ya., 530
Eissner, W., 118
Ekers, J. A., 33
Ekers, R. D., 29
Elgin, R. L., 439, 440
Elias, J. H., 209
Ellis, D. G., 48
Ellis, G. F. R., 417
Elmergreen, B. S., 160
Elsässer, H., 229
Elsmore, B., 25
Endt, P. M., 70
Eneev, T. M., 17
Enkenberg, J., 375
Epps, H. W., 115, 119, 123
Epstein, R. I., 129, 150, 528
Ergma, E., 48
Evans, R. D., 438
Evans, W., 433

F

Faber, S. M., 238, 239, 241-44, 248
Fabian, A. C., 409, 424, 431, 434, 463
Fackerell, E. D., 386
Fahr, H. J., 139, 161
Fano, U., 455
Faraggiana, R., 148
Fassio-Canuto, L., 49
Faulkner, D. J., 115
Favier, J., 449
Feenberg, E., 361
Feierman, B. H., 303, 304
Feinberg, E. L., 375, 376
Feix, M. R., 18
Fejes, I., 260, 262, 268, 281, 282, 287
Felten, J., 436
Ferguson, D. C., 519, 520
Fermi, E., 375
Festa, G. C., 49, 61
Feynman, R. P., 45, 47
Fichtel, C. E., 156, 436
Field, G. B., 140, 144, 146, 148, 155, 157
Fillius, R., 440, 441, 443
Finzi, A., 48, 62
Fireman, E. L., 140
Fischel, D., 138
Fishbone, L. G., 396
Fisher, J. R., 521
Fisher, P. C., 461
Fishman, G. J., 457, 474
Fishman, P. C., 480, 494
Fitch, L. T., 35
Flannery, B., 404
Flowers, E., 49, 64
Fomalont, E. B., 35-39, 41
Ford, H. C., 115, 119, 123, 227
Ford, W. K. Jr., 121, 125
Foreman, W., 411, 433
Forrest, D. J., 474
Forrest, W. J., 207
Forrester, W. T., 225, 239
FOWLER, W. A., 69-112; 46, 48, 57, 58, 69, 70, 88-92, 96, 100, 101, 103, 104, 109, 110, 129, 150
Fox, G., 107
Fraley, G., 58, 62
Franz, O. G., 308, 326
Frautschi, S., 372
Frautschi, S. C., 50
Freedman, D. Z., 51, 64
Freeman, K. C., 223, 227
Fricke, K. J., 386
Friedman, H., 424, 437, 438, 447, 449
Fritz, G., 424, 437, 438, 447, 449
Frogel, J. A., 125, 191, 198
Frost, K. J., 450, 457, 480
Frue, G. M., 528
Fulbright, H. W., 166, 176
Fusi-Pecci, F., 227, 248

G

Gallagher, J. S., 17
Galt, J. A., 35
Gammon, R. H., 209
Gamow, G., 45, 47
Gandelman, G. M., 49
Gaposchkin, S., 229
Garde, V. K., 341-43, 356
Gardner, F. F., 29, 35, 115
Garmire, G., 423, 428, 431-33, 449, 468, 469
Gascoigne, S. C. B., 225, 226, 231
Gasteyer, C., 321
Gatewood, G., 314, 324
Gatley, I., 209
Geisel, S. L., 195
Gell-Mann, M., 45, 47
Genrels, T., 519
Gerard, F., 276
Gerlach, U. H., 413
Gerola, H., 158, 160
Gezari, D. Y., 209, 210
Giacconi, R., 402, 409, 411, 423-25, 427, 428, 433, 435, 444, 445, 448, 466, 468, 499
Gibbons, G. W., 412, 414, 418
Gibson, J., 225
Gilbert, I. H., 4
Gillett, F. C., 195, 207, 213
Gilra, D. P., 146
GINZBURG, V. L., 511-35; 512, 513, 515, 517, 518, 528, 530-33
Giovanelli, R., 261, 262, 268, 269, 272, 279
Glaspey, J., 411
Glassgold, A. E., 142, 153, 156, 157
Glenn, S. W., 457, 474
Goad, J. W., 123
Goddard, R., 459
Gold, T., 64, 386, 515
Goldreich, P., 387, 513, 529, 530
Goldsmith, D., 145
Goldsmith, D. W., 116, 146, 155
Goldstein, S., 19
Gomez-Gonzales, J., 152
Gorenstein, P., 140, 423-25, 428, 431, 432, 445, 449, 466, 468, 469, 485
Gott, J. R., 14, 15, 19, 249
Gott, J. R. III, 128, 412
Gottlieb, C. A., 205, 206, 214
Gould, R. J., 122, 140, 438
Goulding, F. S., 455
Gower, J. F. R., 25-28
Graboske, H. C., 70
Grader, R. J., 480, 490
Graham, B. A., 526
Grands, S., 532
GRASDALEN, G. L., 187-216; 190-92, 195, 198, 201-3, 205-12
Gratton, L., 423
Grayzeck, E. J., 138
Greenberg, J. M., 146
Greisen, E. W., 161
Greisen, K., 438
Grewing, M., 149
Griffiths, D., 31, 32
Griffiths, R. E., 434
Grossman, A. S., 70, 305
Gruber, D. E., 457, 466, 469, 494
Gryzinski, M., 71
Gunn, J. E., 15, 49, 59, 119, 128, 239, 249, 415
Gurr, H. S., 51
Gursky, H., 402, 409, 423-25, 428, 429, 431, 433-36, 445, 448, 449, 459, 461, 466, 468, 469, 499, 505
Guseynov, O. M., 411
Guyer, R. A., 360

Habing, H. J., 138, 145, 146, 155, 262, 277, 279, 282
Hack, M., 148
Hagan, P. J., 178
Hagedorn, R., 372
Hagen, G. L., 228
Hall, R. G. Jr., 303
Hamilton, P. A., 533
Hansen, C. J., 47-50, 62, 64
Hansen, J. P., 361
Harm, R., 10, 53, 59
Harnden, F. R., 457, 474
Haro, G., 190
Harper, D. A., 213
Harri, J., 452
Harrington, R. S., 5, 325
Harris, B., 424
Harris, D. L. III, 330
Harris, W. E., 226, 236, 237
Hart, M. H., 3, 5, 11, 386
Harten, A., 19
Hartle, J. B., 382, 391
Hartman, R. C., 436
Hartwick, F. D. A., 220, 234, 237
Harvey, G. A., 39
Harvey, J., 182
Harwood, J., 177, 182
Hatchett, S., 158, 160
Hawking, S. W., 391, 395, 412, 414-18
Hawkins, F. J., 410
Hayakawa, S., 424, 437, 438
Hayashi, C., 56-58, 188
Hayli, A., 2, 7, 9
Haymes, R. C., 428, 457, 474
Hearnshaw, J. B., 221
Heggie, D. C., 2, 6, 10, 12
Heiles, C., 138, 262, 265, 268, 280
Heintz, W. D., 305, 314, 322, 325
Heisenberg, W., 378
Helfer, H. L., 220
Henderson, A. P., 282
Henke, B. L., 439, 440
Hénon, M., 3, 7, 10-12, 14, 18
Henry, R. C., 139, 218, 237, 424, 437, 438, 447, 449
Henyey, L. G., 187
Herbig, G. H., 145, 146, 156, 190, 192, 193, 197-203, 207
Hershey, J. L., 300, 303, 315, 320, 325, 332
Hesse, K. N., 520
Hesser, J. E., 237
Hetherington, J. H., 358
Hewish, A., 28, 511, 512
Hicks, D. B., 505
Higgs, L. A., 180
Hill, E. R., 24
Hill, J. K., 158
Hill, R. W., 480, 490
Hill, S. J., 183
Hillier, R. R., 528
Hills, J. B., 160
Hills, J. G., 6
Hiltner, W. A., 219, 227, 411
Hindman, J. V., 233
Hirshfeld, A., 135, 137, 140, 144, 153
Hjellming, R. M., 409, 429
Hoag, A. A., 178, 181
Hobbs, L. M., 133, 135, 145
Hodge, P. W., 225, 233, 237, 240, 243, 246
Hoekstra, R., 133
Høg, E. H., 181
Hogg, A. R., 229
Hogg, H. S., 219
Hohl, F., 18
Holden, D. J., 29, 30, 36
Hollenbach, D. J., 141, 142, 144
Hollis, J. M., 166, 171, 178, 183
Holloway, N. J., 514
Holmes, J. A., 70
Holt, S. S., 409, 428, 430, 434, 435, 446, 480, 486, 487
Hooton, I. N., 178
Hoover, P. S., 123, 232
Hoshi, R., 56-59, 188
Howard, A. J., 103
Howard, H. T., 522
Howard, W. E. III, 145
Hoyle, F., 46, 48, 57, 58, 88, 89, 92, 96, 104, 129, 150
Hubbard, W. B., 119
Hubbell, J. H., 438, 442
Hubble, E., 217
Huchtmeier, W., 114
Hughes, M. P., 161, 272
Huguenin, G. R., 525
Hulsbosch, A. N. M., 260-62, 268, 279, 287
Humblet, J., 107
Hunger, K., 197
Hunstead, R. W., 32, 39
Hutchings, J. B., 411
Hyland, A. R., 207

I

Ianna, P. A., 320
Iben, I. Jr., 56, 234
Ichimaru, S., 532
Iddings, C. K., 337, 342, 353
Ikeuchi, S., 56-58
Illingworth, G., 11, 221, 249
Imhof, W. L., 457, 480
Imshennik, V. S., 62
Infeld, L., 378
Ingber, L., 340
Innanen, K. A., 225, 287
Ipser, J. R., 386
Isaacs, J. D., 415
Israel, W., 391
Itoh, N., 49, 64
Ivanova, L. N., 62
Iyengar, V. S., 433

J

Jackson, A. A., 415
Jackson, J. D., 438
Jackson, W. R., 528
Jacobson, A. S., 428, 431, 455, 457, 458, 467, 480, 505
Jagada, N., 459
Janes, K. A., 119, 220
Janin, G., 2, 10, 18
JAUNCEY, D. L., 23-44; 24-33, 35, 39, 41
Jenkins, E., 265
JENKINS, E. B., 133-64; 133, 134, 138, 139, 144-46, 154
Jenner, D. C., 115, 119, 123
Jensen, H. P., 103
Jerde, R. L., 452, 457, 467
Johnson, H., 435
Johnson, H. M., 229
Johnson, M. B., 348-51, 354, 355
Johnson, M. W., 156
Johnson, R. G., 457, 480
Johnson, W. N., 428
Johnston, M., 413
Joly, M., 238
Jones, B. F., 10
Jones, C., 409, 411, 433
Jones, S. G., 180
Jones, T. W., 425, 435, 465
Joseph, G., 474
Joss, P., 429
Joss, P. C., 56
Joyce, R. R., 210
Julian, W. H., 513
Jura, M., 142-44, 156, 158

K

Kafatos, M., 158, 160
Kaler, J. B., 125
Kaliberda, V. S., 8, 10
Kalinina, E. P., 9
Kalman, G., 368
Kalos, M., 361
Kamper, K. W., 321
Kamperman, T. M., 133
Kaplan, S. A., 7, 400, 529, 531-33

Kapteyn, J. C., 299
Kasturirangan, K., 452, 474
Katgert, P., 35-37
Katgert-Merkelijn, J. K., 35-37
Katz, J. I., 409
Keenan, D. W., 225, 287
Kellermann, K. I., 23, 29, 33, 37-39, 41, 42
Kellman, S. A., 279
Kellogg, E., 409, 424, 425, 428, 435, 445, 465, 466, 470
Kenderdine, S., 31, 33
Kennedy, J. E. B., 35
Kepner, M., 279, 283
Kerr, F. J., 138, 140, 258, 273, 277, 280, 282, 283
Kerr, R. P., 383, 389, 393
Keyes, C. D., 119, 123
Kholopov, P. N., 9
Kinahan, B. F., 137, 159
King, I. R., 9-11, 217
Kinman, T. D., 203, 206
Kinter, E. C., 15
Klebesadel, R., 433
Klemola, A. R., 174, 182
Knacke, R. F., 200, 201
Knapp, G. R., 140, 209, 273
Kniffen, D. A., 436
Koch, D., 470
Kojoian, G., 213
Komesaroff, M. M., 519, 533
Koonin, S. E., 107
Kovach, W. S., 138
Kozlov, N. N., 17
Kozlovsky, B.-Z., 92
Kraus, J. D., 35, 262, 268, 279, 288
Kraushaar, W. L., 154, 427, 438, 487, 489, 490
Krishna-Swamy, K. S., 212
Kristian, J., 525, 528
Kron, G. E., 197
Kronberg, P. P., 251
Krzeminski, W., 429
Kuhi, L. V., 199, 205
Kulsrud, R. M., 155, 409
Kumar, C. K., 125
Kunkel, W. E., 229
Kurfess, J. D., 528
Kurz, R. J., 450
Kutter, G. S., 57

L

Labuhn, F., 423
Lacy, J. L., 449
Lada, C. J., 205, 206, 214
Lamers, H. J., 133, 149
Lampton, M., 467, 470
Landau, L. D., 375
Lande, K., 66
Landman, D. A., 181
Langer, W. D., 142, 153, 156, 157, 339, 340, 343
Laplace, P. S., 382
Larach, D. R., 145
Large, M. I., 30
Laros, J., 431, 432, 463, 467
Larson, R. B., 188, 222, 223, 240, 385
Lasker, B. M., 172, 178, 180
Lassila, K. E., 375
Lattimer, J. M., 383, 396
Lauberts, A., 17
Lauritsen, T., 70
Lavakare, P. J., 474
Lavine, J. P., 478
Lazcano-Araujo, A., 160
LECAR, M., 1-21; 1-3, 16, 18, 19
Leckrone, D. S., 148
Ledoux, P., 58
Lee, R. H., 181
Lee, T. A., 198
Lefevre, F., 11
LeLevier, R., 187
Le Poole, R. S., 23, 35-37
Lequeux, J., 35-37, 39, 152
Lerche, I., 515, 517, 529
Lesh, J. R., 152
Leung, Y. C., 371
Levee, R. D., 187
Levesque, D., 361
Lewin, W. H. G., 452, 467
Li, F. K., 410
Liden, K., 443
Lightman, A. P., 407, 411
Liller, W., 113
Lilley, A. E., 205, 206, 214
Limber, D. N., 9
Lindsey, R. S., 449
Ling, J. C., 457, 474, 476, 481, 494
Linnell, A. P., 183
Linsky, J. L., 139
Lippincott, S. L., 300, 301, 305, 314, 315, 319, 320, 322, 323
Little, G. A., 23, 29, 31, 32
Litvak, M. M., 205, 206, 214
Livingston, W., 182
Lloyd-Evans, T., 227, 232
Lo, K. Y., 213
Lock, J. L., 39
Lockhart, I. A., 239
Lockhart, P., 161
Lodenquai, J., 64, 360, 361, 363, 366, 374, 377
Loh, E. D., 433
Long, R. J., 29, 30
Longair, M. S., 27, 28
Loren, R. B., 192, 210
Lucaroni, L., 355
Lüst, R., 400, 423
Lynden-Bell, D., 11, 18, 386, 401
Lyne, A. G., 526
Lyutyi, V. M., 411

M

MacConnell, D. J., 123, 232
Machalski, J., 34
Mack, D. A., 178
MacKay, H. B., 459, 481
Mackey, M. B., 29
Macklin, R. L., 455
MacLeod, J. M., 30
Madore, B. F., 220, 231
Magnan, C., 203
Mahoney, W., 467
Manchanda, R. K., 452
Manchester, R. N., 519, 525
Mankin, W. G., 181
Maraschi, L., 529
Margon, B., 404, 410, 433, 438, 467, 470
Marion, J. B., 110
Marks, G. H., 95
Marschall, L. A., 133
Martin, A. R., 324
Martin, G. E., 320
Martins, P. de A. P., 118
Martynov, D. Ya., 403
Marx, G., 368
Mashoon, B., 396, 412
Maslowski, J., 31, 33-37
Mason, K. O., 410
Mathews, W. G., 211, 249
Mathewson, D. S., 262, 267, 274, 275, 277, 281, 292
Mathis, J. S., 115, 119
Matteson, J. L., 423, 431, 432, 454, 457, 462, 463, 467, 490, 494
Matthews, H., 525, 528
Mayall, N. U., 218, 237
Mazurek, T. J., 47, 48, 61, 64
McCammon, D., 154, 437, 438, 487
McClintock, J. E., 467, 487
McClintock, W., 139
McClure, R. D., 119, 120, 221, 223, 225, 230, 237, 239, 241-43, 246, 249
McCord, T. B., 182
McCray, R. A., 146, 155, 158, 160
McCrea, W. H., 520
McFee, R. W., 48, 49
McGee, M., 452
McGee, R. X., 145, 233
McGoff, C. W., 490
McIlwain, C. E., 472

McNall, J. F., 183
McQueen, R. M., 181
Meaburn, J., 280
Medd, W. J., 39
Meier, R. R., 139, 161
Meloy, D. A., 154
Mendoza, E. E., 197
Menefee, J., 450
Meng, S. Y., 262, 268, 279, 288
Merkelijn, J. K., 23, 38, 40, 41
Merrill, K. M., 207
Mészaros, P., 146, 158, 160
Metzger, A. E., 478
Metzger, P. G., 114, 115
Meyer, P., 471, 472
Meyerott, A., 461
Michalski, D. E., 183
Michaud, G., 90, 96, 101, 103
Michel, F. C., 514
Michie, R. W., 1, 5, 10
Mickiewicz, S., 449, 459, 485
Miedaner, T. L., 183
Milekhin, G. A., 378
Miley, G. K., 409
Milkey, R. W., 196
Miller, J. S., 522, 523
Miller, M. D., 366
Miller, R. H., 2, 3, 6
Mills, B. Y., 23-25, 29, 31, 32, 34, 42
Milne, D. K., 119
Milton, J. A., 233
Minkowski, R., 251
Mintz, B. F., 308
Miranda, L. C. M., 532
Misner, C. W., 383, 389, 396, 412-14
Monnet, G., 121
Moore, C. H., 171
Moore, E. P., 183
Moos, H. W., 139
Morfill, G., 480, 494
Morgan, W. W., 218, 220, 237, 242
Morris, D., 35
Morris, M., 191, 192, 205, 209
Morrison, N. D., 411
Morton, D. C., 133-38, 145, 146, 148, 151-53, 156, 157, 238
Morton, W. A., 134, 144
Moszkowski, S., 343, 345, 347
Muller, C. A., 258
Muller zum Hagen, H., 418
Mullin, W. J., 358
Muncaster, G. W., 519
Munsuk, C., 227
Murai, T., 56-58
Murdin, P., 409, 410
Murdoch, H. S., 24-26, 28-30
Murray, A., 528
Murray, J. D., 161
Murray, S., 435

N

Nachman, P., 135
Nadezhin, D. K., 47, 62
Nakano, G. H., 457, 480
Nakazawa, K., 56-58
Naranan, S., 433
Nassau, J. J., 218
Nather, R. E., 173, 179, 182
Nauenberg, M., 382
Nduka, A., 396
Ne'eman, Y., 417
Nelms, A. T., 440, 443
Nemeth, J., 368
Neugebauer, G., 191, 208, 213, 525, 528
Neville, A. C., 25, 27, 28
Newell, E. B., 235
Newman, E. T., 413
Nichols, D. B., 108, 110
Nicholson, G., 38
Niell, A. E., 30-32, 35
Nielsen, R. F., 180, 181
Nijgh, G. J., 443
Nobili, L., 397
Nörlund, N. E., 297, 322
Nosanow, L. H., 357, 358, 362
Novick, R., 423, 428
Novikov, I. D., 46, 59, 383, 385, 397, 399, 405-7, 414, 415, 417, 418
Nussbaumer, H., 118

O

Oda, M., 409, 423, 430-32, 468
O'Dell, C. R., 119, 212
O'Donnell, E. J., 144, 156
Oemler, A., 15
Ögelman, H., 423, 467
Oke, J. B., 525, 528
Olowin, R. P., 412
Olson, D. W., 413
Olson, R., 433
Olthof, H., 148
O'Mongain, E., 469
O'Neil, E. J., 235
O'Neill, T. G., 180
Oort, J. H., 9, 10, 149, 218, 219, 260, 272, 276, 277, 279, 282, 287
Oppenheimer, J. R., 382
Orrall, F. Q., 181
Osmer, P. S., 233, 411
Osterbrock, D. E., 113, 220, 237, 242
Østgaard, E., 358
Ostriker, J. P., 12, 15, 16, 19, 49, 59, 221-23, 239, 245, 248, 382, 387, 402, 404, 405, 410, 429
Osvalds, V., 314, 324
Osvalds, Z., 314
Otsuki, S., 354
Overbeck, J., 430

P

Paavola, S., 182
Pacholczyk, A. G., 397, 398, 401, 407, 520
Pacini, F., 528-30
Paczynski, B., 403, 404, 409, 410
Paczynski, B. E., 48, 53, 55-57, 60, 61
Page, D. N., 407
Page, T. L., 15
Pagel, B. E. J., 145
Pal, Y., 435
Palmer, P., 191, 192, 205, 209
Palmer, R. G., 356, 357
Panagia, N., 152
Pandharipande, V. R., 341-43, 349-52, 360, 363, 364
Parish, L., 360, 361, 363
Parish, L. J., 362
Parker, E. N., 156
Parker, P. D., 95
Parsons, S. B., 231
Pasachoff, J. M., 149
Patrick, T. J., 424
Pauliny-Toth, I. I. K., 23, 29, 33, 37-39, 41, 42
Payne-Gaposchkin, C., 229, 230
Peach, J. V., 221
Peacock, A., 434
Peebles, P. J. E., 14-16, 19, 221-23, 239, 248, 383, 387, 401, 402
PEIMBERT, M., 113-31; 113-29, 160, 220, 230, 232-34, 245
Pels, G., 9
Pels-Kluyver, H. A., 9
Penrose, R., 412, 417, 418
Penston, M. V., 207, 241
Pequignot, D., 159
Perry, M. E., 411
Persson, S. E., 125, 191, 198
Peters, C. F., 2
Peters, W. L., 210
Peterson, B. A., 411
PETERSON, L. E., 423-509; 423, 425, 428, 431, 433, 435, 450, 452, 454, 457, 459, 460, 462, 465-67, 473-75, 478, 480, 481, 494, 497-500, 504
Peterson, V., 47
Petrosian, V., 48, 49, 51, 62, 528

Petrovskaya, I. V., 8, 10
Petterson, J. A., 408
Pikel'ner, S. B., 400
Pilkington, J. D. H., 25, 26, 30, 511
Pinaev, V. S., 49
Pines, D., 64
Pisano, D. J., 95
Plagemann, S., 197
Planck, M., 414
Plaut, L., 219
Plavec, M., 403
Podurets, M. A., 389
Polnarev, A. G., 397
Pomeranchuk, I. Ya., 375
Pooley, G. G., 29-31, 33
Potenza, R. M., 340
Pottasch, S. R., 122, 133, 137, 143, 148, 149, 404
Pounds, K., 424, 431
Poveda, A., 3, 8, 10
Praderie, F., 148
Prata, S. W., 10
Prendergast, K. H., 400, 404
PRESS, W. H., 381-422; 16, 391, 392, 396, 411-13, 415
Price, R. H., 391, 418
Price, R. M., 411
Price, S. D., 212
Pringle, J. E., 397, 398, 401, 407, 409
Pritchet, C. J., 251
Przyblyski, A., 233

R

Racine, R., 237
Radeka, V., 455
Radhakrishnan, V., 161, 515, 519, 520, 526
Raimond, E., 258, 260-62
Rakavy, G., 52, 53, 55-58, 62, 385
Rananbaum, H., 435
Rank, D. M., 145
Rankin, J. M., 533
Rao, N. K., 190
Rao, U. R., 474
Rappaport, S., 424, 430, 431
Rather, E. D., 171
Ray, E. C., 386
Raychaudhuri, P., 46
Reagan, J. B., 457, 480
Recillas-Cruz, E., 117
Redfern, R. M., 528
Rees, M. J., 383, 386, 392, 397, 398, 401, 407, 414, 529, 530
Refsdal, S., 27
Reidy, W. P., 423, 424
Reines, F., 51
Renzini, A., 227, 248
Reuyl, D., 320
Reynolds, R. J., 160
Rhoades, C. E., 374, 382
Rickard, J. J., 268, 272, 281
Ricker, G., 432, 493
Ricker, G. R., 467
Riegler, G., 431
Righini, G., 210
Ring, J., 181
Rios, M., 103
Roberts, D. H., 513
Roberts, M. S., 113, 239, 248, 275
Roberts, R., 179, 183
Robertson, J. G., 29, 31-34, 42
Robinson, D. C., 391
ROBINSON, L. B., 165-85; 174, 182, 198
Rodgers, A. W., 179, 183
Rodney, W. S., 96, 106, 110
Rodriguez, L. F., 116, 117, 121, 125
Roesler, F. L., 160
Rogerson, J. B., 133, 158
Rogerson, J. B. Jr., 128, 134, 149
Rogstad, D. H., 239
Rolfs, C., 95, 106, 110
Rood, H. J., 15, 239, 243, 251
Rood, R. T., 227, 234, 246
Rose, W. K., 60, 410
Rosen, L., 339, 340
Rosenberg, F., 434
Rosenbluth, M. N., 409
Rosenwald, R. D., 388
Ross, J. E., 233
Rossi, B. B., 438, 445
Rothe, E. D., 450, 457
Rothschild, R. E., 409, 428, 430
Rots, A. H., 239, 248
Rubin, V. C., 121, 125
Ruderman, M., 64, 512
Ruderman, M. A., 46, 49, 61, 513, 514, 517, 531, 533
Rudge, P. T., 179, 183
Ruffini, R., 370, 374, 382, 387, 391, 413
Rugge, H. R., 148
Ryan, M. P., 386, 415
Rybakov, A. I., 9
Ryle, M., 25, 27-30, 33
Ryter, C., 140

S

Saakyan, G. S., 336, 338, 339
Sabbadini, A. G., 382
Sack, N., 58
Saggion, A., 49
Sale, R. G., 528
Salmi, D. S., 490
Salpeter, E. E., 47-49, 51, 56, 62, 141, 142, 144, 149, 222, 336, 383, 397
Sandage, A. R., 127, 232, 234, 236, 237, 243, 251, 252
Sanders, R. H., 386
Sanders, W. T., 154
Sanduleak, N., 123, 232
Sanford, P. W., 410
Saraph, H. E., 118
Sargent, W. L. W., 222, 229, 241
Sarkady, A. A., 474
Saslaw, W. C., 16, 19
Sastry, K. S., 16
Satz, H., 376
Savage, A., 410
Savage, B. D., 133, 138, 139, 142, 146, 149
Savedoff, M. P., 56, 57
Sawyer, R. F., 353, 371, 372
Scargle, J. D., 523
Scarinci, C., 355
Schatzman, E., 203
Schechter, P., 16, 411
Scheer, D. J., 35
Scherb, F., 160
Scheuer, P. A. G., 37, 41
Schiff, D., 360, 361
Schiffer, F. H. III, 119
Schlitt, D., 48, 49
Schlosser, W., 219
Schmidt, M., 222, 223
Schmidt-Kaler, T., 219
Schnopper, H. W., 423
Schönberg, M., 45, 47
Schramm, D. N., 64, 128, 129, 149, 150, 222, 383, 385, 386
Schreier, E., 409, 426, 429, 433
Schubert, G., 387
Schwartz, D. A., 423, 435, 436, 446, 459, 461, 466, 474, 480, 486, 487, 505
Schwartz, R., 203, 205, 210
Schwarz, J., 160
Schwarz, R. A., 62
Schwarz, W. J., 258
Schwarzschild, M., 53, 59
Schwinger, J. S., 414
Scott, P. F., 25, 26, 30, 511
Searle, L., 114-17, 119-21, 124, 126, 220, 222, 223, 229, 230, 240, 241
Sears, R. L., 46, 51
Seaton, M. J., 113, 114, 118
Seeliger, H., 297, 321
Seifert, H. J., 418
Serkowski, K., 197
Serlemitsos, P. J., 409, 428, 430, 446, 486, 487
Seward, F., 490
Seward, F. D., 480

Shaffer, D. S., 38, 39, 41
Shakeshaft, J. R., 25
Shakura, N. I., 406, 407
Shapiro, M. H., 95, 106, 110
Shapiro, S. L., 11, 398, 411
Shapley, H., 246
Shaposhnikov, V. E., 519, 520, 526, 528
Sharpe, J., 445, 449, 454
Shaviv, G., 52, 53, 55-58, 62
Shaw, G. L., 357
Shaw, P. B., 92
Sheather, P. H., 424
Shemming, J., 118
Shen, L., 351
Sher, D., 9, 10, 249
Shields, G. A., 120, 240
Shimmins, A. J., 23, 29, 38, 40, 41
Shklovskii, I. S., 59, 528
Shulman, S. D., 424, 437, 438, 447, 449
Shvartsman, V. F., 397, 398
Sieber, W., 520
Siedentopf, H., 17
Silk, J., 155, 158, 161, 436, 438
Silvan, J. P., 152
Silvestro, M. L., 47
Simon, M., 210
Simpson, J. P., 119
Simpson, M., 417
Sinha, R. P., 533
Slaughter, C., 182
Slaughter, C. D., 181
Slee, O. B., 24, 25
Slettebak, A., 182
Smart, N. C., 434
Smith, A. M., 133, 148
Smith, B. W., 154, 160
Smith, F. G., 519, 520, 526
Smith, H., 3
Smith, H. E., 119, 121, 224, 239, 244
Smith, L. F., 230
Smith, M., 200
Smith, M. A., 29, 30
Smith, M. G., 123
Smith, W. B., 452
Smith, W. H., 135, 151, 153
Snell, C. M., 229
Snellen, G., 182
Sobel, H. W., 51
Sofia, S., 59
Sofue, Y., 281, 291
Solf, J., 180, 182
Solinger, A., 430, 431
Somerville, W. B., 140
Spada, G., 423, 431, 432, 468
Spinrad, H., 113, 115-17, 119, 198, 217, 239-42, 244, 252
Spitzer, L., 3, 5, 7, 8, 10-12, 14, 15, 133-35, 137, 139, 140, 142-44, 146, 153-57
SPITZER, L. JR., 133-64; 386, 402
Sreekantan, B. V., 423, 431-33, 468
Standish, E. M., 2, 3, 7, 9
Stapinski, T., 179, 183
Starfelt, N., 443
Starobinsky, A. A., 413, 415
Stassinopoulos, E. G., 478
Staub, H. H., 445
Stecher, T. P., 138, 148
Stegun, I. A., 477
Steigman, G., 143, 150, 158, 159, 372
Stein, W., 452
Stein, W. A., 195
Stephens, C. L., 178, 181, 183
Stephens, S. A., 474
Stephens, T. L., 142
Stevens, J., 428, 431
Stewart, J., 396, 413
Stewart, P., 29, 30
Stoering, J. P., 490
Stone, R. P. S., 410
Stothers, R., 386
Stothers, R. B., 46, 53, 55, 56, 59
Straka, W. C., 305
Strand, K. Aa., 298, 330
Strand, R. C., 166
Strittmatter, P. A., 143
STROM, K. M., 187-216; 190-92, 195, 198, 200-3, 206-10, 212
STROM, S. E., 187-216; 190-92, 195, 198, 200, 201, 203, 205-10, 212, 213
Strömgren, B., 135, 145
Strong, I., 433
Sturch, C. R., 220
Sturrock, P. A., 513, 515, 519, 521
Sugimoto, D., 53, 57, 188
Suhonen, E., 375, 376
Sullivan, W. J., 258, 277, 280, 293
Suntantyo, W., 404, 412
Sunyaev, R. A., 17, 405, 407, 408, 411
Suri, A. N., 457, 480
Sutherland, P. G., 513, 514, 517, 531, 533
Suvorov, E. V., 518, 530, 532
Swarup, G., 533
Swenson, G. W. Jr., 30
Swinehart, C., 450
Swope, H. H., 246
Szebehely, V., 2

T

Taam, R. E., 404
Tademaru, E., 517, 525
Talbot, R. J., 145, 223
Talpaert, Y., 11
Tamagaki, R., 354
Tammann, G. A., 229
Tanaka, Y., 487
Tananbaum, H., 409, 411, 424, 425, 428, 430, 433, 445, 466
Tarter, C. B., 158
Taylor, B. J., 119, 239, 241
Taylor, B. N., 69
Taylor, D. J., 119, 239, 244
Taylor, J. H., 525
Teague, M. J., 478
Teitelboim, C., 391, 396, 412, 413
ter Haar, D., 512, 520, 530, 532
Terzian, Y., 160
Teukolsky, S. A., 413
Thompson, A. R., 161, 272
Thompson, J. H., 25
Thompson, R. I., 423
Thorne, K. S., 383, 385, 386, 389, 396, 397, 399, 405-8, 411, 414, 418
Thuan, T. X., 7, 11, 150, 153, 238, 386
Thüring, B., 6
Tifft, W. G., 225-27, 229, 243, 246
Tinsley, B. M., 128, 222
Tolbert, C. R., 258, 260, 262
Tombrello, T. A., 101, 107
Toomre, A., 17, 291
Toomre, J., 17, 291
Toor, A., 452
Torres-Peimbert, S., 114-17, 119, 121, 123-29, 160, 232, 234
Tosa, M., 281, 291
Townes, C. H., 145
Tremaine, S. O., 402
Treves, A., 529
Trimble, V., 374, 410
Trimble, V. L., 411
Trombka, J. I., 436, 454, 470, 478, 480, 482
Truran, J. W., 61, 92, 150, 222, 224
Tscharnuter, W., 190
Tsuneto, T., 49, 64
Tsuruta, S., 47, 48, 61, 64, 65
Tsytovich, V. N., 529, 531-33
Tucker, W. H., 402
Turner, B. E., 191, 192, 205, 209

U

Ulmer, M. P., 425, 433, 459, 481
Ulrich, R. K., 109
Unruh, W. G., 413
Usov, V. V., 513

V

Vaiana, G. S., 423, 424
Valtonen, M. J., 3
van Agt, S. L. T. J., 246
van Albada, T. S., 4, 5, 8, 9
van Altena, M. F., 10
Van Breda, I. G., 178, 183
van de Hulst, H. C., 146, 149
VAN DE KAMP, P., 295-333; 297, 307, 308, 314, 320-22, 324, 326, 327, 330
VAN DEN BERGH, S., 217-55; 9, 10, 119, 125, 218-23, 228, 230-37, 239-44, 249, 251
Vanden Bout, P. A., 192, 210
van den Heuvel, E. P. J., 404, 410, 414
Vanderhill, M., 489, 490
van der Hucht, K. A., 133
van der Laan, H., 23, 35-37
Van der Leun, C., 70
Van Dilla, M. A., 478
Van Horn, H., 56
van Kampen, N. G., 363
van Kuilenburg, J., 262, 274
Van Lieshaut, R., 443
Van Speybroeck, L. P., 423, 424, 445, 448, 466, 468
Vasilevskis, S., 174, 182
Verlet, L., 361
VERSCHUUR, G. L., 257-93; 258, 259, 261-63, 265, 266, 268, 269, 271, 272, 277, 279-82, 284-87, 291
Vetešnik, M., 219
Vette, J. I., 478, 494
Vidal, N. V., 411, 429
Viehmann, W., 450
Vila, S. C., 56
Vogt, E., 96, 101
Volkoff, G., 382
von Hoerner, S., 5, 7, 9, 28
von Neumann, J., 4, 245
Vrba, F. J., 190, 191, 198, 206, 208, 210

W

Wada, M., 354
Wade, C. M., 409
Wagman, N. E., 323
Wagoner, R. V., 88, 109, 126-29, 149, 150, 383, 388
Walborn, N., 245
Walborn, N. R., 410
Wald, R. M., 391, 412, 418
Walecka, J. D., 368, 369, 377
Walker, A. B. C., 148
Walker, G. A. H., 411
Walker, M., 396
Walker, M. F., 115, 119, 188, 197, 199, 209, 220, 225-27, 229, 245
Walker, R., 212
Wall, J. V., 23, 38-41
Wallerstein, G., 145, 236
Wallis, M. K., 161
Walmsley, C. M., 149
Walraven, J. H., 229
Walraven, T., 229
Wampler, E. J., 174, 182, 198, 522, 523
Wang, C. G., 371
Wannier, P., 261, 262, 273, 274
Wapstra, A. H., 443
Wares, G. W., 233
Warner, J. W., 123, 125
Watson, W. D., 144, 146, 156, 157
Watt, S., 423
Wayland, J. R., 423, 467
Wayte, R. C., 181
Weaver, T. A., 150, 399
Weber, J., 410
Webster, B. L., 409
Wehner, H., 176, 180
Weigert, A., 59
Weinberg, S., 50, 378
Weisheit, J. C., 142, 158
Weiss, K., 148
Welch, G. A., 123, 239, 241
Welch, W. J., 145
Weller, C. S., 139, 161
Wells, D. C., 172, 182
Wentzel, D. G., 145, 155
Werner, M. W., 141, 142, 155, 209
Wesselius, P. R., 260, 262, 268, 281, 282, 287
Westbrook, W. E., 209
Westerhout, G., 282
Westerlund, B. E., 230
Wheaton, W. A., 425, 433, 435, 459, 460, 465
Wheeler, J. A., 383, 387, 389, 396, 412, 417
Wheeler, J. C., 47, 60-62, 372, 385, 388
Whelan, J. A. J., 409
White, N. M., 182
White, R. H., 62, 63
White-Grodstein, G., 438
Whitehurst, R. N., 275
Whiteoak, J. B., 35
Whitford, A. E., 114, 238
Whittle, R. P. J., 161
Wick, G. L., 415
Wickramasinghe, D. T., 411
Wielebinski, R., 180, 520
Wielen, R., 1-4, 7-10, 12
Wildey, R., 234
Williams, P. J. S., 29, 30, 33, 37, 39
Williams, R. E., 143
Williamson, F. O., 154, 489, 490
Wills, D., 25
Willson, M. A. G., 31
Wilson, J. R., 51, 62, 63
Wilson, R. E., 412
Wilson, R. W., 261, 262, 273, 274
Wilson, T. L., 115
Winkler, H., 95, 106, 110
Winnberg, A., 183
Withbroe, G. D., 145
Witt, A. N., 156
Wolf, A. M., 386
Wolf, B., 233
Wolf, R. A., 48-50, 62, 64, 151
Wolfe, J. L., 180
Wolff, R., 177, 182
Wolff, S. C., 411
Woltjer, L., 125, 374, 512
Womack, E. A. Jr., 455, 457
Woo, C. W., 351, 366
Wood, D. B., 198
Wood, R., 11, 386
Woolf, N. J., 10
Woosley, S. E., 70
Worley, C. E., 330
Worth, M. D., 326
Wright, A. E., 17
Wright, E. L., 143
Wright, M. C. H., 239, 275, 291
Wrixon, G. T., 261, 262, 273, 274
Wu, F. Y., 361
Wulf-Mathies, C., 149
Wyckoff, S., 319
Wyllie, D. V., 31, 33
Wynn-Williams, C. G., 191, 208, 212, 213

Y

Yabushita, S., 17
Yamashita, K., 487
Yang, C. Y., 314
Yang, K. S., 30
Yenicay, G., 452, 467
Yodzis, P., 418
York, D. G., 128, 133, 134, 136, 137, 140, 142, 146, 149, 153, 154, 158, 159, 161
Yost, J., 190-92, 195, 201

Young, E., 200, 201
Young, J. W., 119, 239

Z

Zaitsev, V. V., 515, 518, 530-33
Zane, R., 181
Zappala, R. R., 201, 410
Zare, K., 2
Zarnecki, J. C., 424, 431
Zaumen, W., 435
Zehnpfenning, T. F., 423, 424
Zel'dovich, Ya. B., 46, 59, 368, 397, 411, 412, 415, 418
Zerilli, F., 413
ZHELEZNYAKOV, V. V., 511-35; 515, 517-20, 526, 528, 530-33
Zieba, S., 34
ZIMMERMAN, B. A., 69-112; 69, 70, 103, 107
Zimmerman, R. L., 370
Zinamon, Z., 52, 55-58
Zuckerman, B., 191, 192, 205, 209, 211
Zulliger, H. R., 452, 467
Zupancic, C., 449
Zweibel, E. G., 144
Zych, A. D., 528

SUBJECT INDEX

A

Aluminum
  low-lying states of, 90
  target of thermonuclear reactions, 85, 103
Argon
  interstellar content of, 135, 136, 148
Astrometry
  dynamical interpretation
    mass function, 309, 310
  frequency of dark companions, 330
  long-focus photographic, 297-305
    accuracy of, 297-300
    parallax, 300, 301, 305
    photocenter concept of, 303
    resolved binaries, 301, 302
    unresolved binaries, 303-5
  perturbations in
    Alpha Ophiuchi, 322, 323
    analysis of, 306-9
    Barnard's star, 323, 324
    BD+6° 398, 319, 320
    BD+27° 4120, 325, 326
    BD+43° 4305, 326, 327
    BD+66° 34A, 311, 317, 319
    BD+67° 552, 322
    BD+68° 946, 323
    Chi Draconis, 324
    Delta Aquilae, 324
    density function, 327, 328
    discovery of, 305, 306
    dynamical elements in, 306, 307
    Epsilon Eridani, 320
    Gamma Geminorum, 321
    general formulas, 308, 309
    general survey, 327-30
    geometric elements, 307, 308
    GV24-16, 325
    luminosity function, 328, 329
    mass-luminosity relation, 328-30
    Mu Cassiopeiae, 319
    observations of, 299, 300
    properties of unresolved binaries, 310-13
    Ross 614, 316, 320, 321
    stars with uncertain perturbations, 311, 314, 315
    VW Cephei, 325
    Xi Ursae Majoris A, 322
    Zeta Aquarii, 326
    Zeta Cancri C, 321, 322
  unseen astrometric companions
    individual objects found, 310-27

B

Beryllium
  neutrons produced from, 103
  target of thermonuclear reactions, 75-77
Black holes
  astrophysical processes near, 381-418
    black hole formation events, 382, 391
    black hole masses and effects, 384
    black hole maturation, 383
    collapse of galactic nuclei, 386
    collapse of nonrotating stars, 383-85
    collapse of rotating galactic nuclei, 388
    collapse of rotating stars, 386-88
    collapse of rotating supermassive stars, 388
    collapse of supermassive stars, 385, 386, 388
    "collapse, pursuit, and plunge" scenario, 392
    description of spherical collapse, 388, 389
    diagram of black hole formation, 390
    emission of gravitational radiation, 389
    event horizons and, 381, 382, 389, 391, 393
    gravitational collapse, 383-95
    Kerr black hole, 393-95
    luminosity decay in collapse, 389
    "no-hair" theorem, 389, 390
    nonspherical collapse, 389-92
    onset of collapse, 383, 385
    overview of, 381-83
    radiation types from collapse, 392
    reviews on, 383
    rotation of, 393
    second law of black-hole

dynamics, 391
space-time geometry of, 393
static limit and, 393, 394
surface gravity of, 395
time relations in collapse, 389
view of exterior of Kerr black hole, 394
binary systems and, 402-12
accretion disks in, 406-9
accretion-powered X-ray sources and, 404-6
black holes in optical binaries, 411, 412
evolution of the hole, 408
explosions in, 404
fluctuations in, 407, 408
Lagrange point in, 403-5
mass flow into black hole, 405
nonalignment of plane and axis, 408, 409
optical identification of, 409-12
radial accretion disk structure, 406, 407
Roche lobes, 403, 405
stellar evolution in close binaries, 403, 404
vertical accretion disk structure, 407
X-ray sources possibly black holes, 409-11
exotic possibilities for, 412-18
black holes inside stars, 416, 417
caveats and naked singularities, 418
charge on black holes, 413, 414
energy extraction from rotating holes, 412, 413
evaporation of black holes, 416
"Kerr-Newman" black holes, 413
primordial black holes, 414, 415
quantum processes of microscopic holes, 415, 416
Reissner-Nordstrom black hole, 413
superradiant scattering, 413
white holes and wormholes, 417, 418
interaction with their environment, 395-402
accretion disks and, 399, 406-9
accretion of matter into hole, 397-99
black-hole generated gamma rays, 399
equatorial motion and, 396
hole rotation and, 396
light from orbiting body, 397
magnetic field dragged into, 398
orbits around a black hole, 395-97
radiation from matter accretion, 398
surrounding star cluster or galaxy, 401, 402
tidal breakup of bodies, 395, 396
X-ray emission from, 430, 431
Boron
interstellar gas content of, 148

C

Calcium
disk of M31 content of, 239
interstellar content of, 135, 145, 157
stellar spectra and, 272, 273
Carbon
burning of, 110
interstellar gas content of, 134, 136, 140, 146, 149, 151, 153, 156-59
nebular content of, 113
neutrons produced from, 103
target of thermonuclear reactions, 77, 78, 85, 102, 103
thermonuclear reactions involving, 101-3
Carbon monoxide
interstellar gas content of, 144, 145
stellar content of, 238
Chlorine
interstellar gas content of, 158
Clouds, galactic
intermediate velocity, 268, 270, 271
maps of, 263-69
velocities of, 258
Clouds, interstellar
velocities of, 133
Computers, astronomical, 165-84
categories of on-line systems, 177, 178
computer data acquisition
advantages of, 175
requirements for, 175, 176
computer hardware, 169, 170
computer interfacing:
CAMAC, 178, 179
computer software, 170-72
assembly languages, 170
compiler languages, 170, 171
interpretative languages, 171
special languages, 171, 172
general comments, 179-84
historical summary of use of, 166
list of those active, 179, 180
problems with, 179
software modules, 172, 173
telescopes controlled by, 165-84
terminology of, 166-69
utility programs, 172
Cyanogen
giant stars and, 218, 220
star spectra and, 237, 239, 242

D

Densities, ultrahigh
equation of state at, 335-78
Deuterium
cosmological evolution and, 129
interstellar gas content of, 149, 150

E

Electrons
capture of
thermonuclear reactions of, 87, 88
Equation of state at ultrahigh densities, 335-78
hyperonic liquids and, 336-53
baryonic density and, 338, 339
baryonic density vs concentration, 343, 346, 347, 350
baryonic liquid compostion, 341
concluding remarks on, 352, 353
energy for various particles, 349
energy per baryon, 345, 348, 351
formulation of problem, 336, 337
hyperonic potentials and, 337, 338

hyperons and, 353
many-body theory and, 351
matter density vs concentration of, 340
models for, 341, 342, 349-52
Moszkowski's models, 345-47
multicomponent many-body theory, 337-39
NN potentials and, 348
scattering cross sections in, 342
work of Ambartsumyan and Soakyan, 339
work of Bethe and Johnson, 348-52
work of Buchler and Ingber, 340
work of Langer and Rosen, 339, 340
work of Moszkowski, 343-48
work of Pandharipande and Garde, 341-43
hyperons and, 335
Landau hydrodynamic model of, 336
neutron states and, 335
region $\rho > 8 \times 10^{15}$ g cm^{-3}, 366-73
baryonic states, 371, 372
concluding remarks concerning, 373
energy per particle (MeV), 367
equation of state for a pure neutron gas, 369
generalities concerning, 366, 367
Hagedorn-type formulations, 371-73
mean-field approximation, 368-70
perturbation approach, 370, 371
relativistic hadronic Lagrangian, 367, 368
review of results, 368-73
relation $P = c_s^2\epsilon$ in the limit $\epsilon \rightarrow \infty$, 373-78
"best" composite $P = P(\epsilon)$ relation, 377
conclusions concering, 378
generalities concerning, 373
high energy p-p collisions, 374-78
hydrodynamic model of, 374-78
neutron stars and, 374
transverse momentum vs E_1, 376
solid core and, 353-66
boson system and, 361, 362
classical lattice-dynamics analysis and, 356
elastic deformation vs density, 361
equation of state for hyperonic matter, 355
equation of state in solid region, 366
law of corresponding states and, 356
liquid-crystal computation, 357
liquid-solid transitions, 362, 363
neutron solidification and, 354-66
pure neutron gases and, 354
solidification densities, 364
work of Banerjee, Chitre, and Garde, 356
work of Canuto and Chitre, 358-61, 365, 366
work of Cazzola, Lucaroni, and Scarinci, 355, 356
work of Coldwell, 357, 358, 365
work of Guyer, 366
work of Nosanow and Parish, 362-64
work of Pandharipande, 363-65
work of Schiff and, 361, 362, 365
velocity of sound and, 335
Extragalactic flux
model of, 438
Extragalactic objects
X-ray emission by, 435

F

Fluorine
target of thermonuclear reactions, 81

G

Galactic clusters
X-ray emission from, 434, 435
Galaxies
evolution of, 252
globular clusters and, 234, 235
infall into
high-velocity hydrogen and, 277-80
matter into galactic plane, 260
masses of
distribution of, 248, 249
nuclei of
black hole accretion disks in, 401, 402
gravitational collapse of, 386, 388
outer arms of
velocities of, 259
rotating nuclei of
gravitational collapse of, 388
spiral arms of
high-velocity hydrogen and, 278, 279
spurs of
high-velocity clouds and, 280, 281
stellar populations in, 217-52
age gradients, 220, 221
chemically inhomogeneous models, 223
color-magnitude relations in, 218
composition gradients in, 220, 221
dwarf spheroidal galaxies, 246-48
elliptical galaxy evolution, 249-52
enrichment in disk or halo?, 223
evolution of, 125-27, 221-23
gas inflow models, 222, 223
globular clusters and, 234, 235
luminosity of galaxy and, 221
Magellanic Clouds and, 224-34
metal content of, 221-24, 234-40, 245, 246, 248
metal-dwarfs in, 223
missing masses in, 248, 249
models of, 218, 221-24
M31 and its companions, 235-46
population types and, 218-20, 223
rotation of, 217
time-dependent mass spectrum of, 222
see also M31 and companions
Galaxies, dwarf elliptical
star formation in, 241, 242
Galaxies, dwarf spheroidal, 246-48
Galaxies, elliptical
evolution of, 243, 249-52
Galaxies, Local Group
high-velocity clouds and, 280
Galaxies, spiral
chemical gradients across, 120-22
disks of, 239
metal distribution in, 240

nuclear radio components
of, 122
Galaxy
bending disk of, 282-84
center of gravity of
galactocentric distance
and, 285
coronal material in, 154
diffuse X-rays in, 436,
437
gaseous halo of, 279
high-velocity clouds in
map of, 289
Hubble type of, 219, 220
maps of
black-hole fraction of,
385
noncircular motions of, 285-
88
nuclear bulge in, 217
spiral structure of
map of, 286, 290
Galaxy NGC 5253
N/O abundance in, 123
Gamma radiation
see Radiation, gamma
Gas, hot hadronic
model of, 375
Gas, interstellar
abundance in H I regions,
145-50
deuterium, 149, 150
general abundances, 145-
49
atomic hydrogen in, 137-40
Copernicus spectrometry
of, 133-35
equilibrium condensation
hypothesis of, 148
general picture of, 155
interstellar molecules, 140-
45
H_2 molecules, 140-44
other molecules, 144,
145
ionization and thermal equilibrium, 155-61
alternatives, 160, 161
cool H I clouds, 155-58
intercloud medium, 158,
159
ionization by energetic
radiation, 155
summary of, 161
pattern of depletion of elements, 146-48
photomultiplier studies of,
133, 134
properties of ionized gases,
150-54
high-temperature low
density gas, 153, 154
normal H II regions, 150-
53
radial velocities in, 136,
137
ultraviolet studies of, 133-
61

H

Helium
burning of, 110
content of
globular clusters and,
234
extragalactic nebular content of, 113-17, 123-28
neutrons produced from,
103
target of thermonuclear
reactions, 74, 75
Helium nuclei
cosmic rays and, 471
Helium, solid
studies of, 359-64
High Energy Astronomical
Observatory, 499, 502-
5
Hydrogen
"big bang" nucleosynthesis
and, 107
extragalactic nebular content of, 113-20, 123
formation and disruption of,
140-44
galactic content of, 251
interstellar gas content of,
133-44, 146, 149-57,
160, 161
neutrons produced from,
103
stellar spectra and, 272,
273
target of thermonuclear
reactions, 72, 73
Hydrogen, high-velocity
neutral, 257-92
anomalous nature of, 258
definitions of, 265, 268
models for, 258-60, 276-
92
anticenter clouds as galactic tail, 291
Davies' model, 287-91
distant spiral arms and,
284-87, 292
extragalactic models, 280
galactic disk bending, 282-
84
galactic spurs and, 280,
281
high-Z extensions to spiral
arms, 283
infall hypothesis, 277-80
intermediate-velocity
clouds, 281, 282
Magnellanic Stream and,
291
overview of, 276, 277
passage of Large Magel-
lanic Cloud, 288
satellites of Galaxy, 280
shell of the Galaxy, 283,
284
sky distribution and, 277-
79
supernova explosion, 281-
83
Verschuur's model, 284-
87, 292
observations of, 260-76
distance of high-velocity
hydrogen, 272, 273
distribution of, 261-68
histograms of cloud mass-
es, 270
Magellanic Stream and,
273-75
maps of, 263-69
mass estimates and,
268
Milky Way anomalous
velocities and, 260, 261
near other galaxies, 275,
276
summary of cloud masses,
271, 272
tabulation of recent, 261
properties of anomalous-
velocity clouds, 268-72
velocity-latitude map, 288,
289
Hydrogen, solid
studies of, 358
Hydroxyl
interstellar gas content of,
145
Hyperons
discovery of, 336
high densities and, 335

I

Interstellar gas
see Gas, Interstellar
Iron
disk of M31 content of,
239
interstellar gas content of,
133, 136, 157
low-lying states of, 90

L

Lagrange point
black holes and, 403-5
Liquids, hyperonic, 336-53
equation of state of, 336
neutron stars and, 336
Lithium
"big bang" synthesis of,
108, 109
neutrons produced from,
103
target of thermonuclear
reactions, 74, 75

M

Magellanic Cloud, Large

passage near Galaxy, 288
Magellanic Clouds
abundance determination in, 233
brightest cluster giants in, 226-29
cosmic structure and, 291
evolution of, 224, 227-29
gas and dust in, 233, 234
globular clusters in
color-magnitude diagrams and, 228, 229
high-velocity clouds in, 261, 273
Large Cloud, 229, 230, 288
LMC Bar and, 229, 230
planetary nebulae in, 232, 233
population distribution in, 229, 233
horizontal branch and, 225-27
reddest giant stars in, 225, 228
Small Cloud, 229
star clusters in, 224-29
stellar populations in, 224-34
variable stars in, 230-34
Cepheids, 230, 231
M-type supergiants, 232
novae, 231, 232
Magellanic Stream
cloud velocities and, 259, 273-75
map of, 267
theories concerning, 275, 276
Magnesium
disk of M31 content of, 239
interstellar content of, 133, 135, 136, 149, 156
low-lying state of, 90
magnesium-aluminum cycle, 110
neutrons produced from, 103-6
target of thermonuclear reactions, 85, 103
Manganese
interstellar content of, 133
Medium, interstellar
enrichment of, 125
M31 and companions
central bulge of M31, 237
composition gradient in, 239, 240
dwarf ellipticals
star formation in, 241, 242
E galaxies and
spectra and luminosity of, 242-44
halo of M31, 239
M32 = NGC 221, 241
NGC 205 and NGC 185, 240, 241
nuclei of galaxies, 244-46
Cn + Mg index of, 244
population model for, 238, 239
spectral studies of, 237
star formation in, 245
stellar population of, 235-46
brightest stars, 235-37
globular cluster resolution in, 236
integrated light studies of, 237-46

N

Naked singularities
unpredictable behavior and, 418
Nebula, Andromeda
cyanogen giants in, 218
X-ray emission by, 431-33
Nebulae, extragalactic gaseous
chemical composition of, 113-29
abundance determinations, 114-23
chemical abundance gradients, 119-21, 124, 125
heavy elements, 117-19
helium-hydrogen ratios, 114-17
nitrogen-sulfur gradients, 122
normal H II regions and, 114-21
nuclear H II regions, 121-23
oxygen-helium ionization structure, 116
planetary nebulae, 123
reviews on, 113
Nebulae, planetary
Magellanic Clouds and, 232, 233
Neon
extragalactic nebular content of, 113, 119, 124
low-lying states of, 90
neon-sodium cycle, 106, 110
neutrons produced from, 105, 106
target of thermonuclear reactions, 81-83
Neon-to-sodium
astrophysical evolution and, 106, 110
Neutrinos
neutrino-producing processes, 71
Neutron gas, pure
equation of state of, 309
Neutron stars
crust composition of, 336
high pressure studies and, 374, 375
vibrational energy of, 343
Nitrogen
extragalactic nebular content of, 113, 119-25
interstellar gas content of, 135, 136, 140, 148-54, 158
target of thermonuclear reactions, 78, 79
thermonuclear evolution and, 109
Novae
Magellanic Clouds and, 230, 231
Novae, X-ray
observation of, 433

O

OSO-7
X-ray astronomy by, 481-84, 486
Oxygen
astrophysical evolution and, 106, 107
burning of, 110
extragalactic nebular content of, 113, 116-24
interstellar gas content of, 140, 148, 153, 154, 158
neutrons produced from, 103, 105, 106
thermonuclear reactions involving, 79-81, 85, 101-3

P

Penrose's "cosmic censorship hypothesis"
naked singularities and, 418
Positrons
capture of
thermonuclear reactions and, 87
Protons
proton-proton chain, 109
Pulsar emission mechanisms, 511-34
Pulsars
beaming mechanism of, 518, 521, 526
maser effect in, 518
models of, 519-23, 526
optical and radio emission centers in, 521, 522
pulsar pulse profile, 521-26
conclusions, 533, 534
electromagnetic radiation mechanisms, 515-18
antenna mechanism in, 516, 517
coherent synchrotron mechanisms and, 518

coherent vs incoherent radiation, 515, 516
emission brightness temperature, 515-17
magnetobremsstrahlung, 517
maser mechanism of, 516, 517
types of pulsars, 516
pulsar magnetosphere, 512-15
energy sources for, 515
model of, 512-14
plasma envelope and, 513-15
radiation from, 515
pulsar's radio emission mechanisms, 530-33
plasma membranes and, 530-33
radiation from pulsar in the Crab, 526-29
frequency spectrum of, 526-28
models of, 528, 529
optical, X-ray and gamma ray, 526-29
synchrotron radiation spectra of, 528, 529
radiation polar diagram formation, 518-25
optical polarization plane rotation, 522-24

R

Radiation, gamma
source model for, 474
Radiation, gamma atmosphere
source strengths of, 476
spectra of, 474, 475
X-ray astronomy and, 474-77
Radiation, trapped
world distribution of, 479
X-ray astronomy and, 478-80
Radioactivity, induced
X-ray astronomy and, 480, 481
Rays, cosmic
energy spectrum of, 472
geomagnetic field and, 471
variations with magnetic latitude, 473, 474
X-ray astronomy and, 471-74
Reaction rates, thermonuclear
see Thermonuclear reaction rates
Roche lobes
black holes and, 403, 405

S

Silicon
burning of, 110
interstellar content of, 136, 137, 149, 151, 153
target of thermonuclear reactions and, 103
Sodium
disk of M31 content of, 239
interstellar content of, 135, 145
target of thermonuclear reactions and, 83
South Pole Cloud
see Magellanic Stream
Spacecraft
cosmic ray outoff by, 472, 474
gamma rays produced by, 474, 477, 478
Spaceflights
modern instruments for, 498-505
Star 09.5V
high-resolution scan of, 134
Stars
black holes inside, 416, 417
evolution of
black holes and, 403, 404, 416, 417
galactic winds, 249
galaxies and, 221, 240-45, 249-52
new star formation, 250-52
time-dependent mass spectrum and, 222
frequency of dark companions, 330
globular clusters of
black holes in, 402
populations in galaxies, 217-52
types of
chemical composition of, 119
unseen astrometric companions of, 295-332
historical aspects of, 296, 297
perturbation and, 295
spectroscopic methods for, 295
see also Astrometry
Stars, binary
observation of, 301, 302
X-ray binaries, 404-6
X-ray studies of, 428-31, 433
see also Black holes, binary systems
Stars, degenerate dwarf
origin of, 403
Stars, nonrotating
gravitational collapse of, 383-85
Stars, red giants
M31 and, 240
Stars, rotating
disintegration of, 387
pancake swarms from, 387, 388
gravitational collapse of, 386-88
Stars, rotating neutron
magnetosphere dynamics of, 511-15
pulsars and, 511-15
X-ray emission by, 431
Stars, rotating supermassive
gravitational collapse of, 388
Stars, supergiant
evolution of, 404
HDE 226868, 410
Stars, supermassive
gravitational collapse of, 385, 386
Stars, variable
distribution of, 217
Stellar evolution
black hole formation and, 385-88
Sulfur
extragalactic nebular content of, 122-24
interstellar gas content of, 135, 136, 148, 154, 156
Sun
interstellar gas near, 138
solar abundance scale, 145
Supergiants
location in galaxies, 220
Supernovae
cloud velocities and, 259
explosions of
deuterium formation and, 150, 161
galactic clouds and, 281-83
interstellar gas ionization and, 160
remnants of
X-ray emission by, 430, 431

T

Telescope control
on-line computers for, 165-84
astronomical data and, 173-75
Hale Observatories and, 173, 174
historical summary of, 166
Lick Observatory, 174, 175
Mauna Kea and, 176, 177
McDonald Observatory and, 173

terminology of, 166-69
see also Computers, astronomical
Thermonuclear reaction rates, 69-111
aluminum target of, 85
atomic masses and, 87
basic ideas underlying, 69
beryllium target of, 75-77
beta decay, 88
carbon target of, 77, 78, 85
changes since FEZ I, 108-11
"big bang" nucleosynthesis, 108, 109
carbon, oxygen, and silicon burning, 110, 111
CNO tri-cycle, 109, 110
controlled thermonuclear fusion, 109
helium burning, 110
NeNa and MgAl cycles, 110
proton-proton chain, 109
data analysis, 92-108
alternative nonresonant reaction rates, 99, 100
continuum reaction rates, 92
contributions of bound states, 106, 107
cutoff for nonresonant reaction rates, 97, 98
deviation from integrated values, 98, 99
direct radiation capture, 100, 101
explosive nucleosynthesis, 92
nonresonant reaction rates, 95-97
reactions involving C^{12} and O^{16} pairs, 101-3
sharp resonance reaction rates, 94, 95
stimulated and induced reaction rates, 91, 92
three-body reactions, 107, 108
wide resonance reaction rates, 95
electron and positron capture, 87, 88
fluorine target of, 81
generalized "black body" model and, 101
helium target of, 74, 75
hydrogen targets of, 72, 73
lithium target of, 74, 75
magnesium target of, 83-85
neon target of, 81-83
neutrino producing reactions, 71
neutron-producing reactions, 103-6
nitrogen target of, 78, 79
oxygen target of, 79-81, 85
partition functions and, 88-91
Q processes in, 71
reactions producing four particles, 87
reverse reaction rates, 88, 90
shock wave effects and, 89
sodium target of, 83
table of, 70-85
instructions for use of, 70, 71
warnings to users of, 92
three-body interactions, 87
two-body interactions, 86
Titanium
interstellar gas content of, 145

U

Universe
evolution of, 127, 128
big bang and, 129
black holes and, 414, 415
white holes and big bang, 417

W

White holes
black holes and, 417
Wind, extragalactic
galactic clouds and, 279
Wormholes
tunnels connecting black holes, 417

X

X-ray astronomy
contemporary instruments, 484-505
High Energy Astronomical Observatory, 499, 502-5
modern photomultipliers, 497
modulation collimators, 494-98
phoswich technique and, 492-94
proportional counter techniques, 485-90
scintillation-counter techniques, 490-94
small scientific satellites, 499-501
spaceflight missions and, 498-505
detector principles and applications, 443-59
electronic and data handling techniques, 457-60
proportional counters, 445-49
scintillation counters, 449-54
scintillation materials' properties, 450, 451
solid-state counters, 454-57
instrumental techniques in, 423-505
reviews on, 423, 424
photon interactions and, 438-43
Compton scattering and, 440, 441, 443
photoelectric attenuation and, 439-43
theory of, 438
radiation environment, 470-84
atmospheric gamma rays, 474-77
cosmic diffuse gamma-ray flux, 482
cosmic rays, 471-74
diffuse sky background, 481
induced radioactivity, 480, 481
OSO-7 background, 481-84
spacecraft production, 477, 478
trapped radiation, 478-80
sensitivity considerations, 459-70
active collimators, 467, 468
background statistical terms, 462
data analysis, 468-70
design of aperture and collimators, 466
diffuse fluxes and, 462, 463
finite background for scanning, 464, 465
flux and position fitting, 468, 469
fundamental statistical relations, 461, 462
modulator collimator, 468
scanning experiments, 463-66
shield criterion, 466
spectral fitting and, 469
statistical limits, 461-66
systematic effects in scanning, 464-66
status of, 424-38

binary X-ray sources, 428-31
black-hole X-ray sources, 430, 431
compact objects as sources, 435
Crab nebula and NPO532 sources, 431-33
diffuse component of, 435
diffuse X-ray spectrum, 435, 436
energy ranges of, 424
extended objects as sources, 434, 435
extragalactic sources, 433-35
galactic sources, 428-33
general considerations, 424-28
mechanism of X-ray emission, 429
observation with collimated counters, 425, 426
photon spectrum and, 428
soft X-rays, 436-38
supernova remnants, 431
transient X-ray sources, 433
UHURU observations and, 424-27, 434-37
unidentified emitters, 435
X-ray sky map from UHURU, 427, 428
see Black holes, binary systems

Z

Zinc
interstellar gas content of, 148

CUMULATIVE INDEXES

CONTRIBUTING AUTHORS VOLUMES 9-13

A

Aannestad, P. A., 11:309-62
Aarseth, S. J., 13:1-21
Acton, L. W., 12:359-81
Anders, E., 9:1-34
Arnett, W. D., 11:73-94
Athay, R. G., 11:187-218

B

Bahcall, J. N., 10:25-44
Barkat, Z., 13:45-68
Beckers, J. M., 10:73-100
Becklin, E., 9:67-102
Bless, R. C., 10:197-226
Blumenthal, G. R., 12:23-46
Burrus, C. A., 11:51-72

C

Canuto, V., 12:167-214; 13:335-80
Caughlan, G. R., 13:69-112
Code, A. D., 10:197-226; 11:239-68
Culhane, J. L., 12:359-81

D

Dalgarno, A., 10:375-426
Duncombe, R. L., 11:135-54

E

Eardley, D. M., 13:381-422
El-Baz, F., 12:135-65

F

Field, G. B., 10:227-60
Findlay, J. W., 9:271-92
Fowler, W. A., 13:69-112
Fricke, K. J., 10:45-72
Fricke, W., 10:101-28

G

Gilman, P. A., 12:47-70
Ginzburg, V. L., 13:511-35
Grasdalen, G. L., 13:187-216

H

Harrison, E. R., 11:155-86
Heiles, C., 9:293-322
Hodge, P. W., 9:35-66
Hummer, D. G., 9:237-70
Hyland, A. R., 9:67-102

I

Iben, I. Jr., 12:215-56
Ingersoll, A. P., 9:147-82

J

Jauncey, D. L., 13:23-44
Jenkins, E. B., 13:133-64
Jokipii, J. R., 11:1-28

K

Kaplan, S. A., 12:113-33
Keenan, P. C., 11:29-50
Kippenhahn, R., 10:45-72
Klepczynski, W. J., 11:135-54

L

Larson, R. B., 11:219-38
Lecar, M., 13:1-21
Leibacher, J., 12:407-35
Leovy, C. B., 9:147-82
Litvak, M. M., 12:97-112
Livingston, W. C., 11:95-114

M

Marsden, B. G., 12:1-21
McCray, R. A., 10:375-426
Mihalas, D., 11:187-218
Miller, J. S., 12:331-58
Morgan, W. W., 11:29-50

N

Neugebauer, G., 9:67-102
Novikov, I. D., 11:387-412
Noyes, R. W., 9:209-36

O

Oke, J. B., 12:315-29
Ostriker, J. P., 9:353-66

P

Paczynski, B., 9:183-208
Palmer, P., 12:279-313
Peimbert, M., 13:113-31
Penzias, A. A., 11:51-72
Peterson, L. E., 13:423-509
Pikel'ner, S. B., 12:113-33
Press, W. H., 10:335-74; 13:381-422
Preston, G. W., 12:257-77
Purcell, E. M., 11:309-62

R

Reeves, H., 12:437-69
Robinson, L. B., 13:165-85
Ruderman, M., 10:427-76
Rybicki, G., 9:237-70

S

Salpeter, E. E., 9:127-46
Schramm, D. N., 12:383-406
Searle, L., 12:315-29
Sears, R. L., 10:25-44
Seidelmann, P. K., 11:135-54
Silk, J., 11:269-308
Smerd, S. F., 10:159-96
Spiegel, E. A., 9:323-52; 10:261-304
Spitzer, L. Jr., 13:133-64
Stein, R. F., 12:407-35
Strom, K. M., 13:187-216
Strom, S. E., 13:187-216
Svestka, Z., 10:1-24

T

Thaddeus, P., 10:305-34
Thorne, K. S., 10:335-74
Tsytovich, V. N., 11:363-86
Tucker, W. H., 12:23-46

V

van de Kamp, P., 9:103-26; 13:295-332
van den Bergh, S., 13:217-55
Verschuur, G. L., 13:257-93

W

Wallerstein, G., 11:115-34

Wentzel, D. G., 12:71-96
Wild, J. P., 10:159-96
Woltjer, L., 10:129-58

Z

Zel'dovich, Ya. B., 11:387-412
Zhelezynakov, V. V., 13:511-35
Zimmerman, B. A., 13:69-112
Zuckerman, B., 12:279-313

CHAPTER TITLES VOLUMES 9-13

THE SOLAR SYSTEM		
Meteorites and the Early Solar System	E. Anders	9:1-34
The Atmospheres of Mars and Venus	A. P. Ingersoll, C. B. Leovy	9:147-82
Ultraviolet Studies of the Solar Atmosphere	R. W. Noyes	9:209-36
Spectra of Solar Flares	Z. Svestka	10:1-24
Solar Neutrinos	J. N. Bahcall, R. L. Sears	10:25-44
Solar Spicules	J. M. Beckers	10:73-100
Radio Bursts from the Solar Corona	J. P. Wild, S. F. Smerd	10:159-96
Turbulence and Scintillations in the Interplanetary Plasma	J. R. Jokipii	11:1-28
Comets	B. G. Marsden	12:1-12
Solar Rotation	P. A. Gilman	12:47-70
Surface Geology of the Moon	F. El-Baz	12:135-65
The Solar X-Ray Spectrum	J. L. Culhane, L. W. Acton	12:359-81
Waves in the Solar Atmosphere	R. F. Stein, J. Leibacher	12:407-35
STELLAR SPECTRA AND ATMOSPHERES		
The Formation of Spectral Lines	D. G. Hummer, G. Rybicki	9:237-70
Spectral Classification	W. W. Morgan, P. C. Keenan	11:29-50
The Effects of Departures from LTE in Stellar Spectra	D. Mihalas, R. G. Athay	11:187-218
Compact X-Ray Sources	G. R. Blumenthal, W. H. Tucker	12:23-46
The Chemically Peculiar Stars of the Upper Main Sequence	G. W. Preston	12:257-77
The Spectra of Supernovae	J. B. Oke, L. Searle	12:315-29
Young Stellar Objects and Dark Interstellar Clouds	S. E. Strom, K. M. Strom, G. L. Grasdalen	13:187-216
PHYSICS OF STARS		
Central Stars of Planetary Nebulae	E. E. Salpeter	9:127-46
Convection in Stars: I. Basic Boussinesq Convection	E. A. Spiegel	9:323-52
Recent Developments in the Theory of Degenerate Dwarfs	J. P. Ostriker	9:353-66
Evolution of Rotating Stars	K. J. Fricke, R. Kippenhahn	10:45-72
Supernova Remnants	L. Woltjer	10:129-58
Ultraviolet Astronomy	R. C. Bless, A. D. Code	10:197-26
Convection in Stars: II. Special Effects	E. A. Spiegel	10:261-304
Pulsars: Structure and Dynamics	M. Ruderman	10:427-76
Explosive Nucleosynthesis in Stars	W. D. Arnett	11:73-94
The Physical Properties of Carbon Stars	G. Wallerstein	11:115-34
Post Main Sequence Evolution of Single Stars	I. Iben Jr.	12:125-56
Neutrino Processes in Stellar Interiors	Z. Barkat	13:45-68
Astrophysical Processes Near Black Holes	D. M. Eardley, W. H. Press	13:381-422

Title	Author(s)	Volume:Pages
On the Pulsar Emission Mechanisms	V. L. Ginzburg, V. V. Zheleznyakov	13:511-35
ASTROMETRY		
Fundamental Systems of Positions and Proper Motions	W. Fricke	10: 101-28
Dynamical Astronomy of the Solar System	R. L. Duncombe, P. K. Seidelmann, W. P. Klepczynski	11:135-54
Unseen Astrometric Companions of Stars	P. van de Kamp	13:295-333
STELLAR ASTRONOMY		
The Nearby Stars	P. van de Kamp	9:103-26
Evolutionary Processes in Close Binary Systems	B. Paczynski	9:183-208
THE INTERSTELLAR MEDIUM		
Physical Conditions and Chemical Constitution of Dark Clouds	C. Heiles	9:293-322
Heating and Ionization of HI Regions	A. Dalgarno, R. A. McCray	10:375-426
Millimeter-Wavelength Radio-Astronomy Techniques	A. A. Penzias, C. A. Burrus	11:51-72
Processes in Collapsing Interstellar Clouds	R. B. Larson	11:219-38
Interstellar Grains	P. A. Aannestad, E. M. Purcell	11:309-62
Cosmic-Ray Propagation in the Galaxy: Collective Effects	D. G. Wentzel	12:71-96
Coherent Molecular Radiation	M. M. Litvak	12:97-112
Large-Scale Dynamics of the Interstellar Medium	S. A. Kaplan, S. B. Pikel'ner	12:113-33
Ultraviolet Studies of the Interstellar Gas	L. Spitzer Jr., E. B. Jenkins	13:133-64
THE GALAXY		
Infrared Sources of Radiation	G. Neugebauer, E. Becklin, A. R. Hyland	9:67-102
Nucleo-Cosmochronology	D. N. Schramm	12:383-406
On the Origin of the Light Elements	H. Reeves	12:437-69
Computer Simulations of Stellar Systems	S. J. Aarseth, M. Lecar	13:1-21
High-Velocity Neutral Hydrogen	G. L. Verschuur	13:257-93
EXTRAGALACTIC ASTRONOMY		
Dwarf Galxies	P. W. Hodge	9:35-66
Diffuse X and Gamma Radiation	J. Silk	11:269-308
Computer Simulations of Stellar Systems	S. J. Aarseth, M. Lecar	13:1-21
Radio Surveys and Source Counts	D. L. Jauncey	13:23-44
Chemical Composition of Extragalactic Gaseous Nebulae	M. Peimbert	13:113-31
Stellar Populations in Galaxies	S. van den Bergh	13:217-55
COSMOLOGY AND COSMOGONY		
Intergalactic Matter	G. B. Field	10:227-60
The Short-Wavelength Spectrum of the Microwave Background	P. Thaddeus	10:305-34
Standard Model of the Early Universe	E. R. Harrison	11:155-86
Physical Processes Near Cosmological Singularities	I. D. Novikov, Ya. B. Zel'dovich	11:387-412
INSTRUMENTATION		
Filled-Aperture Antennas for Radio Astronomy	J. W. Findlay	9:271-92
Image-Tube Systems	W. C. Livingston	11:95-144
New Generation Optical Telescope Systems	A. D. Code	11:239-68
Radio Radiation from Interstellar Molecules	B. Zuckerman, P. Palmer	12:297-313
On-Line Computers for Telescope Control and Data Handling	L. B. Robinson	13:165-85
Instrumental Technique in X-Ray Astronomy	L. E. Peterson	13:423-509
OBSERVATIONAL TECHNIQUES		
Planetary Nebulae	J. S. Miller	12:331-58
PHYSICAL PROCESSES		
Gravitational-Wave Astronomy	W. H. Press, K. S. Thorne	10:335-74
Interaction of Fast Particles with Waves in Cosmic Magnetoactive Plasma	V. N. Tsytovich	11:363-86
Equation of State at Ultrahigh Densities	V. Canuto	12:167-214
Neutrino Processes in Stellar Interiors	Z. Barkat	13:45-68

Thermonuclear Reaction Rates, II	W. A. Fowler, G. R. Caughlan, B. A. Zimmerman	13:69-112
Equation of State at Ultrahigh Densities, II	V. Canuto	13:335-80